AN INTRODUCTION TO COMPARATIVE ARACHNOLOGY

AN INTRODUCTION TO COMPARATIVE ARACHNOLOGY

L. van der Hammen

SPB Academic Publishing bv

ISBN 90-5103-023-1

CIP-DATA KONINKLIJKE BIBLIOTHEEK, DEN HAAG

Hammen, L. van der

An introduction to comparative arachnology / L. van der Hammen. – Den Haag : SPB Academic Publishing. – Ill.
With ref.
ISBN 90-5103-023-1 bound
SISO 597.5 UDC 595.4 NUGI 823
Subject heading: arachnology

TABLE OF CONTENTS

PREFACE

This book represents the final development of my acarological and arachnological views, as a result of comparative studies over a period of some thirty years. It is not a handbook, but a general survey of personal insights, written after the conclusion of my investigations, i.e., after my retirement from the Rijksmuseum van Natuurlijke Historie, Leiden (where I had been curator of Arachnida for a long time). The book provides a general introduction, in which the most important problems are formulated, and which can serve as a framework for continued research. In the course of my studies, my main concern had been with the preparation of detailed comparable descriptions and figures, and with the development of a uniform terminology. Complete descriptions of species, chosen as examples for comparison (and as examples for the demonstration of structural relations), still constitute an important part of this book.

My arachnological studies are scattered over a great number of separate papers. Although my first arachnological papers (published in 1947) dealt with Spiders and Harvestmen, the greater part of them are (since 1949) devoted to acarology. My earliest acarological interests were in Oribatid mites and the soil fauna. As a result of my co-operation with the French acarologist F. Grandjean, I became increasingly interested in morphology and evolution. In 1959 I started the comparative morphological study of other groups of Mites, and in 1977 the comparative morphological study of other groups of Chelicerata. Closely connected with these morphological studies, were my studies of the relationships and higher classification of Chelicerata. In the course of my investigations, most attention was paid to both groups of Mites (Actinotrichida and Anactinotrichida), to the groups which appeared to be related to Mites (Palpigradi, Ricinulei; I discovered too late that it would have been interesting to include a detailed study of Schizomida), and to a group which was erroneously supposed to be closely related to Mites (Opilionida). Because of my retirement, I was obliged to restrict my study of other groups to the most essential morphological characters. Extinct groups are mentioned in passing only; with the exception of Eurypterida, little is known in fossils about those morphological characters which are regarded by me as the most important. Not until the fourth part of my Comparative Studies in Chelicerata (Van der Hammen, 1986a) did I arrive at the final understanding of some of the most important problems, i.e., those with reference to the evolution of prosomatic architecture (the evolution of the coxa; the importance of the association of orifices of coxal glands, sternapophyses and mouthparts in Cryptognomae as well as Arachnida, s. str.).

The results summarized in the present book are influenced by series of studies and projects carried out simultaneously with my comparative studies in Chelicerata, viz., (1) a series of theoretical papers (Van der Hammen, 1978, 1981, 1981a, 1983, 1985b, 1986, 1986d) which led to a structuralist approach in the study of evolution and classification, and of which a revised and adapted version is included in my 'Unfoldment and manifestation. Seven essays on evolution and classification' (Van der Hammen, 1988); (2) the 'Glossary of acarological terminology' (Van der Hammen, 1976, 1980; the third part is in course of preparation) which has been the base for the development of a general arachnological terminology; and (3) the editing (with introductions and a detailed index) of Grandjean's 'Complete acarological works', in seven volumes, published in the period 1972–1976.

The development of comparative acarology and arachnology (including the development of a homogeneous morphological terminology) is not a grateful task. Generally, the introduction of changes in current views and terminology meets with much opposition. The task is, moreover, difficult because it demands a repeated mastering of the morphology of other groups. When the new framework is finally accepted, it can easily be improved by specialists in the fields of the separate groups.

The personal satisfaction with comparative studies (I paid special attention to the architecture of prosoma and gnathosoma, and to segmentation and articulation of the legs) lies particularly in the development of a complete overall view, and in the conviction that this view can be the starting-point for very interesting continued research. Because of their segmented body and the occurrence, in the course of postembryonic development, of separate instars, Chelicerata are excellent subjects for a structuralist approach in the study of evolution; the continued application of this method could considerably advance our insights in the complicated hierarchy of laws underlying Chelicerate evolution. The morphology and postembryonic development of several groups of Chelicerata, and several parts of Chelicerate comparative morphology, are still insufficiently known; continued studies of these subjects will further advance our insights in Chelicerate relationships. I hope that my own pioneer studies have prepared the way for future discoveries in these fields.

Baronielaan 1
2242 RA Wassenaar

L. van der Hammen
30 December 1988

A. GENERAL PART

I. INTRODUCTION

The concept Chelicerata dates back to the classic investigations by Lankester (1881), as a result of which Xiphosura (King crabs or Horseshoe crabs) were regarded as Arachnids. The name Chelicerata was introduced by Heymons (1901); it referred to a group comprising Merostomata (Xiphosura and the extinct Eurypterida) and Arachnida (the name Arachnidae was introduced by Lamarck, 1801). Generally, Chelicerata are classified with the phylum Arthropoda,[1] a group characterized by the possession of a jointed body, jointed appendages, and a central nervous system ancestrally consisting of a ganglionic mass surrounding the oesophagus, and a ventral ganglionic chain. It is not impossible that Arthropods represent a grade of evolution, which has been attained independently by several groups (Crustacea, Chelicerata, Uniramia). Pycnogonida (Sea spiders) could be the nearest relatives of Chelicerata.

Chelicerata are Arthropods the body of which is composed of two parts (or tagmata): prosoma and opisthosoma. The prosoma comprises the precheliceral metameres and the segments I–VI; the opisthosoma comprises ancestrally 10–13 segments and, in several cases, also a so-called 'telson'. The central nervous system lacks a deutocerebrum, and there are no antennae and no mandibles. Ancestrally, there are six pairs of prosomatic appendages (which belong to segments I–VI): a pair of chelicerae, a pair of palpi, and four pairs of legs. In many cases, homologues of opisthosomatic appendages (up to seven pairs; embryologically up to nine pairs) can be recognized. The postembryonic development is characterized by moulting and the occurrence of several immature instars; the first instar, or the first instars, have been subject to regressive evolution. The subphylum Chelicerata takes its name from the first pair of prosomatic appendages (the chelicerae) which are, ancestrally, three-segmented and chelate, and of which the function is, ancestrally, prehensile.

Generally, twelve main groups of extant Chelicerata are distinguished: Palpigradi, Actinotrichida (Actinotrichid mites), Solifugae, Pseudoscorpionida

---

1 The name Arthropoda was introduced by C. Th. von Siebold, in the first part of his *Lehrbuch der vergleichenden Anatomie der Wirbellosen Thiere*. According to Engelmann (1846: 237) this first part was published in 1845, although 1848 is mentioned on the title-page of the book. The concept Arthropoda dates back to Latreille (1829), who subdivided the Articulata into jointed animals with and without jointed legs.

(False scorpions), Anactinotrichida (Anactrinotrichid mites), Ricinulei, Uropygi (Whip scorpions), Amblypygi (Whip spiders), Araneida (Spiders), Opilionida (Harvestmen), Scorpionida (Scorpions) and Xiphosura (King crabs or Horseshoe crabs). Fossil Xiphosura are already known from the Cambrian, fossil Scorpionida already from the Silurian, fossil Actinotrichida already from the Devonian, and fossil Ricinulei, Uropygi, Amblypygi, Araneida and Opilionida already from the Carboniferous. These fossil species present already the typical habitus of the extant representatives. There are six extinct groups of Chelicerata, viz., Kustarachnae, Trigonotarbi, Haptopoda, Anthracomarti, Architarbi and Eurypterida; the last-mentioned group is already known from the Ordovician, the other groups already from the Carboniferous. Evidently, Chelicerata constitute a very old group (they belong also to the earliest terrestrial animals) which has little changed since the Carboniferous. In the present book, data with reference to extinct taxa are hardly mentioned, as my own investigations have been restricted to the twelve extant groups.

II. COMPARATIVE MORPHOLOGY

The present chapter deals with the comparative morphology of those structures which are regarded by me as of fundamental importance to an understanding of relationships; the morphology of these structures is studied here in evolutionary perspective. The chapter is subdivided into nine sections, in which the following subjects are discussed: constitution of the body, prosoma, opisthosoma, prosomatic appendages, mouthparts and ingestion, orifices of coxal glands and associated structures, opisthosomatic appendages, genital region, and respiratory organs. On several occasions I had already prepared concise introductions or compendious surveys of Acarid or Chelicerate morphology. Two of these introductions (Van der Hammen, 1968c, 1972) dealt exclusively with Mites (Actinotrichida and Anactinotrichida), whilst much information on Acarid, and to a less extent Chelicerate, morphology (including references to the main literature) can be found in the general part of the *Glossary of acarological terminology* (Van der Hammen, 1980). In two subsequent introductory papers (Van der Hammen, 1977a, 1985a) I have discussed the comparative morphology of all extant groups of Chelicerata.

1. Constitution of the body

In the past, I have repeatedly dealt with general problems relating to the constitution of the Chelicerate body, particularly in the case of both groups of Mites (cf. Van der Hammen, 1963b: 443–454; 1968: 26–28; 1968c: 407–408; 1969: 196; 1970a: 3–9, fig. 1, tab. 1; 1972: 9–11, fig. 5, tab. 2). Many terms with reference to the constitution of the body are mentioned in the general part of the *Glossary of acarological terminology* (Van der Hammen, 1980; the terms in question are listed on p. 47 of this glossary, under the entry 'Divisions of the body'). A general survey of constitution and segmentation of the body in all extant groups of Chelicerata was recently prepared by me (Van der Hammen, 1985a: 137–139).

In all Chelicerata, the body is ancestrally differentiated into two main divisions or tagmata: prosoma and opisthosoma. Each of these two tagmata was ancestrally composed of more or less similar segments or metameres, and constitutes a distinct unity characterized by its own individuality. As mentioned already in the introduction, the Chelicerate prosoma comprises the precheliceral metameres and the segments I–VI; the opisthosoma comprises (ancestrally) 10–13 segments and, in several groups, a so-called 'telson'. It is supposed (Van der Hammen, 1981a: 37) that the fundamental ('prototactic') numbers of segments (16–19) arose, in the course of evolution, from more primitive, irregular ('primordiotactic') numbers. Prosoma and opisthosoma are either broadly joined (latigastric Chelicerata) or connected by a narrow pedicel (cauligastric Chelicerata). In Arachnida s.str. (Uropygi, Amblypygi, Araneida), the pedicel is constituted by segment VII, the first opisthosomatic segment; in Ricinulei, the pedicel is constituted by the segments VII and VIII. In several groups of latigastric Chelicerata, the anterior segments of the opisthosoma are slightly narrowed, but do not constitute a pedicel (Palpigradi, Solifugae); this is supposed to represent the primitive latigastric condition. In all groups of latigastric Chelicerata, prosoma and opisthosoma are more or less distinctly separated by the disjugal furrow.

In cases where the recognition of the original subdivision of the body has become problematic (as in Actinotrichid mites), comparison with the general laws of Arthropod metamerism, formulated by Lankester (1902, 1904), can be conclusive (cf. Van der Hammen, 1963b). According to one of these laws, regressive evolution always affects the most anterior or most posterior segment of a tagma. In Actinotrichid mites, e.g., regressive evolution affects the posterior segment of the prosoma (leg IV is suppressed in the larva, whilst the number of phaneres is reduced in the protonymph), and the posterior segments of the opisthosoma (the segments XVII–XIII, successively, can have been subject to regressive evolution; see below). The regressive evolution of the Actinotrichid segment VI (the segment of leg IV) demonstrates that the primary division of the Actinotrichid body is indicated by the disjugal furrow (between segments VI and VII) and not by the sejugal furrow (between segments IV and V) which constitutes the border between proterosoma and hysterosoma.

Secondary subdivisions of the body, bordered by furrows, and also constituting more or less distinct functional unities, but which do not represent true tagmata, are called pseudotagmata (Van der Hammen, 1980: 131). Mesosoma and metasoma (separated by a segmental border) in Scorpionida (the pygidium of Palpigradi, Ricinulei and Uropygi can also be regarded as a metasoma), gnathosoma and idiosoma (separated by the circumcapitular furrow) in Actinotrichida, Anactinotrichida and Ricinulei, proterosoma and hysterosoma (separated by the sejugal furrow) in Actinotrichida, and myliosoma in Opilionida, Scorpionida and Xiphosura, are pseudotagmata. Other divisions of the body, not representing tagmata, and also not regarded as pseudotagmata, can have a practical value in descriptions (epiprosoma, stethosoma, aspidosoma, podosoma, propodosoma and metapodosoma in Actinotrichida; see Van der Hammen, 1980).

The prosoma comprises the acron, the precheliceral metameres, and the segments I–VI. Acron and precheliceral metameres must originally, in the ancestors of the modern Chelicerata, have been anterior to segment I; they have been subject to considerable translocation. Data with reference to precheliceral metameres have been obtained by embryological studies and, particularly, by the study of the nervous system (Legendre, 1958, 1959; cf. also Van der Hammen, 1968: 26–27; 1980: 122–123). In several cases it has been possible to recognize the following precheliceral elements: acron (ganglion still recognizable as archicerebrum or central body; the rostral gland also originates from the acron); one or more metameres associated with the eyes; a metamere with the cerebroid ganglion and fugacious appendages of the vesicula laterales; and the metamere of the tritocerebrum (with rostral ganglion and labrum). The last-mentioned metamere is followed (i.e. ancestrally followed) by the segments bearing chelicerae and palpi, between which a fugacious segment has possibly been present. Because of the uncertain number of precheliceral metameres, numbering of the prosomatic segments starts with the segment bearing the chelicerae (segment I). Segment II bears the palpi, and segments III–VI the legs.

The ancestral numbers of opisthosomatic segments, mentioned for the different groups of extant Chelicerata, are: ten (Opilionida, Xiphosura), eleven (Palpigradi, Actinotrichida, Solifugae, ?Pseudoscorpionida), twelve (Uropygi, Amblypygi, Araneida), and thirteen (Anactinotrichida, Ricinulei, Scorpionida). A so-called 'telson' (the true nature of the element is uncertain) is present in Palpigradi (flagellum), Uropygi (flagellum), Scorpionida (sting) and Xiphosura (tail). The genital opening of all Chelicerata is ancestrally in segment VIII.

In many cases, homologues of opisthosomatic appendages (up to seven pairs, embryologically up to nine pairs) can be recognized (cf. Van der Hammen, 1977a: 313–314, tab. 2; 1985a: 143, fig. 7). These appendages can have become associated with the genital opening, or with respiration (branchial appendages in Xiphosura, booklungs in Arachnida s.str. and Scorpionida). As a result of the concentration of segments in the ventral region, several pairs of opisthosomatic appendages can have become associated with the genital region (up to five pairs in Actinotrichida; see below).

True anamorphosis (a type of metamorphosis in which segments, that had not been present in the embryo, are added posteriorly in the course of postembryonic development) is not known from Chelicerata. Many Actinotrichida, however, present a particular type of epimorphosis (it is called hysteromorphosis; see Van der Hammen, 1980: 79), in which the posterior metameres of the embryo are subject to suppression although they can reappear in the course of postembryonic development. In the latter case, segment XIII is larval, segment XIV protonymphal, segment XV deutonymphal, and segment XVI tritonymphal; segment XVII, although not known, must have been present in the adults of the ancestors (it could still be present in the adults of unknown primitive species). In most cases, the postembryonic addition is restricted to one, two, or three segments (see Van der Hammen, 1970a: 9, fig. 1A, tab. 1; 1982a:

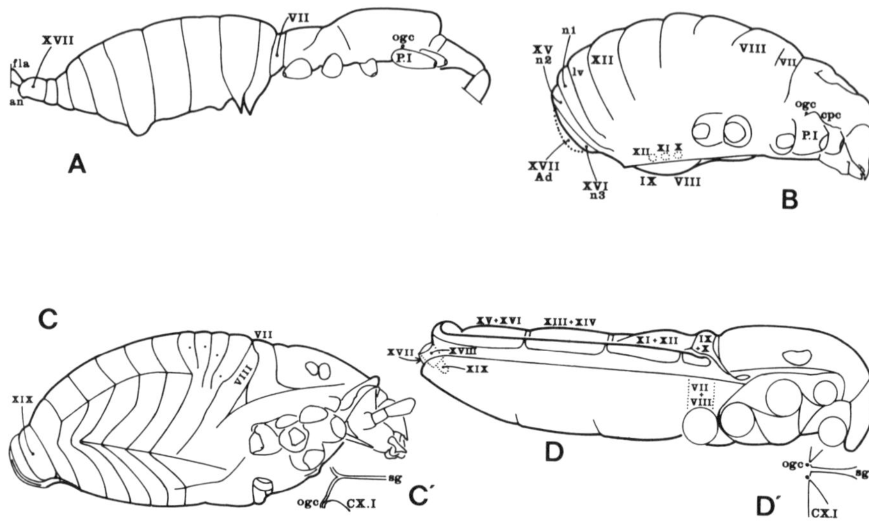

Fig. 1. Segmentation of the body in four groups of Chelicerata (lateral view). A, Palpigradi (*Eukoenenia*): seventeen body segments and flagellum; orifice of coxal gland above the base of leg I. B, Primitive Actinotrichida (*Alycus*): the posterior opisthosomatic segments are suppressed during embryonic development, but can reappear in the course of postembryonic development, in which case segment XIII is larval, XIV protonymphal, XV deutonymphal, XVI tritonymphal, whilst the unknown segment XVII is supposed to have been present, ancestrally, in the adult; ventrally, the segments VIII–XII are concentrated in the genital region; the orifice of the coxal gland is above the base of leg I, and is associated with a podocephalic canal. C, C′, Primitive Anactinotrichida (*Opilioacarus*): nineteen body segments; C′, ventral view of anterior sternal region with orifice of coxal gland (posterior to coxa I; assocation with subcapitular gutter). D, D′, Ricinulei (*Cryptocellus*): nineteen body segments; segments VII and VIII constitute a pedicel which is concealed by the locking device of prosoma and opisthosoma; IX and X, XI and XII, XIII and XIV, and XV and XVI have fused; the segments XVIII and XIX of the pygidium can be retracted; D′, ventral view of anterior sternal region with orifice of coxal gland (posterolateral to coxa I; association with subcapitular gutter). Notations: *Ad*, adult base level; *cpc*, podocephalic canal; *CX.I*, coxa of leg I; *lv*, larval base level; *n1*, protonymphal base level; *n2*, deutonymphal base level; *n3*, tritonymphal base level; *ogc*, orifice of coxal gland; *P.I*, leg I; *sg*, subcapitular gutter; *VII–XIX*, body segments. A, × 58; B, × 78; C, C′, × 31; D, D′, × 8.

27–29, figs. 11–12; 1985a: 138–139, fig. 1b). Hysteromorphosis is a rare phenomenon, known from Actinotrichid mites only; it is not known from any other group of Arthropods. It could be associated with the small size of most Actinotrichida, which probably arose, in the course of evolution, by reduction.

A recapitulation of the numbers of segments, and the constitution of the body, in the main groups of Chelicerata, is given below. It may be remarked here that many of these data need confirmation by embryological studies, and by studies of the musculature of the body.

In Palpigradi (fig. 1A; see Van der Hammen, 1982a: 6–13, figs. 1–3), seventeen segments and a long 'telson' (flagellum) are present. Prosoma and opisthosoma are separated by the disjugal furrow; the separation is particularly distinct because of the slight narrowing of segments VII and VIII (although Palpigradi are latigastric). The posterior segments of the body (XIV–XVII) are also narrowed. Segments XV–XVII are movable with respect to each other; they are slightly telescoping, and constitute a pygidium (or metasoma) which can be slightly elevated with respect to segment XIV. The flagellum is long, thin and multiarticulate; its base is dorsal to the anal opening (which has a terminal position).

In Actinotrichida (fig. 1B; see Van der Hammen, 1968a: 275–277; 1969: 181–185, 197–199, figs. 1–3; 1970a: 5–6, fig. 1A, tab. 1; 1982: 240, 250–252, figs. 1, 2, 11; 1982a: 25–29, 31–34, figs. 11–13, 26D–G), traces of segmentation are still distinctly discernable in some primitive groups, particularly primitive Actinedida and Oribatida. As mentioned above, the ancestral number of segments must have been seventeen, but segments XIII–XVII are suppressed in the course of embryonic development, although one or more of them can reappear in the course of postembryonic development. In Tarsonemida the terminal segment is apparently constituted by segment XIII, in many Actinedida also by segment XIII (with larval base level), in several other Actinedida and in Acaridida by segment XIV (with protonymphal base level), in most Oribatida by segment XV (with deutonymphal base level), and only in some primitive Oribatida and Actinedida by segment XVI (with tritonymphal base level). As mentioned above, segment XVII (with adult base level) is not known, but must have been present in the ancestors (the absence of a 'telson' could be connected with the suppression of this segment). Actinotrichida are latigastric; prosoma and opisthosoma are, generally, separated by a dorso-disjugal and a ventro-disjugal furrow; these furrows are, as a rule, not continuous because of the advanced position of the dorsal region of the opisthosoma. Generally, several pseudotagmata (gnathosoma and idiosoma, proterosoma and hysterosoma), bordered by furrows, and constituting functional units, can be distinguished. The anus has, generally, a ventral position, and this translocation is associated with a concentration of segments in the genital region. The last-mentioned region includes, ancestrally, the opisthosomatic appendages of segments VIII and IX (of which the 'endites' border the eugenital opening, whilst the 'exites' constitute the genital covers) and the opisthosomatic appendages of segments X–XII (the genital papillae) (cf. Van der Hammen, 1982a: 31–34, 49–51, figs. 11, 14, 26).

The body of Solifugae (fig. 2A; see Van der Hammen, 1986a: 5–7, fig. 1A) consists of seventeen segments. It is latigastric, although the anterior segments of the opisthosoma are slightly narrowed. The tergite of segment VII has an advanced position, above the base of leg IV. The dorso-disjugal and the ventro-disjugal furrow, separating prosoma and opisthosoma are generally not continuous. The tergites and sternites of segments VIII–XVI have corresponding positions and the intersegmental furrows are more or less parallel and, laterally, vertical. The anus (in segment XVII) is more or less terminal, with the exception of the family Rhagodidae in which it has a more ventral position.

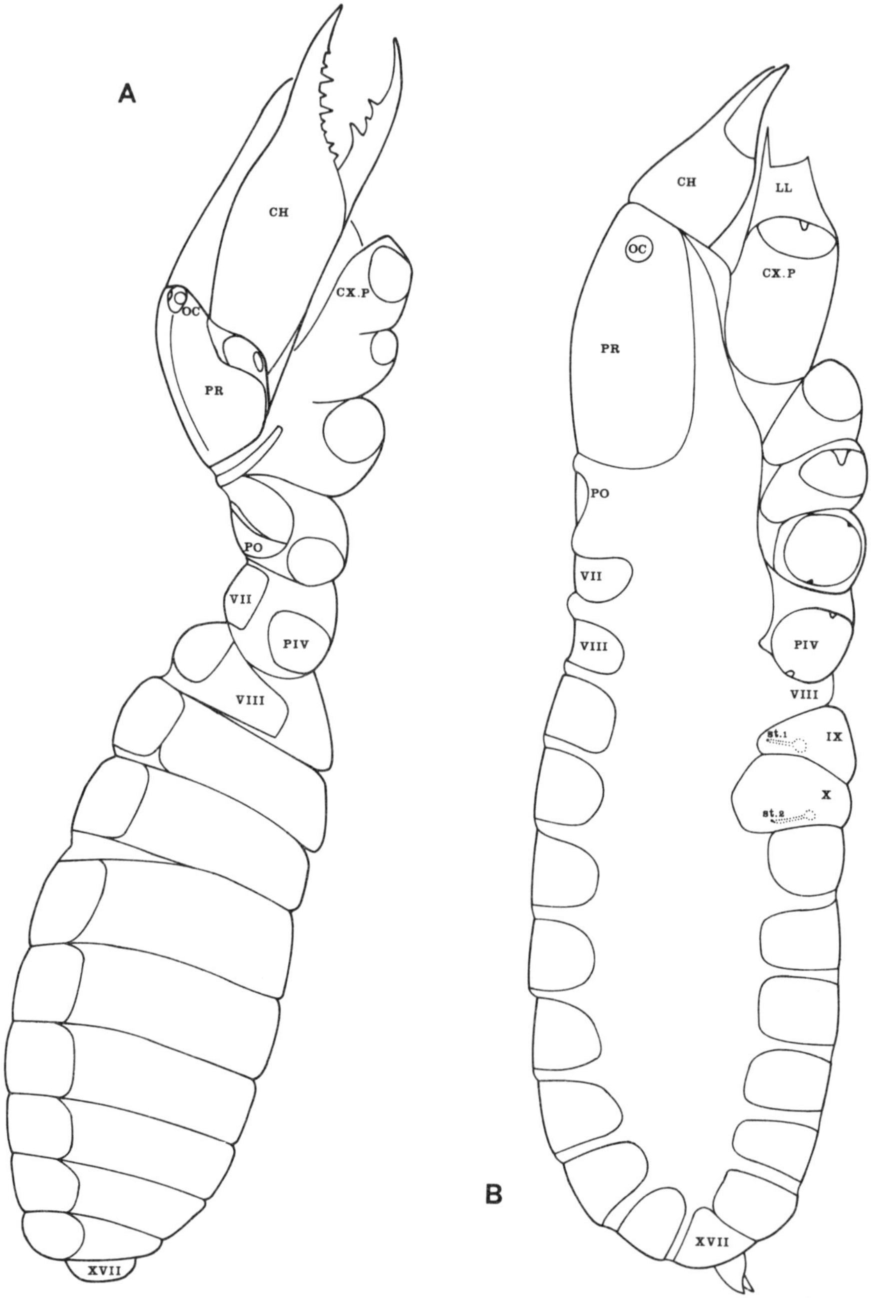
A
CH
OC
CX.P
PR
PO
VII
PIV
VIII
XVII
B
CH
LL
OC
CX.P
PR
PO
VII
VIII
PIV
VIII
st.1
IX
X
st.2
XVII

Pseudoscorpionida (fig. 2B) are, generally, regarded as characterized by the presence of eighteen segments. Comparison of the condition in *Chthonius* (see Van der Hammen, 1986a: 7–8, fig. 1B) with that in the related Solifugae suggests that the tergite currently regarded as representing the tergite of segment VII could in fact represent the dorso-posterior part of the prosoma (the postpeltidium). In that case, the posterior segment bearing the anal opening could represent segment XVII, whilst the evaginable membraneous tegument surrounding the opening could be regarded as not representing a segment. This hypothesis could easily be tested by a study of the musculature of the body and by comparison with representatives of other groups of Pseudoscorpionida (there are important differences in the position of the furrow now supposed to represent the dorsal border between prosoma and opisthosoma).

In Anactinotrichida (fig. 1C, C′) distinct traces of segmentation are known from Opilioacarida only (Van der Hammen, 1966: 8–10, figs. 1–3; 1968c: 407–408; 1969b: 115–119, fig. 2; 1970a: 4–5, fig. 1B, tab. 1; 1979a: 15, fig. 29A; 1985a: 137–138, fig. 1C, C′). In this primitive group, the original segmentation is indicated by furrows and muscle insertions. The number of segments appears to be nineteen. The ventral border between prosoma and opisthosoma is posterior to the base of legs IV, and anterior to the genital opening (with advanced position between the legs). The Opilioacarid anal opening has a terminal position. In other groups of Anactinotrichida, the dorsodisjugal furrow (distinct in Opilioacarida) is supposed to be absent or, in rare cases, to occupy an advanced position (as in *Rhodacarus*; see also Van der Hammen, 1966: 51); it is certainly not identical with the so-called notomeristic scissure (Van der Hammen, 1970a: 9; 1979a: 22, fig. 29; 1980: 104) which, in many Gamasida, subdivides the dorsal shield into two parts. In all Anactinotrichida, two pseudotagmata can be distinguished: gnathosoma and idiosoma. In some Holothyrida (*Australothyrus*; see Van der Hammen, 1983b: 8, figs. 1, 2), dorsal idiosomatic muscle attachments are indicated by large sigilla, arranged in a pattern which presents some traces of the original segmentation, but which does not allow of a reconstruction of the number of segments; in this case, the dorsal part of the prosoma appears indeed to be considerably reduced.

In Ricinulei (fig. 1D, D′; see Van der Hammen, 1979a: 7, 51–52, figs. 1, 29C; 1985a: 137–138, fig. 1d, d′), the body is distinctly subdivided into two tagmata (prosoma and opisthosoma) joined by a pedicel which is constituted by the narrowed segments VII and VIII. The pedicel is completely concealed

←

Fig. 2. Representatives of the two groups of Apatellata, i.e., Solifugae and Pseudoscorpionida, in lateral view (palp and legs removed); A, Solifugae: *Pseudocleobis andinus* (Pocock), adult; B, Pseudoscorpionida: *Chthonius (Chthonius) tenuis* L. Koch, adult. Notations: *CH*, chelicera; *CX.P*, coxal region of palp; *LL*, lateral lip; *OC*, eye; *PO*, postpeltidium; *PR*, propeltidium; *PIV*, leg IV; *st.1*, stigma 1; *st.2*, stigma 2; *VII–XVII*, segments VII–XVII. A, × 11; B, × 73.

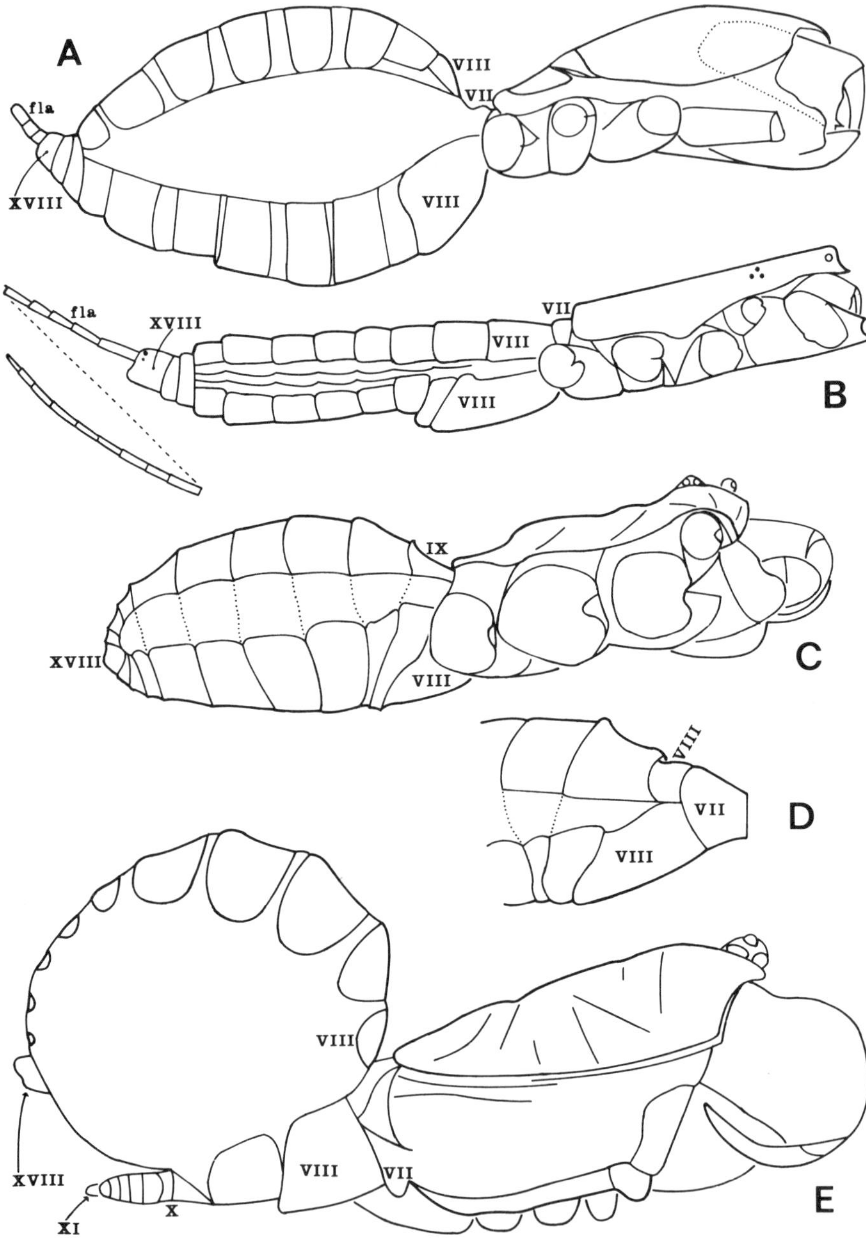
A
fla
XVIII
VIII
VII
VIII
B
fla
XVIII
VII
VIII
VIII
C
IX
XVIII
VIII
D
VIII
VII
VIII
E
VIII
XVIII
XI
X
VIII
VII

when prosoma and opisthosoma are coupled by the locking device. Parts of the anterior prosomatic segments constitute a movable gnathosoma (a pseudotagma) which is located in a camerostome. The segments of the opisthosoma have partly coalesced (four sets of sclerites are supposed to represent fusions of two segments). The posterior part of the opisthosoma is constituted by the pygidium or metasoma, which consists of three narrowed segments which can be withdrawn, each into the preceding segment, like a telescope, and then retracted into the mesosoma. The numbering of the segments depends on the interpretation of the dorsal sclerites immediately following the pedicel; they are supposed to represent a fusion of two segments, by which the total number of segments is nineteen.

The Uropygid body (fig. 3A, B; see Van der Hammen, 1986a: 8–10, fig. 2A, B) consists of prosoma and opisthosoma, distinctly separated because of the narrowing of segment VII. The opisthosoma consists of the segments VII–XVIII and a 'telson' (flagellum). The segments XVI–XVIII constitute a distinctly narrowed pygidium or metasoma. The flagellum is either short (Schizomida) or long (Holopeltida).

The Amblypygid body (fig. 3C, D; see Van der Hammen, 1986a: 10, fig. 2C, D) consists of prosoma and opisthosoma, connected by the narrowed segment VII (tergite VII is hidden by the dorsal part of the prosoma; sternite VII is represented by a triangular sclerite between coxae IV). The opisthosoma consists of the segments VII–XVIII. There is no narrowed metasoma and no 'telson'.

The Araneid body (fig. 3E; see Van der Hammen, 1986a: 10, fig. 2E) consists of prosoma and opisthosoma, connected by a narrow pedicel (segment VII). True segmentation of the opisthosoma is found in the spider family Liphistiidae only; in this group, the opisthosoma distinctly consists of the segments VII–XVIII. There is no narrowed metasoma and no 'telson'.

In Opilionida (fig. 4; see Van der Hammen, 1985: 6–9, figs. 1–4), the body consists of prosoma and opisthosoma; these two tagmata are broadly joined (the body is latigastric). A pseudotagma is constituted by the myliosoma which comprises the chelicerae, the palpi, the coxae of the anterior pairs of legs, and the interjacent mouthparts. The opisthosoma is, generally, more or less distinctly segmented; it consists of the segments VII–XVI. The posterior segments of the opisthosoma have been subject to regression; the tergites and/or sternites of the segments XIII and, particularly, XIV–XVI can have fused in

---

←

Fig. 3. Representatives of the three groups of Arachnida s.str., i.e., Uropygi, Amblypygi and Araneida, in lateral view (palp and legs, with the exception of the coxae, removed); A, Uropygi (Schizomida): *Trithyreus* spec., adult female; B, Uropygi (Holopeltida): *Tetrabalius seticauda* (Doleschall), adult male: C, D, Amblypygi: *Heterophrynus alces* Pocock, adult (D, anterior part of separated opisthosoma; orientation slightly different from C); E, Araneida: *Liphistius desultor* Schiödte, adult female (coxae of the right side also removed). Notations: *fla*, flagellum; *VII–XVIII*, segments VII–XVIII. A, × 32.7; B–E, × 3.8.

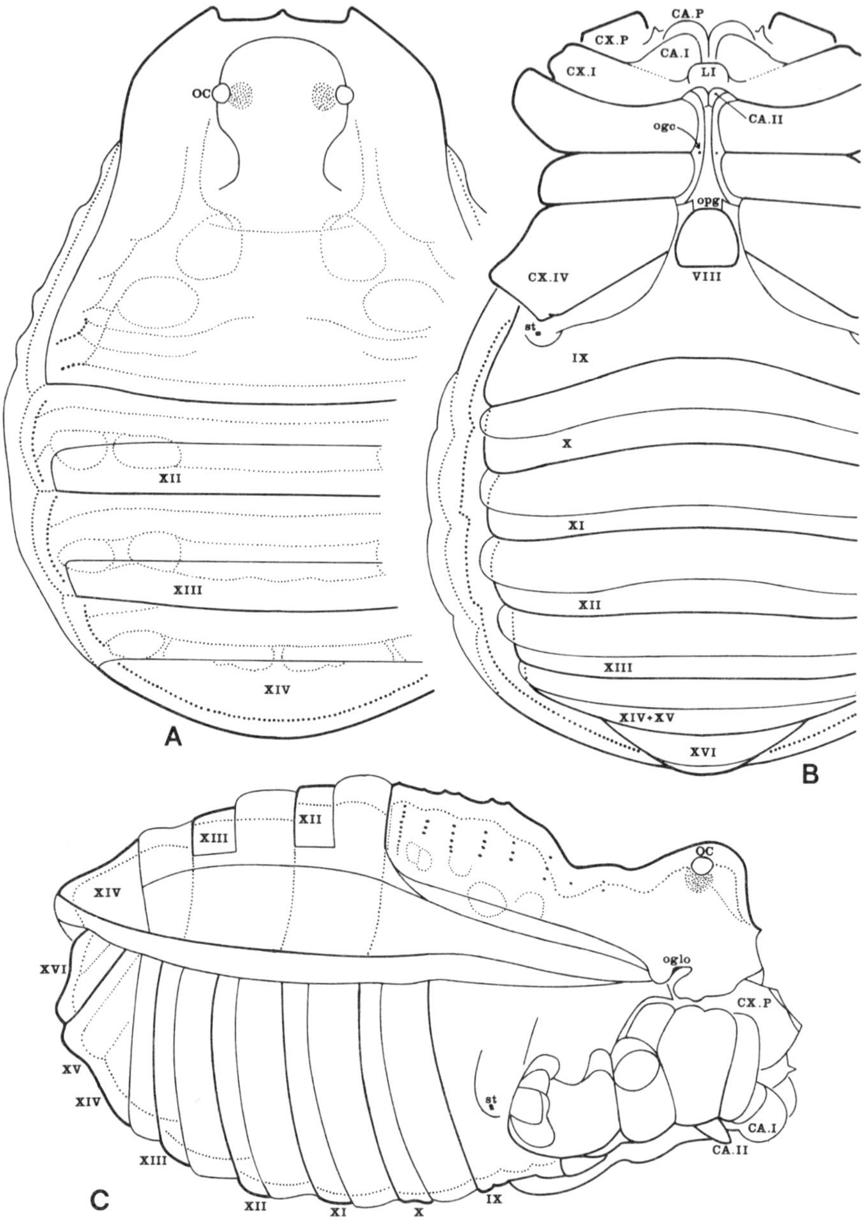

Fig. 4. Opilionida: *Scotolemon lespesi* Lucas; A, dorsal view; B, ventral view; C, lateral view. Notations: *CA.P*, coxapophysis of palp; *CA.I*, coxapophysis of leg I; *CA.II*, coxapophysis of leg II; *CX.P*, coxa of palp; *CX.I*, coxa of leg I; *CX.IV*, coxa of leg IV; *LI*, labium; *OC*, eye; *ogc*, orifice of coxal gland; *oglo*, orifice of odoriferous gland; *opg*, progenital opening; *st*, stigma; *VIII–XVI*, segments VIII–XVI. A–C, × 37.

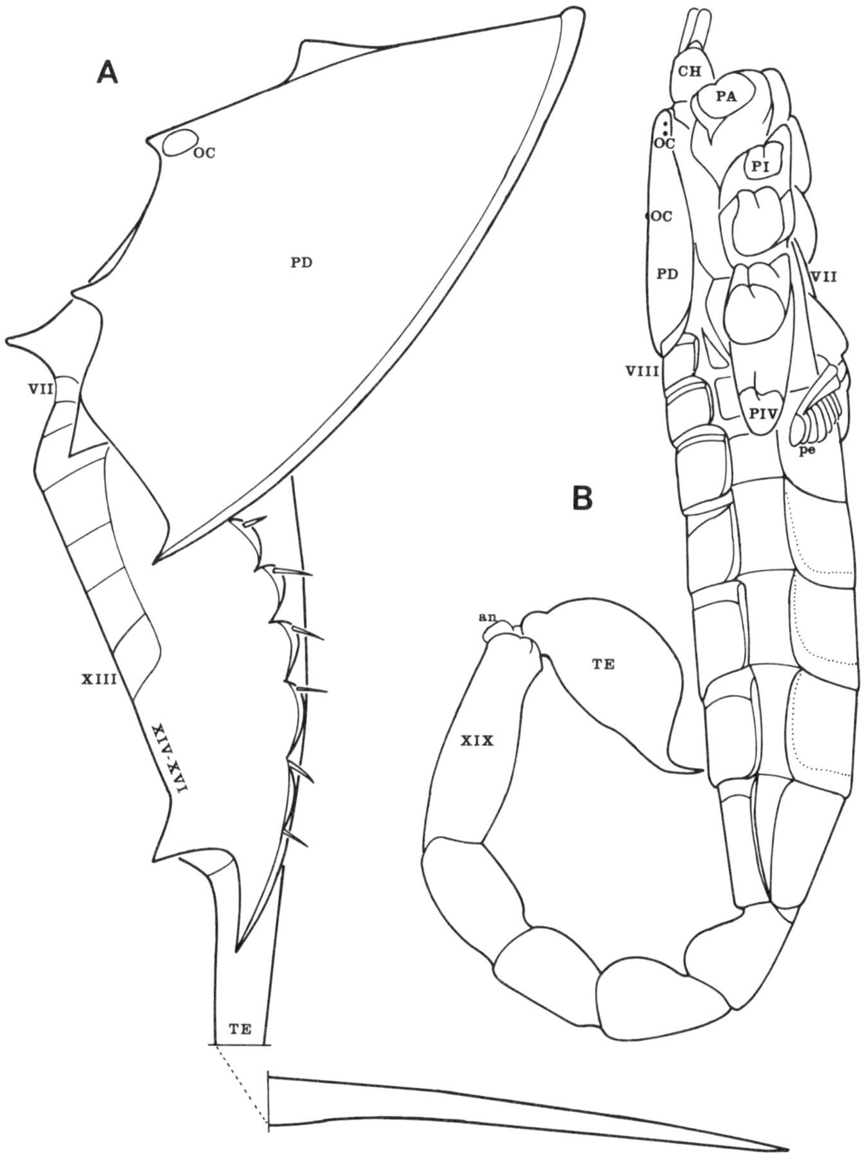

Fig. 5. Representatives of Xiphosura and Scorpionida in lateral view; A, Xiphosura: *Limulus polyphemus* (Linnaeus), juvenile specimen; B, Scorpionida: *Euscorpius carpathicus*(Linnaeus), adult male (palp and legs, with the exception of the coxae, removed). Notations: *an*, anal opening; *CH*, chelicera; *OC*, eye; *PA*, palp; *PD*, prodorsum; *pe*, comb (pecten); *PI*, leg I; *PIV*, leg IV; *TE*, 'telson'; VII– XIX, segments VII–XIX. A, × 3.4; B, × 8.4.

various degrees. The anal cover generally represents segment XVI or a fusion of this segment with tergite XV.

In Scorpionida (fig. 5B; see Van der Hammen, 1986a: 12, fig. 3B), the body is latigastric. The opisthosoma comprises the segments VII–XIX and a 'telson' (the sting); it is subdivided into two pseudotagmata: mesosoma (segments VII–XIV) and metasoma (the distinctly narrowed segments XV–XIX and the 'telson').

In Xiphosura (fig. 5A; see Van der Hammen, 1986a: 12–14, fig. 3A), prosoma and opisthosoma are broadly joined (the body is latigastric). The opisthosoma comprises the segments VII–XVI and a large 'telson' or tail. The segments VII–XIII are still recognizable because of the presence of furrows, muscle insertions, spines and opisthosomatic appendages. Because the posterior part of the opisthosoma of some extinct Xiphosura consists of three segments and a 'telson', the segments XIV–XVI are supposed to have been incorporated in the posterior part of the opisthosoma of extant Xiphosura, without leaving any trace of segmentation.

2. Prosoma

As mentioned above, the Chelicerate prosoma comprises the precheliceral metameres and the segments I–VI. Generally, several prosomatic regions can be distinguished: a dorsal region (or aspidosoma; its dorsal surface is constituted by the prodorsum), a ventral region (or podosoma), and the mouthparts (which can constitute a pseudotagma, viz., the gnathosoma or the myliosoma).

In many Chelicerata, the aspidosoma is covered by a single prodorsal shield; in other groups, there are two or more shields separated by one or more scissures (interruptions of the sclerotized cuticle, constituted by soft skin; see Van der Hammen, 1980: 139); in some primitive groups, the prodorsum is mainly constituted by soft skin. In Palpigradi (Van der Hammen, 1982a: 8, figs. 1A, C, 8A), the prodorsum consists of four pseudosclerites (hard parts of the cuticle without distinctly defined border; see Van der Hammen, 1980: 131) separated by soft skin, viz., a relatively large anterior propeltidium, a smaller posterior meta- or postpeltidium, and one pair of small lateral mesopeltidia. The scissure between pro- and postpeltidium can apparently be extended and contracted, whilst the small mesopeltidia apparently constitute intercalary sclerites. In Actinotrichida (Van der Hammen, 1969: 183, 196–197, fig. 1; 1972: 17–18, fig. 9A; 1982a: 29–31, figs. 10A, 11), the prodorsum can be scleritized (prodorsal shield) or not; it can be indistinctly subdivided by the presence of transverse, internal apodemes. The Solifugid prodorsum (Van der Hammen, 1986a: 5) generally consists of two main sclerites (propeltidium and postpeltidium) whilst smaller arthrodial sclerites can be present in the soft skin between them (they are supposed to be connected with the articulation of the prodorsum). The Pseudoscorpion prodorsum (Van der Hammen, 1986a: 7) is generally regarded as consisting of a single prodorsal sclerite. Comparison with the related Solifugae suggests that, at least in *Chthonius*, the supposed anterior opisthosomatic tergite could represent the posterior sclerite of the prodorsum (the postpeltidium); this hypothesis could easily be tested by an anatomical study of musculature. In Anactinotrichida (Van der Hammen, 1966: 10–12, figs. 1, 3; 1969: 196–197; 1972: 17–18, fig. 1B; 1979a: 15), the prodorsum can be scleritized or not; in many cases the prodorsal shield has fused with the opisthonotal shield in such a way that the posterior border of the prodorsum is indistinguishable (it is not identical with the so-called notomeristic scissure of Gamasida). In Opilioacarida, the faintly sclerotized prodorsum is subdivided by a short transverse furrow into a relatively large anterior part (prostethidium) and a much smaller posterior part (poststethidium). In Ricinulei (Van der Hammen, 1979a: 7, figs. 1A, C, 5A), the aspidosoma is entirely covered by a single prodorsal shield; the extended rostral part of the prodorsum constitutes a movable scleritized hood (cucullus) which can be raised and lowered, and of which the main function is associated with the concealment of the mouthparts and the protection of the camerostome. In Uropygi (Van der Hammen, 1986a: 8–12, figs. 1A, B, 17E), the aspidosoma is either covered by a single prodorsal shield (Holopeltida), or by an anterior propeltidium and a posterior postpeltidium, connected by two arthrodial sclerites (Schizomida). The Amblypygid aspidosoma (Van der Hammen, 1986a: 10, fig. 1C) and the Araneid aspidosoma (Van der Hammen, 1986a: 10, fig. 1E) are covered by a single, large prodorsal shield. In Opilionida (Van der Hammen, 1985: 9, figs. 1A, C, 2A, C, 3A, 4, 13A), most of the aspidosomatic tergites have fused, but tergites V and VI often constitute separate sclerites; the prodorsum can, however, also be integrated into a larger dorsal shield which includes opisthosomatic tergites. In Xiphosura (Van der Hammen, 1986a: 12, fig. 3A) and Scorpionida (Van der Hammen, 1986a: 12, fig. 3B), the aspidosoma is covered by a single prodorsal shield; this shield is relatively very large in Xiphosura.

Eyes are not present in Palpigradi. There is, however, a so-called frontal organ which resembles, on the one hand, the supracoxal setae of the palp (see below) and, on the other hand, the naso of Actinotrichida (also a frontal organ; it can sometimes present an inferior bilobate eye); the true nature of the organ is still unknown (Van der Hammen, 1982a: 8, 51, figs. 1A, C, 3A, 4A, C, 6E, 8A, C). In Actinotrichida (Van der Hammen, 1980: 62; 1982a: 29–31, 51–53) up to two pairs of lateral eyes can be present and (rarely) an unpaired anterior median eye; many Actinotrichid mites (Acaridida, Tarsonemida, most Oribatid mites, several groups of Actinedida) have no eyes. A naso, or frontal protuberance (Van der Hammen, 1980: 102–103), an unpaired organ which is found at (or under) the rostral extremity of the aspidosoma, is present in many Actinedida and some Oribatida; and eye (often bilobate) can be present at its ventral surface. In Solifugae (Millot & Vachon, 1949a: 500), one pair of median eyes and one or two pairs of (regressive) lateral eyes are present. In Pseudoscorpionida (Vachon, 1949: 453–454), there are no median eyes, but there can be one or two pairs of lateral eyes (although these can also be absent). In several Anactinotrichida, lateral eyes are present (Van der Hammen, 1966: 12, figs. 1, 3, 4A; 1968b: 60, fig. 1A, C; 1971: 467, fig. 2A; 1979a: 15–16, 22; 1983b: 8, figs. 1, 2A, 6B). Opilioacarida are characterized by the presence of two or three pairs of lateral eyes; in Holothyrida, one pair of lateral eyes is known from a species of *Australothyrus*; in Ixodida, up to two pairs of lateral eyes can be present; most Anactinotrichida, however, do not possess eyes. In some Ricinulei, light spots are present at the place of lateral eyes, and these are considered vestiges of two pairs of eyes (Van der Hammen, 1979a: 7, 22); Selden (1983) discovered true lateral eyes, at similar places, in fossil Ricinulei. In Uropygi (Millot, 1949b: 545), one pair of median eyes and three pairs of lateral eyes can be present (Holopeltida); in Schizomida, however, there are no median eyes, and up to one pair of more or less vestigial eyes. Amblypygi (Millot, 1949c: 571) typically possess one pair of median eyes and three pairs of lateral eyes; in some species, eyes have been subject to regression. In Araneida (Millot, 1949d: 595–597) there are typically two pairs of median eyes and two pairs of lateral eyes; reduced numbers of eyes (0–3 pairs) are, however, also found in this group. Opilionida (Van der Hammen, 1985: 9) typically present only one pair of median eyes; this pair can, however, be absent. Scorpionida (Millot & Vachon, 1949: 403) present one pair of median eyes and 2–5 pairs of lateral eyes. In Xiphosura (Fage, 1949: 229–231) there are one pair of relatively large lateral eyes, associated with a pair of vestigial eyes, and a pair of small median eyes; in this group, there is also a frontal organ, which ventral position, with vestiges of one pair of eyes (functional in the larva).

In several groups, the anterior or the anteroventral part of the prosoma constitutes a pseudotagma associated with ingestion. This is the case with the gnathosoma of Actinotrichida, Anactinotrichida and Ricinulei, and with the myliosoma of Opilionida, Xiphosura and Scorpionida. These pseudotagmata are dealt with below, in the section on the mouthparts.

In Palpigradi and Actinotrichida, so-called supracoxal or laterocoxal setae are found in the anterolateral part of the prosoma, above the bases of one or more pairs of anterior appendages (Van der Hammen, 1980: 153–154; 1982a: 13, 34, 53, figs. 3A, 4A, 6E, 15A, C, D, E; see also Van der Hammen, 1977b). Laterocoxal setae are also mentioned (although the homology is uncertain) for Ricinulei (Van der Hammen, 1979a: 9, fig. 6D, E; this concerns the laterocoxal setae of the palp, which secondarily occupy a ventral position on the gnathosoma) and for Anactinotrichida (Van der Hammen, 1964a: 12, 16, 47, 51, figs. 5–7, 9C, 10D, 11B, 15G; 1966: 38–40, 57–58, figs. 2, 3, 6D; this concerns the laterocoxal setae of legs I and II in Opilioacarida, and the laterocoxal setae of palp and leg I in some Gamasida). Supra- or laterocoxal setae are supposed to represent sensory organs with unknown function (in Palpigradi and Actinotrichida, they could have taken the place of lateral eyes). Coxal glands and their orifices are found in the prosoma; they are dealt with below, in a separate section.

The ventral and ventrolateral part of the prosoma, which includes the bases of the legs, is sometimes called podosoma; it can be separated from the aspidosoma (the prodorsum) by the abjugal furrow (see Van der Hammen, 1980: 7). The podosomatic region can include the epimera (when coxae are not present), the (neo)sternum (when separate coxae are present), the coxae of the legs, the sternapophyses (or coxapophyses), and the pleural regions. The coxae of the anterior legs can form part of a myliosoma (associated with ingestion). The evolution of coxae and (neo)sternum is dealt with below, in the section on the legs; coxae forming part of a myliosoma are dealt with in the section on the mouthparts. Sternapophyses (see Van der Hammen, 1980: 149–150; 1986a: 18, figs. 7A, B, E, 10B) are regarded as 'endites' of the legs (they are known from Anactinotrichida, Ricinulei, Uropygi and Amblypygi; in these groups they can be associated with the transport of the coxal gland secretions, see below); when they are attached to the coxae of the legs, they are called coxapophyses (Opilionida, Scorpionida, Xiphosura, in which groups they participate in ingestion; see below, and Van der Hammen, 1985: 22–30, figs. 13–19; 1986a: 26, figs. 13–14).

3. Opisthosoma

As mentioned above, the Chelicerate opisthosoma (it can consist of a mesosoma and a narrower metasoma or pygidium) comprises, ancestrally, from ten (VII–XVI) up to thirteen (VII–XIX) segments and sometimes also a so-called 'telson'. Segment VII (and sometimes also segment VIII) can constitute a narrow pedicel. The genital opening is, ancestrally, at the posterior border of segment VIII. The anal opening is more or less terminal, or has secondarily developed a position which is more ventral; when a 'telson' is present, its base is dorsal to the anal opening. Genital region and opisthosomatic appendages are dealt with below, in separate sections.

4. Prosomatic appendages

The Chelicerate appendages and their evolution have repeatedly been the subject of my investigation (see Van der Hammen, 1977a: 308–312, fig. 1; 1977b; 1985: 50–51; 1985a: 139–141, figs. 2–3; 1985b: 400–403, fig. 4; 1986: 26–31, figs. 7–8; 1986c: 217–219, fig. 1), and not until recently (Van der Hammen, 1986a: 24–31, figs. 5, 7–8) have I succeeded in solving the last important problem (that of the homology of epimera and coxae). It has now become possible to give a more or less complete (and partly hypothetical) survey of the evolution of the Chelicerate appendages; in the present section my previous views will be summarized, corrected and completed.

Chelicerata (which the exception of Xiphosura) differ from other Arthropods by the fact that the main promotor-remotor movement is at the coxa-

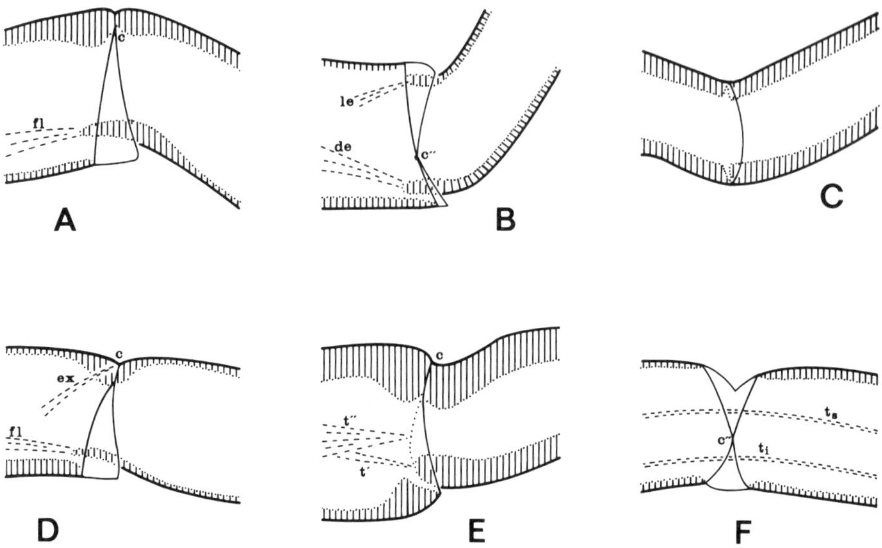

Fig. 6. Types of joints present in Chelicerate legs. A, Hinge joint with flexor muscles (*fl*) and one articulation point (*c*). B, Pivot joint with levator (*le*) and depressor (*de*) muscles, and two articulations points (*c'*, *c''*; *c'* not visible in the drawing). C, Adesmatic joint without traversing tendons. D, Hinge joint with flexor (*fl*) and extensor (*ex*) muscles, and one articulation point (*c*). E, Rocking joint with two sets of tendons and muscles (*t'*, *t''*) and one articulation point (*c*). F, Adesmatic joint with two articulation points (*c'*, *c''*; *c'* not visible in the drawing), levator and depressor movements being operated by traversing tendons (t_s, t_i) not inserted on the adesmatic segments constituting the joint.

trochanter or body-trochanter joint, and not at the body-coxa joint (it is supposed here that the movability of the Xiphosurid coxae first arose in connection with ingestion, as in other Myliosomata, and that participation of the body-coxa joint in locomotion is secondary). It has now become evident that coxae have evolved, relatively late, from epimera (the sternites of the prosoma) by a gradual separation of these from the sternal exoskeleton, a gradual increase in movability, and a subsequent extension in antiaxial direction (associated with the inclusion of pleural regions). A sternum (it is, in fact, a neosternum) evolved simultaneously with the free coxae, and became broader when the position of the coxae became more lateral. Epimera are present in Palpigradi, Actinotrichida, Solifugae and Pseudoscorpionida (in these groups the ancestral condition is perfected by a further development of the ventral prosomatic apodemes). The coxae of Arachnida s.str. (Uropygi, Amblypygi, Araneida) demonstrate the gradual transition from coxae with ventral position and slight movability to coxae with lateral position and increased movability.

Chelicerate appendages consist of a series of separate segments (or podomeres) connected by various types of joints. As in other Arthropods, these segmented appendages probably arose, by sclerotization of the cuticle and the origination of joints, from lobopodia (i.e. muscular-walled limbs, the rigidity of which was controlled by hydrostatic pressure working against the tension of muscles). The appendages present several types of joints: adesmatic joints (i.e. joints without tendons inserted on the base of the distal segment) and various eudesmatic joints (i.e. joints with one or more tendons inserted on the base of the distal segment). Generally, two basic types of eudesmatic joints can be distinguished, and two derived types (fig. 6). The two basic types of eudesmatic joints are hinge joint (with one articulation point and one set of tendons and muscles) and pivot joint (with two articulation points and two sets of antagonistic tendons and muscles). When, in a pivot joint, one of the articulation points disappears, a rocking joint arises. In some cases, extensor muscles can arise in a hinge joint; this is particularly found in the cases of palps. Pivot joints can be associated with levator-depressor, or with promotor-remotor movements. Hinge joints are associated with flexure of the legs.

The problem of the homologization of leg segments is now definitely solved: the main promotor-remotor movement is (with the exception of Xiphosura) at the base of the trochanter, a levator-depressor movement is at the pivot joint between trochanter and femur (in case this joint is repeated, there is a second trochanter), a knee with a hinge joint is present between femur and patella (genu) or tibia (in the case of a second femur, an extra hinge joint is present between the trochanter-femur and the knee-joint); patella, tibia and tarsus can easily be recognized because of the number of true (eudesmatic) articulations in the descending part of the leg, distal to the knee-joint (the articulation between basi- and telotarsus, if present, is adesmatic). This original pattern can be transformed by secondary transformations (the origination of rocking joints from pivot joints, the origination of extensor muscles in hinge joints, etc.), but generally remains recognizable. The following model of the evolution of the Chelicerate appendages has now been prepared by me (Van der Hammen, 1985: 50–51; 1985b: 402–403, fig. 4; 1986: 24–26, fig. 5).

Because the six pairs of Chelicerate appendages (chelicera, palp and legs) are homonomous structures which have developed by parallel evolution from lobopodia (fig. 7A), the ancestral number of the segments of the chelicera (three) could be the ancestral number for all the appendages (archepodia; see fig. 7B, C); these three segments are usually called trochanter, body (of the chelicera), and apotele (or claw-segment). This number of segments could have occurred in short-legged species which stood up on their legs. In Chelicerate embryos (Vachon, 1945), palp and leg first divide into four segments: prototrochanter, protofemur, prototibia and prototarsus. A leg with these four segments and an apotele (i.e. five segments: a so-called protopodium; see fig. 7D, E) could represent the first stage in the evolution of a hanging stance (Chelicerata generally hang down from their legs). All types of Chelicerate palp and legs can be derived from this primitive type. As mentioned above, free coxae arose from epimera; they can include pleural regions. Protopodia arose from archepodia, either by a further subdivision of segments, or by the addition of

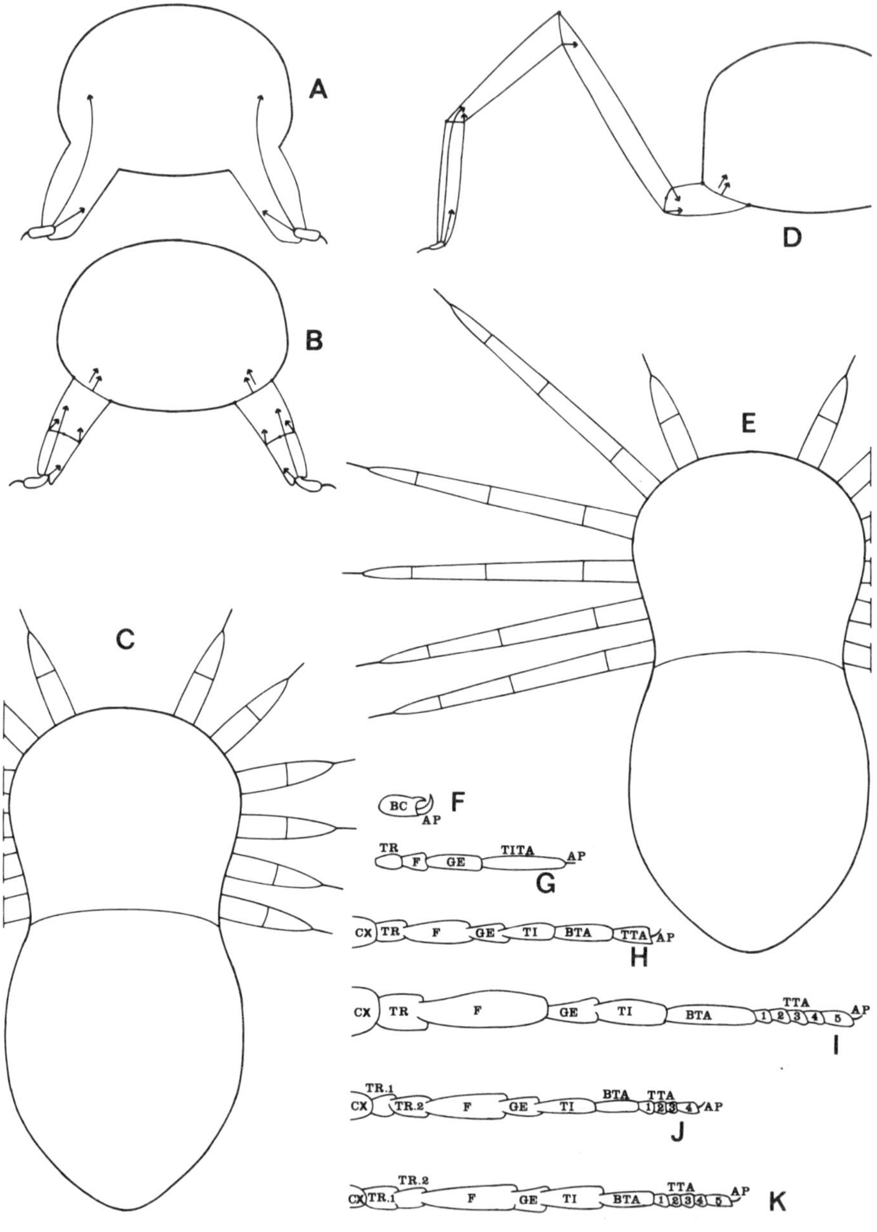

Fig. 7. Stages in the evolution of the Chelicerate appendages: A, B, single body-segment, with pair of limbs, of hypothetical Arthropod in transverse section; A, with lobopodia; B, with archepodia; C, dorsal view of hypothetical Arthropod with archepodia: D, single body segment with left limb, in transverse section, of hypothetical Chelicerate with protopodia; E, dorsal view of hypothetical Chelicerate with three-segmented limb 1 and five-segmented protopodia (limbs 2–6); F–K, ap-

→

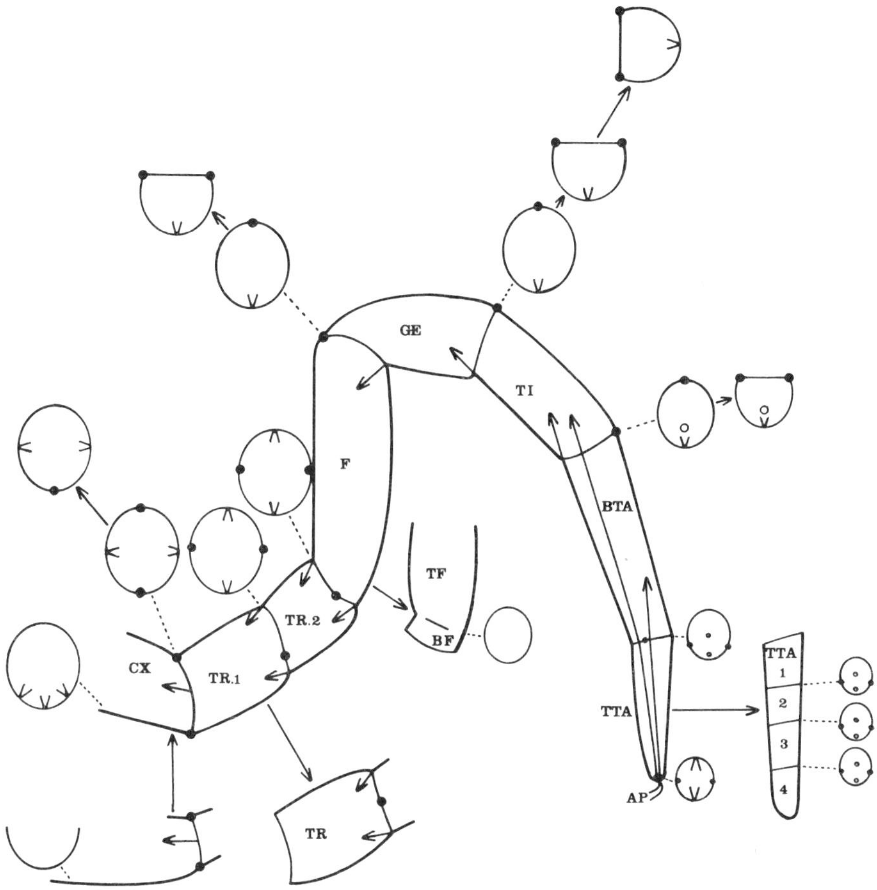

Fig. 8. Model of the evolution of leg III in Cryptognomae (Anactinotrichida and Ricinulei). The ancestral leg III is supposed to be composed of: fixed coxa, trochanter 1, trochanter 2. femur, genu (patella), tibia, basitarsus, telotarsus and apotele. In the figure of the entire leg, articulation points are represented by dots, tendons by arrows. The model demonstrates the evolution of a free coxa, the suppression of trochanter 2, the evolution of a basifemur, and the evolution of four telotarsi. The model demonstrates also the evolution of the separate joints. The schematic sections of the ancestral joints are connected, by dotted lines, with the joints in question (in the figure of the entire leg). The direction of the evolution is indicated by an arrow. In the schematic sections, an articulation point is represented by a dot, a tendon by a v-shaped symbol. The sections demonstrate the evolution of a rocking joint between coxa and trochanter, the evolution of hinge joints with flattened dorsal surface and two main articulation points, and the evolution of torsion in the genu-tibia articulation. Notations: *AP*, apotele; *BF*, basifemur; *BTA*, basitarsus; *CX*, coxa; *F*, femur; *GE*, genu (patella); *TF*, telofemur; *TI*, tibia; *TR*, trochanter; *TR.1*, trochanter 1; *TR.2*, trochanter 2; *TTA*, telotarsus; *TTA.1–4*, telotarsus 1–4.

---

pendages of the adult of a group of extant Chelicerata (Ricinulei) in dorsal view; F, chelicera; G, palp; H, leg I; I, leg II; J, leg III; K, leg IV. Notations: *AP*, apotele; *BC*, body of chelicera; *BTA*, basitarsus; *CX*, coxa; *F*, femur; *GE*, genu; *TI*, tibia; *TITA*, tibiotarsus; *TR*, trochanter; *TR.1*, trochanter 1; *TR.2*, trochanter 2; *TTA*, telotarsus; *TTA.1–5*, telotarsus 1–5.

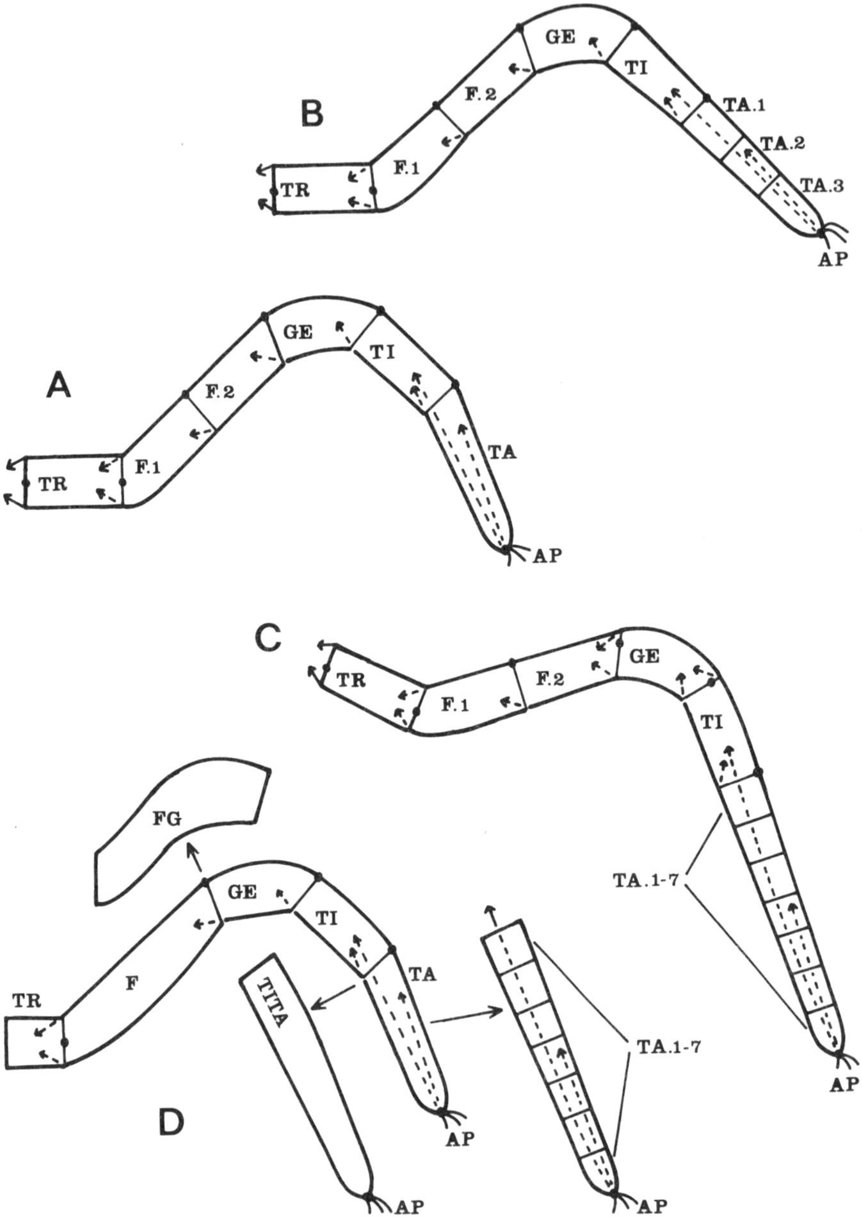
B
TR
F.1
F.2
GE
TI
TA.1
TA.2
TA.3
AP
A
TR
F.1
F.2
GE
TI
TA
AP
C
TR
F.1
F.2
GE
TI
TA.1-7
AP
FG
TR
F
GE
TI
TA
TITA
D
AP
AP
TA.1-7
AP

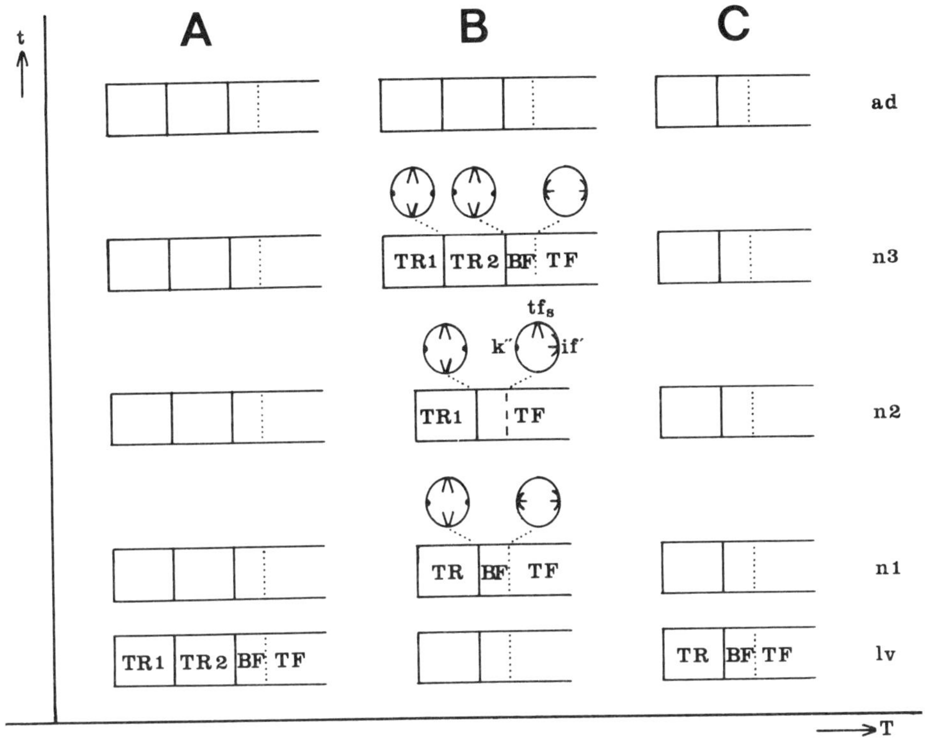

Fig. 10. Ontophylogenetic diagram (not drawn to scale) of the regressive evolution of trochanter 2 in Anactinotrichida. Trochanter 2 disappears by ascendant suppression. In Opilioacarid deutonymphs, the regressive trochanter 2 manifests itself partly (the 'basifemoral ring' presents some characters of the trochanter 2-femur articulation, viz., the superior tendon and the posterior condyle; and some characters of a true basifemoral ring, viz., an anterior lyrifissure and the absence of the inferior tendon). A, ancestral Anactinotrichid stage (two trochanters in all instars are still found in other Cryptognomae, viz., Ricinulei, in which group a basifemoral ring is, however, not present); B, Opilioacarid stage; C, Holothyrid, Gamasid and Ixodid stage. Notations: *ad*, adult; *BF*, basifemur; *if′*, anterior lyrifissure of basifemoral ring; *k″*, posterior condyle of femur; *lv*, larva; *n1*, protonymph; *n2*, deutonymph; *n3*, tritonymph; *t*, ontogenetic time; *T*, phylogenetic time; *TF*, telofemur; tf_s, superior tendon of femur; *TR*, trochanter; *TR1*, trochanter 1; *TR2*, trochanter 2.

←

Fig. 9. Schematic representation of segmentation and articulation of the legs in Epimerata (Palpigradi and Actinotrichida). A, Fundamental type of Epimerate leg (with two femora). B, Palpigradid leg IV (with tarsi 1–3). C, Palpigradid leg I (with tarsi 1–7, and extensor muscles at the femur 2-genu and the genu-tibia articulations). D, Diagram of the evolution of the legs in Actinotrichida, starting from species with integrated femora (the diagram represents the development, by fusion, of femorogenu and tibiotarsus; and the evolution, by subdivision and multiplication, of tarsi 1–7). Notations: *AP*, apotele; *F*, femur; *F.1*, femur 1; *F.2*, femur 2; *FG*, femorogenu; *GE*, genu; *TA*, tarsus; *TA.1–7*, tarsus 1–7; *TI*, tibia; *TITA*, tibiotarsus; *TR*, trochanter.

projecting parts of the body (see Van der Hammen, 1985: 50).

Starting from five-segmented protopodia (similar in all Chelicerate ancestors), evolution has become divergent, by which a number of different fundamental leg types can be distinguished (figs. 8, 9). A coxa evolved in Anactinotrichida, Ricinulei, Uropygi, Amblypygi, Araneida, Opilionida, Scorpionida and Xiphosura. A second trochanter (in legs III and IV) evolved in Solifugae, Anactinotrichida and Ricinulei. A second femur evolved in Palpigradi, Actinotrichida, Solifugae and Pseudoscorpionida. A patella evolved in all groups of Chelicerata with the exception of Solifugae and Pseudoscorpionida. A subdivision (by an adesmatic articulation) of the tarsus, into basi- and telotarsus, arose in all groups of Chelicerata with the exception of Palpigradi, Actinotrichida and Xiphosura.

Because trochanter 2, in Anactinotrichida (see Van der Hammen, 1981: fig. 6; 1985a: fig. 5; 1985b: fig. 5; 1986: fig. 8; see also fig. 10) disappears by suppression, it must originally have evolved by repeated manifestation of the same genotypic information. Because femur 2, in Epimerata (Palpigradi and Actinotrichida), disappears by fusion of femur 1 and femur 2 (see Van der Hammen, 1982a: 38; fusion takes place by the regression and disappearance of the muscles, followed by the gradual disappearance of the articulation), it must originally have evolved by subdivision of the femur. Patella (genu) and tibia originally evolved by subdivision of the prototibia; it is interesting that this subdivision has manifested itself in two different ways: the patella (genu)-tibia articulation is, ancestrally, either a hinge joint or a pivot joint.

A general survey of the fundamental leg-types in the main groups of Chelicerata is given in fig. 11. This figure demonstrates that, in extant Chelicerata, the following eight types can be distinguished (subsequent changes in these types are not considered; cf. figs. 8, 9).

The first type, found in Palpigradi and Actinotrichid mites, is characterized by the absence of a coxa, the presence of two femora, and the absence of a separate basitarsus.

The second type, found in Anactinotrichid mites and Ricinulei, is characterized by the presence of two trochanters in legs III and IV.

The third type, found in Opilionida, is characterized by the presence of rocking joints between trochanter and femur, and between patella and tibia.

The fourth type, found in Solifugae, is characterized by the presence of two trochanters (in legs III–IV) and two femora, and the absence of coxa and patella.

The fifth type, found in Pseudoscorpionida, is characterized by the presence of two femora and the absence of coxa and patella.

The sixth type, found in Uropygi, Amblypygi and Araneida, is characterized by the presence of specialized rocking joints between coxa and trochanter, and between patella and tibia.

The seventh type, found in Xiphosura, is characterized by the presence of an articulation between body and coxa, which (in contradistinction to all other

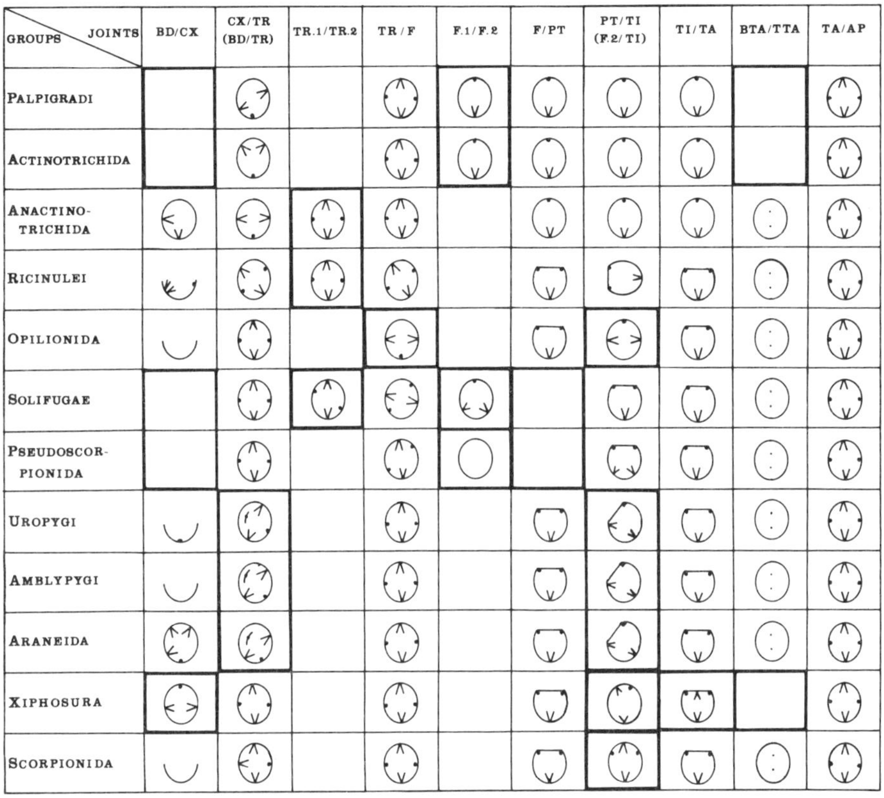

Fig. 11. Schematic representations of the segments and joints in the Chelicerate legs. Each joint is viewed distally, with the anterior face on the right. Fixed (or nearly fixed) coxae are represented by their ventral outline. Dots represent articulation points, v-shaped symbols represent muscle- and tendon-insertions. Compartments of the scheme presenting important characters are shown by a heavy black border. The continuation of the inferior tendon t_i of the apotele, in the tibia, is not drawn. The coxa-trochanter joint of Uropygi, Amblypygi and Araneida presents a sclerite (with two 'hinge-points') which is represented by a short line with a dot. The patella-tibia articulation of Uropygi and Araneida can be much more complicated than represented in this scheme (additional condyle, additional traversing muscle); the joint between patella and tibia in Amblypygi permits very little mobility. The scheme is based on right leg IV, and represents the fundamental leg type of each group. Notations: *AP*, apotele; *BD*, body; *BTA*, basitarsus; *CX*, coxa; *F*, femur; *F.1*, femur 1; *F.2*, femur 2; *PT*, patella (genu); *TA*, tarsus; *TI*, tibia; *TR*, trochanter; *TR.1*, trochanter 1; *TR.2*, trochanter 2; *TTA*, telotarsus.

extant Chelicerata) is associated with the promotor-remotor swing of the leg; in this type there is no separate basitarsus.

The eighth type, found in Scorpionida, is particularly characterized by the presence of a pivot joint between patella and tibia.

Each of these fundamental types must originally have been adapted to one particular type of locomotion. In the case of the first type, found in Palpigradi and Actinotrichid mites, ancestral locomotion could have been associated with life in the interstitial environment and in the deep fissures of the soil, where primitive representatives of these groups are now found (see Van der Hammen, 1985a: 139, 141); a functional analysis of this leg type is, however, still wanting. In the case of the second type and the fourth type, with two trochanters in legs III and IV, ancestral locomotion could have been walking as well as swimming; legs III and IV could have been used, as in the fossil *Baltoeurypterus* (cf. Selden, 1981), for swimming.

5. Mouthparts and ingestion

The chelicerate mouthparts include the mouth and the lips (labrum, lateral lips and labium), the chelicerae, the palpi and their basal parts, and, in some groups, also the basal parts of the anterior or even all pairs of legs. The position of the mouth can be rostral or more or less ventral (associated with the occurrence of a curved pharynx or oesophagus); in this connection, two ways of feeding can be distinguished: rostral feeding and coxisternal feeding (Van der Hammen, 1985: 24). Because coxisternal feeding is associated with a secondary translocation of structural elements, it is regarded as a derived condition.

The marine ancestors of extant Chelicerata were certainly characterized by internal digestion (the aquatic life does not permit of external digestion, unless special provisions have been developed, as in the water spider *Argyroneta*), and by the possession of organs for seizing food and mastication (see Van der Hammen, 1977a: 312; 1985a: 142). Terrestrial habits evidently arose from earlier marine habits.

Ingestion of solid food, and internal digestion, are still found in Xiphosura, in many Actinotrichid mites, in some Anactinotrichid mites (e.g. Opilioacarida) and in Opilionida; this is supposed to represent an ancestral condition. As far as is known, external digestion is found in all other groups of Chelicerata.

The original position of the mouth must have been precheliceral (evidence from embryology). Because of a secondary translocation of structural elements, the mouth is now, generally, situated between the bases of the palpi, whilst the bases of the chelicerae are more dorsal and often more or less postoral. The mouthparts can include a preoral cavity, i.e., a space rostral to the mouth and surrounded by the lips and the bases of some of the appendages.

The shape of the mouth depends upon the number of lips and the number of oral commissures or commissural lines (the places of junction of the lips). There can be two, three or four lips (a labrum, a pair of lateral lips and a labium) and two, three or four commissures; in consequence of this, the mouth can be crescent-shaped, or three- or four-pointed (fig. 12A–C). In Opilionida, where the coxapophyses are also associated with the mouth, the shape of the mouth is six-partite. Generally, the base of the labrum is constituted by the cervix.

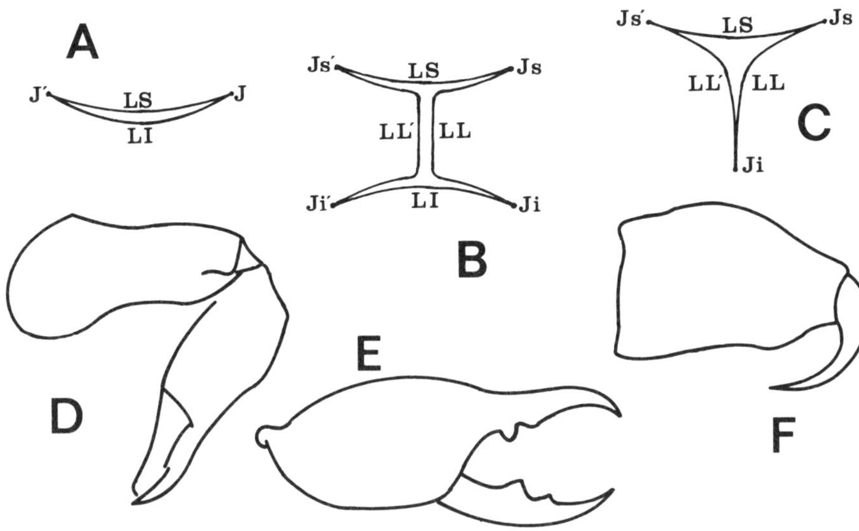

Fig. 12. Schematic representations with references to Chelicerate mouthparts; A–C, types of mouths (frontal view); A, crescent-shaped mouth with two oral commissures (*J, J'*); B, four-pointed mouth with four oral commissures (*Js, Js', Ji, Ji'*); C, three-pointed mouth with three oral commissures (*Js, Js', Ji*); D–F, types of chelicerae; D, prehensile type; E, 'raptorial' type; F, pocket-knife type. Notations: *LI*, labium; *LL, LL'*, lateral lips; *LS*, labrum.

In some groups, such as Palpigradi, Opilioacarida (Anactinotrichid mites) and Opilionida, the chelicerae are three-segmented, have a large trochanter, and a prehensile (and sometimes also masticatory) function (fig. 12D). This prehensile chelicera probably represents the ancestral type. In some other groups, such as Araneida and Amblypygi, the chelicera has the shape of a pocket-knife, lacks a separate trochanter and a fixed jaw, and is particularly adapted to holding and kneading (fig. 12F). In the remaining groups of Chelicerata, the chelicerae belong to intermediate types: they are either three- or two-segmented, generally present two jaws, and are more or less adapted to holding and kneading (in Solifugae, the two-segmented 'raptorial' chelicerae are, however, prehensile; see fig. 12E).

The original function of the palp probably was ambulatory, as still found in Xiphosura and Palpigradi; its ancestral plan of construction must have been similar to that of the walking legs. In many groups of mites, the palp has become an appendage with sensory function, a 'feeler'. In the course of the palpal evolution of the appendages of segment II, extensor muscles can have arisen in the ancestral hinge joints. The segmentation of the palp is more or less similar to that of the legs. It is interesting that, not only in Palpigradi and Actinotrichid mites, but also in Anactinotrichid mites and Ricinulei, the palp has no coxa. In Solifugae and Pseudoscorpionida, the palp has of course no patella;

in Pseudoscorpionida, Uropygi, Amblypygi and Scorpionida, it presents a patellotibia; in Ricinulei, it presents a tibiotarsus. In most groups (with the exception of Palpigradi, Solifugae and Amblypygi), the tarsus presents no subdivisions. In Actinotrichida, Ixodida (Anactinotrichida) and several Opilionida, the palpal apotele has disappeared. In many groups of Chelicerata, the palp has developed a more or less pronounced prehensile function; in several groups chelae are present, and in this case the palp can have taken over the original function of the chelicera. In some groups of Chelicerata, the palpal evolution has also included leg I.

Coxal apophyses of the appendages can also participate in feeding, such as those of appendages 2 in many Araneida (it is possible that these do not represent true coxapophyses); and those of appendages 2, 3 and 4 in Opilionida, appendages 3 and 4 in Scorpionida, and appendages 2–6 in Xiphosura (coxapophyses).

The most simple Chelicerate mouthparts are found in Palpigradi (Van der Hammen, 1969a; 1982a: 13–15, 53, figs. 4, 28A, D; 1986b: 81–82). They comprise the chelicerae, the labrum, the labium, and a crescent-shaped mouth with two oral commissures (situated between the bases of the chelicerae; see fig. 13A). The bases of the two lips have fused; the base of the labrum could represent the cervix, but it is not distinctly differentiated. It may be remarked here that the epimera of the segments of palp and leg I (which do not participate in ingestion) have secondarily fused, as in *Epilohmannia* (an Actinotrichid mite; see Van der Hammen, 1986b); in Palpigradi, this fusion is probably connected with an association of functions of these appendages. The Palpigradid mouthparts could be supposed to represent a regressive gnathosoma (see below), but there is no evidence to support this hypothesis. A capitular apodeme is not mentioned for Palpigradi, but could be present.

The mouthparts of Arachnida s.str. (Uropygi, Amblypygi, Araneida; see fig. 13C) are relatively simple (Van der Hammen, 1986a: 23–26, figs. 10–12). They comprise labrum and cervix, labium, mouth, and the 'coxapophyses' of the palpi (absent in some primitive Araneida; as mentioned above, it is possible that the 'coxapophyses' of Arachnida s.str. are not homologous with the coxapophyses and sternapophyses of other Chelicerata). In many of the higher Araneida, labrum and labium (and the enclosed preoral cavity) are curved downwards, although the mouth has preserved its original position; this condition does not pertain to coxisternal feeding. In Uropygi and Amblypygi, the mouthparts are associated with the transport of coxal gland secretions (see below). The basal apodemes of the Arachnid mouthparts (the paired palpal apodemes and the unpaired median cervical apodeme) are associated with the attachment of intrinsic muscles of the mouthparts.

The mouthparts of Solifugae and Pseudoscorpionida (Van der Hammen, 1986a: 21–23, figs. 8–9) are constituted by the chelicerae and a so-called rostrosoma (consisting of labrum, cervix, rostrosomatic apodeme, lateral lips and labium; see fig. 13B); the base of the rostrosoma has fused with the palpal coxae.

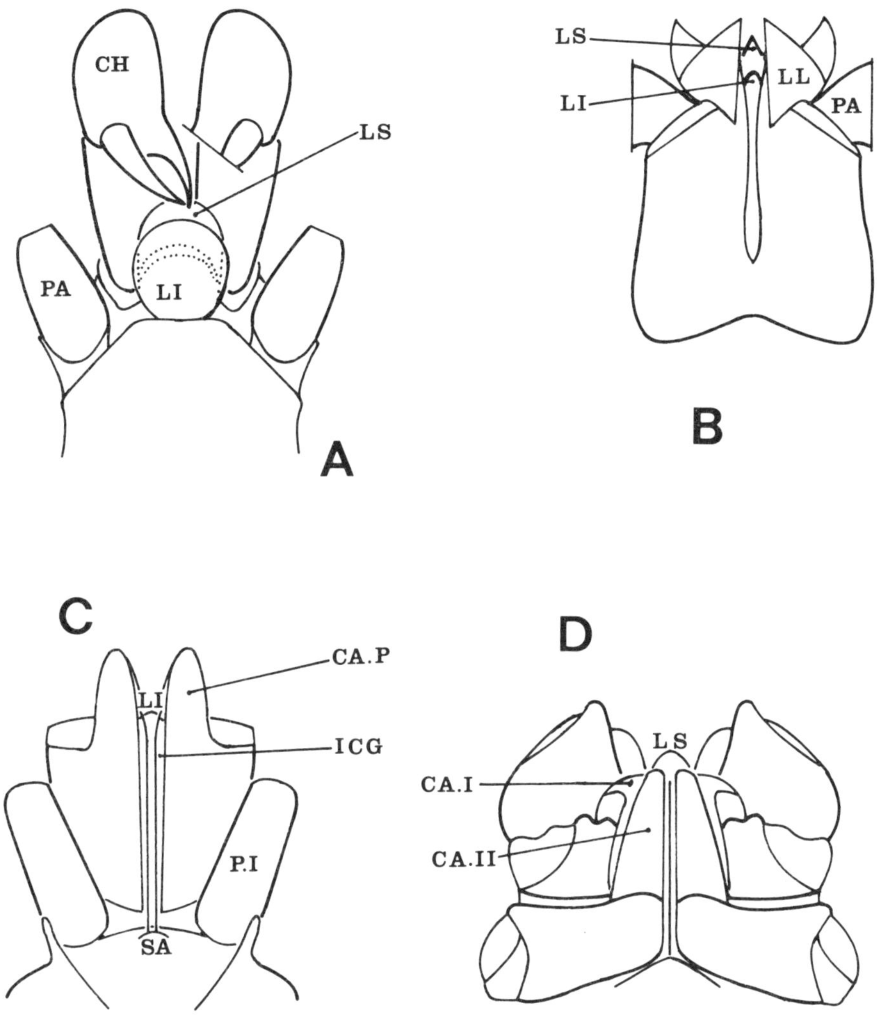

Fig. 13. Schematic representations of Chelicerate mouthparts in ventral view. A, Palpigradi: simple mouthparts with labrum and labium. B, Apatellata (example: Pseudoscorpionida): rostrosoma with labrum, labium and lateral lips. C, Arachnida s.str. (example: Schizomida): mouthparts with labrum, labium, palpal coxapophyses and intercoxal gutter, associated with sternapophysis and orifices of coxal glands. D, Myliosomata (example: Scorpionida): mouthparts which include coxapophyses of the legs. Notations: *CA.P*, coxapophysis of palp; *CA.I*, coxapophysis of leg I; *CA.II*, coxapophysis of leg II; *CH*, chelicera; *ICG*, intercoxal gutter; *LI*, labium; *LL*, lateral lip; *LS*, labrum; *PA*, palp; *P.I*, leg I; *SA*, sternapophysis.

The pharynx is included in the rostrosoma. The rostrosomatic apodeme is constituted by the interior prolongation of the cervix; it is the place of insertion of extrinsic rostrosomatic muscles.

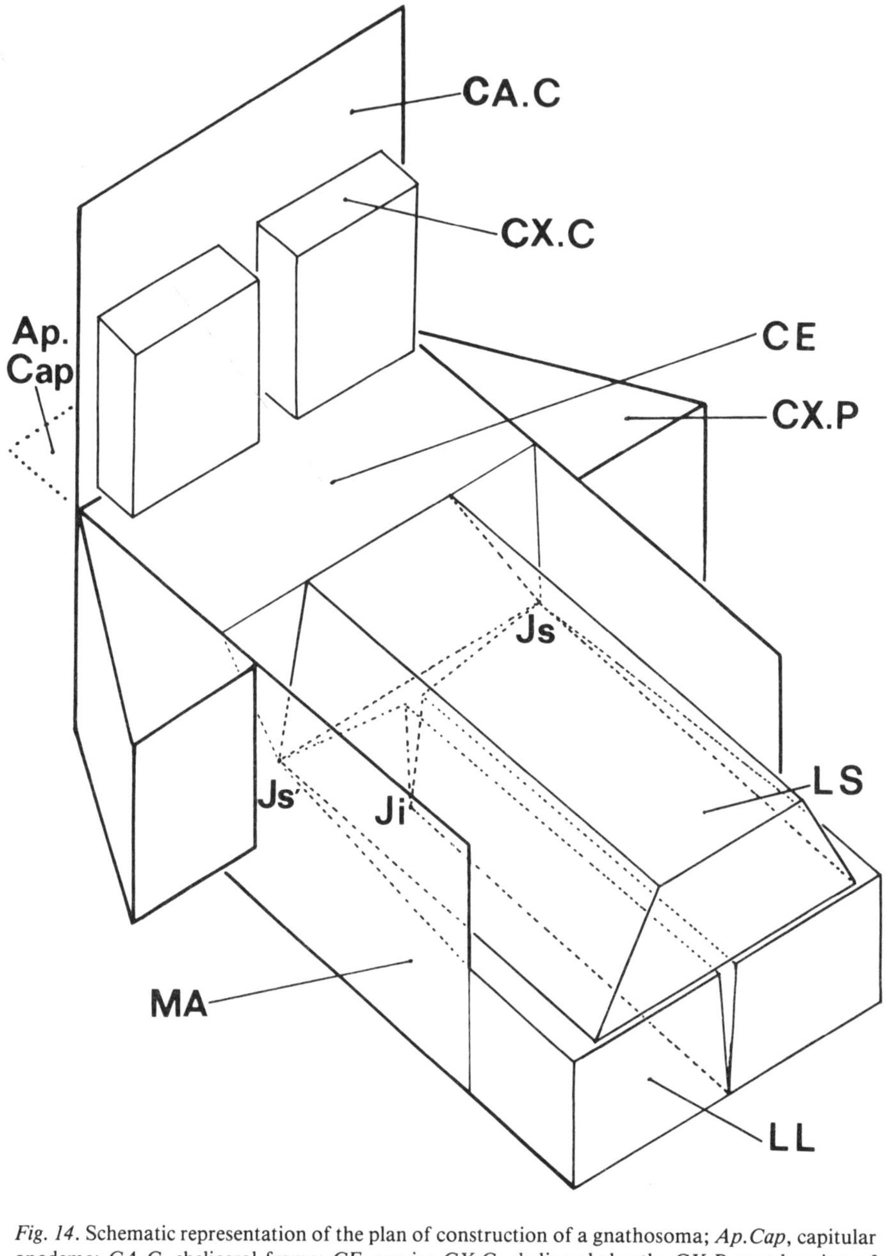

Fig. 14. Schematic representation of the plan of construction of a gnathosoma; *Ap.Cap*, capitular apodeme; *CA.C*, cheliceral frame; *CE*, cervix; *CX.C*, cheliceral sheath; *CX.P*, coxal region of palp; *Ji*, inferior oral commissure; *Js, Js'*, superior oral commissures; *LL*, lateral lips; *LS*, labrum; *MA*, malapophysis.

A so-called gnathosoma (fig. 14) is present in Actinotrichid mites, Anactinotrichid mites and Ricinulei. It is a pseudotagma, constituted by the mouthparts, a division of the body which secondarily has become mobile (see Van der Hammen, 1961: 176–177; 1968; 1968c: 407–408; 1970c; 1980: 70–71; 1986c: 220–223, fig. 2). There is a superficial similarity in the structure of the gnathosoma of these groups, which is, however, mainly attributable to general similarities in the structure of the Chelicerate mouthparts (particularly those found in Solifugae, Pseudoscorpionida, Uropygi, Amblypygi and Araneida). As will be demonstrated below, there are, however, also fundamental differences between its structure in Actinotrichida, on the one hand, and its structure in Anactinotrichida and Ricinulei, on the other.

In a gnathosoma two parts can, generally, be distinguished: cheliceral frame and infracapitulum. The cheliceral frame is constituted by a more or less membraneous part of the tegument, which represents the body wall between the rostral region of the aspidosoma and the infracapitulum; the chelicerae are attached to it, in a movable way, by means of the cheliceral sheaths (the 'coxal' regions of the chelicerae). The superior (dorsal) part of the cheliceral frame constitutes the tegulum; the distal part of the tegulum can develop an anterior extension (the supracheliceral limbus), superior to the free parts of the chelicerae.

The infracapitulum includes part of the palpal segment (segment II) and elements of precheliceral origin (such as the labrum); it constitutes the inferior part of the gnathosoma. The infracapitulum bears the lips and the palpi, and contains the mouth and the pharynx. Generally, the following infracapitular regions can be distinguished: mentum (the unpaired part of the ventral surface, posterior to the mouth), cervix (the dorsal part, extending from the base of the labrum to the line of attachment of the cheliceral frame), large lateral ridges (including, on either side, the base of the palp) and the malapophyses (the paired anterior regions, which are supposed to represent the 'endites' of the palp). The mouth is bordered by two, three of four lips: labrum, labium (present in primitive cases only) and the paired lateral lips (not present in Ricinulei).

In many cases, the gnathosoma (cheliceral frame and infracapitulum) is located in a camerostome, a cavity which is sheltered dorsally and laterally, by the rostral part of the idiosoma. Generally, a gnathosoma is movable by muscles inserted on a gnathosomatic base and/or one or two apodemes.

I have repeatedly dealt with the gnathosoma of the four groups of Actinotrichida (Van der Hammen, 1968; 1969: 187–189, figs. 5D, 6; 1970: 11–12, figs. 4, 7A; 1982: 243–247, figs. 4C, 5–6; 1982a: 34–37, 53–54, figs. 16–18, 20B, C, E, F). The evolution of the Actinotrichid gnathosoma (Van der Hammen, 1971: 472; 1972a: 274–275) is associated with the evolution of a podocephalic canal, with laterodorsal position (see below; see also fig. 1B). The movements of the Actinotrichid gnathosoma are operated by muscles inserted on a capitular apodeme (the internal continuation of the cervix); additional, subcapitular muscles are mentioned for Caeculidae (Coineau, 1971: 130, fig. 29; 1974: 78, fig. 28).

The Ricinuleid gnathosoma (Van der Hammen, 1979a: 9–10, figs. 5C, 6, 7, 9D; see also Van der Hammen, 1986c: 221) is relatively simple. There are no lateral lips, and the labium is incorporated in the mentum (which is considerably extended in anterior direction). The gnathosoma is movable by muscles inserted on the gnathosomatic base (not by muscles inserted on a capitular apodeme). Its evolution is associated with the evolution of a subcapitular gutter and sternapophyses, not with the evolution of a podocephalic canal (see below; see also fig. 1D′).

I have published several detailed studies of the morphology of the gnathosoma in the four groups of Anactinotrichida (Van der Hammen, 1961: 185–189, figs. 2–4, 5a; 1964a: 12–25, figs. 5–12; 1965: 258–269, figs. 2B, C, 3–7; 1966: 23–32, figs. 7–12; 1983a: 218–224, figs. 7–11; 1983b: 11–13, figs. 2C, 6A, 7–8; 1986c: 220–223, fig. 2). The Anactinotrichid gnathosoma (fig. 1C′) differs from the Actinotrichid gnathosoma by the same characters as the Ricinuleid gnathosoma: gnathosoma movements are operated by muscles inserted on the gnathosomatic base, not by muscles inserted on a capitular apodeme; the evolution of the Anactinotrichid gnathosoma is associated with the evolution of a subcapitular gutter and sternapophyses. The Anactinotrichid gnathosoma is, ancestrally, also characterized by the considerable anterior extension of the mentum, and the advanced position of the inferior oral commissure.

In Opilionida (Van der Hammen, 1985: 22–30, figs. 14–19), Xiphosura (Van der Hammen, 1986a: 26, fig. 14A) and Scorpionida (Van der Hammen, 1986a: 26, fig. 13), the mouthparts constitute a so-called myliosoma, a more or less movable division of the body, which is associated with coxisternal feeding, and which includes the coxae and coxapophyses of palp and/or one or more pairs of legs (fig. 13D); in a myliosoma, the mouth had developed a ventral position, whilst the pharynx (in Xiphosura the oesophagus) is curved downwards. A myliosoma can include, besides coxae and coxapophyses: cheliceral frame, cervix, labrum, labium, mouth, pharynx (not present in Xiphosura, in which group the mouth gives entrance to the oesophagus), and myliosomatic apodeme (not known from all groups). The mouth, with ventral position, is bordered by labrum, labium, and the coxapophyses of the appendages. The movability of the coxae is restricted to those (or part of those) associated with ingestion.

6. Orifices of coxal glands and associated structures

Coxal glands (see Van der Hammen, 1985a: 141–142, fig. 1; 1986c: 225) are found in all Chelicerate classes. They originate from segmentally arranged nephridia. In the course of the embryonic development of some Chelicerata, rudiments of several nephridia are still produced. Although six prosomatic pairs must have been present originally, the number of coxal glands in extant Chelicerata is reduced, and those present in the different groups are not in all cases the same homologous glands. There are, moreover, important differences in the position of the orifices, and in the associated structures and the direction of the transport of the coxal gland secretions. As will be demonstrated below, these differences are connected with important differences in prosomatic architecture.

In Palpigradi (Van der Hammen, 1982a: 13, 53, fig. 6E; see also fig. 1A), the single pair of coxal glands belongs to segment III, whilst the orifice is posterodorsal to the base of leg I; a primitive podocephalic canal, extending from this orifice to the mouthparts, is supposed to be present (it could, however, not yet be discerned with certainty). In Actinotrichida (Van der Hammen, 1968: 19–22, 29–30, 38–39, figs. 1, 7, 8B, D, E, 9–11; 1971: 472; 1972a: 274; 1980: 39, 120–121; 1982a: 34, 53, fig. 15A, D, F; see also fig. 1B), the orifice of the coxal gland (the gland belongs to segment III) is also posterodorsal to the base of leg I. It is associated with a podocephalic canal (which is also connected with other podocephalic glands) and extends to the dorsal surface of the infracapitulum. The podocephalic canal can be external (open or partly closed) or internal; in the last-mentioned case, it can be attached to the cuticle, or be free inside the body. In the case of an open, external canal, the function of the canal and the coxal gland secretions is apparently osmoregulation. The architecture of the prosoma, in Palpigradi and Actinotrichida, by which the position of the orifices and the associated structures is determined, is characterized by the presence of epimera (with apodemes) and the absence of coxae and sternum.

In Solifugae (Alberti, 1979a; Van der Hammen, 1986a: 17), the coxal gland probably belongs to segment III, although it occupies a region extending from segment II to segment VI. The orifice is dorsal, near the base of the palpal coxal region and the base of the rostrosoma. The duct of the gland is close to the lateral wall of the prosoma, and resembles the podocephalic canal of those Actinotrichida where it is internal. In Pseudoscorpionida (Van der Hammen, 1986a: 18, fig. 4B), the coxal gland belongs to segment V (the segment bearing leg III) and the orifice is at the posterior paraxial border of epimeron 3. In Solifugae and Pseudoscorpionida, the prosomatic architecture (particularly the architecture of the coxisternal region) is more or less similar to that in Palpigradi and Actinotrichida. The condition of the coxal gland in Solifugae, and the associated way of transport of its secretions, can be derived from the condition in Actinotrichida. The condition in Pseudoscorpionida, however, is quite different, although it will be important to investigate (in primitive species) the possible presence of a gland in segment III, with an internal canal extending to the rostrosoma.

In Anactinotrichida (Van der Hammen, 1966: 21–22, fig. 2; 1971: 472; 1972a: 274–275; 1979a: 16, 24; 1980: 39, 149–151; see also fig. 1C′), the ancestral way of transport of coxal gland secretions is most clearly demonstrated by Opilioacarida. In this group, the single pair of coxal glands belongs to segment III, and the orifice is, on either side, posteroventral to coxa I. The orifices are associated with a pair of sternal taenidia (open canals) in the sternal region of the segment bearing legs I, and these taenidia are connected with the unpaired subcapitular gutter (also a taenidium). The last-mentioned gutter, which is associated with a pair of sternapophyses, is situated at the ventral surface of the infracapitulum; it extends to the base of the lateral lips. In Ricinulei (Van der Hammen, 1979a: 9, figs. 5B, C, 6D; see also fig. 1D′), the condition is very similar to that in Opilioacarida: the paired orifices of the coxal glands are posteroventral to coxa I, and the coxal gland secretions are transported to a subcapitular gutter which is associated with sternapophyses. The architecture of the prosoma, in Anactinotrichida and Ricinulei, by which the position of the orifices and the associated structures is determined, is characterized by the development of coxae and a sternum (the last-mentioned structure is not yet well-developed in Ricinulei).

In Uropygi (Van der Hammen, 1986a: 18, figs. 7A, B, 10B), the coxal glands extend from segment IV to segment VI; it is probably the result of a fusion of two or more glands. The orifice of the gland is paraxial, at the base of coxa I; it is associated with a taenidium. The taenidia of both sides are connected with an unpaired intercoxal gutter (also a taenidium), with ventral position; it belongs to the mouthparts and is situated between the palpal coxae. The intercoxal gutter is associated with a setiferous sternapophysis, of which the long setae are normally lying in the intercoxal gutter (very similar to the ancestral condition in Anactinotrichida and Ricinulei).

Amblypygi (Van der Hammen, 1986a: 18, fig. 7E, F) are characterized by the presence of one or two coxal glands: a large one in segment III, and a smaller one (present in the family Charontidae only) in segment V. The orifice of the last-mentioned gland, if present, is paraxial at the posterior border of coxa III. The orifice of the first-mentioned gland is paraxial at the base of coxa I. It is associated with a taenidium and a sternapophysis. The Amblypygid sternum generally presents three sternapophyses, of which those associated with coxae II and III are more or less vestigial (they could originally have been associated with the coxal glands of segments IV and V).

In Araneida (Van der Hammen, 1986a: 18, fig. 7C, D), two pairs of coxal glands can be present. The anterior pair is present in all Araneida, and belongs to segment III; its orifice is paraxial, at the base of coxa I (it can be associated with a taenidium). The posterior gland, present in Liphistiidae and Mygalomorphae, belongs to segment V; its orifice is at the base of coxa III. Araneid coxal glands are not associated with sternapophyses (because of the close relationships of Araneida and Amblypygi, they are supposed to have disappeared).

The architecture of the prosoma, in Uropygi, Amblypygi and Araneida, by which the position of the orifices of the coxal glands, and the associated structures, is determined, is characterized by the development of coxae and a sternum (as in Anactinotrichida).

In Opilionida (Van der Hammen, 1985: 19, 22, figs. 11D, 12C, 13A; 1985a: 142), the paired orifices of the coxal glands are in the sternal area, close to the paraxial border of coxa III; the gland belongs to segment V. According to Moritz (1959) there is a transient embryonic coxal gland orifice in coxa I.

The Scorpionid coxal gland (Van der Hammen, 1986a: 21, fig. 14B) belongs to segment V; the orifice is paraxial, at the base of coxa III.

The Xiphosurid coxal gland (Van der Hammen, 1986a: 21, fig. 13C) originates from a fusion of the glands of segments II–V. The orifice is at the posterior border of the coxa of limb 5 (homologous with leg III of other Chelicerata); it is associated with a taenidium.

The architecture of the prosoma in Opilionida, Xiphosura and Scorpionida, by which the position of the orifices of the coxal glands is determined, is characterized by the development of a myliosoma (the association of some or all of the large coxae with ingestion).

7. Opisthosomatic appendages

Rudiments of opisthosomatic appendages (up to nine pairs; see Dawydoff, 1949: 347) can be present at a certain stage of the Chelicerate embryonic development. Various numbers of these rudiments can subsequently develop into organs with different function (Van der Hammen, 1977a: 314, tab. 2; 1985a: 143, fig. 7). A survey of these organs in the twelve extant groups of Chelicerata is given in fig. 15.

In Palpigradi (Van der Hammen, 1982a: 10, 49, figs. 2B, 3B, 26C), the opisthosomatic appendages of the segments VIII and IX are associated with the genital opening (the appendages of segment VIII have fused in the case of the female); in *Prokoenenia* (representing the ancestral condition), segments X–XII are moreover characterized by the presence of altogether three pairs of invaginable papillae (the rudiments of these papillae are apparently suppressed in the course of embryonic development and gradually reappear in the course of postembryonic development; see Van der Hammen, 1982a: 17–20, fig. 9).

In Actinotrichida (Van der Hammen, 1982a: 31–34, 49–50, figs. 11, 14, 26D, E), the ancestral condition (represented by Endeostigmata and Oribatida) is the following. The trifid eugenital opening is situated between the segments VIII and IX, and is surrounded by the opisthosomatic appendages (the 'endites') of segment VIII (one pair) and those of segment IX (a fused pair). Three pairs of genital papillae represent the opisthosomatic appendages of segments X–XII; they are included in the progenital chamber, which is covered by the paired progenital valves. The progenital opening between the valves is longitudinal (not transverse); the valves represent, on either side, the fused 'exites' of the opisthosomatic appendages of segments VIII and IX (a suture separating these appendages is still present in some species). The rudi-

Segments / Groups	VII	VIII	IX	X	XI	XII	XIII	XIV	XV	XVI	XVII	XVIII	XIX
PALPIGRADI		○	○	○	○	○						☒	☒
ACTINOTRICHIDA		○	○	○	○	○					(☒)	☒	☒
ANACTINO-TRICHIDA		○											
RICINULEI													
OPILIONIDA		?										☒	☒
SOLIFUGAE		○	○	○	○							☒	☒
PSEUDOSCOR-PIONIDA		?	○	○									☒
UROPYGI		○	○										☒
AMBLYPYGI		○	○										☒
ARANEIDA		○	○	○	○								☒
XIPHOSURA	○	○	○	○	○	○	○				☒	☒	☒
SCORPIONIDA		○	○	○	○	○	○						

Fig. 15. Homologues of opisthosomatic appendages in Chelicerata. For each group, the fundamental condition is represented (in the case of Palpigradi, the figure is based on *Prokoenenia*; in the case of Actinotrichida, the figure is based on Endeostigmata and Oribatida; in the case of Anactinotrichida, the figure is based on Opilioacarida; in the case of Uropygi, the figure is based on Holopeltida). In Palpigradi, the appendages of segments VIII and IX are associated with the genital opening, whilst the appendages of segments X–XII (in *Prokoenenia*) constitute evaginable opisthosomatic papillae. In Actinotrichida, the appendages of segments VIII and IX are, ancestrally, associated with the eugenital opening, whilst those of segments X–XII constitute the genital papillae. In Anactinotrichida (Opilioacarida), the single pair of opisthosomatic appendages, attributed to segment VIII, is represented by a pair of evaginable papillae with verrucae. Homologues of opisthosomatic appendages are supposed to be present in the genital region of Solifugae, Pseudoscorpionida and Opilionida. The homology, with opisthosomatic appendages, of the respiratory organs of Anactinotrichida, Ricinulei, Opilionida, Solifugae and Pseudoscorpionida, is uncertain or doubtful. The respiratory organs of Uropygi, Amblypygi and Araneida are homologous with the opisthosomatic appendages of segments VIII and IX, the spinnerets of Araneida with the appendages of segments X and XI. Xiphosura present seven pairs of opisthosomatic appendages (they belong to segments VII–XIII), of which those of segments VIII are associated with the genital opening, and those of segments IX–XIII carry gills. In Scorpionida, the six pairs of appendages belong to segments VIII–XIII; those of segment VIII are associated with the genital opening, those of segment IX constitute the combs, whilst those of segments X–XII constitute the respiratory organs.

ments of the genital papillae are apparently suppressed in the course of embryonic development and gradually reappear in the course of postembryonic development (the base levels of the three pairs are respectively protonymphal, deutonymphal and tritonymphal). The concentration of segments in the genital region is closely connected with the advanced ventral position of the anal opening. It may be remarked here that, in species with an evaginable ovipositor, the eugenital opening is in the apical part of this organ.

In Solifugae (Van der Hammen, 1986a: 7, 14, fig. 4A), where rudiments of nine pairs are present at a certain stage of embryonic development and subsequently disappear, opisthosomatic appendages could be represented by the genital covers (appendages of segment VIII; the opening between them is longitudinal), and the respiratory organs (paired stigmata in segments IX and X, an unpaired stigma can, in some groups be present in segment XI).

In Pseudoscorpionida (Van der Hammen, 1986a: 7, 14, figs. 1B, 4C), rudiments of four pairs of opisthosomatic appendages are present at a certain stage of embryonic development. In this group, the genital operculum, which belongs to segment VIII and which presents gonopods, could be of appendicular origin; the 'exites' of these appendages (represented by the operculum) have apparently fused, and the genital opening is represented by a transverse slit. The posterior genital plate is sometimes also regarded as the result of fusion of opisthosomatic appendages (those of segment IX). The respiratory organs (the stigmata occupy a lateral position in segments IX and X) could perhaps also have originated from opisthosomatic appendages, although this hypothesis is apparently incompatible with the supposed integration of the opisthosomatic appendages of segment IX into the unpaired posterior genital plate.

In Opilioacarida (Anactinotrichida) (Van der Hammen, 1966: 16–19, figs. 2, 3, 4E, F, 5; 1968b: 64, fig. 2B, C), the paired evaginable genital papillae and verrucae are regarded as homologues of the opisthosomatic appendages of segment VIII. The four pairs of Opilioacarid stigmata occupy a lateral position in the segments IX–XII. The origin of these respiratory organs is unknown (various hypotheses have been introduced); they could originate from glands or from opisthosomatic appendages. The peridium of Allothyridae (Holothyrida) (Van der Hammen, 1968a: 263–265, fig. 1; 1983b: 8, fig. 3A) and the paired stigmata of Ixodina (Ixodida) (Van der Hammen, 1983a: 217, figs. 1A, 2B, D) occupy about the same position as the genital papillae of Opilioacarida. The homology of the single pair of stigmata (with lateral position) in Holothyrida (Van der Hammen, 1961: 184, pl. VI, fig. 1c; 1965: 258, fig. 1D; 1983b: 8, fig. 2A), Argasina (Ixodida) (Van der Hammen, 1983a: 217, fig. 4B) and Gamasida (Van der Hammen, 1964a: 11, fig. 3) is uncertain; it could be homologous with one of the Opilioacarid stigmata.

Homologues of opisthosomatic appendages are not known from Ricinulei. The single pair of Ricinuleid stigmata (Van der Hammen, 1979a: 8, 9, 23, fig. 3A, E) resembles, however, the stigmata of Holothyrida, Gamasida and Argasina (Ixodida), although its position is prosomatic. This position could be the result of a secondary translocation, and in that case the Ricinuleid respiratory organs could either originate from glands or from a pair of opisthosomatic appendages; data in support of this hypothesis are, however, not known.

In all three groups of Arachnida, viz., Uropygi, Amblypygi and Araneida (see Van der Hammen, 1986a: 14–17, fig. 5), respiratory organs are present which are supposed to originate from gill-bearing opisthosomatic appendages. The Uropygid respiratory organs are represented by booklungs and stigmata with ventral position: one pair (stigmata in segment VIII) in Schizomida, two pairs (stigmata in segments VIII and IX) in Holopeltida. The Amblypygid respiratory system is also represented by booklungs and paired stigmata with ventral position (in segments VIII and IX); a pair of small evaginable ventral sacs, with unknown function, can also be present in segment IX (it could represent the paired 'endites' of the appendages of this segment). The respiratory system of Araneida is represented by paired stigmata with ventral position in segments VIII and IX, associated with either booklungs or tracheae. Araneida can, moreover, present up to four pairs of spinnerets; these are homologous with the opisthosomatic appendages of the segments X and XI (the inner pairs are supposed to represent the 'endites').

In Opilionida (Van der Hammen, 1985: 19), the paired lobes in the distal part of the ovipositor, and in the distal part of the Cyphophthalmic penis (spermatopositor; in the penis of others groups, the lobes have apparently fused), could perhaps represent the opisthosomatic appendages of segment VIII; the genital opening between the paired lobes is represented indeed by a longitudinal, not a transverse, slit. Opilionida present respiratory organs in segment VIII (paired stigmata posterior to coxa IV); the hypothesis could be introduced (it is incompatible with the above-mentioned hypothesis) that these respiratory organs originate from opisthosomatic appendages.

In Scorpionida (Van der Hammen, 1986a: 12, 17, fig. 6A), homologues of six pairs of opisthosomatic appendages are present, which belong to the segments VIII–XIII. The appendages of segment VIII constitute the genital operculum, the appendages of segment IX are represented by the paired combs or pectines, those of the segments X–XIII are represented by the paired respiratory organs (stigmata and booklungs, supposed to originate from branchial appendages).

Xiphosura (Van der Hammen, 1986a: 13–14, 17, fig. 6B) present homologues of seven pairs of opisthosomatic appendages, belonging to the segments VII–XIII. Those of segment VII constitute the so-called chilaria, those of segment VIII constitute the genital operculum, and those of segments IX–XII are branchial (gill-bearing) appendages.

8. Genital region

The Chelicerate genital opening is ancestrally represented by a transverse slit at the posterior border of segment VIII. In the course of evolution, this condition can be transformed in various ways, particularly by the association of opisthosomatic appendages with the genital region (dealt with in the previous section, but briefly repeated here), by the origination of genital covers, and by the origination of ovipositor and spermatopositor or penis.

In Palpigradi, the opisthosomatic appendages of segments VIII and IX are associated with the genital opening. In Actinotrichida, the trifid eugenital opening is ancestrally bordered by the 'endites' of the opisthosomatic appendages of segment VIII (paired) and segment IX (fused); the opening is situated in the progenital chamber which is covered by the 'exites' of these appendages (the longitudinal opening between these progenital lips or covers is called progenital opening). The opisthosomatic appendages of the segments X–XII are included in the progenital chamber. As a result of an extension, in the distal region, of the internal reproductive system, an evaginable ovipositor and an evaginable spermatopositor or penis can have evolved; in these organs, the eugenital opening is in the apical part.

In Solifugae, the longitudinal genital opening is supposed to be bordered by homologues of the opisthosomatic appendages of segment VIII. In Pseudoscorpionida, the transverse genital opening is bordered by two sclerites: the genital operculum (which bears a pair of gonopods, and is supposed to represent the opisthosomatic appendages of segment VIII) and the posterior genital plate (belonging to segment IX).

In Anactinotrichida, the genital opening is ancestrally represented by a transverse slit (still present in Opilioacarida and Ixodida). The opening can be secondarily protected by a variable number of genital sclerites (Holothyrida, Gamasida). A primitive ovipositor is known from Opilioacarida (Van der Hammen, 1986b: 64, fig. 2; 1977: 56, fig. 7C, D), Holothyrida (Van der Hammen, 1983b: fig. 2B) and Ixodida (Van der Hammen, 1983a: 238). In Ricinulei (Van der Hammen, 1979a: 8, fig. 4A, B), the female genital opening is represented by a transverse slit; it is bordered anteriorly by the genital shield (belonging to segment VIII). In male Ricinulei, the genital opening is at the posteroventral side of a conical projection of segment VIII. In Uropygi, Amblypygi and Araneida, the genital opening is represented by a transverse slit at the posterior border of sternite VIII (Van der Hammen, 1986a: 8–10).

In Opilionida (Van der Hammen, 1985: 11–19, figs. 1B, 2B, 3B, 4–10), the eugenital opening is in the terminal part of the female ovipositor and the male spermatopositor or penis. These organs originate from extensions of the internal reproductive system. The genital opening, at the anterior border of segment VIII, through which ovipositor and penis are evaginated, is unprotected (Cyphophthalmi) or protected by a genital cover originating from a forward projection of segment VIII (in Phalangiidae it is supposed to include also the sternite of segment VI). The opening between genital cover and prosomatic sternites is named progenital opening. Originally, in the ancestors, eugenital and genital opening must have coincided. Ovipositor and penis can be very long. As mentioned in the previous section, the lobes in the distal part of ovipositor and spermatopositor or penis, could be homologous with a pair of opisthosomatic appendages; the eugenital opening between them is represented by a longitudinal slit.

In Scorpionida (Van der Hammen, 1986a: 14, fig. 6A), the genital opening is covered by a genital operculum which is supposed to originate from a fusion of the opisthosomatic appendages of segment VIII. In Xiphosura (Van der Hammen, 1986a: 14, fig. 6B), the paired genital opening is also covered by the genital operculum, originating from a fusion of the paired opisthosomatic appendages of segment VIII.

9. Respiratory organs

General surveys of respiratory organs in Chelicerata have been prepared by me on two occasions (Van der Hammen, 1977a: 313–314; 1985a: 143). In two other papers I have, moreover, made some remarks on respiratory organs in the two groups of mites (Van der Hammen, 1968a: 270–274, fig. 3; 1983b: 45; in the first-mentioned paper, I have given a provisional classification of respiratory organs in mites). In the present section, my previous views are reconsidered.

Among the adaptations of Chelicerata to terrestrial life, the development of respiratory organs has been of special importance. Many of the marine ancestors, particularly those with heavily sclerotized cuticle, were probably characterized by the possession of gills. In other, less sclerotized, species the tegument was probably associated with the exchange of oxygen and carbon dioxide. Among extant Chelicerata, gills are found in Xiphosura only. In several groups, respiratory organs are supposed to have evolved from opisthosomatic appendages; booklungs are supposed to have evolved from opisthosomatic appendages carrying gills (branchial appendages). Other respiratory organs could have evolved from various types of glands, whilst cutaneous respiration is also still found in some groups of terrestrial Chelicerata.

In Palpigradi (Van der Hammen, 1982a: 8, 51) respiration is apparently cutaneous, as secondary organs are not present in this group. In Actinotrichida (Van der Hammen, 1980: 135–136; 1982a: 42–44, 51), there are many types of respiratory organs, which must have evolved separately and independently in the main groups. Cutaneous respiration is still present in some groups of Actinotrichida, and also in the immature forms of several other groups. Many types of secondary respiratory organs in Actinotrichida, starting from various parts of body and appendages (superficial parts of the cuticle, the base of the appendages, hidden depressions of the surface, such as bothridium, preanal fold and apodemes) are supposed to originate from glands, particularly tegumental glands, although in some cases an origination from excretory glands is also possible. In some species, supposed genital tracheae are present, which could have originated from accessory glands.

The Solifugid respiratory system (Van der Hammen, 1986a: 14, fig. 4A) is represented by a vast system of tracheae and three pairs of stigmata (a posterior unpaired stigma is, moreover, present in the families Galeodidae and Hexisopodidae). The anterior pair of stigmata is prosomatic, the other stigmata are opisthosomatic. The prosomatic stigmata are situated near the posterior border of epimera 2, the two pairs of opisthosomatic stigmata are situated near the posterior border of the sternites of segments IX and X respectively (the unpaired stigma of Galeodidae and Hexisopodidae is near the posterior border of sternite XI). As mentioned in section 7, the Solifugid opisthosomatic respiratory organs could have arisen from opisthosomatic appendages. The origin of the prosomatic respiratory organs in Solifugae is unknown (it will be interesting to compare them with the apodematic tracheae in Oribatid mites). The Pseudoscorpionid respiratory system consists of tracheae and two pairs of stigmata situated near the lateroposterior border of sternites IX and X respectively. The stigmata have a more lateral position than the opisthosomatic stigmata of the Solifugae, but could be homologous with those of the corresponding segments (IX and X) in that group.

Among the Anactinotrichida (Van der Hammen, 1968a: 273; 1979a: 18–19, 22–23; 1980: 135–136), four pairs of opisthosomatic stigmata (and corresponding tracheae) are found in Opilioacarida (Van der Hammen, 1966: 16–17, fig. 3, 4E, F); as mentioned in section 7, they could have evolved from opisthosomatic appendages, although a different hypothesis is mentioned below. A single pair of idiosomatic stigmata, dorsal to the posterior coxae, is present in Holothyrida, Gamasida and Argasina (Ixodida). This pair could be homologous with one of the pairs of Opilioacarid stigmata. In Allothyridae (Holothyrida) (see Van der Hammen, 1983b: 8, 45, fig. 2A), the stigma has an intermediate position between the orifices of two excretory glands, which suggests that stigma and tracheae could have evolved from excretory glands. This could also apply to Gamasida and Argasina (Ixodida), and even to all pairs of stigmata and tracheae in Opilioacarida. In Ixodina (Ixodida) (see Van der Hammen, 1983a: 217, figs. 1A, 2B, D), the single pair of stigmata has a ventral position and could originate from a pair of opisthosomatic appendages. In Ricinulei (Van der Hammen, 1979a: 9, 12, 23, fig. 3A, E), the single pair of stigmata is prosomatic. As mentioned in section 7, it resembles the single pair of stigmata in Holothyrida, Gamasida and Argasina (Ixodida), and the prosomatic position could perhaps be attributable to a secondary translocation. In that case, the pair could have originated either from opisthosomatic appendages or from excretory glands.

Conditions in the three groups of Arachnida s.str. (Uropygi, Amblypygi and Araneida) are very similar (see Van der Hammen, 1986a: 14–16, fig. 5). The Uropygid respiratory organs are represented by booklungs and stigmata with ventral position: one pair (stigmata in segment VIII) in Schizomida, two pairs (stigmata in segments VIII and IX) in Holopeltida. The Amblypygid respiratory system is also represented by booklungs and paired stigmata with ventral position (in segments VIII and IX). The respiratory system of Araneida is represented by paired stigmata with ventral position in segments VIII and IX, associated with either booklungs or tracheae. It is supposed that the Araneid tracheae have replaced the booklungs in the course of evolution. As mentioned above, booklungs are supposed to have evolved from branchiate opisthosomatic appendages.

The Opilionid respiratory system (Van der Hammen, 1985: 11, 45, figs. 2B, 3B, 4, 11B, 12B, 13A, 32B, 33A) consists of tracheae and a pair of stigmata in sternite VIII (either free on the surface of the sternite, or hidden in a fold posterior to coxa IV). Phalangiidae present, moreover, tibial stigmata and pedal tracheae. The origin of the Opilionid opisthosomatic respiratory system is unknown (it could have evolved from opisthosomatic appendages or from glands).

Xiphosura (Van der Hammen, 1986a: 17, fig. 6B) constitute the only extant group of Chelicerata characterized by the presence of gills. Five pairs of gills are carried in this group by the opisthosomatic appendages of the segments IX–XIII.

The Scorpionid respiratory system (Van der Hammen, 1986a: 17, fig. 6B) consists of four pairs of booklungs and the associated four pairs of stigmata with ventral position in the segments X–XIII. It is generally accepted that these booklungs have arisen, in the course of evolution, from branchial appendages.

III. REPRODUCTION

Chelicerate reproduction can be sexual or asexual; it generally includes gametogenesis (spermiogenesis and oogenesis), reproductive behaviour (among which sperm transfer), fertilization (absent in the case of asexual reproduction) and oviposition.

Much attention has been paid of recent years to Chelicerate spermiogenesis and sperm morphology (Baccetti, 1979; Weygoldt & Paulus, 1979: 110–113, fig. 7; Alberti, 1979, 1980, 1980a, 1980b; Alberti & Palacios-Vargas, 1984, 1987; Alberti & Weinmann, 1985). A primitive aquatic sperm type (flagellate, with spherical nucleus and large acrosome) is still present in extant Xiphosura. Three types of spermatozoa are found in terrestrial Chelicerata: (a) a flagellate-filiform type (nearest to the primitive aquatic type) occurs in Scorpionida; (b) a flagellate-coiled type (spermatozoa which are rolled up during spermiogenesis) occurs in Pseudoscorpionida, Uropygi, Amblypygi, Araneida and Ricinulei; and (c) an aflagellate type occurs in Opilionida (a reduced coiling process was observed in Cyphophthalmi), Palpigradi, Solifugae, Actinotrichida and Anactinotrichida. The great diversity in Chelicerate sperm morphology is probably related to the great variety in modes of sperm transfer.

Reproductive behaviour includes the activities of one or two mates, mechanisms of recognition and attraction of one sex to the other, selection of the mate, courtship behaviour (which can include physical contact) and sperm transfer. Fertilization (the union of egg- and sperm-nuclei) can take place inside the maternal body (internal fertilization) or outside (external fertilization). External fertilization is still found in extant Xiphosura. Internal fertilization is associated with active sperm transfer.

Sperm transfer constituted an important problem during the transition, in the course of evolution, from aquatic to terrestrial life (see Schaller, 1965, 1971, 1979; Van der Hammen, 1977a: 314–315; 1985a: 144). Fertilization had certainly been external in the marine ancestors of extant Chelicerata (as mentioned above, it is still external in extant Xiphosura). As in other primitive terrestrial Arthropods, the transition to terrestrial life involved the transfer of sperm cells in spermatophores, the first stage in the evolution of internal fertilization. In many cases, this type of indirect sperm transfer involves courtship behaviour. Indirect sperm transfer by spermatophores (spermatophory; spermatophores can be deposited by means of a spermatopositor) was originally associated with the early soil fauna. The evolution of other modes of internal fertilization created the possibility of colonizing new niches.

Nearly direct sperm transfer with gonopods (gonopody) refers to the implantation or transfer of a spermatophore or a sperm drop to the female system by means of the male appendages; it is found in three groups of Chelicerata. In Solifugae and some groups of Anactinotrichida, the chelicera can function as a gonopod; in Araneida the palp; and in Ricinulei leg III.

Direct sperm transfer through copulation is associated with the presence of a penis; it is found in Opilionida and several groups of Actinotrichida. In some Actinotrichida, the receptaculum seminis or bursa copulatrix can have a separate opening (copulatory orifice).

After sperm transfer, sperm can be relocated in the female reproductive system, mostly by peristalsis of oviducts and ovaries. During storage in the female system, sperm can probably remain viable for extended periods.

Oviposition can take place by means of an ovipositor (several groups of Actinotrichida and Anactinotrichida; Opilionida). Eggs can remain attached to the female genital region till eclosion, either included in a brood pouch or not (Pseudoscorpionida, Uropygi, Amblypygi). Chelicerata are either oviparous or ovoviviparous (in the last-mentioned case, eclosion of the eggs takes place inside the maternal body or during oviposition); ovoviviparity is known from several Actinotrichida and from Scorpionida.

Parthenogenesis (the production of an embryo from a female gamete without any genetic contribution from a male gamete) is known from a Scorpion, very few Opilionida, Amblypygi and Araneida (Bell, 1982: 238) and from both groups of Mites; it can be occasional, obligatory, facultative, cyclical or even artificial. Three types of parthenogenesis (arrhenotoky, deuterotoky and thelytoky; see Van der Hammen, 1980: 19, 44, 159) are known from Mites; thelytoky is also known from a Scorpion. Parthenogenesis in Mites can be gynogenetic, i.e. it can be induced by some part of the act of mating (mechanical effect of copulation, presence of sperm or genital accessory materials, actual penetration of the egg by a spermatozoon without contributing the male genetic material).

Little is known about reproduction in Palpigradi. In some species, males are extremely rare and reproduction could be parthenogenetic; in other species, males seem to be more numerous than females. Eggs are apparently laid throughout the year, in small numbers (two relatively large eggs can be simultaneously present in the female opisthosoma).

In Actinotrichida, reproduction is either sexual or parthenogenetic. Sperm transfer is either indirect (spermatophory), or direct (copulation). Spermatophory can be associated with courtship behaviour. Copulation is found in Acaridida, Tarsonemida and some Prostigmata (Actinedida); it can be associated with the presence of a separate copulatory orifice and a bursa copulatrix. Oviposition can take place by means of an ovipositor (Oribatida, several groups of Actinedida). Several Actinotrichida are known to be ovoviviparous.

Solifugid reproduction is sexual. Sperm is transferred in spermatophores, and the chelicera can function as a gonopod. Courtship behaviour is unknown. Solifugae are oviparous; the female watches over the eggs till eclosion. Pseudoscorpionid reproduction is sexual. Sperm is transferred in spermatophores which are deposited on the substrate; this act can be accompanied by courtship behaviour and physical contact. Pseudoscorpionida are oviparous; eggs are laid in a brood pouch which remains attached to the female genital region till eclosion.

Anactinotrichid reproduction is either sexual or asexual. Parthenogenesis is known from several Gamasida and Ixodida; it is either arrhenotokous or thelytokous (it can be gynogenetic). Sperm transfer is indirect (spermatophory), or nearly direct (gonopody; the chelicera can function as a gonopod). In Dermanyssina (a group of Gamasina), gonopody is associated with the occurrence, in the female, of receptacula seminis of which the orifice is at the base of coxa III (it is called podospermia). Anactinotrichida are oviparous. A primitive ovipositor is known from Opilioacarida, Holothyrida and Ixodida.

Ricinuleid reproduction is sexual. Sperm transfer is nearly direct; after an initial phase of courtship behaviour, the male mounts the female, collects a spermatophore from his genital opening in one of the gonopods (leg III) and introduces the spermatophore (or its contents) into the female genital opening.

Uropygid reproduction is sexual. Sperm is transferred in spermatophores; the transfer is accompanied by courtship behaviour (which can include physical contact). Uropygi are oviparous; eggs remain attached to the genital opening, either in a brood pouch or not. Amblypygid reproduction is generally sexual; parthenogenesis seems to occur in rare cases. Sperm is transferred in spermatophores; sperm transfer is accompanied by courtship behaviour. Amblypygi are oviparous; eggs remain attached to the female genital opening, surrounded by a layer of secretion. Araneid reproduction is generally sexual (parthenogenesis seems to occur in rare cases). Sperm transfer is nearly direct; the male palp can function as a gonopod, and sperm is transported to the female receptacula seminis. Various types of courtship behaviour are known. Araneida are oviparous; eggs are nearly always deposited in a silk cocoon which is guarded (or even carried) by the female.

Opilionid reproduction is generally sexual (parthenogenesis seems to occur in rare cases). Sperm transfer is direct (copulation): sperm is transferred by the male penis to the female genital system (not known from Cyphophthalmi). Opilionida are oviparous; eggs are laid in a suitable place by means of an ovipositor. Scorpionid reproduction is generally sexual; thelytokous parthenogenesis is mentioned for the Brazilian scorpion *Tityus serrulatus*. Sperm transfer is indirect (spermatophory) and is associated with courtship behaviour (which includes physical contact). Scorpionida are ovoviviparous. Xiphosurid reproduction is sexual. Fertilization is external (it is associated with the presence of motile flagellate spermatozoa); males sit on the back of the female while she lays the eggs in a hole on the beach.

IV. POSTEMBRYONIC DEVELOPMENT

The Chelicerate postembryonic development is characterized by the occurrence of instars and moults: because of the presence of an exoskeleton, growth involves a periodical shedding of the outer covering (as in other Arthropods). The change of form and structure (gradual or radical) undergone in the course of postembryonic development is, generally, referred to as metamorphosis. Metamorphosis includes the repeated development of a new instar inside the skin of the preceding instar; during this development (the so-called pupal period of the old instar, a quiescent stage), the developing new form is referred to as the pharate instar. Moulting, consequently, includes a separation of the old cuticle from the hypodermis, and the subsequent formation of a new cuticle, followed by splitting and shedding of the old one (ecdysis, or dehiscence and eclosion). During the process of moulting, appendages which have been partly or completely lost can be reconstituted (regeneration). The dead skin of an instar is referred to as the exuviae. Exuviae nearly always split (dehisce) according to a well-defined line of dehiscense. After eclosion, the exuviae are cast off (shed) or retained (carried on the back).

In previous publications, I have repeatedly dealt with the Chelicerate postembryonic development, and with the life-cycle and its evolution (see Van der Hammen, 1974, 1975, 1978, 1979, 1985a, 1985b). Many generalities and references can also be found in the first part of the Glossary of Acarological Terminology (Van der Hammen, 1980). In the present chapter, the Chelicerate postembryonic development and the evolution of the Chelicerate life-cycle are dealt with in six sections: forms and moults; types of ecdysis; phases; regressive instars; evolutionary phenomena; and a general model of the evolution of the Chelicerate life-cycle.

1. Forms and moults

The Chelicerate life-cycle presents, in the different groups, important differences in numbers of forms and moults. In Actinotrichida and Anactinotrichida, there are, generally, in each group, a fixed number of moults, and consequently a fixed number of (distinctly different) forms. In most groups of Actinotrichida and in Opilioacarida (i.e. primitive Anactinotrichida), the number of forms is six (prelarva, larva, protonymph, deutonymph, tritonymph and adult; see figs. 16, 17). The same number is found in Uropygi (Holopeltida and Schizomida) and Pseudoscorpionida; it could also be present in Palpigradi (four forms are now known in this group; the supposed prelarva and protonymph are still unknown). There are, however, also groups with a less constant number of moults. In Scorpionida, there are apparently 6–8 moults. Numerous moults are mentioned in the case of Amblypygi, where the adult continues moulting. There is also a variable number of moults in Araneida, where, in some groups, the adult also continues moulting. A rather great and variable number of moults is mentioned for Opilionida and Solifugae.

Not every moult has the same effect. In many cases moulting results in a form distinctly different from the preceding form. Sometimes, however, moulting brings no change, or there is only a difference in size; in these cases, the moults are called repetition-moults and growing-moults. Growing-moults are, for instance, present in the case of moulting adults of Amblypygi and Araneida.

Among the forms which are the result of moulting, three types can now be distinguished. Forms which are the result of growing-moults or repetition-moults (type 1) are called isophena. The forms representing types 2 and 3 are characterized by distinct, discontinuous differences from the forms preceding them; they are called instars. Forms (instars) of type 2 are found in the life-cycle of groups with a fixed fundamental number of moults; in this case, the levels of development of species of these groups can be homologized, and the forms (instars) are called stases. Forms (instars) of type 3 are found in the life-cycle of groups in which the life-cycle presents a variable number of moults; in this case the separate levels of development of species of these groups cannot be homologized, and the forms (instars) are called stasoids. In one and the same life-cycle, however, forms of more than one type can be present (see section 3).

2. Types of ecdysis

The process of ecdysis comprises splitting and shedding of the old cuticle, i.e. dehiscence and eclosion. There are various ways in which exuviae dehisce (split), and dehiscence, at least in Actinotrichida, nearly always takes place according to a well-defined line of dehiscence, which is predetermined by its slighter resistance. A comparative study of moulting in Chelicerata is still wanting, and the following survey of types of ecdysis is mainly based on studies in Actinotrichida (fig. 18).

Splitting of the old cuticle can take place in the anterior part of the body (proterodehiscence; in hatching, the animal moves forwards), or in the posterior part (hysterodehiscence; in hatching, the animal moves backwards). Three types of proterodehiscence (fig. 18A–D) are known: prodorsal dehiscence

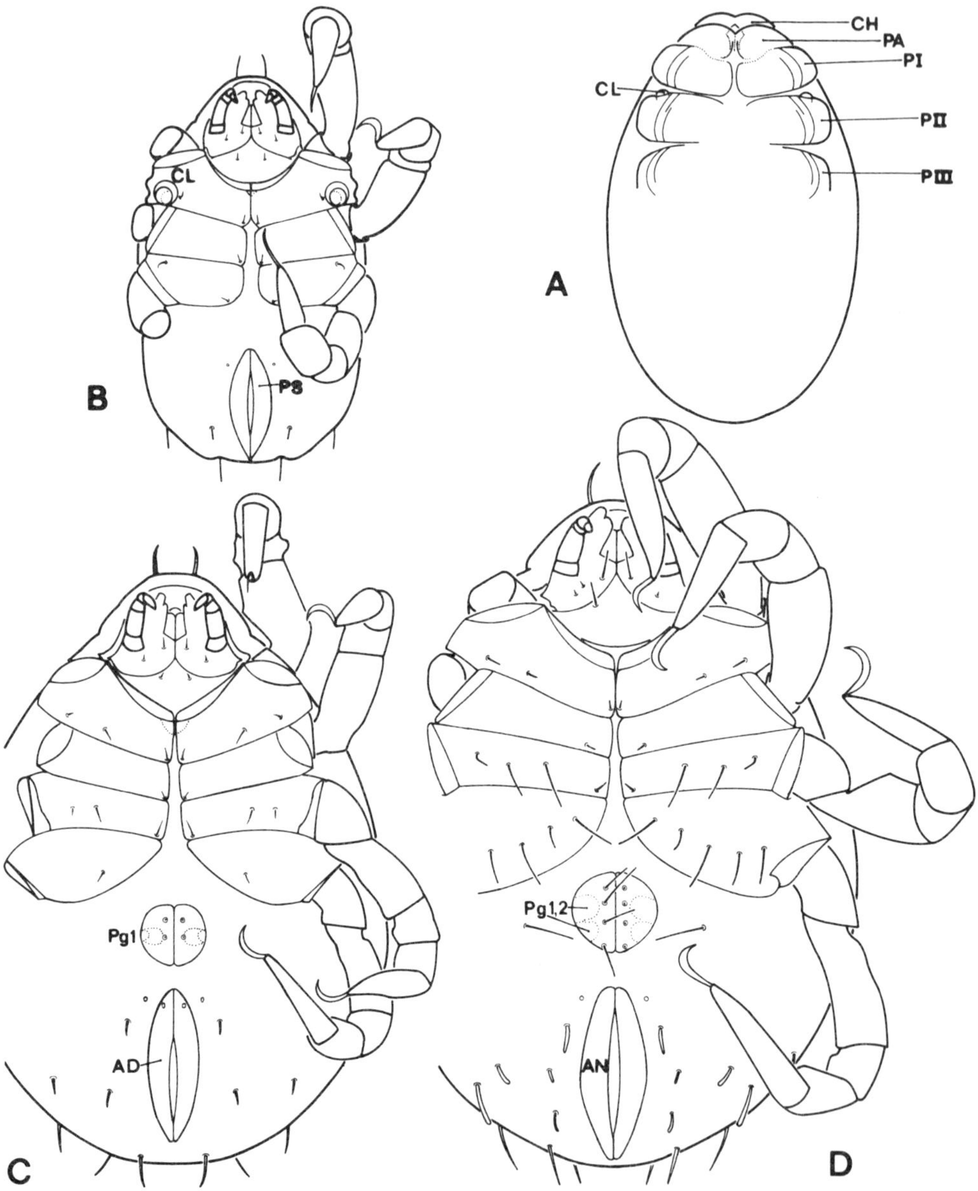

Fig. 16. Postembryonic development of *Hermannia convexa* C.L. Koch (Actinotrichida: Oribatida) (1); ventral views of: A, prelarva; B, larva; C, protonymph; D, deutonymph; *AD*, adanal segment; *AN*, anal segment; *CH*, chelicera; *CL*, Claparède's organ; *PA*, palp; PI–III, legs I–III; *Pg1–2*, genital papillae; *PS*, pseudanal segment; A–D, × 118.

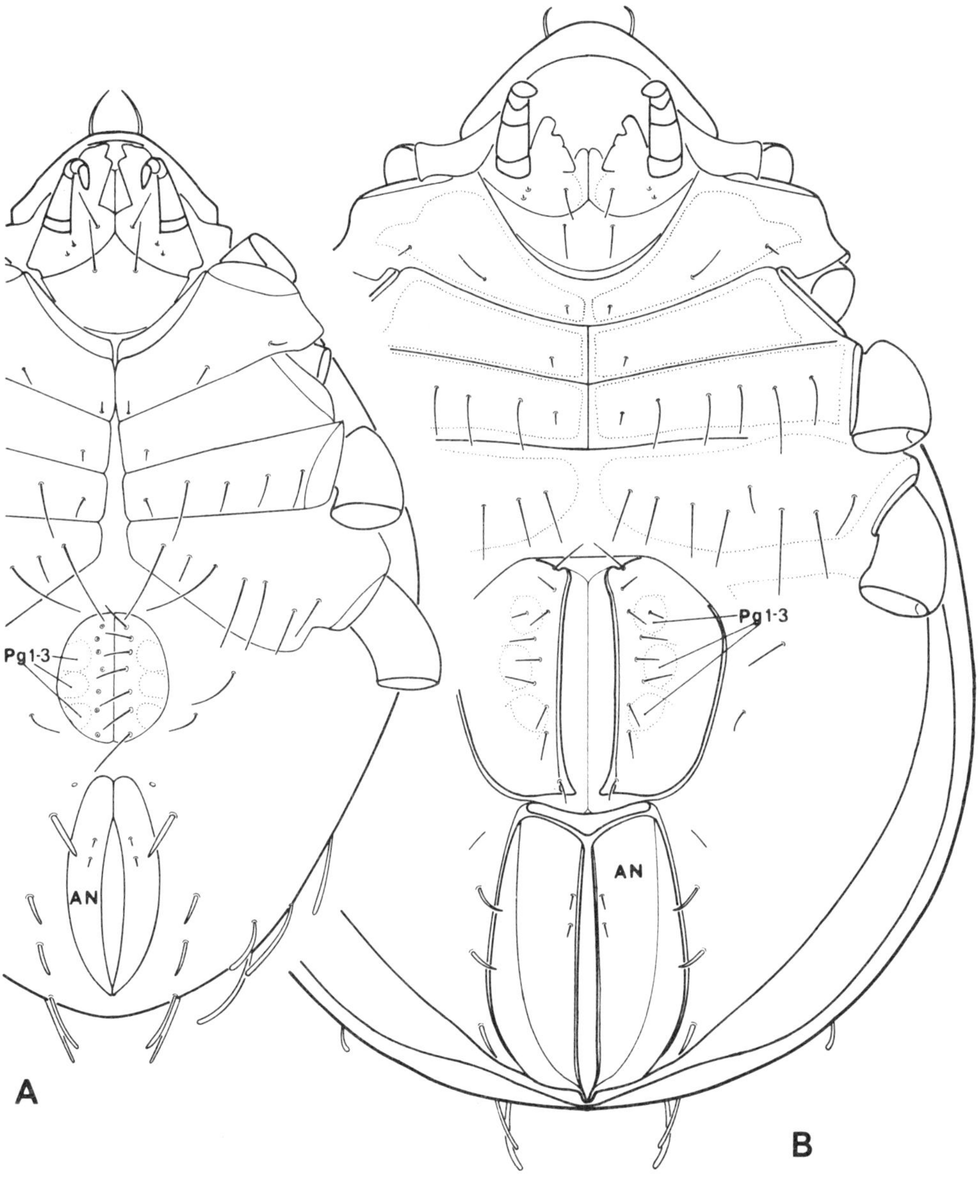

Fig. 17. Postembryonic development of *Hermannia convexa* C.L. Koch (Actinotrichida: Oribatida) (2); ventral views of: A, tritonymph; B, adult female; *AN*, anal segment; *Pg1–3*, genital papillae; A–B, × 118.

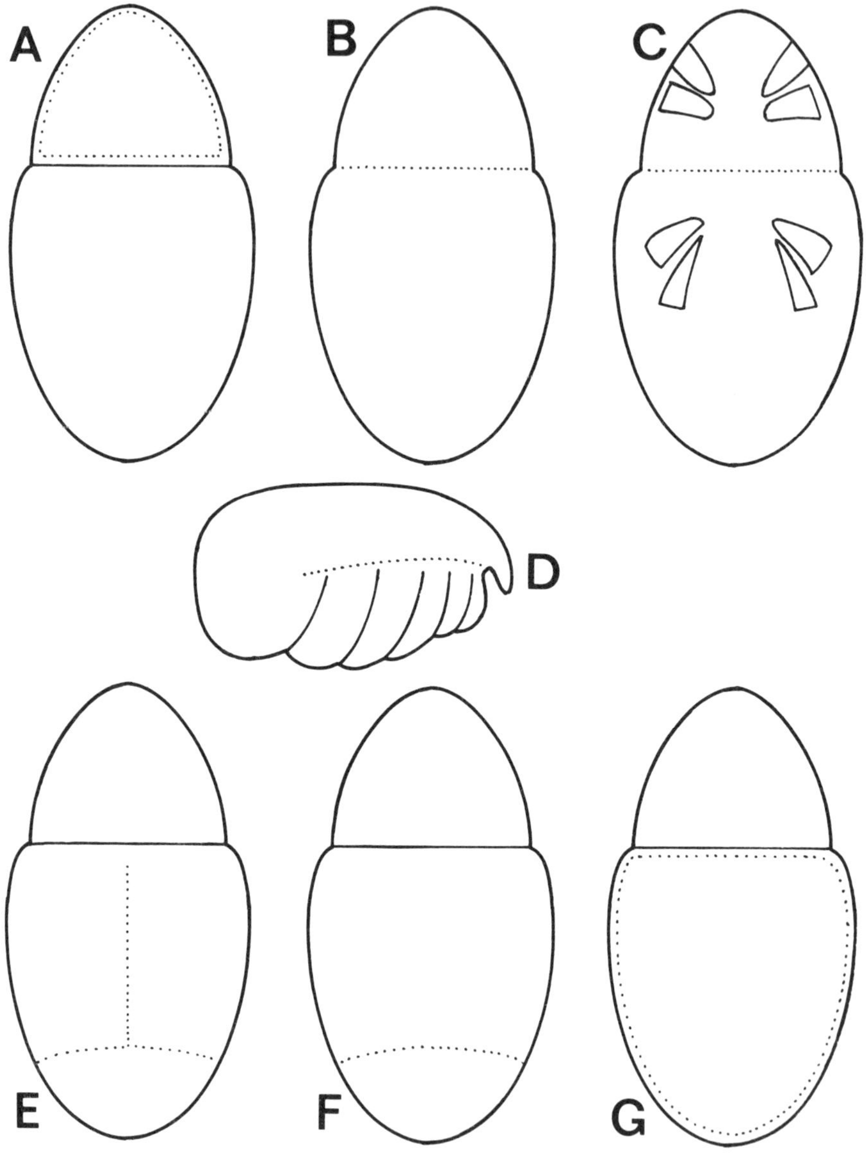

Fig. 18. Schematic representation of types of Chelicerate dehiscence (the dotted lines represent the lines of dehiscence); A–D, proterodehiscence; A, prodorsal dehiscence (dorsal view); B, C, amphidehiscence (B, dorsal view; C, ventral view); D, prelarval prodehiscence (lateral view); E–G, hysterodehiscence (dorsal views); E, dehiscence in the shape of a reversed T; F, transverse dehiscence; G, circumdehiscence.

(the line of dehiscence follows the abjugal furrow; the exuviae of the prodorsum can be completely cut off, or they can remain partly attached), prodehiscence (the line of dehiscence is laterofrontal and passes the frontal protuberance, or just below the naso), and amphidehiscence (the line of dehiscence cuts off part of the prodorsum; ventrally, it passes in front of the sejugal furrow, so that, in hatching, the anterior part of the exuviae is completely separated like a hood). There are also three types of hysterodehiscence (fig. 18E–G): transverse dehiscence (the line of dehiscence, in the opisthosomatic skin, is transverse and remains restricted to the posterodorsal region; the two parts of the exuviae are not completely separated), dehiscence in the shape of a reversed T (the line of dehiscence, in the opisthosomatic skin, has two parts: a transverse posterior line, and a longitudinal, i.e. sagittal, line running from the transverse line in anterior direction), and circumdehiscence (the line of dehiscence encircles the opisthosoma posteriorly, laterally, and often also anteriorly).

The type generally found in most groups of Chelicerata is probably constituted by prodorsal dehiscence, although variations, each entitled to a proper term, are known (in Araneida, e.g., an additional line continues on the opisthosoma; this condition is certainly connected with the presence of a pedicel); among Actinotrichida, prodorsal dehiscence is known from Actinedida. Prodehiscence is known from Actinotrichid calyptostases (regressive instars; see below, section 4).[1] Amphidehiscence is known from Erythraeidae (Actinotrichida: Actinedida). Transverse dehiscence and dehiscence in the shape of a reversed T are known from primitive Oribatida (Actinotrichida); circumdehiscence is known from higher Oribatida.

3. Phases

In a sequence of instars, the forms can be more or less similar, or present important differences from those which immediately precede and succeed them. In the first case they are called homostases or homostasoids, in the second case they are called heterostases or heterostasoids. In many Actinotrichida, we can distinguish a prelarva, a larva, three more or less similar nymphs, and an adult. In this case, the three nymphs are homostases. The life-cycle can be subdivided into phases according to the presence of homostases (or homostasoids) and heterostases (or heterostasoids). In the above-mentioned case of the Actinotrichida, e.g., we can distinguish four phases: a prelarval, a larval, a nymphal, and an adult phase. In Chelicerata, the number of phases varies. Generally,

---

[1] A different, unnamed, type is found in the calyptostasic protonymph and tritonymph of the Erythraeid mite *Balaustium florale* Grandjean (Actinotrichida: Actinedida; see Grandjean, 1959a: 165, 172, fig. 1C). In these stases the line of dehiscence is prodorsal and latero-opisthosomatic; this type could be a variety of amphidehiscence.

there are three or four phases. A primitive life-cycle with two phases (an immature and an adult phase) could have been present in the ancestors of extant Chelicerata.

The phase concept can be applied to stases as well as stasoids. It enables us to standardize the terminology of the subsequent forms of a life-cycle. Until now there have been two criteria for a definition of prelarva, larva, prenymph and nymph (stases), and larvoid and nymphoid (stasoids): the criterion of the level of postembryonic development and the criterion of regressive evolution. Acarologists have used the first criterion: in Actinotrichida, the instar of level 1 is called prelarva, the instar of level 2 larva, etc. Araneologists, on the contrary, have used the criterion of regressive evolution: according to the degree of regression, one or two prelarvae, and one or two larvae can be distinguished. Application of the phase-concept to the terminology of instars of various levels, enables us to standardize the names of the Chelicerate instars. In the theoretical case of two phases, the instars are called nymphs (or nymphoids) and adult; in the case of three phases, we can also distinguish larvae (or larvoids); in the case of four phases, we can moreover distinguish prelarvae. In this terminology, the term prenymph could only be used in the case of an additional (fifth) phase between larval and nymphal phase.

For a standardization of terminology, it is also necessary that in each separate group the basic number of phases (i.e., the number of phases on which the terminology in a group is based) should be that of the primitive condition in that group. In Actinotrichida, where ancestrally three nymphs can be distinguished, which constitute one nymphal phase, the three nymphs of Trombidina (Actinedida) are still named proto-, deuto- and tritonymph, although in reality they represent heterostases constituting three separate phases. In contradistinction to stasoids (which cannot be homologized with corresponding instars in related specimens and species), phases can always be homologized with corresponding sequences of instars in other specimens and species of the same group.

As mentioned above, the terminology of stases (prelarva, larva, nymph) is different from that of stasoids (larvoid, nymphoid). The nymphal phase of Araneida, e.g., apparently comprises 5–10 nymphoids. Isophena preserve the name of the level of postembryonic development of which they form part. We can, e.g., speak of the isophena of nympha 2.

As examples of a standardized terminology, the following life-cycles can be mentioned here. In Scorpionida, there are apparently a larva, several nymphoids and an adult. In Amblypygi, there are apparently a larva, several nymphoids, an adult, and isophena of the adult. In Uropygi, there are a larva, four nymphs and an adult. In Xiphosura there are apparently four phases (a prelarval, a larval, a nymphal and an adult phase); the prelarval phase remains inside the egg (there are two prelarval moults); the larval phase comprises the so-called trilobitic larva; the nymphal phase comprises the so-called prestwickianella larvae (it is possible that it should be subdivided into a prenymphal and a nymphal phase).

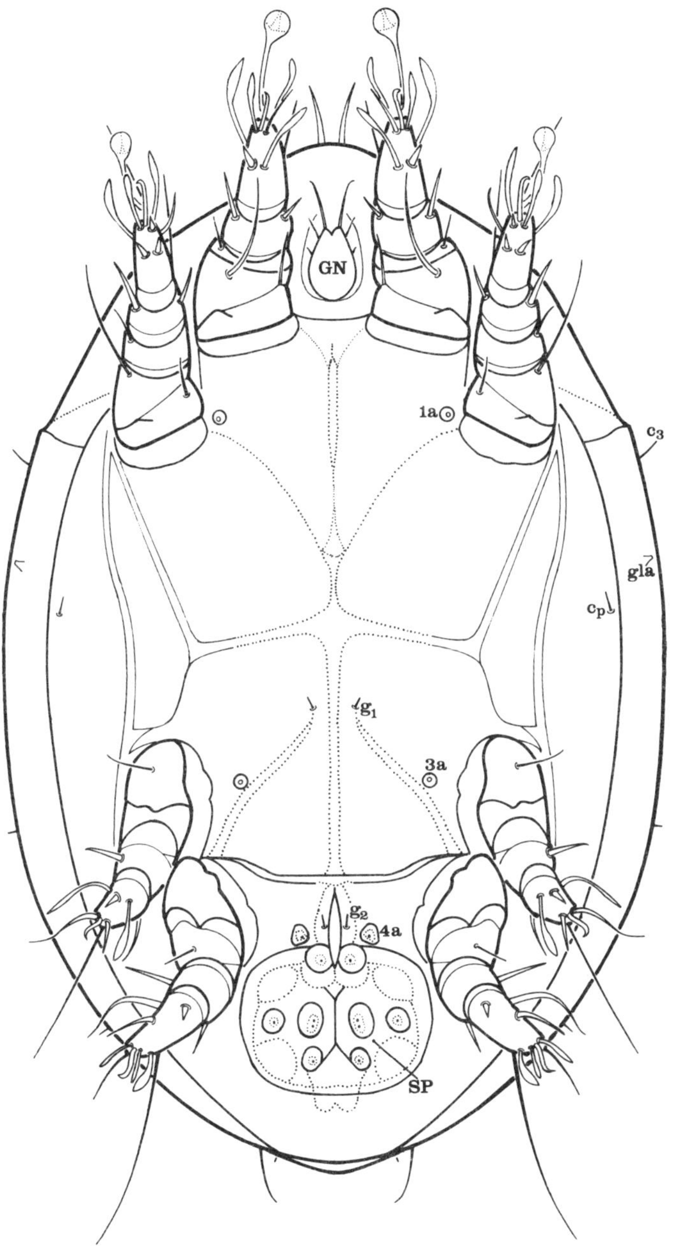

Fig. 19. Rhizoglyphus echinopus (Fumouze & Robin) (Actinotrichida: Acaridida); ventral view of hypopus (elattostasic deutonymph); c_3, c_p, opisthosomatic setae attributed to segments VII / VIII; g_{1-2}, genital setae; *gla*, latero-opisthosomatic gland; *GN*, regressive gnathosoma; *SP*, sucker plate; *1a, 3a, 4a*, epimeral setae; × 368.

4. Regressive instars

In the course of evolution, stases (and perhaps also stasoids) can be subject to various regressive evolutions. They can loose the use of mouthparts (so that they are unable to eat), or they can loose the use of mouthparts and appendages (so that they are also unable to walk). Regressive stases of the first type are called elattostases, those of the second type calyptostases. The prelarva of a species of *Saxidromus* (Actinotrichida: Actinedida), the larva (or larvae) of Araneida, the larva of Scorpionida, Uropygi and Amblypygi, and the deutonymph of many Acaridida (Actinotrichida; see fig. 19) are examples of elattostases. The prelarva of Opilioacarida (Anactinotrichida) is a calyptostase, although it constitutes a transition to an elattostase (fig. 20B). The prelarva of many Actinotrichida (figs. 17A, 20A), the prelarva(e) of Araneida, the larvae of Pseudoscorpionida, and the prelarva, the proto- and the tritonymph of Trombidina (Actinotrichida: Actinedida; see fig. 21) are examples of calyptostases. The examples from Actinotrichida demonstrate that a calyptostase can take the place of an elattostase, or a normal stase, at the same homologous level.

5. Evolutionary phenomena

The Chelicerate life-cycle can be subject to various types of evolution. A very characteristic evolutionary phenomenon is constituted by the regression of the first instar or instars; as far as known, this type of regression is present in all groups of extant Chelicerata. The first instar of Uropygi, Amblypygi, Opilionida and Scorpionida is an elattostase or a calyptostase. In Araneida several of the first instars can be subject to regression. In Actinotrichida, the first instar generally is a calyptostase, but in one primitive group it can be an elattostase. In Opilioacarida (Anactinotrichida) it is a calyptostase with some characters of an elattostase. The larva of Actinotrichida, Anactinotrichida and Ricinulei has only three pairs of legs (there can be a vestige of leg IV); this is also the result of a regressive evolution. In some species of Actinotrichida, the larva has even become an elattostase. The evolutionary phenomenon of the regression of the first levels of postembryonic development is called protelattosis. It comprises the regression of separate structures (such as leg IV), the formation of elattostases, and the formation of calyptostases.

A second evolutionary phenomenon is constituted by the attainment of maturity at a previously immature level. This phenomenon is generally called neoteny. Neoteny is apparently possible, or probable, or even practically certain in a number of cases: for instance in the case of some male Araneida and Scorpionida, and in those Actinotrichida, Pseudoscorpionida and Anactinotrichida, where two nymphs are present instead of three.

A third evolutionary phenomenon pertains to the regression of instars at other levels than at the beginning of postembryonic development. It is called

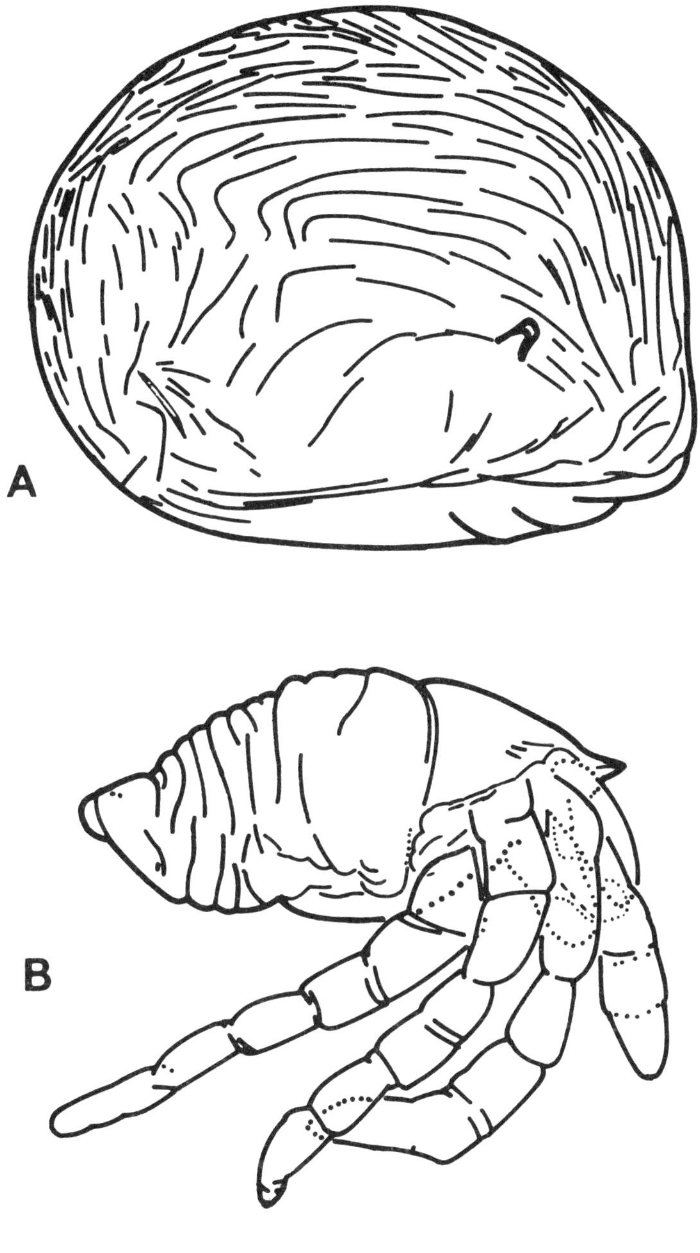

Fig. 20. Prelarvae of mites, in lateral orientation; A, *Hoplophthiracarus* spec. (Actinotrichida: Oribatida); B, *Phalangiacarus brosseti* Coineau & Van der Hammen (Anactinotrichida: Opilioacarida); A, × 600; B, × 132. (Fig. 4B redrawn after a drawing by Coineau, in Van der Hammen, 1976).

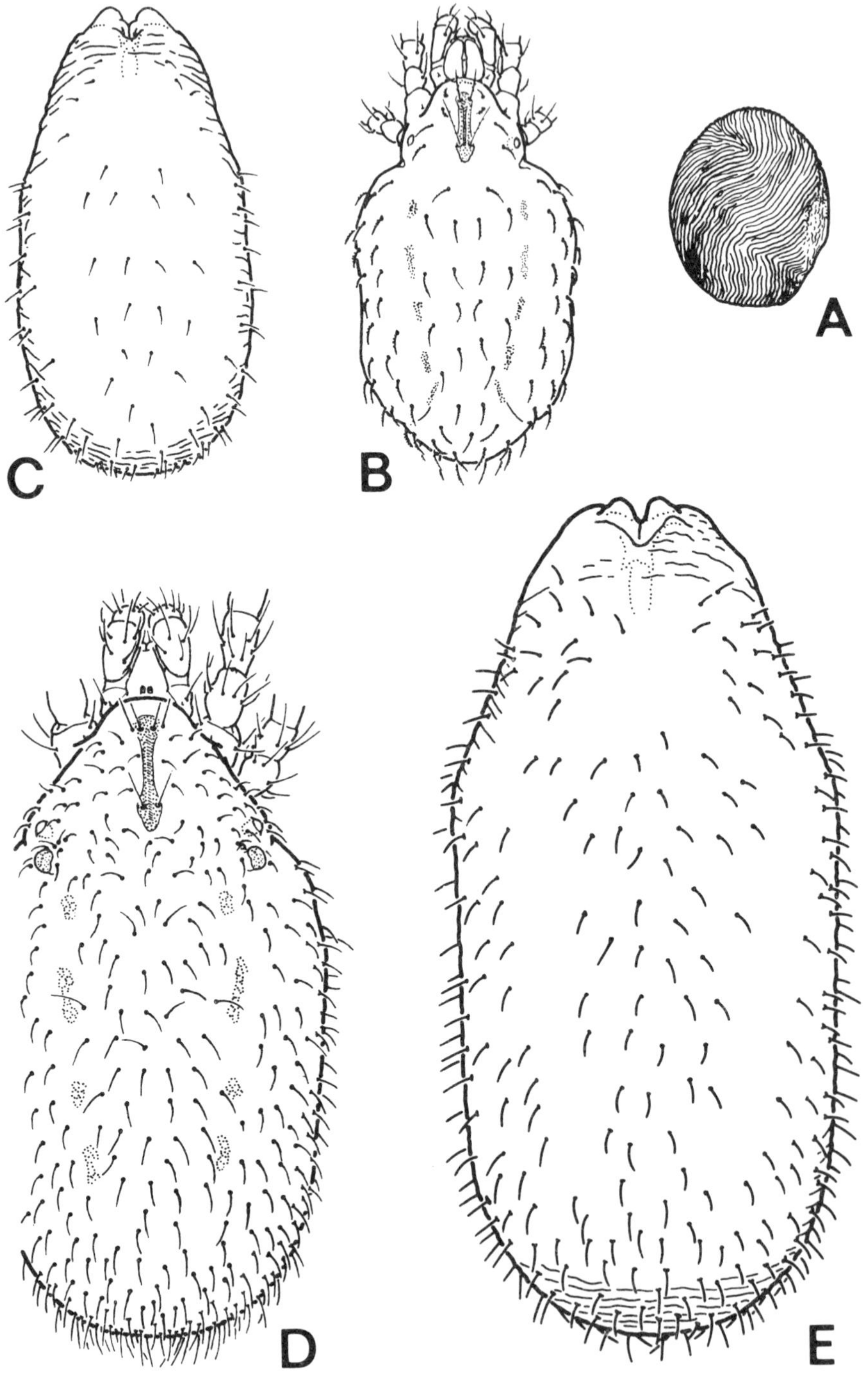
A
B
C
D
E

metelattosis. This phenomenon is well-known from Actinotrichida. I point to the formation of hypopi (elattostases at the deutonymphal level) in many Acaridida (Actinotrichida; see fig. 19), and to the formation of calyptostases at the level of proto- and tritonymph in Trombidina (Actinotrichida: Actinedida; see fig. 21). This phenomenon is apparently very rare in other groups of Chelicerata. I can mention here only one example discovered by Weygoldt (1970): the regression of trichobothria in the third instar of Amblypygi. This case apparently represents the very first beginning of a separation of the third instar from its phase.

The fourth evolutionary phenomenon pertains to the formation of isophena at one of the levels of postembryonic development; it is called plethomorphosis. This phenomenon is found in nymphs of Argasidae (Anactinotrichida: Ixodida), and in the case of moulting adults (as in Amblypygi and some groups of Araneida). It could also be present in nymphs and adults of several other groups, but unmistakable data are lacking (the rare occurrence of adult moults in mites, such as Microthrombidiidae, could represent an abnormal phenomenon; see Robaux, 1974: 95).

6. General model of the evolution of the Chelicerate life-cycle

Starting from the data, collected and classed in the previous sections of the present chapter, we can try to prepare a general model of the evolution of the Chelicerate life-cycle. Such a model can be developed by arranging the life-cycles, the various types of forms and instars, and the evolutionary phenomena, according to the hypothetical interpretation of character states (ancestral or derived), in a chronological diagram in which phylogenetic time is on the horizontal axis, and ontogenetic time on the vertical axis (fig. 22). The hypothesis can be introduced that the evolution of the life-cycle has started with stasoids and presented only one immature phase (consisting of nymphoids). It is supposed that a life-cycle with stases arose from a life-cycle with stasoids. In this way, two evolutionary periods (or stages of evolution) can be distinguished: stasoidy and orthostasy. It is further supposed that the ancestral number of stases has, in all cases, been six (protostasy; as still found in several groups). This number has apparently arisen several times in different groups of Chelicerata, and the potentiality of an evolution of this number is apparently a general character of the Chelicerata. By the divergent phylogeny of the levels of postembryonic development, different immature phases can have arisen (prelarval, larval and nymphal). In the course of evolution, various lev-

←

Fig. 21. Part of the life-cycle of *Balaustium florale* Grandjean (Actinotrichida: Actinedida); A, calyptostasic prelarva; B, larva; C, calyptostasic protonymph; D, deutonymph; E, calyptostasic tritonymph; A–E, × 120. (Redrawn after Grandjean, 1957, 1959).

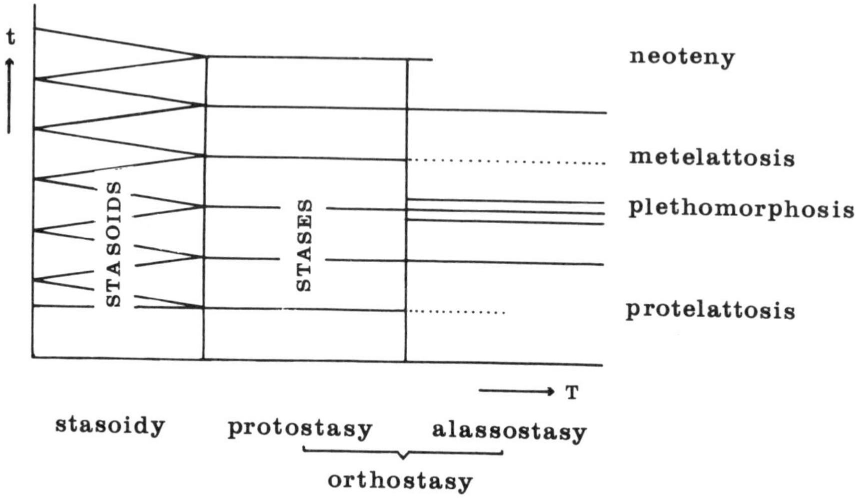

Fig. 22. Model of the evolution of the Chelicerate life-cycle (ontophylogenetic diagram; *t* = ontogenetic time, *T* = phylogenetic time). The hypothesis is that the evolution of the life-cycle has started with stasoids (instars which are morphologically different from the instar which precedes them by discontinuous characters, but which occur in variable numbers, so that they cannot be homologized with instars in other specimens and taxa of the same group) and presented only one immature phase (consisting of nymphoids). It is supposed that a life-cycle with stases (instars which, as in the case of stasoids, are morphologically different from the instar which precedes them, but which occur in fixed number, by which they can be homologized with instars in other specimens and taxa of the same group) arose from a life-cycle with stasoids, by which two evolutionary periods can be distinguished: stasoidy and orthostasy. It is further supposed that the ancestral number of stases has, in all cases, been six (protostasy; as still found in several groups). As a result of the divergent phylogeny of the levels, different phases can have arisen (prelarval, larval, nymphal and adult). In the course of evolution, various levels of the ancestral protostatic life-cycle can have been subject to changes (alassostasy), including regressive evolution (protelattosis and metelattosis), the attainment of maturity at previously immature levels (neoteny), and the formation of isophena, i.e., forms which are the result of growing or repetition moults (plethomorphosis). At a certain stage of evolution, stases and stasoids can occur in one and the same life-cycle. It is probable that, as to phylogeny, the model should be read in both directions, and that a third period (neostasoidy) must be distinguished.

els of the ancestral protostasic life-cycle can have been subject to changes (alassostasy), including regressive evolution (protelattosis and metelattosis), the attainment of maturity at previously immature levels (neoteny), and the formation of isophena (plethomorphosis).

For ease of survey, the model is considerably simplified. It is, for instance, highly probable (or even certain) that, during a period of transition, stases and stasoids are present in one and the same life-cycle (in Amblypygi, Araneida and Scorpionida, the first one or two phases probably consist of stases, whilst the nymphal phase apparently consists of stasoids). The possibility cannot be excluded that, as to phylogeny, the model should be read in both directions and that neostasoids could have secondarily arisen by a chaotic increase in the number of instars (neostasoïdy). It is supposed that, in the case of neostasoidy, a single stase, or a phase, is replaced by an irregular number of instars. Evidently these groups (or phases), consisting of irregular numbers of instars, can only be homologized with similar groups or phases in other Chelicerata (single instars of this type cannot be homologized with similar single forms). It is now also evident that, in the case of stasoidy and neostasoidy, instars evolve as groups or phases; it is only in the case of stases that they evolve by age (by distinctly defined levels of postembryonic development).

The simplified model, prepared in the present section, has the advantage that it has led to a perfection of the terminology, and that it has indicated many gaps in our knowledge; a detailed program of investigations could now be based on it. It is, for instance, now possible to indicate which of the instars in Palpigradi and Ricinulei are probably still unknown. It is now also evident that the nature of many immature forms (stasoids, stases or isophena), in the life-cycle of Opilionida and Solifugae is still uncertain.

It might, finally, be remarked here that the functional aspect of the evolutionary changes is not yet well known. Some of the regressive changes could be associated with greater protection and shortening of the life-cycle. Other phenomena are evidently associated with dispersal: regressive deutonymphs of Acaridida (Actinotrichida) are actively transported by other animals.

V. RELATIONSHIPS, CLASSIFICATION AND EVOLUTION

The problem of the relationships, the evolution and the natural classification of the Chelicerata, has repeatedly been the subject of my papers. I started with studies of the recognition, the definition and the relationships of the main natural groups of mites, and of the relationships of these with some other groups of Chelicerata (Van der Hammen, 1961: 177–179; 1966: 60–64; 1968a: 276–277; 1968c: 402–407; 1971: 471–472; 1972a; 1973). The results of these papers led to studies of the relationships and the natural classification of all main groups of Chelicerata (Van der Hammen, 1977a; 1985: 52–55, fig. 34; 1985a: 145; 1986a: 46–47; 1986c: 224–228, fig. 3). Not until the completion of my comparative studies in Chelicerata have these relationships been fully understood.

Chelicerata constitute a very old group, of which representatives are known since the Cambrian (Xiphosura), the Ordovician (Eurypterida) and the Silurian (Scorpionida). Primitive Actinotrichida, closely related to extant species, are known from the Devonian. Several other Chelicerate groups had already developed their characteristic general habitus in the Carboniferous (Ricinulei, Uropygi, Araneida). Evidently, most of the higher Chelicerate taxa have evolved separately since the Paleozoic era. Because of this long separation in time, the affinities between the groups are indistinct, and several controversial views were introduced. Current taxonomic (cladistic) methods, for instance, have resulted in highly artificial classifications which were evidently based on atomistic views in which the organism as a functional whole had virtually no place.

Terrestrial life was not possible before the Silurian, and extant terrestrial Chelicerata have evidently evolved from aquatic ancestors. The biology of modern terrestrial Chelicerata, consequently, must be regarded as an adaptation of earlier aquatic ways of life to new terrestrial circumstances. For this reason, the earliest separations must have been connected with life in the aquatic environment; these separations could, in some cases, have pertained to characters which subsequently, in the terrestrial period, have been superseded by others.

Because of these considerable difficulties, I have gradually developed a structuralist approach in the study of Chelicerate evolution and classification (Van der Hammen, 1981, 1985b, 1986d, 1987, 1988), in which particular attention was paid to the hierarchy of systems of transformation. In the model of evolution, developed in this way, and which must constitute the base of a natural classification, further attention was paid to the evolutionary potentialities of the genotype. Separations of groups were supposed to pertain to separations of evolutionary programs, which can subsequently manifest themselves discontinuously, at various moments of evolution, by parallelism. Particular attention was further paid to a hypothetical explanation of the hierarchic value attributed to a character (or a character complex).

As a result of this structuralist analysis of the characters of the main Chelicerate groups, it has become evident that the earliest phylogenetic branchings must have been associated with important differences in prosomatic architecture (segmentation and articulation of the legs; position of the mouth; the presence of epimera, or of neosternum and coxae, or of coxae involved in food-intake; the position of the orifices of coxal glands, and differences in associated structures and in ways of transport of coxal gland secretions). Branchings at lower levels must have been associated with different adaptations pertaining to respiration, sperm transfer, new specializations in locomotion, etc.

In this way, three large groups have now been distinguished: (a) Coxisternata (rostral feeding; absence of coxae and sternum, presence of epimera with apodemes; mouthparts with posterior apodeme on which extrinsic muscles are inserted; orifice of coxal gland ancestrally posterodorsal to the base of leg I; coxal gland secretions ancestrally transported to the dorsal surface of the mouthparts); (b) Neosternata (rostral feeding; coxae and neosternum present; mouthparts with posterior apodemes, to which intrinsic muscles are attached; orifices of coxal glands ancestrally posteroventral to coxae I, and ancestrally associated with sternapophyses and subcapitular or intercoxal gutter; coxal gland secretions transported to the ventral surface of the mouthparts); and (c) Myliosomata (coxisternal feeding; coxae large and movable in connection with food intake; orifices of coxal glands not associated with coxae I, at least in the adult). The distinction of two superclasses among these three groups remains more or less arbitrary, but the most obvious solution appears to be a division into Rostrostomata (with rostral feeding) and Myliosomata (with coxisternal feeding). Classification at the next lower level is associated with fundamental differences in locomotion. As a result of these considerations, the following survey of the higher classification of Chelicerata (it is a very simple solution) has now been prepared; phylogenetic relationships are represented in fig. 23.

Subphylum *Chelicerata* Heymons, 1901

Arthropods, the body of which is composed of two tagmata: prosoma (precheliceral metameres and segments I–VI) and opisthosoma (ancestrally 10–13 segments, and sometimes also a so-called 'telson'). Ancestrally, six pairs of prosomatic appendages (chelicerae, palpi, legs I–IV) are present, and up to

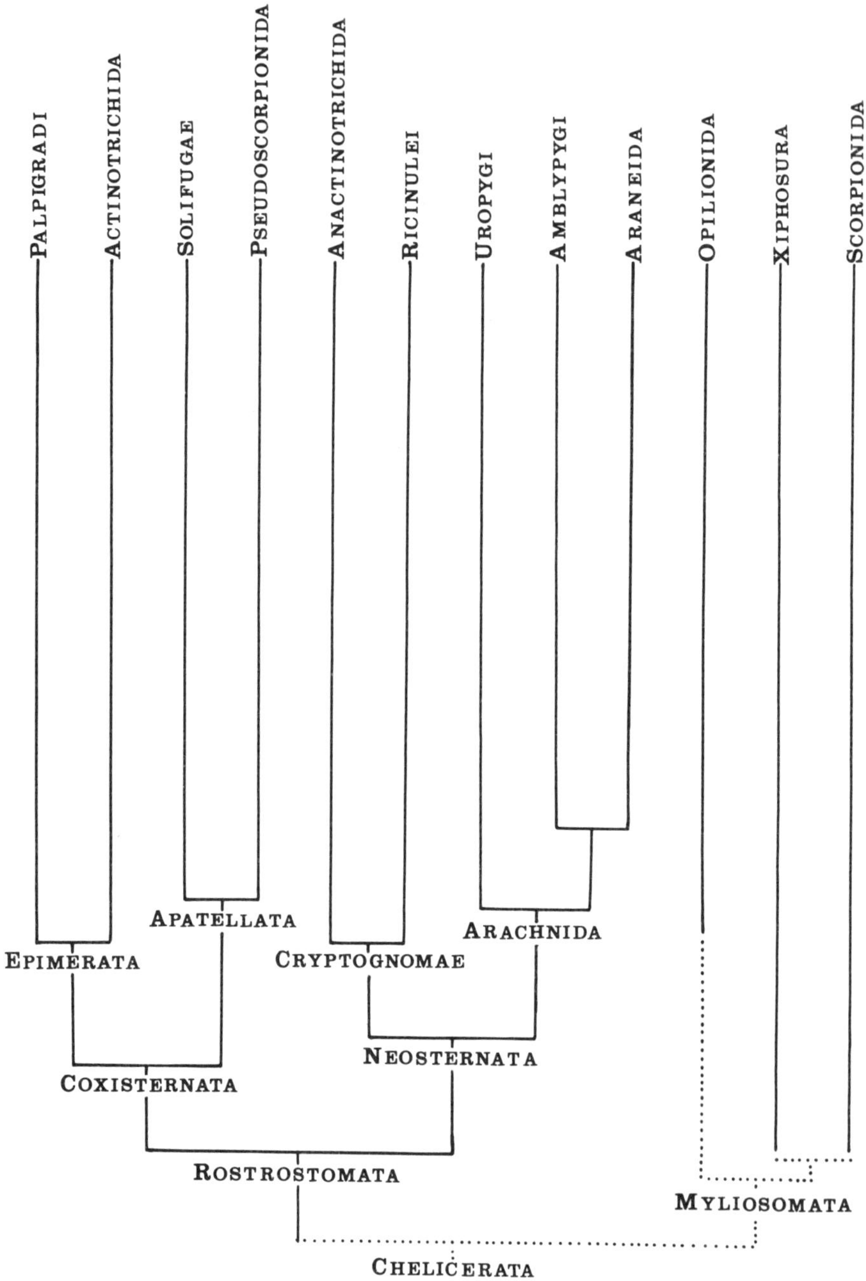

Fig. 23. Diagrammatic representation of the present view of Chelicerate phylogenetic relationships.

seven pairs of homologues of opisthosomatic appendages. Postembryonic development characterized by moulting and the occurrence of a series of instars (the first of which is a regressive form).

Chelicerata are supposed to be related to Pycnogonida. They are usually classified with Arthropoda, although this (super)phylum could represent a grade of evolution, which has been attained independently by several groups.

Superclass 1. *Rostrostomata* nov.

Chelicerata characterized by rostral feeding. Although this type of ingestion could have been the ancestral way of food intake, it has subsequently been perfected by a further specialization of the mouthparts. As mentioned above, the classification of the three main groups of Chelicerata in two superclasses is perhaps arbitrary, although according to the present state of our knowledge it appears to be the most logical solution.

Class 1. *Coxisternata* nov.

Rostrostomata, the body of which is ancestrally composed of seventeen segments (not known with certainty from Pseudoscorpionida). Prosomatic architecture characterized by the presence of epimera and apodemes (a perfection of a primitive condition). Mouthparts characterized by the presence of a capitular or rostrosomatic apodeme which constitutes the place of insertion of extrinsic muscles. Orifice of coxal gland ancestrally posterodorsal to the base of leg 1. Coxal gland secretions are transported to the dorsal region of the mouthparts, ventral to the base of the chelicerae (not known from Pseudoscorpionida). Legs ancestrally with two femora.

Subclass 1. *Epimerata* Van der Hammen, 1977a

Coxisternata, in which supracoxal setae are ancestrally present. Opisthosomatic appendages of segments VIII and IX associated with the genital opening. Ancestrally, the opisthosomatic appendages of segments X–XII are suppressed during embryonic development; they gradually re-appear in the course of postembryonic development. Legs with genu, and ancestrally with tridactyl claws; tarsus not subdivided into basi- and telotarsus, although a subdivision into tarsus 1, 2, etc. can be present. Life-cycle ancestrally composed of six stases (not completely known in the case of Palpigradi).

Superorder 1. *Palpigradi* Thorell, 1888

Epimerata, the prodorsum of which is subdivided into propeltidium, mesopeltidium and metapeltidium. Rostral region with frontal organ. No eyes. Thirteen pairs of prodorsal setae. Supracoxal seta *e* present above the base of the palp; it can be multiplied. Mouth surrounded by two lips: labrum and labium. The epimera of palp and leg I have fused and present epimeral setae. The opisthosoma is subdivided into eleven segments. Terminal flagellum long and filiform, with numerous subdivisions. Anus with terminal position in segment XVII (ventral to the base of the flagellum). Segments X–XII, ancestrally, with opisthosomatic appendages in the shape of evaginable papillae; these can be replaced by verrucae. Chelicerae three-segmented; cheliceral trochanter relatively large. Palp relatively long and tridactyl. The number of adesmatic tarsal segments in palp and legs I–IV (of the adult) is: 5–7–2–2–3. The articulation between femur 2 and genu of leg I, and that between genu and tibia of palp and leg I, are constituted by bidesmatic hinge joints (with extensor muscles).

Superorder 2. *Actinotrichida* Grandjean in Van der Hammen, 1961

Epimerata, the prodorsum of which is scleritized and entire, or transversely subdivided by two furrows. A frontal protuberance (or naso) can be present in the rostral region. An unpaired anterior eye and one or two pairs of lateral eyes can be present. Aspidosoma ancestrally with six pairs of prodorsal setae. Supracoxal setae *e*, *eI*, and *eII* can be present above the base of palp and legs I and II respectively; supracoxal seta *eII* can be multiplied. Mouthparts constituted by a gnathosoma. Mouth originally surrounded by four lips, viz., labrum, lateral lips and labium: the labium has disappeared in higher Actinotrichida. Opisthosomatic segmentation particularly recognizable in some primitive species. Opisthosomatic segments XIII–XVII are suppressed in the course of embryonic development, and can re-appear (with the exception of segment XVII) in the course of postembryonic development (hysteromorphosis). There is no flagellum. The position of the anus is ventral. The eugenital opening is trifid; it is, ancestrally, situated in a progenital chamber which is closed by progenital valves; the opening between the valves is, ancestrally, longitudinal. The opisthosomatic appendages of the segments X–XII constitute the genital papillae which are included in the progenital chamber. Chelicerae with reduced trochanter (or trochanter no more recognizable). Palp relatively small, without apotele (adactyl). Legs exceptionally with tarsal subdivisions (tarsus 1, 2, etc.).

Subclass 2. *Apatellata* Van der Hammen, 1977a

Coxisternata in which supracoxal setae are unknown. Mouthparts constituted by a rostrosoma (with lateral lips). Opisthosomatic appendages of segment VIII associated with the genital opening. The opisthosomatic respiratory organs could have originated from the opisthosomatic appendages of segments IX and X. Chelicerae 'raptorial', i.e. large and two-segmented with well-developed chela. Legs bidactyl, ancestrally without patella; the tarsus consists of basi- and telotarsus. Life-cycle, ancestrally, either with six stases or with a great and apparently irregular number of forms.

Superorder 3. *Solifugae* Sundevall, 1833

Rather large or medium-sized Apatellata. Prodorsum subdivided into pro-and postpeltidium, connected by arthrodial sclerites. Epimera of legs II and III separated by a sejugal interval. Opisthosoma distinctly segmented, with dorsal and ventral sclerites, and pleural soft skin. One pair of median eyes and one or two pairs of (regressive) lateral eyes. Respiratory system comprising one pair of prosomatic stigmata and two pairs of opisthosomatic stigmata (in some groups, an additional unpaired opisthosomatic stigma is also present), connected with a vast system of tracheae. Labrum with long anastomosed setae. Cervix and labium attached to the coxal region of the palpi. The male chelicera can function as a gonopod. A small palpal coxapophysis is present. Palpal apotele transformed into a sucker. Legs III and IV with two trochanters. Epimeron 4 and trochanter 1 of leg IV with malleoli. Life-cycle characterized by the occurrence of a rather great and variable number of moults and forms. Regressive (elattostasic) larva with so-called lateral organ.

Superorder 4. *Pseudoscorpionida* Latreille, 1817

Apatellata of rather small size. One or two pairs of (sometimes regressive) lateral eyes. Respiratory system consisting of tracheae and two pairs of opisthosomatic stigmata, lateroventrally in segments IX and X. Lateral lips with superior and inferior projection. Pharynx with pharyngeal pump. Coxal gland belonging to segment V; orifice at the posterior border of coxa III. Palp with terminal chela (constituted by tarsus and apotele). Femur 1 and 2, and basi- and telotarsus, can have fused in the two anterior pairs or in all pairs of legs. The life-cycle includes two calyptostasic larvae and three nymphs (two nymphs in the case of supposed neoteny).

Class 2. *Neosternata* nov.

Rostrostomata, the body of which is ancestrally composed of nineteen or eighteen segments (and precheliceral metameres). Prosomatic architecture characterized by the presence of a neosternum and coxae. Mouthparts with posterior apodeme, to which intrinsic muscles of the mouthparts are attached. Orifice of coxal gland ancestrally posteroventral to coxa I, and ancestrally associated with sternapophyses and subcapitular or intercoxal gutter; coxal gland secretions transported to the ventral surface of the mouthparts. Genital opening ancestrally represented by a transverse slit. Legs ancestrally with one femur, a patella, and basi- and telotarsus.

Subclass 3. *Cryptognomae* Van der Hammen, 1977a

Small to medium-sized Neosternata, the body of which is ancestrally composed of precheliceral metameres and nineteen segments. The mouthparts constitute a movable gnathosoma. Eyes absent, or one to three pairs of lateral eyes. One to four pairs of stigmata. Trichobothria absent. Palp without free coxa. The palpal tibia and tarsus tend to constitute one segment (tibiotarsus). Legs III and IV ancestrally with two trochanters. Life-cycle ancestrally with six stases: a calyptostasic prelarva, a hexapod larva (a vestige of leg IV is present in Opilioacarida and Ricinulei), three nymphs and an adult.

Superorder 5. *Anactinotrichida* Grandjean in Van der Hammen, 1961

Small or medium-sized Cryptognomae. Prosoma and opisthosoma broadly joined. Opisthosomatic segmentation recognizable in Opilioacarida. Eyes absent or present (one, two or three pairs of lateral eyes). One to four pairs of stigmata. Podosoma with relatively large neosternum or neosternal region. One pair of sternapophyses (often fused; sometimes suppressed). Mouth with three commissures, bordered by the labrum and the paired lateral lips. Infracapitulum (ancestrally) with cervix, mentum, large lateral ridges and malapophyses. Chelicerae ancestrally prehensile, three-segmented and chelate; body of chelicera ancestrally with two lyrifissures. Palp ancestrally with apotelic claw (with paraxial position). Border between palpal tibia and tarsus ancestrally oblique; the two segments together have ancestrally the shape of a single tibiotarsus. Legs with free coxae. Legs III and IV of Opilioacarida with two trochanters. Femora with basifemoral ring (associated with lyrifissures). Telotarsi II–IV subdivided in Opilioacarida. A pretarsus (sometimes with setae or vestiges of setae) is often present.

Superorder 6. *Ricinulei* Thorell, 1892

Medium-sized Cryptognomae. Prosoma with a cucullus and sometimes with vestiges of (two pairs of) eyes. One pair of stigmata with (?secondarily) prosomatic position. Anterior part of opisthosoma (segments VII and VIII) constituting a pedicel. Prosoma and opisthosoma can be coupled by a locking device. The posterior three opisthosomatic segments (XVII–XIX) constitute a retractable pygidium or metasoma. Coxae large, neosternum small. Two or three sternapophyses. Gnathosoma simple, located in a camerostome. Mouth with two commissures; labrum large, labium incorporated in mentum. Chelicerae two-segmented (there is no trochanter), and chelate. Palp with tibiotarsus and apotele; the terminal part of the tibiotarsus and the apotele constitute a chela. Coxa IV movable; coxae I–III fixed. Legs III and IV with two trochanters. Telotarsi II–IV subdivided. Leg III of the male can function as a gonopod (basitarsus, telotarsus 1 and telotarsus 2 are involved in sperm transfer).

Subclass 4. *Arachnida* Lamarck, 1801 (s.str. Van der Hammen, 1977a)

Small, medium-sized or rather large Chelicerata. Body composed of precheliceral metameres and eighteen segments. Segment VII narrowed. Ancestrally, one pair of median and three pairs of lateral eyes (the number of eyes can be reduced to various degrees). Mouthparts rather simple; mouth generally bordered by two lips, viz., labrum and labium (the labium is absent in Amblypygi; there are no lateral lips). Anus with terminal position in segment XVIII. Respiratory organs ancestrally constituted by two pairs of booklungs and ventral stigmata in segments VIII and IX. Trichobothria present. The chelicerae are two-segmented and represent the pocket-knife type. Palp with free coxa, or palpal coxae fused; palp ancestrally with apotele. Legs with one trochanter. Coxae of the legs with anterior internal apodeme. Coxa-trochanter joint of the legs with arthrodial sclerite; patella-tibia joint of the legs associated with antero-posterior movements. Apotele of the legs in many cases with a so-called pseudonychium (not homologous with a central ungues). Life-cycle either with six stases (larva, four nymphs, adult) or with an irregular and sometimes even rather great number of moults and forms (adult moulting is also found in some groups). First instars octopod (not hexapod).

Superorder 7. *Uropygi* Thorell, 1882

Arachnida with narrowed metasoma (segments XVI–XVIII), in which tergites and sternites are not differentiated. 'Telson' represented by a flagellum (either short or long). Prodorsum represented by a large prodorsal sclerite or by propeltidium, postpeltidium and arthrodial sclerites. The mouthparts include the fused palpal coxae, an intercoxal gutter and palpal coxapophyses. Paired orifices of coxal glands associated with sternapophyses and intercoxal gutter. One or two pairs of booklungs. Small neosternum present between the coxae of legs II. The palp presents a large apotele (with prehensile function). The coxae of the legs occupy the ventral surface of the prosoma. Apotele absent in the case of leg I. Telotarsi subdivided into several adesmatic segments. Life-cycle with six stases (elattostasic larva, four nymphs and adult).

Superorder 8. *Araneomorphae* nov.[1]

Arachnida without narrowed metasoma and without 'telson'. Prodorsum represented by a single prodorsal shield. Palpal coxae not fused; coxapophyses present in many groups (absent in several Araneida). Orifices of coxal glands paraxial, at the base of coxa I, and in some cases also at the base of coxa III. Transport of coxal gland secretions either associated with sternapophyses (Am-

---

[1] The name Labellata Petrunkevitch, 1949 does not refer to a character by which this group is particularly distinguished.

blypygi) or not (Araneida). Neosternum generally rather large. Ancestrally two pairs of booklungs which, in Araneida, can be replaced by tracheae. Postembryonic development with a great and irregular number of moults.

There are two Araneomorph orders: Amblypygi (opisthosoma distinctly segmented; mouthparts without labium; sternapophyses present; palp with large apotele and prehensile function; leg I very long, with numerous adesmatic tibial and tarsal segments, and with reduced apotele) and Araneida (opisthosomatic segmentation present in Liphistiidae only; mouthparts with labium; no sternapophyses; opisthosomatic appendages of segment X and XI transformed into spinnerets; chelicerae with poison glands; male palp can function as a gonopod; leg I not exceptionally long).

Superclass 2. *Myliosomata* Van der Hammen, 1985

Chelicerata characterized by the presence of a so-called myliosoma, associated with coxisternal feeding. The anterior part of the alimentary canal (pharynx or oesophagus) presents a downward or even backward curvature. The mouth is surrounded by labrum, labium and the coxapophyses of palp and/or one or more pairs of legs. The orifices of the coxal glands are near coxae III. Prosoma and opisthosoma broadly joined.

The superclass comprises three extant superorders, which must have evolved separately since the Cambrian. The phylogenetic relationships of the three groups are still unknown.

Superorder 9. *Opilionida* Sundevall, 1833

Myliosomata the opisthosoma of which is composed of the segments VII–XVI. Prosoma with one pair of median eyes (sometimes absent) and one pair of odoriferous glands (with lateral orifice). The myliosoma includes the coxapophyses of palp and anterior legs. The respiratory system consists of tracheae and one pair of stigmata with ventral position, in segment VIII, posterior to coxa IV (pedal stigmata and tracheae are present in Phalangiidae). Genital opening with advanced position, between the coxae of the legs. Adult female with ovipositor, adult male with penis or spermatopositor. Chelicerae three-segmented. Palp and legs with coxa, trochanter, femur, patella, tibia, tarsus and apotele (the palpal apotele is absent in many cases). Palpal coxa free. Coxae of the legs with ventral position; posterior coxae immovable, anterior coxae movable in connection with ingestion. Palp and legs with rocking joints between trochanter and femur, and between patella and tibia. The tarsus of the legs is subdivided into basi- and telotarsus; these adesmatic segments can also be subdivided. Trichobothria are not present. Life-cycle with an elattostasic or calyptostasic larva, a nymphal phase with a variable number of instars (probably nymphoids; four nymphs could be present in some cases), and an adult.

Superorder 10. *Scorpionida* Latreille, 1817

Myliosomata of relatively large size, the body of which is composed of nineteen segments and a 'telson' (the sting). The opisthosoma consists of a mesosoma and a distinctly narrowed metasoma (segments XV–XIX and 'telson'). Prosoma with a single prodorsal shield. One pair of median eyes (rarely absent), and 2–5 pairs of lateral eyes. Anus terminal in segment XIX, ventral to the 'telson'. The opisthosomatic appendages of segment VIII constitute the genital operculum; those of segment IX constitute the combs or pectines. Four pairs of booklungs and four pairs of stigmata with ventral position in the segments X–XIII. Orifices of the coxal glands at the base of coxae III. The coxapophyses of legs I and II participate in ingestion (there are no palpal coxapophyses). Chelicerae three-segmented. Palp with relatively large terminal chela. The legs consist of coxa, trochanter, femur, patella, tibia, basitarsus, telotarsus and apotele. Coxae II–IV are not movable (coxa I is movable in connection with ingestion). Pivot joint between patella and tibia. The apotele has two ungues and a pseudonychium. Sperm transfer by spermatophores. Life-cycle apparently with an elattostasic larva, several nymphoids and an adult (there are apparently 6–8 moults).

Superorder 11. *Xiphosura* Latreille, 1802

Large-sized Myliosomata with large prodorsal shield, narrow sternum, and an opisthosoma which is indistinctly composed of the segments VII–XVI and a large 'telson' or tail. There are homologues of seven pairs of opisthosomatic appendages: the chilaria (segment VII, with advanced position), the genital operculum (segment VIII) and five pairs of branchial appendages (segments IX–XIII). Anal opening terminal, ventral to the base of the tail or 'telson'. One pair of relatively large lateral eyes (associated with a pair of vestigial eyes) and one pair of small median eyes. Frontal organ with vestigial eyes. Coxal glands of segments II–V fused; orifice at the posterior border of the coxa of limb 5 (homologous with coxa III of other Chelicerata). Chelicera three-segmented, with terminal chela. The coxapophyses of limbs 2–6 participate in food-intake (the coxapophyses of limbs 2–5 are associated with the mouth). Pharynx not present. The foregut consists of oesophagus and crop; the oesophagus is curved downwards and backwards in such a way that the labrum is ventral to the mouth and the labium dorsal. Limbs 2–5 (homologous with palp and legs I–III of other Chelicerata) consist of coxa, trochanter, femur, patella, tibiotarsus and apotele; limb 6 (homologous with leg IV of other Chelicerata) consists of coxa, trochanter, femur, patella, tibia, tarsus and apotele. All coxae movable (associated with ingestion and with promotor-remotor movements). The coxa of limb 6 presents an 'exite' (the so-called flabellum). Trichobothria are not present. External fertilization. Internal digestion. Life-cycle with four phases (prelarval, larval, nymphal, adult); the prelarval phase remains inside the egg (there are two prelarval moults); the larval phase comprises the so-called trilobitic larva; the nymphal phase comprises the so-called prestwickianella larvae.

Extinct groups are not included in this classification because the morphology of fossil Chelicerata, with the exception of Eurypterida, is not very well known. Trigonotarbi and Kustarachnae can probably be classified as separate superorders with Arachnida s.str. Stethostomata (Haptopoda and Anthracomarti) probably constitute a separate subclass among the Neosternata. Eurypterida are undoubtedly Myliosomata. Architarbi could perhaps represent another group of Myliosomata.

As mentioned above, the earliest evolutionary events in the phylogeny of the Chelicerata are supposed to have pertained to ingestion and locomotion (both closely connected with the entire prosomatic architecture). In this way, three main groups of Chelicerata have been distinguished: Coxisternata, Neosternata and Myliosomata. A few remarks are added here on some of the most primitive representatives of these groups, and on some other key events in Chelicerate evolution (those associated with the transition from aquatic to terrestrial life). Some remarks are also made on the unfoldment of successful groups.

In nearly all groups of Chelicerata, species are found which are much more primitive than other representatives of the same group. Palaeosomata, Endeostigmata (Actinotrichida) and Opilioacarida (Anactinotrichida), for instance, can be characterized, because of the possession of numerous primitive attributes, as living fossils. The most primitive Chelicerate subclass is unquestionably constituted by Palpigradi. In this group, the prosoma presents a number of separate dorsal sclerites, and a number of separate sternites; the endosternite has six pairs of ventral suspensor muscles (the primitive number), four pairs of dorsal suspensors, and five pairs of transverse suspensors; the cuticle is thin, and respiration is apparently cutaneous; the chelicerae are three-segmented; and there is a 'telson' (the flagellum).

Some groups of Chelicerata are characterized by a marked degree of variation in body form, by the adaptation to various types of environment, and by striking numerousness. Other groups of Chelicerata present less diversity and are mainly restricted to the soil fauna. Among the very successful groups, mention must be made here of Araneida, both groups of Mites (Actinotrichida and Anactinotrichida) and Opilionida. The development of a gnathosoma (in Mites), and the development of direct or nearly direct sperm transfer are probably closely connected with this successful evolution. Numerous representatives of these four groups have left the soil fauna, and have colonized many new niches. Araneida have certainly reached the highest stage of Chelicerate development, partly because of visual faculties and spinning behaviour. Actinotrichid mites undoubtedly present the greatest diversity.

As mentioned above, the biology of extant terrestrial Chelicerata is a transformation and an adaptation of earlier marine ways of life. Practically nothing is known with certainty about these marine ancestors, but some hypotheses can be formulated.

Some primitive species of Palpigradi and Actinotrichida belong to the interstitial fauna, viz., representatives of the genus *Leptokoenenia* (Palpigradi) and of the family Nematalycidae (Actinotrichida). Palpigradi were even found in sand in the sea, not far from the coast. This could be regarded as an argument in favour of the marine origin of both groups. Many Palpigradi and Actinotrichida belong to the soil fauna (they present dorsal and ventral prosomatic articulations, by which they are particularly appropriate to life in narrow fissures of the soil); several representatives, among which Palaeosomata (Actinotrichida; known already from the Devonian), are found in the deeper layers of the soil. Ancestral Palpigradi and Actinotrichida (and perhaps also other ancestral Coxisternata) could have colonized the deeper layers of the terrestrial soil from the sandy bottom of shallow coastal waters in an early period (even before the Silurian, when animal life at the surface of the soil was still impossible).

The biology of extant Xiphosura suggests that the ancestors of several other groups of Chelicerata could have lived near the coast in shallow water, walking on the bottom, burrowing, and temporarily leaving the water. This could, for instance, have been the biology of the ancestors of extant Scorpionida, and perhaps also of other Myliosomata and Neosternata.

Further studies in the functional morphology of extant and fossil Chelicerata could perhaps yield further indications. Evidently, the plan of construction of the extant Chelicerate groups includes a hierarchy of levels of adaptation, which can only be understood by studying the morphology in evolutionary perspective.

B. SYSTEMATIC PART

In the systematic part, the higher classification of Chelicerata (as explained in the last chapter of the general part) is not repeated. The superorders (and, in the case of Araneomorphae, the two orders) are, however, dealt with in the same sequence as in the general classification.

I. PALPIGRADI

On four occasions, Palpigradi have been the subject of my investigations (Van der Hammen, 1969a; 1977a: 310, 312, 313, 316, 318, tab. 1–3; 1977b; 1982a: 3–22, 47–59, figs. 1–9, 26A–C, 28A, D, 29A, B, D, F, 30A–C). In the first paper, a detailed study was made of the mouthparts; the second and the third paper dealt with segmentation of the legs, classification, relationships and generalities; the last-mentioned paper constituted the second part of my comparative studies in Chelicerata, devoted to Epimerata. The results of these four studies are summarized in the present chapter. It is subdivided into three parts: an introduction, a descriptive part, and additional remarks.

1. Introduction

A representative of the Palpigradi was first discovered and described by the Italian zoologist Grassi (1885, 1886) who, in 1883, had become Professor of zoology and comparative anatomy at Catania (Sicily). He studied the external morphology, the internal anatomy and the affinities of a species collected under rocks near Catania, and which he named *Koenenia mirabilis*. Grassi compared it with Uropygi, Schizomida, Amblypygi, Solifugae and Opilionida, and regarded it as a representative of a new order (Microthelyphonida).

Thorell (1888: 358) replaced the name Microthelyphonida by Palpigradi, a name considered by him more appropriate. Pocock (1893: 9), in his study of Arachnid morphology and classification, could not add further data on the group; although he discussed its affinities, it could not be included in his classification.

Hansen & Sörensen (1897) published a detailed morphological study of *Koenenia mirabilis*, in which attention was paid also to the systematic position (they pointed out a relationship with Schizomida). Börner (1901, 1902, 1904) and Hansen (1901) continued the study of the morphology of the Palpigradi. Börner (1902, 1904) classified the Palpigradi with the Pedipalpi. Hansen (1930) dealt with the comparative morphology of the walking legs.

In the mean time, Wheeler and Rucker had discovered interesting Palpigradid species in Texas, of which the external morphology, the internal anatomy and part of the postembryonic development were described in a series of papers (Wheeler, 1900; Rucker, 1901, 1903, 1903a).

Versluys & Demoll (1921: 750–751; 1922: 134, 141–148, 158, 163, 165, 372, 375), in their discussion of the relationships of Merostomata and other groups of Chelicerata, also dealt with Palpigradi. They enumerated a series of ancestral and derived character states present in Palpigradi, but not in Scorpionida, and pointed out the fact that Palpigradi cannot be descendants of Scorpion-like ancestors (Arachnid ancestors, moreover, cannot have been Merostomata-like).

Kaestner (1932a, 1955, 1956) and Roewer (1934) published useful summaries of the knowledge on Palpigradi. New details, particularly with reference to internal anatomy, were described and summarized by Millot (1942, 1943, 1949a).

Snodgrass (1948: 19–22, fig. 6) summarized the knowledge with reference to the Palpigradid mouthparts. Petrunkevitch (1955: 116–118, figs. 82, 83) classified the Palpigradi with the Arachnid subclass Cauligastra (including also Uropygi, Amblypygi, Araneida, Solifugae and Ricinulei); in this subclass, they constituted the superorder Latisterna.

Condé (1965) described a species belonging to the interstitial microfauna of the Farasan Islands in the southern Red Sea. Monniot (1966) described a species belonging to the interstitial microfauna of Pointe Noire, Congo Brazzaville (she collected several specimens of the species); she made also a study (Monniot, 1970) of the fine structure of tegument and phaneres in *Eukoenenia mirabilis*.

Savory (1935: 90–95, figs. 43–46; 1964: 142–147, figs. 61–65; 1971: 27–28, 30–32, 35, 38–39, 41, fig. 8; 1974: 43–45; 1977: 125–131, figs. 33–37) published several summaries and notes on phylogeny; he pointed out the possibility of a Palpigradid evolution different and separate from that of other Chelicerates.

Van der Hammen (1969a, 1977a, 1977b) studied the morphology of the mouthparts and the appendages, and classified the Palpigradi, together with the Actinotrichida, in a new Chelicerate subclass, the Epimerata.

Firstman (1973: 6–8, 46, 49–50, figs. 3, 4, 35) investigated the Palpigradid arterial system and endosternite; he characterized the endosternite as the most primitive of all known extant Chelicerata

Weygoldt & Paulus (1979, 1979a) and Weygoldt (1980) studied the phylogeny of the Chelicerata and introduced a cladistic classification. They regarded the presence of a flagellum and three-segmented chelicerae in Palpigradi as primitive characters, and mentioned the following derived characters: reduction of size; subdivision of the prosoma; absence of eyes, lyrifissures and respiratory organs; the occurrence in fissures of the soil. The cladistic classification of these authors is based on a hypothesis about the origin of the Chelicerata, and the interpretation of character states is adapted to this view.

Alberti (1979) investigated the ultrastructure of sperm and spermiocytogenesis in *Prokoenenia wheeleri* (Rucker), which appeared to present special derived characters. He pointed out the isolated position of Palpigradi as far as sperm morphology is concerned.

Rowland & Sissom (1980) described a fossil Palpigrade species from the Tertiary, and prepared a review of the morphology and systematics, as well as a complete bibliography, of the Palpigradi. In the course of nearly a century, about fifty Palpigrade species have been described by some fifteen authors.

The following discussion of Palpigradid morphology is, for the greater part, based on original studies; it is supplemented with data from literature. Some data are added on Palpigradid reproduction and postembryonic development. The original part of the present chapter is based on the study of *Eukoenenia mirabilis* (Grassi) (material from Banyuls-sur-Mer, France). Figs. 24–26 and 28–29 were prepared after one and the same specimen (an adult female). With the help of figs. 24–26 (the specimen in dorsal, ventral and lateral view respectively), a three-dimensional image can be built up. The material used for the present chapter was studied in cavity slides, in diluted lactic acid. In this medium, it becomes very soft and can easily be deformed. The outlines of figs. 24 and 25, consequently, present slight differences to one another, although the figures are based on the same specimen.

2. Descriptive part

Habitus. – Palpigradi are small, elongate, whitish and delicate Chelicerata with a long, narrow, fragile flagellum, and long, thin appendages. The movements of living specimens are rapid and vivid; opisthosoma and flagellum can be actively elevated and even bent to the front.

Divisions of the body, tagmata and segmentation. – The Palpigradid body is subdivided into two tagmata: prosoma and opisthosoma. The dorsal surface of the prosoma (prodorsum) is subdivided into a propeltidium, a small paired mesopeltidium (often difficult to distinguish) and a metapeltidium, separated by soft skin. The border of these sclerites is not distinct, and they apparently represent pseudosclerites (cf. Grandjean, 1965: 557–558). The epimeral regions of palp and leg I have fused, those of legs II–IV are separated by soft skin. The soft skin between pro- and metapeltidium, between metapeltidium and opisthosoma, and between the epimera, permit of opisthosomatic movements in the vertical as well as in the horizontal plane. The movability is not restricted to a single interval of soft skin, as in the case of dichoidy or ptychoidy in Oribatid mites (cf. Van der Hammen, 1980: 45, 131–132), but is associated with two dorsal and probably three or four ventral intervals (figs. 24A, 25A, 26A). The opisthosoma is subdivided into eleven segments (VII–XVII) and a flagellum. The segments are separated by intersegmental furrows, distinct in the dorsal region, partly less distinct in the ventral region.

Cuticle. – The chitinous tegument is thin, flexible and whitish. The borders between pseudoscleritic parts (cf. Van der Hammen, 1980: 131) and soft skin are indistinct. The cuticle presents a papillate microsculpture closely resembling a pubescence (cf. Monniot, 1970: 559, fig. B); it bears various types of

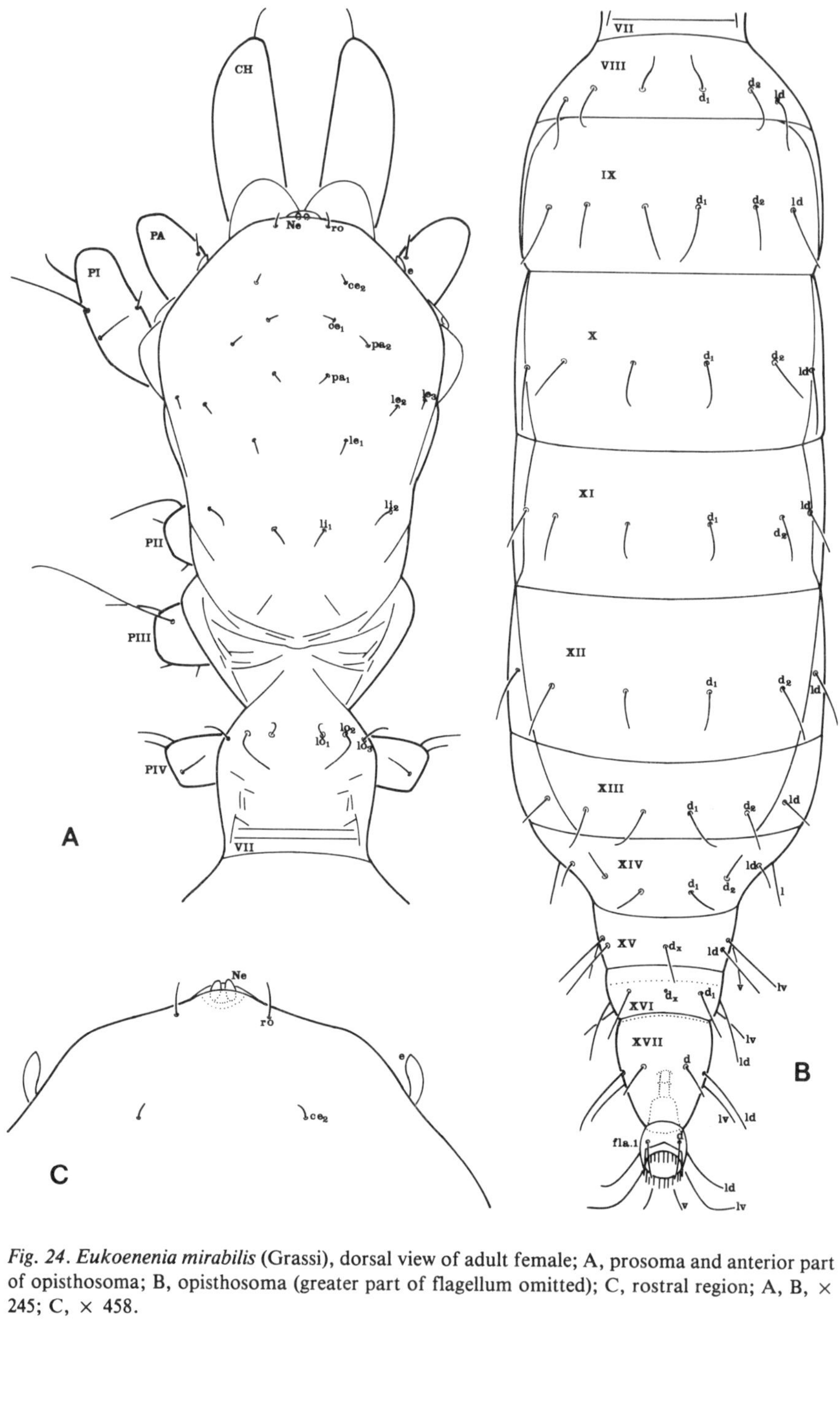

Fig. 24. Eukoenenia mirabilis (Grassi), dorsal view of adult female; A, prosoma and anterior part of opisthosoma; B, opisthosoma (greater part of flagellum omitted); C, rostral region; A, B, × 245; C, × 458.

phaneres. There are no lyrifissures. Respiration is supposed to be cutaneous, and one of the functions of the cuticle must consequently pertain to gaseous interchange.

Aspidosoma. – Four speudosclerites are present, separated by soft skin: a propeltidium, paired mesopeltidia and a metapeltidium; the paired mesopeltidia constitute intercalary pseudosclerites. The soft skin is striate. The borders of the pseudosclerites are not distinct (figs. 24A, C, 26A, 31A). All specimens at my disposal presented an elevated rostral part, which is said to be an artificial condition caused by the preservation medium (alcohol); it could, however, also represent a relaxed condition (cf. Van der Hammen, 1969a: 42). The shape of living specimens appears to correspond with fig. 31A. There are no eyes. The frontal organ has more or less the shape of a pair of partly fused setae (figs. 24A, C, 26A, 27A, C, 29E). In *Leptokoenenia* the two elements are globular, with echinulate microsculpture and terminal point (fig. 31C). In living specimens, the frontal organ is hidden under the rostral part of the aspidosoma (fig. 31A). The frontal organ resembles, on the one hand, the lateral organ (the supracoxal seta *e*), and could, consequently, represent the supracoxal setae of the chelicerae. It is, on the other hand, because of its position, very similar to the Actinotrichid naso (also a frontal organ; it can sometimes present an inferior bilobate eye). The two hypotheses (which could both be valid) are simultaneously expressed in the notation (*Ne*). Apart from the frontal and lateral organ (the latter is dealt with in the paragraph on the lateral aspect of the podosoma), there are thirteen pairs of aspidosomatic (prodorsal) setae: ten pairs on the propeltidium (ro, ce_1, ce_2, pa_1, pa_2, le_1, le_2, le_3, li_1, li_2) and three pairs on the metapeltidium (lo_1, lo_2, lo_3). This number appears to be constant in all extant species of Palpigradi in which it is described in a reliable way (genera *Eukoenenia, Prokoenenia, Leptokoenenia*).

Dorsal and laterodorsal regions of opisthosoma. – The dorsal and laterodorsal regions of the opisthosomatic segments are perhaps partly pseudoscleritic. The intersegmental furrows are distinct. The posterior three segments (XV–XVII) are movable with respect to each other; they are slightly telescoping, and constitute a pygidium which can be slightly elevated with respect to segment XIV. The numbers of dorsal and laterodorsal opisthosomatic setae of the segments VIII–XIV (figs. 24B, 26B) appear to be constant in all species of Palpigradi in which they are described in a reliable way (genera *Eukoenenia, Prokoenenia, Leptokoenenia*); there are, on each of these segments, three pairs of dorsal and laterodorsal setae: d_1, d_2, and *ld*. The number is less constant in the posterior narrowed part of the opisthosoma (the segments XV–XVII, which constitute the pygidium), in which, moreover, an unpaired dorsal seta d_x (or a vestige of it) can be present; the unpaired seta probably represents a fused pair. In the figured specimen (cf. figs. 24B, 26B), the dorsal and laterodorsal setae of segments XV–XVII are the following: XV, d_x, *ld*; XVI, d_x (vestige), d_1; XVII, *d*, *ld*.

Flagellum. – The flagellum is movable: its basal piece articulates with an articular tooth inside segment XVII; it can be actively elevated and depressed by a set of antagonistic muscles (figs. 24B, 25B, 26B). In my specimens, the flagellum was broken off, with the exception of the proximal segment. According to data in literature (fig. 31E) it is composed of a variable number of flagellar segments (often fourteen or fifteen, apart from the basal piece). The flagellar segments present verticils of 6–8 setae; some of the flagellar segments present moreover a terminal whorl of spines. The distalmost segment of the flagellum presents a terminal whorl of some seven setae. The flagellum is generally regarded as representing the 'telson', its presence constituting a primitive character. Although the presence of a 'telson' must indeed be regarded as primitive, its segmented condition certainly constitutes a derived character state.

Ventral and lateroventral region of opisthosoma. – Ventrally, the intersegmental furrows can be less distinct, particularly in the anterior part. The ventral regions of the opisthosomatic segments are of unequal length, particularly because of the ventral shortening of segment IX, and the ventral lengthening of segment X. The segments can present ventral and lateroventral setae, lobes, verrucae and invaginable papillae. Segments VIII and IX of the female can be produced in the shape of one lobe (segment VIII) or two lobes (segment IX) (figs. 25B, 26B, C). In the male (fig. 31D) segments VIII and IX both present two lobes which can be subdivided. The lobes bear setae, some of which contain the ducts and orifices of glands. In *Prokoenenia*, segments X–XII each present a pair of invaginable papillae, reminiscent of the genital papillae of Actinotrichida. In *Eukoenenia mirabilis*, the ventral regions of segments X and XII present an unpaired wart-like structure with setae, here called anterior and posterior verruca (*Va*, respectively *Vp*; cf. figs. 25B, D, E, 26B); some of the setae of the verrucae contain the ducts and orifices of glands. The anterior and posterior verrucae could be homologous with a pair of papillae (they resemble the sternal and genital verrucae of Opilioacarida, which are supposed to be homologous with endites or exites of appendages). The anus has a ventral position in the distal part of segment XVII; it is best visible in lateral view (fig. 26B). The number of ventral and lateroventral opisthosomatic setae is more variable than the number of dorsal and laterodorsal setae; in literature, mention is made of sexual dimorphism pertaining to the number of ventral setae. In the figured specimen (cf. figs. 25B, 26B), the ventral and lateroventral setae are the following (there are no setae inserted on segment VII): VIII, g_{1-4}, v_{1-2}; IX, pg_{1-2}, lv_{1-2}; X–XI, v_{1-2}, lv_{1-2}; XII, vp_x, vp_{1-2}, lv_{1-2}; XIII, v_{1-2}, *lv*; XIV, v_{1-3}, *lv*, *l*; XV, v_x, v_1, *lv*; XVI–XVII, *v*, *lv*.

Ventral aspect of podosoma. – The coxisternal part of the podosoma (the Palpigradid podosoma includes the coxisternal region of the palp) is constituted by epimera, separated by intersegmental intervals of soft skin (fig. 25A). The epimera are partly or completely pseudoscleritized (data published until now are contradictory); a detailed comparative study of the pseudoscleritization in various Palpigradid genera, after staining with Chlorazol Black (cf.

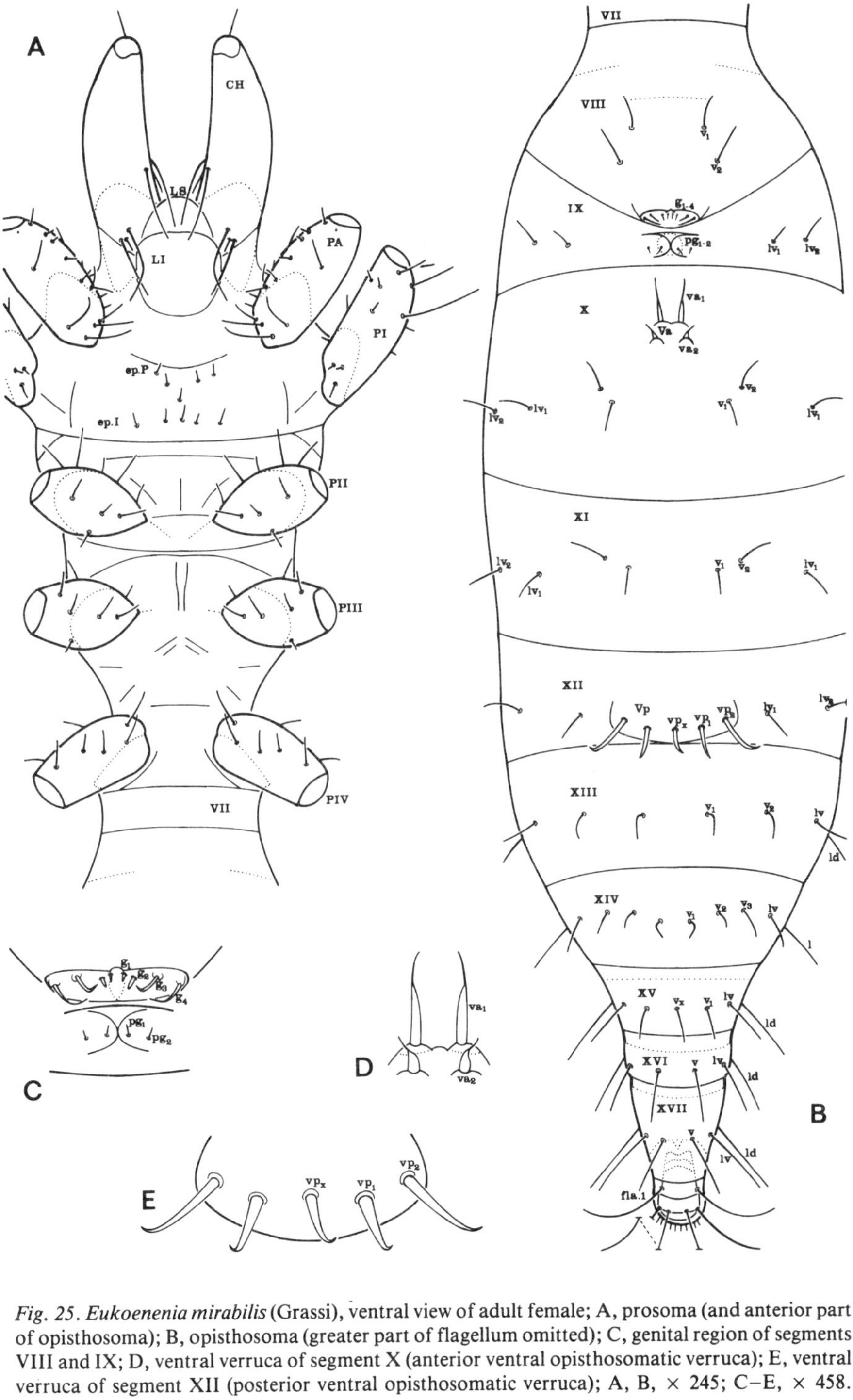

Fig. 25. Eukoenenia mirabilis (Grassi), ventral view of adult female; A, prosoma (and anterior part of opisthosoma); B, opisthosoma (greater part of flagellum omitted); C, genital region of segments VIII and IX; D, ventral verruca of segment X (anterior ventral opisthosomatic verruca); E, ventral verruca of segment XII (posterior ventral opisthosomatic verruca); A, B, × 245; C–E, × 458.

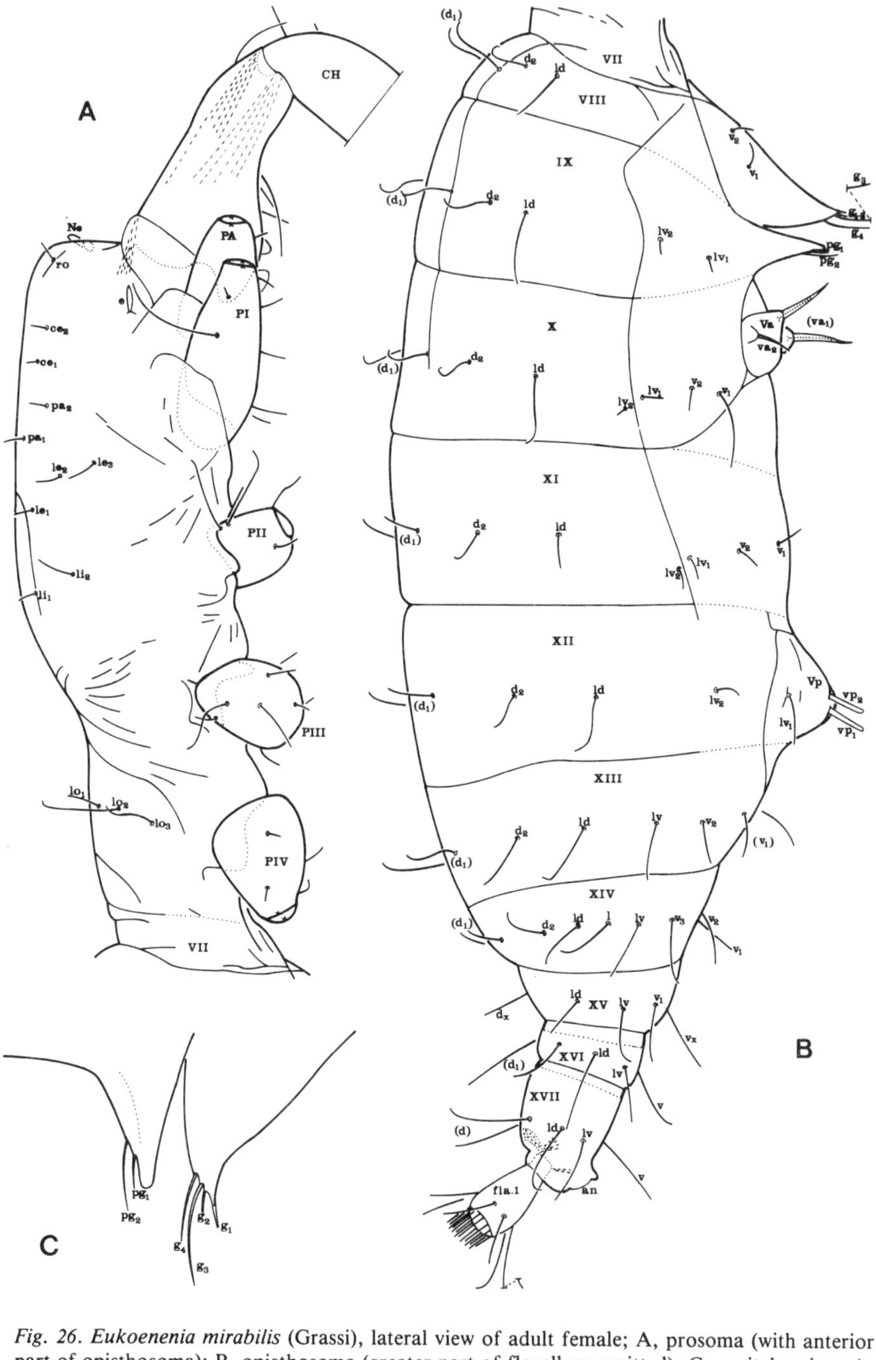

Fig. 26. Eukoenenia mirabilis (Grassi), lateral view of adult female; A, prosoma (with anterior part of opisthosoma); B, opisthosoma (greater part of flagellum omitted); C, genital region; A, B, × 245; C, × 458.

Coineau, 1974a), will be interesting. The epimera of palp and leg I have fused to a single pseudosclerite (figs. 25A, 27B). Epimera II and III of *Leptokoenenia* (cf. Monniot, 1966: 48, fig. 4) each consist of several pseudosclerites; fig. 25A of the present paper suggests that the same condition could be present in *Eukoenenia*. Epimeron IV consists of a single pseudosclerite. The intersegmental intervals are extensile; the normal condition in living specimens is represented in fig. 31B, whilst the specimen of fig. 25A is completely stretched out. The pseudosclerite constituting the epimera of palp and leg I bears epimeral setae, of which the number varies not only with sex and age, but also presents specific differences.

Lateral aspect of podosoma. – The supracoxal setae *e* of the palp are inserted in the lateral region of the podosoma, above the palpal acetabulum (figs. 26A, 27A, 29E); in Actinotrichida, supracoxal setae are also associated with the base of the appendages (cf. Van der Hammen, 1980: 153–154). The supracoxal setae can be multiplied; the numbers recorded in literature, for different species, vary from one to five (cf. fig. 31F–H). Multiplication of supracoxal setae is also known from Actinotrichida (cf. Van der Hammen, 1977b: 13–15, fig. 2F, H–L). In chapter II (section 2) of the general part, it was suggested that supracoxal setae could have taken the place of lateral eyes. The orifice of the coxal gland is above the acetabulum of leg I (cf. fig. 29E), as in Actinotrichida; the coxal gland itself was described by Millot (1942: 42–45, fig. 7). There could be a primitive podocephalic canal, extending from the orifice of the coxal gland to the mouth, although it could not be discerned by me with absolute certainty. At the base of palp and leg I, the prosoma is slightly extending, constituting a coxal region; the soft skin does not represent a coxa, nor does it exclusively represent an arthrodial membrane. The lateral aspect of the Palpigradid prosoma resembles that of Actinotrichida in many respects.

Mouthparts. – In Palpigradi, the elements involved in food-intake comprise the labrum or upper lip, the labium or under lip, the mouth (and pharynx) and the chelicerae (fig. 27). The chelicerae are described below in a separate paragraph. The mouth is situated between the bases of the chelicerae. It is crescent-shaped; there are two oral commissures (*J* and *J'*), at the place of junction of labrum and labium. The bases of the two lips have fused; the base of the labrum (i.e. the part which has fused with the labium) could, in reality, represent the cervix (cf. Van der Hammen, 1980: 28–29). The dorsal surface of the labrum is convex. It overhangs the labium and is capable of closing the preoral cavity. Its anterodorsal surface presents numerous ciliae. The labium is rounded and swollen in its anterior part. According to Millot (1942: 36, fig. 2), it contains many glandular (mucous) cells.

Chelicera. – The first pair of appendages (figs. 27, 28A) is composed of trochanter, body of the chelicera (the terminal part of which constitutes the fixed jaw), and apotele (the movable jaw); fixed jaw and movable jaw constitute the chela. The three segments are well developed and the chelicera is relatively large. The articulations between trochanter and body of the chelicera,

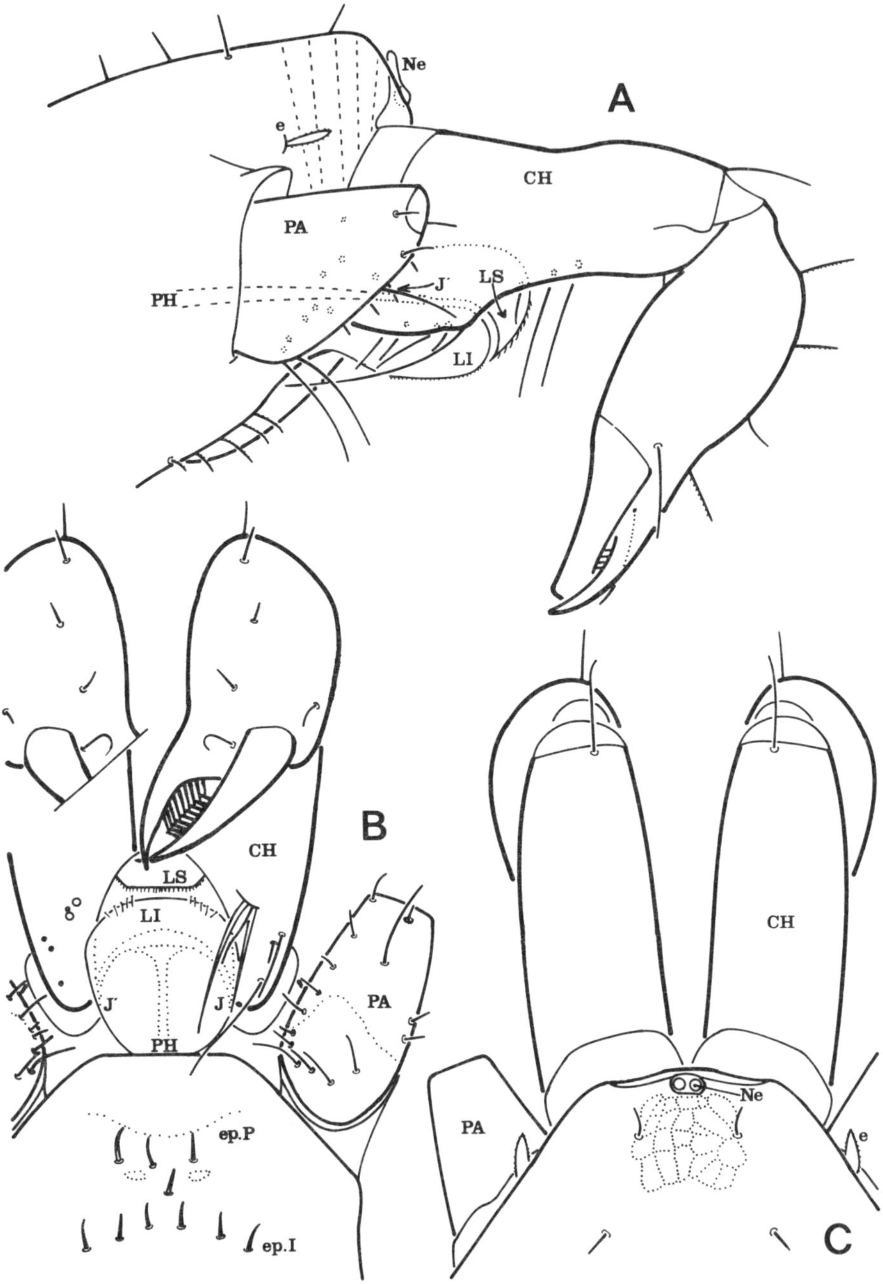

Fig. 27. Eukoenenia mirabilis (Grassi), female, anterior prosomatic region with mouthparts; A, lateral view; B, ventral view; C, dorsal view; A–C, × 368.

and between body of the chelicera and apotele, are bidesmatic and bicondylar. The articulation between prosoma and chelicera (at the base of the trochanter) is apparently tridesmatic. At the base of the chelicera (as mentioned above, a similar condition is found at the bases of palp and leg I), the prosoma is slightly extending, constituting a coxal region; the soft skin does, however, not represent a coxa, nor does it exclusively represent an arthrodial membrane. The plane of pseudosymmetry of the body of the chelicera has been subject to considerable torsion: the chela operates in a plane nearly perpendicular to the axis of the body; of the two jaws of the chela, the apotele (the movable jaw) is the antiaxial one. Both jaws of the chela present some nine oblique, narrow, pointed teeth. The cheliceral trochanter and the body of the chelicera present several setae; in *Eukoenenia mirabilis*, two groups of three ventral setae appear to be associated with the lips.

Palp and legs. – The segmentation and articulation of palp and legs is represented in fig. 28B–F. The segments of the palp are: trochanter, femur, genu (patella), tibia, tarsus 1–5, and apotele. All legs present the following segments: trochanter, femur 1, femur 2, genu (patella), tibia, a variable number of tarsal segments, and apotele. The adesmatic tarsal segments are the following: leg I, tarsus 1–7; legs II–III, tarsus 1–2; leg IV, tarsus 1–3. There is no evidence for an ancestral division of the tarsus into basi- and telotarsus. The articulation between trochanter and body (in the case of palp and legs) apparently constitutes a rocking joint (cf. also fig. 29A). The articulation between trochanter and femur (or femur 1) constitutes a bidesmatic, bicondylar, bivalent pivot joint, with levator and depressor muscles. The articulation between femur 1 and femur 2 of the legs (in the case of legs I and IV, the separation is incomplete at the posterior face) is constituted by a monodesmatic, monocondylar, monovalent hinge joint; the muscles of this articulation are distinctly present, although I did not discern them in 1977 (cf. Van der Hammen, 1977b: 13, fig. 1F, G). The articulation between femur and genu of the palp, and between femur 2 and genu of legs II–IV are also constituted by normal hinge joints. The articulations between femur 2 and genu of leg I, and between genu and tibia of palp and leg I, are constituted by bidesmatic, monocondylar hinge joints; the two sets of muscles of these joints are flexor and extensor muscles respectively. The extensor muscles are inserted just above the articulation point; in lateral view, they seem to pass through the condyle (fig. 29B, D). The occurrence of this type of joint in Palpigradid palp and leg I is apparently not only associated with the palpal function of leg I, but is also an indication of the palpal (not ambulatory) origin of the palp. The articulation between genu (patella) and tibia of legs II–IV is constituted by a hinge joint with flexor muscles only. The articulation between tibia and tarsus 1 of palp and legs is constituted by a monocondylar, monovalent hinge joint; one tendon is inserted on the base of the tarsus (*tta*), whilst the inferior tendon of the apotele (t_i) passes on into the tibia. The articulation between the tarsal segments is adesmatic; they are moved under the influence of the tendons of the

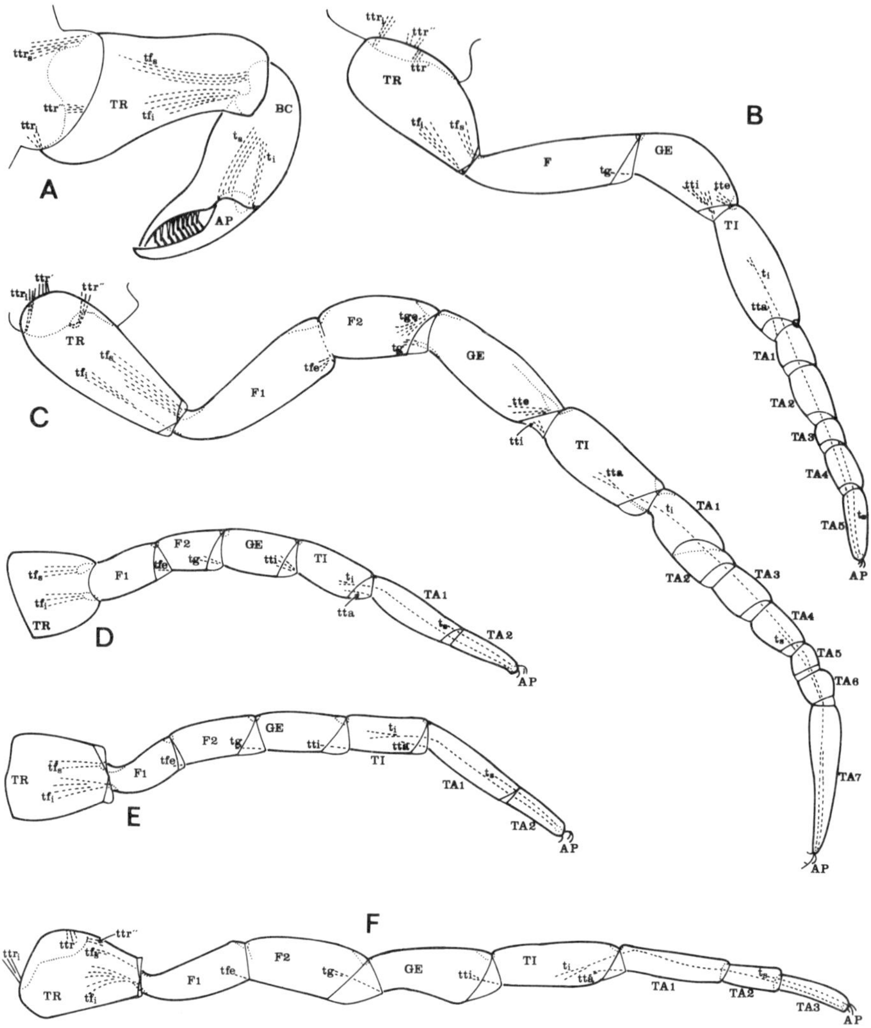

Fig. 28. Eukoenenia mirabilis (Grassi), adult female, segmentation and articulation of the appendages; A, lateral (antiaxial) view of right chelicera; B, lateral (antiaxial = posterior) view of right palp; C, lateral (antiaxial = posterior) view of right leg I; D, lateral (antiaxial = posterior) view of right leg II; E, lateral (paraxial = posterior) view of right leg III; F, lateral (antiaxial = anterior) view of left leg IV; A–F, × 245.

Fig. 29. Eukoenenia mirabilis (Grassi), adult female; A, base of the appendages of the left side, viewed from the interior (and slightly from below); B, distal part of femur 1, femur 2 and proximal part of genu of right leg I, lateral (antiaxial) view; C, distal part of tarsus 6, and apotele of right leg I, lateral view; D, distal part of genu, and proximal part of tibia of right leg I, lateral (antiaxial) view; E, anterior part of prosoma, lateral view (obliquely from above); A, × 368; B–D, × 687; E, × 470.

→

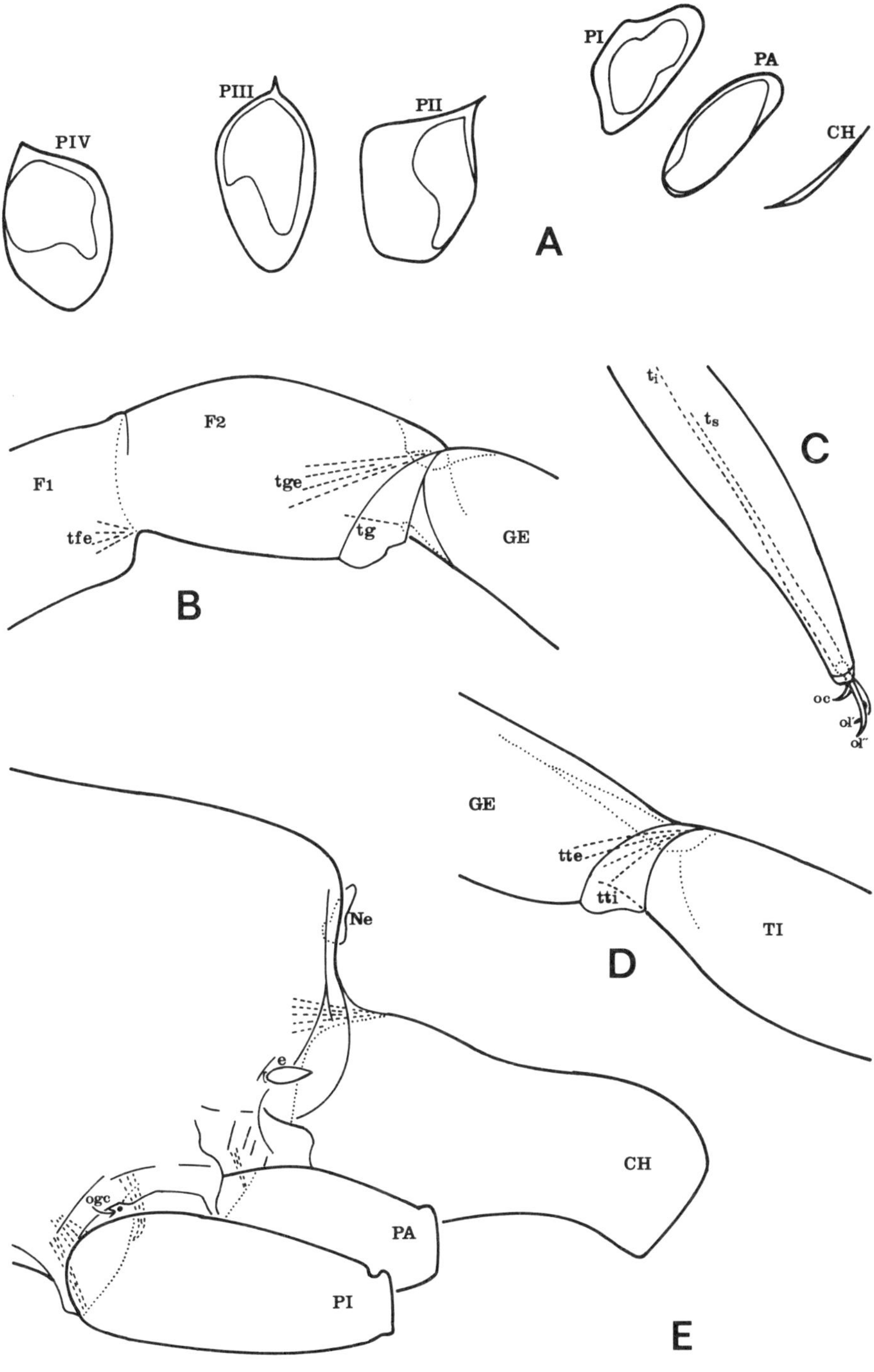

PIV
PIII
PII
PI
PA
CH
A
F1
F2
tfe
tge
tg
GE
B
C
oc
D
GE
tte
tti
TI
Ne
e
ogc
PA
PI
CH
E

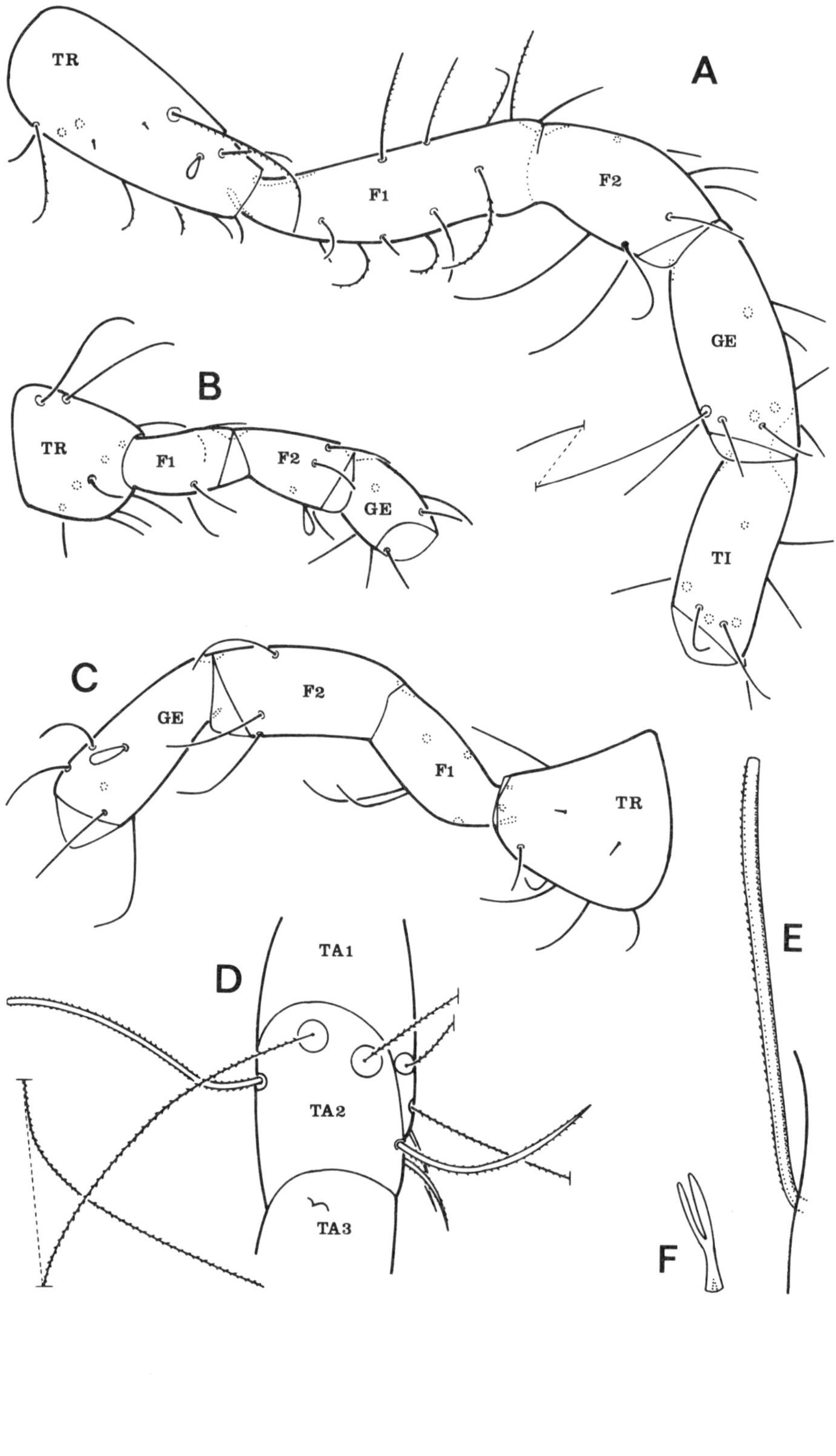
TR
F1
F2
GE
TI
A
B
C
D
TA1
TA2
TA3
E
F

apotele (t_s and t_i). The articulation between tarsus and apotele of palp and legs is bidesmatic, bicondylar and bivalent (fig. 29C). Palp and legs present various types of phaneres (fig. 30). Ordinary setae are present in all segments (the three ungues of the claws of palp and legs I–IV are considered homonomous with ordinary setae). Trichobothria (also homonomous with ordinary setae) are found on genu (patella) and tarsi 1, 2, 4 and 6 of leg I (cf. fig. 30A, D). The number of trichobothria on tarsi 1–7 of leg I in *Eukoenenia mirabilis* and *Leptokoenenia scurra* can be represented by the formula 2–2–0–1–0–1–0. Tarsi 2, 3, 5 and 7 of leg I (and tarsi 1 and 3 of leg IV in *Leptokoenenia scurra*) present moreover a different type of phanere (figs. 30D, E, 31I, J), which could be homologous with a solenidion (cf. Van der Hammen, 1980: 144); this type is long, stiff, flat, hollow, more or less reclinate (i.e. bent downwards upon the surface of the appendage), and often more or less blunt (i.e. without point). These stiff phaneres have a rectangular section, whilst each of the angles presents a longitudinal row of ciliae. A fourth type of phanere is constituted by the bifurcate phaneres occurring, for instance, on tarsi 4, 6 and 7 of leg I in *Eukoenenia mirabilis*, and apparently also on tarsus 5 of the palp in *Leptokoenenia scurra* (cf. figs. 30F, 31I, J); these bifurcate phaneres could be homologous with the Actinotrichid famulus (cf. Van der Hammen, 1980: 62–63). The claws of palp and legs are tridactyl; the central ungues represents a true ungues (not an empodium).

Internal anatomy. – Important contributions to our knowledge of the internal anatomy were published by Rucker (1901, 1903), Börner (1904), Millot (1942, 1943), Firstman (1973) and Alberti (1979). Summaries were prepared by Kaestner (1932a), Roewer (1934), Millot (1949a) and Rowland & Sissom (1980). The following is a compendious survey. The alimentary canal consists of mouth, pharynx, oesophagus, midgut and caeca (one pair in the prosoma, six in the opisthosoma), hindgut (rectum) and anus. The excretory system comprises the coxal glands, and absorption cells in the gut; there are no Malpighian tubes, nor any other excretory organ. The Palpigradid endosternite is more primitive than that of any other extant Chelicerate: there are six pairs of ventral suspensors, four pairs of dorsal suspensors, and five pairs of lateral suspensors. The circulatory system is very simple. The central nervous system consists of a relatively large supraoesophageal ganglionic mass and a smaller suboesophageal ganglionic mass; from the posterior end of the last-mentioned mass, a nerve cord passes into the opisthosoma, in the anterior part of which

←

Fig. 30. Eukoenenia mirabilis (Grassi), adult female; A, trochanter, femur 1, femur 2, genu and tibia of right leg I, lateral (antiaxial) view; B, trochanter, femur 1, femur 2 and genu of right leg II, lateral (antiaxial) view; C, trochanter, femur 1, femur 2 and genu of right leg IV, lateral (antiaxial = anterior) view; D, distal part of tarsus 1, tarsus 2, and proximal part of tarsus 3 of right leg I, lateral (antiaxial) view; E, reclinate phanere (supposed solenidion) of tarsus 7 of right leg I, ventral view; F, right 'famulus' of tarsus 7 of right leg I, ventral view; A–C, × 368; D, × 707; E–F, × 1312.

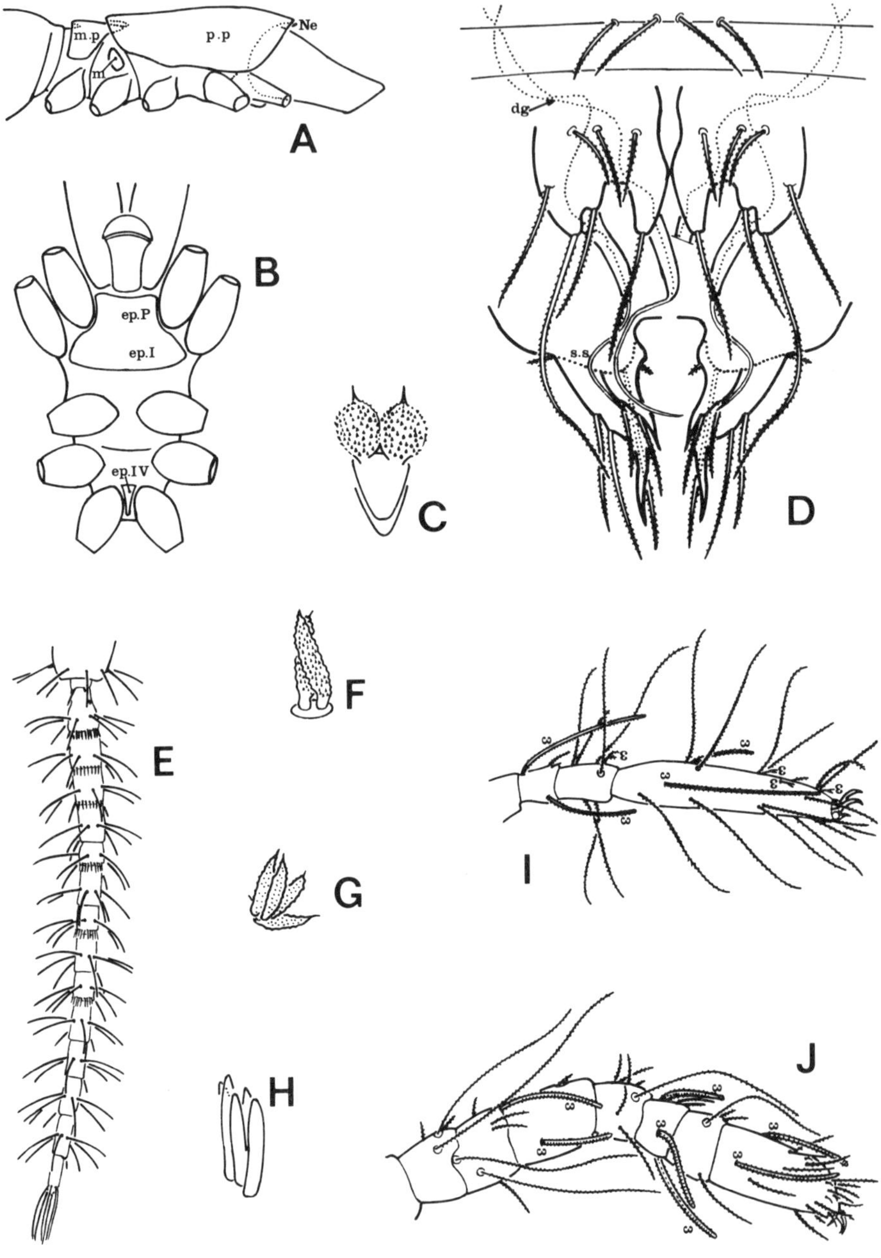
m.p
p.p
Ne
m
A
B
ep.P
ep.I
ep.IV
C
dg
s.s
D
E
F
G
H
I
J

there is a distinct ganglionic swelling. The male reproductive system comprises paired testes; Palpigradid sperm and spermiocytogenesis present derived characters and are strongly different from other groups of Chelicerata. The female reproductive system comprises two ovaries and two oviducts which converge towards the uterus. A pair of prosomatic glands of unknown function debouches posterior to leg II. Several opisthosomatic glands are associated with the genital lobes and the opisthosomatic verrucae; the orifice of these glands is in the terminal part of large setae. The opisthosomatic musculature consists of longitudinal (dorsal, lateral, ventral), oblique and dorsoventral muscles; there are five pairs of dorsoventrals (in segments VIII–XII).

Reproduction. – Little is known about reproduction in Palpigradi. In some species (e.g., *Eukoenenia mirabilis*), males are extremely rare, and reproduction could be parthenogenetic; in other species (e.g., *Prokoenenia wheeleri*), males seem to be even more numerous than females. Eggs are apparently laid throughout the year in small numbers (two relatively large eggs can be simultaneously present in the female opisthosoma).

Postembryonic development. – Our knowledge of the Palpigradid life-cycle is based on the study by Rucker (1903) who described three immature instars and the adult of *Prokoenenia wheeleri*. The instars are, among others, distinctly different in size, chaetotaxy, development of the genital region, characters of the flagellum, and the number of opisthosomatic papillae (in segments X–XII). The behaviour of the papillae (they are known from *Prokoenenia* only), in the course of postembryonic development, is interesting because it reminds us of the postembryonic development of the genital papillae in Actinotrichida (the normal development of the Actinotrichid papillae is the following: absent in prelarva and larva, one pair in the protonymph, two pairs in the deutonymph, three pairs in the tritonymph and the adult). Rucker found no papillae in the first known instar of *Prokoenenia*, two pairs in the second known, and three pairs in the third known instar and the adult. There is an important gap between the sizes of Rucker's first and second known instar, and the occurrence of an instar of intermediate size, with one pair of papillae, seems highly probable. Judging from conditions known in Actinotrichida, the Palpigradid life-cycle could include the following instars (stases): prelarva (probably an

---

←

Fig. 31. Details of various Eukoeneniid species. A, B, *Eukoenenia mirabilis* (Grassi); A, lateral view of prosoma and anterior part of opisthosoma; B, ventral view of prosoma. C–F, *Leptokoenenia scurra* Monniot; C, frontal organ *Ne*; D, ventral view of male genital region; E, flagellum of female; F, supracoxal setae *e*. G, *Prokoenenia millotorum* Rémy, supracoxal setae *e*. H, *Eukoenenia pyrenaica* (Hansen), supracoxal setae *e*. I, *Eukoenenia mirabilis* (Grassi), lateral view of tarsi 5–7 and apotele of leg I. J, *Leptokoenenia scurra* Monniot, lateral view of tarsi 1–7 and apotele of leg I. A, B, about 75 ×; C, F, G, H, much enlarged; D, about 310 ×; E, about 125 ×; I, about 350 ×; J, about 300 ×. (A, B, redrawn after Millot, 1949a; C–F, J, redrawn after Monniot, 1966; G, redrawn after Rémy, 1950; H, redrawn after Hansen, 1926; I, redrawn after Börner, 1901).

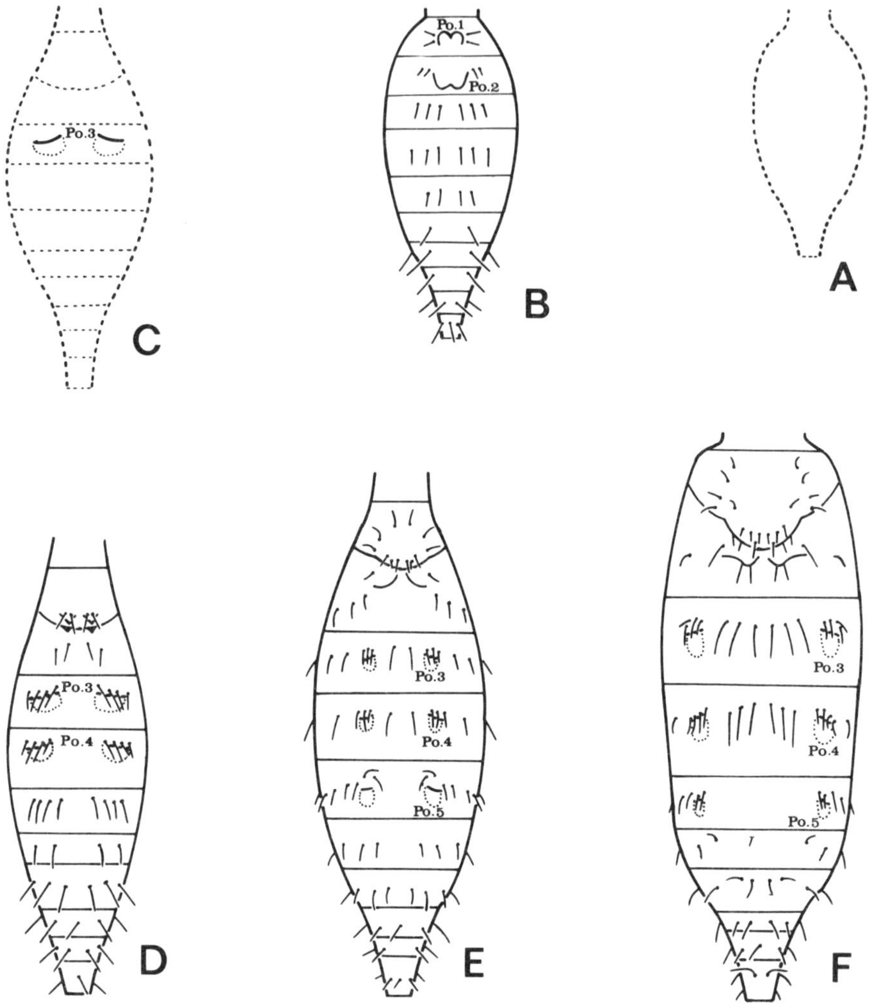

Fig. 32. Schematic representation of the postembryonic development of the opisthosoma (ventral view; flagellum omitted) in *Prokoenenia wheeleri* (Rucker). In this species (as in other Palpigradi) the life-cycle is still incompletely known, and the present representation is partly hypothetical. A, hypothetical first instar (prelarva); B, second instar or larva (representing the first instar now known to us), with evaginated papillae *Po1* and *Po2*; C, hypothetical third instar (protonymph), with invaginated opisthosomatic papillae *Po3*; D, deutonymph (representing the second instar now known to us), with invaginated opisthosomatic papillae *Po3* and *Po4*; E, female tritonymph (representing the third instar now known to us), with invaginated opisthosomatic papillae *Po3*, *Po4* and *Po5*; F, adult female, with invaginated opisthosomatic papillae *Po3*, *Po4* and *Po5*. A–F, about 90 ×. (B, D, E, F, based on Rucker 1903).

elattostase or a calyptostase), larva (Rucker's first known instar), protonymph, deutonymph (Rucker's second instar), tritonymph (Rucker's third instar), and adult. A schematic representation of the postembryonic development of the opisthosoma in the stases of this hypothetical life-cycle is given in fig. 32.

3. Additional remarks

Species of Palpigradi are known from the tropical, the subtropical or the warmer regions of Europe, Asia, Africa and America. It is interesting to mention, in this connection, the living Palpigradid specimen collected by Grandjean near Périgueux (Dordogne), France, and recorded by me in the introduction to the second volume of his Complete Acarological Works; this specimen now constitutes one of the northernmost records for the group.

An extinct species, supposed to constitute a separate Palpigradid family, was described from the Jurassic of Germany (Haase, 1890: 11; 1890a: 653–657, pl. 31). A fossil Eukoeneniid is now known from the Tertiary of Arizona, U.S.A. (Rowland & Sissom, 1980). Extant Palpigradid species apparently prefer the deeper layers of the soil, although they have also been found in litter and on the surface of the soil (after heavy rain). Two species are known from the interstitial environment. According to Monniot (1966: 42), the behaviour of her interstitial Palpigradi in water was normal. It has been suggested that Palpigradi are of marine origin, and that, in the course of evolution, members of this group colonized the terrestrial soil by way of the interstitial environment (Monniot, 1966; Van der Hammen, 1977a). It is generally assumed that terrestrial life was not possible before the Silurian (because of the penetration of harmful radiation, owing to the low oxygen content of the atmosphere). Life in the deeper layers of the soil, however, must have been possible at an earlier moment in the history of the earth, and Palpigradi (together with Actinotrichida) could have been among the earliest terrestrial animals.

Although Palpigradi present a great number of primitive characters (they probably constitute the Chelicerate group with the most important number of ancestral characters), they have also developed many derived characters, of which the following are mentioned here: the development of the 'telson' into a many-segmented flagellum; the development of the opisthosomatic appendages into genital lobes, evaginable papillae and opisthosomatic verrucae; the specialized morphology of the sperm (possibly connected with the life habits in an environment close to the groundwater); the fusion of the epimera of palp and leg I; the multiplication of the supracoxal seta *e*, in many species; the association of setae of the cheliceral trochanter with the lips; the torsion in the body of the chelicera; the subdivision of the tarsus of palp and legs; the occurrence of extensor muscles in some hinge joints of palp and leg I; and the presence of trichobothria.

Important contributions to the study of Palpigradid evolution (and the interpretation of character states) could result from a comparative study of the postembryonic development, as soon as the life-cycle of several species will be completely known. A study of the embryonic development will also be of considerable importance.

List of the notations of figs. 24–32

an	anus.
AP	apotele.
BC	body of the chelicera.
ce_{1-2}	prodorsal setae attributed to segment I.
CH	chelicera.
d_{1-2}	dorsal setae of opisthosoma.
d_x	unpaired dorsal seta.
dg	duct of gland.
e	supracoxal seta of palp.
ep.P	epimeron of segment II (palpal segment).
ep.I	epimeron of segment III (segment bearing leg I).
ep.IV	epimeron of segment VI (segment bearing leg IV).
F	femur.
fla.1	first segment of flagellum.
F1	femur 1.
F2	femur 2.
g_{1-4}	genital setae.
GE	genu.
J, J'	oral commissures.
l	lateral seta.
ld	laterodorsal seta.
le_{1-3}	prodorsal setae attributed to segment III.
LI	labium.
li_{1-2}	prodorsal setae attributed to segment IV.
lo_{1-3}	prodorsal setae of metapeltidium.
LS	labrum.
lv, lv_{1-2}	lateroventral setae.
m	mesopeltidium.
m.p	metapeltidium.
Ne	frontal organ.
ogc	orifice of coxal gland.
PA	palp.

pa_{1-2}	prodorsal setae attributed to segment II.
pg_{1-2}	postgenital setae.
PH	pharynx.
Po.1–5	opisthosomatic appendages.
p.p	propeltidium.
PI–IV	legs I–IV.
ro	rostral seta.
ss	secretory seta.
t_i	inferior tendon of apotele.
t_s	superior tendon of apotele.
TA1–7	tarsus 1–7.
tf_i	inferior tendon of femur.
tf_s	superior tendon of femur.
tfe	tendon of femur 2.
tg	tendon of genu.
tge	tendon of extensor muscle of genu.
TI	tibia.
TR	trochanter.
tta	tendon of tarsus.
tte	tendon of extensor muscle of tibia.
tti	tendon of tibia.
ttr_i	inferior tendon of trochanter.
ttr_s	superior tendon of trochanter.
ttr'	anterior tendon of trochanter.
ttr''	posterior tendon of trochanter.
v, v_{1-3}	ventral setae.
v_x	unpaired ventral seta.
Va	anterior opisthosomatic verruca (verruca of segment X).
va_{1-2}	setae of anterior opisthosomatic verruca.
Vp	posterior opisthosomatic verruca (verruca of segment XII).
vp_x	unpaired seta of posterior opisthosomatic verruca.
vp_{1-2}	setae of posterior opisthosomatic verruca.
ϵ (epsilon)	supposed famulus.
ω (omega)	supposed solenidion of tarsus.
′ (prime)	anterior (pertaining to the anterior face of an appendage); the right one of two symmetrical structures.
″ (double prime)	posterior (pertaining to the posterior face of an appendage).

II. ACTINOTRICHIDA

Actinotrichid mites have repeatedly been the subject of my investigations, since I started my work as an acarologist. In the course of time, I studied the general morphology of the superorder Actinotrichida (Van der Hammen, 1963b; 1968a: 273–274; 1968c; 1970a; 1970b; 1970c; 1972; 1972a; 1973; 1977b; 1979; 1982a: 22–47, figs. 10–25), as well as the morphology of the separate orders, viz., Actinedida (Van der Hammen, 1968a: 275–276; 1969; 1986c: 218–220, fig. 1), Oribatida (e.g., Van der Hammen, 1953; 1955; 1955a; 1956; 1963; 1963a; 1963c; 1968; 1986b), Tarsonemida (Van der Hammen, 1970; correction in Van der Hammen, 1972: fig. 36), and Acaridida (Van der Hammen, 1968a: 270–273; 1982). In the present chapter these studies are summarized and reconsidered; the chapter is subdivided into eight sections: an introduction, a general descriptive part, a discussion of the four orders, a discussion of the relationships of the four orders, and additional remarks.

1. Introduction

The mites, classified here with the superorder Actinotrichida, were first recognized as constituting a separate group (Trombidi-Sarcoptiformes) by Oudemans (1931: 318). Oudemans (1936: 3; 1940: 295) subsequently restricted the use of the name Acari to his Trombidi-Sarcoptiformes, and classified this group, together with Scorpionida, Pedipalpi, Anthracomarti, Pseudoscorpionida, Solifugae, Opilionida and Ricinulei, in his Fixicoxata.

Grandjean (1935a: 123–126; 1936: 84–85, 88–89) recognized the same group of mites (which he named Actinochitinosi), and mentioned several of its characters. A more detailed study of the distinguishing characters and the affinities (Grandjean pointed out a certain resemblance to Solifugae) was published by him shortly afterwards (Grandjean, 1936b: 438–442). Later on (Grandjean, 1970: 800–807, 809–814), he reinvestigated and rediscussed the characters and affinities of the group.

The name Actinotrichida, suggested by Grandjean to replace his Actinochitinosi, was introduced in one of my papers (Van der Hammen, 1961: 179), and the distinguishing characters and affinities were subsequently also studied by me (Van der Hammen, 1966: 63; 1968a: 277; 1968c: 403, 404; 1969: 194–

196; 1972: 3, 36; 1972a: 274–279, 280–281, 283, 286–289; 1973: 275–278; 1977a: 316; 1985: 52–55, fig. 34; 1986a: 46–47; 1986c: 224–228, fig. 3). Among many other characters, the fundamental differences between the Actinotrichid and Anactinotrichid gnathosoma (and associated structures) were pointed out by me.

Weygoldt & Paulus (1979, 1979a) and Weygoldt (1980) returned to a classification of Actinotrichida with Acari (together with Anactinotrichida), although they were apparently convinced of the polyphyletic origin of the group. They united the Acari with the Ricinulei as Acarinomorpha, because of the presence of a hexapod larva, but erroneously supposed that the Acari were different from Ricinulei by the possession of a gnathosoma.

Alberti (1980a) investigated the ultrastructure of spermatozoa and spermiocytogenesis of Actinotrichida. In this connection, he pointed out the fundamental differences between Actinotrichida and Anactinotrichida; he also discovered interesting similarities between Actinotrichida and Solifugae, and established the absence of striking similarities between Actinotrichida and Palpigradi (as far as sperm morphology and spermiocytogenesis are concerned).

Lindquist (1984) made an attempt to reintroduce a monophyletic view of the mites. For that purpose he tried, on the one hand, to invalidate characters supporting the different relationships of Actinotrichida and Anactinotrichida, on the other hand, to find new shared derived characters for the mites as a single group (the characters previously used, viz., the presence of a gnathosoma and the occurrence of a hexapod larva, had proved invalid). Lindquist's paper, which is mainly based on an interpretation of characters from literature (several important data escaped from his notice), is characterized by a rigid application of the current methods of phylogenetic systematics, an unconvincing argumentation, and the development of atomistic views. The new shared derived characters of the mites as a single group, introduced by Lindquist, refer to small characters, the homology of which is often uncertain, and the connection of which with the adaptation to a new way of life (a prerequisite for the successful evolution of a large group) is not evident. I confine myself here to the discussion of a few examples. It is, for instance, not quite clear why the highly specialized association of coxal glands, subcapitular gutter and sternapophyses in Anactinotrichida is characterized as primitive (it is, in fact, not the association which is discussed by him: the separate characters are regarded as primitive and the association is not analysed). As to Lindquist's discussion of the segmentation of the Chelicerate legs, it is sufficient to compare his argumentation with chapter II, section 4 of the general part of the present book. The following remarks on the supposed shared derived characters of Actinotrichida and Anactinotrichida may be added here. Lateral lips are also present in Solifugae and Pseudoscorpionida. Three setae on the body of the chelicera is not the maximum number for mites (four setae are known from *Paracarus*, five setae from the male of *Opilioacarus vanderhammeni* Juvara-Bals & Baltac). The homology of the 'rutella' in both groups of mites is uncertain (in

Opilioacarida, three of the four pairs of paralabial setae can be more or less hypertrophied; it is even uncertain which of these is homologous with the corniculus of Gamasida and Holothyrida). Several of Lindquist's formulations are misleading, such as in the case of prosomatic lyrifissures (prodorsal and sternal); these are generally present in Anactinotrichida and absent in Actinotrichida (they are also absent in Palpigradi, and it is unknown if they have been present in Actinotrichid ancestors; the Anactinotrichid neosternum is, moreover, not homologous with the Actinotrichid coxisternal region); Lindquist transforms this important difference into a shared derived character (reduction to a fundamental maximum number which includes absence; it is evident that conclusions of this type are logically inadmissible). The maximum number of sternal lyrifissures mentioned by Lindquist is even invalid in the case of Anactinotrichida: ten sternal lyrifissures were described by me for *Opilioacarus texanus* (Chamberlin & Mulaik) (Opilioacarida), numerous cupuloids (supposed to be homonomous with lyrifissures) for the case of *Australothyrus ocellatus* Van der Hammen (Holothyrida). As a result of my analysis, not a single one of Lindquist's supposed shared derived characters of the mites has remained valid.

2. General descriptive part

Habitus (fig. 33). – Small or very small Chelicerata, rarely medium-sized (the length of the body of *Dinothrombium tinctorium* (Linnaeus), the largest known Actinotrichid mite, is up to 1.6 cm), exhibiting a great variety in habitus, varying from narrow to semi-globular, from soft-skinned to heavily sclerotized, from hairy to nearly glabrous, and from pale to bright-coloured. The length of the legs varies from relatively short to relatively very long.

Divisions of the body, tagmata and segmentation. – The original division of the Actinotrichid body is into a prosoma and an opisthosoma, the two being separated by the disjugal furrow. Incomplete segmentation of the opisthosoma is present in various species. In the course of evolution, the constitution of the Actinotrichid body has been subject to important changes. Fusion of segments, shifting of the paraproctal region from a posterior terminal to a ventral position, changes in the genital region by the evolution of a progenital chamber (see below), and the evolution of pseudotagmata and other subdivisions of the body (fig. 34A) considerably transformed the ancestral condition. Two furrows (circumcapitular and sejugal) can divide the body into four pseudotagmata: gnathosoma and idiosoma (separated by the circumcapitular furrow), and proterosoma and hysterosoma (separated by the sejugal furrow). The abjugal furrow separates a podosoma from an aspidosoma. Other divisions which can be distinguished are: epiprosoma (the prosoma without the podosoma), propodosoma and metapodosoma (anterior and posterior part of the podosoma, separated by the sejugal furrow), and stethosoma (the proso-

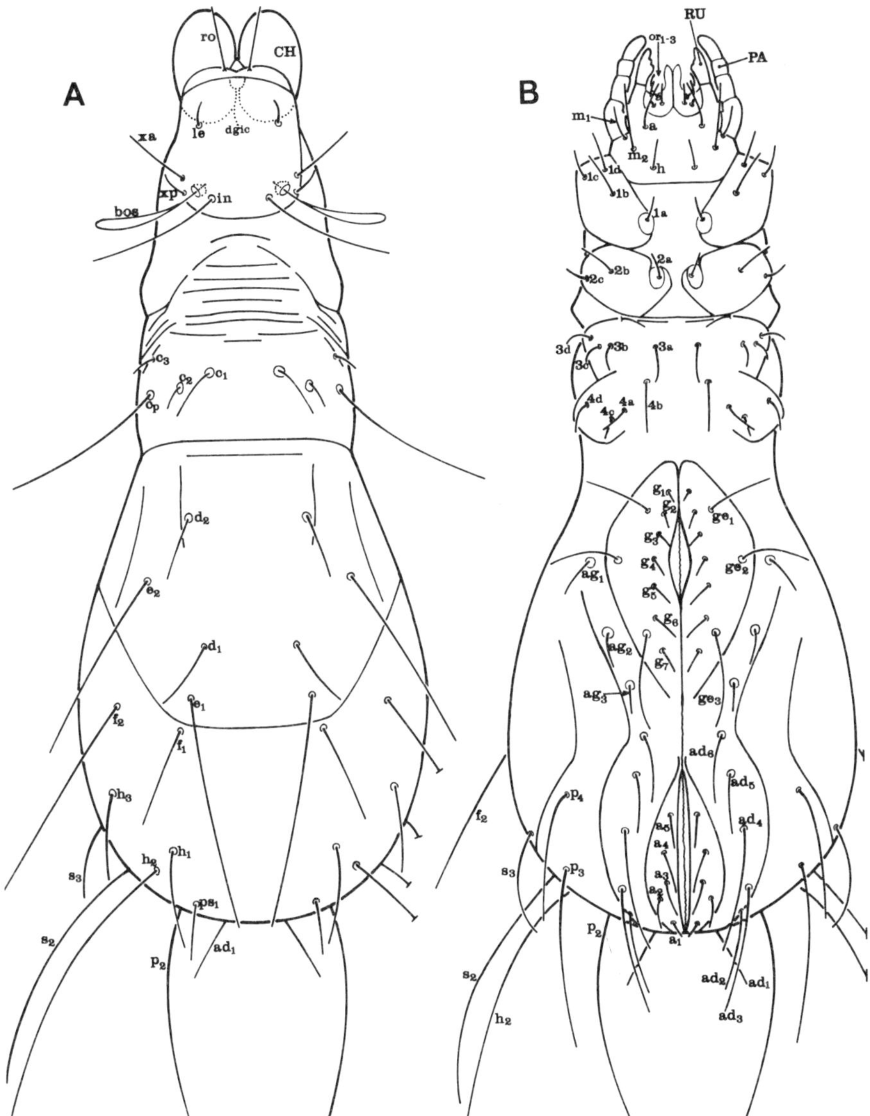

Fig. 33. Aphelacarus acarinus (Berlese) (Oribatida: Palaeosomata), adult female; A, dorsal view; B, ventral view; A, B, × 295.

ma without the gnathosoma). Modern soma-terminology is the result of studies by Van der Hammen (1963b, 1968c, 1969) and Grandjean (1970). References to the older literature, including the first introductions of the terms, are

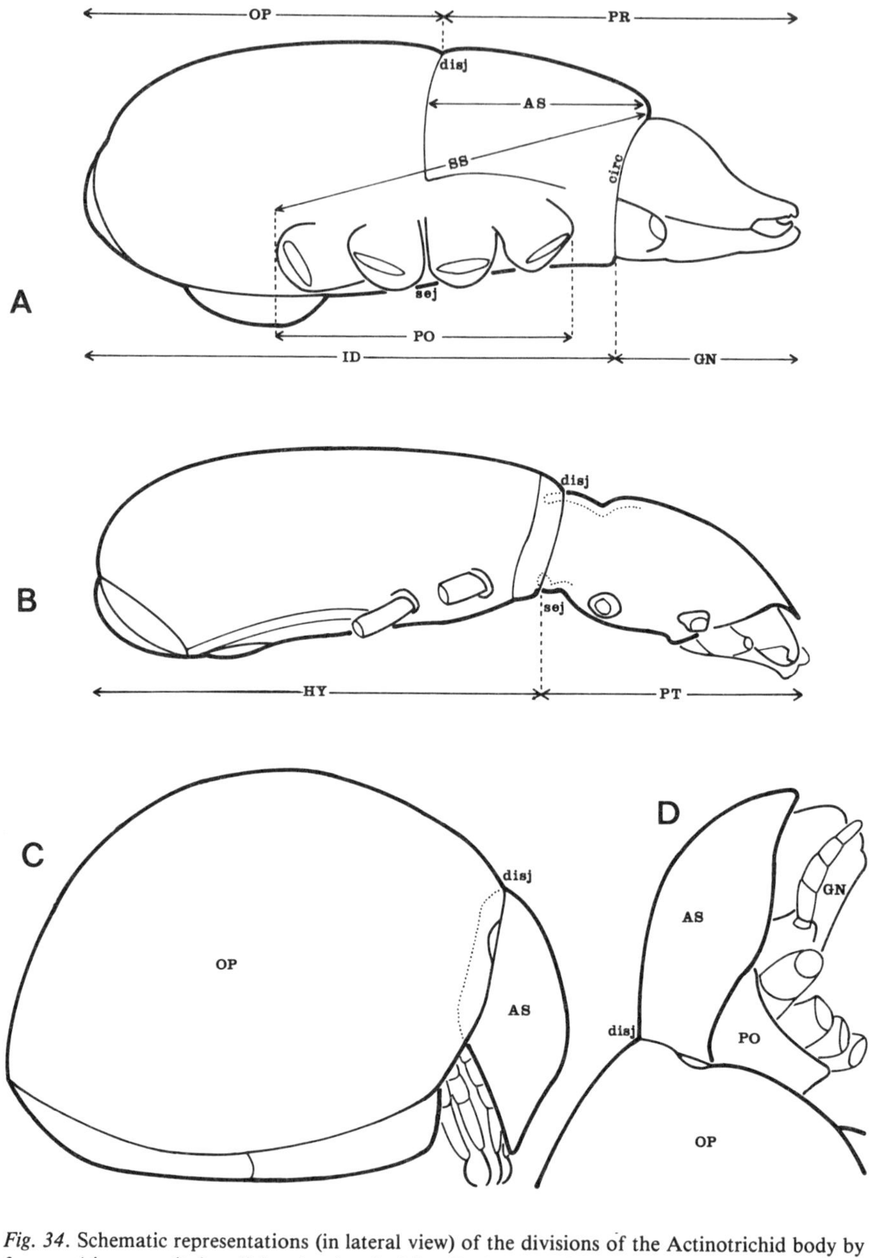

Fig. 34. Schematic representations (in lateral view) of the divisions of the Actinotrichid body by furrows (circumcapitular, disjugal, sejugal, abjugal). A, normal condition, in which two tagmata (prosoma, opisthosoma), four pseudotagmata (gnathosoma, idiosoma, proterosoma, hysterosoma) and five further divisions (stethosoma, aspidosoma, podosoma, propodosoma, metapodosoma) can be distinguished. B, division into protero- and hysterosoma in the case of protero-hysterosomatic articulation (dichoidy), in which dorso-disjugal and ventro-sejugal intervals are

given in the Glossary of Acarological Terminology (Van der Hammen, 1980). Generally, gnathosoma and idiosoma are movably connected (gnatho-idiosomatic articulation). The divisions of the Actinotrichid idiosoma can be immovably connected (holoidy) or some can be movably joined. In the last-mentioned case, we can distinguish protero-hysterosomatic articulation or dichoidy (fig. 34B, 37F), known from many primitive Oribatid mites; and ptychoidy (the presence of a dorsal hinge between prosoma and opisthosoma, permitting the prosoma to be folded down and the legs to be concealed; cf. figs. 34C, D, 37G), known from several other groups of primitive Oribatid mites (Mesoplophoroidea, Protoplophoroidea, Euphthiracaroidea, Phthiracaroidea). Ptychoidy presupposes: (1) a complete separation, by soft skin, of the podosoma from the remaining part of the exoskeleton, (2) a completely sclerotized opisthosoma, and (3) an articulated coxisternal region, in the ancestors, at the levels of development at which the mites are now ptychoid (cf. Grandjean, 1969: 132–139).

Studies by Van der Hammen (1969, 1970a) have demonstrated that the body of primitive segmented Actinotrichid mites consists of a number of precheliceral metameres and the segments I–VI (prosoma) and the segments VII–XVI (opisthosoma) (cf. fig. 48). In the course of embryonic development, the paraproctal segments (i.e. the segments XIII–XVI) are suppressed and can reappear in the course of postembryonic development (hysteromorphosis, cf. fig. 35). Each of these segments is eustasic and can reappear at a fixed level of the postembryonic development. The base levels of segments XIII–XVI are respectively: larval, protonymphal, deutonymphal and tritonymphal. When a certain paraproctal segment does not appear at this level, it will not appear later on in ontogeny, nor will any of the segments which, in the plan of construction of the group, are posterior to it. As mentioned in chapter II, section 1 of the general part, it is logical to suppose that, in the Actinotrichid ancestors, segment XVII, with adult base level, must have been present; this segment is not known from extant Actinotrichida, but could still be present in some primitive species (particularly primitive Actinedida). The presence of segment XVI (the peranal segment) is known from: Parhypochthoniidae, Brachychthoniidae and Pterochthoniidae (primitive Oribatida); and Terpnacaridae and some Pachygnathidae (primitive Actinedida). The anal segment (segment XV) remains the terminal paraproctal segment in most Oribatida. The adanal segment (segment XIV) remains the terminal paraproctal segment in several groups of Actinedida (Caeculidae, Anystidae) and Acaridida. The pseudanal segment (segment XIII) remains the paraproctal segment in many

---

←

continuous. C, D, division into prosoma and opisthosoma in the case of ptychoidy (in Euptyctima); the podosoma can be completely withdrawn, and the aspidosoma folded down, by which the mite is closed up like an armadillo; C, specimen in partly closed up condition; D, anterior part of specimen in normal 'opened' condition.

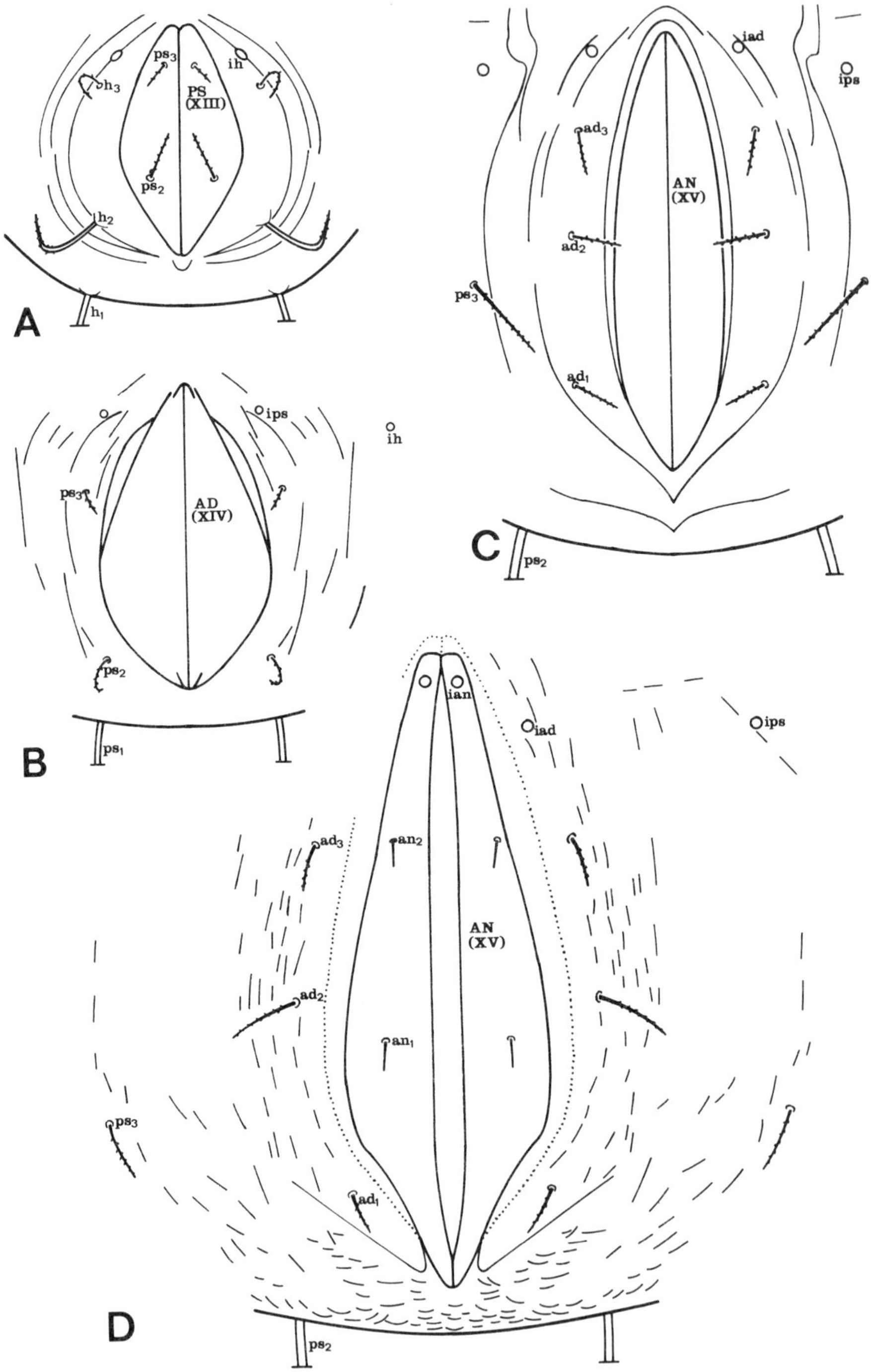
ps3
ih
h3
PS
(XIII)
ps2
h2
h1
A
iad
ips
ad3
AN
(XV)
ad2
ps3
ad1
C
ps2
ips
ih
ps3
AD
(XIV)
ps2
B
ps1
ian
iad
ips
ad3
an2
AN
(XV)
ad2
an1
ps3
ad1
D
ps2

other Actinedida (Tydeidae, Raphignathoidea, Tetranychoidea, Cheyletoidea), and in Tarsonemida.

Tegument. – The Actinotrichid tegument (or skin) generally consists of a living layer or hypodermis, and three non-living layers, viz., chitonostracum (procuticle), epiostracum (epicuticle) and cerotegument (cement and lipid layers, and wax blooms). The chitonostracum generally consists of three layers: Schmidt's layer, endostracum and ectostracum. Chitonostracum and epiostracum constitute the cuticle. The skin can be soft, or sclerotized and melanized; the highest degree of sclerotization is found in the ectostracum. Sclerites, when present, can be segmentally arranged. The cuticle can present various types of microsculpture: aciculate, alveolate, areolate, corniculate, costate, costulate, foveolate, granulate, plicate, punctate, reticulate, rugose, scabrous, shagreened, striate, striolate, sulcate, tuberculate, vermiculate, verrucose, etc. The cerotegument develops by exudation through pores in the cuticle; it can present various types of microsculpture. In many Actinotrichida, respiration is cutaneous, and the cuticle not heavily sclerotized. It is supposed here that, in Actinotrichid ancestors, the skin was soft.

Aspidosoma. – The aspidosoma is bordered laterally by the abjugal furrow, anteriorly by the circumcapitular furrow, and posteriorly by the disjugal furrow. The surface of the aspidosoma is constituted by the prodorsum; it can be scleritized (prodorsal shield) or not. The prodorsum can be transversely subdivided, by two furrows, into prostethidium, mesostethidium and metastethidium (fig. 48). The ancestral number of prodorsal setae is six pairs (figs. 33A, 48): rostral setae (*ro*), lamellar setae or anterior trichobothria (*le*), interlamellar setae (*in*), trichobothria or sensilli, or posterior trichobothria (*bos*), anterior exobothridial setae (*exa*), and posterior exobothridial setae (*exp*). A naso, or frontal protuberance (of precheliceral origin), is present in many Actinedida and some Oribatida (e.g., Archeonothridae, Brachychthoniidae). It is unpaired and, when present, it is found at (or under) the rostral extremity of the aspidosoma, overhanging the chelicerae (fig. 36B, D, G); an eye (often bilobate) can be present at its ventral surface. The prodorsum can present an unpaired anterior, and one or two pairs of lateral eyes, but many Actinotrichid mites (Acaridida, Tarsonemida, most Oribatid mites, several groups of Actinedida) have no eyes.

Opisthosoma (dorsal, laterodorsal and paraproctal regions). – The opisthosomatic cuticle can be soft or sclerotized. Segmentation can be indicated by furrows in the soft skin, or by scissures in the scleritized cuticle. The paraproctal region has developed a ventral position; external traces of segmentation (when present) are, as a result of this dislocation, not continued in lateroventral direction. The ancestral number of opisthosomatic lyrifissures or cupules

---

←

Fig. 35. Postembryonic development of the paraproctal region of *Archegozetes magna* (Sellnick) (Oribatida: Nothrina); A, larva; B, protonymph; C, deutonymph; D, tritonymph; A–D, × 472.

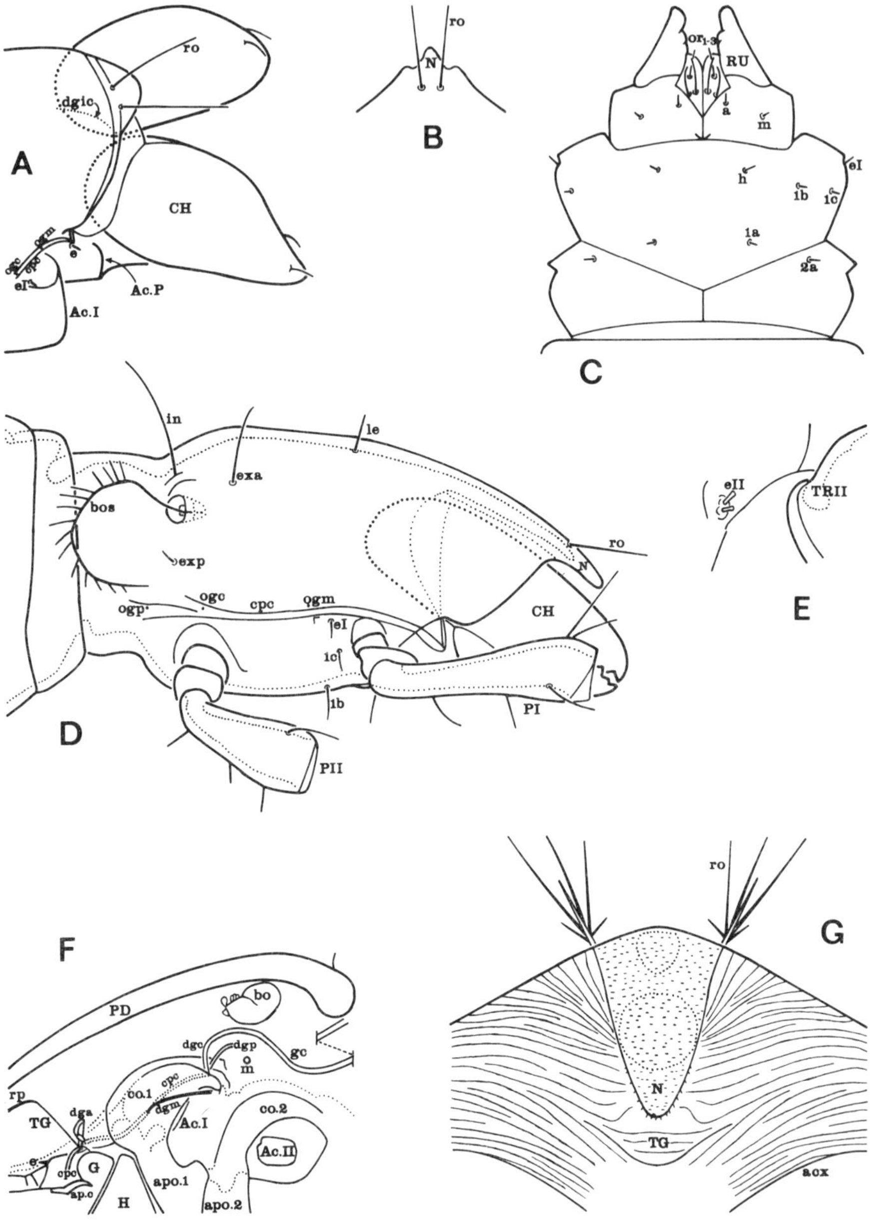

Fig. 36. Prosomatic regions of various species of Actinotrichida. A, *Aphelacarus acarinus* (Berlese) (Oribatida: Palaeosomata), adult, laterodorsal view of anterior part of prosoma. B, *Eulohmannia ribagai* (Berlese) (Oribatida: Mixonomata), adult, dorsal view of rostrum. C, *Epilohmannia cylindrica* (Berlese) (Oribatida: Mixonomata), adult, ventral view of anterior part of prosoma. D, *Eulohmannia ribagai* (Berlese), adult, lateral view of anterior part of prosoma. E, *Oecosmaris callitricha* Grandjean (Actinedida: Trombidina), adult, laterodorsal view of

is seven pairs, attributed to the segments IX–XV (cf. Van der Hammen, 1969: 185, 198). Chaetotaxy can be primordiotrichous (primitive, adelonymous and chaotic), hypertrichous (primitive, idionymous and supernumery), holotrichous (normal, idionymous), meritrichous (idionymous, reduced in number), or neotrichous (increased in number, idionymous or adelonymous).

Genital region. – The Actinotrichid genital region is generally close to the anal region (the two regions are often even contiguous), not separated by a distance proportionate to the distance between segments VIII and XVI (figs. 33B, 48); it is also relatively very large. It is laterally bordered by the aggenital region which can present one or more pairs of setae. The ancestral condition is already highly complicated by the development of a secondary genital opening, and can be characterized in the following way. There is a secondary, longitudinal progenital opening, closed by a pair of progenital valves, giving entrance into a progenital chamber. This chamber contains (in the adult) three pairs of evaginable progenital papillae, a trifid eugenital opening, and eugenital setae (fig. 37A). The genital papillae represent opisthosomatic appendages; they can be segmented (fig. 37B). The postembryonic development of the genital papillae is interesting: there are one pair in the protonymph, two pairs in the deutonymph, and three pairs in the tritonymph and the adult (a development very similar to the postembryonic development of the opisthosomatic papillae of segments X–XII in *Prokoenenia*). In Oribatida and several Actinedida, an evaginable ovipositor has developed as an outgrowth of the vagina (figs. 37C, E). The trifid eugenital opening is found, in these cases, at the distal end of the ovipositor (fig. 37D); it is surrounded by three terminal lobes. An ovipositor is not present in Tarsonemida, Acaridida, and several groups of Actinedida (in the last-mentioned group, the absence can be an ancestral or a derived character state). In Tetranychidae, genital region and ovipositor are quite different (fig. 37F, G): there is a single progenital valve, whilst an ovipositor can develop mainly by extension of the plicate tegument of the genital region.

The following interpretation of the ancestral Actinotrichid genital region is given here. The trifid eugenital opening is situated in, or between, the segments VIII and IX, and is surrounded by the opisthosomatic appendages of segment VIII (one pair) and the opisthosomatic appendages of segment IX (a fused pair); it may be repeated here that a similar condition, although reversed, is found in female Palpigradi. The three pairs of genital papillae represent the opisthosomatic appendages of segments X–XII, which are included in the progenital chamber. Fig. 48 demonstrates that the posterior bor-

---

←

supracoxal setae *eII* (redrawn after Grandjean, 1947). F, *Hermannia convexa* (C.L. Koch) (Oribatida: Nothrina), adult, internal view of part of the lateral prosomatic region; G, *Alycus roseus* C.L. Koch (Actinedida: Endeostigmata), adult female, ventral view of naso and adjoining part of prosoma. A, × 550; B, D, × 295; C, × 376; E, × 522; F, × 118; G, × 1050.

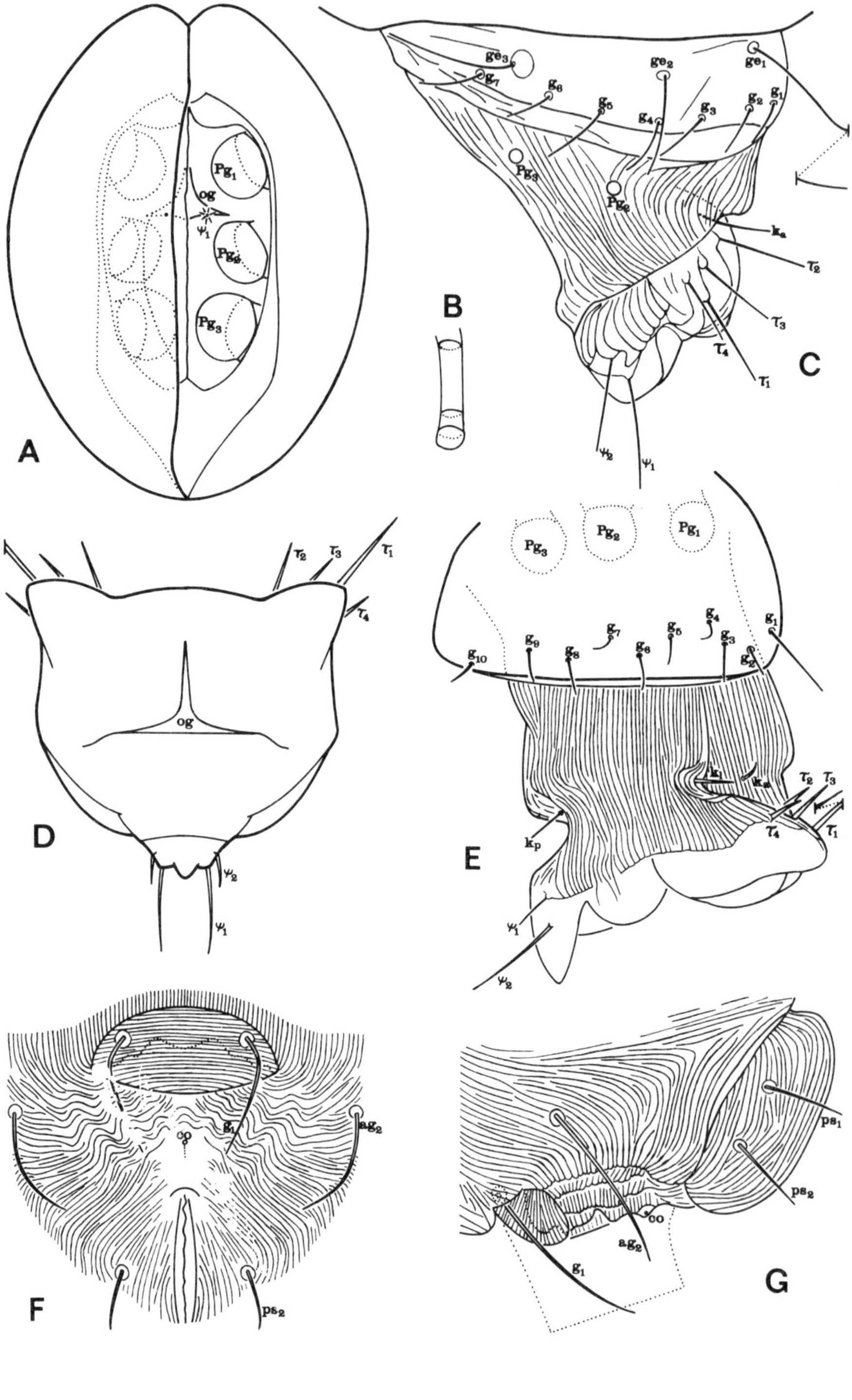
A
Pg_1
og
ψ_1
Pg_2
Pg_3
B
C
ge_3
g_7
g_6
g_5
ge_2
g_4
g_3
g_2
g_1
ge_1
Pg_3
Pg_2
k_a
τ_2
τ_3
τ_4
τ_1
ψ_2
ψ_1
D
τ_2
τ_3
τ_1
τ_4
og
ψ_2
ψ_1
E
Pg_3
Pg_2
Pg_1
g_{10}
g_9
g_8
g_7
g_6
g_5
g_4
g_3
g_2
g_1
k_1
k_a
τ_2
τ_3
τ_4
τ_1
k_p
ψ_1
ψ_2
F
g_1
co
ag_2
ps_2
G
ps_1
ps_2
co
ag_2
g_1

der of the progenital opening indeed coincides with the posterior border of segment XII. The progenital valves are supposed to represent the exites of opisthosomatic appendages of segments VIII and IX (a subdivision into two parts, by a suture, is still present in some species, e.g., in the Oribatid genus *Hypochthonius*; the presence, in Chelicerata, of a longitudinal genital opening instead of a transverse one, is always the result of an association with opisthosomatic appendages). This interpretation (see also fig. 38), which has gradually developed by way of previous hypotheses (Van der Hammen, 1963b: 448–449; 1969: 198–199, fig. 7A; 1977a: tab. 2; 1979: 429–430; 1981a: 28–31, fig. 11), finally explains all problems posed by the complicated morphology of the Actinotrichid genital region.

Coxisternal region. – The ventral region of the podosoma can be partly or completely scleritized (see, e.g., figs. 17B, 33B, 54B, 64A, 68D, 75B, 107). In *Epilohmannia cylindrica* (Berlese), epimeron 1 has fused with the epimeral region of the palp (fig. 36C), as in Palpigradi. A sejugal furrow or interval is present between epimera 2 and 3. Epimeral chaetotaxy can be subject to numerical changes (multiplications and regressions); the number of epimeral setae generally increases in the course of postembryonic development, as in Palpigradi.

Lateral aspect of podosoma. – Supracoxal setae (*eI* and *eII*; cf. Van der Hammen, 1977b: 13–15, fig. 2B, C, F–L; 1980: 153–154; 1981a: 26) can be present above the bases of legs I and II. Seta *eII* is the first supracoxal seta to disappear; it can also be subject to duplication (cf. fig. 36E). A podocephalic canal (fig. 36A, D, F) can be present, originally extending from above the base of leg I to the gnathosoma. The orifices of four podocephalic glands can be associated with the podocephalic canal, viz., those of anterior, median, coxal and posterior (lateral) gland. The number of glands can, however, be reduced, and the podocephalic canal can become internal (several Actinedida) (cf. Van der Hammen, 1968: 19–22, 29–30, figs. 1, 7–11; 1980: 120–121). In *Eulohmannia ribagai* (Berlese) (fig. 36B), the podocephalic canal extends from the base of leg II to the gnathosoma, and the orifice of the coxal gland is above the base of leg II; this could be a secondary condition, although it has not yet been explained.

Gnathosoma. – The division of the body, anterior to the circumcapitular furrow, is called gnathosoma. It can be situated in a camerostome, a cavity

---

←

Fig. 37. Genital region of various species of Actinotrichida. A, *Alycus roseus* C.L. Koch (Actinedida: Endeostigmata), ventral view (left progenital lip removed). B, *Epilohmannia cylindrica* (Berlese) (Oribatida: Mixonomata), genital papilla *Pg3*, lateroventral view. C, *Aphelacarus acarinus* (Berlese) (Oribatida: Palaeosomata), evaginated ovipositor, lateral view (right face). D, E, *Hermannia convexa* (C.L. Koch) (Oribatida: Nothrina), evaginated ovipositor; D, distal view; E, lateral view (right face). F, G, *Tetranychus urticae* C.L. Koch (Actinedida); F, ventral view; G, lateral view (left face; the position of the protruded genital region, functioning as an ovipositor is represented by a dotted line). A–C, F–G, × 550; D–E, × 186.

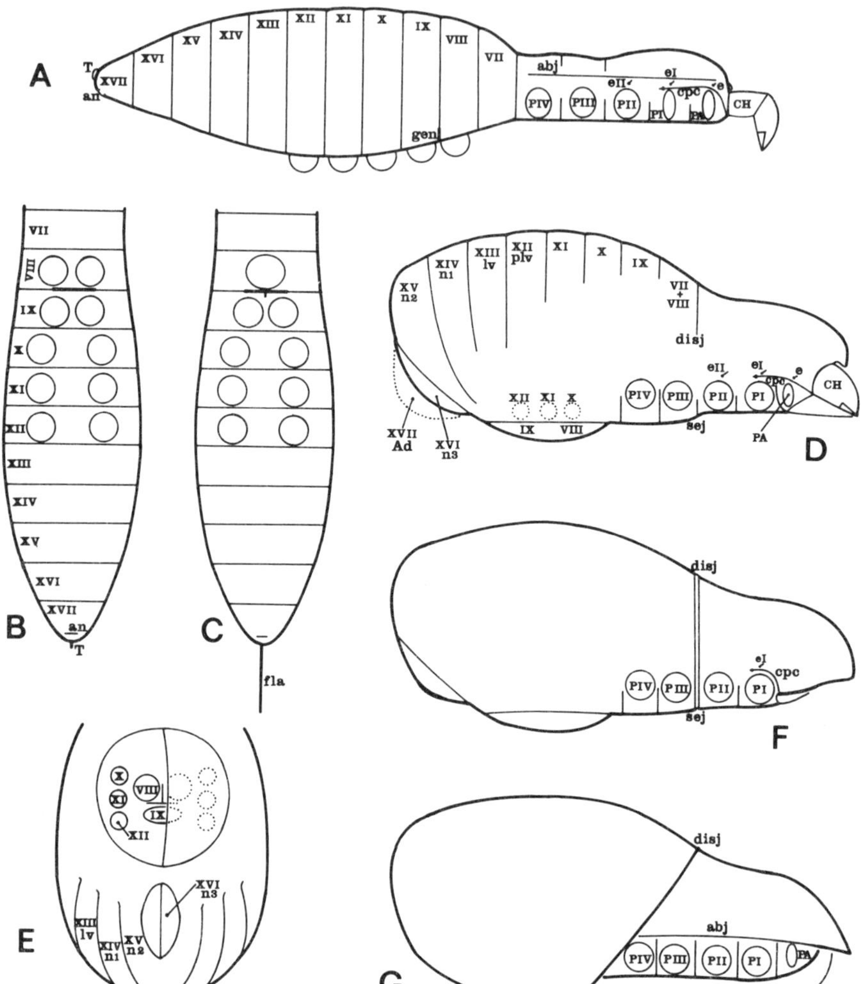

Fig. 38. Schematic representation of the evolutionary development of the architecture of the Actinotrichid body; A, B, ancestral Epimerate type; A, lateral view; B, ventral view of opisthosoma; C, ancestral Palpigradid type, ventral view of opisthosoma; D, E, ancestral Actinotrichid type; D, lateral view; E, ventral view of opisthosoma; F, Actinotrichid dichoid type, lateral view: G, Actinotrichid ptychoid type, lateral view.

sheltered dorsally and laterally by the aspidosoma (fig. 93). The gnathosoma bears two pairs of appendages: chelicerae and palps. Two principal parts can be distinguished: cheliceral frame and infracapitulum (figs. 51A, 93). The cheliceral frame consists of the membraneous tegument which constitutes the body-wall between the rostral region of the aspidosoma and the infracapitulum. The chelicerae are movably attached to the cheliceral frame by means of the cheliceral sheaths. The superior part of the cheliceral frame constitutes the tegulum.

The infracapitulum consists of a dorsal region or cervix (fig. 88), a ventral region or mentum (figs. 51B, 89C), a pair of large lateral ridges, and a pair of malapophyses, probably representing palpal 'endites' (of which the ventral surface, or gena, is a continuation of the mentum). The cervix presents an axial convexity, highest near the base of the labrum, which is called capitular saddle; it separates the two cheliceral grooves. The ventral surface of the infracapitulum presents infracapitular setae. The infracapitulum contains mouth and pharynx. The mouth is surrounded by three or four lips: labrum, lateral lips, and labium. The lateral lips bear the adoral setae. The labium has disappeared in most Actinotrichid mites. When it is present, the mouth has four commissures and a quadrangular section; when the labium is absent, there are three oral commissures, and the section is triangular. The exoskeleton of the infracapitulum generally presents internal projections, or apodemes, on which several tendons and muscles can be inserted. A capitular apodeme, constituting the internal continuation of the cervix, is generally present (fig. 88); it separates, internally, the cheliceral frame and the infracapitulum (segments I and II). A subcapitular apodeme, constituting the internal continuation of the mentum, is moreover known from Caeculidae (Coineau, 1974: 78, fig. 28); it separates, internally, the infracapitulum and the coxisternal region of segment III (epimera 1). The simultaneous presence of capitular and subcapitular apodeme could constitute a primitive character.

The evolution of the gnathosoma (gnathosomatization) in Actinotrichida is closely connected with the evolution of the podocephalic canal which extends to the cervix (fig. 88; cf. the previous section, which deals with the lateral aspect of the podosoma).

Papers of fundamental importance, dealing with the gnathosoma, have been published by Grandjean (1936b, 1957a, 1957b) and Van der Hammen (1968, 1970c). Further literature on the gnathosoma is mentioned, under various entries, in the Glossary of Acarological Terminology (Van der Hammen, 1980).

Chelicera. – The Actinotrichid chelicera is ancestrally composed of three segments: trochanter, body of the chelicera, and apotele (fig. 39B–D). Generally, the Actinotrichid trochanter is reduced to a ventral or lateroventral element. The body of the chelicera takes the place of femur, genu, tibia and tarsus. Its distal part constitutes the fixed jaw of the chela. Conditions in Endeostigmata (Actinedida) and Enarthronota (Oribatida) (cf. Grandjean, 1947b) demonstrate that the fixed jaw represents a hypertrophied seta or a group of coalesced setae (fig. 39A). The apotele constitutes the movable jaw of the chela. The body of the chelicera can present two setae (*cha*, *chb*) and, in rare cases, one antiaxial lyrifissure ($i\alpha$). Various oncophyses, i.e. extensions of an arthrodial membrane, often in the shape of a hyaline intumescence, have been described from Actinotrichid chelicerae. The ancestral function of the Actinotrichid chelicera probably was partly prehensile; it could also have participated

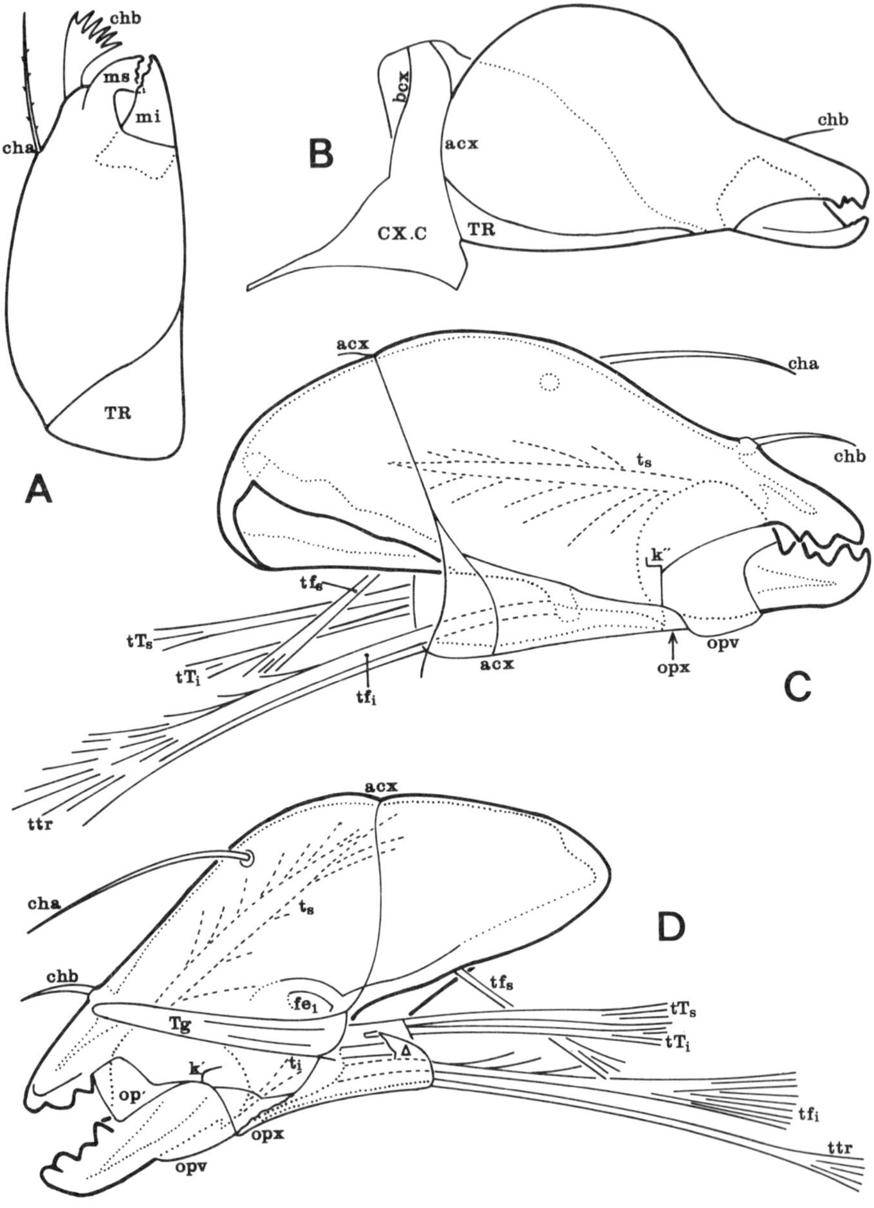

Fig. 39. Actinotrichid chelicerae. A, *Cosmochthonius lanatus* (Michael) (Oribatida: Enarthronota), lateral (antiaxial) view of right chelicera. B, *Alycus roseus* C.L. Koch (Actinedida: Endeostigmata), lateral (antiaxial) view of right chelicera. C, D, *Hermannia convexa* (C.L. Koch) (Oribatida: Nothrina); C, lateral (antiaxial) view of right chelicera; D, lateral (paraxial) view of right chelicera. A, × 1312; B, × 707; C, D, × 368.

in the fragmentation of food (ingestion of solid food and internal digestion are known from extant Oribatida and Endeostigmata). In the course of evolution, the chelicera has developed several specialized functions (e.g., suctorial), by which, in several groups, the shape has been radically transformed. In Tarsonemida, e.g., the greater part of the chelicera is supposed to have fused with the cervix, whilst the apotele has remained free (it constitutes a stylet, situated inside the infracapitulum) (cf. Van der Hammen, 1970: 11–12, fig. 7A, B).

Palp. – The ancestral number of Actinotrichid palpal segments is five (fig. 40A): trochanter, femur, genu, tibia and tarsus (a palpal apotele is not present in Actinotrichida). The articulations are originally similar to those in the corresponding joints of the legs (the genu-tibia articulation could, in some cases, perhaps also present extensor muscles, although I have not been able to discern these with certainty; cf. fig. 40A). The number of palpal segments is reduced in many Actinotrichid species; there are, for instance, four segments in *Eulohmannia ribagai* (Berlese) and *Cheyletus eruditus* (Schrank) (fig. 40C), three segments in *Phthiracarus laevigatus* (C.L. Koch) (fig. 40B), two segments in *Epilohmannia* (cf. fig. 56B). In many families of Actinedida (the so-called Dactylognatha) a thumb-claw complex is present (fig. 40C), consisting of a hypertrophied seta of the tibia (the tibial claw) and the palpal tarsus (the thumb); in these cases, the tarsus is reduced in shape and size, and has shifted to a ventrobasal position on the tibia. The main function of the Actinotrichid palp is now sensory; in Dactylognatha it is, apparently, also prehensile. The palpal segments can present setae, the palpal tarsus also a solenidion and a proximal lyrifissure; several of the tarsal setae can have evolved into eupathidia. A supracoxal seta *e* is generally inserted above the base of the palp. It may be remarked here that, in comparison with the legs, the Actinotrichid palp (generally) is relatively small.

Legs. – The ancestral number of leg-segments is seven (fig. 41A–D): trochanter, femur 1, femur 2, genu, tibia, tarsus, and apotele (cf. Van der Hammen, 1970b). Two femora (fig. 41E) are present in some primitive Oribatida and in many Actinedida. In most Oribatida (fig. 41F) femur 1 and 2 have been integrated (by anergastic union) into a single femur (cf. Grandjean, 1954). In *Fusacarus* (fig. 42E–F), femur and genu of leg II have been integrated, by anchylosis, into a single femorogenu (cf. Grandjean, 1953). A femorogenu is also present in leg III of some Tarsonemida (fig. 42D) (cf. Van der Hammen, 1970). Tibia and tarsus of the legs can be integrated into a single tibiotarsus (fig. 42C) (cf. Grandjean, 1954b: 356). The apotele can disappear as a result of suppression. The number of tarsal segments can increase (fig. 42A, B), first by subdivision, thereupon by cosmiotaxy (tarsus 1, 2, etc.); these adesmatic subdivisions are rare in Actinotrichida (they are known from some Actinedida). The Actinotrichid tarsus does not consist of a basi- and a telotarsus.

The ancestral types of articulation in the Actinotrichid legs are the following (fig. 41E–G): the body-trochanter, trochanter-femur, and tarsus-apotele articulations are constituted by pivot joints; the femur 1-femur 2, femur-genu,

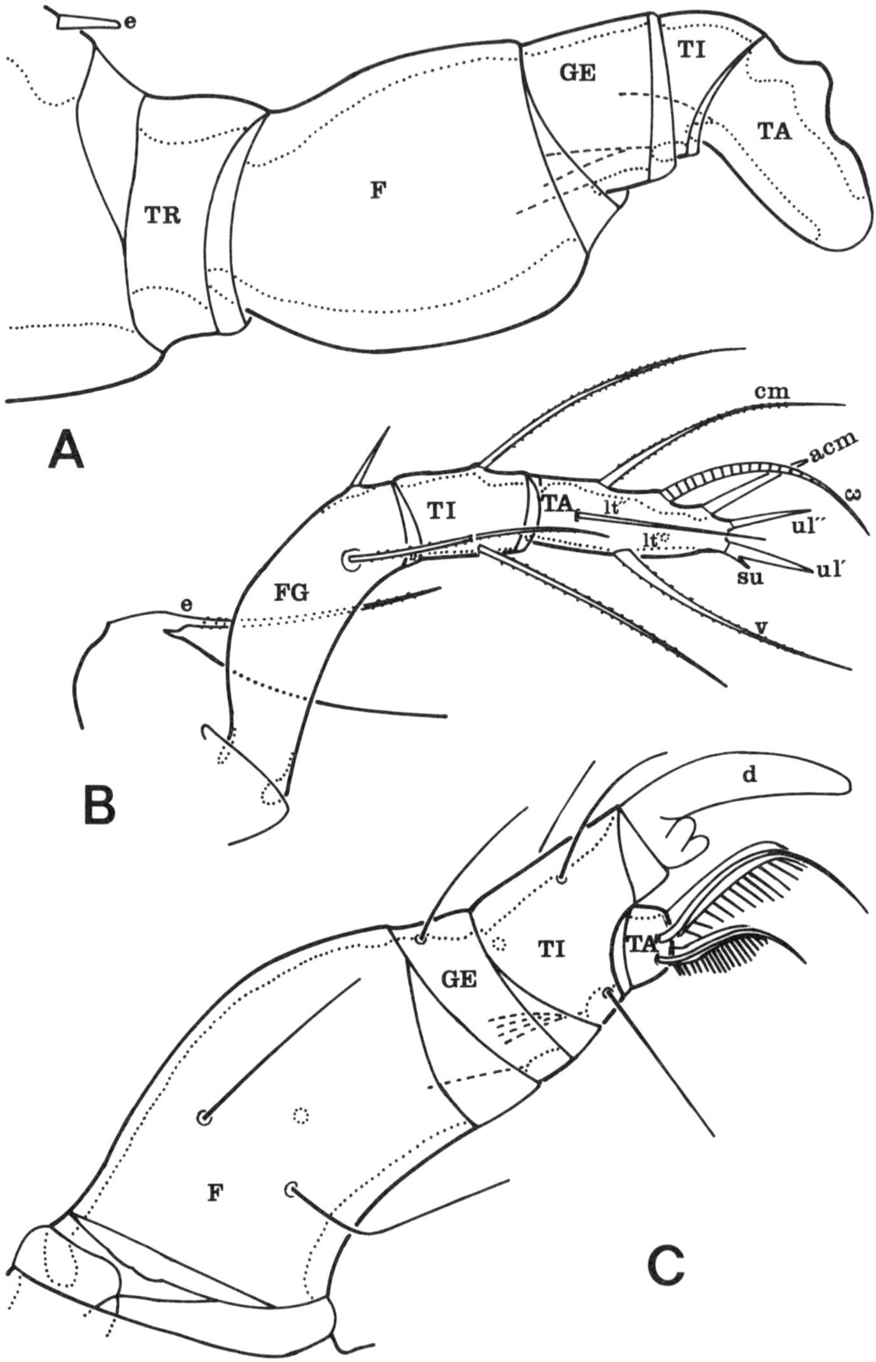
e
TI
GE
TA
F
TR
A
cm
acm
ω
TI
TA
lt″
lt′
ul″
ul′
su
FG
e
v
B
d
TI
TA
GE
F
C

genu-tibia, and tibia-tarsus articulations are constituted by hinge joints. In higher Oribatid mites, a rocking joint has developed between body and trochanter. Leg movements are complicated here by the pseudosymmetrical arrangement of the legs (with respect to the sejugal furrow or interval): legs I and II are directed to the front and the femur can be elbowed; legs III and IV are directed backwards and the trochanter can be elbowed (cf. Grandjean, 1952: 24–27); leg-rocking has not been analysed in detail in these cases. The legs generally present a rising part, a knee-joint or a knee-segment (genu), and a descending part. In higher Oribatid mites, the knee-segment is relatively small.

In the course of evolution, the main function of leg I can change from mainly ambulatory to mainly sensory. This evolution can involve suppression of the apotelic claw, and the development of hexapodous stepping. It is, for instance, known from Tydeidae and Cheyletidae (Actinedida), and Staurobatidae (Oribatida) (cf. Grandjean, 1967: 720–721).

The Actinotrichid tarsus generally presents one proximal, dorsal lyrifissure; other pedal lyrifissures are never present. The legs can present various types of phaneres (figs. 43, 44): ordinary setae (which can be present on all segments; cf. fig. 44A), trichobothria (compound elements, consisting of a small cavity and a specialized seta, and constituting vibro- and anemoreceptors; cf. fig. 44C, D), eupathidia (modified setae of which the solid axis and the root are pierced by a canal; cf. fig. 44B), famulus (a hollow phanere, inserted in the dorsal or antiaxial region of tarsus I and, rarely, tarsus II; cf. 44E–G), and solenidia (hollow phaneres with thin wall, numerous pores, and a short, largely open root, which are supposed to be chemosensory, and which can be present on femur, genu, tibia and tarsus; cf. fig. 44I–K). Actinotrichid setae are characterized by the presence of an actinopilinous axis (actinopilin is an anisotropic matter exhibiting strong birefringence; cf. fig. 44A, and Van der Hammen, 1980: 10).

In many groups of Actinotrichida, the larva and the prelarva present a variously shaped organ of unknown function, situated on either side between epimera 1 and 2 (Claparède's organ). In its most primitive shape, it has the aspect of a relatively long, segmented appendage, with hemispherical distal part (fig. 42G). Its shape is similar to that of the genital papillae, and it is supposed to represent the 'endite' of leg I or leg II.

Internal anatomy. – The main literature on Actinotrichid internal anatomy is summarized in the Glossary of Acarological Terminology (Van der Hammen, 1980), where detailed references are also given. The alimentary canal comprises: mouth (and gullet), pharynx, oesophagus, midgut and caeca, hind-

←

Fig. 40. Lateral (antiaxial) view of the right palp in the adult of three species of Actinotrichida; A, *Hermannia convexa* (C.L. Koch) (Oribatida: Nothrina) (all palpal phaneres omitted); B, *Phthiracarus laevigatus* (C.L. Koch) (Oribatida: Mixonomata); C, *Cheyletus eruditus* (Schrank) (Actinedida); A, × 917; B, × 487; C, × 627.

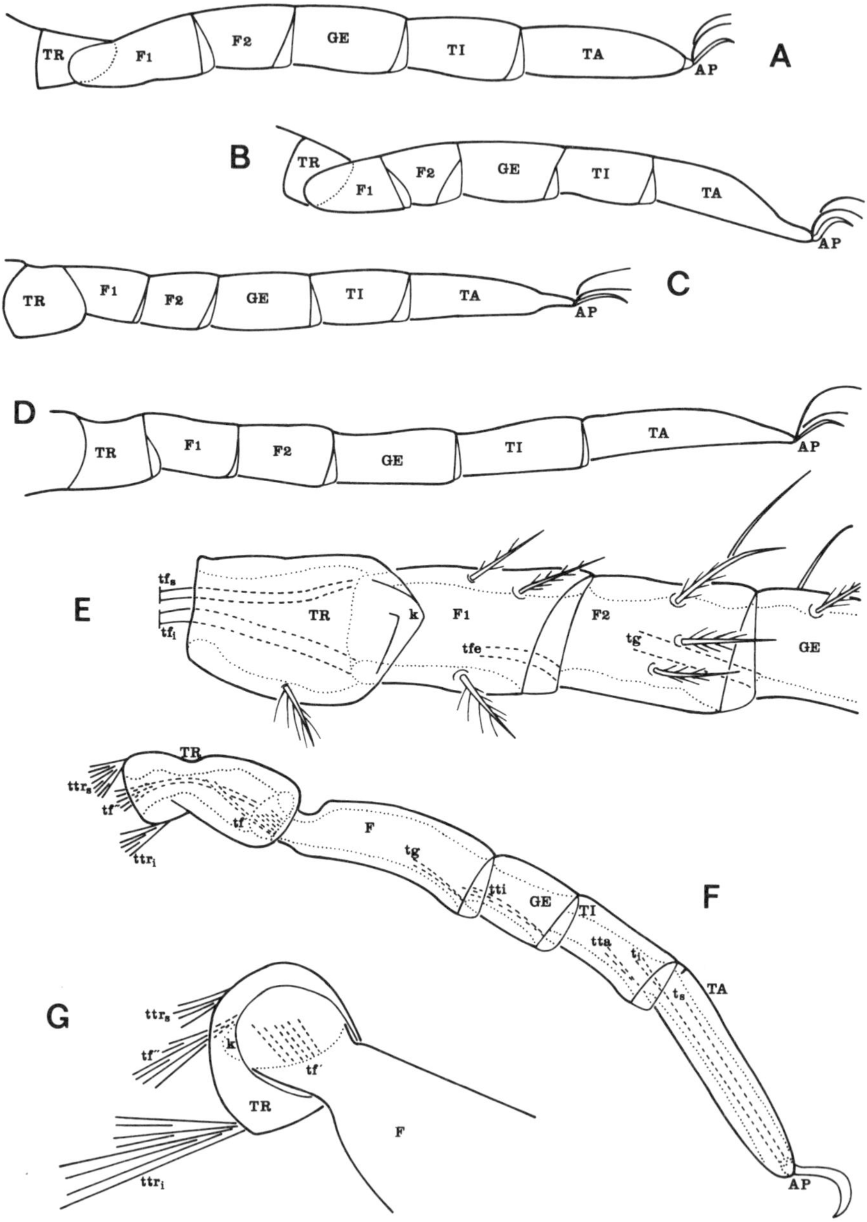

Fig. 41. Segmentation and articulation of Actinotrichid legs (1). A–D, *Aphelacarus acarinus* (Berlese) (Oribatida: Palaeosomata), lateral (antiaxial) view of legs I–IV; A, right leg I; B, right leg II; C, left leg III; D, left leg IV. E, *Alycus roseus* C.L. Koch (Actinedida: Endeostigmata), lateral (antiaxial) view of trochanter, femur 1, femur 2, and proximal part of genu of left leg IV. F, G, *Hermannia convexa* (C.L. Koch) (Oribatida: Nothrina); F, lateral (antiaxial) view of left leg IV; G, lateral (antiaxial) view of trochanter and proximal part of femur of right leg I. A–D, × 550; E, × 566; F, G, × 295.

gut and anus. The excretory system can comprise: Malpighian tubes, and unpaired excretory tube, a uroporus (an excretory pore at the place of the anus, in cases where the gut ends blindly), and coxal glands (originally debouching, dorsal to leg I, into the podocephalic canal). An endosternite is present in Actinotrichida, sometimes as a transverse band, sometimes more extensively developed (cf. Van der Hammen, 1969: 187, fig. 4A; Firstman, 1973: 12–14); it has not yet been studied in detail. The circulatory system is open; a dorsal vessel (heart), or its vestige, is apparently often present. Respiration is cutaneous, or by means of secondary respiratory organs (tracheae, etc.). Secondary respiratory organs can start from various superficial parts of the idiosomatic cuticle, from the progenital chamber, from the base of the appendages, and from hidden depressions of the surface (bothridium, preanal fold, apodemes). The central nervous system is concentrated to a solid mass in the prosoma, pierced by the oesophagus. Actinotrichid sensory organs (or supposed sensory organs) include: eyes and other photoreceptors, various sensory phaneres, genital papillae and Claparède's organ, lyrifissures, etc. The male reproductive system can comprise: accessory glands, ductus ejaculatorius, penis or spermatopositor, testes, vasa deferentia and vesicula seminalis. Fine structure of Actinotrichid spermatozoa, and spermiocytogenesis have been studied by Alberti (1980a); his results confirm the monophyletic character of the Actinotrichida, and demonstrate the absence of striking similarities between Actinotrichida, on the one hand, and Anactinotrichida (incl. Opilioacarida) and Palpigradi, on the other hand, whilst Actinotrichida and Solifugae appear to have some characters in common. The female reproductive system can comprise accessory glands (rarely), bursa copulatrix, ovarium, oviduct, ovipositor, receptaculum seminis, vagina and vulva. The glandular system comprises, among others, various prosomatic glands, and various tegumental glands. The opisthosomatic musculature consists, among others, of longitudinal and dorsoventral muscles; a detailed study of opisthosomatic musculature in a primitive species would be an important contribution to our knowledge.

Reproduction. – Actinotrichid reproduction can be sexual or parthenogenetic. Reproductive behaviour involves sperm transfer (spermatophory or copulation), often associated with courtship behaviour. Oviposition can take place by means of an ovipositor. Actinotrichida are oviparous or ovoviviparous.

Postembryonic development. – The ancestral Actinotrichid life-cycle comprises six stases: prelarva, larva, protonymph, deutonymph, tritonymph and adult (figs. 16, 17). The prelarva, when present, is a calyptostase or, rarely, an elattostase (cf. Coineau, 1977); in several groups, a prelarva is not known. Generally, the larva is characterized by hexapody. In Trombidina, the proto- and tritonymph are calyptostasic; in Acaridida, the deutonymph can be elattostasic (hypopus). Neoteny could be present in groups with a reduced number of stases (several groups of Actinedida, Tarsonemida). Adult moults (and isophena of the adult) are mentioned for some species of Actinedida (Trombidina) and Acaridida (*Dermatophagoides*) (cf. Van der Hammen, 1980: 12).

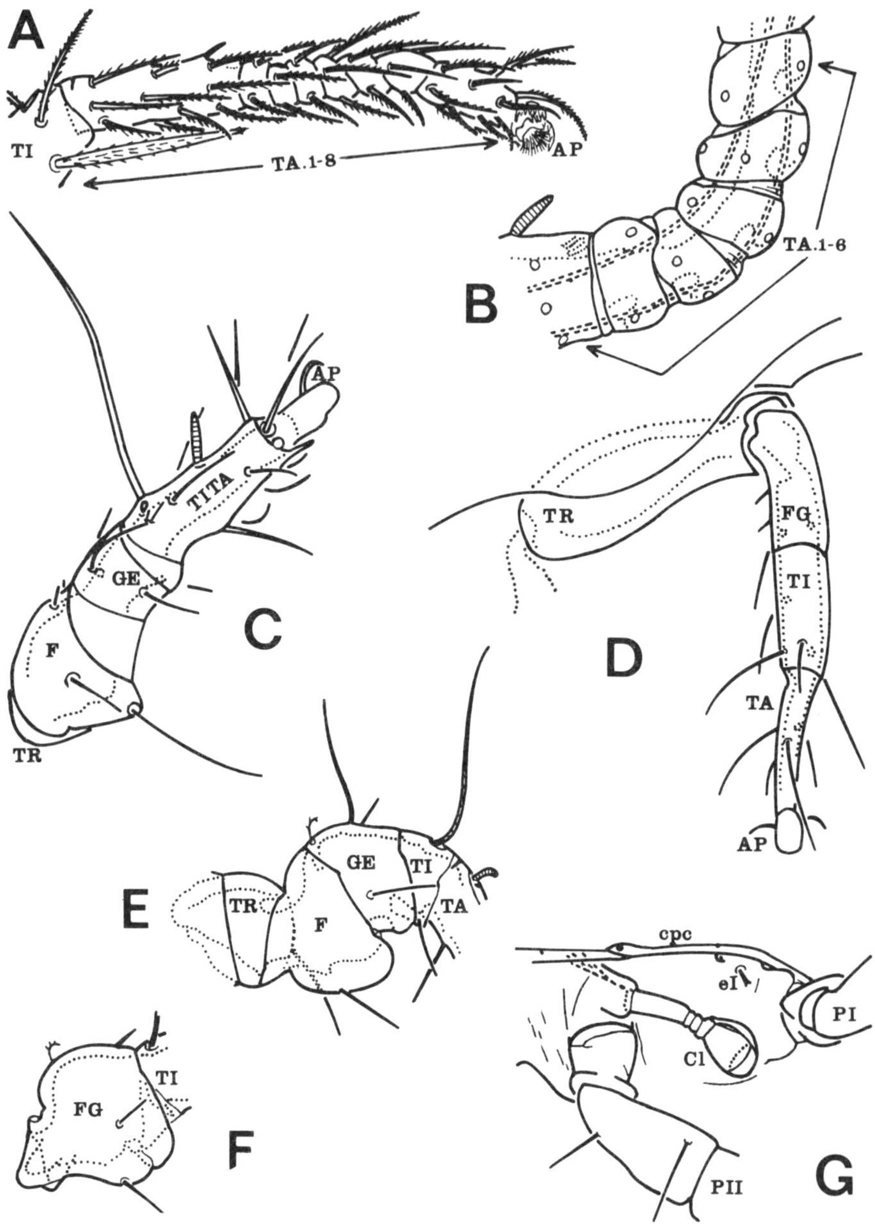

Fig. 42. Segmentation and articulation of Actinotrichid legs (2). A, B, *Tarsolarkus* spec. (Actinedida), lateral (antiaxial) view of right leg I (redrawn after Grandjean, 1952a); A, tarsus and ambulacrum; B, tarsus 1–7. C, D, *Tarsonemoides limbatus* Van der Hammen (Tarsonemida), female; C, lateral (antiaxial) view of right leg I; D, lateral (antiaxial) view of left leg III. E, F, *Fusacarus* spec. (Acaridida) (redrawn after Grandjean, 1953); E, lateral (antiaxial) view of right leg I (from

→

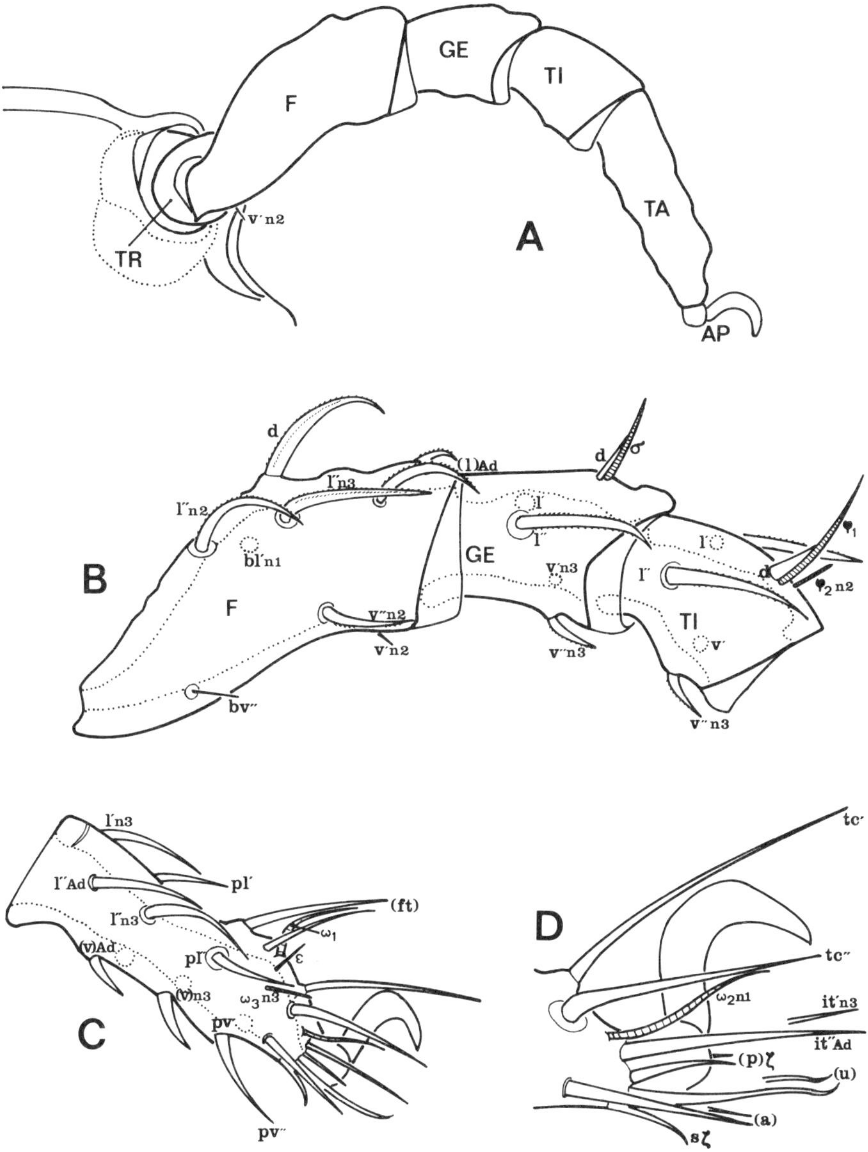

Fig. 43. Platynothrus peltifer (C.L. Koch) (Oribatida: Nothrina), adult, lateral (antiaxial) view of right leg I; A, leg and acetabulum; B, femur, genu and tibia; C, tarsus and apotele; D, terminal part of tarsus, and apotele; A, × 232; B, C, × 368; D, × 687.

←

the trochanter up to the base of the tarsus); F, lateral (antiaxial) view of the femorogenu of right leg II. G, *Eulohmannia ribagai* (Berlese) (Oribatida: Mixonomata), lateral view of the anterolateral part of the podosoma with Claparède's organ and the bases of legs I and II (redrawn after Grandjean, 1939b). A, × 323; B, × 700; C, D, × 943; E, F, × 457; G, × 470.

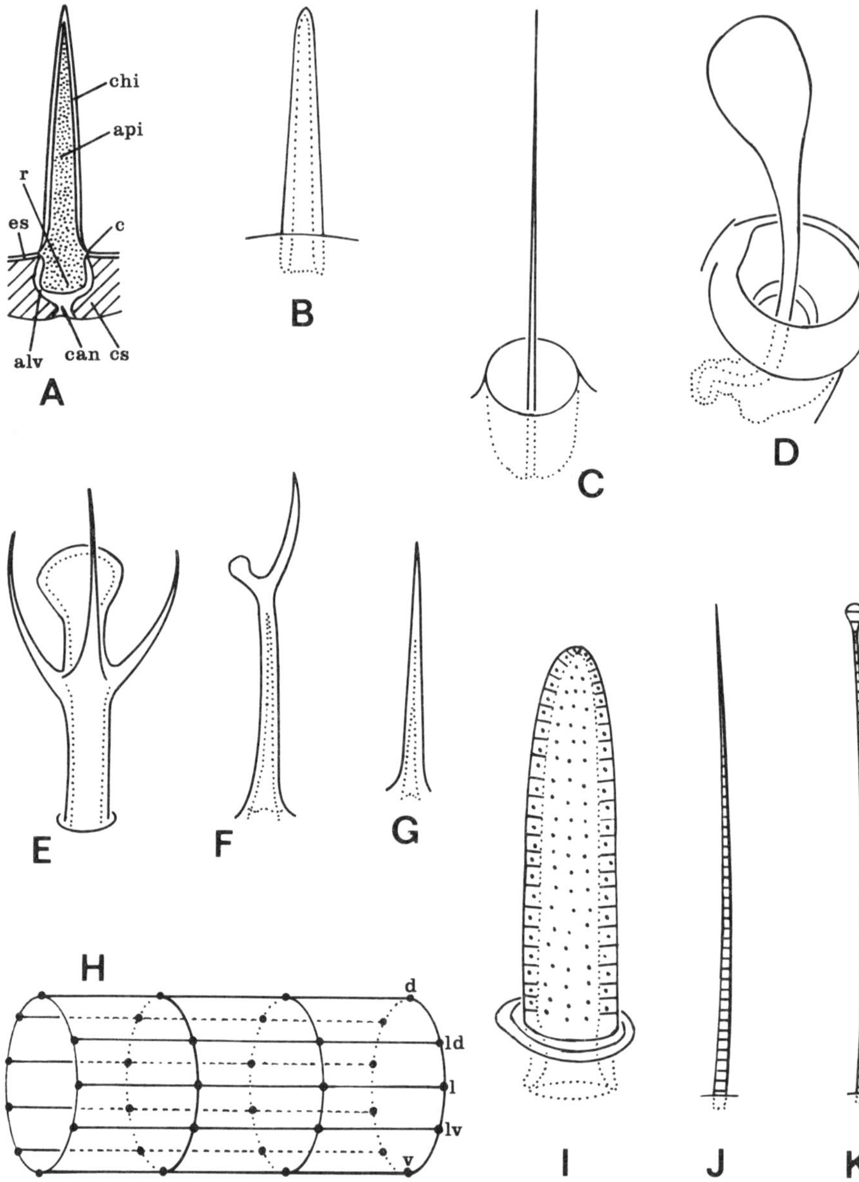

Fig. 44. Schematic representation of phaneres and phanerotaxy in Actinotrichida; A, homotype of Actinotrichid seta (with actinopilinous axis); B, eupathidium; C, D, trichobothria; E–G, famuli; H, arrangement of setae in verticils and files; I–K, solenidia (I, based on Vercammen-Grandjean, 1965). Eupathidia and trichobothria represent modified setae; they have evolved from ordinary setae by: (1) the development of a canal in the solid setal axis (eupathidia); (2) the development of a cup-shaped bothridium at the place of insertion, and of various specialized setal shapes (bothridial setae). The ancestral shape of a famulus (sensory function unknown) is that of a globular head surrounded by bracteae, borne on a stalk. The function of solenidia could be chemosensory.

Ontophylogenetic diagrams of the various types of Actinotrichid postembryonic development are given in fig. 45.

List of the notations of figs. 33–45

a	anterior infracapitular seta.
a_{1-5}	anal setae.
(*a*)	antelateral setae of tarsus.
abj	abjugal furrow.
acm	anteroculminal seta.
Ac.P	acetabulum of palp.
acx	line of attachment of cheliceral sheath to chelicera.
Ac.I	acetabulum of leg I.
Ac.II	acetabulum of leg II.
ad	adult.
Ad	adult base level.
AD	adanal segment (segment XIV).
ad_{1-6}	adanal setae.
ag_{1-3}	aggenital setae.
alv	alveolus.
an	anus.
AN	anal segment (segment XV).
an_{1-2}	anal setae.
AP	apotele.
ap.c	capitular apodeme.
api	actinopilinous core.
apo.1–2	apodemata 1 and 2.
AS	aspidosoma.
bcx	base of cheliceral sheath.
bl'	anterior basilateral seta of femur.
bo	bothridium.
bos	bothridial seta.
bv''	posterior basiventral seta of femur.
c	collar.
c_{1-3}, c_p	notogastral setae of 'segment' *C*.
can	canal.
CH	chelicera.
cha	posterior seta of chelicera.
chb	anterior seta of chelicera.
chi	chitinous part of setae.
circ	circumcapitular furrow.
Cl	Claparède's organ.
cm	culminal seta of palpal tarsus.
co	copulatory orifice.
co.1	cotyloid wall of acetabulum I.
co.2	cotyloid wall of acetabulum II.
cpc	podocephalic canal.
cs	chitonostracum (procuticle).
CX.C	cheliceral sheath.

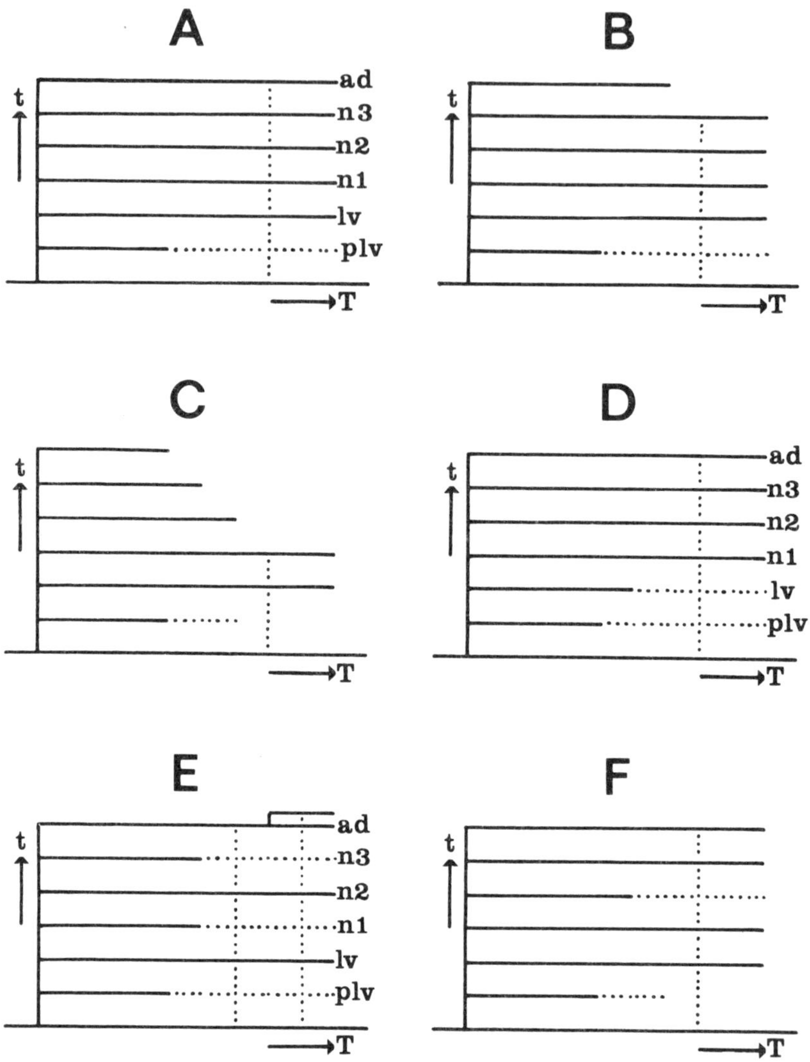

Fig. 45. Ontophylogenetic diagrams of the different types of postembryonic development found in Actinotrichida; the horizontal dotted lines represent regressive stases (elattostases or calyptostases); the vertical dotted lines represent the postembryonic development of the group for which the type is characteristic. A, ancestral type, with six stases, of which the prelarva is elattostasic or calyptostasic (this type is present in Oribatida and many Actinedida). B, neotenous type, with two nymphs, in which maturity is supposed to have been attained at the tritonymphal level (this type is present in several groups of Actinedida). C, neotenous type with two stases (larva and adult); it is supposed that maturity is attained at the level of the protonymph (this type is found in Tarsonemida). D, type with progressive protelattosis, in which the larva has become elattostasic (it is present in some species of Actinedida). E, type with metelattosis at the level of proto- and tritonymph; in this type, plethomorphosis (the secondary formation of isophena by growing moults

d	dorsal seta.
d_{1-2}	notogastral setae attributed to segment IX.
dga	duct of anterior podocephalic gland.
dgc	duct of coxal gland.
dgic	duct of intercheliceral gland.
dgm	duct of median podocephalic gland.
dgp	duct of posterior podocephalic gland.
disj	disjugal furrow.
e	supracoxal seta of palp.
e_{1-2}	notogastral setae attributed to segment X.
es	epiostracum (epicuticle).
exa	anterior exobothridial seta.
exp	posterior exobothridial seta.
eI	supracoxal seta of leg I.
eII	supracoxal seta of leg II.
F	femur.
f_{1-2}	notogastral setae attributed to segment XI.
fe_1	fenestrate area of body of chelicera.
FG	femorogenu.
fla	flagellum.
(ft)	fastigial setae of tarsus.
F1	femur 1.
F2	femur 2.
G	gena.
g_{1-10}	genital setae.
gc	coxal gland.
GE	genu.
ge_{1-3}	antiaxial genital setae.
gen	genital opening.
GN	gnathosoma.
h	posterior infracapitular seta.
H	mentum.
h_{1-3}	notogastral setae attributed to segment XII.
HY	hysterosoma.
iad	lyrifissure (cupule) of adanal segment (segment XIV).
ian	lyrifissure (cupule) of anal segment (segment XV).
ID	idiosoma.
ih	lyrifissure (cupule) attributed to segment XII.
in	interlamellar seta.
ips	lyrifissure (cupule) of pseudanal segment (segment XIII).
it', *it''*	iteral setae.

←
or repetition moults) can occasionally be found at the level of the adult (the type is characteristic for Trombidina). F, type with metelattosis at the level of the deutonymph (this type is present in Acaridida with hypopus-formation).

k, k', k'',	condyles.
ka, kl, kp	setae of constricted part of ovipositor.
l, l', l'', (*l*)	lateral setae.
ld	laterodorsal seta.
le	lamellar seta.
lt', lt''	lateral setae of palpal tarsus.
lv (fig. 45)	larva.
lv	lateroventral seta.
m, m_{1-3}	median infracapitular setae.
m (fig. 36F)	lateral prosomatic mark.
mi	movable jaw of chelicera.
ms	fixed jaw of chelicera.
N	naso.
n1	protonymph, protonymphal base level.
n2	deutonymph, deutonymphal base level.
n3	tritonymph, tritonymphal base level.
og	eugenital opening.
ogc	orifice of coxal gland.
ogm	orifice of median podocephalic gland.
ogp	orifice of posterior podocephalic gland.
op'	paraxial oncophysis of body of chelicera.
OP	opisthosoma.
opv	ventral oncophysis of body of chelicera.
opx	coxal oncophysis of chelicera.
or_{1-3}	adoral setae.
(*p*)	proral setae.
p_{2-4}	setae of segment XIII.
PA	palp.
PD	prodorsum.
Pg_{1-3}	genital papillae.
pl', pl''	primilateral setae.
plv	prelarva.
PO	podosoma.
PR	prosoma.
PS	pseudanal segment (segment XIII).
ps_{1-3}	setae of pseudanal segment.
PT	proterosoma.
pv', pv''	primiventral setae.
PI–IV	legs I–IV.
r	root of seta.
ro	rostral seta.
rp	rostrophragma.
RU	rutellum.
s	subungual seta.
s_{2-3}	setae of segment XIII.
sej	sejugal interval or furrow.
SS	stethosoma.
su	subultimal seta.

t	ontogenetic time.
T (fig. 45)	phylogenetic time.
T	telson.
t_i	inferior tendon of apotele.
t_s	superior tendon of apotele.
TA	tarsus.
TA1–8	tarsi 1–8.
tc', tc''	tectal setae of tarsus.
tf_i	inferior tendon of femur.
tf_s	superior tendon of femur.
tf'	anterior tendon of femur.
tf''	posterior tendon of femur.
tfe	tendon of femur 2.
tg	tendon of genu.
Tg	Trägårdh's organ.
TG	tegulum.
TI	tibia.
TITA	tibiotarsus.
TR	trochanter.
TRII	trochanter of leg II.
tT_i	inferior tendon of Trägårdh's organ.
tT_s	superior tendon of Trägårdh's organ.
tta	tendon of tarsus.
tti	tendon of tibia.
ttr	tendon of trochanter.
ttr_i	inferior tendon of trochanter.
ttr_s	superior tendon of trochanter.
(*u*)	ungual setae.
ul', ul''	ultimal setae.
v, v', v'', (v),	ventral setae.
xa	anterior exobothridial seta.
xp	posterior exobothridial seta.
ϵ (epsilon)	famulus.
ζ (zeta)	eupathidium
σ (sigma)	solenidion of genu.
τ, τ_{1-4} (tau)	anterior eugenital setae.
φ_{1-2} (phi)	solenidia of tibia.
ψ_{1-2} (psi)	posterior eugenital setae.
ω, ω_{1-3} (omega),	solenidia of tarsus.
1a–d	epimeral setae of segment III.
2a–c	epimeral setae of segment IV.
3a–d	epimeral setae of segment V.
4a–d	epimeral setae of segment VI.
VII–XVII	segments of opisthosoma.
' (prime)	anterior face of an appendage; the right one of two symmetrical structures.
" (double prime)	posterior face of an appendage.
()	a pair.

3. Actinedida

In five papers I have, previously, paid particular attention to Actinedida. I have first published some notes on the relationships of the main Actinedid groups (Van der Hammen, 1968a: 275–276). Thereupon a detailed study was made of the morphology of a primitive species (Van der Hammen, 1969), in which study attention was also paid to generalities. In two publications, I summarized some of the general Actinedid characters (Van der Hammen, 1972: 54–56, figs. 1–2; 1972a: 287). In a recent paper (Van der Hammen, 1986c: 218–220, fig. 1), I have made a study of the functional morphology of the legs in a species of *Anystis*.

The diagnostic characters of the order Actinedida are the following. The body of the adult is composed of 13–16 segments, which have partly or nearly completely fused. In numerous cases, the cuticle of the adult is little sclerітized; small or larger shields (microsclerites or sclerites) are often found, but a complete, scleritized armour is rare (Labidostommidae). As a rule, there is no rostral tectum. In case tracheae are present, these start from the base of the chelicerae (most so-called Prostigmata), and (in rare cases) from the progenital chamber (respiratory function not certain); tracheae are absent in many primitive Actinedida (Endeostigmata) and in Phytoptina. The ancestral number of prodorsal setae is six pairs: rostrals, anterior bothridials, interbothridials, posterior bothridials, anterior exobothridials, and posterior exobothridials (consequently, two pairs of prodorsal setae can be represented by trichobothria). In higher Actinedida, the number of prodorsal setae is often reduced, and homologization of setae can be difficult. Trichobothria can also be present on the legs. A latero-opisthosomatic gland is never present. The anus is sometimes replaced by a uroporus. The podocephalic canal is external or internal. The ungues of the legs can present tenent hairs; the central unguis can be replaced by an empodium. Solenidia can be present on femur, genu, tibia, and tarsus. Besides a calyptostasic prelarva (it is an elattostase in rare cases), other calyptostases (protonymphal and tritonymphal) are found in Trombidina. The larva of *Labidostomma* and *Rhagidia* is an elattostase. In cases of supposed neoteny, one of the three nymphal stases can be lacking (some Endeostigmata, Raphignathoidea, Tetranychoidea, Cheyletoidea); many Phytoptina appear to have only one nymphal stase.

Actinedida and Oribatida are distinctly different as far as the higher developed groups are concerned; they are apparently closely related in the case of their primitive representatives. Few diagnostic characters apply to all species. Actinedida and Oribatida are apparently characterized by different evolutionary programs and different evolutionary potentialities which, in several cases (primitive representatives), have not yet become manifest. The two groups can generally be distinguished by the following combination of characters: a rostral tectum is absent in Actinedida, generally present in adult Oribatida; two pairs of prodorsal setae can be bothridial in Actinedida, one pair in Oribatida;

a large opisthosomatic sclerite is generally absent in the case of Actinedida (present in Labidostommidae), generally present in Oribatida; pedal trichobothria can be present in Actinedida and are always absent in the case of Oribatida; some of the Actinedid femora can present solenidia, whilst femoral solenidia are not present in Oribatida; the Actinedid ungues can present tenent hairs (not present in Oribatida); the central unguis can be replaced by an empodium in Actinedida (not in Oribatida).

Several groups (suborders) of Actinedida have been distinguished, of which the phylogenetic relationships are not yet clear (cf. Cunliffe, 1955; Van der Hammen, 1968a: 275–276; 1972: 54–57; Krantz, 1978: 228). Some of these groups can be the following: Endeostigmata (Alycina); Bdellina (Eupodina); Labidostommina; Anystina and Raphignathina (which could constitute one suborder); Phytoptina (Tetrapodili; also regarded as related to Tetranychoidea, and classified with Raphignathina); and Trombidina (Parasitengona).

The present section is subdivided into two subsections, in which the morphology of *Alycus roseus* (subsection 3.1), and the functional morphology of the legs in *Anystis* (subsection 3.2) are successively dealt with.

3.1. *The morphology of* Alycus roseus *C.L. Koch*

The species described in the present subsection is one of the most primitive Actinedida. Mention must be made here, among other characters, of the fact that the opisthosoma is distinctly segmented and consists of the highest number of segments known from Actinotrichida. An early study of the segmentation of the species was already made by Kramer (1882). The following description is based on material collected at Arnhem (The Netherlands); these specimens are identical with topotypic material of *Alycus roseus* C.L. Koch (from Regensburg) and with topotypic material of *Pachygnathus dugesi* Grandjean (from Coulounieix, France).

Measurements. – The length of the specimen represented in figs. 46–48 is 0.535 mm (measured from the naso up to the posterior border), the breadth is 0.320 mm; the ratio length: breadth is 1.67. According to Grandjean (1937: 56) the length varies from 0.430–0.560 mm, the ratio length: breadth from 1.65–2. In some of my specimens the 'shoulders' are more distinctly developed than in the remaining material. In some females, the opisthosoma has become strikingly plicate, by which condition a description is thwarted. The specimens described here are females which probably had recently moulted, and in which eggs had not yet developed; in this condition, they are best suited for a detailed description.

Colour. – The colour of the specimens is caused by the light brown colour of the cuticle and the violet (slightly reddish) colour of the pigment grains in the hypodermis. Together they are the cause of the slightly pinkish colour of a specimen observed in reflected light. In alcohol (under a dissecting microscope) and in a slide (under a high power microscope), the violet colour is particularly visible. The narrow intersegmental areas and the muscle attachments lack the pigment grains; this is especially visible under a dissecting microscope. In this way, segmentation is already easily discernable. A detailed description of the colour of the prodorsum was given by Grandjean (1937: 56–58, fig. 1A).

Tegument. – Although a thin layer of cerotegument is probably present, it is difficult to observe because observations on a carbon block are thwarted by folding of the cuticle of drying specimens. The greater part of the cuticle is not scleritized; its ornamentation consists of a great many, more or less parallel, winding ridges. The crests of the ridges each bear a row of pointed granules.

The prodorsum presents a central sclerite (cf. fig. 46), which bears the two pairs of bothridial setae and the interbothridials. In the anterior part, the sclerite presents a narrow median prolongation which continues on the naso. The genital covers and the coxal regions of the legs are scleritized (fig. 47); the same applies to the chelicerae, the segments of palp and legs, and parts of the infracapitulum. Small microsclerites are present at the bases of setae not inserted on larger sclerites, especially in the dorsal region of the opisthosoma (fig. 49F); these microsclerites often constitute the wall of small, funnel-shaped hollows, at the bottom of which the setae are inserted. The surface of all sclerites (including the microsclerites) is more or less granulate; the 'granules' have the shape of small ridges, for which reason they were named costules by Grandjean (1936a: 400). In the case of the larger sclerites, these costules are arranged in winding rows. In lateral view, the costules have more or less the appearance of small cones. The borders of the larger sclerites are not always distinct; gradual transitions occur between the soft skin and the scleritized parts.

As mentioned above, the hypodermis contains numerous violet pigment grains; the occurrence of this pigment could constitute an ancestral character.

Subdivision of the body. – Although the gnathosoma is movably attached to the idiosoma, the former is still distinctly the normal prolongation of the latter; chelicerae and palps have the ordinary Chelicerate position (cf. figs. 47, 48). This clearly demonstrates once more that the infracapitulum (the mentum) is the coxisternal region of the palp.

The idiosoma is subdivided into stethosoma and opisthosoma. Dorsally, this subdivision is represented by the disjugal furrow; laterally and ventrally, the subdivision passes posterior to leg IV.

Prodorsum. – There are six pairs of prodorsal setae (figs. 46, 48): rostral (*ro*), anterior bothridial (= lamellar *le*), interbothridial (= interlamellar *in*), posterior bothridial (= sensillus *bo*), anterior exobothridial (*xa*), and posterior exobothridial setae (*xp*). Consequently, there are two pairs of bothridial setae, viz., *le* and *bo*; their bothridia are simple, but relatively deep (fig. 49B). Rostral, interbothridial, and anterior and posterior exobothridial setae are dendroid; the two pairs of bothridials are long and ciliate.

In the anterior part, the prodorsum presents a distinct frontal protuberance, the naso, which is directed downwards, between the chelicerae (fig. 50A–C). This naso (*N*) has a circular base; its shape is more or less conical. Its dorsal

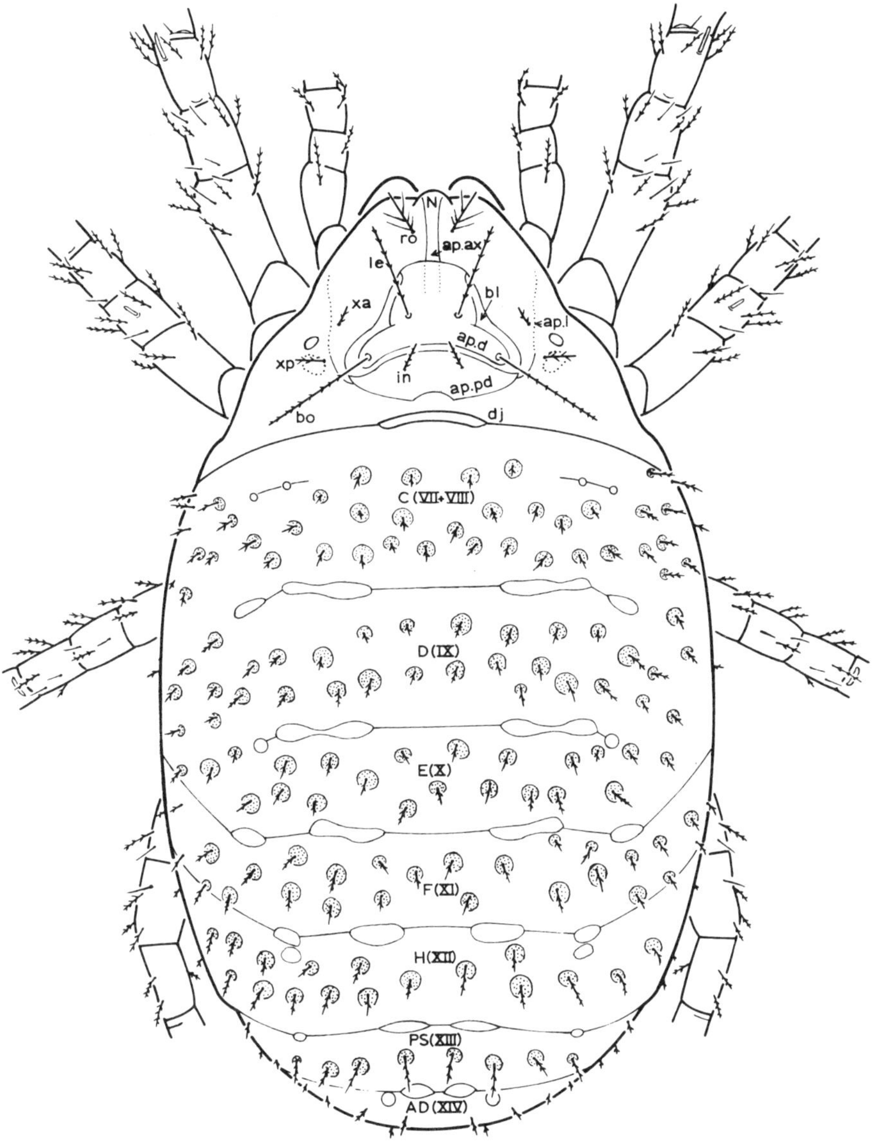

Fig. 46. Alycus roseus C.L. Koch, dorsal view of female; × 235.

surface is granulate ('costulate'), its ventral surface is smooth (fig. 50B); there is no inferior eye. A rostral tectum is not present.

There are one pair of eyes and one pair of vestigial eyes (fig. 40C). The vestigial eye is situated on either side posterior to (and close to) the anterior eye;

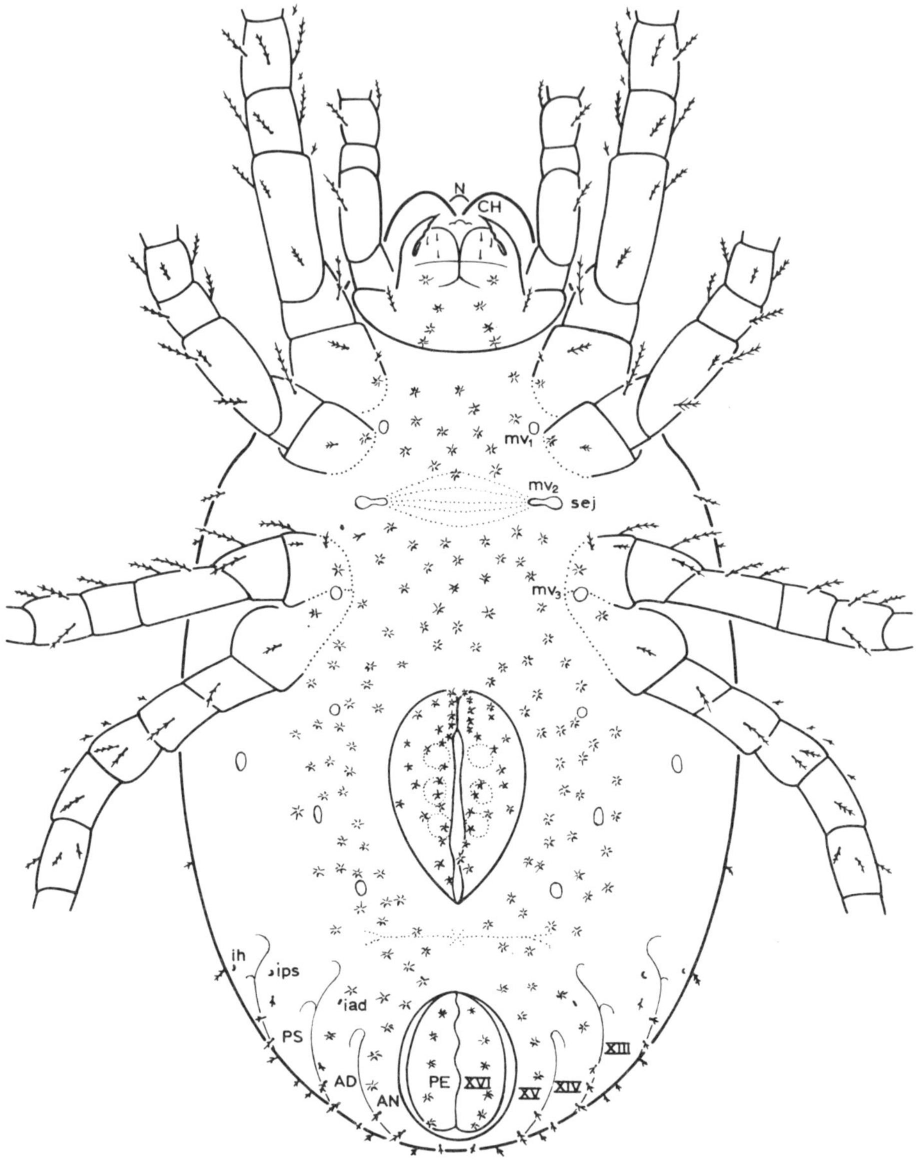

Fig. 47. Alycus roseus C.L. Koch, ventral view of female; × 235.

the two still constitute a distinct group. The vestigial eye, recognizable as a light spot, lacks a lens; its surface is granulate, and its outline is distinctly circular.

The prodorsum presents a central sclerite (figs. 46, 49B) which bears the anterior and posterior bothridials, and the interbothridials. The central sclerite

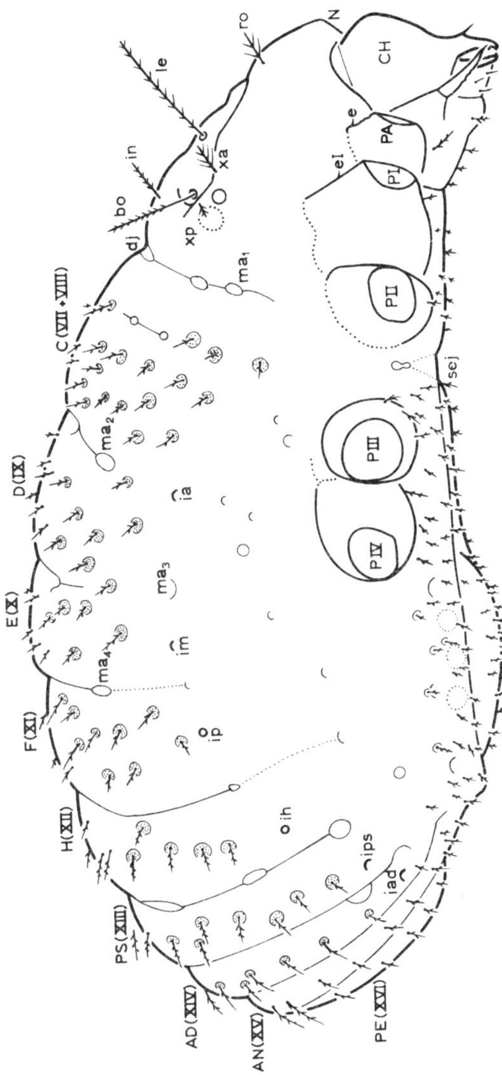

Fig. 48. Alycus roseus C.L. Koch, lateral view of female; × 235.

continues on the dorsal surface of the naso. The rostral setae are inserted at the border of this anterior continuation. The prodorsum presents five internal, chitinous ridges which serve as places of muscle attachment: axial, dorsal, posterodorsal, and a pair of lateral apodemes; they are indicated by light bands without pigment grains. The axial apodeme (*ap.ax*) presents an anterior continuation in the naso, and a posterior continuation under the central prodorsal sclerite. The dorsal apodeme (*ap.d*) is situated posterior to the posterior both-

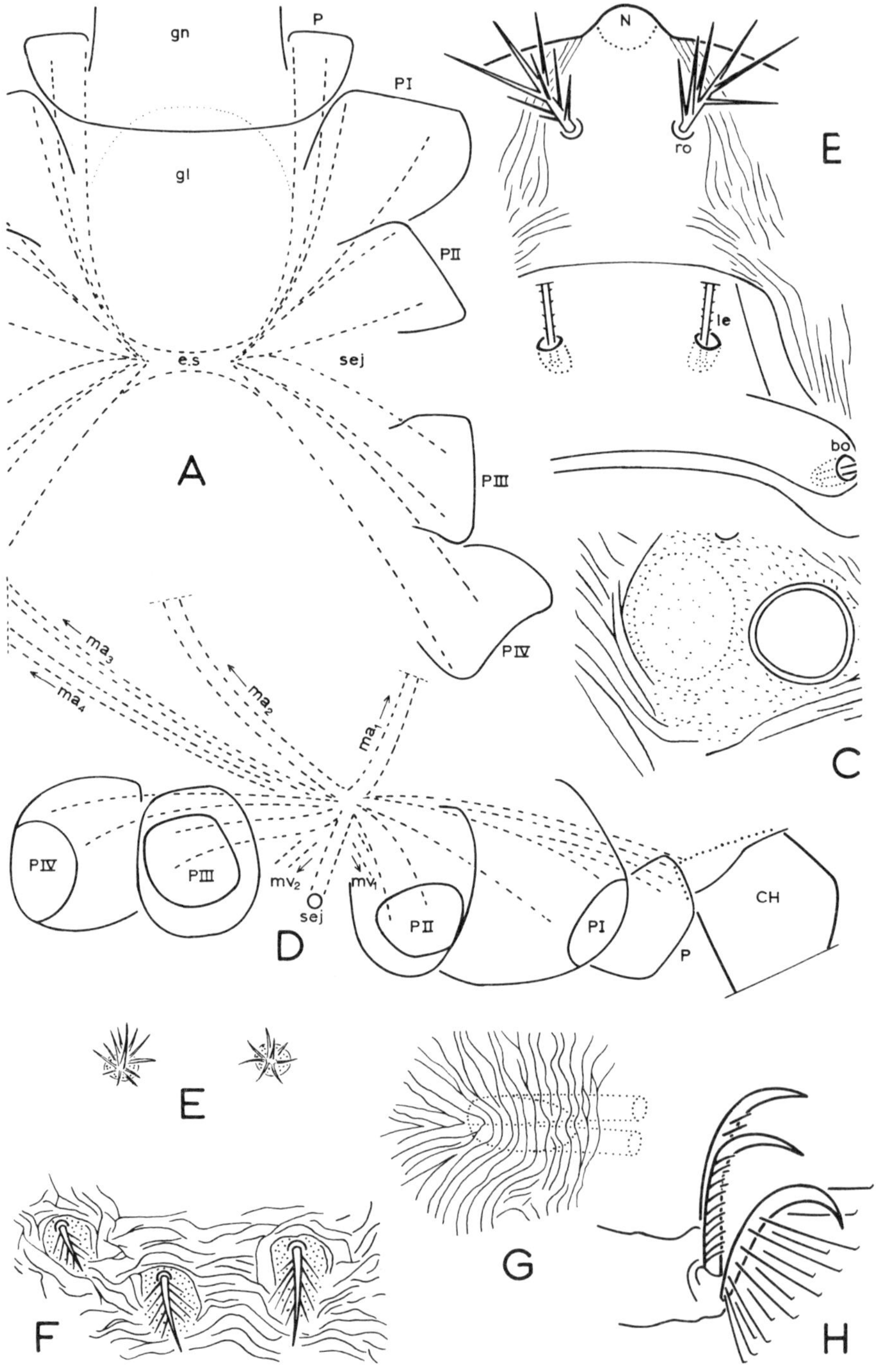
gn
P
PI
gl
PII
e.s
sej
A
PIII
PIV
ma_3
ma_4
ma_2
ma_1
mv_2
mv_1
sej
PIV
PIII
PII
PI
P
CH
D
N
ro
E
le
bo
C
E
G
F
H

ridia, the posterodorsal (*ap.pd*) posterior to the interbothridial setae. The lateral apodeme (*ap.l*) extends from the fusion of posterodorsal and dorsal apodeme up to the anterolateral part of the prodorsum. A pair of colourless bands (*bl*) laterally border the central prodorsal sclerite; they do not correspond with an internal apodeme. Because of the presence of these apodemes, the prodorsum can be regarded as consisting of pro-, meso-, and metastethidium.

Dorsal and lateral aspect of opisthosoma, and anal region. – Prosoma and opisthosoma are dorsally separated by the disjugal furrow (figs. 46, 48); various muscles are attached to the internal side of this furrow. The opisthosoma is markedly different from the prosoma because of its neotrichy.

At first sight, the opisthosoma of *Alycus roseus* seems to be composed of nine segments: *C, D, E, F, H, PS, AD, AN*, and *PE*; these segments are separated by colourless bands which internally correspond with intersegmental muscle attachements (fig. 49G). Segment *C*, however, appears to be composed of two segments: muscle attachments are present in the anterior part, lateral to the anterior row of setae (figs. 46, 48). This condition reminds one of Opilioacarida, in which group the anterior dorsal opisthosomatic division is also composed of two segments. 'Segment' *C*, consequently, is composed of segments VII (generally reduced in Chelicerata) and VIII. The peranal segment (*PE*), consequently, represents segment XVI. This number constitutes the base for the study of segmentation in Actinotrichida.

The opisthosoma presents a distinct neotrichy, particularly developed in the anterior segments. All dorsal setae are inserted on microsclerites (fig. 49F). The setae are ciliate or dendroid. 'Segment' *C* (= VII + VIII) presents some three rows of setae, segments *D–H* (IX–XII) some two rows; segments *PS–PE* (XIII–XVI) present only one row. The paraproctal lips (segment *PE*) each bear 4–6 often asymmetrically arranged setae.

There are six pairs of lyrifissures which have the shape of cupules: *ia, im, ip, ih, ips, iad*; there is no anal lyrifissure *ian*. Consequently, the two posterior segments (*AN, PE*) and apparently the two anterior (*C*) lack lyrifissures. This means that the lyrifissures *ia, im*, and *ip* belong to the segments *D, E* and *F* respectively. This conclusion is of fundamental importance to the comparative morphology of all groups of Actinotrichida.

The shape of the opisthosoma is slightly variable. In the specimen represented here, it is nearly parallel-sided; the shoulders, however, can be more pronounced. Although the peranal segment is, in the ancestral plan of con-

←

Fig. 49. Alycus roseus C.L. Koch, female; A, ventral view of podosoma (with endosternite); B, dorsal view of median part of prodorsum; C, lateral view of lateral eyes; D, lateral view of podosoma with endosternite; E, ventral view of two coxisternal setae of the segment bearing leg II; F, dorsal view of three dorsal setae of segment XI; G, dorsal view of right dorsal muscle attachment between segments IX and X; H, lateral (paraxial) view of the ungues of left leg II; A, × 370; B, × 690; C, × 1315; D, × 370; E, F, G, × 690; H, × 1315.

struction, the posterior segment, its position is now ventral; this means that the posterior part of the opisthosoma is curved downwards.

Genital region. – The progenital lips (genital covers) (figs. 47, 48) are relatively large, as in other Endeostigmata; they are separated by a distinct interval from the anal opening. The lips are scleritized. There is a great number of (pro)genital setae; these are, generally, asymmetrically arranged. In the figured specimen the number is 21–23; according to Grandjean (1937: 60), it varies from 18–24. The longitudinal opening between the progenital lips (the progenital opening) gives entry into the progenital chamber, as in other Actinotrichida which present the ancestral conditon. The progenital chamber (fig. 52A) contains three pairs of genital papillae (Pg_{1-3}). Between the anterior pair of papillae, the true genital (eugenital) opening (*og*) of the female is visible; it is triangular (trifid). One pair of eugenital setae (ψ_1) is present near the posterior border of the opening. In Chelicerata, the genital opening is generally at the posterior border of segment VIII (between segments VIII and IX). As mentioned above, the anterior genital papilla belongs to segment X, the median and posterior papillae to segments XI and XII respectively. This interpretation of the genital region is very important for a clear apprehension of the opisthosomatic architecture. A progenital chamber including five segments constitutes a unique character of the Actinotrichida.

Ventral and aggenital region. – This area is characterized by a strongly developed neotrichy (fig. 47). A series of muscle attachments is present in the aggenital region.

Coxisternal region. – This region is also neotrichous (fig. 47). There is a large sejugal interval. It appears that the endosternite is restricted to this area (fig. 49A, D); it has the shape of a narrow, transverse band, to which extrinsic muscles of chelicerae, palps, and all legs are attached. Ventral muscles extend from the endosternite to the pair of muscle attachments in the sejugal interval (mv_2), and to the muscle attachments between legs I and II (mv_1), and legs III and IV (mv_3) (fig. 47). Dorsally, muscles extend from the endosternite to the muscle attachments in the disjugal fold (fig. 48: ma_1), between segments *C* and *D* (ma_2), *D* and *E* (ma_3), and *E* and *F* (ma_4).

Lateral region of podosoma. – The pleural regions are glabrous. The coxal regions of the legs are distinctly differentiated. The coxal region of leg I presents a small supracoxal seta *eI* (fig. 52C), similar to the supracoxal seta *e* of the palp; the coxal region of leg II has no supracoxal seta. The podocephalic canal is represented by a superficial taenidium.

Gnathosoma. – The general position of the gnathosoma is represented in figs. 47, 48; it is rostral and slightly curved downwards. There is no rostral tectum and no camerostome. Chelicerae and palps have more or less the same position as in other groups of Chelicerata; the presence of an infracapitulum, however, remains an important difference in comparison with the general Chelicerate condition. The gnathosoma represents, moreover, a distinct pseudotagma: it can easily be separated as a whole, whilst the chelicerae remain attached to the infracapitulum.

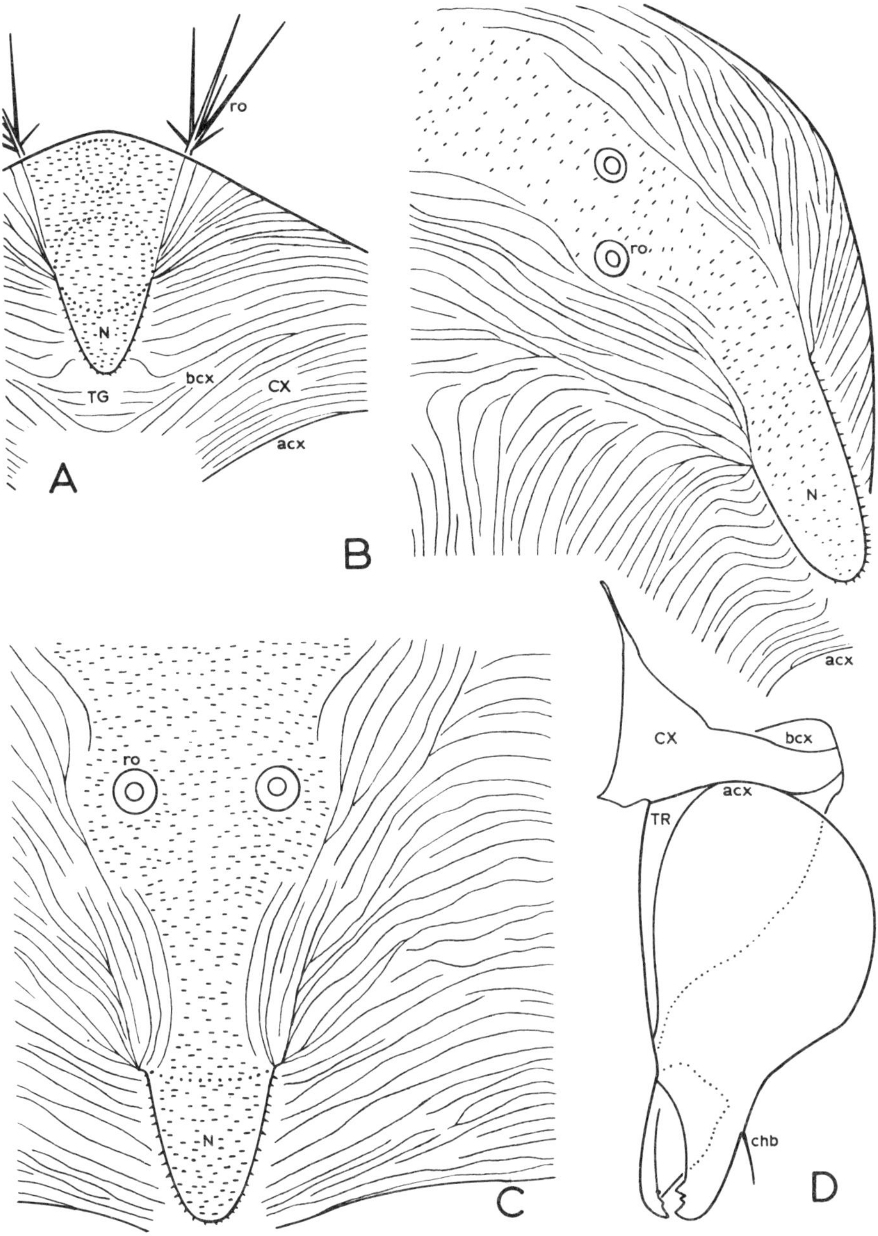

Fig. 50. Alycus roseus C.L. Koch, female; A–C, naso and adjoining parts of prosoma; A, ventral view; B, lateral view; C, frontal view; D, lateral (antiaxial) view of right chelicera with coxal region; A–C, × 1315; D, × 710.

Alycus roseus clearly demonstrates the primitive condition of the tegulum: it is still nothing but part of the rostral body wall (cf. fig. 50A–C: *TG*). Apart from the simple tegulum, the cheliceral frame consists of a rather large interspace of soft skin between the chelicerae, and of the membraneous attachment of the chelicerae to the infracapitulum. The three parts of the cheliceral frame, together with the naso, represent the complete rostral body wall (cf. Van der Hammen, 1968: 7–9). There is a distinct coxal region of the chelicera, which is attached to the tegulum according to the line *bcx* (fig. 50A–D). The gnathosoma as a whole is represented in lateral view in fig. 51A; in this figure, the chelicera is slightly raised, in order to show the labrum. The primitive condition of the cheliceral frame is clearly evident from this figure.

The infracapitulum is anarthric: there is no labiogenal articulation; mentum and genae are not separated. In contradistinction to the genae, the lateral lips are not scleritized. There is a distinct difference in level between the base of the lateral lips and the anterior border of the genae.

There are three pairs of simple, strongly curved adoral setae. The rutellum belongs to the narrow, primitive type. Its shape is represented in lateral and ventral view in fig. 51A and B respectively. In the Dutch specimens, the rutellum ends in a single point; it presents some three rounded ventral teeth. Apart from the lateral lips, there are a three-lobed labrum, and a small, rounded, unpaired underlip. Consequently, the mouth has four commissures (fig. 51B: *Js*, *Js'*, *Ji*, *Ji'*). There are a variable number of infracapitular setae: one pair of anteriors (*a*), one to three pairs of medians (m_{1-3}), and one pair of hysterostomatics (*h*); as a rule, they are asymmetrically arranged. The pharynx has a dorsal sclerite (fig. 51B: *sc.ph*) which is pointed in the anterior part; this pointed part continues in the labrum as labral sclerite (*sc.ls*).

Chelicera. – The chelicera (fig. 50D, 51A) presents only one seta (*chb*) with an anterior, dorsal position; a comparison with other species of the genus demonstrates that the posterior seta (*cha*) has disappeared. The trochanter is small and narrow; its position is particularly ventral and paraxial.

The coxal regions of the chelicerae can function as sheaths: they can be invaginated and evaginated. They are attached to the base of the chelicerae (fig. 50D: *acx*). The fixed jaw of the chelicerae presents three teeth (including a large one), the movable jaw two teeth.

Palp. – A left palp in lateral (antiaxial) view was described and figured in detail by Grandjean (1937: 58, fig. 1B). The formula of the palp is 0–2–1–3–14; the tarsus bears one small solenidion.

Legs. – Parts of the legs are represented in figs. 46 and 47; leg I is represented in detail in fig. 52B, D, E, part of leg IV in fig. 52F, the ambulacrum of leg II in fig. 49H. Legs I–III consist of six eudesmatic segments: trochanter, femur, genu, tibia, tarsus and apotele. Leg IV consists of seven eudesmatic segments: there are two femora (femur 1 and femur 2). The articulation between the two femora (cf. fig. 52F) is constituted by a monodesmatic hinge joint.

The formulae of the chaetotaxy of the legs are the following (the formula

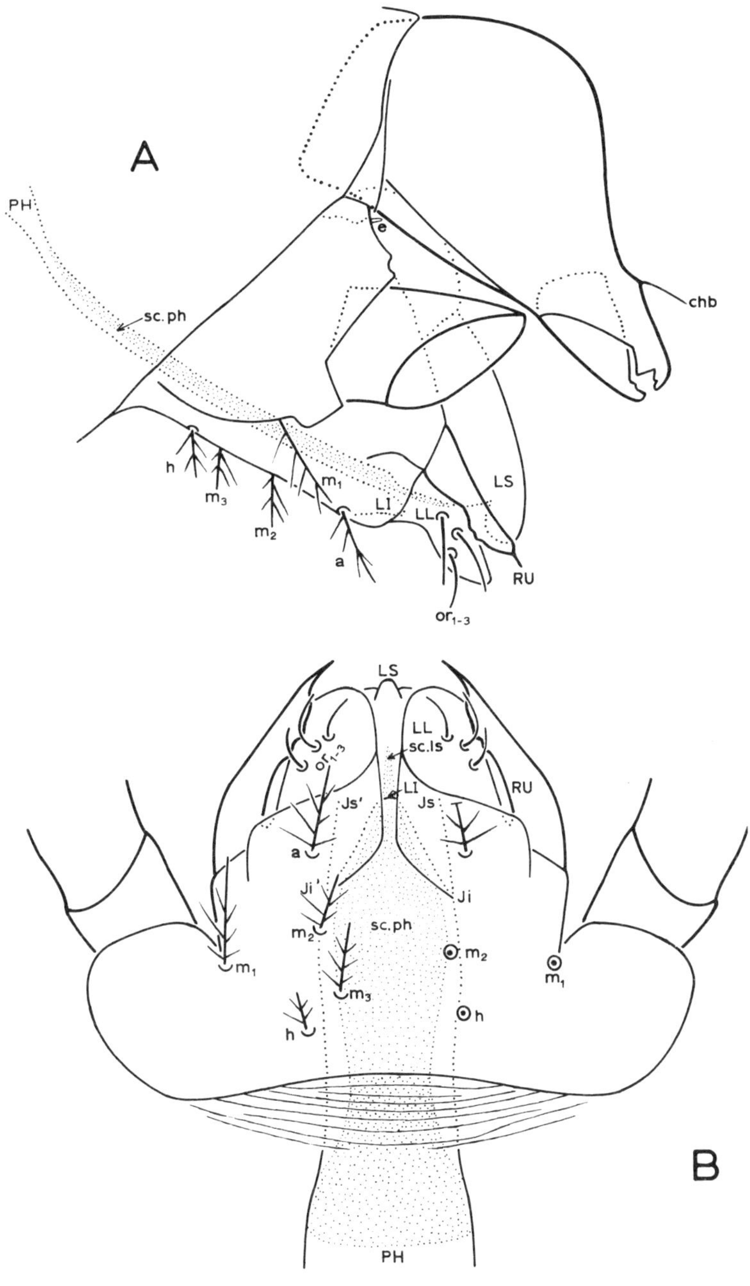

Fig. 51. Alycus roseus C.L. Koch, gnathosoma of adult female; A, lateral view (chelicera slightly raised); B, ventral view; A–B, × 710.

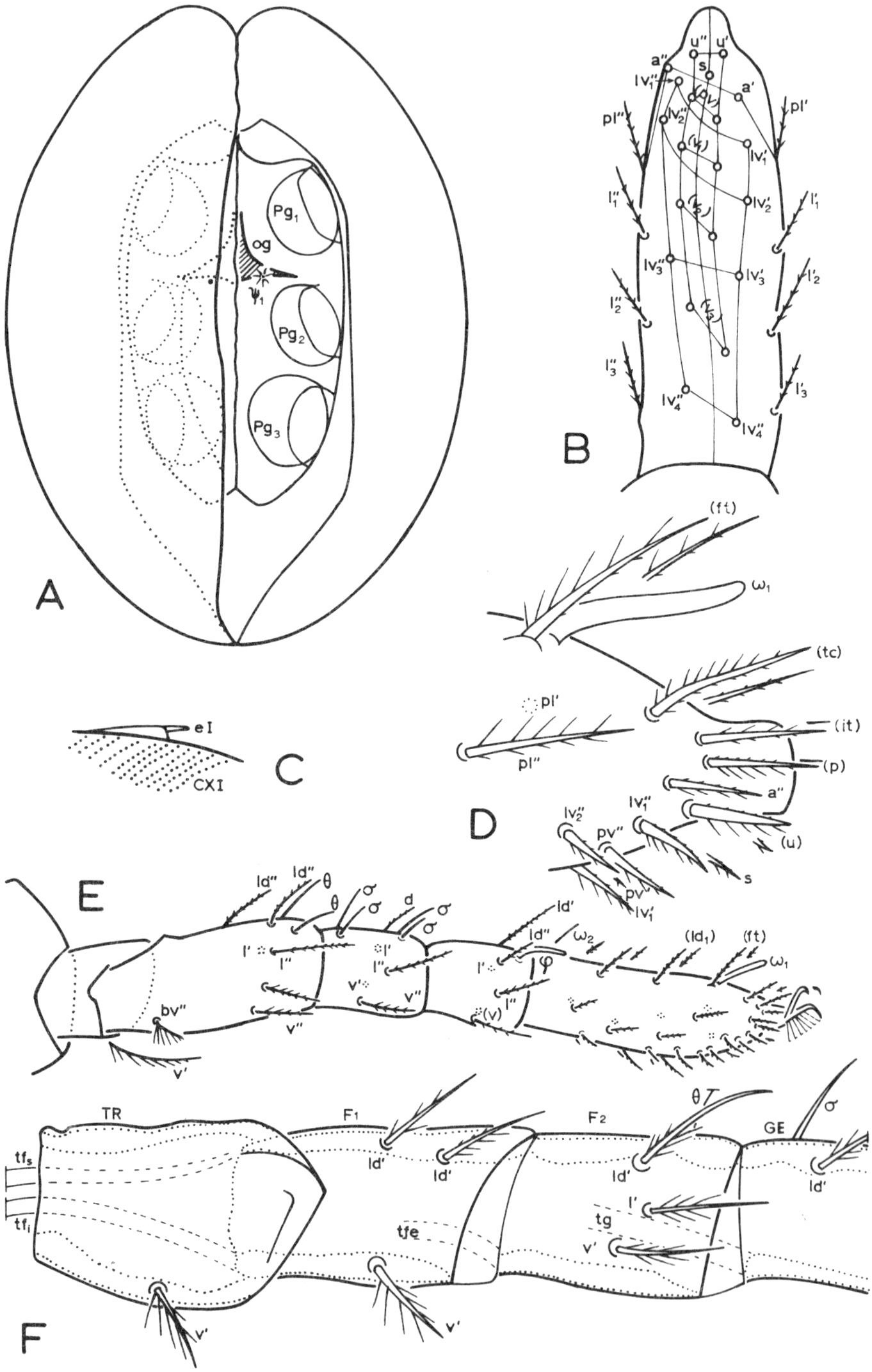
A
Pg1
Pg2
Pg3
og
B
u''
u'
a''
s
a'
pl''
pl'
lv1''
lv2''
lv1'
lv2'
lv3''
lv3'
lv4''
l1''
l1'
l2''
l2'
l3''
l3'
C
eI
CXI
D
(ft)
ω1
(tc)
pl'
pl''
(it)
(p)
a''
(u)
lv2''
pv''
lv1''
pv'
lv1'
s
E
ld''
θ
σ
d
ld'
ω2
φ
(ld1)
bv''
v'
v''
l'
l''
(v)
F
TR
F1
F2
GE
tfs
tfi
tfe
tg
ld'
l'
v'

of leg IV has seven numbers instead of six, because of the presence of two femora; the last number of each formula pertains to the three ungues of the claw): I (1–6–5–6–42–3); II (1–5–4–5–29–3); III (2–3–3–4–25–3); IV (1–3–3–4–5–23–3). A very small famulus (not drawn) is present on the tarsi of the legs I and II. The formulae of the solenidia are the following (the formulae have four numbers because solenidia can occur on femur, genu, tibia and tarsus): I (2–4–1–2); II (0–2–2–1); III (0–2–2–0); IV (1–1–0–0). The number of solenidia of femur IV refers to femur 2. The presence of solenidia on some femora distinctly distinguishes Endeostigmata from Oribatida.

In fig. 52E, F I have used the notation *d* for the unpaired dorsal seta; the paired dorsals have received the notations *ld'* and *ld''*. The setae of tarsus I (fig. 52B, D, E) are arranged in rows, viz., a pair of laterodorsal rows: (*p*), (*it*), (*tc*), (*ft*), (ld_1), (ld_2), ld_3'; a pair of lateral rows: (*a*), (*pl*), (l_1), (l_2), (l_3); a pair of lateroventral rows: (lv_1), (lv_2), (lv_3), (lv_4); a pair of ventral rows: (*u*), (*pv*), (v_1), (v_2), (v_3); and an unpaired medioventral seta *s*.

In the Dutch specimens, femur 1 IV has three setae; Grandjean (1937: 61) mentioned two setae for the specimens from Coulounieix (France).

The apotele presents three ungues which bears hairs (fig. 49H). The hairs of the central unguis appear to be hooked in the terminal part; they are supposed to represent tenent hairs. The possession of ungues of this type also constitutes a character that profoundly separates *Alycus* from the Oribatida.

3.2 *The functional morphology of the legs in* Anystis

Locomotion in mites of the Actinedid family Anystidae can generally be characterized as fast running (as in the related family Erythracaridae). In these mites (see Grandjean, 1943: 34, 51), the orientation of the legs with respect to the sagittal plane of the body is more or less transverse, and in this connection the interesting problem arose whether this transverse orientation is ancestral or represents a derived condition (an adaptation to fast running). In order to solve this problem, the morphology of the Anystid legs is studied in an unidentified species of the genus *Anystis*, collected from hawthorn in The Netherlands (see Van der Hammen, 1986c). Leg I of this species is represented in fig. 53.

The legs of *Anystis* present the ancestral Actinotrichid number of seven leg segments, i.e., trochanter, femur 1, femur 2, genu, tibia, tarsus and apotele

---

←

Fig. 52. Alycus roseus C.L. Koch, female; A, ventral view of genital region (the left progenital lip is removed); B, ventral view of right tarsus I (the ventral setae are represented by their place of insertion only); C, lateral view of supracoxal seta *el*; D, lateral (antiaxial) view of distal part of right tarsus I (the famulus, with exceptional distal position among the terminal setae, is extremely small and is not represented); E, lateral (antiaxial) view of right leg I; F, lateral (antiaxial) view of trochanter, femur 1, femur 2, and proximal part of genu of left leg IV; A, B, × 690; C, D, × 1315; E, × 370; F, × 710.

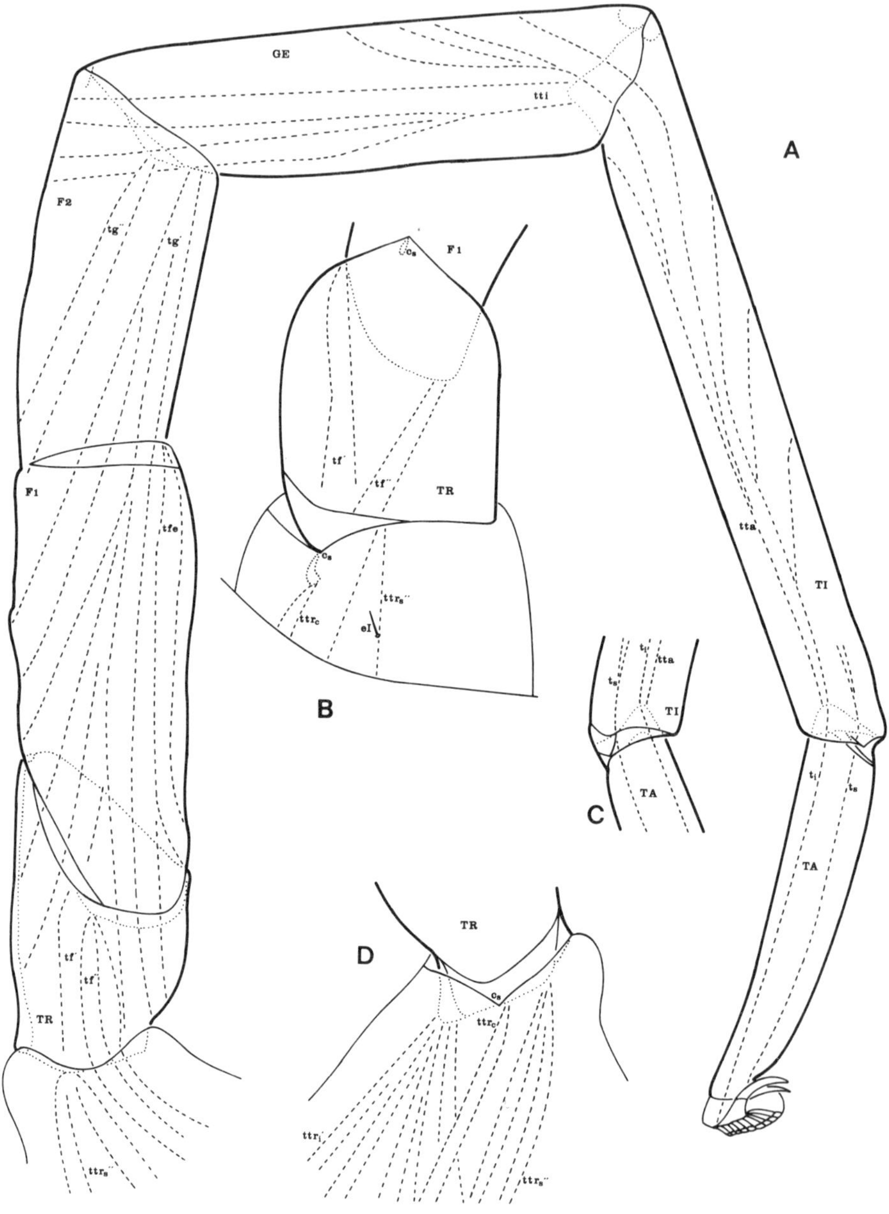

Fig. 53. Anystis spec., right of leg I of adult female; A, lateral (posterior) view (basal part separated from body); B, dorsal view of basal part of leg attached to body; C, lateral (anterior) view of tibia-tarsus joint; D, lateral (anterior) view of base of leg (separated from body); A–D, × 221.

(fig. 53A). The articulation between body and trochanter (fig. 53A, B, D) is constituted by a rocking joint with superior condyle, a set of antero-inferior muscles, a set of postero-superior muscles and a single superior tendon inserted close to the condyle. This specialized joint is associated with promotor-remotor movements. In comparison with the normal condition in other Actinotrichid mites (and many other groups of Chelicerata), the Anystid trochanter-femur jointed represents a derived condition: the articulation-points are superior and inferior, instead of anterior and posterior, whilst the superior articulation-point occupies a more distal position than the inferior (the axis of articulation is oblique). As a consequence of this, the trochanter-femur joint is not associated with the normal Chelicerate levator-depressor movements: the femoral part of the legs is not raised (Anystid and Erythracarid mites do not hang down from their legs). For this reason, the more or less transverse position of the legs cannot be characterized as a primitive condition, but as an adaptation to fast running. The femur 1-femur 2 articulation is a hinge joint with a single inferior muscle; the articulation-point is represented by the superior cuticle which is uninterrupted (there is no condyle). The femur 2-genu articulation is constituted by a hinge joint with superior broadened hinge and two sets of muscles (*tg'* and *tg''*) of which the powerful *tg'* continues as far as the trochanter. The genu-tibia joint is operated by a set of muscles which continue into femur 2. The tibia-tarsus articulation (fig. 53A, C) is bicondylar; it is operated by the tendon and muscles of the tarsus and by the inferior and superior tendons of the apotele. The superior tendon of the apotele (t_s) can function as an extensor as well as a levator, although levator movements are restricted (the tarsus can be raised to a slight extent only) by the presence of small superior protuberances and corresponding small cavities. The inferior tendon of the apotele (t_i) passes dorsal to the place of insertion of the tendon of the tarsus (*tta*) (which represents the normal condition), and not ventral (as mentioned for the Erythracarid genus *Tarsolarkus* by Grandjean, 1952a: 119, fig. F). As far as now known, the continuation of the superior tendon of the apotele into the tibia is unusual (a derived condition associated with extensor and levator movements). The tarsus-apotele articulation is bicondylar; it is operated, as usual, by the tendons t_s and t_i.

It is interesting that in Solifugae (which are also fast runners) the trochanter-femur articulation (trochanter-femur 1 in legs I–II, trochanter 2-femur 1 in legs III–IV) presents similar characters: the articulation axis is oblique and the superior articulation-point is dorsal (legs I–II) or anterodorsal (legs III–IV). Because of the distant relationships between Apatellata (Solifugae and Pseudoscorpionida) and Epimerata (Palpigradi and Actinotrichida), the hypothesis could be introduced that the common ancestors were also fast runners. Because of the highly specialized character of the trochanter-femur articulation, this hypothesis is unlikely. It is more probable that the evolution of fast running, in the two groups, represents a case of parallelism (founded on a common evolutionary potentiality).

List of the notations of figs. 46–53

a	anterior infracapitular seta.
a', *a''*	antelateral setae of tarsus.
acx	line of attachment of cheliceral sheath to chelicera.
AD	adanal segment (segment XIV).
AN	anal segment (segment XV).
ap.ax	axial apodeme of prodorsum.
ap.d	dorsal apodeme of prodorsum.
ap.l	lateral apodeme of prodorsum.
ap.pd	posterodorsal apodeme of prodorsum.
bcx	base of cheliceral sheath.
bl	colourless lateral prodorsal band.
bo	bothridial seta.
bv''	basiventral seta of femur.
C	anterior opisthosomatic 'segment' (segments VII + VIII).
c_s	superior condyle.
CH	chelicera.
chb	(anterior) seta of chelicera.
CX	cheliceral sheath (coxal region of chelicera).
CXI	coxal region of leg I.
d	unpaired dorsal seta of legs.
D	segment IX.
dj	disjugal furrow.
e	supracoxal seta of palp.
E	segment X.
e.s	endosternite.
eI	supracoxal seta of leg I.
F	segment XI.
(ft)	fastigial setae.
F1	femur 1.
F2	femur 2.
GE	genu.
gl	gland.
gn	gnathosoma.
h	posterior infracapitular seta.
H	segment XII.
ia	cupule of segment *D* (IX).
iad	cupule of segment *AD* (XIV).
ih	cupule of segment *H* (XII).
im	cupule of segment *E* (X).
ip	cupule of segment *F* (XI).
ips	cupule of segment *PS* (XIII).
(it)	iteral setae.

Ji, Ji'	inferior oral commissures.
Js, Js'	superior oral commissures.
$l', l'', l_{1-3}', l_{1-3}''$,	lateral setae of legs.
ld', ld'', (ld)	laterodorsal setae of legs.
le	anterior bothridial seta.
LI	labium.
LL	lateral lip.
LS	labrum.
lv_{1-4}', lv_{1-4}''	lateroventral setae of tarsus.
m_{1-3}	median infracapitular setae.
ma_{1-4}	attachments of dorsal muscles of endosternite.
mv_{1-3}	attachments of ventral muscles of endosternite.
N	naso.
og	eugenital opening.
or_{1-3}	adoral setae.
(p)	proral setae.
P	palp.
PE	peranal segment (segment XVI).
Pg_{1-3}	genital papillae.
PH	pharynx.
pl', pl''	primilateral setae.
PS	pseudanal segment (segment XIII).
(pv)	primiventral setae.
PI–IV	legs I–IV.
ro	rostral seta.
RU	rutellum.
s	subungual seta.
sc.ls	labral sclerite.
sej	sejugal interval.
t_i	inferior tendon of apotele.
t_s	superior tendon of apotele.
TA	tarsus.
(tc)	tectal setae.
tf_i	inferior tendon of femur (or femur 1).
tf_s	superior tendon of femur (or femur 1).
tf', tf''	tendons and muscles of femur (or femur 1).
tfe	tendon of femur 2.
tg, tg', tg''	tendons and muscles of genu.
TG	tegulum.
TI	tibia.
TR	trochanter.
tta	tendon and muscles of tarsus.
tti	tendon and muscles of tibia.
ttr_c	mediosuperior tendon of trochanter.
ttr_i'	antero-inferior tendons and muscles of trochanter.

ttr_s''	postero-superior tendons and muscles of trochanter.
u', u'', (u)	ungual setae.
v', v'', (v)	ventral setae.
xa	anterior exobothridial seta.
xp	posterior exobothridial seta.
θ (theta)	solenidion of femur.
σ (sigma)	solenidion of genu.
φ (phi)	solenidion of tibia.
ψ (psi)	eugenital seta.
ω (omega)	solenidion of tarsus.
VII–XVI	opisthosomatic segments.
′ (prime)	anterior; the right one of two symmetrical structures.
″ (double prime)	posterior.
()	a pair.

4. Oribatida

Many of my papers, particularly the early ones, are devoted to Oribatid mites. Mention can be made here of those dealing with the morphology of Northrina (Van der Hammen, 1953, 1955, 1955a, 1956), Fortuyniidae (Van der Hammen, 1960, 1963), Phthiracaridae (Van der Hammen, 1963a, 1963c, 1964b, 1965a), and Epilohmanniidae (Van der Hammen, 1986b), and of the detailed study of the gnathosoma of *Hermannia convexa* (C.L. Koch) (Van der Hammen, 1968).

The diagnostic characters of the order Oribatida are the following. The body of the adults is composed of precheliceral metameres and 14–16 (mostly 15) segments; these have partly or nearly completely fused. The cuticle of the adults (and sometimes also of the immature forms) is partly or completely scleritized. There is nearly always a rostral tectum (in the case of the adults). Tracheae are absent in many primitive Oribatid mites; the adults of higher Oribatid mites are characterized by the presence of a tracheal system starting from acetabula and apodemes; besides these, various other types of tracheae are found. One pair of prodorsal trichobothria (never two pairs) is nearly always present; the trichobothria generally consist of a bothridial seta (of specialized shape) and a highly developed bothridium. Basic chaetotactic patterns of the body are represented in fig. 54, basic phanerotactic patterns of the legs in fig. 43. The legs never present trichobothria. The podocephalic canal is external. In many cases (many primitive and all higher Oribatid mites) there is a pair of latero-opisthosomatic glands. The ungues of the legs never bear tenent hairs; empodia are not found, and pulvilli are rarely present. Solenidia occur

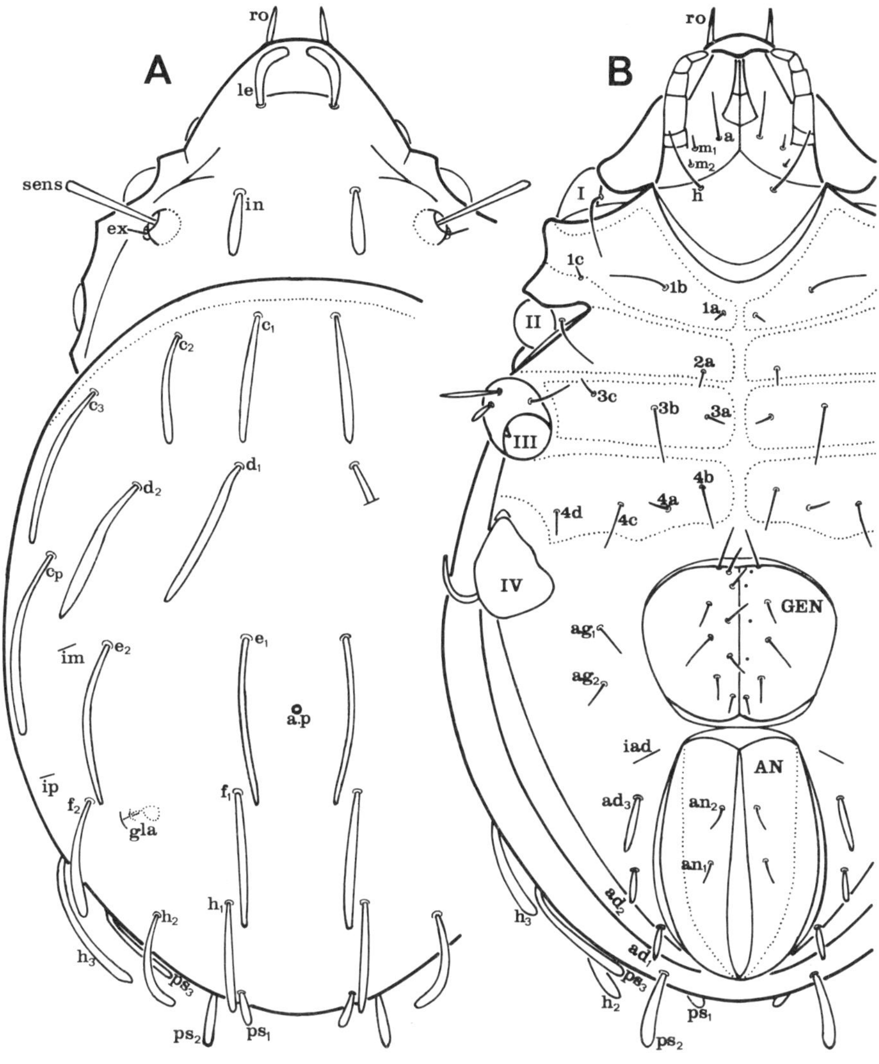

Fig. 54. Hermannia gibba (C.L. Koch), adult male; A, dorsal view; B, ventral view (legs, with the exception of the trochanter, removed); A, B, × 161.

on genu, tibia and tarsus only. There are nearly always three nymphal stases. Besides the calyptostasic prelarva, no other calyptostases are found; there are no elattostases.

According to Grandjean (1954, 1965a, 1969), Oribatid mites comprise six main groups, which are here regarded as suborders: Palaeosomata, Enarthronota, Parhypochthonina, Mixonomata (previously Perlohmannoidea and

Phthiracaroidea), Nothrina and Circumdehiscentiae. The first five suborders comprise the isolated groups of primitive Oribatid mites; the last-mentioned suborder comprises the numerous groups of higher Oribatid mites.

The present section is divided into five subsections, in which the morphology of *Epilohmannia* (subsection 4.1), the morphology of *Hoplophthiracarus* (subsection 4.2), the morphology and postembryonic development of *Archegozetes* (subsection 4.3), the morphology and postembryonic development of *Fortuynia* (subsection 4.4) and the gnathosoma of *Hermannia* (subsection 4.5) are successively dealt with.

4.1. *The morphology of* Epilohmannia zwarti *Van der Hammen*

As mentioned already in the general part of this chapter, species of the genus *Epilohmannia* are interesting because of the fusion of the epimera of palp and leg I (a phenomenon also observed in Palpigradi, but not known from other Chelicerata). The differences between the mouthparts of the two taxa are important, and the fusion of the epimera must be related to quite different functional adaptations (in Palpigradi, it could pertain to an association of functions of palp and leg I; in *Epilohmannia*, it is apparently associated with a change in infracapitular articulation).

The genus *Epilohmannia* is classified with the family Epilohmanniidae and the suborder Mixonomata. Although it belongs to a group of primitive Oribatid mites, it presents many derived characters. The following description is based on material from former Netherlands New Guinea (cf. Van der Hammen, 1986b), which included males and females, but no immature stases.

Measurements. – Males: length 0.688–0.765 mm; breadth 0.376–0.417 mm; ratio length: breadth 1.82–1.83. Females: length 0.832 mm; breadth 0.454–0.457 mm; ratio length: breadth 1.82–1.83. Females are not only characterized by the larger size, but also by the relatively larger genital opening.

Colour. – The idiosoma is for the greater part dark chestnut-brown. The transparent rostral tectum and the legs are lighter brown. The dark colour of the idiosoma is partly caused by the contents of the body; in three of the four specimens studied here, the colour is of irregular intensity, whilst a female, represented by an exoskeleton only, is lighter brown.

Tegument. – The surface of the cuticle is very glossy; there is no cerotegument. In most parts of the body, the superficial microsculpture consists of small, shallow, rather irregularly distributed pits. The microsculpture of epimera 2 is reticulate, with meshes broader than long; this reticulation is partly constituted by low, irregular, transverse ridges, of which the profile is particularly visible at the lateral border. In transparent light, the cuticle is characterized by a dense porosity, caused by the presence of numerous narrow canals in the cuticle (not in the epiostracum). In oblique view, the porosity has the appearance of an internal striation; in a view perpendicular to the surface, the porosity has the appearance of a punctuation. The punctuations are distinctly larger in the cuticle of the epimera and part of the prodorsum.

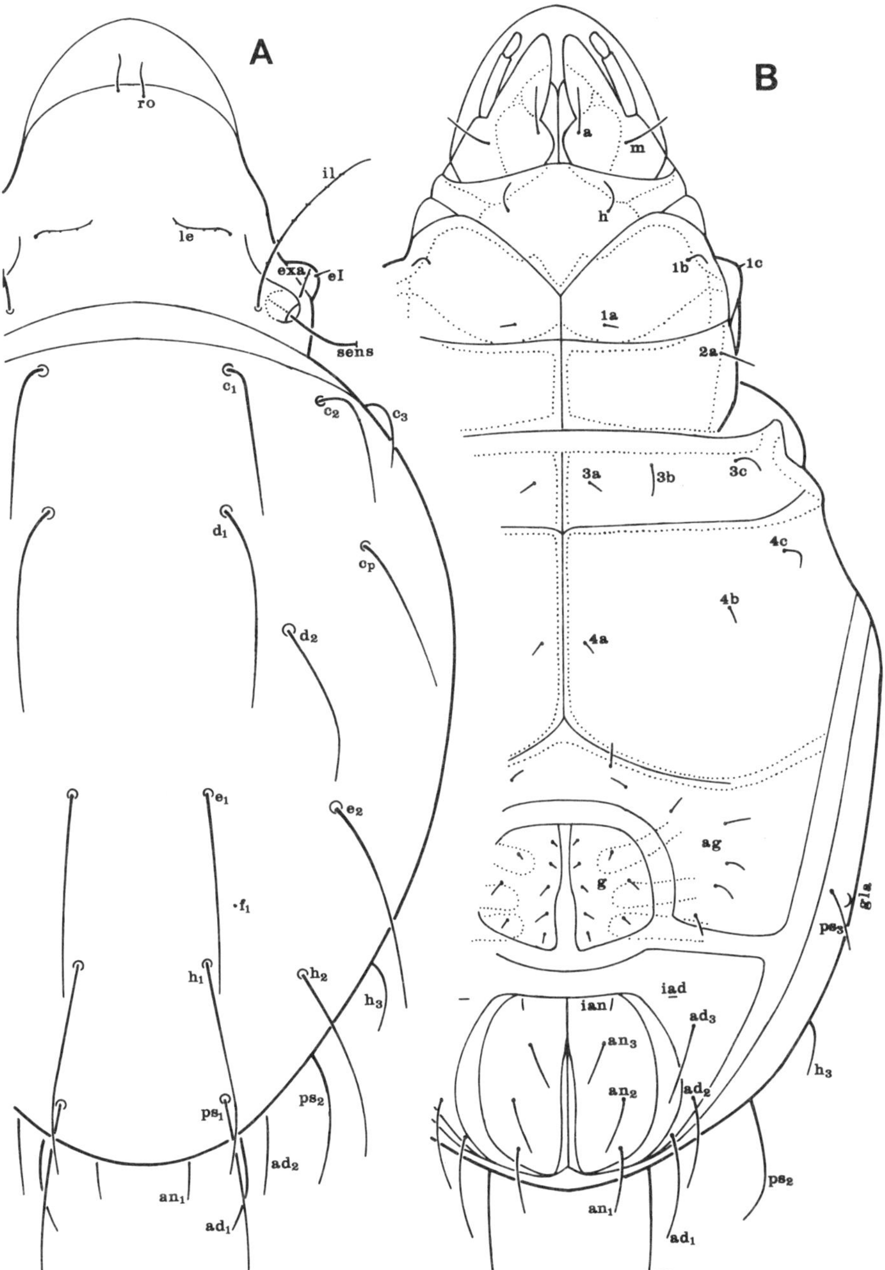

Fig. 55. Epilohmannia zwarti Van der Hammen, adult male; A, dorsal view; B, ventral view; A, B, × 200.

Prodorsum. – The prodorsal shield (figs. 55A, 56A) is laterally contiguous with the lateral region of the propodosoma. The posterior border of the prodorsum is connected with the notogaster by a disjugal area of soft skin, by which it can be retracted under the anterior border of the notogaster (fig. 56A). The anterior part of the prodorsal shield is constituted by a large rostral tectum with its limbus (fig. 57D, E); a naso with an inferior convexity is present at the ventral surface of the rostral base. The rostral setae are rather small and thin, with a few minute ciliae (as in the other setae); their position is more or less variable, but one of the paired setae (the left in both figured specimens) is generally more advanced than the other. The lamellar, interlamellar, bothridial and anterior exobothridial setae (and the supracoxal seta *el*) are also represented separately in fig. 56B; in all specimens investigated by me, the posterior exobothridial seta (fig. 56A) was reduced to its alveolus. The lamellar setae are relatively short and thin. The interlamellar setae are inserted close to the bothridium; they are long (much more than half as long as their mutual distance). The sensillus is setiform and ciliated. The anterior exobothridial seta is small.

Notogaster. – The notogastral setae (fig. 55A, B) are generally very long, and several of them (c_1, c_p, e_1, e_2, h_1, h_2) surpass the place of insertion of the seta immediately posterior to them. Notogastral setae c_3, h_3, and ps_3, however, are much shorter. All notogastral setae are slightly ciliated. Setae f_1 and f_2 are vestigial; f_2 (fig. 56A) is close to the orifice of the latero-opisthosomatic gland. (The problem of the homology of the notogastral setae and lyrifissures in the family Epilohmanniidae was recently solved by Norton, Metz & Sharma, 1978). Notogastral lyrifissures are best visible in lateral view (fig. 56A; this figure was prepared after an exoskeleton in which setal bases, lyrifissures and glands are easier observable than in the dark specimens which have preserved their internal organs); *ia* and *im* are more or less ventral to c_p; *ips* and *ih* are anterior to ps_3. The position of the orifice of the latero-opisthosomatic gland is anterior to h_3 and dorsal to ps_3.

Ano-adanal region (fig. 55B). – The anal opening is relatively large and occupies a postero-ventral position. There are three anal setae (an_{1-3}) and an anal lyrifissure *ian* (with more or less longitudinal orientation). There are also three adanal setae (ad_{1-3}) and an adanal lyrifissure *iad* (with more or less transverse orientation).

Genito-aggenital region (fig. 55B). – The genito-aggenital region is separated from the ano-adanal region by a transverse scissure, an interruption of the sclerotized cuticle, in the shape of a narrow band of soft skin (this type of ventral scissure is referred to as schizogastric). There are eight genital setae, arranged in two rows: one row of five setae near the paraxial border, and a row of three setae with a more antiaxial position. The adults present the normal number of three genital papillae. The female genital opening (fig. 56C) is relatively larger, but presents the same number of setae. The aggenital region is neotrichous; in my specimens, there are seven or eight aggenital setae with slightly asymmetrical and irregular positions.

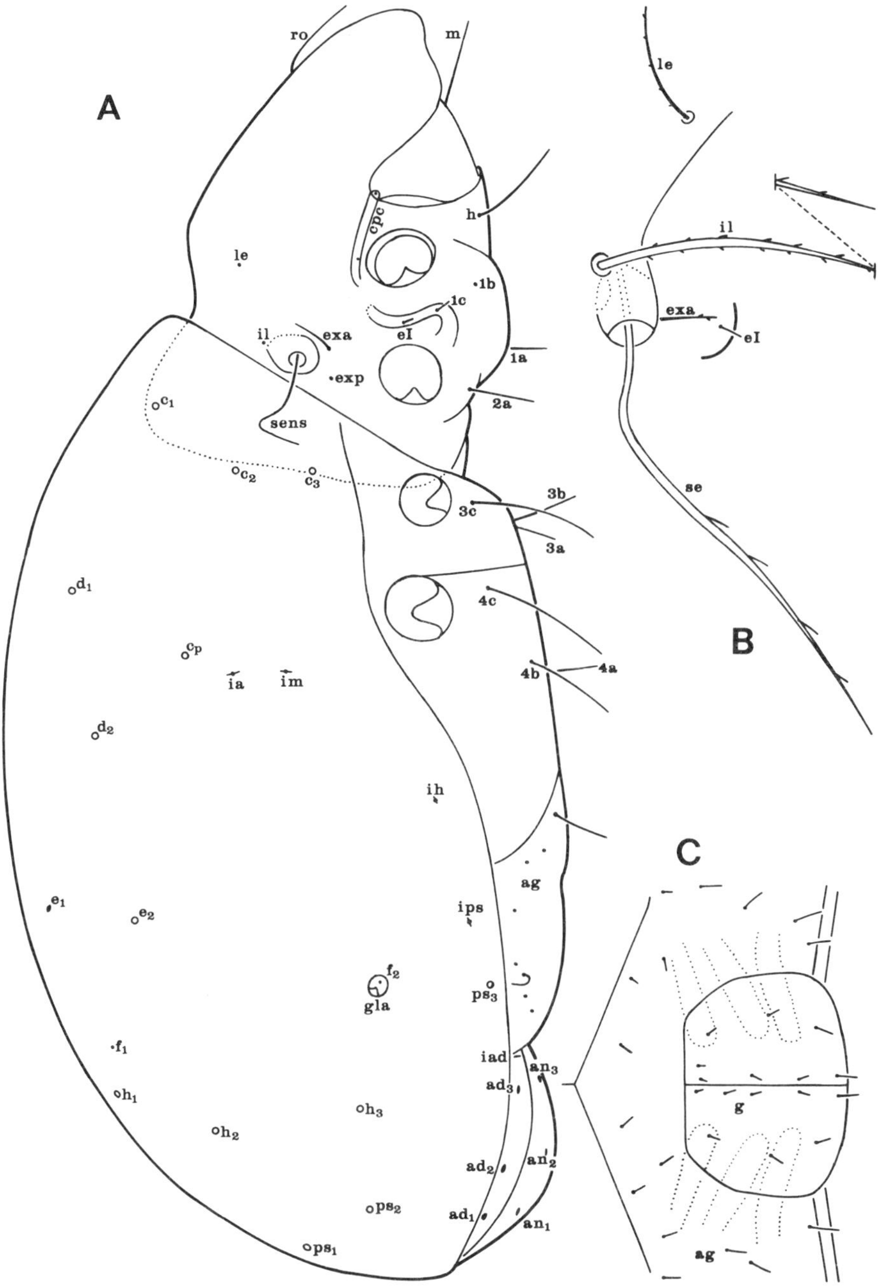

Fig. 56. Epilohmannia zwarti Van der Hammen; A, lateral view of exoskeleton of adult female; B, dorsal view of right bothridium, sensillus, lamellar seta, exobothridial seta and supracoxal seta *eI* of adult male; C, ventral view of genital and aggenital region of adult female; A, C, × 200; B, × 404.

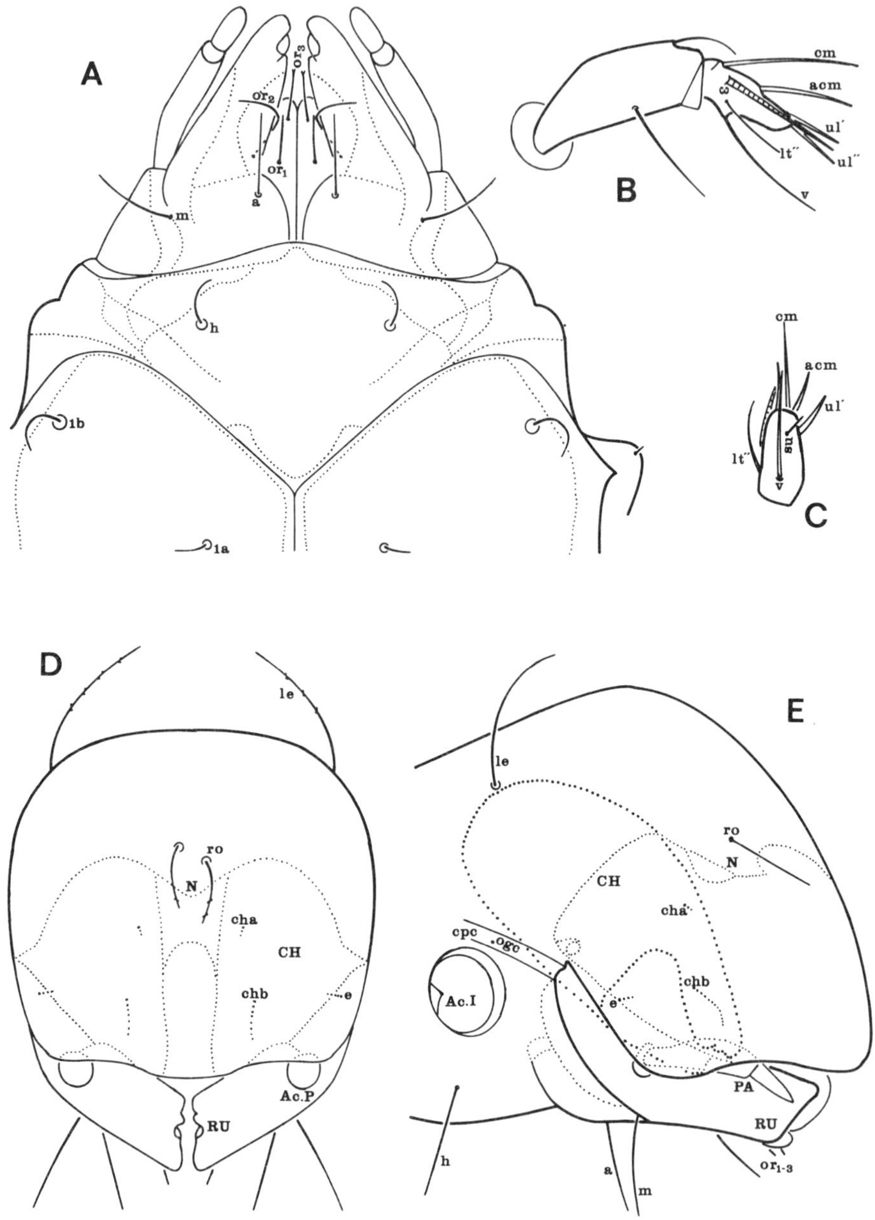

Fig. 57. Epilohmannia zwarti Van der Hammen, adult male; A, ventral view of anterior part of prosoma (infracapitulum and epimera 1); B, lateral (antiaxial) view of right palp; C, ventral view of right palpal tarsus; D, frontal view of anterior part of prosoma; E, lateral (oblique) view of anterior part of prosoma (the rostral part is at the same level as acetabulum I); A, × 376; B, C, × 550; D, E, × 295.

Ventral aspect of podosoma (fig. 55B). – The podosoma is characterized by the occurrrence of a protero-hysterosomatic articulation (dichoidy) at the place of the ventro-sejugal interval. The epimera are separated and bordered by apodemes; there is also a sternal apodeme in the sagittal plane, internally separating the apodemes of the left and the right side. Epimera 1 are separated from the infracapitulum by apodemes 1 (there is no gnatho-idiosomatic articulation and no mentotectum); epimera 4, which are relatively very large, are separated from the aggenital region by apodemes 4. The formula of the epimeral setae is 3–1–3–3.

Lateral aspect of podosoma (fig. 56A). – There is a rounded projection between the acetabula of legs I and II; it bears the supracoxal seta *eI* and epimeral seta *1c*. The idiosomatic part of the podocephalic canal extends from above the base of leg I to the base of the rostral tectum. The orifice of the coxal gland is above the base of leg I. Another podocephalic orifice (homologous with the orifice og_1 of the anterior podocephalic gland in *Hermannia convexa* (C.L. Koch); see Van der Hammen, 1968: 19–22, fig. 11) is present near the base of the rostral tectum. There is probably also an orifice of a median podocephalic gland (homologous with og_2 in *Hermannia*), although it could not be discerned with absolute certainty.

Infracapitulum. – The infracapitulum is represented, in ventral view, in fig. 57A. As mentioned above, the mentum has fused with epimera 1 (apodemes 1 are, however, present), and there is no mentotectum. The infracapitulum is diarthrous and the transverse labiogenal articulation is well-developed, presenting a distinct tectum; the mentum is relatively large, but the genae are much smaller. Part of the role of the gnatho-idiosomatic articulation is apparently taken over by the labiogenal articulation. There are three pairs of infracapitular setae (*h, m, a*). The rutellum is atelebasic and its separation from the genae is apparently incomplete. There are three pairs of adoral setae (or_{1-3}); or_3 is bifurcate in the terminal part.

Chelicera. – The chelicera (fig. 57D, E) is relatively large, with robust jaws. Cheliceral seta *cha* is small; *chb* is much longer.

Palp (fig. 57B, C). – As a result of fusion, the palp consists of two segments. The basal segment represents trochanter, femur, genu and tibia; it bears two setae: a dorsal and an antilateroventral. The tarsus presents a solenidion and seven setae (*cm, acm, ul', ul'', su, lt''* and *v*; *ul'* and *ul''* are eupathidic).

Legs (figs. 58, 59). – The legs consist of six segments: trochanter, femur, genu, tibia, tarsus and apotele. The femur of legs I and II is exceptionally small, the genu much larger. All legs are monodactyl. The numbers of solenidia on genu, tibia and tarsus of legs I–IV are the following: I (2–1–3); II (1–1–2); III (1–1–0); IV (1–1–0). On tarsus I, the famulus is reduced to a small, rounded projection. The solenidia of genu (one of the two in the case of genu I) and tibia of legs I–IV are coupled with the dorsal seta *d*, which is very small; the true size of seta *d* can only be discerned after extraction of the soleni-

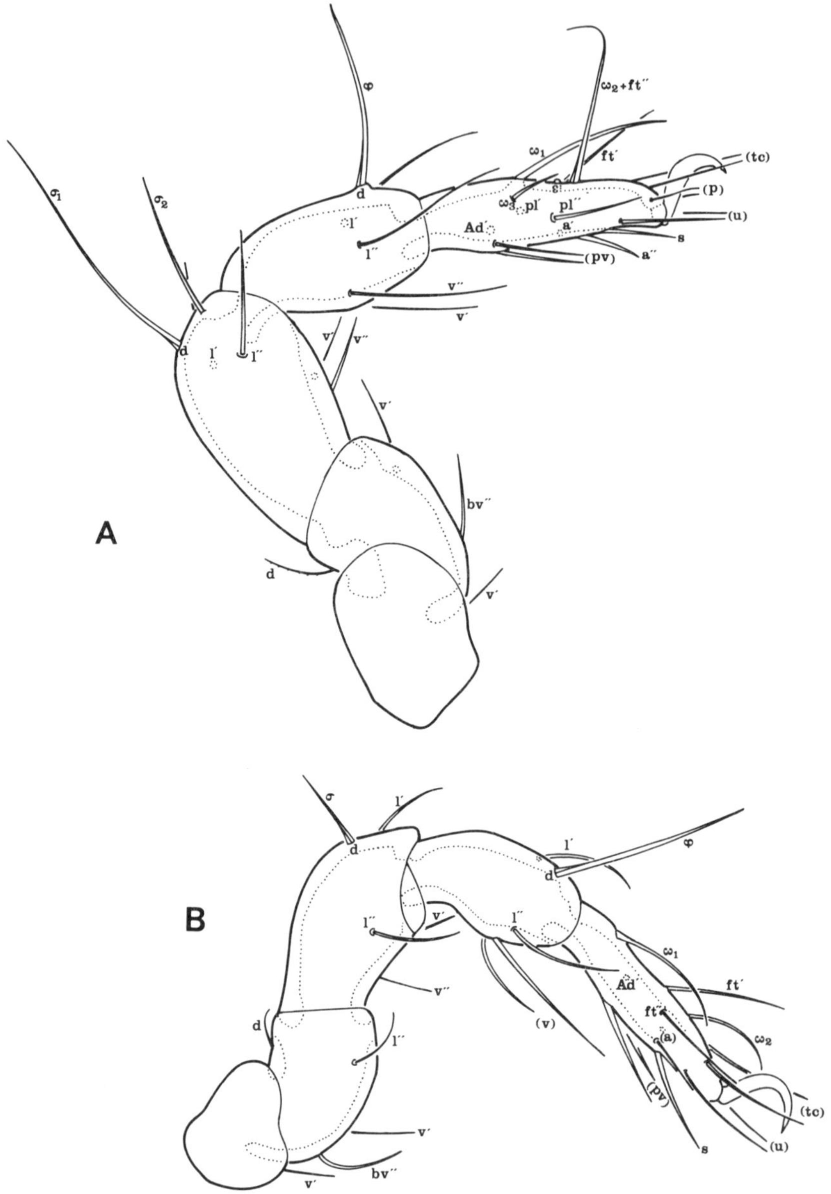

Fig. 58. Epilohmannia zwarti Van der Hammen, adult male; A, right leg I, lateral (antiaxial = posterior) view; B, right leg II, lateral (antiaxial = posterior) view; A, B, × 295.

dion. The second solenidion of the tarsus is coupled with seta *ft″*. The formulae of the setae on trochanter, femur, genu, tibia and tarsus of legs I–IV are

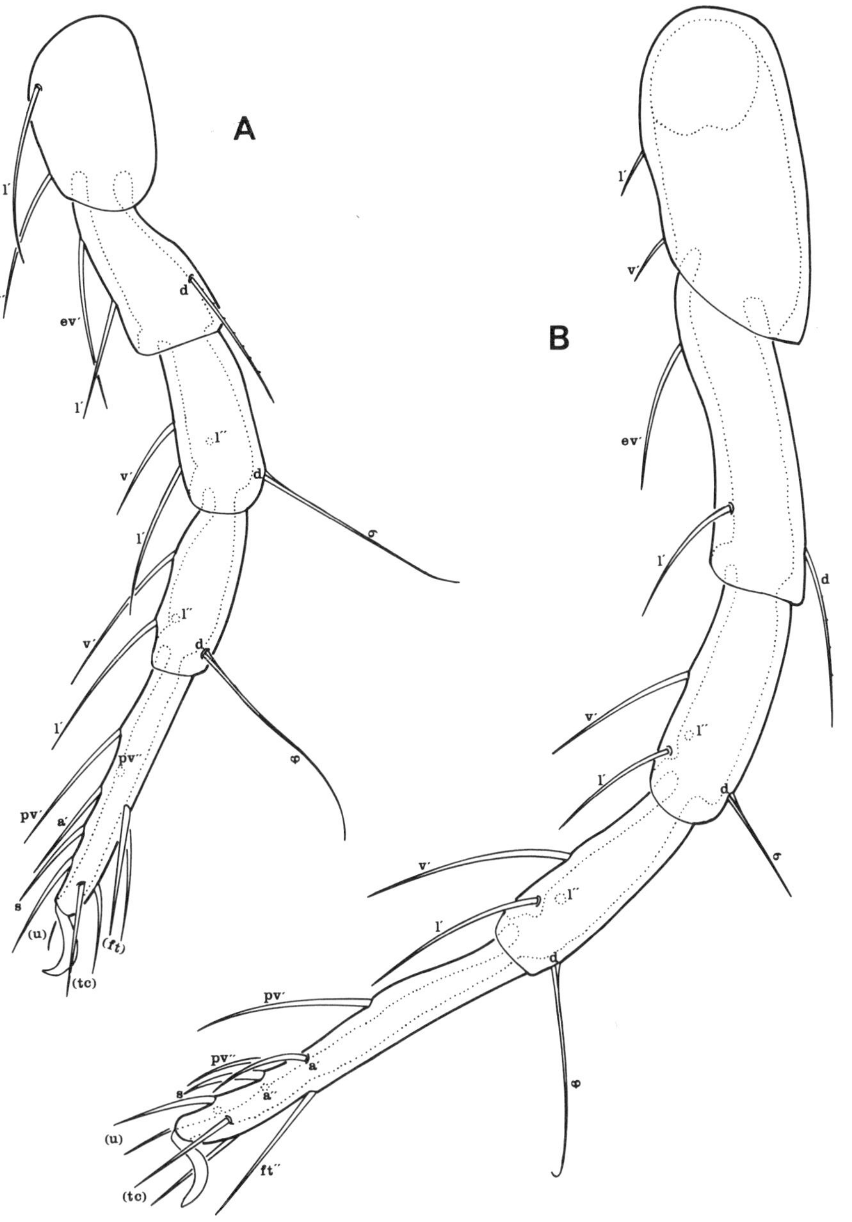

Fig. 59. Epilohmannia zwarti Van der Hammen, adult male; A, left leg III, lateral (antiaxial = anterior) view; B, left leg IV, lateral (antiaxial = anterior) view; A, B, × 295.

the following: I (1–3–5–5–17); II (1–4–5–5–12); III (2–3–4–4–10); IV (2–3–4–4–9). Seta *s* of tarsus IV is the only thickened spiniform seta (it is thickened in both sexes).

Remarks. – The differences between the simple mouthparts of Palpigradi and the specialized gnathosoma of *Epilohmannia* are considerable. For the present comparison it is of particular interest that the long Palpigradid palp (limb 2) has an ambulatory function, whilst the function of the reduced Actinotrichid palp is sensory. The fusion of the epimera of the segments bearing palp and leg I, in the two groups, must indeed be associated with completely different adaptations. The hypothesis could be introduced that the Palpigradid mouthparts represent a regressive gnathosoma, but there is no evidence to support it. An important element in the comparison of the Palpigradid and Actinotrichid mouthparts is constituted by the uncertain homology of the labium. A Chelicerate labium is generally homologized with the sternite of the palpa segment (sternite 2), but that of the Palpigradi (situated anterior to sternite 2) was supposed to be homologous with the sternite of the cheliceral segment (see Millot, 1949a: 521). In Actinotrichida, a labium is present in primitive species only; it is no more present (or no more recognizable) in Epilohmanniidae. The Actinotrichid labium could be homologous with the labium in Palpigradi, and that in Solifugae and Pseudoscorpionida.

4.2. *The morphology of* Hoplophthiracarus *spec.*

The *Hoplophthiracarus* species described in the present subsection is a representative of the family Phthiracaridae, one of a group of those families of the suborder Mixonomata, which are characterized by ptychoidy (the presence of a dorsal hinge between prosoma and opisthosoma, permitting the prosoma to be folded down and the legs to be concealed). The species was originally described by me (Van der Hammen, 1963a) as *Hoplophthiracarus pavidus* (Berlese); a subsequent comparison with specimens from the type-locality of *pavidus* has demonstrated that it represents a different, unnamed species of the genus. The present description is based on numerous specimens collected from *Sphagnum* in one of the northern provinces of The Netherlands.

Measurements. – Length of prodorsum 0.180–0.225 mm (average 0.200); length of notogaster 0.295–0.395 (average 0.360), height 0.220–0.395 (average 0.250), breadth 0.250–0.295 mm (average 0.270).

Habitus and colour. – The species is particularly characterized by the curved notogastral setae and the erect interlamellar setae. The colour of the notogaster is rather dark chestnut-brown; the anterior and ventral notogastral limbus is darker brown. The anterior border of the genital covers and the border between genital and anal covers are also darker. The prodorsum or aspis is lighter brown, with a posteromedian and two anterolateral dark spots. The legs are light brown. All setae are whitish.

Tegument. – The cerotegument layer is particularly developed in older

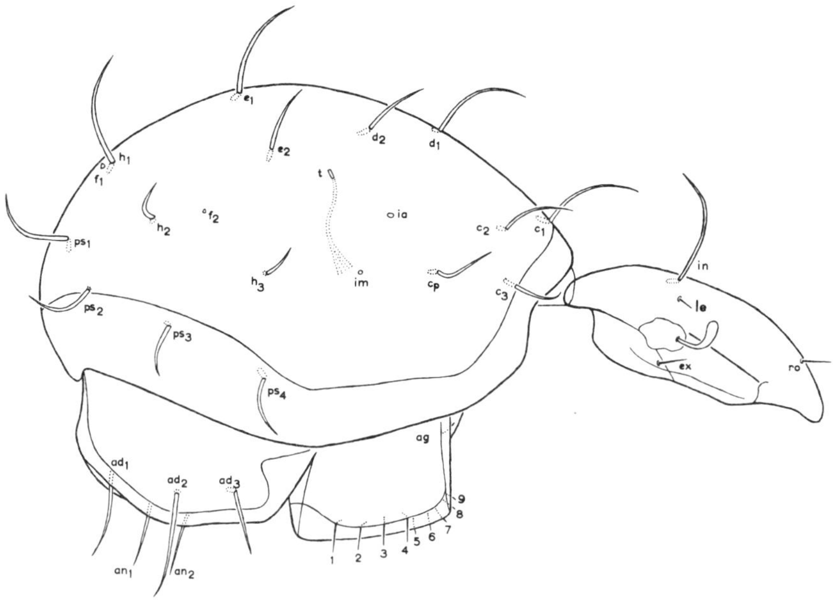

Fig. 60. Hoplophthiracarus spec., female; lateral view, × 217.

specimens. It consists of irregular masses of granules which are interconnected in an irregular way. The layer is especially distinct in the lateral regions of the notogaster. The cuticle is shining. The microsculpture is for the greater part shagreened (as in *Phthiracarus*), but the anterior limbus and the anteroventral part of the notogaster close to the ventral limbus (not the ventral limbus itself) present a punctate microsculpture (as in *Steganacarus*). The punctate areas are easily discernable when the animals are studied on a carbon block.

Prodorsum. – A lateral view of the prodorsum or aspis is represented in fig. 60, a dorsal view in fig. 61A. In both views, the lateral ridge is distinct; there is no dorsal ridge. The three brachytracheae attached to the bothridium are easily visible (fig. 62B). The sensillus is directed sidewards; it has a thickened distal part with blunt tip. The bothridium is protected by the usual scale. Apart from the paired sensillus, there are four other pairs of prodorsal setae: interlamellar, lamellar, rostral, and exobothridial setae. The interlamellar setae are erect and slightly curved backwards in the distal part. The small lamellar setae are close to the bothridium. The rostral setae are pointing forwards; a long canal (in the cuticle) extends from the alveolus to the root (the rostral setae are apobasic).

Notogaster. – A lateral view of the notogaster is represented in fig. 60, a dorsal view in fig. 61A, and a ventral view in fig. 61B. The notogastral limbus is darker than the remaining part of the notogaster. As mentioned above, the microsculpture of the notogaster is mainly shagreened, but the anterior limbus

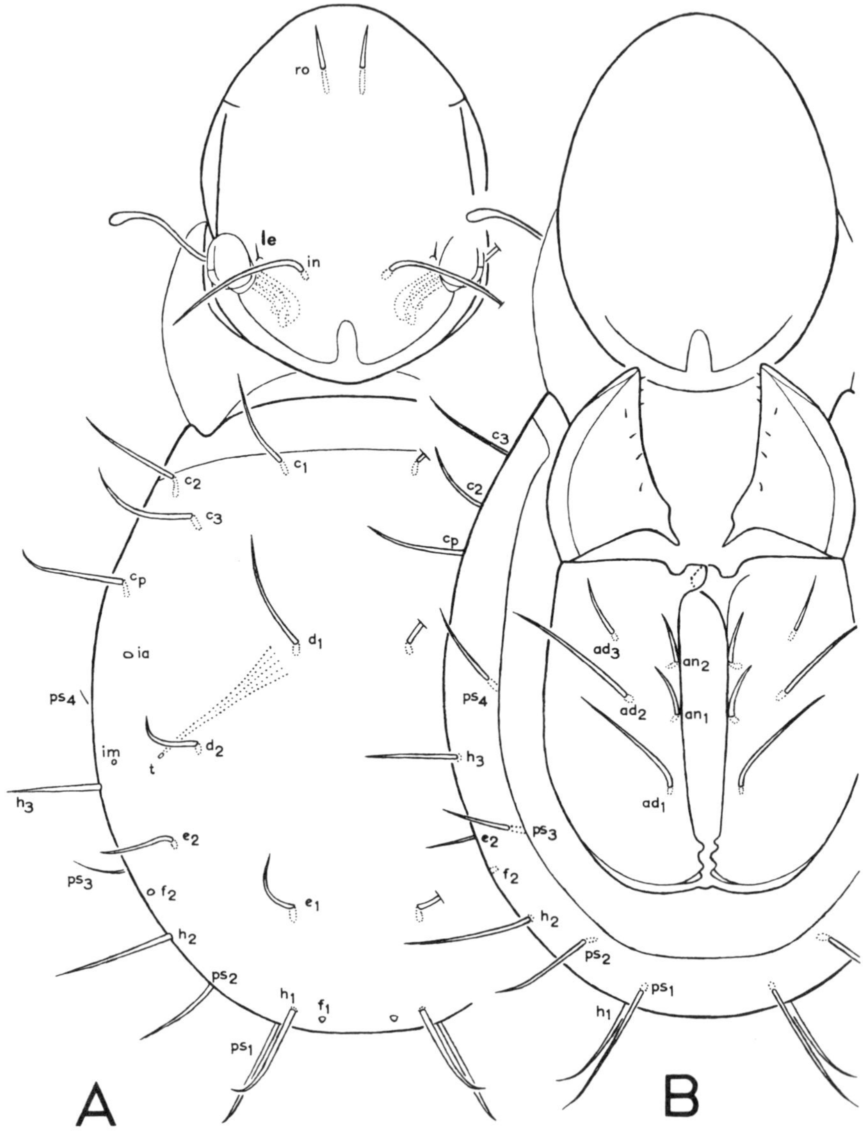

Fig. 61. Hoplophthiracarus spec., female; A, dorsal view; B, ventral view; A, B, × 290.

and the anteroventral part, close to the ventral limbus, are punctate. There are fifteen pairs of notogastral setae and two pairs of vestiges. In accordance with Grandjean (1950: 73–78, fig. 1) the setae have received the following notations: c_1, c_2, c_3, c_p, d_1, d_2, e_1, e_2, h_1, h_2, h_3, ps_1, ps_2, ps_3, ps_4. The vestiges of setae f_1 and f_2 are easily recognizable; one of them (f_1), together with seta h_1,

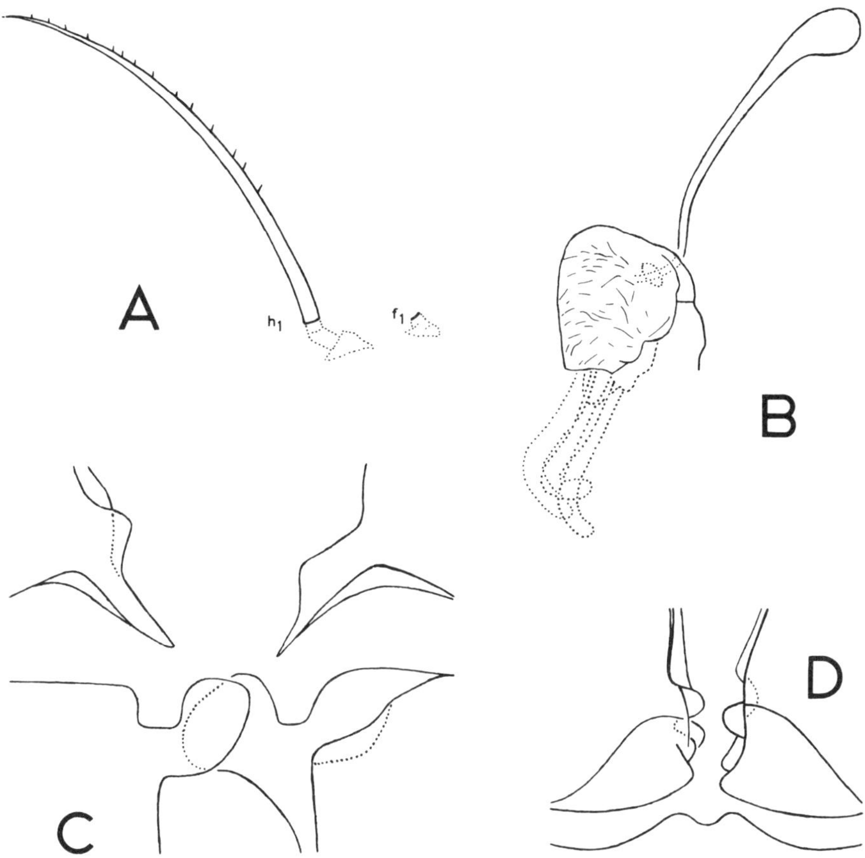

Fig. 62. Hoplophthiracarus spec., female; A, lateral view of notogastral seta h_1 and vestige of seta f_1; B, dorsal view of bothridium and sensillus; C, posterior lock of genital covers, and anterior lock of anal covers; D, posterior lock of anal covers; A–D, × 866.

is represented in fig. 62A; seta h_1 is situated slightly anterior to f_1, whilst it is posterior to f_1 in *Phthiracarus anonymum* Grandjean (see Grandjean, 1950). The notogastral setae are rather strong and slightly rough; they are more or less erect and curved (mostly forwards). There are two pairs of lyrifissures, viz., *ia* and *im*, which have the shape of cupules; *ia* is larger than *im*. The paired opisthosomatic muscle-attachment *t* is also visible.

Ano-genital region (fig. 61B). – Although the ano-genital region must be regarded as a fusion of anal and adanal, and of genital and aggenital shields, the covers or valves are simply indicated here as anal and genital covers (or valves). On each of the two anal covers, there are two anal setae (inserted at the paraxial border) and three adanal setae (inserted in line with each other; ad_1 is close to the paraxial border). The anal covers present a particular lock

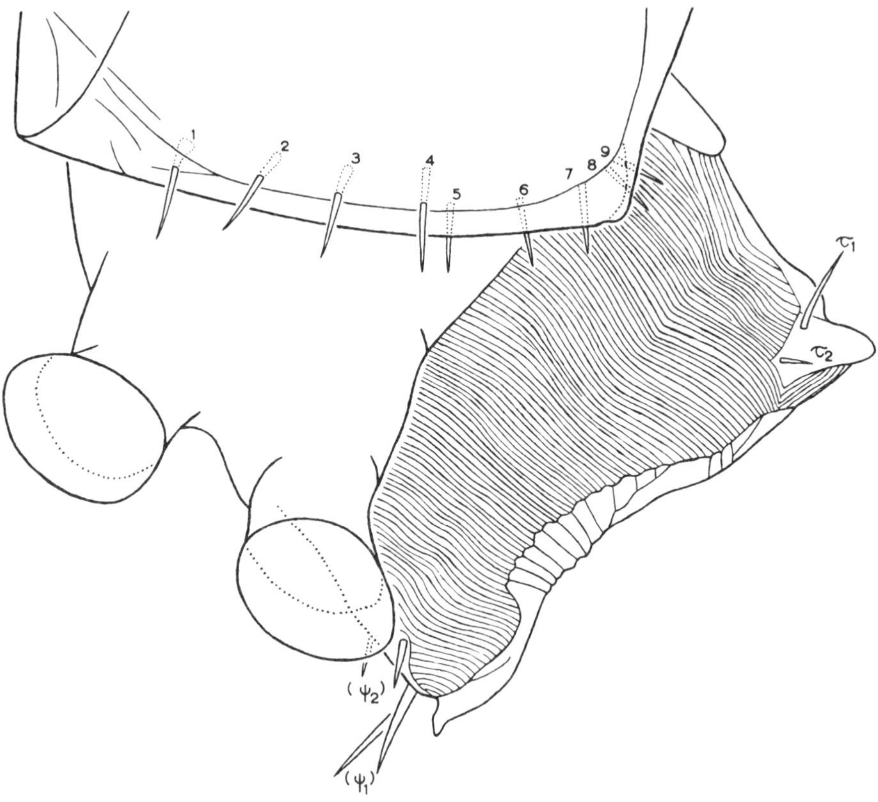

Fig. 63. Hoplophthiracarus spec., female; lateral view of ovipositor, genital papillae, and part of right genital cover; × 866.

(fig. 62C, D); in the anterior part, this lock consists of two lobes of which the right one covers the left (it fits into a corresponding lock) when the anal opening is closed (this condition is characterized as left-fitting; the right-fitting condition is found in different species of Phthiracaridae); the posterior anal lock consists, on either side, of some three small lobes, corresponding with small cavities at the opposite side.

There are nine pairs of genital setae, of which the anterior five are smaller. The last-mentioned setae are inserted on the inner border of the cover; they have a long canal that runs from the alveolus to the root (they are apobasic). The aggenital seta is small; it is difficult to discern because of its insertion in an anterolateral groove of the cover; it is represented in fig. 60.

The ovipositor (fig. 63) is short and relatively broad. It presents a posterior lobe with two pairs of setae (on either side ψ_1 and ψ_2) and two anterior lobes, each with two setae (τ_1 and τ_2). The surface of the ovipositor is striate, with the exception of the lobes and the distal border. There are only two pairs of genital papillae.

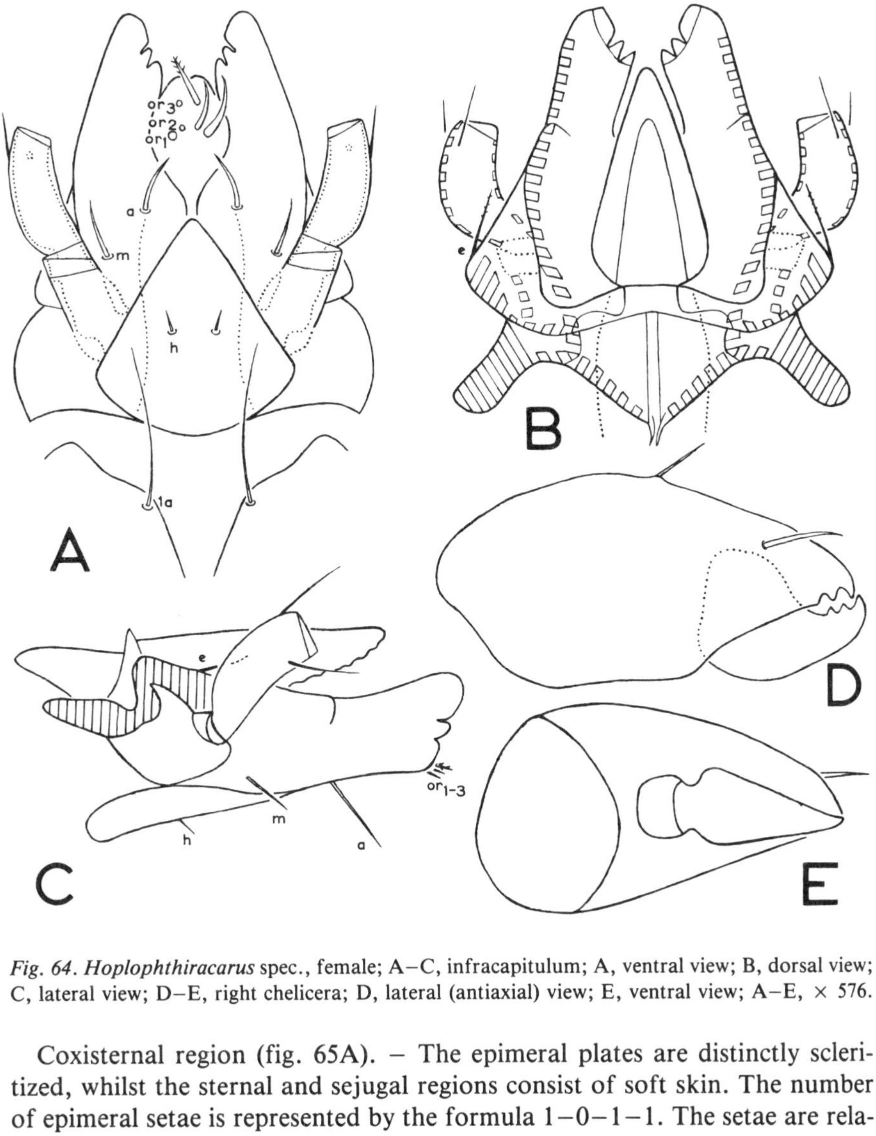

Fig. 64. Hoplophthiracarus spec., female; A–C, infracapitulum; A, ventral view; B, dorsal view; C, lateral view; D–E, right chelicera; D, lateral (antiaxial) view; E, ventral view; A–E, × 576.

Coxisternal region (fig. 65A). – The epimeral plates are distinctly scleritized, whilst the sternal and sejugal regions consist of soft skin. The number of epimeral setae is represented by the formula 1–0–1–1. The setae are relatively very long.

Gnathosoma. – The infracapitulum is stenarthric, and the rutellum is atelebasic (fig. 64A); the usual infracapitular setae *h*, *m* and *a* are present. The lateral lips present three pairs of adoral setae (or_{1-3}), of which or_1 is distinctly pectinate. The dorsal surface of the infracapitulum (fig. 64B) is distinctly scleritized; the labrum and the commissures of the mouth are easily discernable. The supracoxal seta *e* is long and pectinate. In lateral view (fig. 64C), the ventral outline of the labrum appears to be undulate.

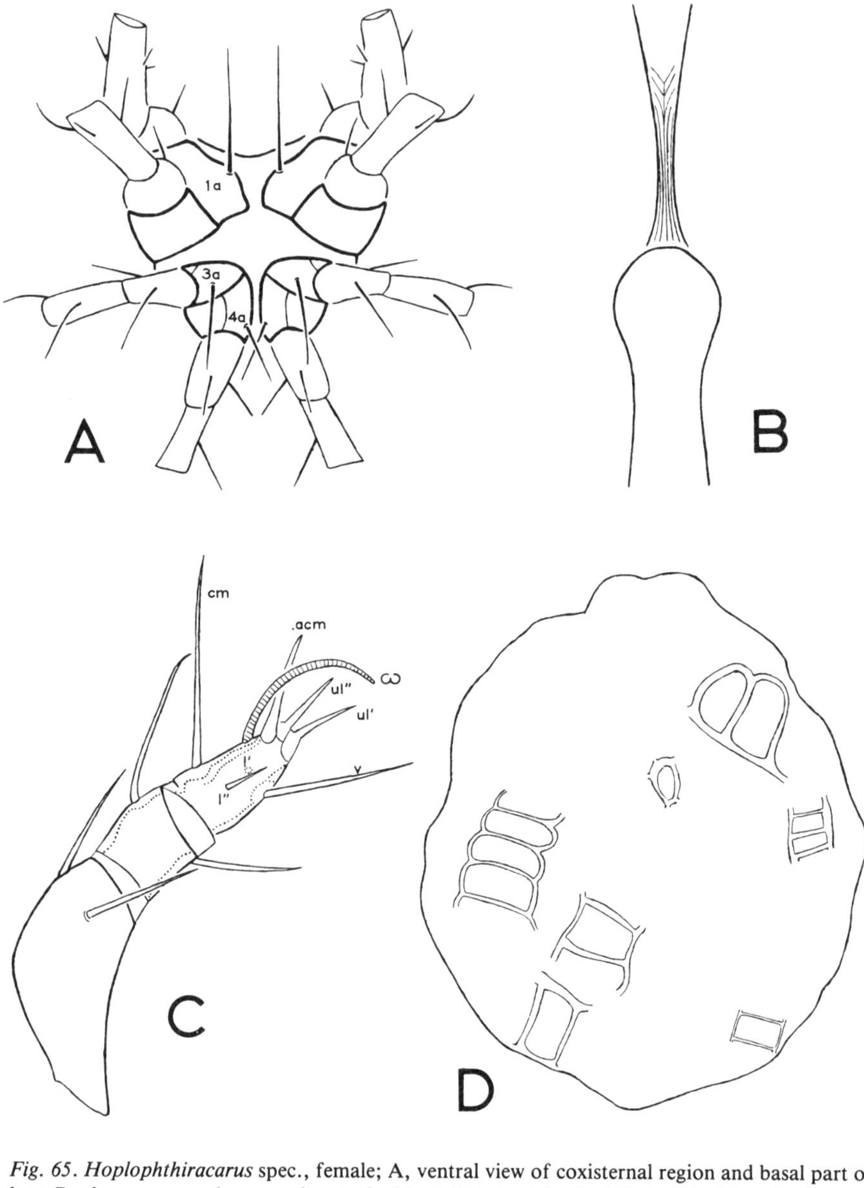

Fig. 65. Hoplophthiracarus spec., female; A, ventral view of coxisternal region and basal part of legs; B, pharynx, oesophagus and part of midgut; C, lateral (antiaxial) view of right palp; D, contents of the midgut (globular mass of food with fragments of *Sphagnum*); A–D, × 290; C–D, × 866.

In transparent light, part of the alimentary canal is easily visible. Pharynx, oesophagus, and part of the midgut are represented in fig. 65B; in dorsal view, part of the foregut is hidden from view by a lobe of the midgut. In most speci-

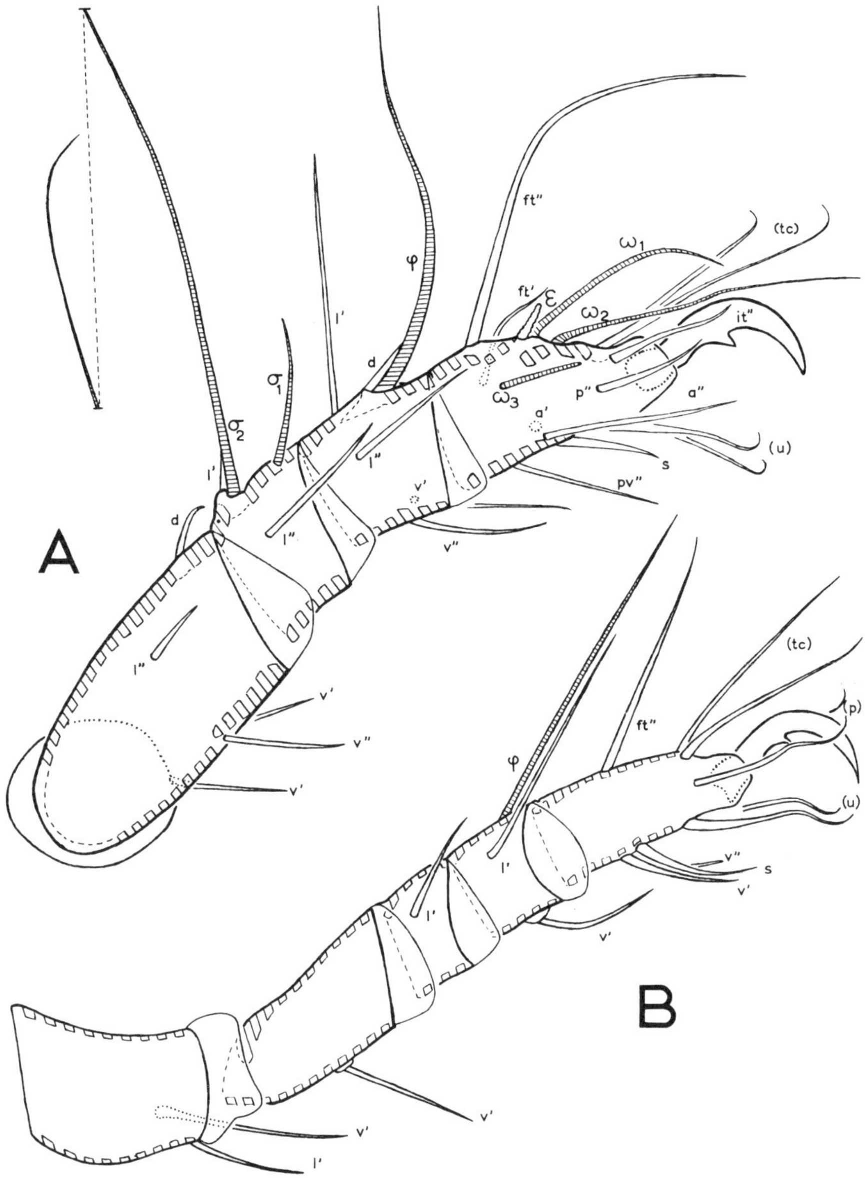

Fig. 66. Hoplophthiracarus spec., female; A, lateral (antiaxial) view of right leg I; B, lateral (antiaxial) view of left leg IV; A–B, × 866.

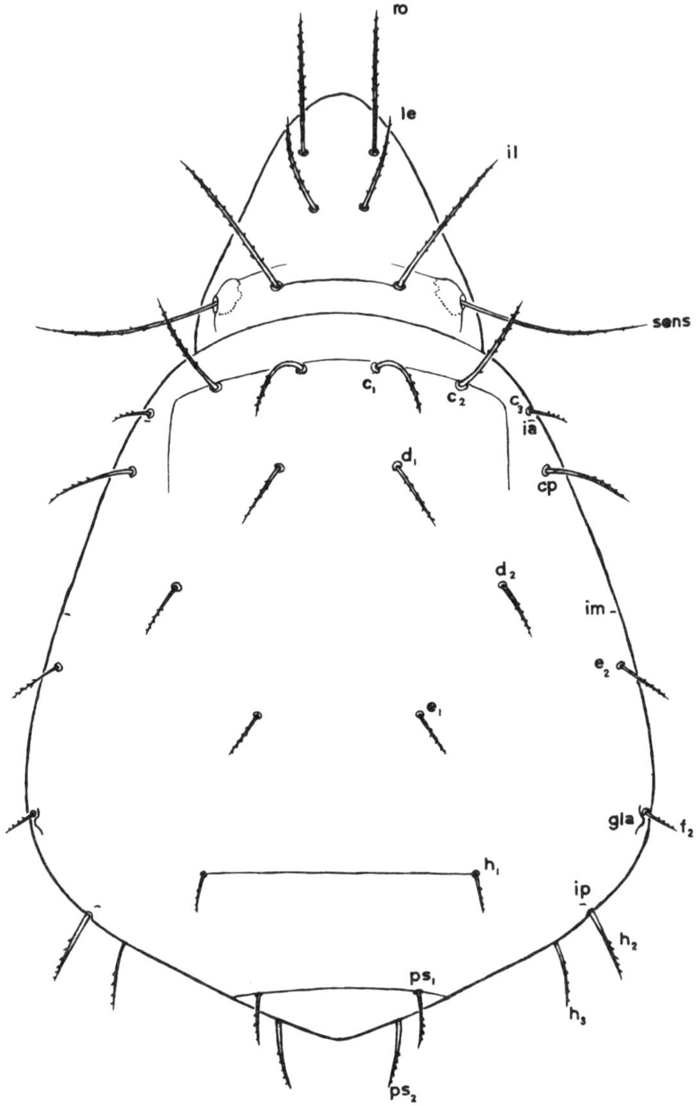

Fig. 67. Archegozetes magna (Sellnick), dorsal view of adult female, × 116.

mens investigated by me, the midgut contained globular masses of food, in which *Sphagnum* fragments were apparently present (fig. 65D).

Chelicera. – The chelicerae (fig. 64D, E) are relatively large; they present a lateral (antiaxial) and a dorsal seta. The two jaws of the chela each present three teeth.

Palp (fig. 65C). – The palp consists of four segments; a separate trochanter

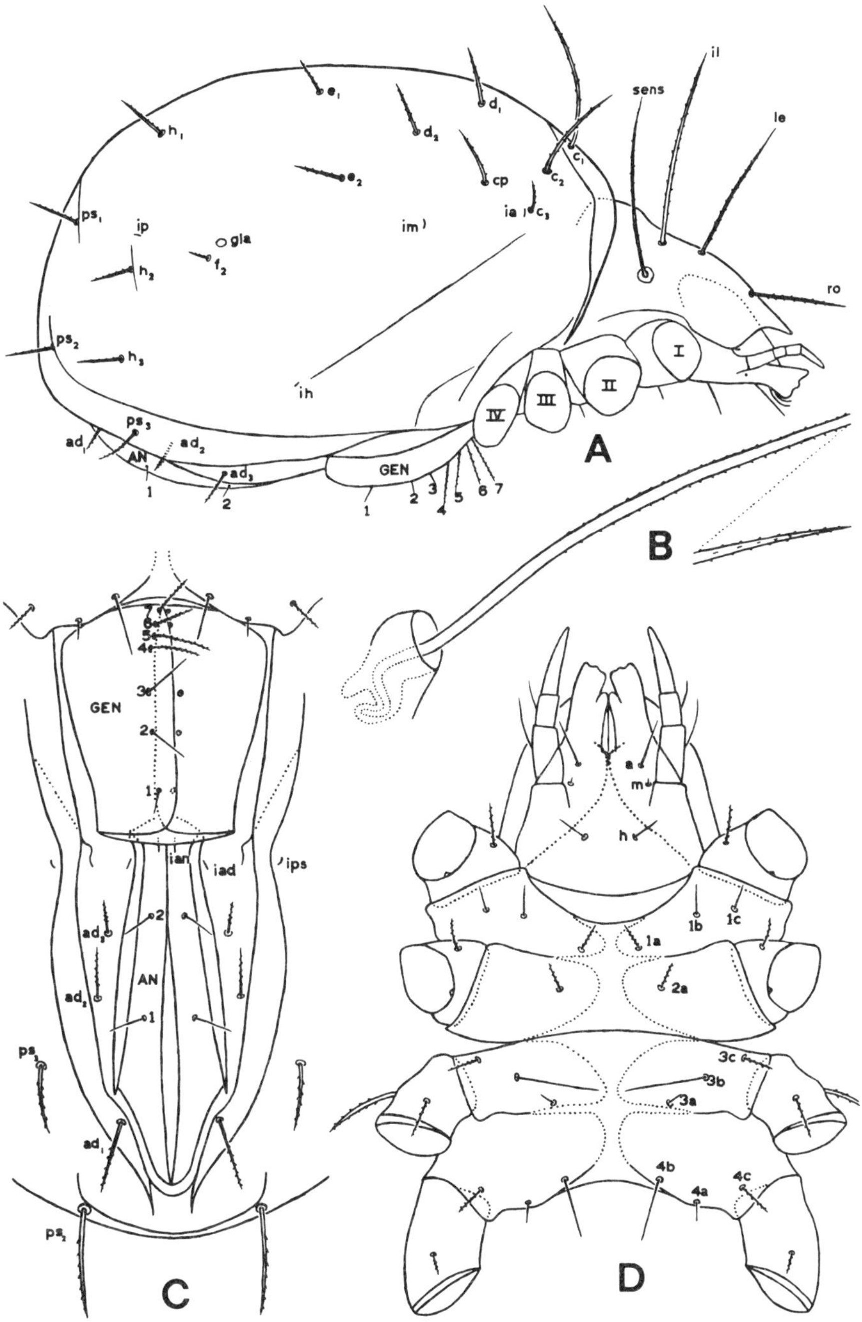

Fig. 68. Archegozetes magna (Sellnick), adult female; A, lateral view; B, bothridium and sensillus; C, ano-genital region; D, ventral view of prosoma; A, × 105; B, × 451; C, D, × 152.

is present. There is one solenidion. The chaetotactic formula is 0–2–2–7; setae *acm*, *ul'* and *ul''* are eupathidic. The subultimal seta *sul* is completely absent, and there is no small denticle at the base of *ul'*. Instead of one pair of ventral setae, there is only a single seta *v*.

Legs. – The number of setae on the legs is reduced, particularly on legs III and IV. The lyrifissure of tarsus IV is very small. Genu I has a distinct pore; it is best visible in dorsal view. Leg I is represented in fig. 66A, leg IV in fig. 66B. The chaetotactic and solenidiotactic formulae of legs I–IV are the following: Setae: I (1–4–2–4–15–1); II (1–3–2–3–14–1); III (2–2–1–3–10–1); IV (2–1–1–2–10–1); the last number of each formula refers to the single unguis of the claw. Solenidia: I (2–1–3); II (1–1–2); III (1–1–0); IV (0–1–0). The single unguis of the claws presents a distinct ventral tooth, and an indistinct basal tubercle. The notation of the setae of legs I and IV is added to fig. 66. There are some difficulties in the notation of the setae of tarsi III and IV. It appears that *ft'* is still present on tarsus III, but that there are only four ventral setae, viz., (*u*), *s* and an antiaxial seta that probably represents *pv'*. In the case of leg IV, the fastigial seta *ft'* is absent, but there are five ventral setae, viz., (*u*), *s*, and a ventral pair (fig. 66B: *v'* and *v''*) that probably represents (*pv*). The iteral setae (*it*) are absent in the case of legs III and IV. In the case of tarsus I, (*it*), (*p*) and *s* are eupathidic. The famulus is rather large; it has a characteristic shape that is represented in fig. 66A. On genu I, the solenidion σ_2 is coupled with seta *l'*; *d* is no more present, and σ_1 is free. The solenidion φ of tibia I is coupled with *d*. The solenidia of the tarsus are free.

Remark. – Numerous females in the material studied by me contained calyptostasic prelarvae (as in many other species of Phthiracaridae), but no eggs; eggs probably develop rather rapidly. The number of prelarvae inside the female body varies from zero to four. The prelarva of *Hoplophthiracarus* is represented in fig. 20A. It is evident that the species is ovoviviparous.

4.3. *Morphology and postembryonic development is* Archegozetes magna *(Sellnick)*

Archegozetes magna is a species of the family Trhypochthoniidae (suborder Nothrina), a group of primitive Oribatid mites. The species is known from several tropical countries, and the material on which the following description is based was collected in former Netherlands New Guinea (see Van der Hammen, 1955; 1955a: 195–200, figs. 1–4).

Measurements. – Length of the female adults 0.900–1.090 mm, breadth 0.660–0.740 mm. Ratio length: breadth 1.36–1.50.

Habitus and colour. – At first sight, the adult reminds one of a nymph, because it is little sclerotized (a faint pressure is sufficient for deformation); females are, however, easily recognizable as adult because of the presence of ovipositor and eggs. The colour is pale yellowish, sometimes light-brown; gen-

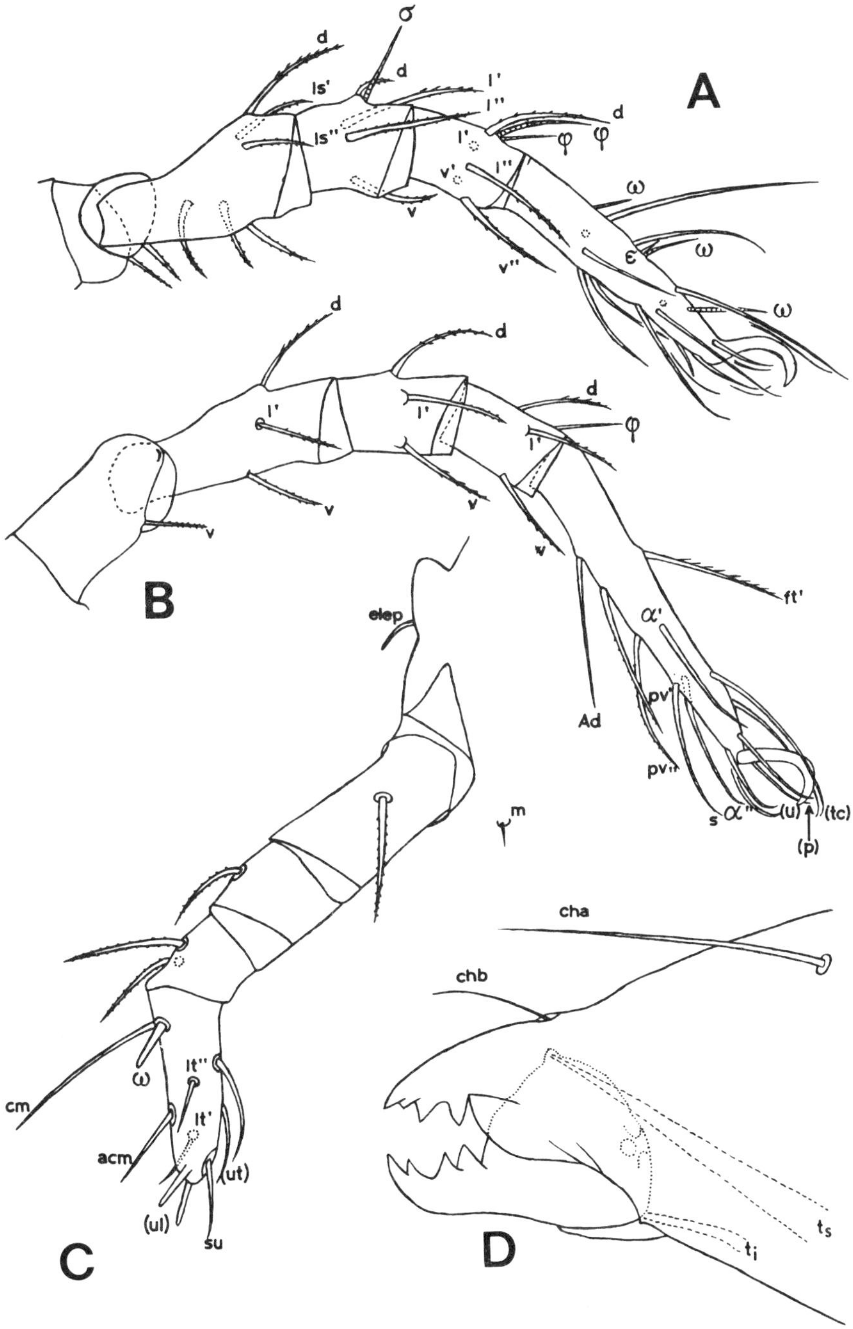

Fig. 69. Archegozetes magna (Sellnick), adult female; A, right leg I, lateral (antiaxial) view; B, left leg IV, lateral (antiaxial) view; C, left palp, lateral (antiaxial) view; D, distal part of right chelicera, lateral (paraxial) view; A, B, × 190; C, D, × 570. (N.B. for *elcp* read *e*).

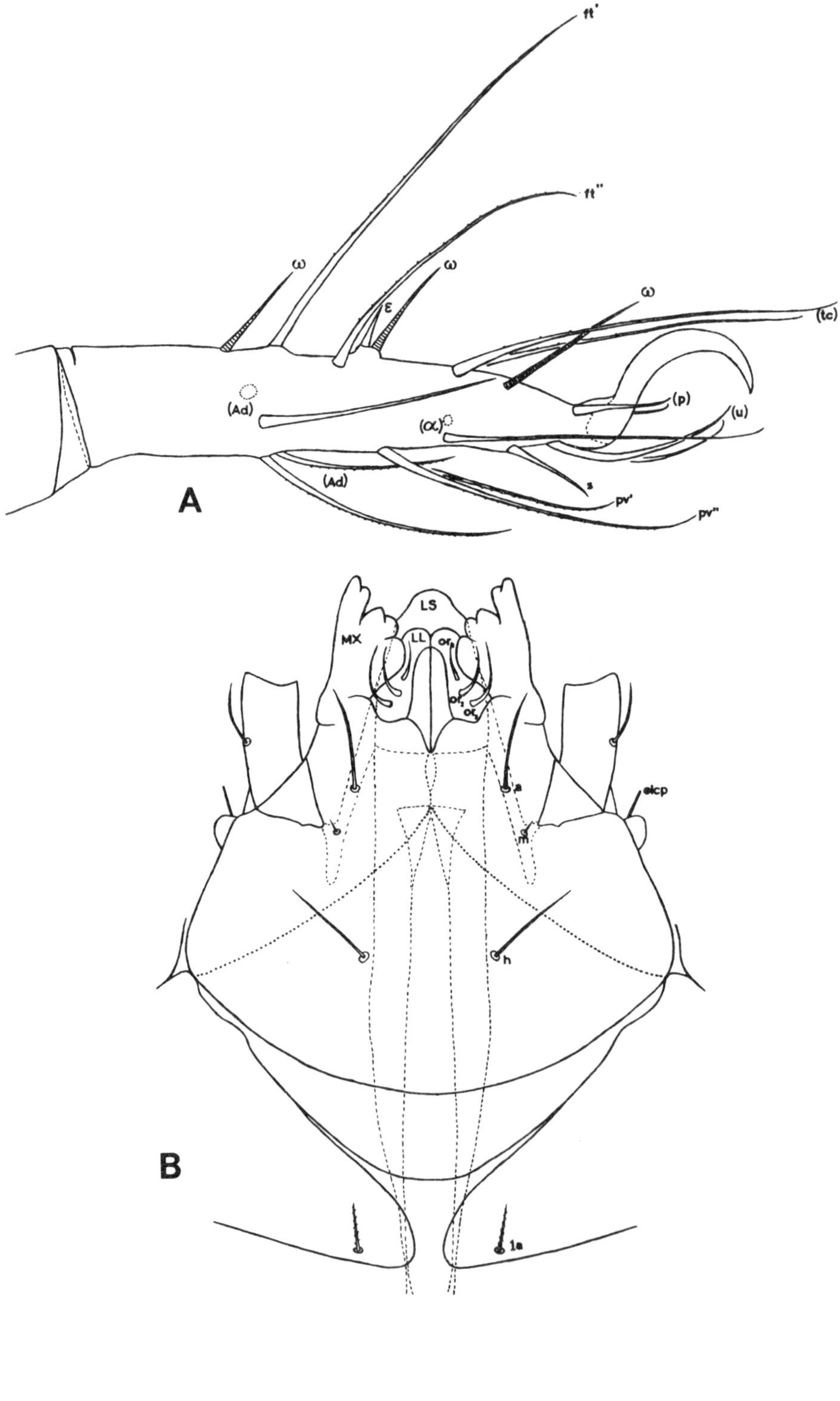
ft'
ft''
ω
ε
ω
ω
(tc)
(Ad)
(p)
(α)
(u)
(Ad)
s
pv'
pv''
A
LS
MX
LL
a
elcp
m
h
B
1a

erally, the contents of the paired latero-opisthosomatic glands are visible as more or less rounded, dark-brown spots.

Tegument. – The cuticle is very glossy, and the cerotegument layer is apparently little developed. The notogaster presents some transverse and longitudinal folds. The microsculpture of the prodorsum is densely puntate; the notogaster is marked with larger dots (best visible at the posterior border).

Prodorsum (figs. 67, 68A). – The prodorsal shield is laterally bordered by soft skin. Anteriorly, it constitutes a rostral tectum which protects the gnathosoma. The posterior border is covered by the notogaster. There are four pairs of prodorsal setae: rostrals, lamellars, interlamellars and sensilli; there are no exobothridial setae. Sensillus and bothridium are represented separately in fig. 68B; the bothridium is simple and cup-shaped, and the sensillus setiform. All prodorsal setae are ciliate.

Notogaster. – In figs. 67 and 68A, the setae have received the usual notation. The notogaster is unideficient: f_1 is lacking. All setae are ciliate. Lyrifissures are not easily discernable: *ia* is close to c_3; *im*, *ip* and *ih* are best visible in lateral view (fig. 68A); *ip* is near h_2, and is often indiscernable when the contents of the latero-opisthosomatic glands are dark; *ips* is visible in ventral view (fig. 68C). The orifice of the latero-opisthosomatic gland is close to f_2.

Ano-genital region (fig. 68C). – There are three pairs of ciliate adanal setae (ad_{1-3}) and two pairs of anal setae. The lyrifissures *ian* and *iad* occupy the usual positions. There are seven pairs of genital setae, of which the anterior four are ciliate.

Coxisternal region (fig. 68D). – The number of epimeral setae is represented by the formula 3–1–3–3.

Gnathosoma. – The infracapitulum (fig. 70B) is stenarthic. There are three pairs of infracapitular setae: *h*, *m* and *a*; *m* is very small. The rutella are atelebasic and lobate. The lateral lips present three setae (or_{1-3}). The supracoxal seta *e* is spiniform.

Chelicera (fig. 69D). – The chelicera is well-developed and the chela presents strong teeth. The cheliceral setae *cha* and *chb* are both present; *cha* is long.

Palp (fig. 69C). – The chaetotaxy of the palp is represented by the formula 0–1–1–2–9; *ul'* and *ul''* of the tarsus are eupathidic. The solenidion of the tarsus is free.

Legs. – Leg I and leg IV are represented in fig. 69A, B; tarsus I also in fig. 70A. All legs are monodactyl. The chaetotaxy of leg I is represented by the formula 1–7–4–5–18 (famulus included), the chaetotaxy of leg IV by the formula 1–3–3–3–12 (13). The solenidiotaxy of legs I and IV is represented by the formulae: I (1–2–3); IV (0–1–0).

←

Fig. 70. Archegozetes magna (Sellnick), adult female; A, tarsus of right leg I, lateral (antiaxial) view; B, ventral view of infracapitulum; A, B, × 428.

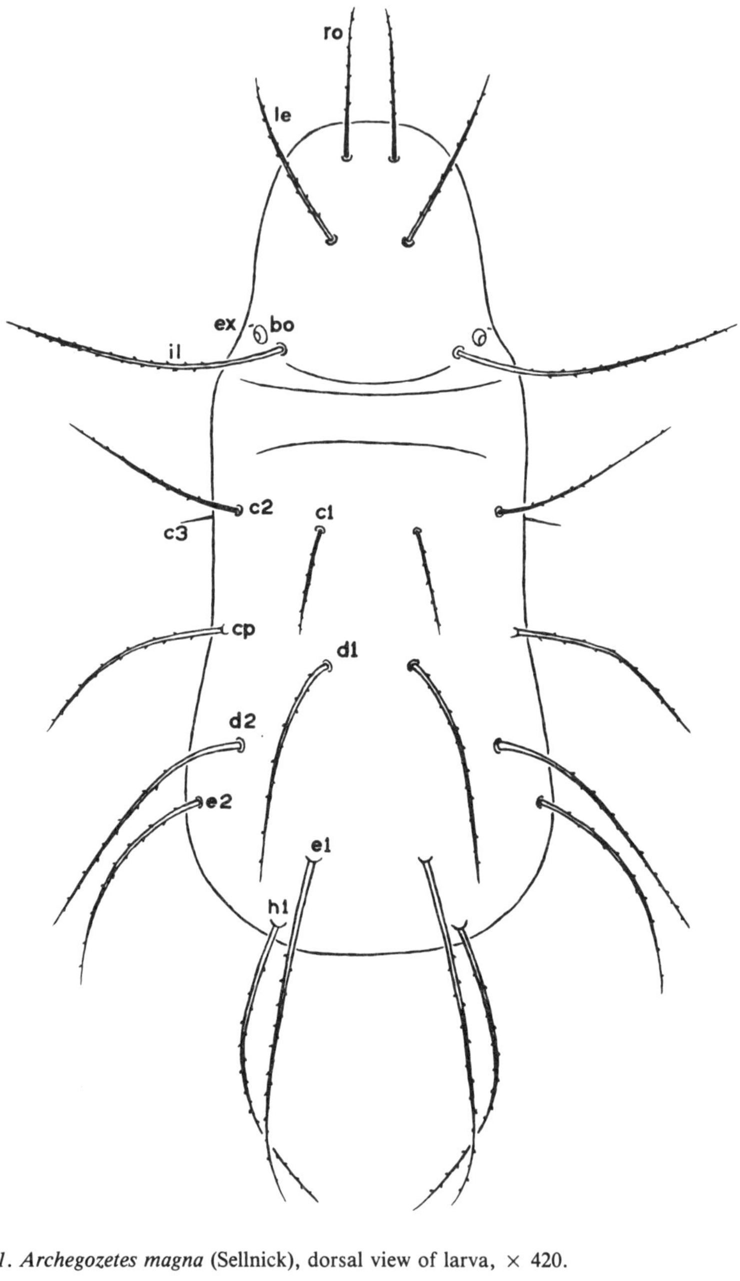

Fig. 71. Archegozetes magna (Sellnick), dorsal view of larva, × 420.

Postembryonic development. – At first sight, the immature stases differ from the adults by the smaller size, the paler colour, and the relatively smaller breadth. Larvae, proto-, deuto- and tritonymphs presen distinct mutual differences in measurements, as well as in numbers of genital papillae (0. 1, 2, 3 pairs respectively) and legs (larvae are hexapod). The larvae are white; the nymphs, particularly the tritonymphs, are yellowish, and present a pair of brown spots caused by the dark contents of the latero-opisthosomatic glands. The larvae are relatively narrow; the relative breadth increases in the course of postembryonic development, but even the tritonymphs are narrow in comparison with the adult. In the larva prodorsum, infracapitulum, epimera and paraproctal lips are more or less scleritized.

The prodorsum of the larva (fig. 71) presenats rostral, lamellar and interlamellar setae, as in subsequent stases. The sensillus, however, is vestigial and the bothridium little developed; close to the bothridium, the larva presents a very small exobothridial seta which is absent in subsequent stases. Starting from the protonymph (fig. 72), the sensillus is present in its normal shape; evidently, the trichobothridial regression is not at an advanced stage.

The notogastral setae f_2 and ps_1 are absent in the larva, but appear in the protonymph; f_1 is absent in all stases. The development of the paraproctal setae (cf. fig. 73A–D) is represented by the following formulae: pseudanal setae, from larva to adult 2–3–3–3–3; adanal setae, from protonymph to adult 0–3–3–3; anal setae, from deutonymph to adult 0–2–2. Anal and adanal segments do not present setae at the base level: *Archegozetes* is characterized by atrichosy at two levels. The development of the genital setae (cf. fig. 73A–D), from protonymph to adult, is represented by the formula 1–4–6–7. The development of the epimeral setae is represented by the following formulae: larva (2–1–2), protonymph (3–1–2–1), deutonymph (3–1–3(2)–3), tritonymph and adult (3–1–3–3); there are vertitional absences in the case of seta 3c of the deutonymph.

In the course of postembryonic development, the phanerotaxy of the legs changes in the following way. Tarsus I of the larva (fig. 74C) presents: (*ft*), ϵ, ω_1, (*tc*), (*p*), (*u*), *s*, (*pv*) and (*a*); the protonymph adds the solenidion ω_2; solenidion ω_3 is of deutonymphal origin; the adult adds two pairs of additional setae. On tibia I, φ_1 is of larval, φ_2 of deutonymphal origin. On tarsus IV, *ft'*, (*p*), (*u*) and (*pv*) are protonymphal; (*tc*), *s*, and (*a*) are of deutonymphal origin (fig. 74A); an additional seta can be vertitionally present in the adult (figs. 69B, 74A).

4.4. *Morphology and postembryonic development of* Fortuynia yunkeri *Van der Hammen*

Fortuynia yunkeri (family Fortuyniidae) is a representative of the suborder Circumdehiscentiae (higher Oribatid mites). Species of the genus occur in the tropical and subtropical marine littoral, and the first known species was de-

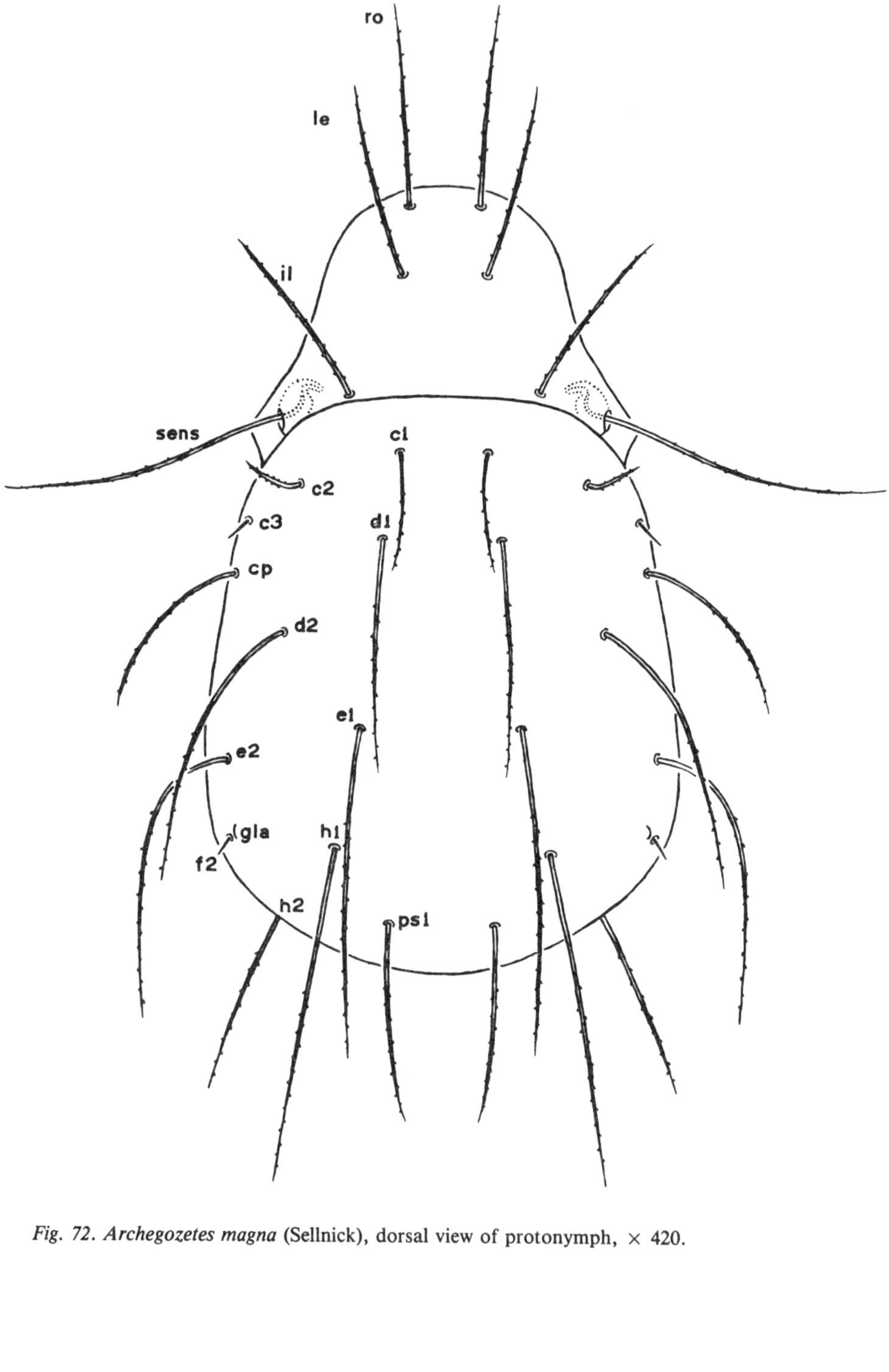

Fig. 72. Archegozetes magna (Sellnick), dorsal view of protonymph, × 420.

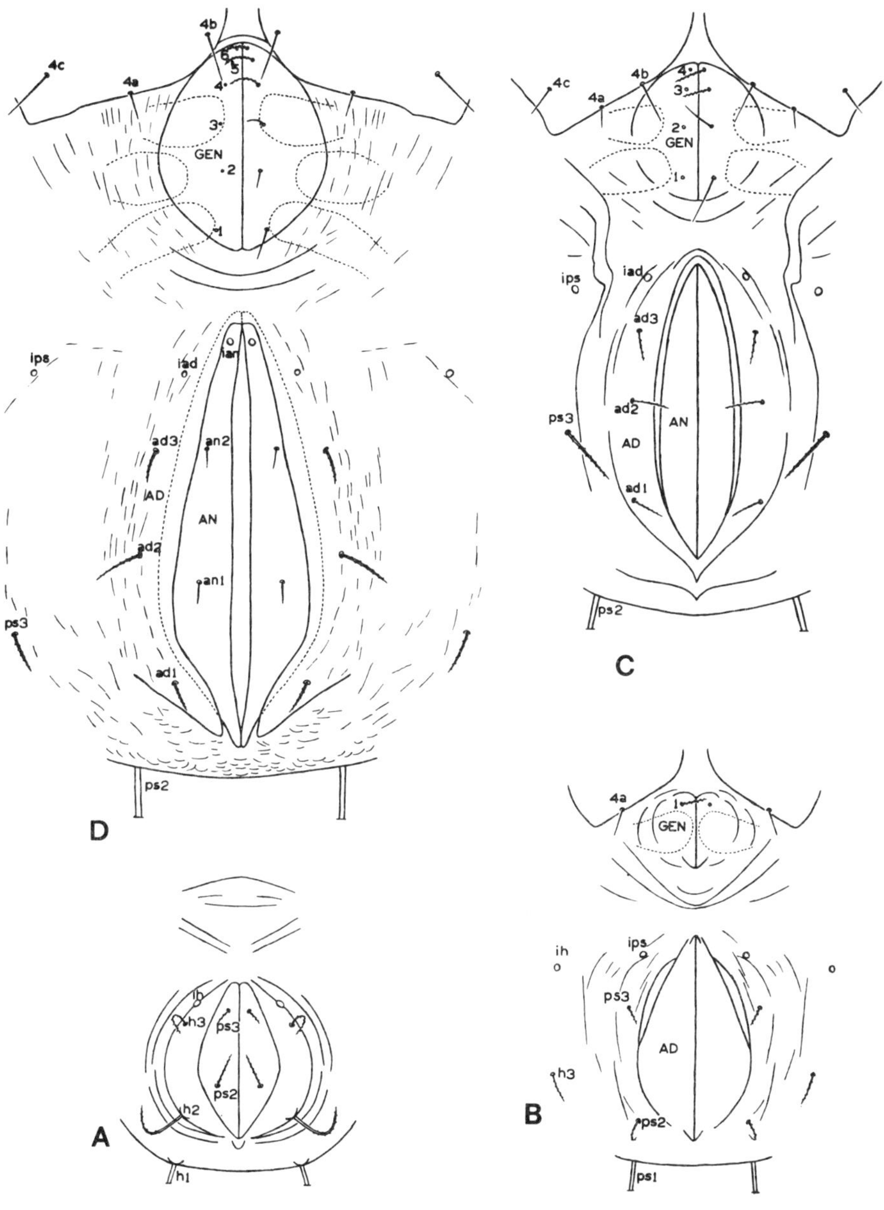

Fig. 73. Archegozetes magna (Sellnick), postembryonic development of ano-genital region; A, larva; B, protonymph; C, deutonymph; D, tritonymph; A–D, × 315.

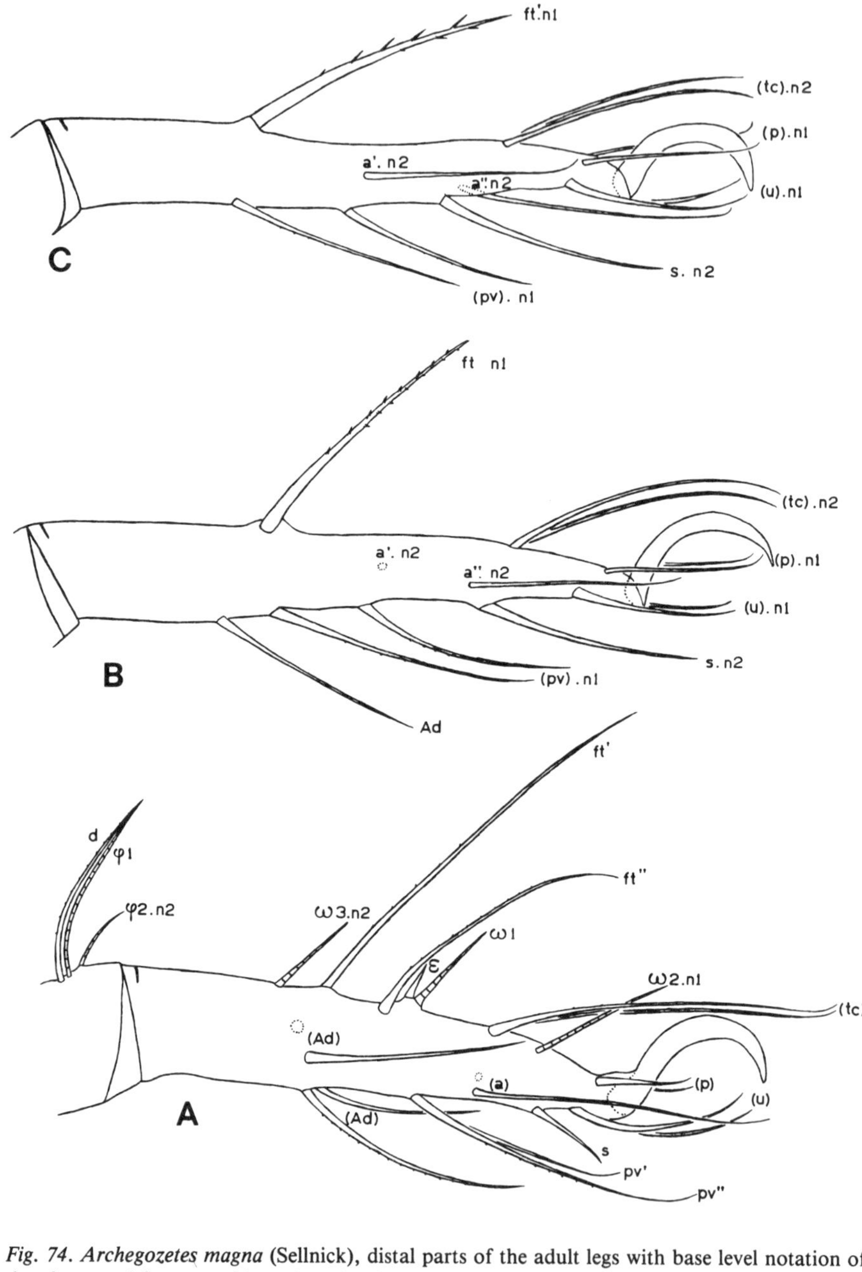

Fig. 74. Archegozetes magna (Sellnick), distal parts of the adult legs with base level notation of the phaneres (phaneres without base level notation are larval); A, right leg I, lateral (antiaxial) view; B, right leg IV, lateral (paraxial) view; C, left leg IV, lateral (antiaxial) view; A–C, × 380.

scribed by me from New Guinea (Van der Hammen, 1960). The present subsection deals with another species of the genus, of which a more numerous material (including immature specimens) was available; it was described by me from the littoral of the Pacific coast of Panama (Van der Hammen, 1963). Species of the genus are particularly interesting because of the adaptation to life in the intertidal zone, i.e., the evolutionary development of an air reservoir, termed 'Van der Hammen's organ' by Luxton (1967: 80, 83; cf. also Luxton, 1986).

Measurements. – Length of the female 0.450–0.480 mm (average 0.460), breadth 0.270–0.300 mm (average 0.290). Length of the male 0.435–0.465 mm (average 0.450), breadth 0.260–0.275 mm (average 0.265).

Habitus and colour. – There is a distinct secondary sexual difference between adult males and females: the notogaster of the male is more flattened than in the female, and distinctly extends beyond the anal opening, whilst the outline of the male ventral plate is triangular (it is rounded in the female).

In reflected light, the specimens are glossy and brown, with a light spot in the anterior central part of the notogaster. The white setae, particularly the posteriors, are relatively large. In transmitted light, local differences in saturation of the brown colour become evident. Genital and anal covers (with anterior locking piece), rutellum, trochanter III and IV, femur (with the exception of a small basal part), genu, and a small basal part of the tarsus, are dark brown. The posterior part of the ventral plate is distinctly darker than the anterior; in figs. 75B and 78A these two parts are separated by a dotted line. The light spot in the anterior part of the notogaster is slightly irregular in shape; two dark pigment spots are visible under this area. Generally, the notogaster is slightly darker in the posterior part; the notogastral border is nearly colourless, although in the male this light border is not visible in the latero-posterior part. The central part of the prodorsum is lighter than the lateral parts. The coxisternal region is as light as the anterior part of the ventral plate; the mentotectum is slightly darker.

Tegument. – The body is covered by a continuous cerotegument layer of irregular thickness, which is difficult to separate. It is colourless, not birefringent, and for the greater part only faintly granulate. In the lateral parts of the idiosoma, viz., the lateral border of the prodorsum, the lateral part of the podosoma, and the lateral part of the ventral plate (partly covered by the notogastral tectum), the cerotegument layer is thicker and more granulate; I succeeded in separating, as a whole, the cerotegument layer covering the granulate border of the prodorsum.

The cuticle is thin, and deformation of the notogaster by slight pressure is easy. In transmitted light, a densely punctate structure is visible, attributable to a porosity of the cuticle under the epiostracum. In a well-defined area of the idiosoma, the cuticle is granulate (also after removal of the cerotegument); this area comprises the lateral border of the prodorsum, the lateral part of the podosoma (the precise extent is represented in figs. 76A, 77A), and the posteri-

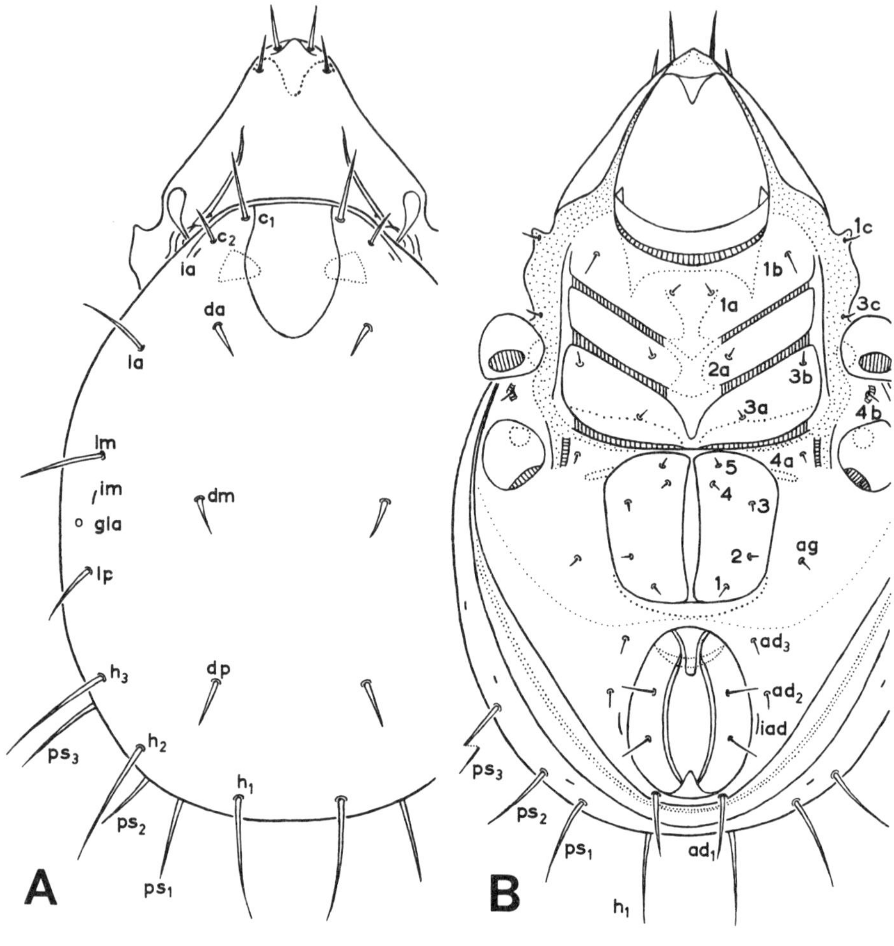

Fig. 75. Fortuynia yunkeri Van der Hammen, female; A, dorsal view; B, ventral view; A, B, × 220.

or border of the third epimeral furrow (cf. fig. 75B). The lateral and posterior border of the notogaster is nearly colourless, very thin and flexible, and constitutes a special type of tectum that will be dealt with below.

Prodorsum. – The prodorsum presents a pair of converging lamellar ridges (figs. 75A, 76A: *cl*) which are slightly curved inwards at the end, and extend only up to half-way the lamellar setae (the faint inner ridge *ci*, described for *Fortuynia marina*, is not visible in the present species). The sensilli are directed upwards; they are also slightly curved inwards. The exobothridial setae are extremely small, and about reduced to their alveolus; the interlamellar setae are reduced to the same degree. The rostral setae are relatively long and strong, inserted in small depressions of the rostrum; the lamellar setae are smaller. The lateral border of the prodorsum is granulate, with the exception of the rostrum.

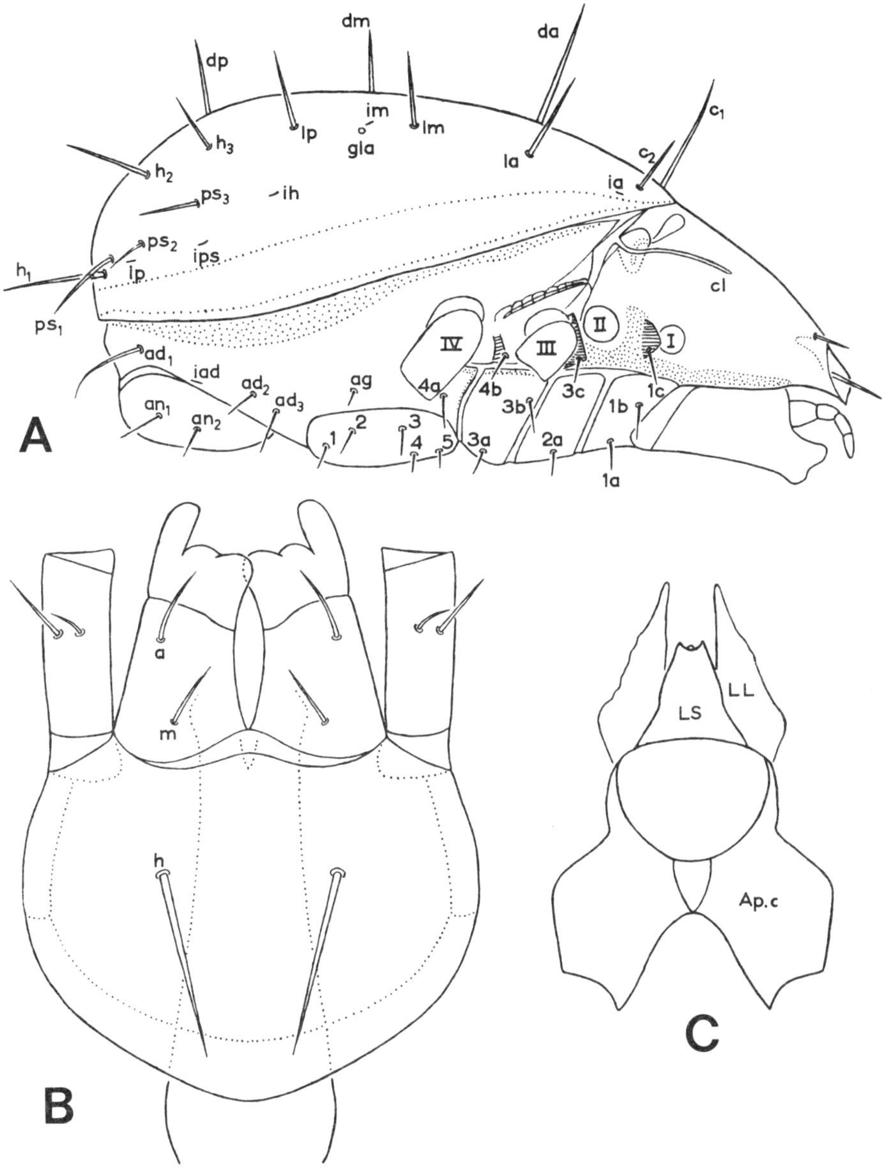

Fig. 76. Fortuynia yunkeri Van der Hammen, female; A, lateral view; B, ventral view of gnathosoma (lateral lips omitted); C, dorsal view of gnathosoma (chelicerae removed; the lateral lips have moved to a more antiaxial position); A, × 220; B, C, × 630.

Notogaster. – The notogaster does not present areae porosae. There is a remarkable lateral and posterior notogastral tectum. The nearly colourless cuticle, bordering the notogaster, is very thin and flexible (but scleritized and shining), and constitutes a fold; cross-sections of this fold are represented in figs. 77B–E, an oblique section in fig. 77F. The thin, colourless tectum is firmly connected with the ventral plate, so that the notogaster can only be separated (even after heating with lactic acid) by tearing of the cuticle. At the place of attachment of notogaster and ventral plate, the border of the ventral plate constitutes a dark tube with thickened wall, which is lengthwise open; the tube is hidden by the tectum. Because the tectum is not bordered by a limbus, it differs from the lateral and posterior tecta found in Achipteriidae and other families. The granulate cerotegument of the lateral region continues under the tectum and extends up to the border of the dark tube. The tube is probably connected with the system of air reservoirs described below.

The shape of the notogaster is represented in figs. 75, 76A and 78. There is a remarkable secondary sexual difference. In ventral view, the enclosure of the female ventral plate is rounded (fig. 75B), whilst it is triangular and much more narrow in the male ventral plate (fig. 78A). In lateral view, the notogaster of the male appears to be flattened and distinctly extended posteriorly (fig. 78B). The surface of the notogaster is smooth; it presents no ridges. There is no lenticula, but a simple light spot which is slightly variable in shape, particularly in the posterior part.

Notogaster seta c_3, of which the alveolus is still distinctly present in *Fortuynia marina*, is absent in *F. yunkeri* (in one or two specimens, I have observed a very small light spot at the ancestral place of insertion, near *ia*, but at one side only). The adult, consequently, is bideficient (28 notogastral setae), but because it is certain that c_3 is one of the lacking setae, the notation of unideficiency can be applied. The posterior setae are distinctly longer than in *Fortuynia marina*. Seta ps_1 occupies a position which is more lateral than in the last-mentioned species: it is not inserted ventral to h_1, as in many other Oribatid mites, but lateral to it. The places of insertion of the notogastral setae are constituted by distinct light spots in the brown notogaster. Lyrifissures *ia*, *im*, *ip*, *ih*, and *ips* are easily visible, just as the latero-opisthosomatic gland and its orifice (*gla*); their arrangement is represented in figs. 75, 76A, 78. The disposition, particularly that of *im*, is slightly variable: *im* can be anterior to the orifice of the latero-opisthosomatic gland, or lateral to it; its orientation can be lenghtwise, oblique or transverse. The position of *ip*, *ih* and *ips* is also slightly variable because of the secondary sexual differences in the shape of the notogaster.

Ventral plate. – Genital and anal openings are large and also rather close to each other. The genital opening of the male appears to be relatively smaller (especially shorter) than that of the female. There is one pair of aggenital setae. As mentioned above, there is a striking secondary sexual difference in the shape of the ventral plate. In the female (fig. 75B), the ventral plate is rounded, the lateral borders are remote from the anal opening, whilst the posterior border is closely parallel to it. In the male (fig. 78A), the ventral plate is nearly triangular, the lateral borders are close to the anal opening, whilst the plate is distinctly elongated posterior to the anal opening. In ventral view, the ventral

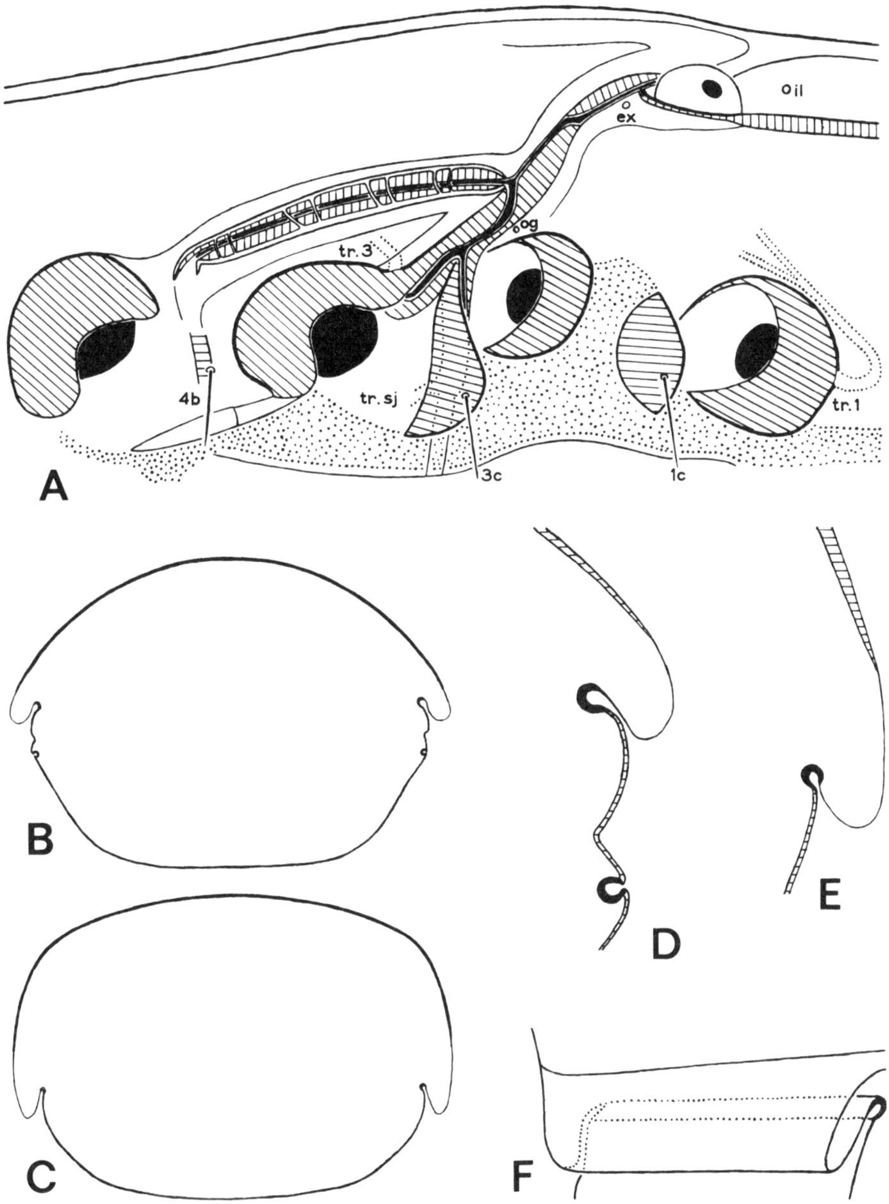

Fig. 77. Fortuynia yunkeri Van der Hammen, female; A, lateral region of the podosoma of the right side (legs removed), with acetabula, orifices of tracheae (stigmata), and Van der Hammen's organ; B, transverse section of idiosoma between legs III and IV; C, transverse section of idiosoma, posterior to legs IV; D, detail of the right side of fig. B with notogastral tectum and section of Van der Hammen's organ; E, detail of the right side of fig. C; F, posterior part of opisthosoma, showing notogastral tectum; A, D–F, × 503; B, C, × 202.

plate of the male presents also a pair of distinct ridges, between the lateral border and the adanal setae; these ridges constitute the borders of a steep declivity, and the adanal setae are inserted on an elevated area. In lateral view (fig. 78B), it is also distinct that the male ventral plate is posteriorly produced. As a result of the different shape of the ventral plate, the male adanal setae occupy a distinctly different position.

Anal region. – There are two pairs of anal setae. The anal region is particularly interesting because of the large, dark anterior locking-piece; in figs. 75B, 80B, D, this piece is represented in different views, together with the smaller, nearly colourless (not sclerotized) posterior locking-piece.

Genital region. – Each genital cover presents five setae, a number which is not common among Oribatid mites. In fig. 75B, the setae are denoted by the numbers 1–5; setae 4 and 5 occupy a marginal position. Females with extended ovipositor were not present among the material; judging from the invaginated condition, the ovipositor is short.

Coxisternal region. – Epimeral furrow 2 and sejugal furrow are less distinct in the median part. The sternal furrow is indistinct in the anterior part. The region of the furrows is distinctly bordered by a granulate area (as represented in fig. 75B). The chaetotaxy of the epimera is represented by the formula 3–1–3–2. The numbers recorded here for epimera 3 and 4 (3–2, and not 2–3) are based on a study of the postembryonic development (see below).

Lateral region of podosoma. – This region is particularly interesting because of the presence of highly specialized structures, unknown in other groups of Oribatid mites. It is represented in figs. 76A and 77. Pedotecta I and II are both present and well-developed; they bear the places of insertion of the epimeral setae *1c* and *3c* respectively. There is a discidial ridge (bearing epimeral seta *4b*) between acetabula III and IV. The distinctly bordered granulate area is represented in figs. 76A and 77A. The orifice of a gland is present above acetabulum II.

The lateral region is characterized by the presence of a chitinous tube above acetabulum III, connected with the disjugal furrow; this complicated structure, first described by me in 1960 (for the case of *Fortuynia marina*), was named 'Van der Hammen's organ' by Luxton (1967). This chitinous tube, reinforced by an irregular system of chitinous rings, appears to be lengthwise open, so that it can be regarded as covered by two tecta. A section of the tube is represented in fig. 77D. The tube is connected with the disjugal furrow that has developed into a canal, covered by tecta. A section through the disjugal canal resembles more or less a section through the chitinous tube, although the disjugal tecta are alternating. The main tectum of the superior part extends from the posterior border, the tectum of the inferior part from the anterior border of the canal. In the inferior part, the disjugal canal divides into two branches. The posterior branch enters acetabulum III and reaches the stigma of trachea 3; the anterior branch continues under pedotectum II and reaches the stigma of the sejugal trachea. It is evident that these complicated structures

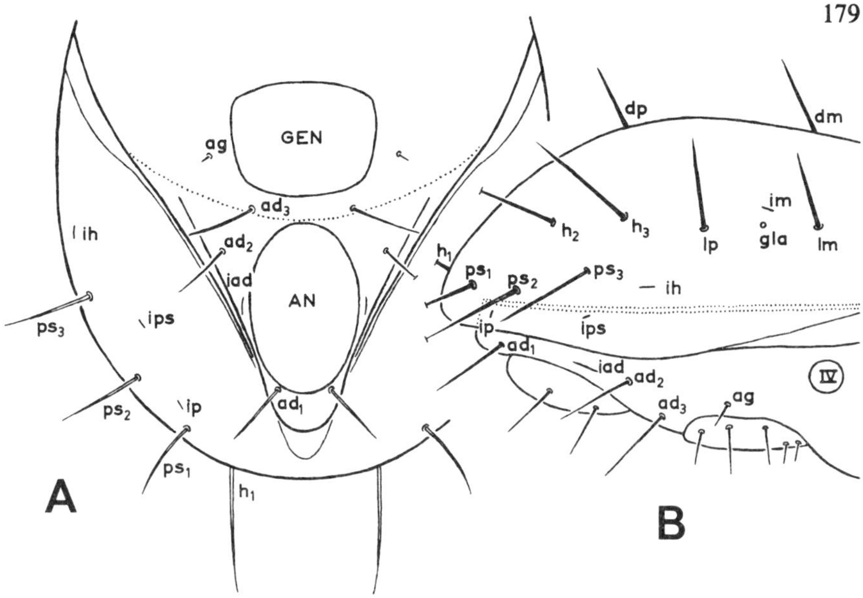

Fig. 78. Fortuynia yunkeri Van der Hammen, male, posterior part of opisthosoma; A, ventral view; B, lateral view; A, B, × 252.

function as an air reservoir during high tide (it may be repeated here that species of the genus *Fortuynia* live in the intertidal zone). It is probable that during high tide the air in the disjugal canal is connected with the air in the circumgastric tube under the notogastral tectum (cf. figs. 77A–F).

Gnathosoma. – The infracapitulum presents the usual ventral setae (*h*, *m* and *a*) which are rather long. It is diarthric: the labiogenal articulation reaches to the base of the palp. The rutellum is broad and dark; the paraxial lobes overlap each other (fig. 76B). A dorsal view of the infracapitulum, with labrum and capitular apodeme, is represented in fig. 76C. The cervix is only faintly sclerotized and the cheliceral grooves are indiscernible.

Palp. – The palp is represented in fig. 80C; its chaetotaxy is represented by the normal formula 0–2–1–3–9. There is no double horn: the solenidion of the tarsus is free. The eupathidia (*acm*, *ul'*, *ul''* and *su*) are relatively large; *su* is slightly curved; of the two ultimal setae, the antiaxial one (*ul''*) is closest to the subultimal seta *su*.

Legs. – There is a distinct secondary sexual difference in the fourth leg, pertaining to the shape of the tibia, and to some setae of genu, tibia and tarsus. In tibia IV of the male, the basal part is distinctly shorter than in the female, whilst it is narrow and suddenly broadens at the paraxial side, near *v''* (fig. 79D). In lateral view, this sudden broadening looks like a ridge (fig. 79C, E). The following setae are differently shaped in the male: *l'* of genu IV, *l'*, *v''*, *v'* of tibia IV, and *pv'* of tarsus IV; they are large and sickle-shaped, with the exception of *v''* of the tibia, which is only larger than in the female.

Leg I is represented in fig. 79A, femur II in fig. 79B, parts of male leg IV

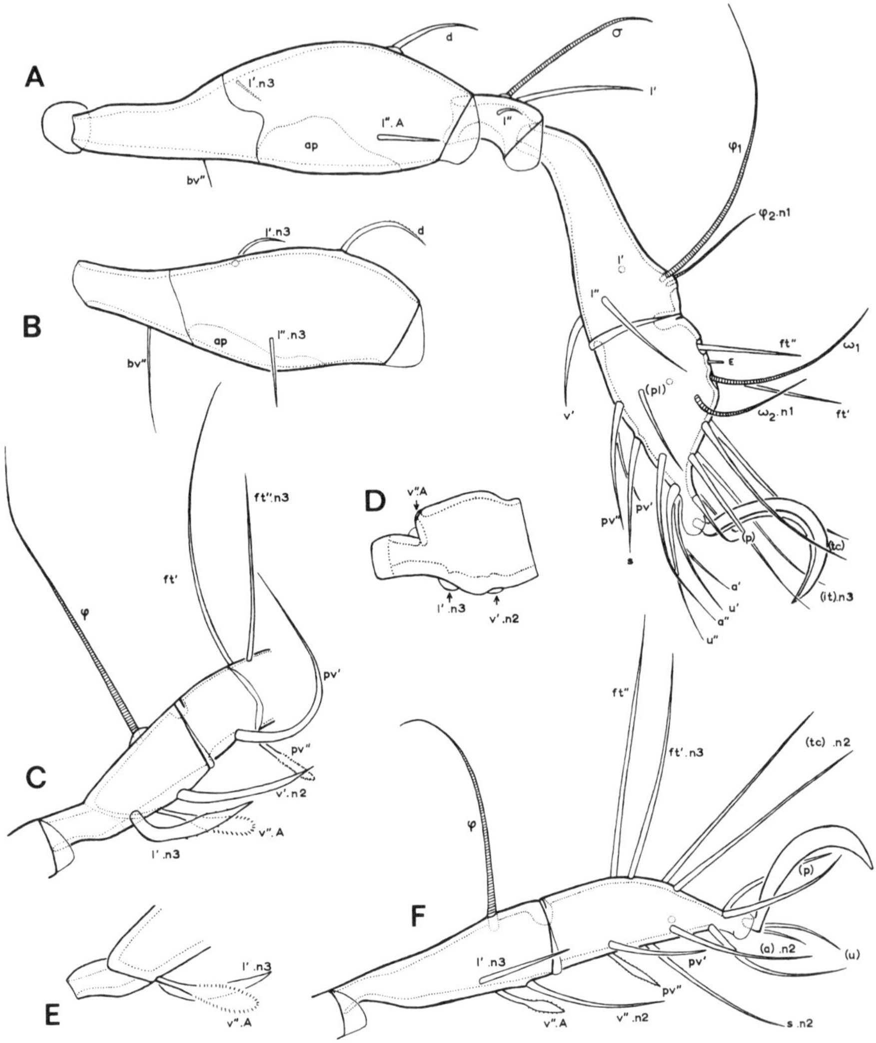

Fig. 79. Fortuynia yunkeri Van der Hammen; A, lateral (antiaxial) view of right leg I of female; B, lateral (antiaxial) view of femur of right leg II of female; C, lateral (antiaxial) view of tibia and part of tarsus of left leg IV of male; D, ventral view of right leg IV of male; E, lateral (paraxial) view of part of tibia of right leg IV of male; F, lateral (antiaxial) view of tibia, tarsus and apotele of left leg IV of female; A–F, × 397.

in fig. 79C–E, and the distal part of female leg IV in fig. 79F; in the figures, the setae are accompanied by the usual notations. The femora present two areae porosae; in the anterior legs, one area porosa is antiventral, whilst the other is paradorsal. Leg chaetotaxy is represented by the following formulae:

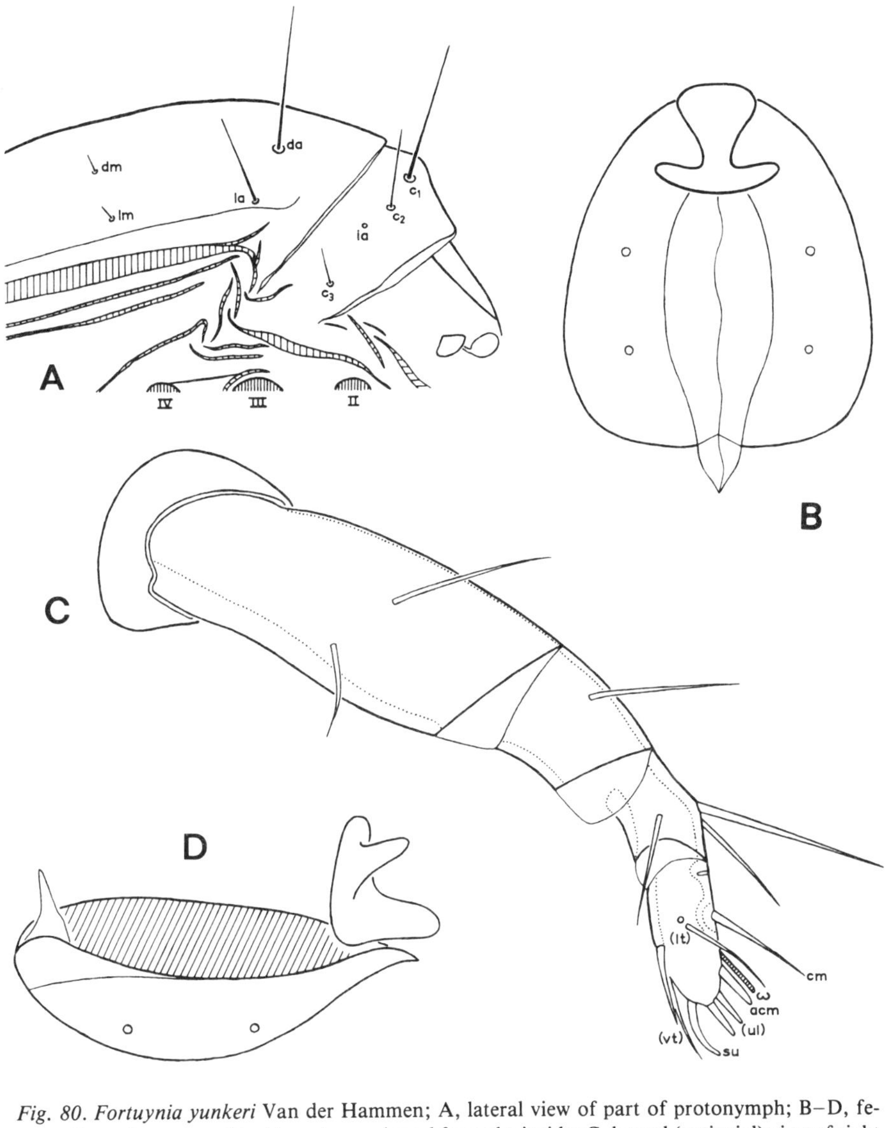

Fig. 80. Fortuynia yunkeri Van der Hammen; A, lateral view of part of protonymph; B–D, female; B, anal covers and locking pieces, viewed from the inside; C, lateral (antiaxial) view of right palp; D, lateral view of anal covers and locking pieces, separated from idiosoma; A, × 348; B, D, × 461; C, × 1196.

I (1–4–2–3–18–1), II (1–4–2–3–15–1), III (2–3–2–3–15–1), IV (1–2–2–3–13–1); solenidiotaxy is represented by the following formulae: I (1–2–2), II (1–1–2), III (1–1–0), IV (0–1–0). All solenidia are free (not coupled with a dorsal seta). Seta *bv''* of femur I is faintly developed; this seta is much more strongly developed in the case of femur II. The proral setae (*p*) of tarsus I are the only eupathidia of the legs. Tarsus IV lacks the iteral setae (*it*). The large unguis of the claws presents a thin inner border, particularly in the case of leg I (as represented in fig. 75A).

Postembryonic development. – The lengths of larva, protonymph, deutonymph and tritonymph (based on one specimen of each stase) are: 0.240, 0.275, 0.345, 0.450 mm. The nymphs are represented in figs. 80A and 81; in the single larva of the material studied by me, the central region of the notogaster was damaged, and could not be represented. The immature stases belong to the plicate type. The folds are, however, less numerous than in related groups (Achipteriidae, etc.); they are mainly present in the lateral (fig. 80A) and ventral (fig. 81B–D) regions. A distinct dorsal area, with the setae *da*, *la*, *dm*, *lm*, *dp*, and *lp* is, however, smooth. The colour is brown (the paraproctal lips are hardly different in colour); the large folds and the narrow border of the progenital opening are distinctly darker. The genu and the greater part of the femur of the legs are darker than the other segments.

There is a continuous layer of colourless cerotegument, which is for the greater part thin (although slightly thicker in the lateral region). The cuticle is soft and thin. As mentioned above, an important dorsal part of the notogastral region is smooth; this area does not stain with Unna's polychrome blue, but because the lateral border (cf. fig. 80A) is rather distinct, it probably represents a transition to a dorsal sclerite. The microsculpture of this dorsal area consists of small dots. Genital sclerites are absent or indistinct; the paraproctal lips are not distinctly bordered in the posterior part; there are no areae porosae.

The prodorsal region of the immature stases is posteriorly bordered by the disjugal fold. Other folds are present between the bothridia, anterior to the lamellar setae, and in the lateral region. A pair of lamellar ridges run from the bothridia in rostral direction, curve inwards, and slightly surpass the lamellar setae. In the larva, the lamellar ridges are interconnected between the lamellar setae; posteriorly, they nearly reach the disjugal fold. The bothridium is simple, the sensillus club-shaped. The rostral setae are rather long, the lamellar setae shorter; a vestige of the interlamellar seta is present, on either side, in the fold between the bothridia.

The notogastral region presents only a few folds. Anteriorly, it is bordered by the disjugal fold. There is also a fold posterior to setae c_{1-3}, which is more or less connected with lateral folds that continue ventrally; there are several other lateral folds and a fold posterior to *dp*. There is also a large fold posterior to the paraproctal lips. As mentioned above, the central dorsal area is smooth. The nymphs are unideficient: there are thirty notogastral setae. In the larva ps_1, ps_2 and ps_3 are lacking, whilst h_3 is represented by a vestige, so that the larval number is twenty-two. The notogastral setae are simple in shape; c_1, *da* and h_2 are long (strikingly longer than the remaining setae). In all nymphal stases, ps_1 occupies a lateral position (not ventral to h_1). The cupules and the orifice of the latero-opisthosomatic gland are difficult to discern because

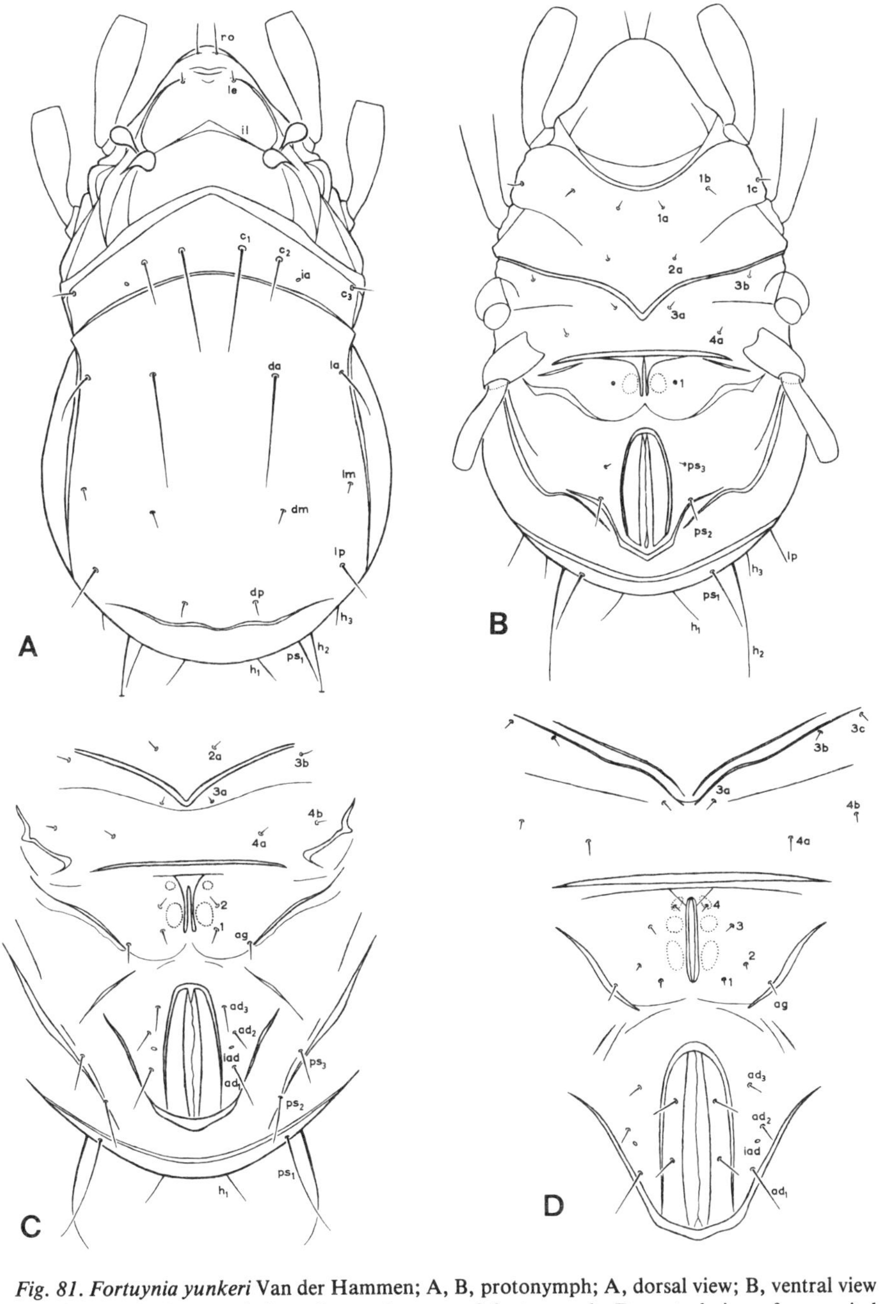

Fig. 81. Fortuynia yunkeri Van der Hammen; A, B, protonymph; A, dorsal view; B, ventral view (less flattened); C, ventral view of posterior part of deutonymph; D, ventral view of anogenital region of tritonymph; A–D, × 288.

of the folds in the lateral and posterior parts of the notogastral region; *ia* (situated between c_2 and c_3) is, however, easily visible.

The paraproctal lips are hardly darker than the surrounding cuticle; the posterior border is indistinct. The lips bear no setae in larva, protonymph and deutonymph, so that there is paraproctal atrichosy at three levels. The anal and adanal region of the three nymphal stases are represented in fig. 81B–D. The cupule *iad* is easily visible; its base level is deutonymphal.

The genital lips are not bordered in the lateral part, whilst the posterior border is indistinct. As usual, there are 1–2–3 pairs of genital papillae in proto-, deuto- and tritonymph respectively. The development of the genital setae, from protonymph to adult, is represented by the formula 1–2–4–5. The development of the paired aggenital seta, from protonymph to adult, is represented by the formula 0–1–1–1.

The ventral region of the podosoma presents a large sejugal fold (which is represented by a furrow, bordered by two folds), a large fold anterior to the progenital opening, and smaller, incomplete separations between epimera 1 and 2, and 3 and 4. The development of epimeral setae is represented by the following formulae: larva (2–1–2), protonymph (3–1–2–1), deutonymph (3–1–2–2), tritonymph (3–1–3–2).

The development of leg chaetotaxy from larva to adult (or from protonymph to adult, in the case of leg IV) is represented by the following formulae. Trochanter: I and II (0–0–0–0–1); III (0–0–1–1–2); IV (0–1–1–1). Femur: I (2–2–2–3–4); II (2–2–3–4–4); III (2–2–2–3–3); IV (0–2–2–2). Genu: I and II (2–2–2–2–2); III (1–1–1–1–1); IV (0–2–2–2). Tibia: I (3–3–3–3–3); II (2–2–2–2–3); III (1–1–1–2–3); IV (0–1–2–3). Tarsus: I (16–16–16–18–18); II and III (13–13–13–15–15); IV (7–12–13–13). The development of the solenidiotaxy of the legs, from larva to adult (or from protonymph to adult, in the case of leg IV) is represented by the following formulae. Genu: I, II and III (1–1–1–1–1); IV (0–0–0–0). Tibia: I (1–1–2–2–2); II and III (1–1–1–1–1); IV (0–1–1–1). Tarsus: I (1–2–2–2–2); II (1–1–2–2–2); III (0–0–0–0–0); IV (0–0–0–0).

4.5. *The morphology of the gnathosoma of* Hermannia convexa (*C.L. Koch*)

The gnathosoma of both groups of mites is relatively small, whilst its structure can often be extremely complicated; as a result of this, a complete understanding of gnathosomatic architecture is often difficult. In the present subsection, an attempt is made to study the Oribatid gnathosoma as a whole, together with its relations to the adjacent parts of the idiosoma (rostrum, camerostome, podocephalic canal). By means of a detailed description, and of detailed illustrations in various projections, I have tried to prepare a model of the gnathosoma, and to demonstrate which parts are external and which internal.

Material for this kind of study must be relatively large (dissection must not be too difficult), numerous specimens must be available, and the gnathosoma

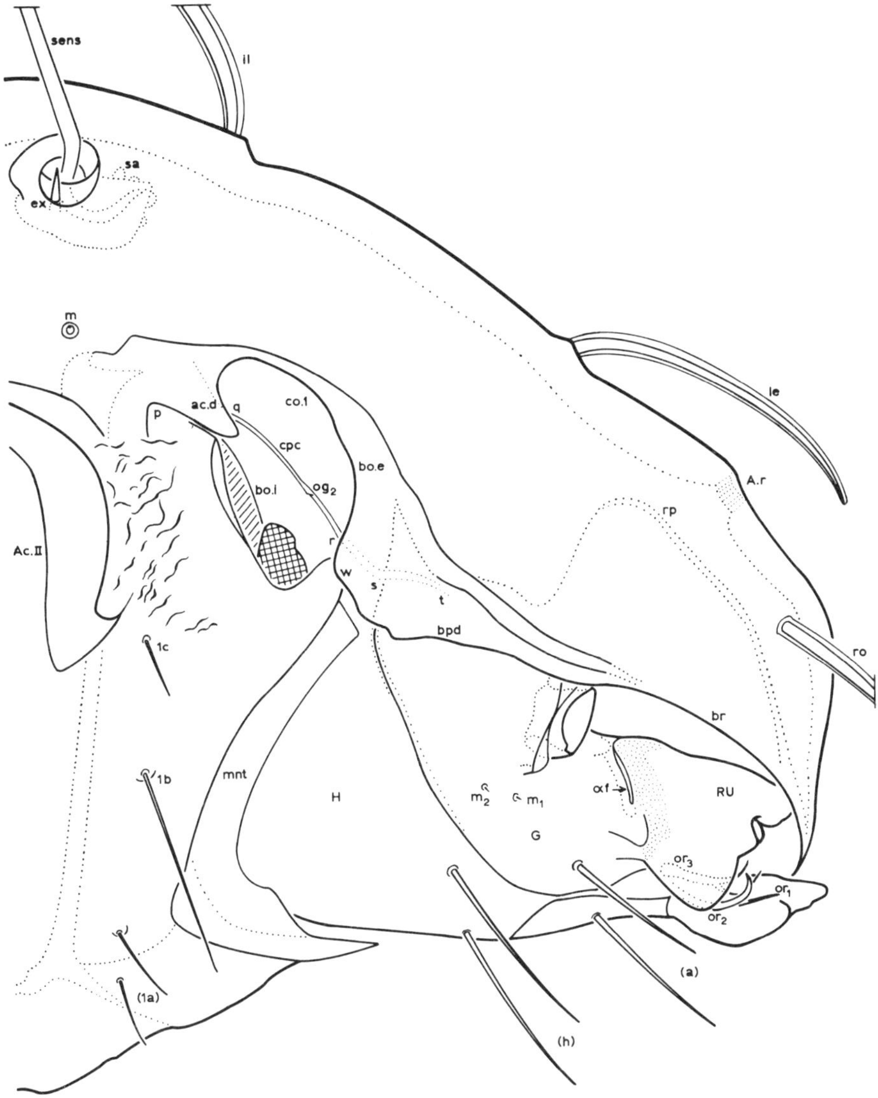

Fig. 82. Hermannia convexa (C.L. Koch), adult; lateral (oblique) view of anterior part of prosoma (chelicerae, greater part of palp, and legs removed; the figure is prepared after a specimen of which the left part was separated by a longitudinal section); × 295.

should preferably represent a more or less normal or typical condition (not too primitive and not too specialized). For this reason, I have selected a large species of the suborder Nothrina, viz., *Hermannia convexa* (C.L. Koch), which is about 1.5 mm in length, and of which I had collected numerous specimens at the type-locality in Bavaria. Since the original publication of this study (Van der Hammen, 1968), the morphological terminology has been further developed by me (e.g., Van der Hammen, 1970c: 20–21n; 1972: 14; 1976: 41; 1980: 95), whilst the internal anatomy of the prosomatic glandular system was studied by Woodring (1973) and Alberti & Storch (1974).

Topography. – A general survey of the region described in this subsection is given in fig. 82, in which the gnathosoma is represented in a natural position, partly hidden by the camerostome. The rostrophragma (to which the cheliceral frame is attached) is visible through the rostral cuticle. Because the specimen of fig. 82 is viewed slightly from below, the podocephalic canal is also visible; it is partly inside acetabulum I, and its further course can be followed, thanks to the slightly transparent condition of the cuticle, under the prodorsum. For ease of survey, the description of these complicated structures is subdivided into the following paragraphs: rostrum, camerostome, cheliceral frame, chelicerae, internal and external parts of the gnathosoma, infracapitulum, podocephalic canal, and spatial relations.

Rostrum. – The cheliceral frame is dorsally attached to a structure named 'cloison rostrale' by Grandjean (1934: 355); this term is translated here by rostrophragma. Rostral tectum (cf. Grandjean, 1934: 355; 1952: 33; 1958: 430, fig. 2) is replaced here by rostrum, because this structure is much more complicated than a simple tectum; it consists of the anterior part of the prodorsum, of the rostrophragma, and of the cavity enclosed by them (this cavity probably contains elements of precheliceral origin).

In fig. 83C, the rostrum is represented in dorsal view (slightly flattened). It presents a marginal limbus, of which the base is indicated by λ. The rostrum presents a dorsal fenestra (*fr*), constituted by a distinctly vaulted light spot. It is supposed that, under this light spot, elements of precheliceral origin are present (homologous with the naso of some primitive Oribatid mites and Endeostigmata). Posterior to the fenestra, a very small, median, rostral area porosa is distinctly visible (fig. 83B, C: *A.r*). Because the greater part of the prodorsum is already constituted by an area porosa with smaller 'pores', the rostral area with distinctly larger pores is possibly composed of sacculi; the structure is not studied here in detail. The posterior border of the rostrophragma (fig. 83C: *brp*) is visible through the cuticle. Because it is important to establish the exact extension of the rostrophragma, its posterior border (to which the cheliceral frame is attached) is also represented in two different orientations in fig. 83A, B; in both figures, the space between prodorsum and rostrophragma is indicated as rostral cavity (*rc*). Fig. 83A represents a ventral view of the camerostome; in fig. 83B, the rostrum is separated and represented in posterior view. Fig. 83A–C illustrates that the rostrophragma is constituted by a thin lamella (it is chitinous) that starts from the base of the rostral limbus (λ) and extends in posterior direction (cf. also figs. 82, 93: *rp*). Its position is nearly parallel to the prodorsum, and only at a small distance from it. Laterally, it is attached to the lateral border of the rostrum.

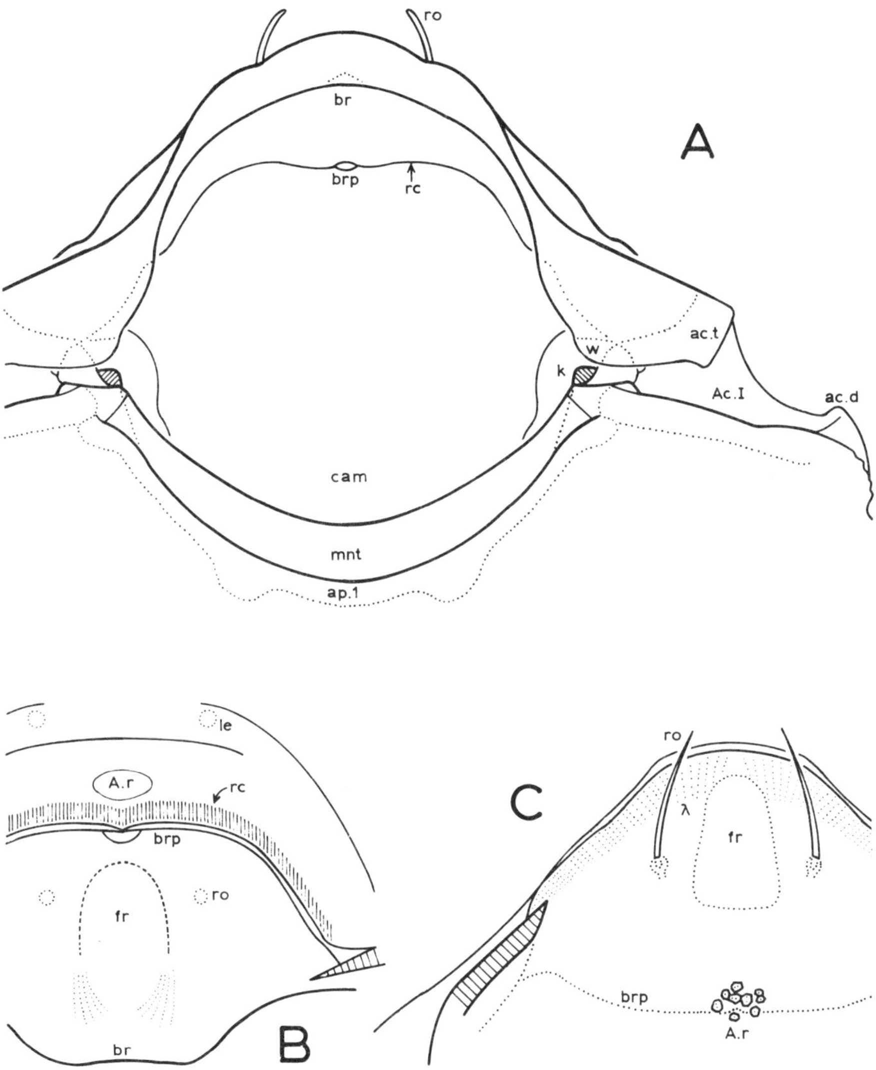

Fig. 83. Hermannia convexa (C.L. Koch), adult; A, ventral view of camerostome; B, rostral part of camerostome (inner side), viewed from behind; C, dorsal view of slightly flattened rostrum (orientation nearly horizontal); A–C, × 235.

Camerostome. – The camerostome is best visible in ventral view, after separation (removal) of the gnathosoma from the idiosoma. It is represented in this orientation in fig. 83A. It contains the rostrophragma. Its anterior and lateral borders are constituted by the borders of rostrum and prodorsum. Its posterior border is constituted by the mentotectum (*mnt*) which protects the posteri-

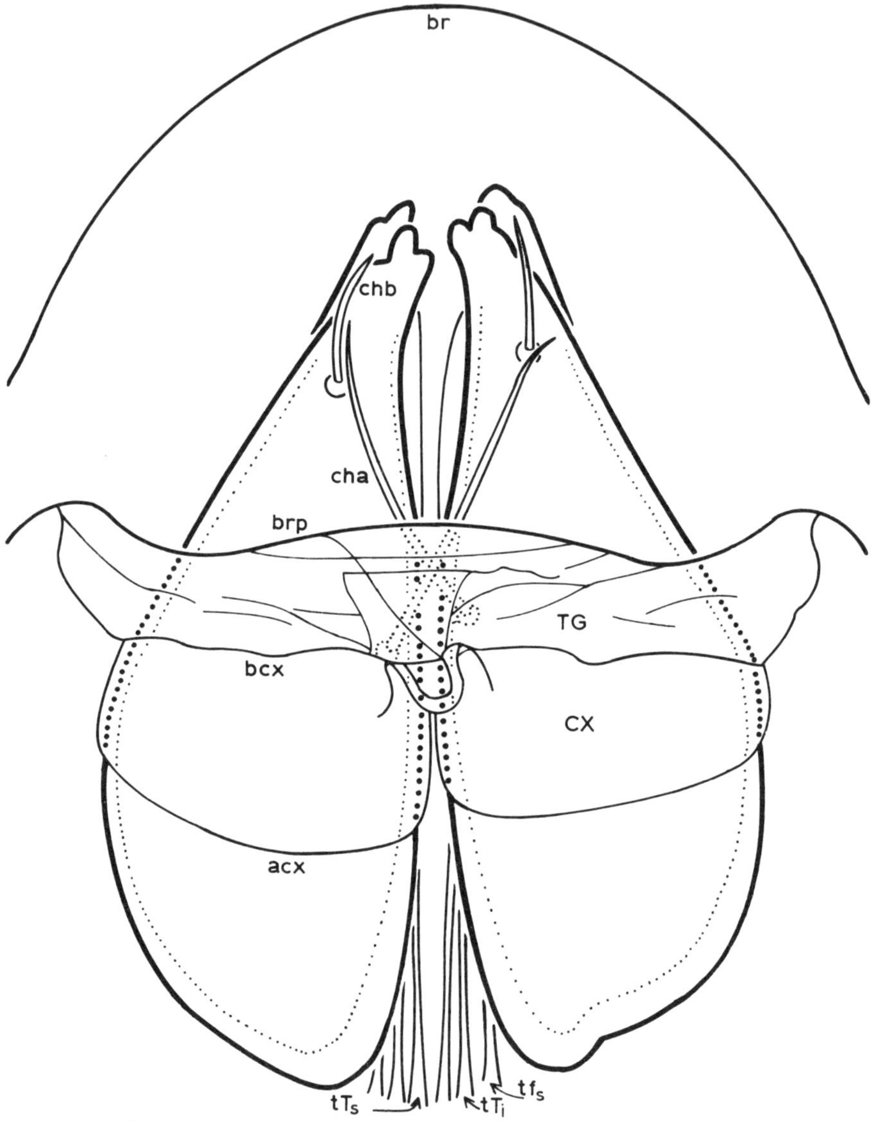

Fig. 84. Hermannia convexa (C.L. Koch), adult; dorsal view of gnathosoma (observed through the cuticle of the prodorsum) with completely retracted chelicerae (although the outline of the rostrum and the posterior border of the rostrophragma are drawn, the gnathosoma is supposed to be uncovered); × 490.

or border of the infracapitulum. Posterolaterally, on either side, a condyle for the articulation of the gnathosoma is present; the transverse axis of articulation of the gnathosoma passes through these condyles.

Cheliceral frame. – One of the two main parts of the gnathosoma is constituted by the cheliceral frame (Grandjean's 'cadre mandibulaire'; cf. Grandjean, 1936b: 416; 1957b: 239). Because of its membraneous nature and hidden position, the cheliceral frame is generally not studied in detail; it is represented here, in different views, in figs. 84–85. In my opinion, the cheliceral frame comprises the epimeral (coxisternal) region of the chelicerae, and the frontal wall of the prosoma; it extends from the rostrum (or, in the present case, the rostrophragma) to the infracapitulum (line of attachment: *at*). The coxal regions of the chelicerae constitute the cheliceral sheaths. The frontal part of the cheliceral frame (with dorsal or superior position) is named tegulum; as mentioned above, it is not regarded here as part of the cheliceral epimeron, but as part of the body-wall superior to it.

The cheliceral frame of *Hermannia convexa* is represented in figs. 84 and 85 (in dorsal, lateral and frontal view). The figures were prepared after a specimen with completely retracted chelicerae; only in this position, tegulum and cheliceral sheaths are sufficiently visible through the prodorsal cuticle. Although the prodorsum was still present, it is supposed in the figures that the gnathosoma is uncovered.

In retracted position, the tegulum (figs. 84, 85: *TG*) is recognizable as an unpaired membrane, attached to the posterior border (*brp*) of the rostrophragma; this unpaired condition can be verified by studying transverse folds of the membrane. The lateral fold drawn, on either side, in fig. 84, does not represent the lateral border; because the cheliceral frame completely separates the body-cavity from the exterior, it extends up to the lateral side of the prosoma.

The bases of the cheliceral sheaths (figs. 84, 85: *bcx*) are in the sternal part of the cheliceral frame. The sheaths represent the coxal regions of the chelicerae (figs. 84, 85: *CX*). The distal part of the sheaths is attached to the chelicerae; the line of attachment is indicated as *acx* (a further description of the attachment is given in the section on the chelicerae). The sternal part of the cheliceral frame between the cheliceral sheaths is very narrow (cf. fig. 84). In the mediodorsal part, the transition from tegulum to sheaths is distinctly reinforced. This reinforcement is certainly connected with evagination and invagination of chelicerae and sheaths. The chelicerae are represented in figs. 84 and 85 in retracted position. When they are protracted, the sheaths evaginate. The protracted condition must be regarded as the original position of a coxal region; the capacity of evagination is supposed to constitute a derived character state.

Chelicerae. – The Oribatid chelicerae (cf. Grandjean, 1947b: 314) comprise three cheliceral segments, viz., trochanter, body of the chelicera, and apotele. The trochanter, which still has a proximal position in some primitive Oribatid mites, has in most Oribatida developed a ventral and mainly paraxial position.

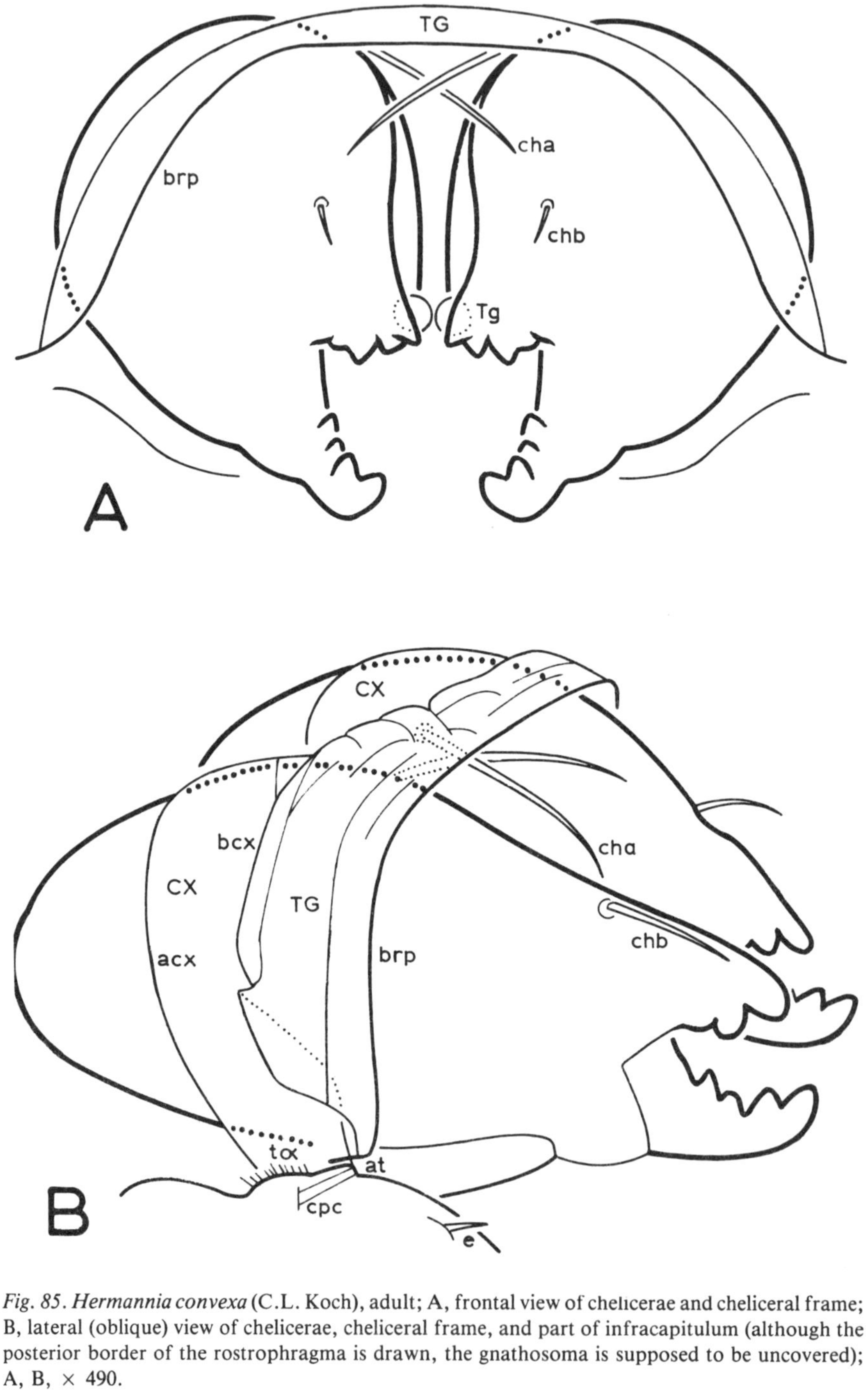

Fig. 85. Hermannia convexa (C.L. Koch), adult; A, frontal view of chelicerae and cheliceral frame; B, lateral (oblique) view of chelicerae, cheliceral frame, and part of infracapitulum (although the posterior border of the rostrophragma is drawn, the gnathosoma is supposed to be uncovered); A, B, × 490.

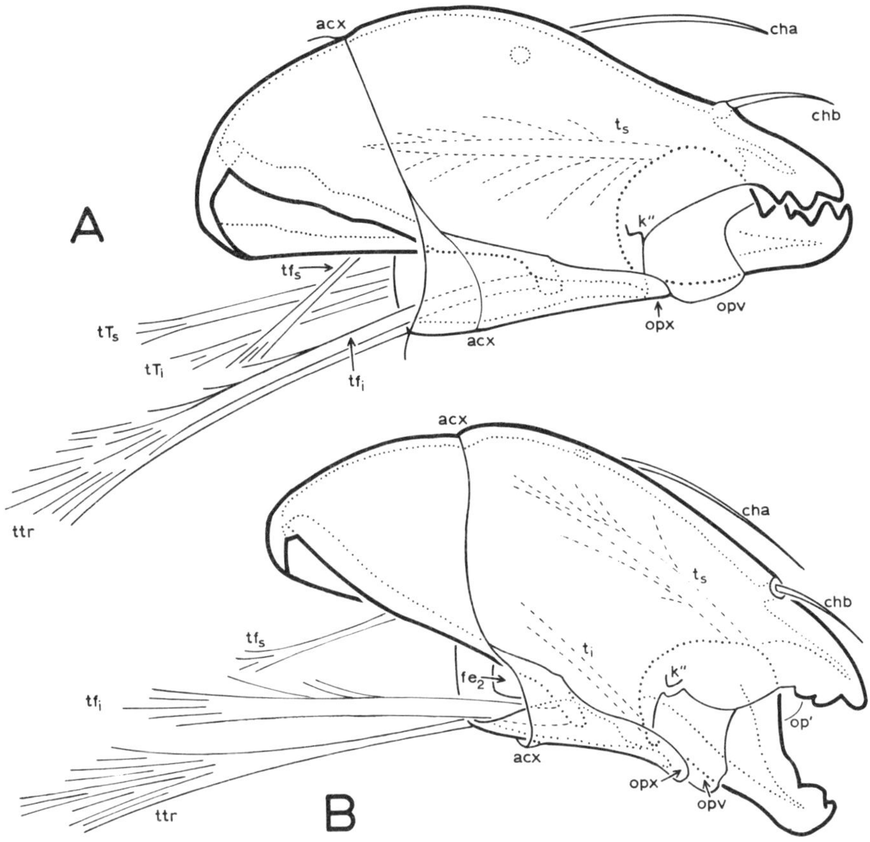

Fig. 86. Hermannia convexa (C.L. Koch), adult; A, lateral (antiaxial) face of right chelicera; B, idem, but oblique view of a chelicera of which the coxal region (the cheliceral sheath) is nearly completely removed; A, B, × 370.

The chelicera of *Hermannia convexa* is represented in figs. 86 and 87. The apotele, or movable jaw, is slightly longer than the fixed jaw; it turns on a transverse axis passing through the condyles *k'* and *k''*. The body of the chelicera presents the usual setae *cha* and *chb*; there is no lyrifissure. The trochanter is reduced to a thin paralateroventral chitinous plate with indistinct borders. The cheliceral sheath is attached to the chelicera according to the line *acx*; part of the body of the chelicera is, consequently, internal. The external part presents a granulate structure which is nearly completely absent in the case of the internal part (some granules are found internally, close to *acx*).

The paraxial surface of the chelicera (fig. 87A) presents a number of interesting characters. Trägårdh's organ (*Tg*) is present in its usual shape and position. Its cuticle appears to be a continuation of the cheliceral sheath. It

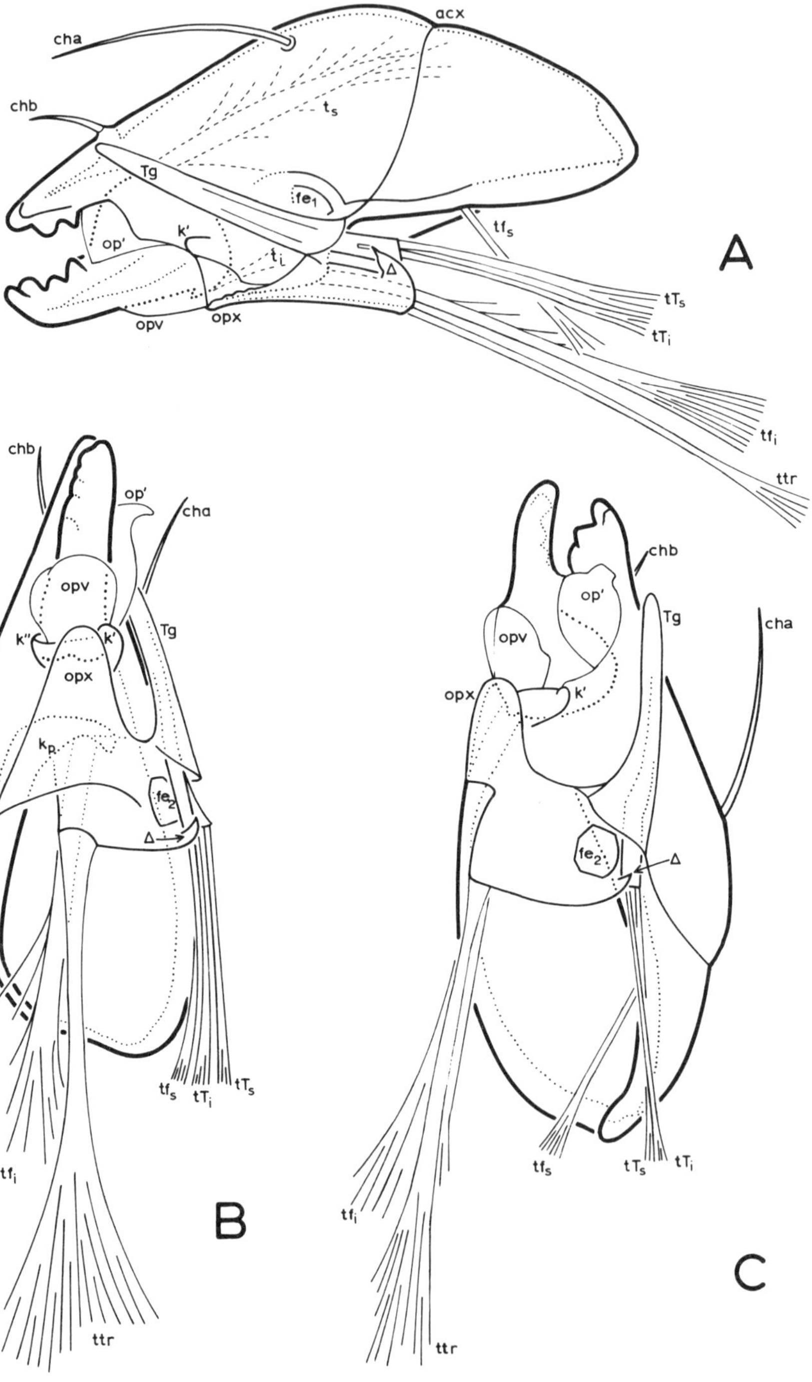
cha
acx
chb
t_s
Tg
fe_1
tf_s
op'
k'
t_i
A
opv
opx
tT_s
tT_i
tf_i
ttr
chb
op'
cha
opv
k''
k'
Tg
opx
k_p
fe_2
acx
tf_s
tT_i
tT_s
tf_i
B
ttr
chb
op'
Tg
cha
opv
opx
k'
fe_2
tf_s
tT_s
tT_i
tf_i
C
ttr

presents an internal sclerite which articulates with a tooth (fig. 87A–C: Δ) in the trochanter; this tooth is difficult to discern, and must be studied in various orientations. The sclerite presents two tendons, inserted on its base (tT_s and tT_i); as a result of this, the organ will certainly be movable. The body of the chelicera presents a paraxial fenestrate area (fe_1), dorsal to the base of Trägårdh's organ. Another fenestrate area (fe_2) is present in the paraxial region of the trochanter, ventral to the base of Trägårdh's organ.

The following tendons are present, besides tT_s and tT_i. The apotele presents the usual two tendons t_s and t_i. Two tendons are attached to the body of the chelicera (tf_s and tf_i); they are considered homologous with the tendons of the femur of palp and legs. It is uncertain whether k_p (at the border between the body of the chelicera and the trochanter; cf. fig. 87B) represents a condyle for the articulation of these two segments. One long tendon is inserted on the trochanter (*ttr*). Operation of the tendons (as well as cooperation) is difficult to reconstruct. I return to this problem in the paragraph on spatial relations. The notations of the tendons are those introduced and standardized by me (Van der Hammen, 1966). They are partly different from those introduced by Grandjean (1947b: 314–316, fig. 2); his t_s, t_i, $t\pi$, tv_s, tv_i correspond with my t_s, t_i, tf_s, tf_i, *ttr* respectively.

The ventral as well as the paraxial surfaces of the chelicerae present membraneous swellings which are difficult to study. They apparently represent continuations of the arthrodial membranes, and they are named oncophyses. There are at least three cheliceral oncophyses (cf. figs. 86, 87), which are named here: coxal oncophysis of the chelicera (*opx*), ventral oncophysis of the body of the chelicera (*opv*), and paraxial oncophysis of the body of the chelicera (*op'*); these terms can be shortened to coxal, ventral and paraxial oncophysis. Grandjean (1959: 360) introduced the name 'bouffissure paraxiale' in the case of *op'*, and 'intumescence ventrale' in the case of *opv*. The exact shape of the three oncophyses is represented in figs. 86 and 87. It is not impossible that Trägårdh's organ represents an oncophysis with an internal sclerite; its cuticle is also a continuation of an arthrodial membrane (of the cheliceral sheath).

Internal and external parts of the gnathosoma. – As mentioned above, the cheliceral frame constitutes the body-wall between rostrum and infracapitulum, and consequently separates internal and external parts. Because dissection of the gnathosoma results in destruction of the cheliceral frame, the exact border between internal and external parts is, as a rule, not sufficiently studied. For this reason, data with reference to the topography of this border and the elements concerned in it, are summarized and repeated once more in

←

Fig. 87. Hermannia convexa (C.L. Koch), adult, right chelicera; A, lateral (paraxial) face; B, antilateroventral face (cheliceral sheath partly removed); C, paralateroventral face (cheliceral sheath removed); A–C, × 370.

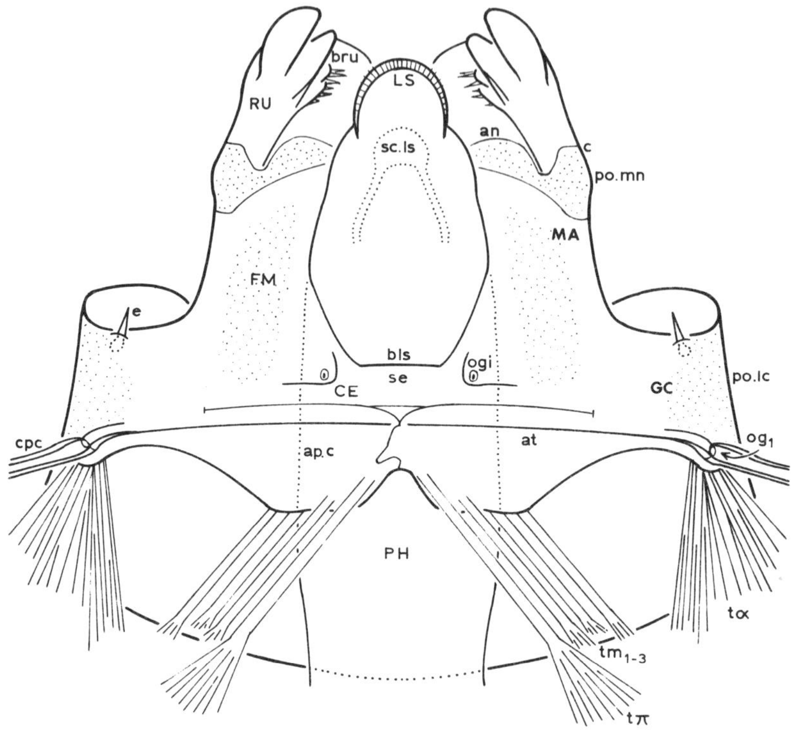

Fig. 88. Hermannia convexa (C.L. Koch), adult; dorsal view of infracapitulum (chelicerae removed); × 370.

the present paragraph. Starting from the base of the rostral limbus (λ), the wall is constituted by rostrophragma and cheliceral frame (cf. fig. 93). The internal part of the chelicerae is bordered by the line *acx* (cf. fig. 86); the body-cavity continues into cheliceral sheaths and chelicerae, and these certainly can be evaginated by hydrostatic pressure. The cheliceral frame is attached to the infracapitulum according to the line *at*, at the posterior border of the podocephalic canal (the canal is external). As mentioned below, the internal part of the chitinous infracapitulum consists of the capitular apodeme. The body-cavity continues of course into the infracapitulum. Between the infracapitulum and the coxisternal region of the podosoma, the body-wall is constituted by an arthrodial membrane, protected by the mentotectum.

Infracapitulum. – The infracapitulum is represented in figs. 88–91. Its main parts are mentum, malapophyses, large lateral ridges and cervix. According to Grandjean's classification (Grandjean, 1957), the infracapitulum of *Hermannia* is stenarthric (the labiogenal articulation is present, and is directed obliquely backwards). In ventral view (fig. 89), the mentum is approximately

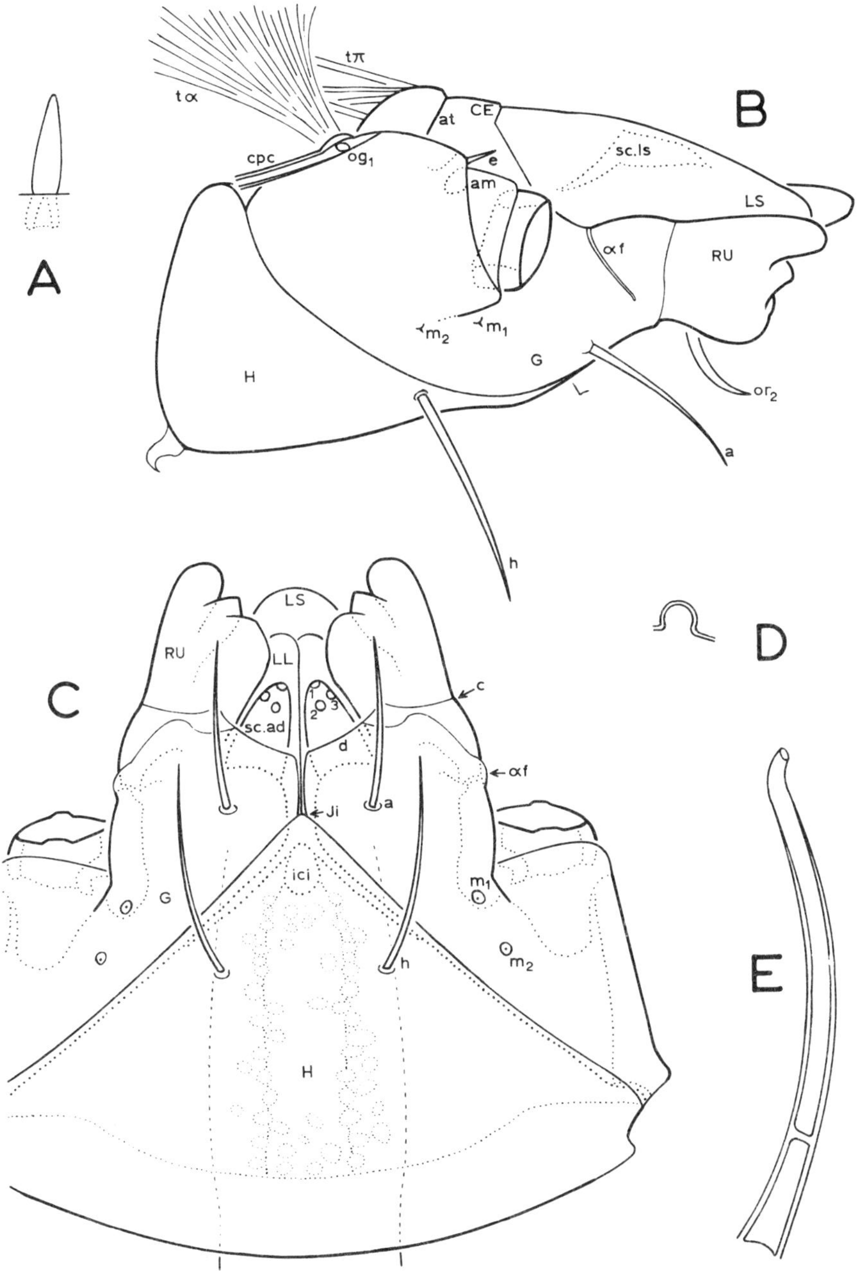

Fig. 89. Hermannia convexa (C.L. Koch), adult; A, supracoxal seta (*e*) of palp; B, lateral view of infracapitulum; C, ventral view of infracapitulum; D, transverse section of podocephalic canal in the region of acetabulum I; E, duct (dg_1) of anterior podocephalic gland; A, D, × 1315; B, D, E, × 370.

triangular; in posterior view (fig. 90B), its surface appears to be strongly curved (convex). The mentum bears the usual pair of setae *h*; numerous round spots, in the median part, represent the attachments of pharyngeal muscles.

The paired region, anterior to mentum and cervix, and antilateral to the lateral lips, represents the malapophyses (ventral sclerite termed gena). These have a dorsal and a ventral surface. Because a section through the pair of malapophyses is V-shaped, the dorsal surface is paraxial, the ventral surface antiaxial. The ventral surface (genae) bears one pair of anterior infracapitular setae (*a*), and two pairs of medians (m_1, m_2); m_1 and m_2 are reduced to very small and thin setae, although the places of insertion are distinct. The surface of the rutella (hypertrophied setae with manducatory function) is a continuation of the surface of the malapophyses. The border between the two (collum) is discernable because the surface of the malapophyses is for the greater part porous. The border can also be discerned in polarized light. A manubrial zone, belonging to the malapophyses, is present at the base of the rutella; it is bordered by the collum (fig. 88, 89C, 90A: *c*) and by an antiaxial fissure (figs. 89A, C, 90A: *αf*); in dorsal view it is easily recognizable because it represents an area porosa. The rutellum belongs to the ordinary atelebasic type: it is large, but the base does not extend to the infrabuccal fissure. Dorsally, it presents the usual brush (figs. 88, 90A: *bru*).

The lateral lips present the usual adoral sclerite which articulates with the gena. This articulation was termed postadoral by Grandjean (1957b: 257). The base of the adoral sclerite is dorsal to the distal border of the malapophyses. This is illustrated in fig. 90A, representing an infracapitulum in dorsal view, from which one lateral lip was removed. The area of articulation is bordered by *d* (the distal or anterior border of the malapophyses) and *b.sc* (the base of the adoral sclerite). In frontal view (fig. 91A), the lateral lips are oval in outline; a section near the base (fig. 91B) demonstrates that the lips are less high at that level, and that the sclerite *sc.ad* is ventral and antiaxial. The usual three adoral setae are present; or_1 and or_3 are bifurcate, or_2 is simple.

The cervix is short and broad. Posteriorly, it is bordered by the line *at*, from which the capitular apodeme is starting. Anteriorly, it is partly bordered by the base of the labrum (figs. 88, 90A: *bls*). Cervix and malapophyses pass into one another, and the transitional part is steeply sloping. Posterolateral to the labral base (i.e., lateral to the capitular saddle: *se*) a pair of foramina is found, with the paired orifices of the infracapitular glands. Part of the posterior border of the cervix is occupied by the podocephalic canal which will be discussed below; its posterior border coincides with the line of attachment of the cheliceral frame. The podocephalic canal is distinct only in the lateral part of the infracapitulum; its course soon becomes indistinct. The dorsal surface of the infracapitulum presents three areae porosae, viz., manubrial (fig. 88: *po.mn*) and laterocoxal (*po.lc*), and the area porosae of the cheliceral groove (*FM*). The last-mentioned groove is a depression in which the chelicerae can move. The supracoxal seta (figs. 88, 89A, B, 90A, 91: *e*) has a dorsal position. It is represented separately in fig. 89A.

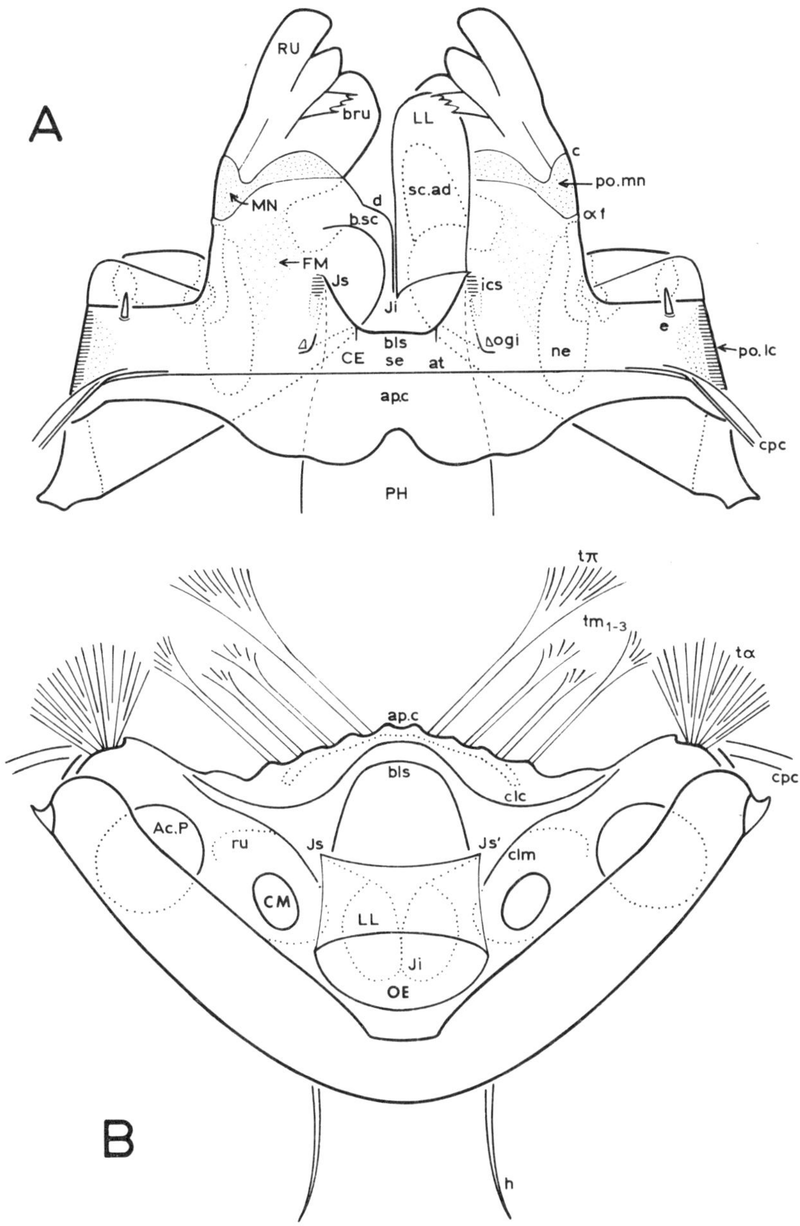

Fig. 90. Hermannia convexa (C.L. Koch), adult; A, dorsal view of infracapitulum (chelicerae, labrum, and left lateral lip removed; mentum omitted; the soft parts of the right lateral lip are slightly schematized); B, posterior view of infracapitulum; A, B, × 370.

The capitular apodeme presents a posterior indentation in the median part. The apodeme is constituted by an internal lamella, on which a number of tendons are inserted, viz., on either side, a large antiaxial tendon (figs. 88–91: $t\alpha$), three small median tendons (tm_{1-3}), and a paraxial tendon ($t\pi$).

The mouth is trifid in shape (cf. figs. 90–91); the commissures are *Js*, *Js'* and *Ji*. The commissural indurations are represented in fig. 90A (the two superior indurations *ics*) and in fig. 89C (the unpaired inferior induration *ici*). The labrum (*LS*) is relatively large. Its base can best be studied after its separation from the infracapitulum. This base (fig. 90A: *bls*) is situated in the anterior declivity of the cervix: its dorsal border is situated posterior to the superior commissures (*Js*, *Js'*). These superior commissures are situated at the border of cervix and malapophyses. The distal part of the labrum presents very small teeth. The labrum has an internal sclerite (*sc.ls*) from which a pair of tendons (fig. 93: *tls*) extends in posterior direction. The sclerite has two posterior branches which probably articulate with the superior indurations. The labrum is operated by the tendons, and its axis of articulation passes through the commissures.

When the infracapitulum is studied in posterior view, the internal cavities and various external contour lines can be discerned. In fig. 90B, *clc* represents the dorsal contour line at the level of the cervix, *clm* that at the level of the manubrium. It is evident that the transition from cervix to malapophyses is steeply sloping. The cavity enclosed by cervix and mentum (the pharyngeal chamber) is connected with the internal parts of the labrum; the trapezoid base of the labrum is represented in fig. 90B (*bls*). The central cavity (the pharyngeal chamber) continues into the malapophyses as malar cavity (fig. 90B: *cg*); it is also connected, on either side, with the coxal cavity and the acetabulum of the palp. The malar cavity is, moreover, connected with the central cavity of the rutellum; a section at the level of the rutellar base is represented in fig. 91B.

Podocephalic canal. – As mentioned above, the podocephalic canal extends from acetabulum I up to the gnathosoma. Its median part (fig. 82) is visible through the cuticle of the prodorsum. The anterior part of the canal, in the laterodorsal part of the infracapitulum, is visible after separation of the gnathosoma from the idiosoma and the camerostome, and after removal of the chelicerae; after separation, an arthrodial membrane with part of the canal remains generally attached to the lateral border of the infracapitulum. It is evident that the canal has an acetabular, a prodorsal, an arthrodial, and an infracapitular part. The full course of the podocephalic canal (including the connections between the above-mentioned parts), and the glands debouching into it, is described in the following paragraph.

The course of the podocephalic canal in acetabulum I is represented in fig. 82. Its posterior extremity *p* is under the acetabular tooth *ac.d*. Its course from *q* to *r* can be discerned when the acetabulum is studied slightly from below. A small orifice og_2 is present in the anterior half of the acetabular part. A trans-

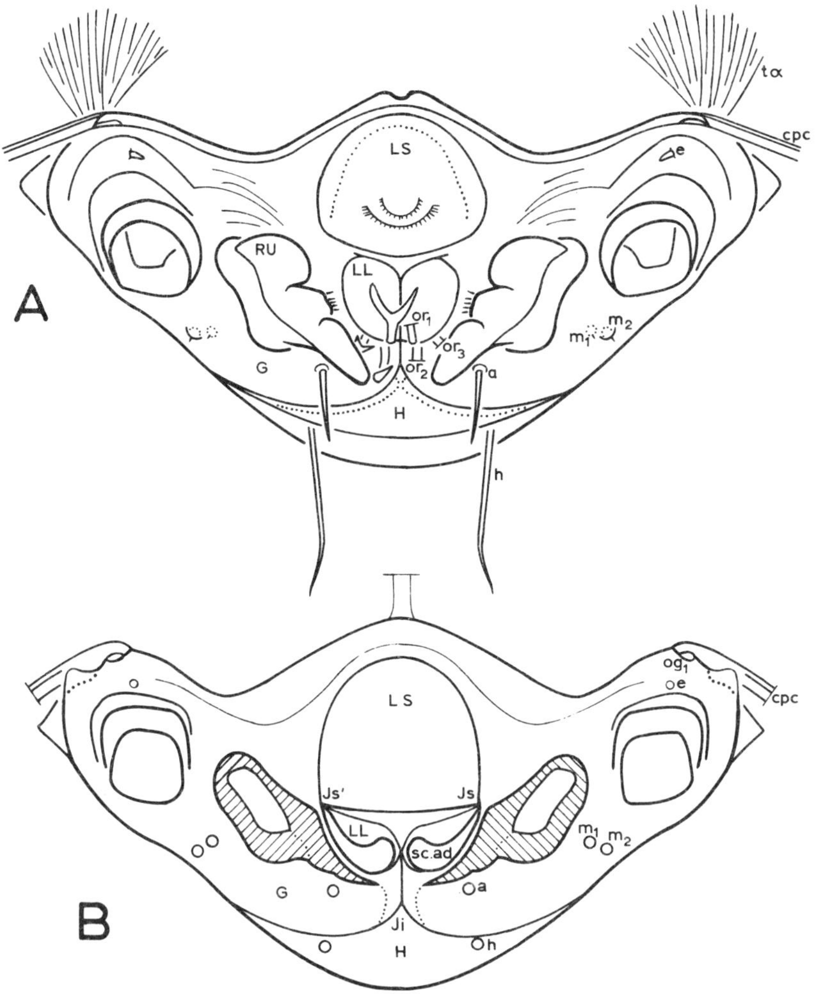

Fig. 91. Hermannia convexa (C.L. Koch), adult; A, frontal view of infracapitulum; B, idem, but greater part of labrum, lateral lips and rutella removed by transverse section; A, B, × 370.

verse section of the canal (it is open) is represented in fig. 89D. At *r* the canal disappears under the capitular angle *w*, and runs to *s* and *t*, following the paraxial border of the angle; this course can be discerned, although with some difficulty, through the prodorsal cuticle. At *t* the podocephalic canal reaches the anterior border of its prodorsal part. From this point to the infracapitulum, it is incorporated in an arthrodial membrane. This is easy to understand: the gnathosoma is movable, and the canal must be adapted to this condition. When a gnathosoma is removed from the camerostome, part of the podocephalic canal, and of the arthrodial membrane remain attached to the infraca-

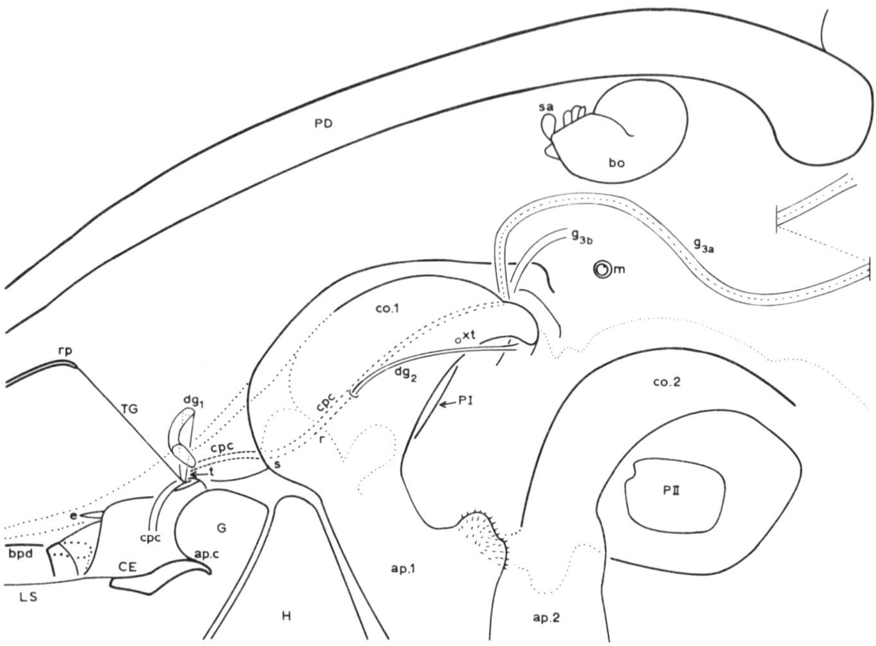

Fig. 92. Hermannia convexa (C.L. Koch), adult; internal structure of lateral part of prosoma (the paraxial face of the right side is represented; chelicerae and legs removed); × 295.

pitulum. A small fragment of the cheliceral frame remains attached to the line *at*; it partly represents the posterior border of the canal. A small orifice og_1 in the cheliceral frame is present at a small distance from the lateral border of the infracapitulum. As a rule a long, chitinous, colourless tube remains attached to og_1, either at one side or at both. This tube is represented separately in fig. 89E; it presents a transverse ridge in the posterior part, and it is much broader proximally than near its distal orifice og_1. The tube can be studied in various orientations; fig. 92 demonstrates that it is indeed hollow. The course of the canal in the dorsal region of the infracapitulum can be followed for a short distance only; it soon becomes indistinct when it approaches the cheliceral groove. It is possible that the gland secretions, transported by it, afterwards follow the groove; the same applies to the secretions of the infracapitular glands (cf. fig. 88: *ogi*). Four glands debouch into the podocephalic canal; they are named here podocephalic glands. The glands or their ducts can be discerned when a lateral part of the prosoma is separated by a longitudinal section, and subsequently studied from the paraxial side. An ideal section is represented in fig. 92; the cotyloid wall (*co.1*) of acetabulum I is visible as well as its connection with apodeme 1 (*ap.1*). The acetabular part of the podocephalic canal can be discerned through the cuticle. A coxal gland (g_{3a}) and a posterior podocephalic gland (g_{3b}) debouch into the posterior part of the canal, at its posterior extremity. In order to study these glands in their original shape, special care must be taken not to heat the preparation too much, because the glands are easily destroyed. The coxal gland has a narrow lumen; the part represented in fig. 92 is long and extends beyond the posterior border of the prodorsum into the opisthosoma. The posterior opisthosomatic gland (g_{3b}) is shorter and more difficult to discern. A very narrow, chitinous duct (dg_2) is attached to the anterior part of the cotyloid wall. Its orifice (og_2) in the podocephalic canal is visible at the external side (fig. 82). The duct is slightly wider in the basal part, where the gland has been destroyed by the action of lactic acid. The prodorsal part of the podocephalic canal (from *s* to *t*) is also represented in fig. 92. The duct dg_1, which debouches into the infracapitular part, is represented in its original position; fig. 92 demonstrates that its orifice is in the cheliceral frame just above the canal. The canal is everywhere external, but the glands and the ducts are of course internal. It is not immediately evident in what direction the secretions of the glands are transported. There are two outlets, a posterior and an anterior. The orifice of posterior podocephalic and coxal gland is close to the posterior end; og_1 is close to the anterior outlet, whilst og_2 is about halfway. Capillary attraction will certainly be an important agent in transport (the secretions will fill the canal in this way); the secretions can, however, only run out from the canal through a different agent.

Spatial relations. – Spatial relations must be studied in a longitudinal (sagittal) section of a gnathosoma in its original position in the camerostome. An ideal section of a gnathosoma, in which the chelicerae are in a retracted position, is represented in fig. 93. In this specimen, the cheliceral sheath is in an invaginated position: the surface represented in fig. 93 is the internal one, whilst air from the outside was originally present under the surface, from *bcx* to *acx*.

It is rather difficult to reconstruct the mechanism operating the movements of the chelicerae. All cheliceral tendons inserted on the body of the chelicera and on the trochanter probably contribute to retraction (invagination); tf_s probably readjusts the chelicera to a position parallel to the plane of the cervix; tf_1 and *ttr* are certainly retractors. Retracted chelicerae extend posteriorly far beyond the capitular apodeme.

When a gnathosoma starts functioning, the infracapitulum is lowered, and partly emerges from the camerostome, whilst the chelicerae are evaginated, following the course of the cheliceral grooves. It is probable that the tegulum can also be moved forwards, so that the surface, which is dorsal in fig. 93, then becomes ventral; in that position it is easily comparable with the tegulum in various other groups of mites. It is supposed that hydrostatic pressure is one of the main agents in evagination, protraction and extension of the gnathosoma.

Fig. 93 clearly demonstrates the position of the lips with respect to each other and to the pharynx; the figure demonstrates also the position of the com-

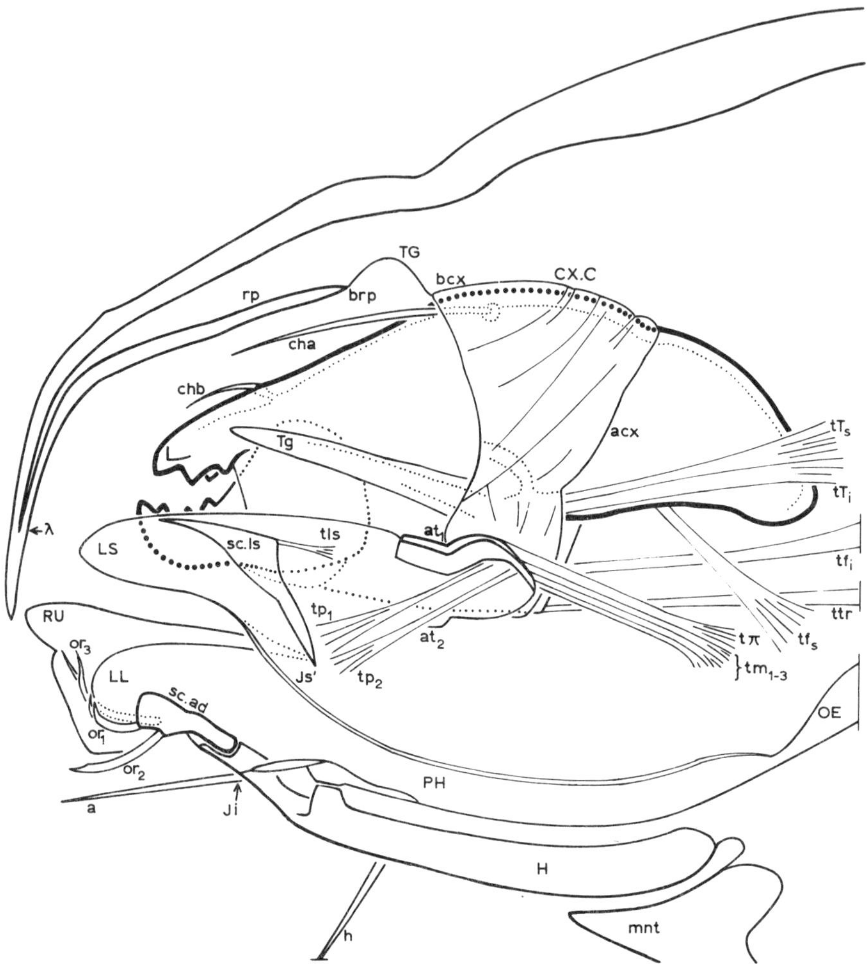

Fig. 93. Hermannia convexa (C.L. Koch), adult; longitudinal (sagittal) section of gnathosoma in camerostome (chelicerae nearly completely retracted); × 370.

missures and articulations. The pharynx is recognizable by its well-developed wall; posteriorly, it passes into the oesophagus. Tendons inserted on the cervix run in anteroventral direction. The tendons inserted on the capitular apodeme can retract the gnathosoma as a whole; during that operation, it turns on an axis passing through the condyles of the camerostome.

List of the notations of figs. 54–93

a	anterior infracapitular seta.
A	adult base level.
a', *a''*, (*a*)	antelateral setae of tarsus.
ac.d	acetabular tooth.
acm	anteroculminal seta of palpal tarsus.
Ac.P	acetabulum of palp.
ac.t	acetabular tectum.
acx	line of attachment of cheliceral sheath to chelicera.
Ac.I	acetabulum I.
Ac.II	acetabulum II.
Ad	adult base level.
Ad, *Ad'*, (*Ad*)	accessory setae of tarsus.
ad_{1-3}	adanal setae.
ag	aggenital seta(e).
am	arthrodial membrane.
an	anterior paraxial border of gena.
AN	anal cover.
an_{1-3}	anal setae.
ap	area porosa.
ap.c, *Ap.c*	capitular apodeme.
ap.1	apodeme 1.
ap.2	apodeme 2.
A.r	rostral porose area.
at, at_1, at_2	line of attachment of cheliceral frame to infracapitulum.
bcx	base of cheliceral sheath.
bls	base of labrum.
bo	bothridium.
bo.e	border of external acetabular opening.
bo.i	border of internal acetabular opening.
bpd	lateral border of prodorsum.
br	lateral border of rostrum.
brp	posterior border of rostrophragma.
bru	brush of rutellum.
b.sc	base of adoral sclerite.
bv''	basiventral seta of femur I–II.
c	collum of rutellum.
c_{1-3}, c_p	notogastral setae attributed to segments VII–VIII.
cam	camerostome.
CC	coxal cavity of infracapitulum.
CE	cervix.
CH	chelicera.
cha	posterior seta of chelicera.
chb	anterior seta of chelicera.
cl	lamellar ridge.
Cl	Claparède's organ.
clc	dorsal contour line of infracapitulum at level of cervix.
clm	dorsal contour line of infracapitulum at level of manubrium.
cm	culminal seta of palp.
CM	malar cavity of infracapitulum.

co.1	cotyloid wall of acetabulum I.
co.2	cotyloid wall of acetabulum II.
CP	pharyngeal chamber.
cpc	podocephalic canal.
CX	cheliceral sheath.
d	dorsal seta of legs.
d (figs. 89C, 96A)	postadoral articular difference in level.
d_{1-2}	notogastral setae attributed to segment IX.
da	anterodorsal notogastral seta.
dg_1	duct of anterior podocephalic gland.
dg_2	duct of median podocephalic gland.
dm	mediodorsal notogastral seta.
dp	posterodorsal notogastral seta.
e	supracoxal seta of palp.
e_{1-2}	notogastral setae attributed to segment X.
ev'	basiventral seta of femur III and IV.
ex	exobothridial seta.
exa	anterior exobothridial seta.
exp	posterior exobothridial seta.
eI	supracoxal seta of leg I.
f_{1-2}	notogastral setae attributed to segment XI.
fe_1	fenestrate area of body of chelicera.
fe_2	fenestrate area of cheliceral trochanter.
FM	cheliceral groove (in cervix).
fr	fenestrate area of rostrum.
ft', *ft''*	fastigial setae.
g	genital setae.
G	gena.
g_{3a}	coxal gland.
g_{3b}	posterior podocephalic gland.
GC	large lateral ridge of infracapitulum.
GEN	genital cover(s).
gla	latero-opisthosomatic gland.
h	seta of mentum (posterior infracapitular seta).
H	mentum.
h_{1-3}	notogastral setae attributed to segment XII.
ia	anterior opisthosomatic lyrifissure.
iad	adanal lyrifissure.
ian	anal lyrifissure.
ici	inferior commissural induration.
ics	superior commissural induration.
ih	opisthosomatic lyrifissure attributed to segment XII.
il	interlamellar seta.
im	median opisthosomatic lyrifissure.
in	interlamellar seta.
ip	posterior opisthosomatic lyrifissure.
ips	lyrifissure of pseudanal segment.

it″, (*it*)	iteral setae.
Ji	inferior oral commissure.
Js, *Js′*	superior oral commissures.
k	condyle for the articulation of the gnathosoma.
k′, *k″*	condyles for the articulation of the cheliceral apotele.
k_p	supposed condyle for the articulation of trochanter and body of the chelicera.
l′, *l″*	lateral setae.
la	anterolateral notogastral seta.
le	lamellar seta.
LL	lateral lip.
lm	mediolateral notogastral seta.
lp	posterolateral notogastral seta.
LS	labrum.
ls′, *ls″*	setae of femur.
lt″, (*lt*)	posterolateral setae of palpal tarsus.
lv″	posterior lateroventral seta.
m (figs. 82, 93)	lateral prosomatic mark.
m, m_1, m_2	median infracapitular setae.
MA	malapophysis.
MN	manubrium.
mnt	mentotectum.
N	naso.
ne	internal rib of dorsal infracapitular wall.
n1	protonymphal base level.
n2	deutonymphal base level.
n3	tritonymphal base level.
OE	oesophagus.
og	orifice of gland.
og_1	orifice of anterior podocephalic gland.
og_2	orifice of median podocephalic gland.
ogi	orifice of infracapitular gland.
op′	paraxial oncophysis of body of chelicera.
opv	ventral oncophysis of body of chelicera.
opx	coxal oncophysis of chelicera.
or_{1-3}	adoral setae.
p (fig. 82)	posterior border of podocephalic canal.
p″, (*p*)	proral setae.
PA	palp.
PD	prodorsum.
PH	pharynx.
pl′, *pl″*	primilateral setae.
po.lc	laterocoxal porose area of infracapitulum.
po.mn	porose area of manubrium.
ps_{1-4}	pseudanal setae.
pv′, *pv″*, (*pv*)	primiventral setae.
PI	leg I.

PII	leg II.
q (fig. 82)	posterior acetabular part of podocephalic canal.
r (fig. 82)	anterior acetabular part of podocephalic canal.
rc	rostral cavity.
ro	rostral seta.
rp	rostrophragma.
ru	basal contour line of rutellum.
RU	rutellum.
s (fig. 82)	transition of acetabular part of podocephalic canal into prodorsal part.
s	subungual seta of tarsus.
sa	bothridial sacculi.
sc.ad	adoral sclerite.
sc.ls	internal sclerite of labrum.
se	capitular saddle.
sens	sensillus.
su	subultimal seta of palpal tarsus.
t (fig. 82)	anterior border of prodorsal part of podocephalic canal.
t (figs. 60, 61A)	opisthosomatic muscle attachment.
t_i	inferior tendon of apotele.
t_s	superior tendon of apotele.
(*tc*)	tectal setae.
tf_i	inferior tendon of body of chelicera.
tf_s	superior tendon of body of chelicera.
Tg	Trägårdh's organ.
TG	tegulum.
tls	tendon inserted on labral sclerite.
tm_{1-3}	median tendons inserted on capitular apodeme.
tp_{1-2}	tendons associated with intrinsic muscles, and inserted (probably) on capitular apodeme.
tr.sj	sejugal trachea.
tr.1	trachea 1.
tr.3	trachea 3.
tT_i	inferior tendon of Trägårdh's organ.
tT_s	superior tendon of Trägårdh's organ.
ttr	tendon of cheliceral trochanter.
$t\alpha$	posterior antiaxial tendon of infracapitulum.
$t\pi$	paraxial tendon inserted on capitular apodeme.
u', u'', (u)	ungual setae of tarsus.
ul', ul''	ultimal setae of palpal tarsus.
v', v'', (v)	ventral setae.
vt	ventral seta of palpal tarsus.
w	capitular angle.
xt	place of attachment (or insertion) of tendon to cotyloid wall 1.
α (alpha)	antiaxial.
αf	antiaxial manubrial fissure.

Δ (delta)	articular tooth of cheliceral trochanter.
λ (labda)	base of rostral limbus.
σ, σ_{1-2} (sigma)	solenidia of genu.
τ_{1-2} (tau)	anterior eugenital setae.
φ, φ_{1-2} (phi)	solenidia of tibia.
ψ_{1-2} (psi)	posterior eugenital setae.
ω, ω_{1-3} (omega)	solenidia of tarsus.
1–3 (fig. 89C)	adoral setae.
1–9 (figs. 60, 63, 68, 73, 75, 76)	genital setae.
I–IV	legs I–IV
1a–c	setae of epimeron 1.
2a	seta of epimeron 2.
3a–c	setae of epimeron 3.
4a–c	setae of epimeron 4.
′ (prime)	anterior face of an appendage.
″ (double prime)	posterior face of an appendage.
( )	a pair.

5. Acaridida

In four papers, I have previously paid attention to Acaridida. I have first published some notes on the so-called genital tracheae of the female of *Gohieria fusca* (Oudemans) (Van der Hammen, 1968a: 270–273, fig. 3). Thereupon I have summarized, in two papers, some of the general Acaridid characters (Van der Hammen, 1972: 37, 61–63, figs. 22B, D, 34, 35; 1972a: 288). In a recent paper (Van der Hammen, 1982), I studied the morphology and the post-embryonic development of a *Rhizoglyphus* species.

The diagnostic characters of the order Acaridida are the following. The body of the adults comprises fourteen segments which have nearly completely fused. As a rule, the cuticle is soft and faintly sclerotized, although single sclerites can be present; a rostral tectum, however, is never present. Generally, tracheae are absent; the function of the so-called genital tracheae, found in the female of *Gohieria fusca* (Oudemans), is unknown. There are no prodorsal trichobothria. A paired latero-opisthosomatic gland is present. The posterior eugenital lip is, generally, only partly covered by the progenital lips. Generally, the female presents a bursa copulatrix. The podocephalic canal is external. The palp has been subject to regression (it is small and more or less flattened, and consists of two segments only); the basal palpal segments have fused with the malapophyses. As a rule, the legs present one apotelic unguis and a pulvillus, although both can be lacking. Generally, there are three nymphal stases. As a rule, the deutonymph is represented by an elattostase (the hypopus), or is lacking; in rare cases, the deutonymph is represented by a calyptostase. In deutonymph and male, several setae are replaced by suckers.

The order Acaridida is, generally, subdivided into two suborders: Acaridina and Psoroptina. Representatives of the first-mentioned suborder are predominantly free-living, those of the second predominantly parasitic.

The present section is subdivided into two subsections, in which the morphology and postembryonic development of *Rhizoglyphus* (subsection 5.1) and the so-called genital tracheae of *Gohieria* (subsection 5.2) are successively dealt with.

5.1. *Morphology and postembryonic development of* Rhizoglyphus echinopus *(Fumouze & Robin)*

The *Rhizoglyphus* species described in the present subsection is a free-living representative of the suborder Acaridina (family Acaridae). The description is based on material collected from bulbs (mainly hyacinth, but also other bulbs), originating from The Netherlands (see Van der Hammen, 1982). The study has contributed to the solution of problems with reference to seta homology (particularly those with reference to the paraproctal setae and their postembryonic development) and the morphology of genital region and gnathosoma. The study has also revealed details of the structure of the bursa copulatrix, which are not in accordance with current views.

Grandjean (1937a, 1937b, 1938a, 1938b, 1939, 1953) published several papers of fundamental importance to Acaridid morphology; he particularly dealt with podocephalic canal and podocephalic glands, genital region, suckers, leg segmentation and leg chaetotaxy, postembryonic development, and the relationships between Acaridida and other groups of Actinotrichida. He studied the leg chaetotaxy of Acaridida (Grandjean, 1939) before he had developed his general system of notations for the use in Actinotrichid mites; afterwards (Grandjean, 1953: 392–393) he published some remarks on possible homologies. Although Grandjean's earlier system of notation was adopted by several specialists of Acaridida, his later remarks remained practically unobserved. Knülle (1958, 1959) studied the morphology and postembryonic development of Acaridida, with particular reference to *Acarus siro* L. He was the first to apply Grandjean's system of notation for Actinotrichid notogastral and epimeral setae to Acaridida. He did not solve the problem with reference to the homology of the tarsal setae, and his interpretation of the postembryonic development of the paraproctal setae is different from that introduced by me. The postembryonic development of the idiosomatic setae was recently reinvestigated by Griffiths (1977); his results are discussed below (he did not apply Grandjean's system of notation).

Measurements. – The measurements of the species are variable. For the present study one or a few specimens only of all stases were measured. My data (in mm) are the following. Adult female: length 0.649–0.781, breadth 0.354–0.469. Adult homomorphic male: length 0.493, breadth 0.266. Larva: length 0.244, breadth 0.141. Protonymph: length 0.313, breadth 0.207. Deutonymph (hypopus): length 0.316–0.366, breadth 0.198–0.241. Tritonymph: length 0.354–0.443, breadth 0.186–0.245.

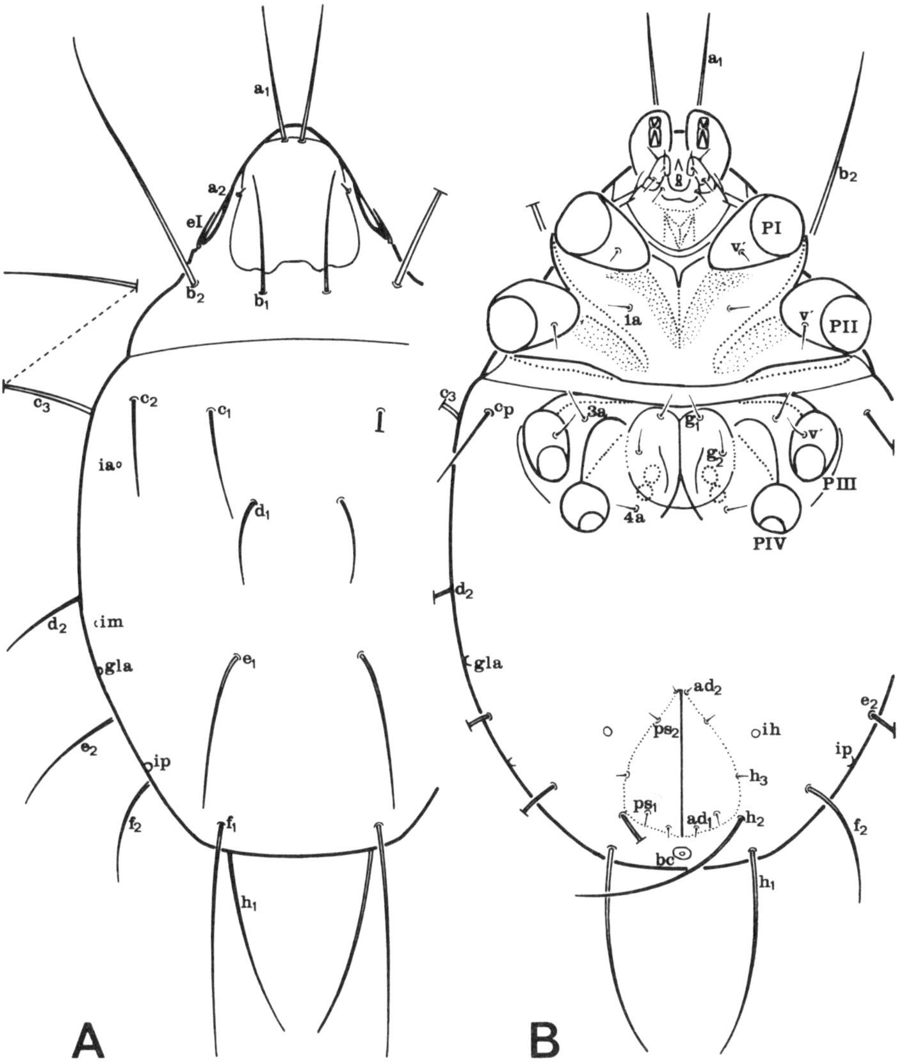

Fig. 94. Rhizoglyphus echinopus (Fumouze & Robin), adult female; A, dorsal view; B, ventral view; A, B, × 138.

Divisions of the body. – The dorsodisjugal furrow separates the prodorsum from the opisthosoma (fig. 94A). The ventrosejugal furrow and interval separate the propodosoma from metapodosoma and opisthosoma (the genital region has an advanced position between legs III and IV; see fig. 94B). Although both furrows are more or less in line with each other, they are not continuous (fig. 95). The gnathosoma is not protected by a rostral tectum (the species is astegasimous).

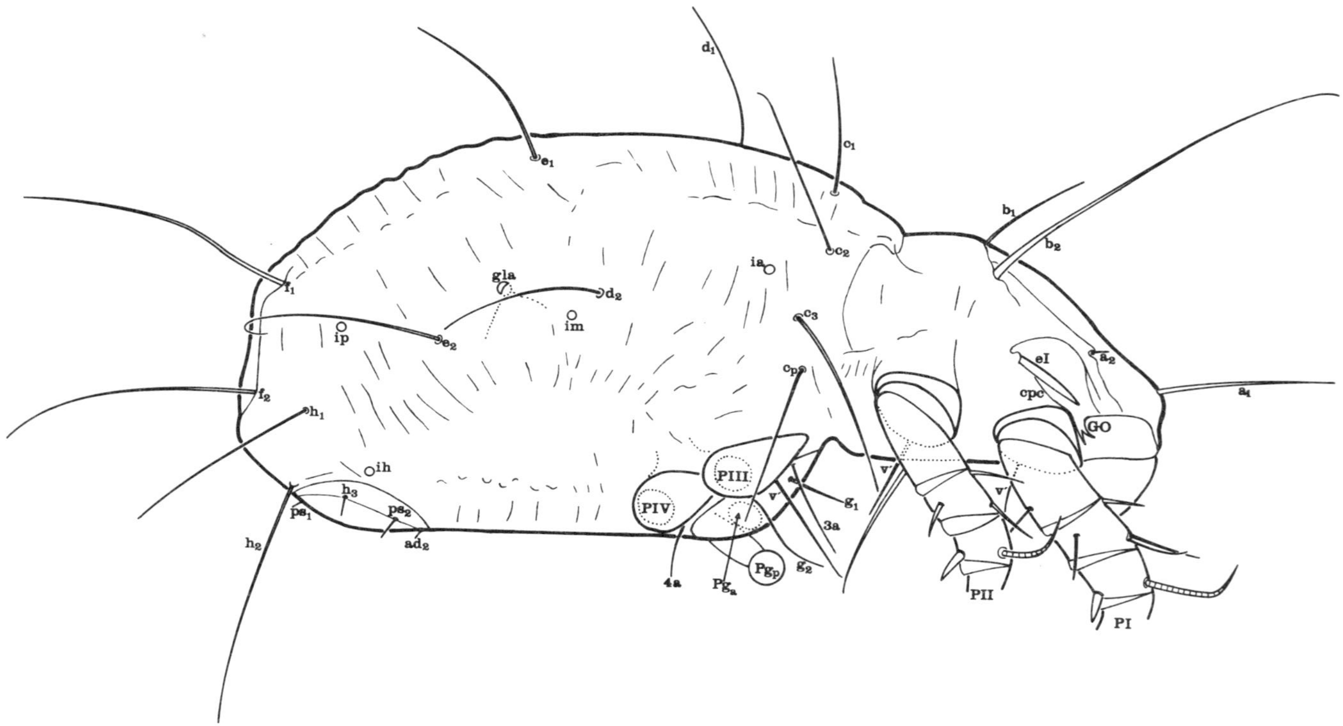

Fig. 95. Rhizoglyphus echinopus (Fumouze & Robin), adult female, lateral view; × 186.

Tegument. – The cuticle is for the greater part soft and pale in colour; the cuticle of the appendages, and particularly of the chelicerae, is more distinctly sclerotized. There is a prodorsal sclerite (fig. 94A) in the anterior part of the aspidosoma. Sclerotization is more pronounced in the heteromorphic male.

Aspidosoma. – There are four pairs of prodorsal setae: a_1 (vertical internals), a_2 (vertical externals), b_1 (scapular internals) and b_2 (scapular externals); the new notation is introduced in order to be consistent with the notation of the notogastral setae (it is not possible to homologize the four pairs of prodorsal setae of Acaridida with some of the six pairs of prodorsal setae ancestrally present in Oribatida and Actinedida).

Dorsal and laterodorsal aspects of opisthosoma. – The chaetotactic condition of this region is easily comparable with the condition in Oribatida and Actinedida. This is demonstrated by a study of the postembryonic development (which will be discussed below, in the paragraph in question). The (fused) opisthosomatic segments present in the adult are *VII/VIII(C), IX(D), X(E), XI(F), XII(H), XIII(PS)* and *XIV(AD)*. The notation of the notogastral setae (figs. 94, 95), according to the system currently used in the study of Oribatida and Actinedida, is the following (the notations and the names of the setae currently used by specialists of Acaridida are added between brackets): c_1 (d_1 = dorsal 1st), c_2 (*hi* or l_1 = internal humeral or lateral 1st), c_3 (*h* = external humeral or humeral), c_p (*hv* or *sh* = ventral humeral or sub-humeral), d_1 (d_2 = dorsal 2nd), d_2 (*la* or l_2 = anterior or 1st lateral, or lateral 2nd), e_1 (d_3 = dorsal 3rd), e_2 (*lp* or l_3 = posterior or 2nd lateral, or lateral 3rd), f_1 (d_4 = dorsal 4th), f_2 (l_3 or l_4 = 3rd lateral or lateral 4th), h_1 (*sai* or d_5 = internal sacral or dorsal 5th), h_2 (*sae* or l_5 = external sacral of lateral 5th), h_3, ps_1, ps_2; the adanal setae, together with the pseudanals, are discussed in the paragraph on the paraproctal region (a comparative study of the notation of the *h*, *ps* and *ad* setae is made in the paragraph on the postembryonic development). There are four pairs of cupules (lyricupules): *ia*, *im*, *ip* and *ih* (cf. figs. 94, 95); *ih* is arranged in the same alignment as h_1, h_2 and h_3. The bursa copulatrix of the female, posterodorsal to the anal opening, is described in the paragraph on the genital region.

Paraproctal region. – The two paraproctal segments of *Rhizoglyphus* (and other Acaridida) are: the pseudanal segment (*PS* = *XIII*) and the adanal segment (*AD* = *XIV*). In female *Rhizoglyphus* these segments present four pairs of setae, the notation of which is: ps_1, ps_2, ad_1, ad_2. The pseudanal seta ps_3, normally present in holotrichous Actinotrichida, is absent. In the adult male (fig. 97B) the adanal setae are replaced by a pair of suckers (homologized here with ad_1), whilst ps_1 has a more posterior position (on a transverse line with h_2). The pseudanal and anal (lyri)cupules are not present (not discernable) in both sexes.

Genital region. – The longitudinal progenital opening is bordered by the progenital lips (figs. 94B, 96A). There are two pairs of progenital setae: g_1 (g_a or c_1 = anterior genital or first central) and g_2 (g_m or c_2 = median genital or

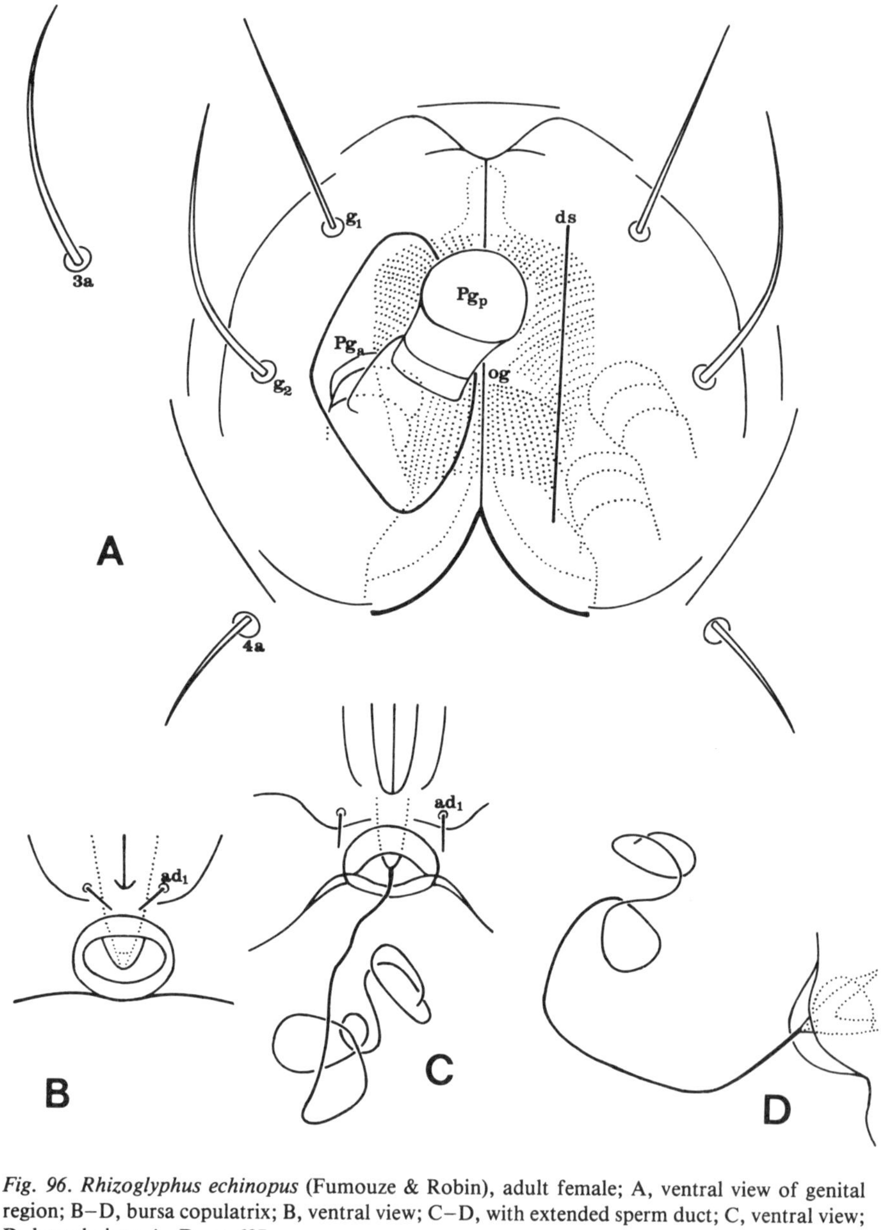

Fig. 96. Rhizoglyphus echinopus (Fumouze & Robin), adult female; A, ventral view of genital region; B–D, bursa copulatrix; B, ventral view; C–D, with extended sperm duct; C, ventral view; D, lateral view; A–D, × 687.

second central). The eugenital opening (fig. 96A: *og*) is trifid. There are two pairs of genital papillae (Pg_a and Pg_p); they are situated (in rest) in cavities of the progenital lips, which open at the so-called diachilous slit (*ds*). Grandjean

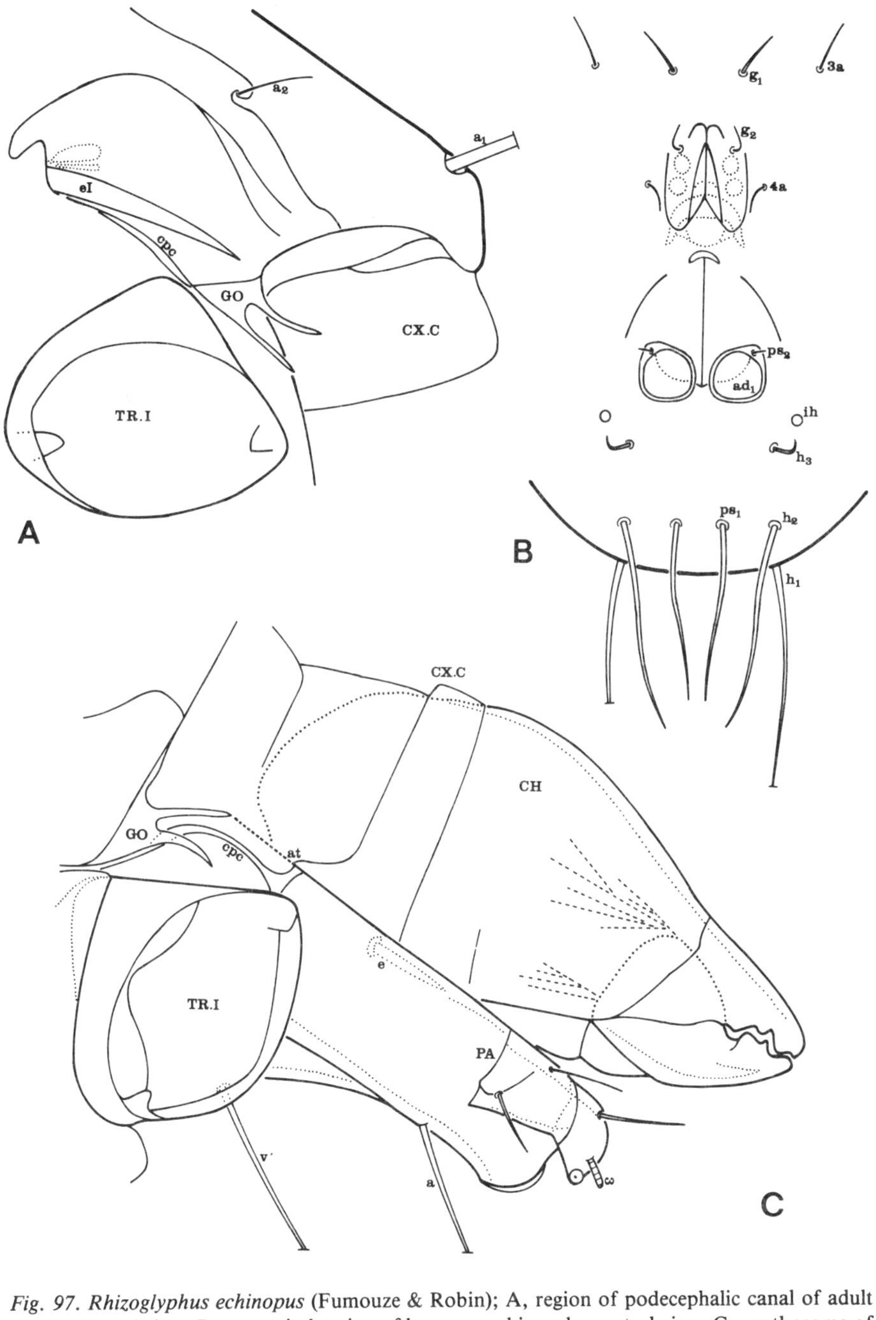

Fig. 97. Rhizoglyphus echinopus (Fumouze & Robin); A, region of podecephalic canal of adult female, lateral view; B, anogenital region of homomorphic male, ventral view; C, gnathosoma of adult female, lateral view; A, C, × 550; B, × 186.

(1938b: 284–285, fig. A–F) described the progenital lips as eugenital. The position of the eugenital opening (*og*) demonstrates that the posterior eugenital lip is indeed only partly covered by the progenital lips, but that the anterior pair of eugenital lips is completely covered. This condition is a particular one, known from Acaridida only. It is now generally assumed (cf. Baker & Krantz, 1985) that the bursa copulatrix of the female, situated posterodorsal to the anal opening (fig. 96B), has internal connections (by means of openings) with the ovaries. In one of my specimens (studied in lactic acid), of which I had just removed the legs, a long, narrow, chitinous duct (fig. 96C, D) was protruding; this duct was not visible in rest. It could represent the duct by which sperm, stored in the bursa copulatrix, is transported to the eugenital opening. The supposed internal openings to the ovaries could indeed represent the chitinous parts of glands (they are very similar to a type of uncellular gland with sclerotized collar at the transition between gland and duct). The male copulatory organ or penis (fig. 97B) is broad and rounded.

Epimeral region. – Epimera 1, 3 and 4 each present one seta (fig. 94B): *1a*, *3a* and *4a* respectively (*CXI* or *s*, *CXIII* or cx_3, and g_p or c_3 in current literature on Acaridida). The genital opening has an advanced position between the pairs of epimera 3 and 4.

Lateral region of podosoma. – The podocephalic canal (figs. 95, 97A, C) extends from posterodorsal to leg I up to the cervix (close to the line of attachment of the cheliceral frame to the infracapitulum). The canal is protected by a tectum of which the anterior part extends as Grandjean's organ (*GO*); the last-mentioned organ is bifurcate is *Rhizoglyphus echinopus*. The supracoxal seta *eI* of leg I is inserted close to the posterior part of the podocephalic canal; it is relatively large.

Gnathosoma. – The Acaridid gnathosoma consists (as in other Actinotrichida) of two parts: cheliceral frame and infracapitulum. The cheliceral frame extends from the rostral region of the aspidosoma to the infracapitulum. It is attached to the infracapitulum according to the line *at* (fig. 99A). The dorsal part constitutes the tegulum (fig. 99C: *TG*). The infracapitulum consists of a dorsal part (the cervix), a ventral part (mentum) and the malapophyses (with which the palps have partly fused). The cervix presents a median elevated part, named capitular saddle (fig. 99D: *se*). The internal continuation of the cervix is constituted by the capitular apodeme (fig. 99A: *ap.c*). Mentum and genae (the ventral region of the malapophyses) are continuous, not separated by an articulation. There is one pair of infracapitular setae, viz., the anterior pair *a* (belonging to the genae). There are three lips (figs. 98, 99A): the labrum (*LS*) and the pair of lateral lips (*LL*). The lateral lips have fused. There is, on either side of the base of the labrum, a distinctly sclerotized thickening (the superior commissural induration), along the inner part of the superior commissural line, close to the superior oral commissure. As mentioned above, the palps have partly fused with the malapophyses. The malapophyses constitute the paired anterior region of the infracapitulum, in front of the base of the labrum

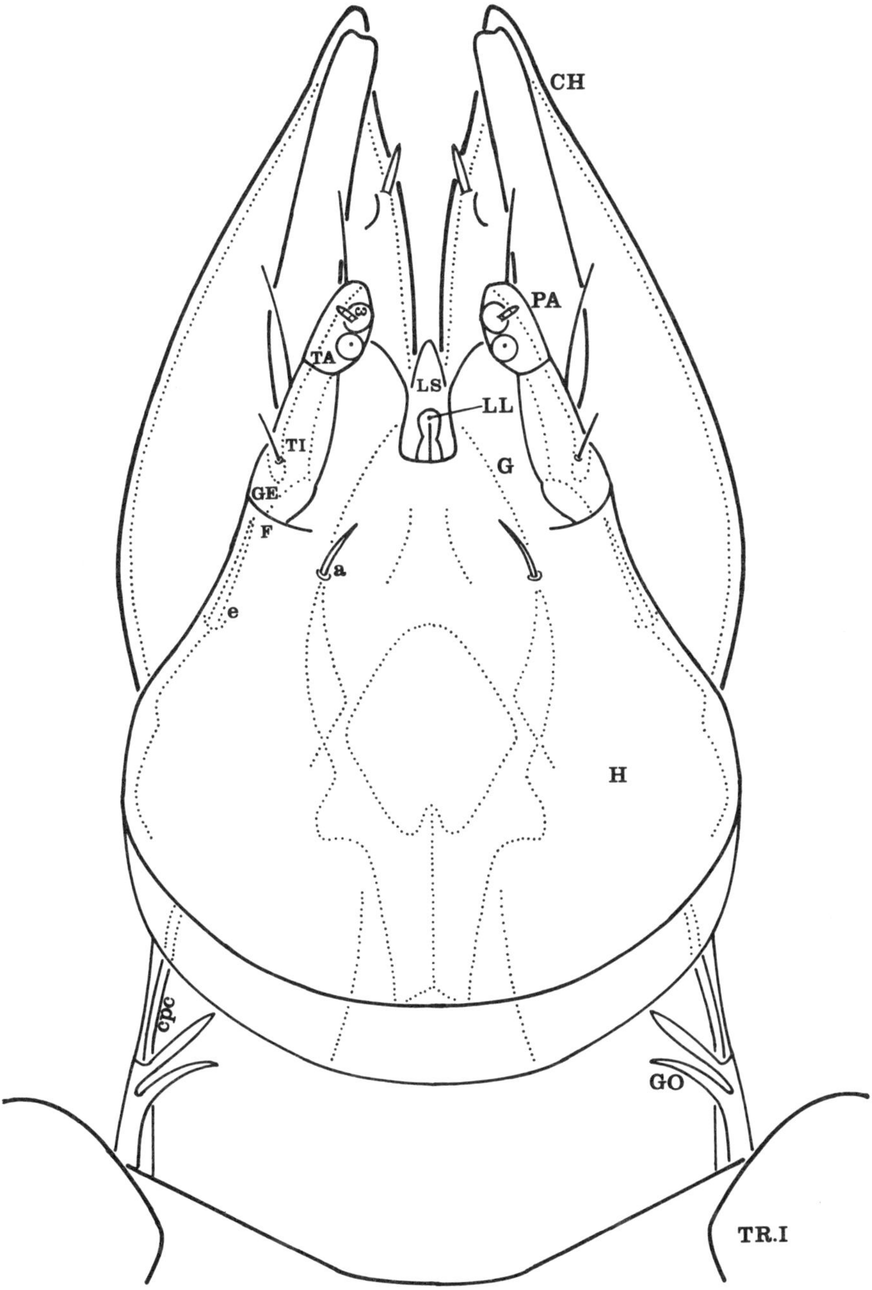

Fig. 98. Rhizoglyphus echinopus (Fumouze & Robin), gnathosoma of adult female, ventral view; × 825.

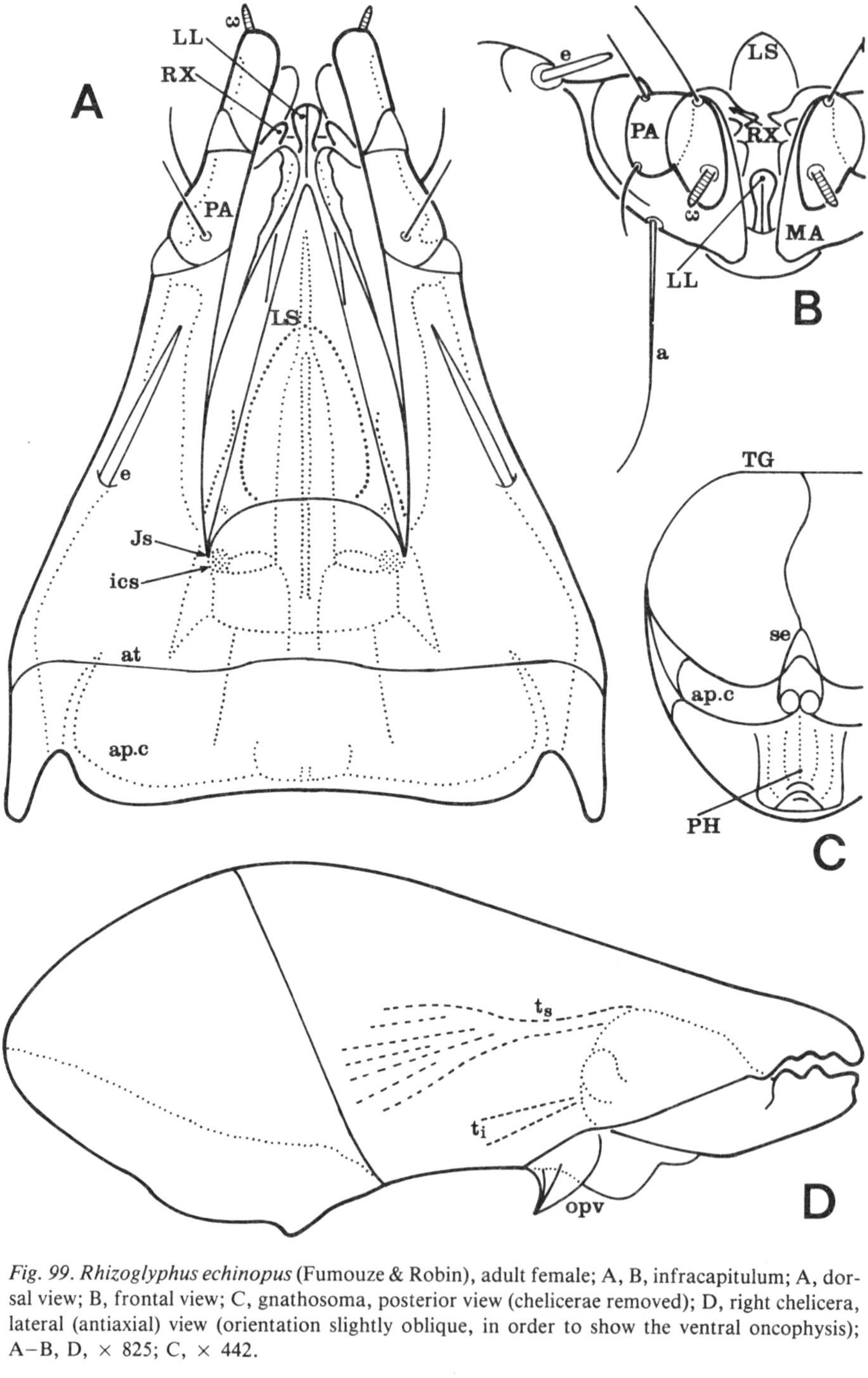

Fig. 99. Rhizoglyphus echinopus (Fumouze & Robin), adult female; A, B, infracapitulum; A, dorsal view; B, frontal view; C, gnathosoma, posterior view (chelicerae removed); D, right chelicera, lateral (antiaxial) view (orientation slightly oblique, in order to show the ventral oncophysis); A–B, D, × 825; C, × 442.

and the bases of the palps. The two distal palpal segments are more or less free; they probably represent a genotibia and a tarsus. The tarsus presents the solenidion ω, a supposed vestige of a eupathidium, and one seta; the genotibia presents two setae (figs. 97C, 98, 99A). The supracoxal seta *e* is relatively large; in Actinotrichid mites, it is usually inserted above the base of the palp, and its position in *Rhizoglyphus* is an indication of the extension of the trochanterofemur, which is, on both sides, completely incorporated into the malapophyses (figs. 97A, 98, 99A). On both sides of the infracapitulum, mentum and cervix together constitute a large lateral ridge (cf. fig. 99C) which is continuous with the malapophyses (fig. 99B). The malapophyses present a compound paralaterodorsal protuberance (fig. 99A, B: *RX*), which was named pseudorutellum by Akimov (1979) and rutellum by Griffiths (1977). The rutellar nature of the protuberance is not proved. I have not been able to discerne a root, nor an actinopilinous core. The position of the protuberance on the malapophyses (paralaterodorsal) differs from that of a rutellum (which is terminal).

Cheliceral. – The chelicera consists of two segments: the body of the chelicera and the apotele (figs. 97C, 98, 99D). A cheliceral trochanter is not discernable. The anterior part of the body of the chelicera (the fixed jaw) and the apotele (the movable jaw) consitute the chela; both jaws present several more or less obtuse teeth. A voluminous ventral oncophysis (fig. 99D: *opv*) is present at the articulation between body of chelicera and apotele. The paraxial face of the chelicera (fig. 98) presents an seta and a small protuberance. The cheliceral sheath is attached to the body of the chelicera; the part posterior to the line of attachment (fig. 99D) is internal. The cheliceral sheath (fig. 97C: *CX.C*) is inserted on the cheliceral frame.

Legs. – All legs consist of six segments: trochanter, femur, genu, tibia, tarsus, and apotele (fig. 100). The numerical aspects of leg chaetotaxy can be represented by the following formulae (in which the number of setae of each leg is enumerated for trochanter, femur, genu, tibia, tarsus, and apotele respectively): leg I (1–1–2–2–13–1), leg II (1–1–2–2–12–1), leg III (1–0–1–1–10–1), leg IV (0–1–0–1–10–1). All legs are monodactyl (the single unguis is represented by the last number in the formulae). Solenidia can be present on genu, tibia and tarsus; the solenidial formulae are: leg I (2–1–3), leg II (1–1–1), leg III (1–1–0), leg IV (0–1–0). The setae of the tarsus (fig. 101) represent the so-called fundamental setae: they are all of larval origin (with the exception of those of leg IV, as will be demonstrated in the paragraph on the postembryonic development). The homology of the famulus (ϵ), the proral (*p*), ungual (*u*) and subungual (*s*) setae is evident. The remaining setae certainly represent (*ft*), (*tc*), (*a*) and *pv*. Difficulties arise in the case of the anterodorsal setae, which are here indicated with the notations tc', tc'' and ft''; it is possible that the notations of the last-mentioned two setae must be interchanged. The case of *Rhizoglyphus* is a simple one, and the case of *Forcellinia*, studied by Grandjean, presented more difficulties. Grandjean (1953: 393) supposed

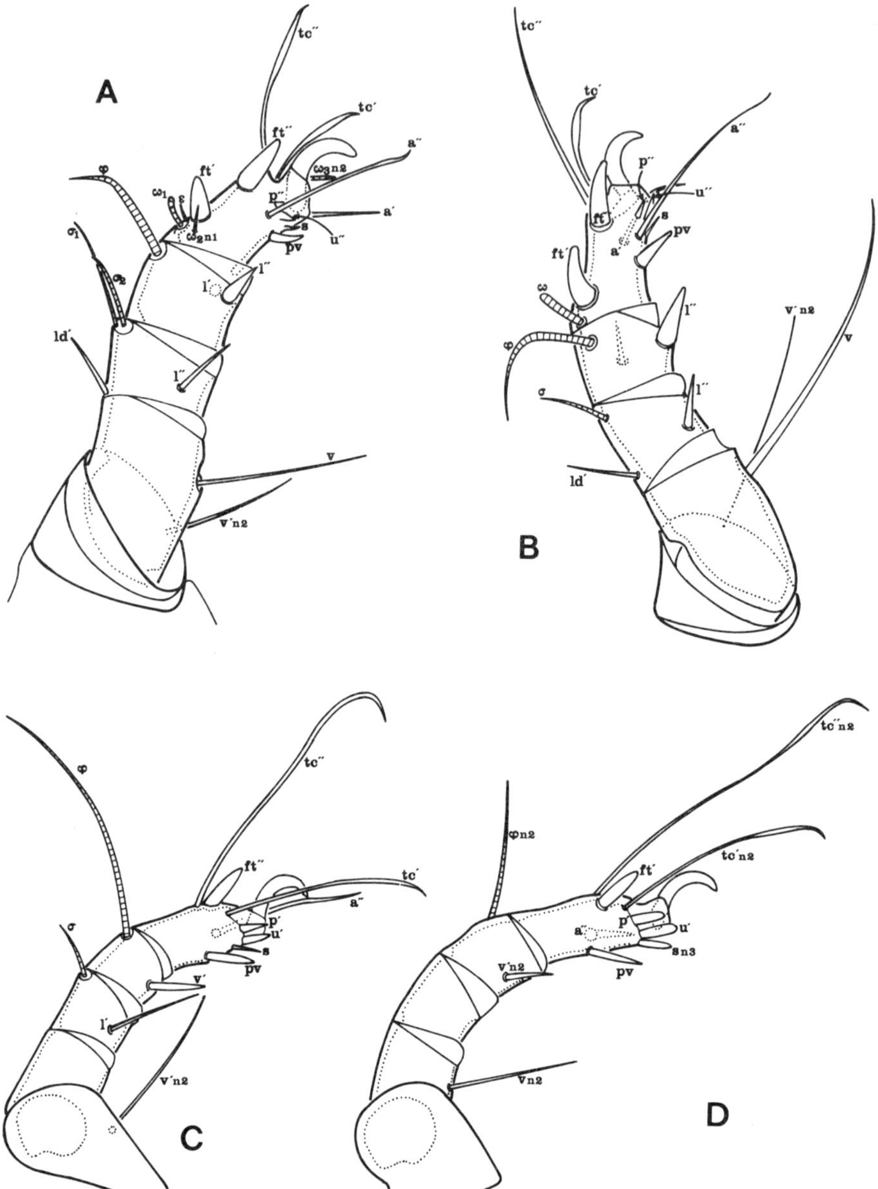

Fig. 100. Rhizoglyphus echinopus (Fumouze & Robin), legs of adult female, lateral (antiaxial) view; A, right leg I; B, right leg II; C, left leg III; D, left leg IV; A–D, × 295.

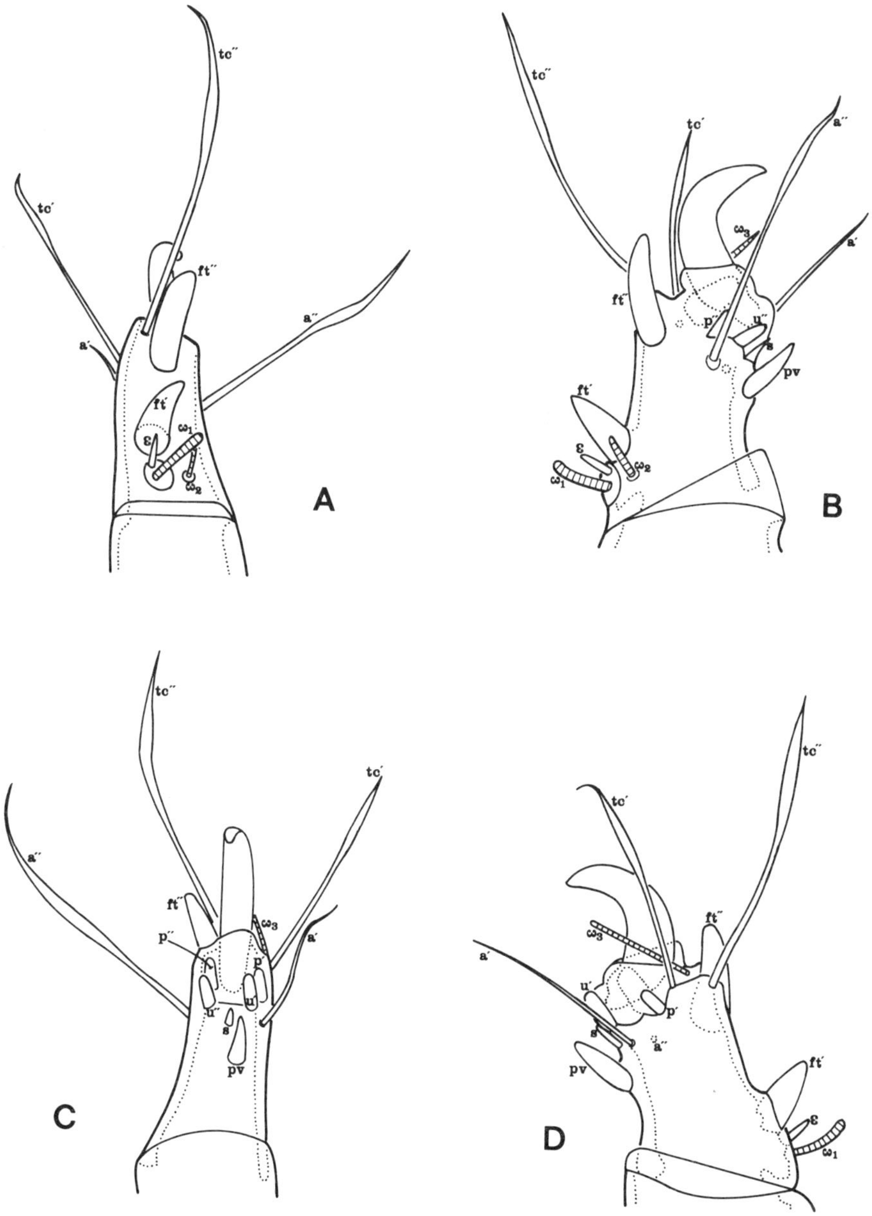

Fig. 101. Rhizoglyphus echinopus (Fumouze & Robin), terminal part of right leg I of adult female; A, dorsal view; B, lateral (antiaxial) view; C, ventral view; D, lateral (paraxial) view; A–D, × 550.

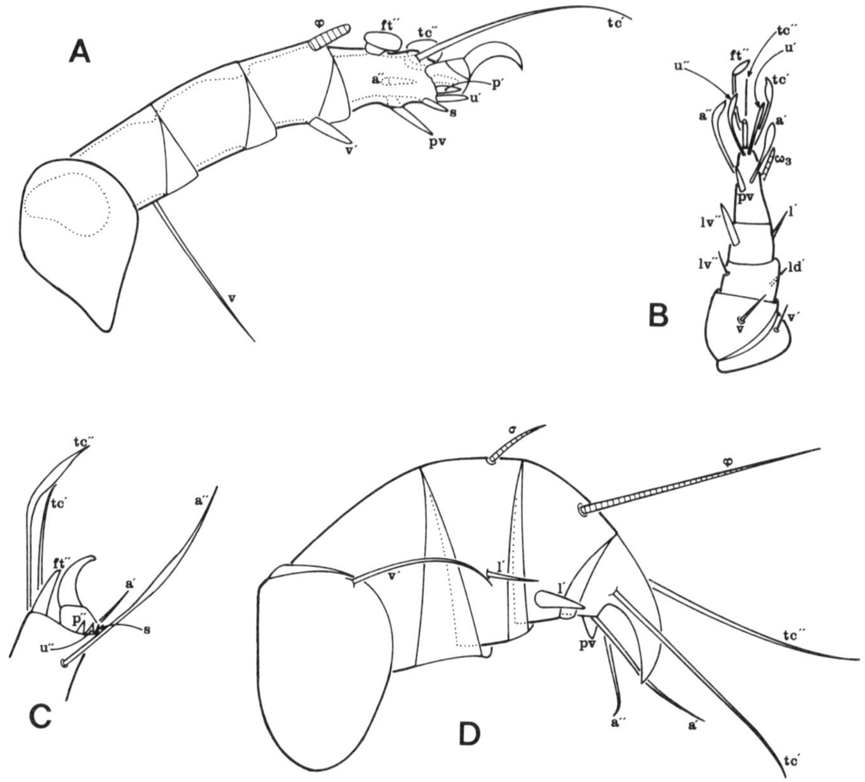

Fig. 102. Rhizoglyphus echinopus (Fumouze & Robin); A, left leg IV of homomorphic male, lateral (antiaxial) view; B, right leg I of hypopus, ventral view; C, terminal part of right leg II of heteromorphic male, lateral (antiaxial) view; D, left leg III of heteromorphic male, lateral (antiaxial) view; A–D, × 295.

that his setae *aa* and *ba* (Grandjean, 1939: 55–60, figs. 1–2) could represent the primilateral setae. The chaetotaxy of *Rhizoglyphus* demonstrates that *aa* represents *a''*, and that *ba* represents *ft'*. In leg IV of the male (fig. 102A), *ft''* and *tc''* are replaced by suckers. In the heteromorphic male (fig. 101D), leg III underwent important transformations: the segments are higher and shorter, and tarsus and apotele constitute one segments with reduced chaetotaxy. In comparison with tarsus III of the homomorphic male (fig. 102C), the proral, ungual and subungual setae have disappeared. The claw-like terminal part could represent the fastigial seta *ft''*, the apotelic claw, or a fusion of both (possibly together with (*p*), (*u*) and *s*).

→

Fig. 103. Rhizoglyphus echinopus (Fumouze & Robin); A, opisthosoma of larva, dorsal view; B, posterior part of opisthosoma of protonymph, dorsal view; C, dorsal view of deutonymph (hypopus); A–C, × 268.

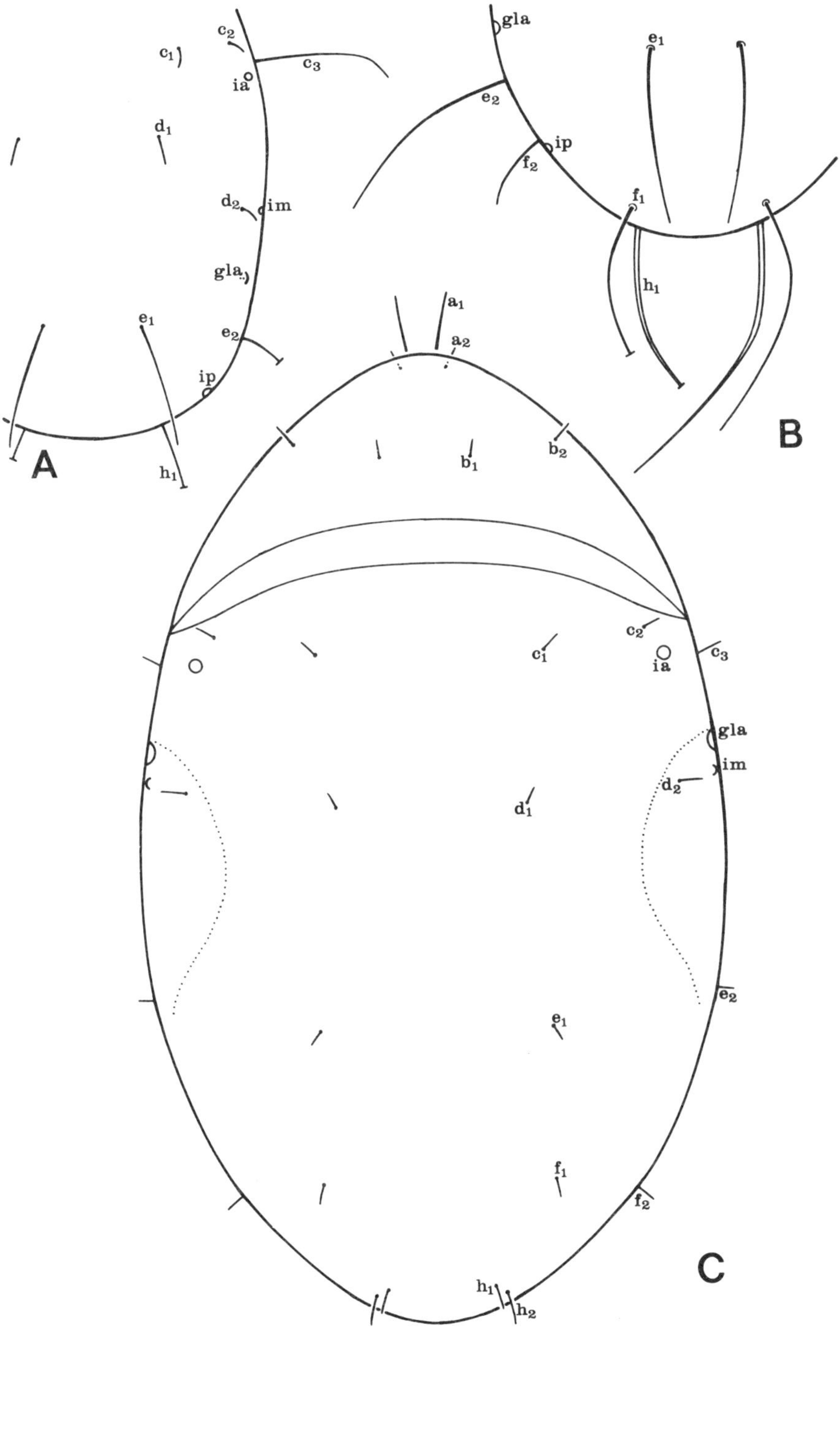
c₁
c₂
c₃
ia
d₁
d₂
im
gla
e₁
e₂
ip
h₁
A
gla
e₁
e₂
ip
f₂
f₁
h₁
B
a₁
a₂
b₁
b₂
c₂
c₁
ia
c₃
gla
im
d₂
d₁
e₂
e₁
f₁
f₂
h₁
h₂
C

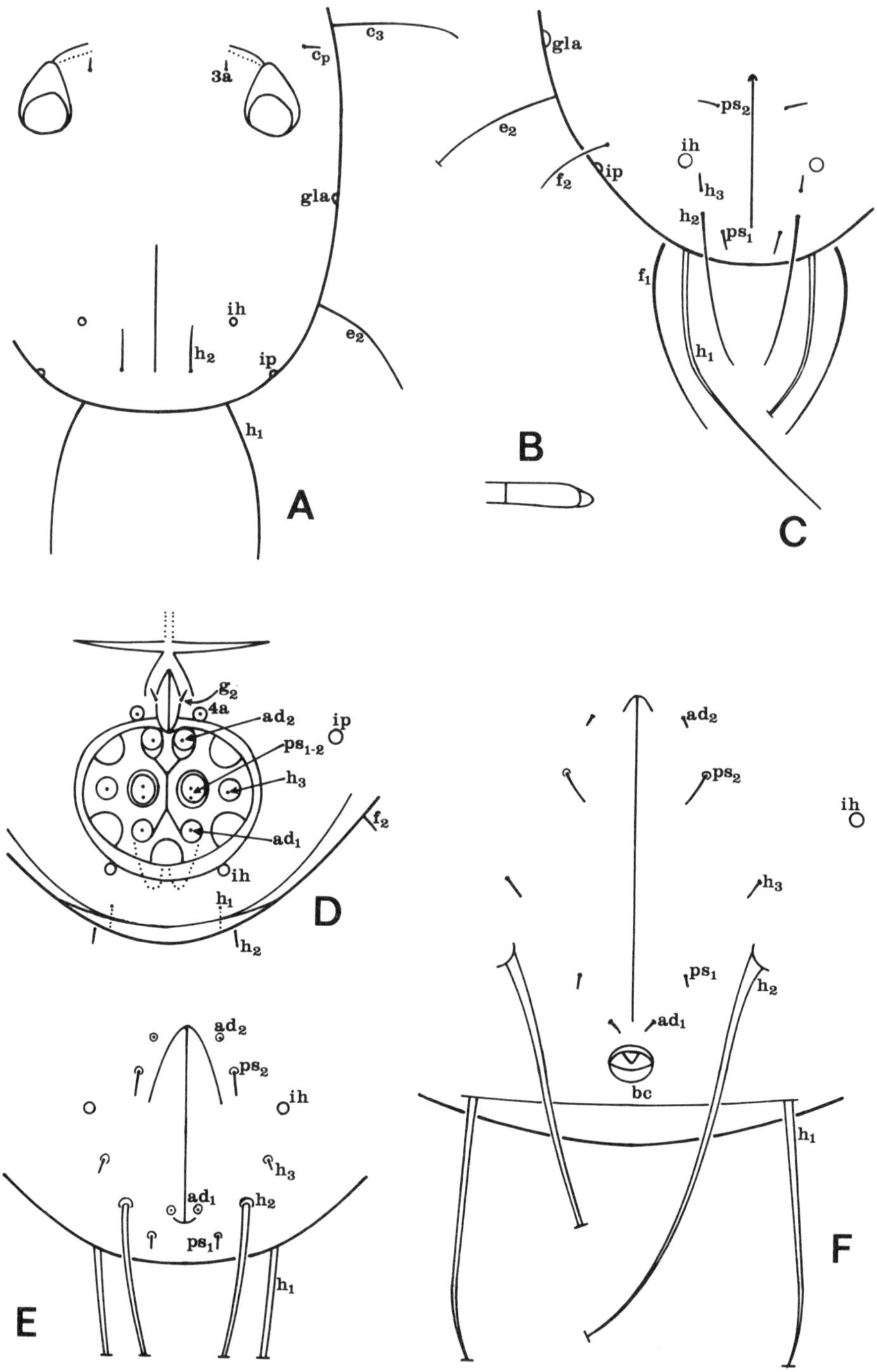
c3
cp
3a
gla
ih
h2
ip
e2
h1
A
gla
ps2
e2
ih
f2
ip
h3
h2
ps1
f1
h1
B
C
g2
4a
ad2
ip
ps1-2
h3
ad1
f2
ih
h1
h2
D
ad2
ps2
ih
h3
ps1
h2
ad1
bc
h1
F
ad2
ps2
ih
h3
ad1
h2
ps1
h1
E

Postembryonic development. – The life-cycle of *Rhizoglyphus* is characterized by the presence of five stases: larva, protonymph, deutonymph, tritonymph and adult. A prelarva is not known. The deutonymph has been subject to regression; it has become an elattostase (hypopus). The elattostasic regression at the level of the deutonymph is an example of metelattosis (cf. Van der Hammen, 1975: 378; 1978: 52, 56; 1980: 100). The postembryonic development of the paraproctal region (fig. 104) is important for the homologization of the setae, and for the recognition of the number of segments. It appears that the anus is situated, in the larva, in segment *XIII*, and, in the protonymph (and subsequent stases), in segment XIV; no segments are added in the deutonymphal, tritonymphal and adult stases (cf. Van der Hammen, 1970a). In the larva (fig. 104A), the paraproctal segment (*XIII*) is glabrous (ps_1 and ps_2 are absent), whilst h_3 is also absent (suppressed, i.e., retarded by one stase). In the protonymph (fig. 104C), the paraproctal segment (*XIV*) is also glabrous (ad_1 and ad_2 are absent), whilst ps_1, ps_2 and h_3 are present. In the deutonymph (hypopus) (fig. 104D), ad_1 and ad_2 are present (they constitute suckers), whilst ps_1, ps_2 and h_3 are replaced by suckers. In tritonymph and adult, ad_1, ad_2, ps_1, ps_2 and h_3 are all present. The case of *Rhizoglyphus* is a simple and normal example of atrichosy at two levels. In the larva (fig. 103A), two other opisthosomatic setae are lacking; evidently, these are f_1 and f_2, which appear in the protonymph (fig. 103B) and are present in deutonymph (fig. 103C), tritonymph and adult. The elattostasic regression of the deutonymph (fig. 19) pertains to the gnathosoma (which is not functional). Important deutonymphal transformations include regression in size of the notogastral setae, an advanced position of the orifice of the latero-opisthosomatic gland, a considerable development of the epimeral region, and changes in the shape of the setae of the legs (fig. 102B).

Knülle (1959: 369, figs. 22, 24–26) supposed that, in *Acarus siro* L., f_1 and f_2 are of larval origin, h_1, h_2, h_3, ps_1 and ps_2 of protonymphal origin, and ad_1 and ad_2 of adult origin. Several years ago (Van der Hammen, 1969: 198), I pointed out the abnormal character of this development. It now appears that Knülle has not understood the absence of f_1 and f_2 in the larva, and probably overlooked vestiges of ad_1 and ad_2 in the nymphs.

Griffiths (1977) studied the postembryonic development of the idiosomatic setae in *Glycacarus* (Glycacaridae); he, however, could not include the study of the deutonymph. His results correspond more or less with my data, although he applied a particular system of notation. His 'pore' represents the

←

Fig. 104. Rhizoglyphus echinopus (Fumouze & Robin); A, B, larva; A, posterior part of idiosoma, ventral view; B, Claparède's organ, lateral view; C, posterior part of opisthosoma of protonymph, ventral view; D, anogenital region of deutonymph (hypopus), ventral view; E, posterior part of opisthosoma of tritonymph, ventral view; F, posterior part of opisthosoma of adult female, ventral view; A, C–F, × 268; B, × 687.

(lyri)cupule *ih*. His larval seta l_5 representes h_2; his protonymphal setae *aP3*, *aP1* and *aP2* represent ps_1, ps_2 and h_3, respectively; his tritonymphal setae *aT2* and *aT1* (deutonymph not studied) represent ad_1 and ad_2, respectively. His tritonymphal 'vestige' *aT3* does not apparently represent a seta, but is included in the notation of the adult (which notation is probably not correct). It appears that he has not understood the larval suppression of f_1 and f_2.

All phaneres of the legs are larval in the case of legs I–III, and protonymphal in the case of leg IV, with the exception of the following phaneres, of which the base level is mentioned between brackets (fig. 100): leg I, v' trochanter (*n2*); ω_2 (*n1*), ω_3 (*n2*); leg II, v' trochanter (*n2*); leg III, v' trochanter (*n2*); leg IV, v femur (*n2*), v' tibia (*n2*), φ (*n2*), tc' and tc'' (*n2*), s (*n3*).

5.2. *The so-called genital tracheae of the female of* Gohieria fusca *(Oudemans)*

Oudemans (1928: 348) was the first to mention the occurrence of tracheae in *Ferminia fuscus* (now *Gohieria fusca*), a species of the family Glycyphagidae (suborder Acaridina). According to his concise note, he could discern two unbranched trunks. In his opinion, these did not originate from the 'pseudostigmatic organ'; he assumed that they represented genital tracheae. From the text it is not evident whether the two trunks represent two tracheae of one pair, or two tracheae of one side. According to one of his unpublished drawings, he supposed to have seen two pairs of tracheae: one running in anterior direction, another in posterior (the latter extending to the genital region, although the exact relation to that region is not drawn).

Previously, Oudemans (1906) had already discovered the presence of genital 'tracheae' in the Actinedid genus *Cyta* (Bdellidae). They are now known from various species of Bdellidae (cf. Grandjean, 1938: 6), from the Actinedid suborder Endeostigmata (Grandjean, 1939a: 45; mentioned for the genera *Speleorchestes* and *Sphaerolichus*), from the Oribatid suborder Enarthronota, viz., the genera *Cosmochthonius* (Grandjean, 1962a: 408, fig. 3D), *Haplochthonius* (Grandjean, 1947: 231, figs. 2B, 3A, F) and *Amnemochthonius* (Grandjean, 1948: 33, figs. 1B, C, D, 2A), and from the Acaridid genus *Gohieria*, which is the subject of the present subsection (see also Van der Hammen, 1968a: 270–274, fig. 3). There can be two pairs (an anterior and a posterior) or one pair. The conditions in Bdellidae and Enarthronota and related to each other: when one pair is reduced or has disappeared, it is always the posterior pair. In Endeostigmata, conditions are different: the single pair in the males and females of *Speleorchestes* represent the anterior pair; the single pair in the male (not present in the female) of *Sphaerolichus* is the posterior pair.

Grandjean (1937a: 280) had also mentioned the supposed occurrence of genital tracheae in *Ferminia fuscus*, although he himself had not studied this species. A.M. Hughes (1948: 70, fig. 95) was the first to confirm, with certainty, the occurrence of genital 'tracheae' in the female (not in the male) of *Gohieria fusca*. Later on (A.M. Hughes, 1961: 137, fig. 186), she repeated her descrip-

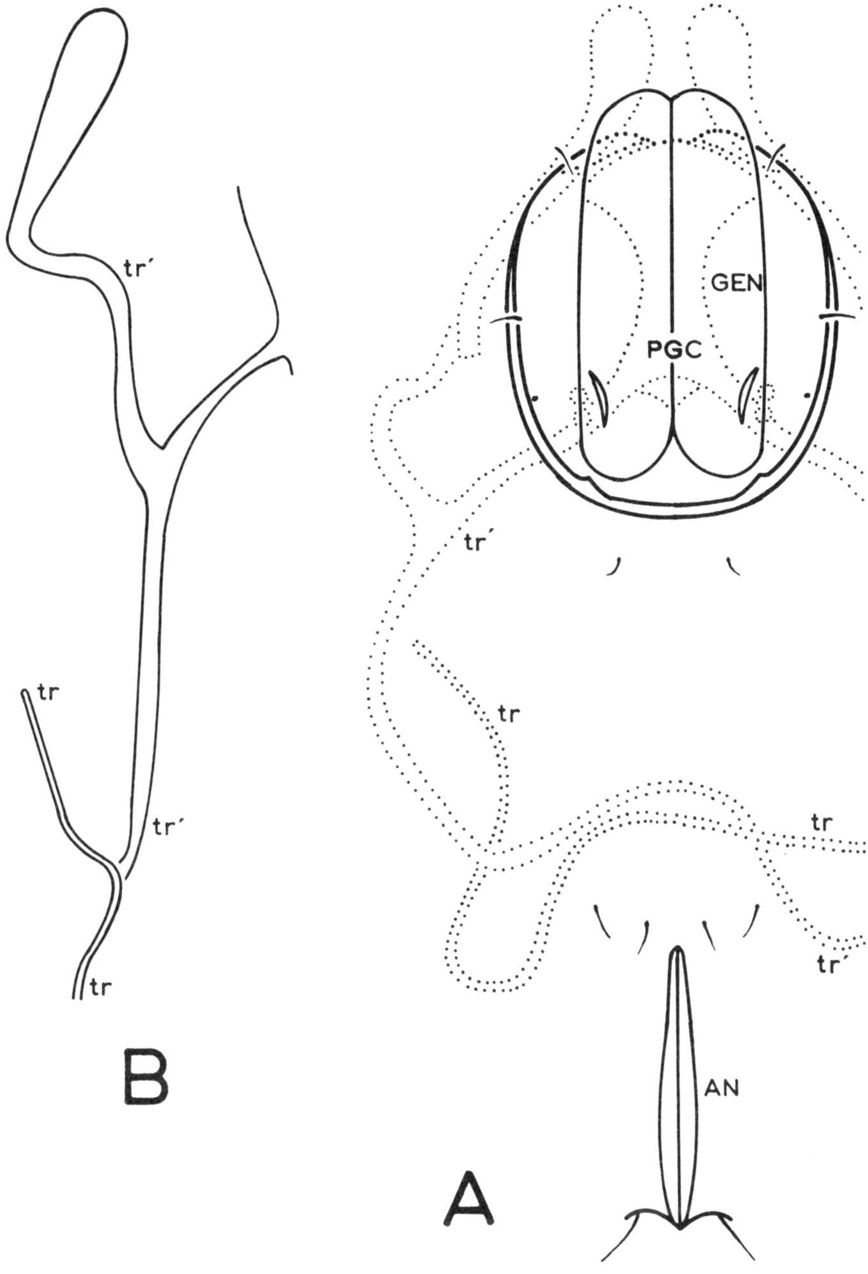

Fig. 105. Gohieria fusca (Oudemans), adult female; A, ventral view of genital region (with genital 'tracheae') and anal opening; B, lateral view of genital 'tracheae'; A, B, × 350.

tion and figure, and added a new observation: in fresh specimens, the 'tracheae' are filled with air. This observation was already published shortly before by T.E. Hughes (1959: 200, pl. 52 fig. 5). Descriptions and figures by A.M. Hughes and T.E. Hughes are not very distinct as to the orifice of the 'tracheae'. In the present subsection the 'tracheae' of *Gohieria* are described and represented according to the standards of this book. In this way, we are in a position to make a comparison with the condition in primitive Actinedida and Oribatida. The subsection is based on laboratory-reared material from England.

The genital 'tracheae' of the female of *Gohieria fusca* are represented in fig. 105. Fig. 105A (the genital region in ventral view) demonstrates that there is one pair of genital 'tracheae', which debouches into the progenital chamber. Each 'trachea' branches into two parts: an anterior part (which ends in a large caecum) and a long winding posterior part (which gradually narrows towards the end). The posterior parts cross, and this probably induced A.M. Hughes to represent them as connected with one another. The crossing of the two 'tracheae' (*tr*, *tr'*) is represented in fig. 105A as well as in fig. 105B (a lateral view of the 'tracheae'). Fig. 105B illustrates also the characteristic curve of anterior branch.

The figure demonstrates that the genital 'tracheae' of *Gohieria* correspond with the posterior pair of 'tracheae' (*tr.p*) of *Haplochthonius* and *Cosmochthonius*. In *Gohieria* no trace of an anterior 'trachea' is present, although the progenital chamber is widened in the anterior part. The genital 'tracheae' apparently constitute segmentally arranged organs. They are supposed to originate from accessory glands (the tracheal nature is, however, not yet unequivocally demonstrated). In Pseudoscorpionida, segmentally arranged accessory glands can also develop into trachea-like organs (cf. Vachon, 1938: 58, 108; Grandjean, 1944: 168).

List of the notations of figs. 94–105

a	anterior infracapitular seta.
a_1, a_2	anterior prodorsal setae (= internal and external vertical setae).
a', a''	antelateral tarsal setae.
ad_1, ad_2	adanal setae.
AN	anal opening.
ap.c	capitular apodeme.
at	line of attachment of cheliceral frame to infracapitulum.
b_1, b_2	posterior prodorsal setae (= internal and external scapular setae).
bc	bursa copulatrix.
c_1, c_2, c_3, c_p	opisthosomatic setae attributed to segments VII and VIII.
CH	chelicera.
cpc	podocephalic canal.
CX.C	cheliceral sheath.

d_1, d_2	opisthosomatic setae attributed to segment IX.
ds	diachilous slit.
e	supracoxal seta of palp.
e_1, e_2	opisthosomatic setae attributed to segment X.
eI	supracoxal seta of leg I.
F	femur.
f_1, f_2	opisthosomatic setae attributed to segment XI.
ft', ft''	fastigial setae of tarsus.
G	gena.
g_1, g_2	genital setae.
GE	genu.
GEN	genital covers.
gla	latero-opisthosomatic gland.
GO	Grandjean's organ.
H	mentum.
h_1, h_2, h_3	opisthosomatic setae attributed to segment XII.
ia	anterior (lyri)cupule of opisthosoma.
ics	superior commissural induration.
ih	(lyri)cupule attributed to segment XII.
im	median (lyri)cupule of opisthosoma.
ip	posterior (lyri)cupule of opisthosoma.
Js	superior oral commissure.
l	lateral seta of legs.
ld	laterodorsal seta of legs.
LL	lateral lips.
LS	labrum.
MA	malapophysis.
n1	protonymphal base level.
n2	deutonymphal base level.
n3	tritonymphal base level.
og	eugenital opening.
opv	ventral oncophysis of chelicera.
p', p''	proral setae of tarsus.
PA	palp.
Pg_a	anterior genital papilla.
Pg_p	posterior genital papilla.
PGC	progenital chamber.
PH	pharynx.
ps_1, ps_2	pseudanal setae.
pv	primiventral seta of tarsus.
PI, *PII*, *PIII*, *PIV*	legs I–IV.
RX	pseudorutellum.

s	subungual seta of tarsus.
se	capitular saddle.
t_i	inferior tendon of apotele.
t_s	superior tendon of apotele.
TA	tarsus.
tc', *tc''*	tectal setae of tarsus.
TG	tegulum.
tr, *tr'*	genital 'tracheae'.
TR.I	trochanter of leg I.
u', *u''*	ungual setae of tarsus.
v	ventral seta of legs.
ϵ (epsilon)	famulus.
σ, σ_1, σ_2 (sigma)	solenidia of genu.
φ (phi)	solenidion of tibia.
ω, ω_1, ω_2, ω_3 (omega)	solenidia of tarsus.
1a	epimeral seta of segment III.
3a	epimeral seta of segment V.
4a	epimeral seta of segment VI.
' (prime)	anterior element of the legs (paraxial in the case of legs I and II, antiaxial in the case of legs III and IV); the right one of two symmetrical structures.
'' (double prime)	posterior element of the legs (antiaxial in the case of legs I and II, paraxial in the case of legs III and IV).

6. Tarsonemidia

In my comparative studies, I have paid little attention to Tarsonemida. These mites are usually extremely small and application of the standard methods (dissection, decription and representation in various orientations) can be very difficult. Unfortunately, I had selected a species of the genus *Tarsonemoides* as an example of this group (Van der Hammen, 1970). Although a very numerous material of that species was at my disposal, the study was particularly thwarted (apart from the extremely small size) by the presence of a relatively very large limbus at the posterior border of the dorsal sclerites. Each limbus covers part of the next sclerite (i.e., the sclerite posterior to the sclerite of which it is an extension), by which condition several details are difficult to locate. The actual extension of each limbus was not discerned by me till after the preparation of some scanning electron micrographs. As a result of the last-mentioned study, a corrected drawing of the dorsal view was published by me, together with some generalities (Van der Hammen, 1972: 37, 63, figs. 36, 37). A survey of Tarsonemid characters, and a short statement of my views of Tarsonemid relationships (as developed by me at that time) was prepared by me shortly afterwards (Van der Hammen, 1972a: 288–289, fig. 1; see also Van der Hammen, 1973: fig. 1). In a short remark (Van der Hammen, 1982: 255), I corrected my earlier views of Tarsonemid relationships.

The main object of my 1970 study of a species of Tarsonemidae was the application of the general acarological terminology to this group. Up to the time of my study, various terminologies had been in use, which were all very different from those introduced by Grandjean. In order to facilitate the introduction of this terminology, I had added, to the text of my paper, an alphabetic list of synonyms (Van der Hammen, 1970: 29–33). It has been a great surprise to me that, soon after its introduction, the terminology proposed by me found nearly general acceptance.

I had proceeded with caution in the designation of the dorsal and paraproctal opisthosomatic setae. Grandjean's system of chaetotactic notation is based on principles of homology; in cases of deficiencies, it can only be applied when it is demonstrated which stetae have disappeared (application becomes increasingly difficult when the degree of deficiency increases; Grandjean introduced several other systems of notation for these deficient chaetotaxies). In the case of Tarsonemida, it is unknown which of the *c, d, e, f* and *h* setae have disappeared, for which reason I introduced a notation preceded by the letter *p*. It is even uncertain whether the paraproctal setae represent indeed the pseudanal setae *ps* (theoretically it is also possible that the paraproctal segment is constituted by segment *H* (= *XII*) with prelarval base level). The fact is that the Tarsonemid paraproctal setae are partly suppressed after the larval stase. Descendant suppression of Actinotrichid opisthosomatic setae is known from the inguinal setae (h_4, ps_4) and from seta f_1; setae ps_{1-3} in Oribatid mites can be subject to retardation (the reverse of descendant suppression). For all these reasons, I have tried to avoid the introduction of notations of which the homology is not proved.

As mentioned above I had made, in the original description of the species, a mistake as to the position of one of the setae (it is inserted on a limbus of which the true extent was afterwards demonstrated by scanning electron micrographs); a corrected drawing was subsequently published by me (Van der Hammen, 1972: fig. 36). It may be remarked here that Lindquist (1976, 1977) loosely applied Grandjean's system of holotrichid notation to the multideficient chaetotaxy of Tarsonemida, in paying attention to segmental arrangement but not to the homology of the individual setae; he criticized, moreover, Grandjean's carefulness (which necessitates the introduction of different systems of notation when the idionymy of the setae is unknown).

Originally, I had also the intention to study a primitive species of Pyemotoidea (e.g., Dolichocybidae). Because of my retirement, this part of my program has remained unaccomplished. For this reason, the present section is only illustrated with the more or less schematic drawings of the species of *Tarsonemoides* previously prepared by me. It may be remarked here that the generic name *Tarsonemoides* is sometimes regarded as a synonym of *Tarsonemus*. It is still mentioned as a separate genus by Krantz (1978: 329), and as such it is also treated here.

The taxon Tarsonemida, as dealt with here, includes the heterostigmatic mites. It is now generally regarded as related to the prostigmatic Raphignathae. Alberti (1980a), in his comparative study of Actinotrichid sperm morphology and spermiocytogenesis, could not establish any particular relationship (based on shared derived character states) of Tarsonemida with other groups of Actinotrichida. The diagnostic characters of the Tarsonemida are the following. Small Actinotrichid mites, the body of which is supposed to include precheliceral metameres and thirteen segments (no paraproctal segments are added in the course of postembryonic development). Part of the original opisthosomatic segmentation is supposed to be indicated by the presence of some four dorsal sclerites. The female generally presents one pair of stigmata, originally associated with apodeme 1; stigmata and tracheae are, generally, not present in the male (Tarsonemida are heterostigmatic). The female generally also presents a latero-ventral, originally prodorsal, trichobothrium with capitate sensillus (rarely present in the male). Eyes are absent. The mouthparts have been subject to regression and the gnathosoma is capsulate; the palps can be greatly reduced. The gut can end blindly, and in that case a uropore (the outlet of an excretory organ) is found at the place of the anus. The claws of legs II–III generally present two ungues and a pulvilloid structure. In the case of leg IV the claw may be lost, and in the case of leg I one of the ungues may be lost. Legs can be reduced in parasitic species of the family Podapolipidae. The life-cycle can be reduced to one or two instars (adult only, or larva and adult); these reductions are supposed to represent cases of neoteny.

The present section includes only one subsection which deals with the morphology of a *Tarsonemoides* species.

6.1. *The morphology of* Tarsonemoides limbatus *Van der Hammen*

The material of the species described in the present subsection was collected by me in the surroundings of Paris, France (numerous specimens of the species were found in wet moss). *T. limbatus* is a representative of the family Tarsonemidae. The following description is based on the description published by me in 1970, but includes several corrections; it deals with the female only.

Measurements. – Length of the female 0.195–0.225 mm (average 0.200); breadth 0.130–0.140 mm (average 0.135). Differences in length can be associated with differences in the degree of extension of the segmented body.

Habitus and colour. – The species has the usual habitus of a Tarsonemid mite, with rather broadly oval outline and convex dorsal surface. The colour is very light yellowish brown, sometimes nearly whitish.

Tegument. – The cuticle is rather hard. In reflected light, it appears to have a polished and shining surface.

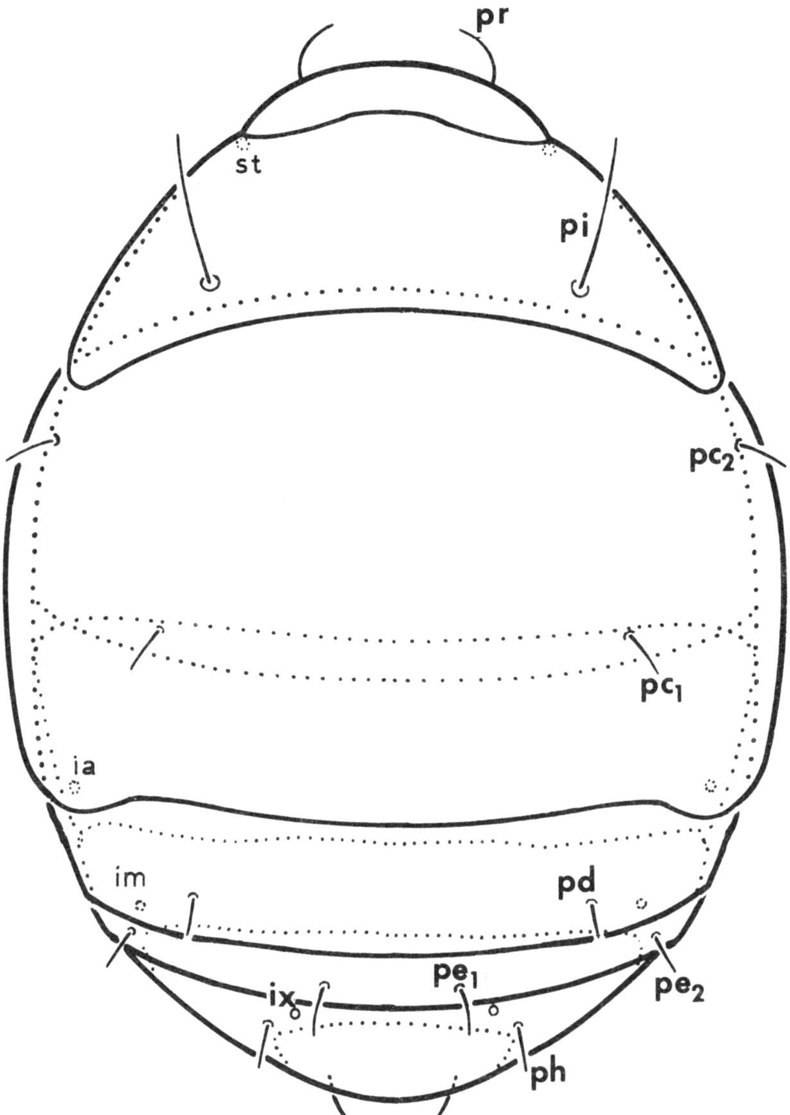

Fig. 106. Tarsonemoides limbatus Van der Hammen, dorsal view of female; × 737.

Subdivision of the body. – The gnathosoma is movably attached to the idiosoma; it can be extended in forward direction. The body is incompletely subdivided into prosoma and opisthosoma by the dorso-disjugal furrow (fig. 109: *dj*); this furrow is covered by the posterior prodorsal limbus (figs. 108, 109:

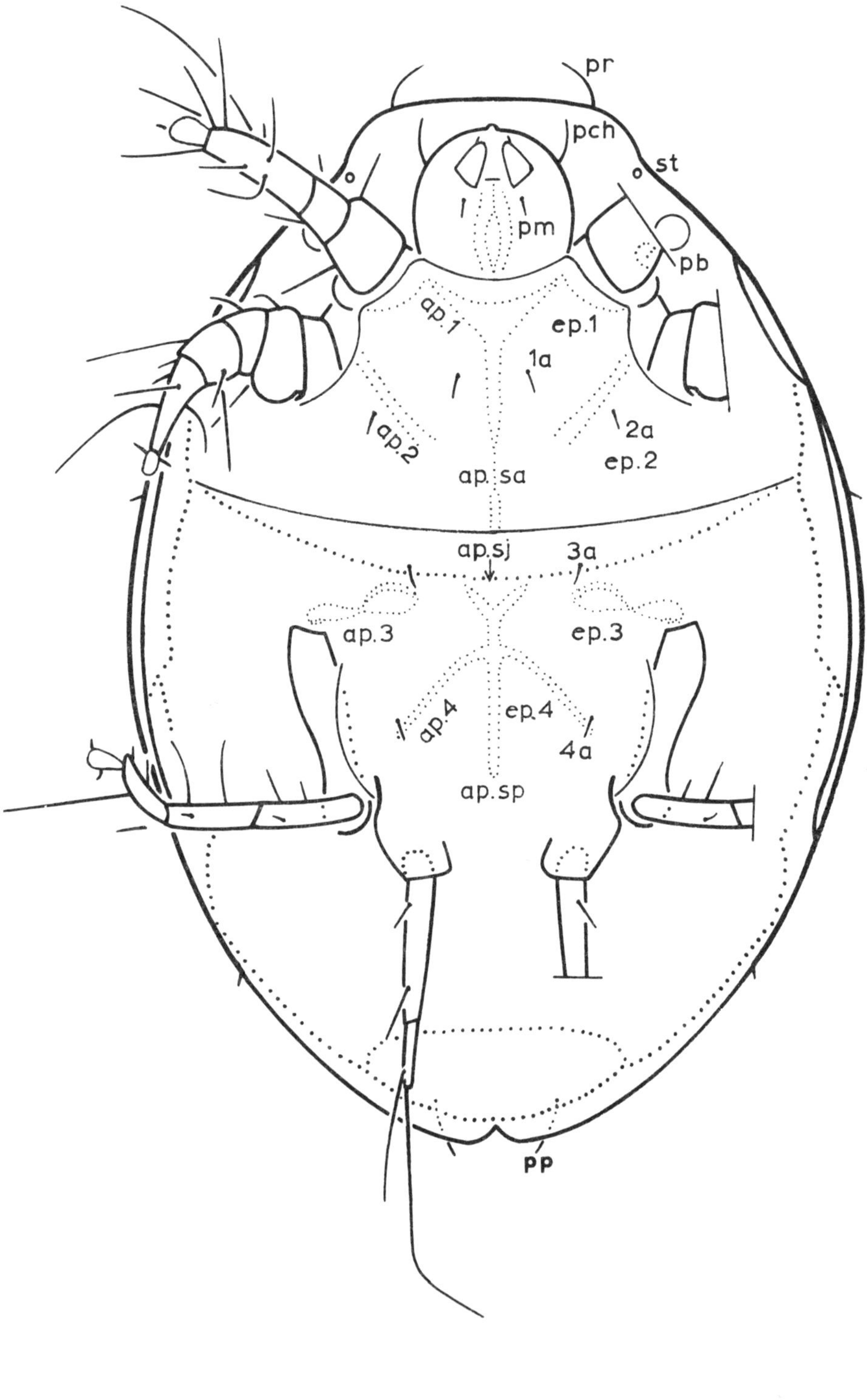
pr
pch
st
pm
pb
ap.1
ep.1
1a
ap.2
2a
ep.2
ap.sa
ap.sj
3a
ap.3
ep.3
ap.4
ep.4
4a
ap.sp
pp

ppl). Epimera 2 and 3 (the coxisternal regions of legs II and III) are separated by a large sejugal interval; this interval is often folded in such a way that the sejugal interval is covered. The region posterior to dorso-disjugal and sejugal furrows is named hysterosoma (a pseudotagma).

Prodorsum. – The prodorsum extends from the rostrum to the dorso-disjugal furrow, and includes a large prodorsal limbus. It presents two pairs of prodorsal setae: the rostral setae (figs. 106, 108, 109: *pr*; their homology with the rostral setae of other Actinotrichida is uncertain) and the posterior prodorsal setae (figs. 106, 108, 109: *pi*; homology uncertain). The third pair of aspidosomatic setae, viz., the capitate bothridials or sensilli (figs. 107, 108, 109: *pb*), occupies a lateral position and is covered by the prodorsal limbus. In primitive Actinotrichida, the ancestral number of prodorsal setae is six pairs; three of these pairs have been lost by Tarsonemidae (one of these is still present in several Pyemotidae). The rostrum presents a rostral limbus. The prodorsal limbus is particularly well-developed in the postero-lateral part. The stigmata occupy a lateral position, and will be dealt with below.

Dorsal and lateral aspect of opisthosoma. – The opisthosoma (figs. 106, 108) is covered dorsally and posteriorly by five sclerites which are supposed to represent the tergites, or fused tergites, of segments. Homologization is partly possible because of the presence of (lyri)cupules in sclerites 2, 3 and 4. The cupules of sclerites 2 and 3 certainly represent *ia* and *im*, and from this it can be deduced that sclerite 1 represents the dorsal part of 'segment' *C* (the fused segments VII and VIII), and sclerite 2 the dorsal part of segment *D* (IX). Sclerite 3 is supposed by Lindquist (1977) to represent a fusion of the dorsal sclerites of segments *E* (X) and *F* (XI), in consequence of which sclerite 4 would represent the dorsal part of *H* (XII) and the paraproctal sclerite segment *PS* (XIII); these homologies are, however, uncertain. Because of the advanced degree of deficiency, none of the dorsal and paraproctal setae can be homologized with setae present in ancestral holotrichous Actinotrichida. It is indeed unknown which of the setae are missing (as a result of regression) and application of Grandjean's system of notation (based on the holotrichous condition) would deprive this system of its scientific value (it is based on the principle of seta homology, and on the idionymy of the individual setae). For this reason, the seven pairs of dorsal and paraproctal setae have received the following notations: pc_1, pc_2, *pd*, pe_1, pe_2 (in Actinotrichida, *f* setae are generally weaker than *e* setae), *ph* and *pp*. The third pair of cupules, of which the homology is also uncertain, has received the notation *ix*.

Sclerite 1 is bordered, in the lateral and posterior part, by a large limbus. When the opisthosoma is in a retracted condition, the limbus covers the greater part of sclerite 2 (in extended condition the greater part of sclerite 2 is free).

←

Fig. 107. Tarsonemoides limbatus Van der Hammen, ventral view of female; × 737.

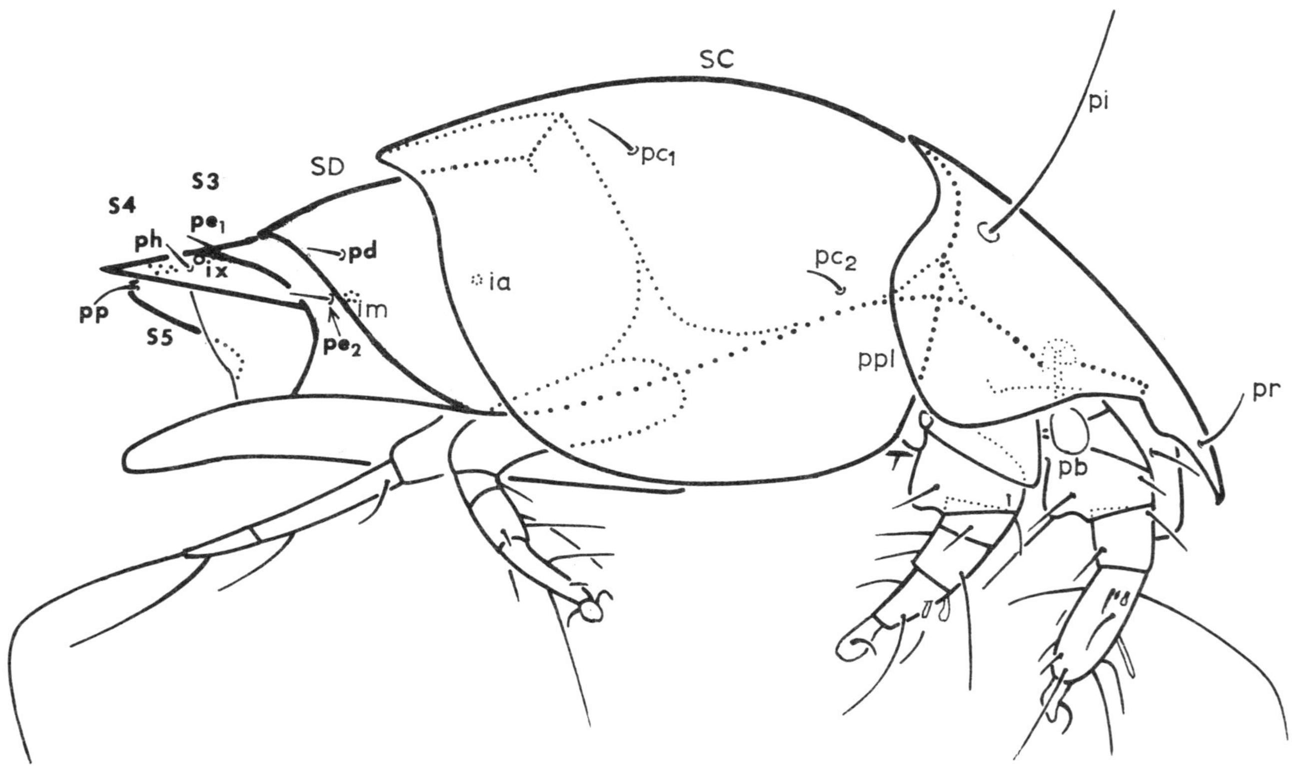

Fig. 108. Tarsonemoides limbatus Van der Hammen, lateral view of female; × 737.

Sclerites 2 and 3 are bordered, in the posterior part, by a smaller limbus.

Paraproctal region. – Sclerite 5 (fig. 110A, B) represents the dorsal part of the posterior opisthosomatic segment (the paraproctal segment). The segment has the shape of a tubercle or a cone; it is covered dorsally by the posterior limbus of sclerite 4, ventrally by the large limbus which covers the genital opening. The two limbi together could be regarded as constituting a secondary genital opening; when they are widely apart, the paraproctal segment is easily visible. This paraproctal segment presents the pair of setae *pp*, and the uropore (*u.p*) which occupies a more or less terminal position. The uropore is easily visible when the paraproctal segment is studied in posterior view. The uropore takes the place of the anal opening (the gut ends blindly); it is the orifice of an excretory canal.

Genital region. – As mentioned above, the genital opening is covered by a ventral limbus. Together with the posterior limbus of sclerite 4, this limbus constitutes a secondary genital opening. This opening is terminal; the two sclerites can be widely apart (cf. figs. 108, 110).

Ventral region of opisthosoma. – The ventral region, posterior to the legs, is entirely occupied by the large limbus which covers the genital opening. This limbus is very thin and transparent, particularly in the lateral and posterior part, and presents a posterior indentation.

Coxisternal region. – This region comprises the epimera, the sejugal interval, and the apodemes (cf. fig. 107). There are four pairs of epimeral setae: *1a, 2a, 3a* and *4a*. Legs II and III are separated by a large sejugal interval. There are several transverse and longitudinal (sagittal) apodemes (internal lamellae or ridges, which form part of the chitinous exoskeleton); these are named here: apodeme 1, apodeme 2, sejugal apodeme, apodeme 3, and apodeme 4 (the transverse apodemes); and anterior and posterior sternal apodeme (the sagittal apodemes). As mentioned above, the sejugal interval can be folded. In a folded condition, the region of the sejugal apodeme is covered by the posterior part of the sejugal interval; this condition is represented in fig. 107. The disposition of epimeral setae and apodemes is also represented in fig. 107.

Lateral region of podosoma. – The lateral region of the podosoma (figs. 108, 109) is partly covered by the posterior prodorsal limbus and the limbus of opisthosomatic sclerite 1. I could not discern supracoxal setae. As mentioned above, the bothridial seta or sensillus (which is one of the prodorsal setae) is inserted in the lateral part of the prosoma; it presents a pyriform head with scabrous surface. The bothridium is cup-shaped and well-developed. The paired stigmata occupy, on both sides, a lateral position, just under the place where the lateral parts of rostral and posterior limbi meet (cf. fig. 111C, D: *st*). There is an extremely small podocephalic canal (fig. 111C, D: *cpc*) extending from the dorsal part of the coxal region of leg I to the dorsal surface of the infracapitulum, just anterior to the line of attachment (*at*) of the cheliceral frame. I could discern the orifices of two glands (og_1 and og_2). It is very difficult to study the canal in detail; it is probably external, i.e., constituted by a taenidium.

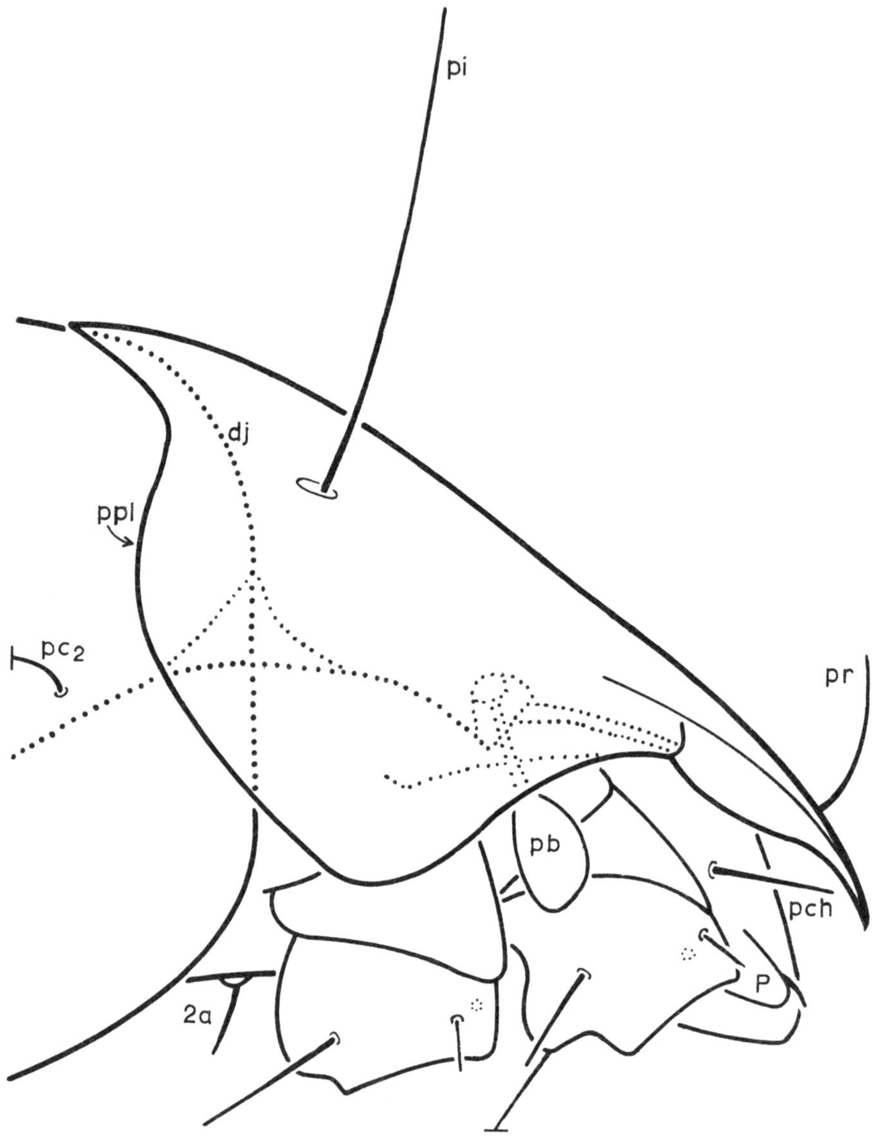

Fig. 109. Tarsonemoides limbatus Van der Hammen, female; lateral view of anterior part; × 1415.

Gnathosoma. – The gnathosoma has been subject to complicated coalescences. The cheliceral frame consists of the tegulum only (fig. 111C, D: *TG*); it extends from the rostrum to the infracapitulum, where it is attached according to the line *at*. In most mites, the chelicerae are attached to the cheliceral frame by means of the cheliceral sheaths. In Tarsonemida, the chelicerae are

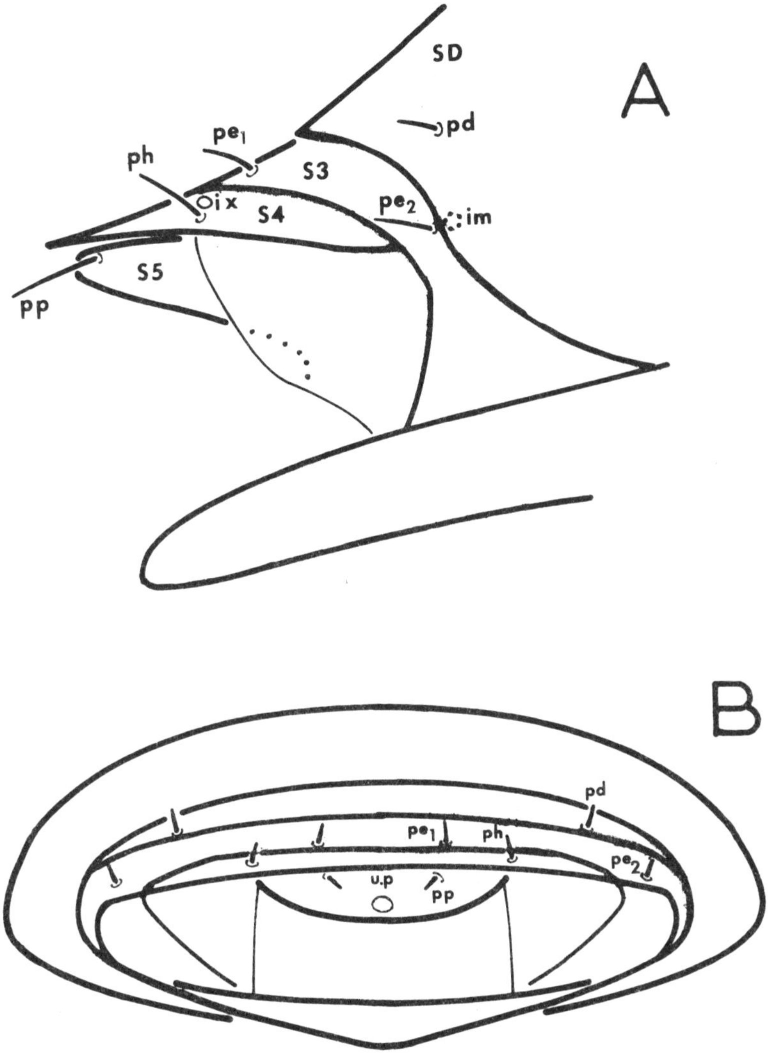

Fig. 110. Tarsonemoides limbatus Van der Hammen, posterior part of opisthosoma of female; A, lateral view; B, posterior view; A, × 1415; B, × 737.

represented by the so-called stylets, which are situated inside the infracapitulum. This exceptional position can be explained by the following hypothesis. The stylets represent the apotele (or movable jaw) of the chelicerae; they present indeed two basal knobs (cf. fig. 111B) to which the two tendons (t_s and t_i) of the apotele are probably attached. The remaining parts of the chelicerae have apparently coalesced with the dorsal surface of the infracapit-

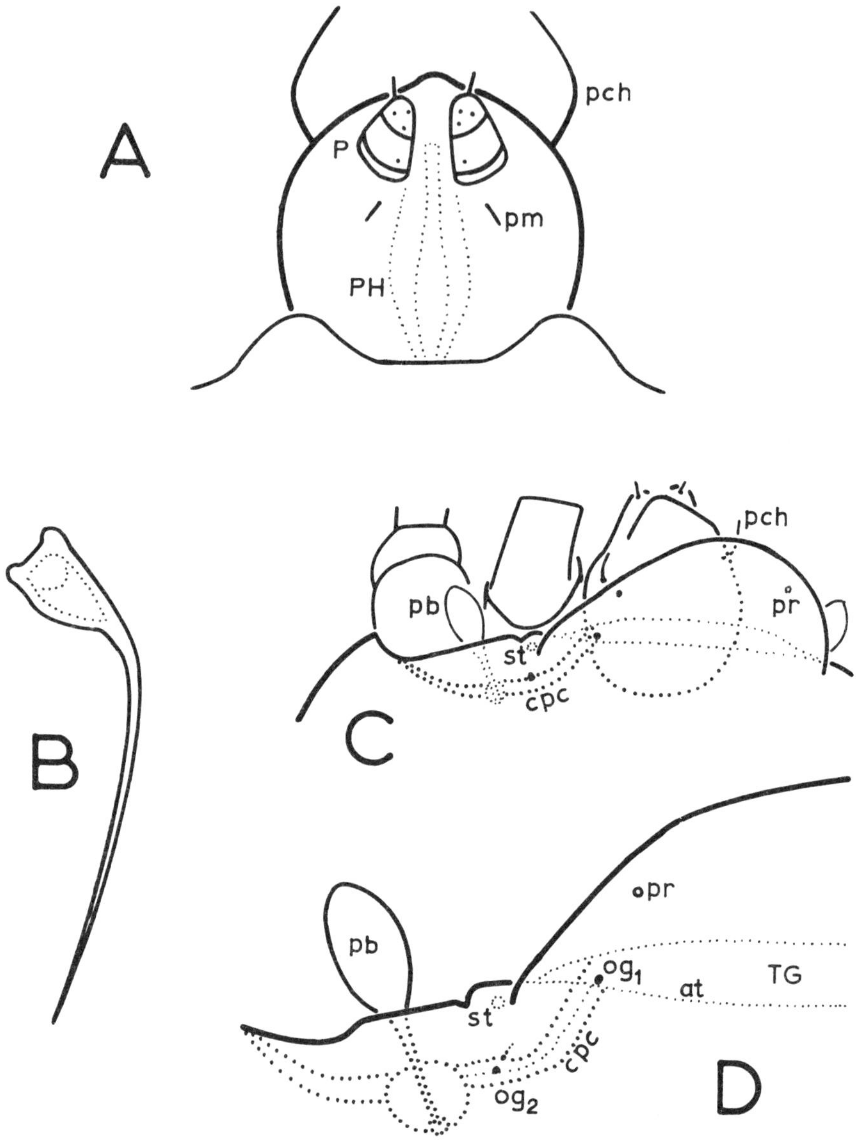

Fig. 111. Tarsonemoides limbatus Van der Hammen, female; A, ventral view of gnathosoma; B, lateral (antiaxial) view of right stylet; C, dorsal (slightly oblique) view of left podocephalic canal and adjacent parts; D, detail of fig, C; A, D, × 1415; B, × 2625; C, × 737.

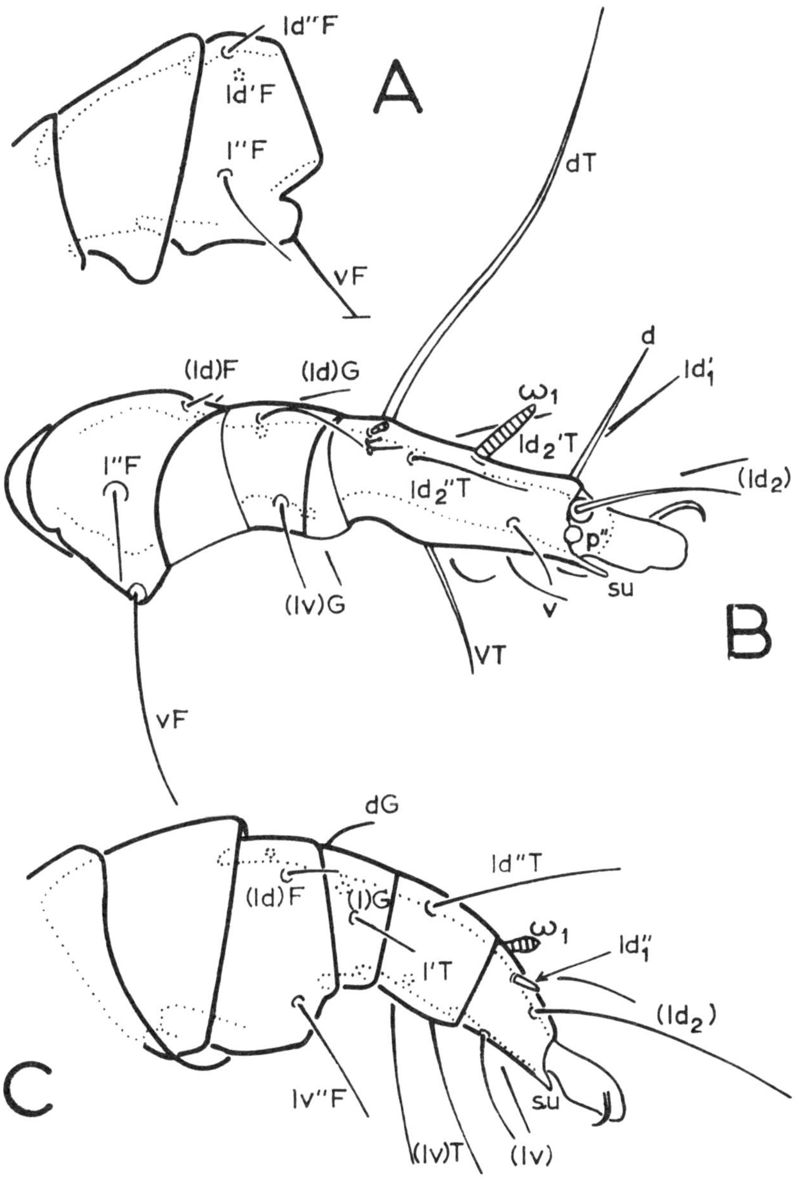

Fig. 112. Tarsonemoides limbatus Van der Hammen, female; A, lateral (antiaxial) view of trochanter and femur of right leg I; B, lateral (antiaxial) view of right leg I (trochanter nearly invisible; leg completely extended); C, lateral (antiaxial) view of right leg II; A–C; × 1415.

ulum, i.e., with the cervix. The dorsal surface of the Tarsonemid infracapitulum presents indeed one pair of setae which is unknown from the infracapitulum of groups with free chelicerae. In some Pyemotidae, there are even two pairs of dorsal setae; these certainly represent the cheliceral setae *cha* and *chb*. In comparison with Pyemotidae, the paraxial seta has disappeared (this could represent *cha*, although it occupies a more advanced position). Because of the uncertain homology, the Tarsonemid cheliceral seta has received here the notation *pch* (those of Pyemotidae could receive the notations *pch'* and *pch''*).

The ventral surface of the infracapitulum (fig. 111A) presents one pair of infracapitular setae. It is unknown whether this pair is homologous with the *a*, *m* or *h* setae of other Actinotrichida; because of its median position, it has received the notation *pm*. The pharynx has a relatively heavy wall, and has apparently developed into a highly specialized suctorial organ.

The palp is very small. It appears to consist of three segments. Segments 2 and 3 present vestiges of setae; segment 3 presents, moreover, a distinct dorsal, subterminal seta. The last-mentioned seta is birefringent when studied in polarized light; consequently, it represents indeed a true seta (no solenidion). Supracoxal setae (*e*) are not discernible; these, however, are present in Pyemotidae (they were named postpalpal setae by Cross, 1965).

Legs. – The terminology of the leg segments is based on the types of articulation: the trochanter-femur articulation is ancestrally constituted by a pivot joint; the femur-genu, genu-tibia and tibia-tarsus articulations are ancestrally constituted by hinge joints. A study of the articulations in the legs of *Tarsonemoides* results in the following terminology. Leg I consists of trochanter, femur, genu, tibiotarsus and apotele. Leg II consists of trochanter, femur, genu, tibia, tarsus and apotele. Leg III consists of trochanter, femorogenu, tibia, tarsus and apotele. Leg IV consists of segment 1, segment 2, and segment 3 (homologies uncertain). This terminology corresponds for the greater part with that of Schaarschmidt (1959) and Suski (1966, 1967, 1968), but differs from that of Cross (1965); these authors, however, did not mention the apotele as a separate segment.

Solenidia are present on legs I and II. Tibiotarsus I presents three solenidia, viz., two on the tibial part (φ_1, φ_2) and one on the tarsal part (ω_1). Tarsus II presents one solenidion (ω_1). No solenidia are found on legs III and IV.

It is not possible to homologize the setae of the legs with those of other Actinotrichida. The notation applied to the setae of the present species (figs. 112–114) is topographic. The formulae of the setae of the legs are the following (the last number of the formulae of legs I–III refers to the number of ungues): I (0–4–4–[6–8]–1); II (0–3–3–4–6–2); III (0–[1–3]–4–5–2); IV (0–2–2).

Legs I and II are represented in fig. 112, legs III and IV in fig. 114; details

→

Fig. 113. Tarsonemoides limbatus Van der Hammen, tibiotarsus and apotele of right leg I of female; A, ventral view of distal part; B, lateral (antiaxial) view; A, B, × 2625.

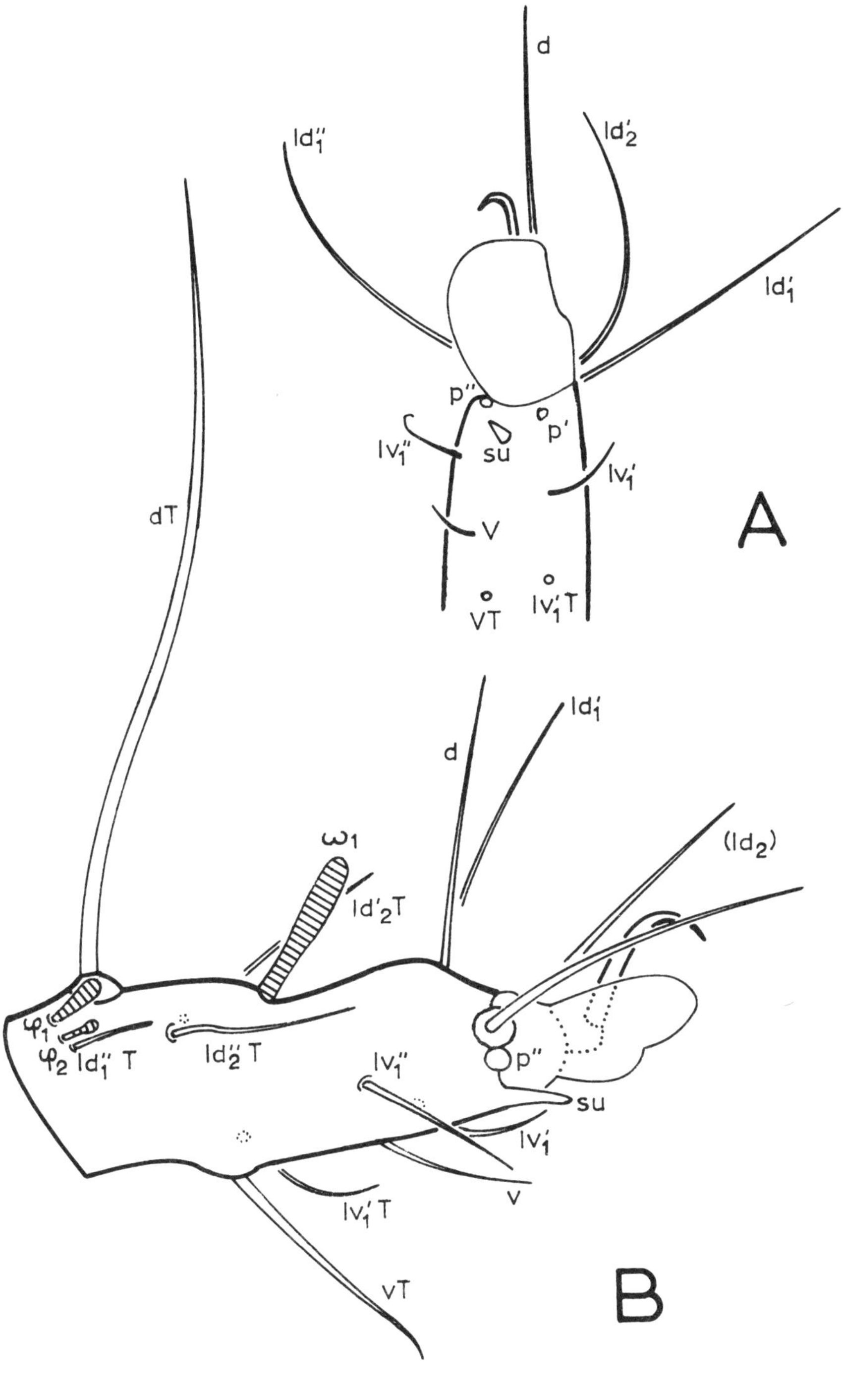
d
ld″1
ld′2
ld′1
p″
p′
su
lv″1
lv′1
v
vT
lv′1T
A
dT
d
ld′1
ω1
ld′2T
(ld2)
φ1
φ2
ld″1 T
ld″2 T
lv″1
p″
su
lv′1
v
lv′1T
vT
B

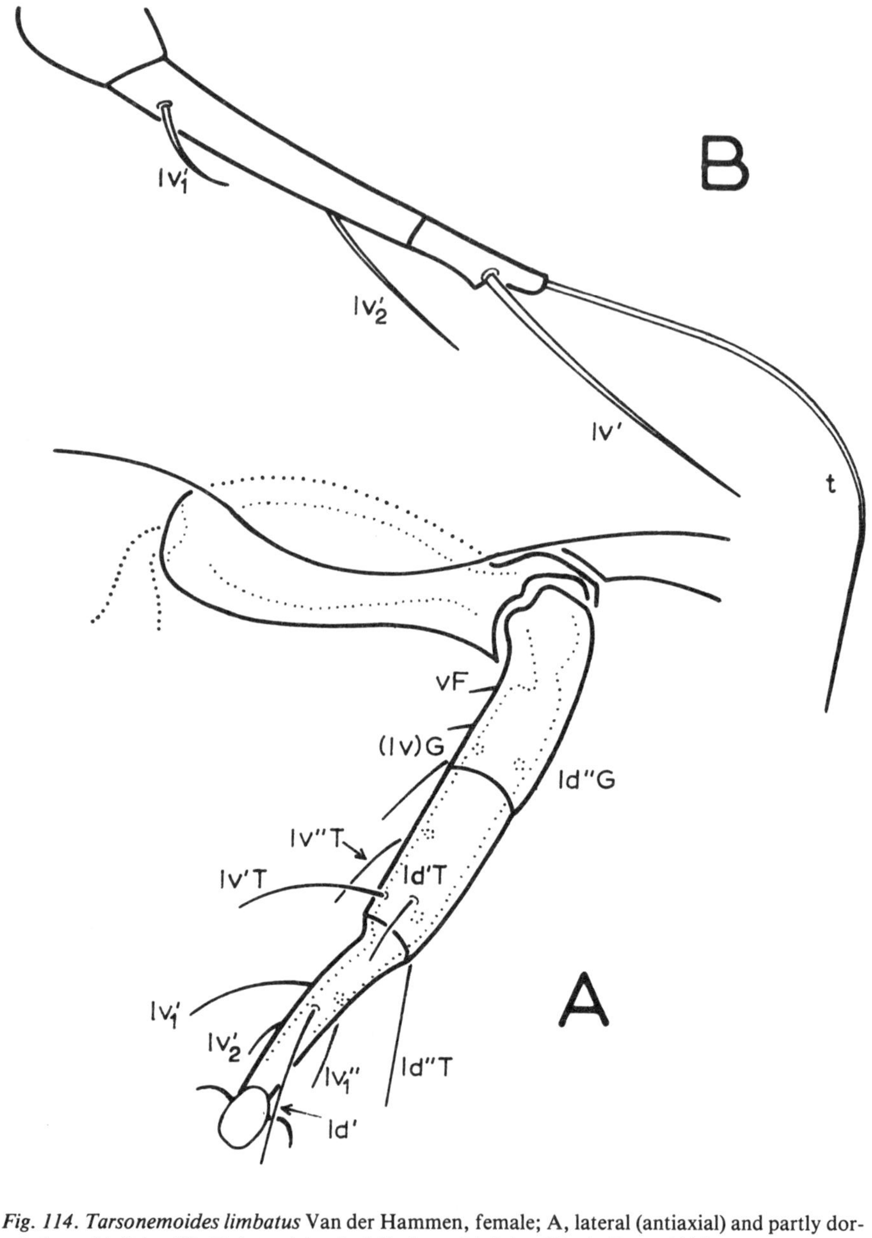

Fig. 114. Tarsonemoides limbatus Van der Hammen, female; A, lateral (antiaxial) and partly dorsal view of left leg III; B, lateral (antiaxial) view of left leg IV; A, B, × 1415.

of leg I moreover in fig. 113. The ventral terminal seta *su* of tarsus I–III is a spine. In the case of tarsus I, I discovered a pair of small, lateral, terminal structures, indicated as *p′* and *p″*; I do not know whether these represent indeed the proral setae of other Actinotrichida (they are not included in the formulae). I could not discern a famulus. It is not impossible that *d* and ld_1' of leg I represent the fastigials (*ft*), ld_2' and ld_2'' the tectals (*tc*).

Legs I, II and III present a pulvilloid structure (which, according to Lindquist, 1976, includes the empodium). The claw of leg I bears one unguis (apparently *ol′*), that of legs II and III two ungues (*ol′* and *ol″*). Leg IV presents a terminal seta *t*, of which the homology is uncertain.

List of the notations of figs. 106–114

ap.sa	anterior sternal apodeme.
ap.sj	sejugal apodeme.
ap.sp	posterior sternal apodeme.
ap.1	apodeme 1.
ap.2	apodeme 2.
ap.3	apodeme 3.
ap.4	apodeme 4.
at	line of attachment of cheliceral frame to infracapitulum.
cpc	podocephalic canal.
d	dorsal seta of appendages.
dj	disjugal furrow.
ep.1	epimeron 1.
ep.2	epimeron 2.
ep.3	epimeron 3.
ep.4	epimeron 4.
F	femur.
G	genu.
ia	anterior cupule of opisthosoma.
im	median cupule of opisthosoma.
ix	posterior cupule of opisthosoma.
l	lateral seta of the appendages.
lv	lateroventral seta of the appendages.
og_{1-2}	orifices of glands debouching into the podocephalic canal.
p	supposed homologue of proral seta.
P	palp.
pb	bothridial seta.
pc_{1-2}	setae of opisthosomatic sclerite *SC*.

pch	cheliceral seta.
pd	seta of opisthosomatic sclerite *SD*.
pe_{1-2}	setae of opisthosomatic sclerite *S3*.
ph	seta of opisthosomatic sclerite *S4*.
PH	pharynx.
pi	posterior prodorsal seta.
pm	infracapitular seta.
pp	paraproctal seta.
ppl	posterior prodorsal limbus.
pr	rostral seta.
SC	first (anterior) opisthosomatic sclerite.
SD	second opisthosomatic sclerite.
S3	third opisthosomatic sclerite.
S4	fourth opisthosomatic sclerite.
S5	fifth opisthosomatic sclerite.
st	stigma.
su	subungual spine of tarsus.
t	terminal seta of segment 3 of leg IV.
T	tibia.
TG	tegulum.
u.p	uropore.
v	ventral seta of the appendages.
φ_{1-2} (phi)	solenidia of tibia.
ω_1 (omega)	solenidion of tarsus.
1a	seta of epimeron 1.
2a	seta of epimeron 2.
3a	seta of epimeron 3.
4a	seta of epimeron 4.
′ (prime)	anterior element of the appendages.
″ (double prime)	posterior element of the appendages.
( )	a pair (of setae).

7. Relationships of the higher Actinotrichid taxa

Little is known about the phylogeny of the main Actinotrichid groups. All of them appear to be related to Endeostigmata (primitive Actinedida), which group apparently closely resembles the Actinotrichid ancestors. Investigations of sperm morphology and spermiocytogenesis, by Alberti (1980a), have confirmed the distinction of the main groups, but did not disclose special relationships (shared derived characters of some of the groups). Fundamental differences, mentioned up to now, appear to refer to evolutionary programs, realized in different ways by the different representatives. Actinedida and Oribatida represent groups which are, for the greater part, very different from each other, although the primitive representatives are closely related.

In his study of the Palaeacaroid mites, Grandjean (1954a: 262–266) discussed also the systematic position of this group (it presents affinities with Oribatid mites as well as with Endeostigmata). Among the fundamental Oribatid characters, Grandjean mentioned the scleritization of the cuticle, and the presence of a rostral tectum (at least in the adults). In a continuation of this discussion (Van der Hammen, 1969: 193–194), I mentioned the following characters which never occur in (adult) Oribatida, but which can occur in Actinedida: the absence of a rostral tectum; the presence of two pairs of prodorsal bothridial setae; the absence of a notogastral sclerite (which character does not apply to all higher Actinedida); the presence of femoral solenidia; the presence of ungues with tenent hairs; the replacement of the median unguis of the claws by an empodium. Other characters, which could be of systematic importance, pertain to the regression in the number of instars (not found in Oribatida) and the occurrence of a pulvillus (in some groups, pulvilli are regularly found; in Oribatid mites, the occurrence of a pulvilloid structure is rare, although it is known from Zetorchestidae).

In essay II of my book on Unfoldment and Manifestation (Van der Hammen, 1988), which deals with numerical changes in postembryonic development and evolution in Actinotrichid mites, I have collected data which appear to indicate that the structural genes for Actinotrichid setae are generally present in all representatives, but that differences in chaetotactic patterns are mainly the result of differences in epigenetic coding. Taxa, consequently, are characterized in particular by evolutionary potentialities and evolutionary programs (Grandjean's priority lists are part of these programs). In essay VI of the same book, which deals with a structuralist approach in the study of evolution and classification, I pointed out that branching points in a scheme underlying phylogenetic classification should correspond with the separation of the evolutionary programs of the deep structure underlying relationship, and not with the manifestation of the evolutionary potentialities at the level of the surface structure. Current taxonomic methods tend to over-estimate the last-mentioned manifestations. Procedures of this type risk the development of atomistic views in which the organism as a functional whole has no place, and in which the hierarchic value of characters is not sufficiently analysed and discussed. Rigid application of current methods could indeed lead to highly artificial classifications. In a study of the relationships of the Pediculochelidae by Norton, O'Connor & Johnson (1983), e.g., much attention is paid to the suppression of individual setae, and in this way the authors arrived at the conclusion that the family belongs to the Cosmochthonioidea. It will be interesting to reconsider this supposed relationship by a study of chaetotaxy as a whole, paying special attention to the priority lists of the Cosmochthonioidea.

The increased interest in chaetotaxy and phanerotaxy has sometimes also led to an incorrect interpretation of character states. Grandjean (1935: 25; 1942) has demonstrated that the association of solenidia and setae is secondary. In contradistinction to the evidence presented by Grandjean, and without mentioning it, Lindquist (1985: 65) regarded the separate occurrence of the phaneres as secondary and the association as ancestral.

8. Additional remarks

Actinotrichida represent one of the oldest extant groups of Chelicerata. Representatives (*Protacarus crani* Hirst, supposedly a species of the Actinedid family Eupodidae; and an unnamed species of the Oribatid suborder Palaeosomata, family Ctenacaridae) are known from the Devonian (cf. Hirst, 1923: 456–459, figs. 1–2, pl. 11 figs. a–b; Petrunkevitch, 1955: 97, fig. 1a–b; Shear, Bonamo, Grierson, Rolfe, Smith & Norton, 1984: 493, fig. 1D). These fossil species do not present important differences from extant representatives of the same taxa.

Actinotrichida constitute also one of the most successful groups of Chelicerata: the diversity and numerousness of extant species is considerable. Actinotrichid mites can be free-living or parasitic, aquatic or terrestrial, epigeous or hypogeous, xerophilous or hygrophilous, arboricolous, arenicolous, cavernicolous, epizoic, halophilous, muscicolous, saxicolous, etc.

Several species of Actinotrichid mites can be found at considerable depths in the soil. Representatives of the Actinedid family Nematalycidae are known from the interstitial environment (cf. Coineau, Fize & Delamare Deboutteville, 1976). Actinotrichid ancestors (as in the case of Palpigradid ancestors) could indeed have colonized the terrestrial soil, long before the Silurian, by way of the interstitial environment. For this reason, they could be among the oldest terrestrial animals.

III. SOLIFUGAE

In the fourth part of my Comparative Studies in Chelicerata (Van der Hammen, 1986a: 5–7, 14, 17, 21, 30, figs. 1A, 4A, 8, 15), I have dealt with some of the elements of Solifugid morphology, viz., segmentation of the body, respiratory system, orifices of coxal glands, mouthparts, and segmentation and articulation of the legs. Previously, I had mentioned Solifugae only in passing, in some general papers dealing with Chelicerate morphology, postembryonic development and classification (Van der Hammen, 1963b: 445, 448; 1969: 197; 1974: 386, 387, 388; 1977a: 310, 312, 313, 314, 316–317, tab. 1–3; 1978: 45, 55; 1985: 55, fig. 34; 1985a: 137, 139, 142–145, fig. 3; 1986c: 224–225, 227, fig. 3). Recently, moreover, I compared the functional morphology of the legs of *Anystis* (Actinotrichida: Actinedida) with that of Solifugae (Van der Hammen, 1986c: 220). The results of these studies, supplemented with some data from literature, are summarized in the present chapter. It is subdivided into three sections: an introduction, a descriptive part, and some additional remarks.

1. Introduction

Data with reference to the comparative morphology of Solifugae have been summarized by Roewer (1932–1934), Kaestner (1933a; 1955: 66–68; 1956: 598–603), and Millot & Vachon (1949a). Snodgrass (1948: 22–26, fig. 7) summarized data with reference to the Solifugid mouthparts. Of the more recent studies, mention must be made here of the study by Manton (1977: 468–469, fig. 10.3) on the functional morphology of the Solifugid legs, the study by Firstman (1973: 18–19) on the Solifugid circulatory system (and the absence of an endosternite), and the studies by Alberti (1979a, 1980b) on the Solifugid coxal glands and on the fine structure of Solifugid sperm. Interesting data with reference to the Solifugid postembryonic development are mentioned by Lawrence (1947) and Muma (1966). The systematic position of Solifugae was recently studied by Weygoldt & Paulus (1979, 1979a) and Weygoldt (1980).

The following discussion of Solifugid comparative morphology and postembryonic development is based on an original study of the morphology of *Pseudocleobis andinus* (Pocock) supplemented with data from literature.

2. Descriptive part

Habitus. – Medium-sized or rather large Chelicerata with long setae. Colour generally more or less greyish brown. Anterodorsal part of prosoma with a strongly curved and sclerotized propeltidium. Opisthosoma cylindrical and segmented. Chelicerae considerably developed and directed forward. Palp large; leg I slender. Ventral surface of the coxisternal region of leg IV, and of trochanter 1 and 2 of leg IV with so-called racked organs (malleoli). Movements rapid. Prosoma and opisthosoma can be raised.

Divisions of the body, tagmata and segmentation. – The Solifugid body is divided into two tagmata (prosoma and opisthosoma) which are broadly joined. Data with reference to Solifugid segmentation have been summarized by Kaestner (1933a: 195–204, figs. 271–279), Roewer (1932: 29–44, 93–101, figs. 13–33, 83–91), and Millot & Vachon (1949a: 483–491, figs. 243–247). The Solifugid prodorsum (the dorsal part of the prosoma) generally consists of two main sclerites (propeltidium and postpeltidium), whilst smaller sclerites can be present in the soft skin between them. These smaller sclerites are, for instance, particularly distinct in *Solpuga* (see Roewer, 1932: figs. 25–26); they are generally named anterior and posterior arcs, and the arrangement in *Solpuga* suggests that they constitute arthrodial sclerites, and must be connected with the articulation of the prodorsum. The subdivided condition of the prodorsum is generally regarded as primitive; Weygoldt & Paulus (1979: 89) suggested that it could also be secondary. The ventral region of the prosoma is occupied by the coxisternal regions of the appendages. The Solifugid opisthosoma comprises the segments VII–XVII. Each opisthosomatic segment generally consists of a tergite, a sternite, and a large pleural region of soft skin; the successive tergites and sternites are also connected by soft skin. The longitudinal genital opening is supposed to be bordered by homologues of the opisthosomatic appendages of segment VIII. Segment XVII (the anal segment) is constituted by a single sclerite. According to Millot & Vachon (1949a: 486), the sclerite generally named metapeltidium (and attributed to the prodorsum) constitutes the tergite of segment VII, whilst the small sternite anterior to the genital segment (and between the coxisternal region of leg IV) is regarded by these authors as the sternite of segment VII (Millot & Vachon, 1949a: 491). The anal opening (in segment XVII) is more or less terminal, with the exception of species of the family Rhagodidae, in which it occupies a more ventral position. The segmentation of *Pseudocleobis andinus* is represented in fig. 2A.

Cuticle. – The chitinous tegument consists of sclerites and soft skin. The sclerites include pro- and postpeltidium (and the associated arthrodial sclerites), the coxisternal regions (epimera), and the tergites and sternites of the opisthosoma. The cuticle of palp and legs is relatively thin. There are various types of phaneres on body and appendages, among which hollow phaneres with internal and external striation (on palp and legs; they could represent solenidia), and variously shaped solid phaneres with external striation only (setae). Lyrifissures are found in the cuticle of the chelicerae.

Aspidosoma. – As mentioned above, the aspidosoma (or prodorsum) generally comprises two sclerites, propeltidium and postpeltidium, connected by a pair of arthrodial sclerites. The propeltidium is strongly curved (it is distinctly convex, and the rostal part is raised). A pair of median eyes is situated on a small elevation, in the rostral part of the propeltidium. Vestiges of one or two pairs of lateral eyes (hardly discernible) are present at the anterior border of the propeltidium, near the place of insertion of the chelicerae. Generally, on either side, a lateral lobe, a lamina major, and a lamina minor can be distinguished.

Dorsal and laterodorsal region of opisthosoma. – This region comprises the opisthosomatic sclerites, and the intersegmental and pleural regions of soft skin. Segmentation can be obscured by the presence of numerous setae (in sclerotized as well as soft parts).

Anal region. – As mentioned above, segment XVII (the anal segment) is constituted by a single sclerite (representing tergite and sternite). The anal opening is more or less terminal, with the exception of species of the family Rhagodidae (in which it has a more ventral position). Generally, the anus has the shape of a vertical slit; it is closed by muscles.

Ventral and lateroventral region of opisthosoma. – This region comprises the opisthosomatic sternites, and the intersegmental and pleural regions of soft skin. The sternite of segment VII is supposed to be represented by the small sternite anterior to the genital segment (between epimera 4). The longitudinal genital opening is supposed to be bordered by homologues of the opisthosomatic appendages of segment VIII (the genital plates). Generally, there are, between the sexes, only small differences in the shapes of the genital plates. Two pairs of opisthosomatic stigmata (fig. 115A) are situated near the posterior border of the sternites of segments IX and X respectively (the unpaired stigma of Galeodidae and Hexisopodidae is near the posterior border of sternite XI). The Solifugid opisthosomatic respiratory organs could have arisen from homologues of opisthosomatic appendages, although, in the course of embryonic development, the rudiments of these appendages disappear before the development of the stigmata.

Ventral aspect of podosoma. – The coxisternal part of the podosoma is represented by the epimera; these are constituted by sclerites, separated in the sagittal plane by a suture. The epimera are associated with large internal apodemes. The coxisternal regions of palp and legs I and II (*CX.P, CX.I, CX.II*) are contiguous. The coxisternal regions of legs II and III (epimera 2 and 3) are transversely separated by the sejugal interval (an intersegmental area of soft skin, which allows of prosomatic articulation). The coxisternal regions of legs III and IV (epimera 3 and 4) are also separated by an interval of soft skin. Paired prosomatic stigmata are present at the posterior border of the coxisternal region of legs II (fig. 115A).

Lateral aspect of podosoma. – The coxal regions of palp and legs (consti-

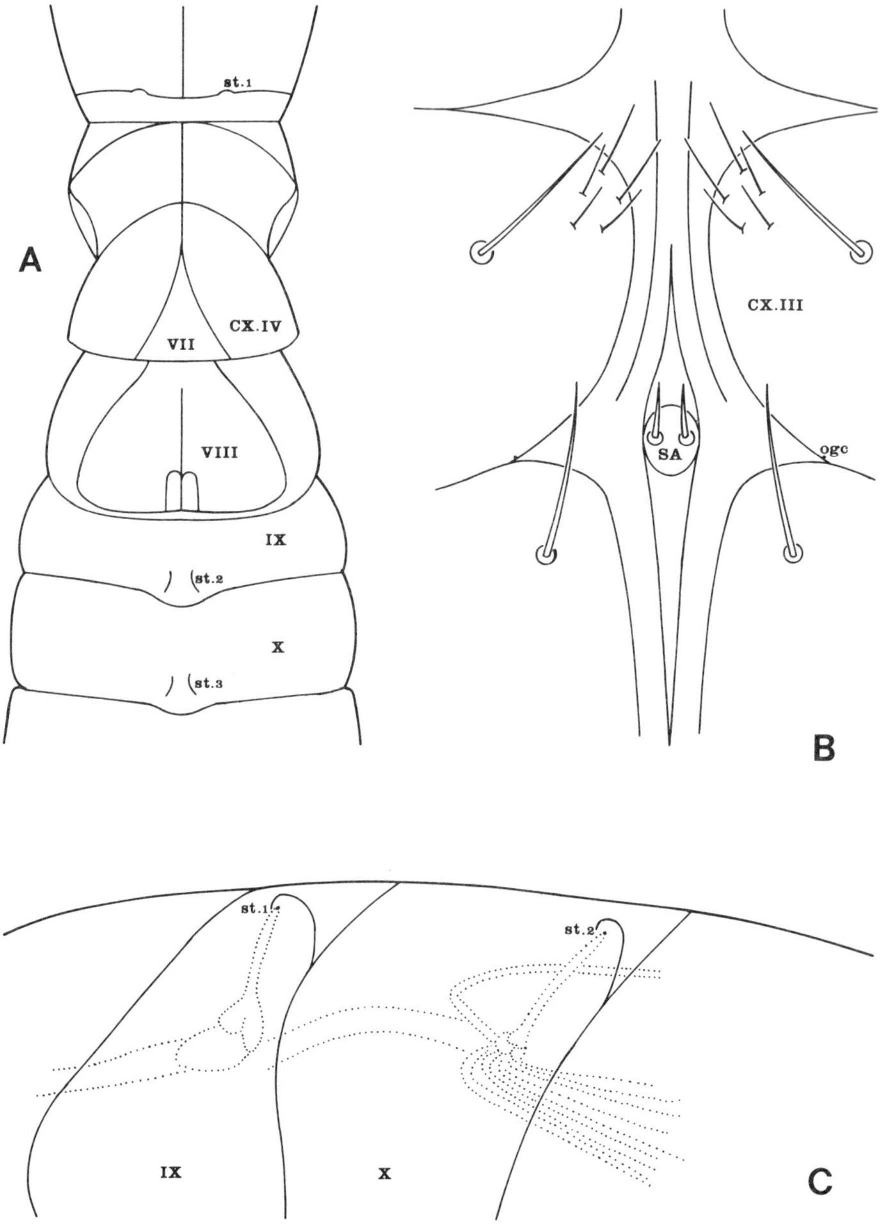

Fig. 115. Representatives of the two groups of Apatellata in ventral view; A, Solifugae: *Pseudocleobis andinus* (Pocock), posterior part of prosoma and anterior part of opisthosoma, ventral view; B, C, Pseudoscorpionida: *Chthonius (Chthonius) tenuis* L. Koch; B, ventral view of part of coxisternal region; C, ventral (oblique) view of left part of opisthosoma with segments IX and X and stigmata 1 and 2; A, × 14; B, × 591; C, × 318.

tuted by epimera and pleural regions) contain the acetabula. Supracoxal setae are apparently not present. The orifice of the coxal gland is dorsal, at the base of the coxal region of the palp.

Mouthparts. – The Solifugid mouthparts (see Roewer, 1932: 44–58, figs. 33–51) comprise the chelicerae and the so-called rostrum (here termed rostrosoma). The last-mentioned structure consists of the labrum and the lateral lips, and includes the mouth and the pharynx. A dorsal view of the Solifugid rostrosoma is represented in fig. 116B, a ventral view in fig. 116A, and a lateral view in fig. 116D. The labrum is characterized by the presence, in the distal part, of anastomosed setae (fig. 116C); its dorsal surface is a prolongation of the cervix which has fused with the coxal regions of the palps. The cheliceral frame is attached to the posterior border of the cervix, according to the line *at*. The interior continuation of the cervix (posterior to *at*) is constituted by the rostrosomatic apodeme, the place of insertion of extrinsic muscles. The labium, which probably represents the sternal part of segment II, has a ventral position between the coxal regions of the palps (to which it is attached). The basal part of the lateral lips has fused with labrum and labium; these lips present many large setae, among which a pair of plumose setae (fig. 116E). The coxal region of the palp presents a small coxapophysis (*CA.P*). As mentioned above, the orifice of the coxal glands is basidorsal, in the coxal region of the palp.

Chelicera. – The huge Solifugid chelicerae are two-segmented; they consist of the body of the chelicera (including the fixed jaw of the chela) and the apotele (the movable jaw of the chela). They are directed straight forward from the rostral part of the body. The chela presents teeth; it operates in a vertical plane. The chelicerae can be characterized as raptorial; they are prehensile as well as adapted to holding and fragmenting. They present numerous setae (of various types), lyrifissures, and a stridulation organ. The male chelicera can also function as a gonopod; it is characterized by the presence (on the fixed jaw) of a so-called flagellum.

Palp. – Apart from the coxal region, the palp consists of trochanter, femur 1, femur 2, tibia, tarsus and apotele; the tarsus is not subdivided into adesmatic segments. The apotele is transformed into an adhesive (or tactile) organ. The palp resembles the legs. It can participate in food-intake, and sometimes has a prehensile function.

Legs. – Apart from the coxal region, the Solifugid legs consist of trochanter 1, trochanter 2 (a single trochanter is present in the case of legs I and II), femur 1, femur 2, tibia, basitarsus, telotarsus and apotele (sometimes reduced or lacking in leg I). Leg IV of *Pseudocleobis andinus* is represented in fig. 117A, leg II in fig. 117B. The coxal regions are not movable. The promotor-remotor movements take place at the body-trochanter articulation, which is constituted by a pivot joint. The articulation between trochanter 1 and trochanter 2 (in legs III and IV) is constituted by a pivot joint with levator and depressor muscles. In the trochanter-femur 1 (legs I and II) or trochanter 2-femur 1 (legs III and IV) articulation, the axis of movement is nearly longitu-

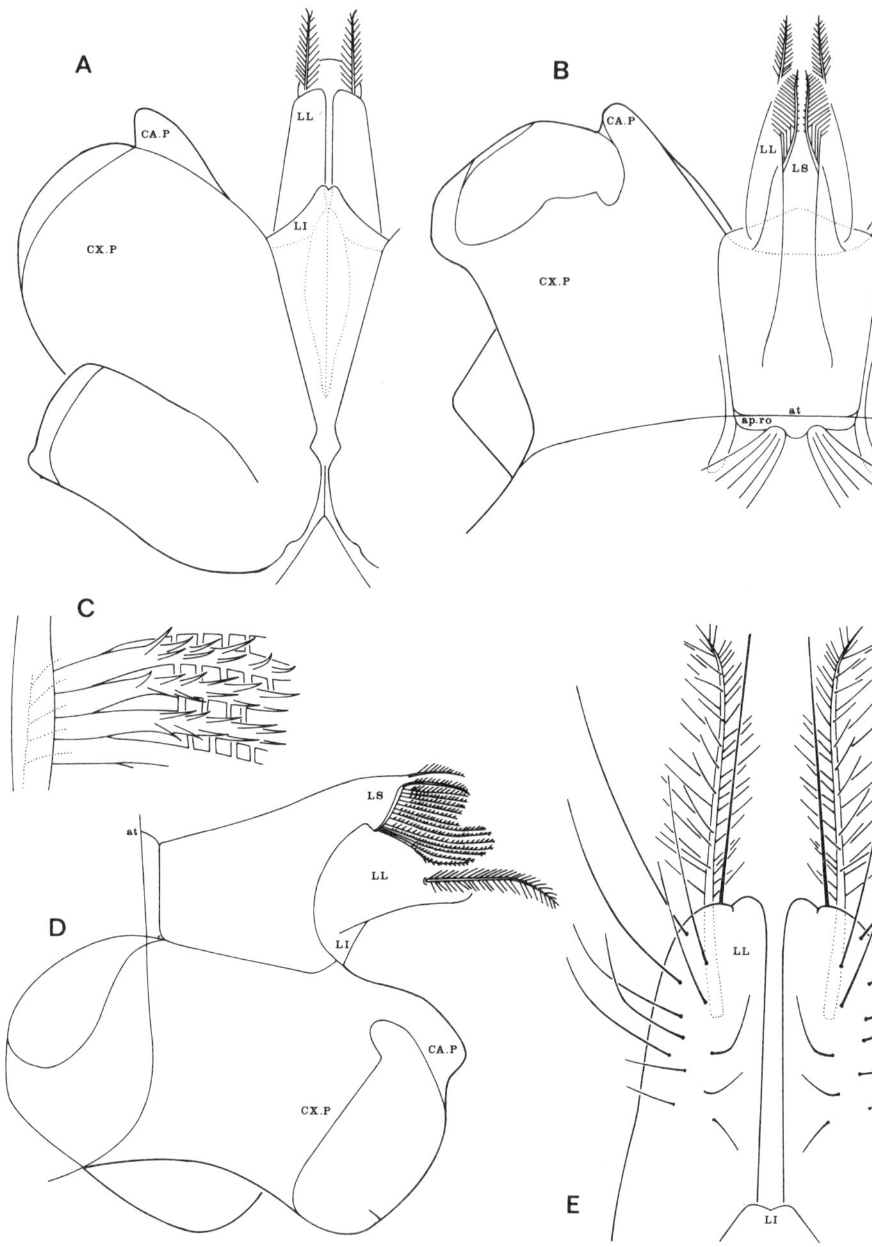

Fig. 116. Pseudocleobis andinus (Pocock), rostrosoma of adult; A, ventral view; B, dorsal view; C, basal part of some of the anastomosed setae of the labrum, lateral view; D, lateral (oblique) view; E, ventral view of lateral lips (in A, B, and D, all setae, with the exception of the large distal setae of the labrum, are omitted); A, B, D, × 33; C, × 198; E, × 106.

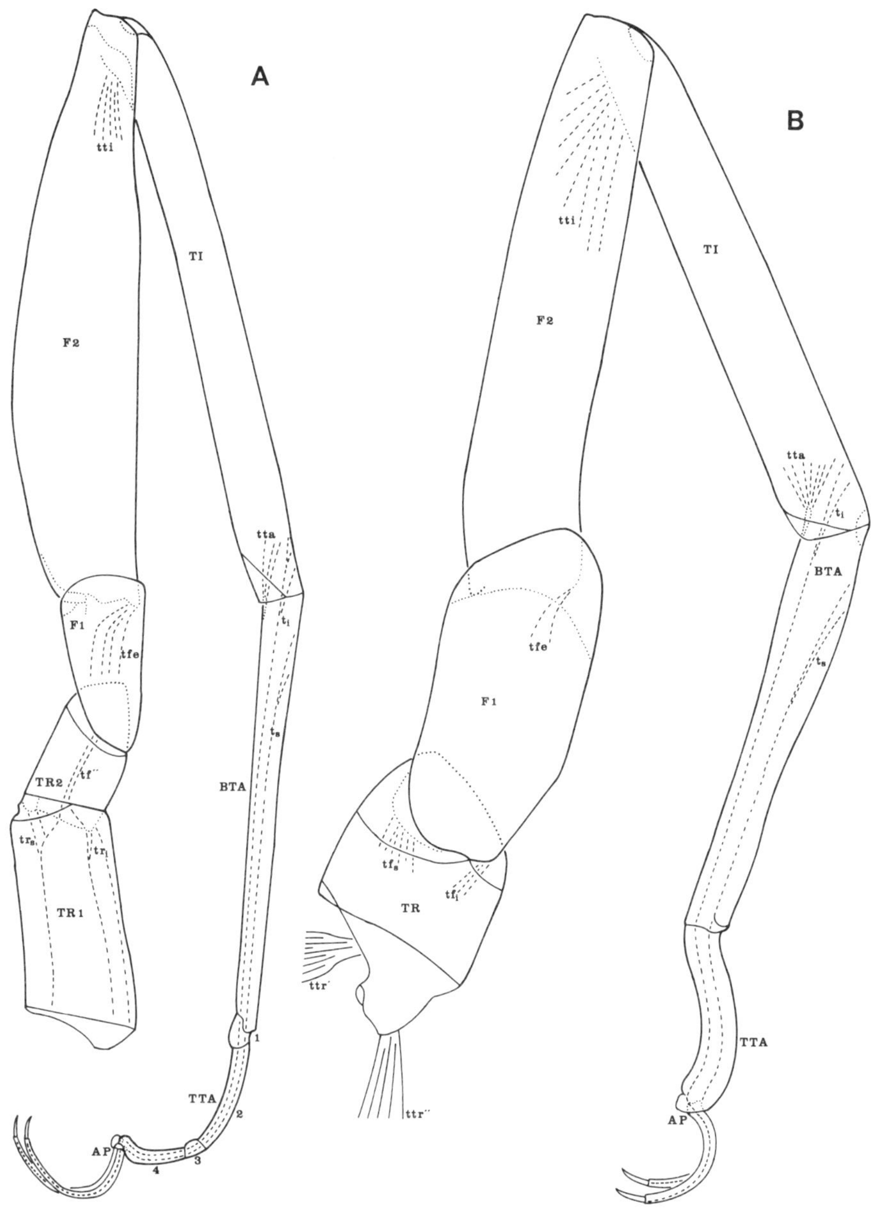

Fig. 117. Pseudocleobis andinus (Pocock); A, lateral (posterior) view of right leg IV; B, lateral (posterior) view of right leg II; A, × 17.4; B, × 36.6.

dinal (because of the advanced position of one of the condyles) and permits of leg rocking. The articulation between femur 1 and femur 2 is constituted by a hinge joint with flexor muscles. The hinge joint between femur 2 and tibia (there is no patella) permits of considerable flexion. A hinge joint with flexor muscles is also present between tibia and basitarsus. The adesmatic articulation between basi- and telotarsus is operated by the tendons of the apotele. The telotarsus-apotele articulation is constituted by the usual bicondylar articulation with levator and depressor muscles. The telotarsus can be subdivided into a variable number of adesmatic segments (leg IV of *Pseudocleobis andinus* presents four adesmatic telotarsal segments: the telotarsi 1–4). The Solifugid ungues are bipartite; the distal part is separately movable (the inferior tendon of the apotele, subdivided in the basal part, is inserted on the distal ungual division).

Internal anatomy. – A summary of the data with reference to the Solifugid internal anatomy was prepared by Millot & Vachon (1949a: 495–511, figs. 253–269). Important additional data were, e.g., published by Firstman (1973: 18–19, fig. 12) and Alberti (1979a, 1980b).

The alimentary canal comprises the mouth, the pharynx, the oesophagus, the midgut with numerous prosomatic and opisthosomatic caeca, the hindgut and the anus. The excretory system includes the coxal glands, the Malpighian tubes and the pericardial cells. The coxal glands (cf. Alberti, 1979a), which present similarities with the coxal glands of the Palpigradi, consist of a saccule, an initial segment, a glandular segment, a distal tubule (possibly a reservoir), a cuticular-lined excretory duct, and a specialized excretory pore in the dorsal part of the coxal region of the palp; form and extension of glandular segment, distal tubule and excretory pore differ considerably in various species. The Malpighian tubes debouch into the midgut (at the level of segment XII). A mesodermal endosternite is lacking. The circulatory system is reduced and comprises a heart (dorsal vessel), an anterior aorta (periganglionic arterial sinus), a posterior aorta, and a lacunar system. The respiratory system is represented by a vast system of tracheae and three pairs of stigmata (a posterior unpaired stigma is, moreover, present in the families Galeodidae and Hexisopodidae). As mentioned above, the anterior pair of stigmata is prosomatic, whilst the other stigmata are opisthosomatic. The prosomatic stigmata are situated near the posterior border of the coxisternal region of legs II, the two pairs of opisthosomatic stigmata are situated near the posterior border of the sternites of segments IX and X (the unpaired stigma of Galeodidae and Hexisopodidae is near the posterior border of sternite XI). The central nervous system comprises a supra-oesophageal syncerebrum, an important suboesophageal ganglionic mass, and a small opisthosomatic ganglionic nodule. The male reproductive system comprises four long testicular tubes (with testicular epithelium, as in Actinotrichida; cf. Alberti, 1980b), vasa deferentia, and a single vesicula seminalis (debouching into the genital chamber). Solifugid sperm presents strongly derived and simplified characters (cf. Alberti, 1980b).

The female reproductive system comprises ovaria, oviducts, a uterus (debouching into the genital chamber) and receptacula seminis. An important part of the prosoma is occupied by the extrinsic muscles of the appendages; besides these, longitudinal, dorso-ventral and oblique prosomatic muscles are present. The opisthosomatic musculature consists of dorsal, ventral and lateral longitudinal muscles, and dorso-ventral muscles (in the segments VIII–XIII).

Reproduction. – Solifugid reproduction is bisexual. The reproductive behaviour (it takes only a few minutes) involves sperm transfer by a gonopod (gonopody). The Solifugid chelicera is functioning as a gonopod, whilst sperm is transferred in spermatophores. Solifugae are oviparous.

Postembryonic development. – The number of Solifugid moults is rather great and variable. The first instar (the larva) is represented by an elattostase (nearly a calyptostase). The subsequent instars probably represent nymphoids, although the occurrence of some nymphs and isophena is not precluded; an elattostasic protonymph, for instance, could be present. The larva is characterized by the occurrence of a so-called lateral organ at the base of leg II; this organ (it varies in size in the different species) could represent a vestige of an 'exite'.

3. Additional remarks

About 600 species of Solifugae are now known; they are generally restricted to dry and warm regions. They live on the ground, burrow or hide under stones, and can move very rapidly; they are carnivorous and very ferocious, and digest their prey externally. Two families are found in America only; the other families are restricted to Africa, Asia and Southern Europe. All species are more or less characterized by the same general habitus. Some ten families are now distinguished. A fossil species of Solifugae, which already presents the general habitus of the extant representatives, is known from the Carboniferous.

List of the notations of figs. 115–117

AP	apotele.
ap.ro	rostrosomatic apodeme.
at	line of attachment of cheliceral frame.
BTA	basitarsus.
CA.P	coxapophysis of palp.
CX.P	coxal region of palp.
CX.III	coxisternal region of leg III (epimeron 3).
CX.IV	coxisternal region of leg IV (epimeron 4).

F1	femur 1.
F2	femur 2.
LI	labium.
LL	lateral lip.
LS	labrum.
ogc	orifice of coxal gland.
SA	sternapophysis.
st.1–3	stigmata.
t_i	inferior tendon of apotele.
t_s	superior tendon of apotele.
tf''	posterior tendons and muscles of femur.
tf_i	inferior tendons and muscles of femur.
tf_s	superior tendons and muscles of femur.
tfe	tendons and muscles of femur 2.
TI	tibia.
tr_i	inferior tendons and muscles of trochanter 2.
tr_s	superior tendons and muscles of trochanter 2.
TR	trochanter.
TR1	trochanter 1.
TR2	trochanter 2.
TTA	telotarsus.
TTA1–4	telotarsus 1–4.
tta	tendons and muscles of tarsus.
tti	tendons and muscles of tibia.
ttr', ttr''	anterior and posterior tendons and muscles of trochanter.
VII–X	segments VII–X.
' (prime)	anterior (pertaining to the anterior side of an appendage).
'' (double prime)	posterior (pertaining to the posterior side of an appendage).

IV. PSEUDOSCORPIONIDA

In the fourth part of my Comparative Studies in Chelicerata (Van der Hammen, 1986a: 7–8, 14, 18, 21–23, 30–32, figs. 1B, 4B, 4C, 9, 16), I have also dealt with some aspects of Pseudoscorpionid morphology, viz., segmentation of the body, respiratory system, orifices of coxal glands, mouthparts, and segmentation and articulation of the legs. Previously, I had mentioned Pseudoscorpionida only in passing, in some general papers dealing with Chelicerate morphology, postembryonic development and classification (Van der Hammen, 1974: 386; 1977a: 310, 312, 313, 317, tab. 1–3; 1978: 45, 49, 52; 1985: 55, fig. 34; 1985a: 137, 139, 142, 143, 144, 145, figs. 3, 7; 1986c: 224–225, 227, fig. 3). The results of these studies, supplemented with some data from literature, are summarized in the present chapter. It is again subdivided into three parts: an introduction, a descriptive part, and some additional remarks.

1. Introduction

Data with reference to the comparative morphology of Pseudoscorpionida have been summarized, among others, by Chamberlin (1931), Beier (1932), Kaestner (1955: 64–66, fig. 43; 1956: 588–597, figs. 582–586) and Vachon (1949). Snodgrass (1948: 33–39, figs. 11–12) summarized data with reference to the Pseudoscorpionid mouthparts. Firstman (1973: 14–16, fig. 10) studied the endosternite and the circulatory system. Important data with reference to the development were published by Weygoldt (e.g., 1968). The systematic position was recently studied by Weygoldt & Paulus (1979, 1979a) and Weygoldt (1980).

The following discussion of Pseudoscorpionid comparative morphology and postembryonic development is based on an original study of the morphology of *Chthonius (Chthonius) tenuis* L. Koch, supplemented with data from literature.

2. Descriptive part

Habitus. – Chelicerata of small size, with rather flattened body and distinctly segmented opisthosoma. The palp presents a well-developed chela. The colour is generally more or less brown. Pseudoscorpionida can sometimes move rather rapidly and are also capable of backward movements.

Divisions of the body, tagmata and segmentation. – In Pseudoscorpionida, the juncture between prosoma and opisthosoma is broad. An important study of Pseudoscorpionid morphology, including segmentation, was published by Chamberlin (1931; see particularly pp. 27–40 and figs. 2–7). Data with reference to Pseudoscorpionid segmentation have been summarized by Beier (1932: 119–124, figs. 159–165) and Vachon (1949: 438–442, figs. 197–201). According to these studies, Pseudoscorpionida are characterized by the presence of the segments I–XVIII (one segment more than in the related Solifugae). It may be remarked here that Chamberlin (1931: 42) supposed that Pseudoscorpionida were related to Araneida (which are characterized by the presence of the segments I–XVIII), and this could have influenced his interpretation of segmentation in Pseudoscorpionida. On the other hand, the great differences in the supposed position of the dorsodisjugal interval (the dorsal interval of soft skin between prosoma and opisthosoma) and the seemingly arbitrary interpretation of the anal opening (sometimes regarded as a segment, although there are no anal covers) suggest that the problem of Pseudoscorpionid segmentation has no yet been solved. A hypothetical solution of the problem, starting from the supposed relationship to Solifugae, and based on the condition in *Chthonius (Chthonius) tenuis*, is represented in fig. 2B. The last-mentioned species belongs to the family Chthoniidae, which is generally regarded as one of the more primitive Pseudoscorpionid families. My hypothetical interpretation could easily be tested by a study of musculature.

Although the Pseudoscorpionid prodorsum is generally described as undivided and presenting only one or two furrows, the condition in *Chthonius (Chthonius) tenuis* is here hypothetically regarded as distinctly consisting of two sclerites: propeltidium and postpeltidium. The postpeltidium is regarded in literature as the first opisthosomatic tergite. In dorsal view, however, propeltidium and postpeltidium distinctly constitute a unity, separated from the opisthosoma by a narrowing. The subdivision of the prodorsum in Chthoniidae (and possibly also in some other families), and the consequent misinterpretation of the border between prosoma and opisthosoma, could account for the supposed differences in the position of this border with respect to the coxal regions of the legs. The ventral surface of the prosoma is nearly completely occupied by the coxal regions of the appendages. The Pseudoscorpionid opisthosoma comprises (according to my hypothetical interpretation) the segments VII– XVII. As in Solifugae, each opisthosomatic segment generally consists of a tergite, a sternite, and a large pleural region of soft skin (the successive tergites and sternites are also connected by soft skin). Segment VII is represented dorsally by a distinct tergite, and ventrally by a triangular sternite situated between the coxal regions of legs IV. The dorsal and ventral parts of the segments VIII–XVI have corresponding positions; in my opinion, this condition demonstrates that the postpeltidium belongs to the prosoma. Vachon (1949: 442) supposed that the ventral region of segment VIII includes homologues of opisthosomatic appendages. The sternite and tergite of segment XVII can have fused and include the circum-anal ring (the anal opening is represented by a transverse slit surrounded by membraneous parts). The position of the anal opening is more or less terminal. A comparison of fig. 2A and 2B distinctly demonstrates the similarities in segmentation of the two groups of Apatellata (Solifugae and Pseudoscorpionida).

Cuticle. – The chitinous tegument consists of sclerites and soft skin. According to the present interpretation, the sclerites include pro- and postpeltidium, the coxisternal regions (epimera), and the tergites and sternites of the opisthosoma. In the sclerotized parts, the cuticle is generally thick and hard. There are several types of phaneres, among which trichobothria (which are often numerous). The cuticle of body and appendages generally presents many lyrifissures.

Aspidosoma. – As mentioned above, the extent of the aspidosoma in various groups is still uncertain. It could, in several cases, include a propeltidium and a postpeltidium. The anterior sclerite (or the aspidosoma when it is entire) can present one or two transverse furrows, and generally presents one or two pairs of eyes (which can be reduced).

Dorsal and laterodorsal region of opisthosoma. – This region comprises the opisthosomatic sclerites, and the intersegmental and pleural regions of soft skin. Segmentation is always distinct. The tergites can be subdivided in the sagittal plane, and consequently be replaced by paired sclerites.

Anal region. – The anal opening generally occupies a terminal position in the posterior opisthosomatic segment; it is surrounded by the circumanal ring. The tergite and sternite of segment XVII are either separate or fused. In literature, the circum-anal ring is often regarded as representing a separate segment (segment XVIII).

Ventral and lateroventral region of opisthosoma. – This region comprises the opisthosomatic sternites, and the intersegmental and pleural regions of soft skin. The sternites can be longitudinally subdivided in the sagittal plane. The sternite of segment VII is represented by a triangular sclerite between the coxal regions of leg IV. The genital opening is represented by a transverse slit at the posterior border of segment VIII; the genital region is supposed to include homologues of opisthosomatic appendages. The stigmata (fig. 115C) occupy a lateroventral position in segments IX and X, near the posterior border of the sternite. The respiratory organs could have originated form opisthosomatic appendages.

Ventral aspect of podosoma. – The coxisternal region of the podosoma is occupied by the paired epimera (the coxisternal regions of legs I–IV). They are separated, in the axial part of the body, by an area of soft skin. The orifice of the coxal glands is, on both sides, at the posterior paraxial border of the coxisternal region of leg III (epimeron 3). In *Chthonius* (fig. 115B), a small

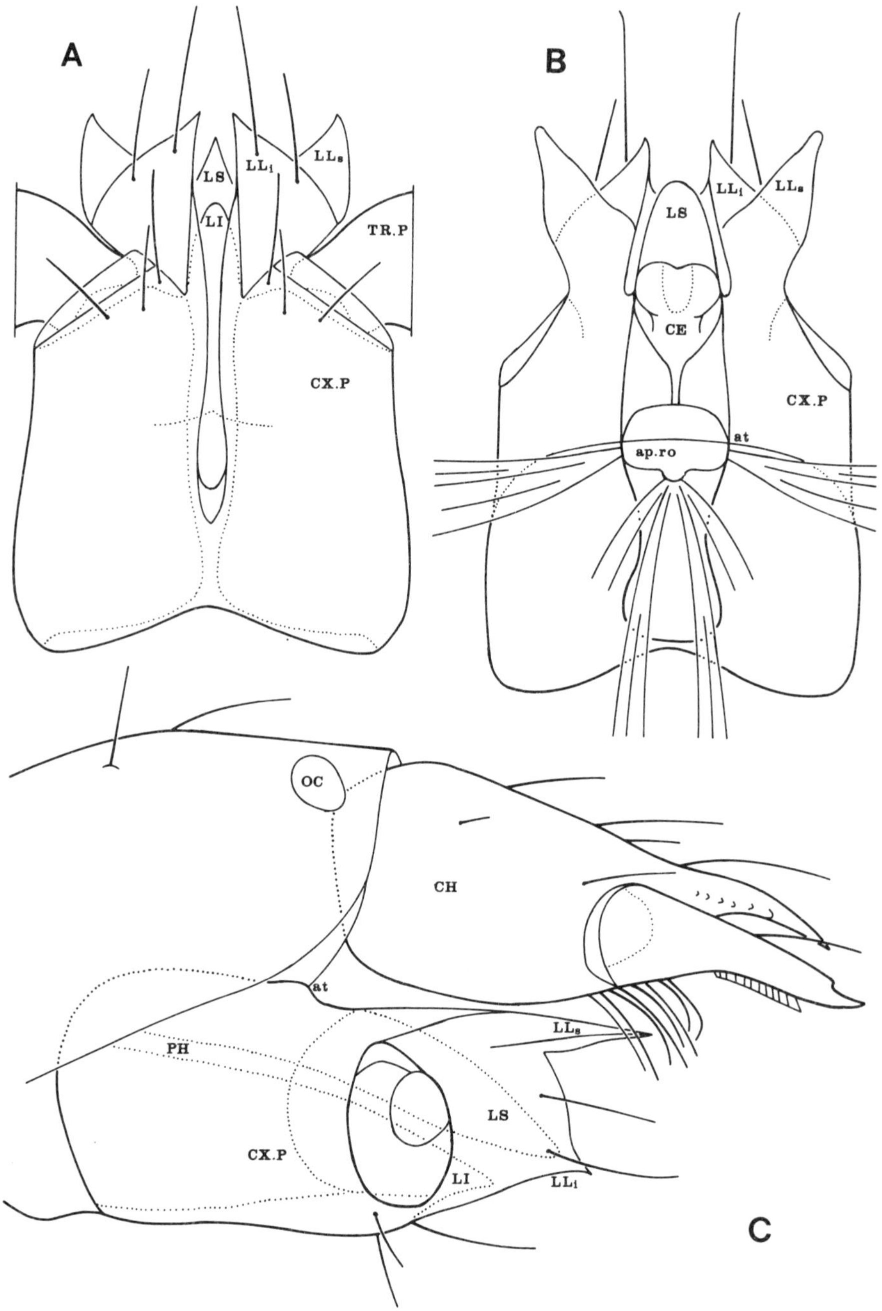
A
LS
LL$_i$
LL$_s$
LI
TR.P
CX.P
B
LL$_i$
LL$_s$
LS
CE
CX.P
at
ap.ro
OC
CH
at
LL$_s$
PH
LS
CX.P
LI
LL$_i$
C

setiferous tubercle is present in the area of soft skin between the orifices; it probably represents a vestigial sternapophysis. In several groups of Chelicerata, sternapophyses are associated with the taenidia of the coxal glands.

Lateral aspect of podosoma. – The coxal regions of the legs (constituted by epimera and pleural regions) contain the acetabula. Supracoxal setae are apparently not present.

Mouthparts. – The Pseudoscorpionid mouthparts (see Chamberlin, 1931: 95–112, figs. 22–26) comprise the chelicerae and the rostrosoma (as in Solifugae). The rostrosoma consists of the labrum, the labium and the lateral lips, and includes the mouth and the pharynx. A dorsal view of the Pseudoscorpionid rostrosoma is represented in fig. 118B, a ventral view in fig. 118A; a lateral view of the mouthparts is represented in fig. 118C. The dorsal surface of the labrum is a continuation of the cervix (which has fused with the palpal coxae). The cheliceral frame is attached to the posterior border of the cervix according to the line *at*. The interior prolongation of the cervix (posterior to *at*) is constituted by the relatively long and narrow rostrosomatic apodeme (the place of insertion of extrinsic muscles). The labium, which could represent the sternal part of segment II, has a ventral position between the coxal regions of the palps. The large lateral lips present a superior and an inferior projection; their basal parts are attached to the coxal regions of the palps, and to labrum and labium. The pharynx is characterized by the presence of a pharyngeal pump.

Chelicera. – The chelicerae are two-segmented; they consist of the body of the chelicera (including the fixed jaw) and the apotele (or movable jaw). They are directed straight forward. The two jaws constitute the chela which can be characterized by the presence of various organs (flagellum, galea, serrulae, setae).

Palp. – Apart from the coxal region, the palp consists of trochanter, femur, tibia, tarsus and apotele. The last-mentioned two segments constitute a chela. The prehensile palp participates in food-intake.

Legs. – Apart from the coxal region, the Pseudoscorpionid legs (see Chamberlin, 1931: 146–165, figs. 40–44) consist of seven eudesmatic and adesmatic segments at the most, viz., trochanter, femur 1, femur 2, tibia, basitarsus, telotarsus and apotele. In several groups there is, however, only one femur in the two anterior or in all legs; the same applies to the tarsus (a single tarsus instead of basi- and telotarsus). In *Chthonius (Chthonius) tenuis*, leg I (fig. 119B) consists (apart from the coxal region) of trochanter, femur, tibia, basitarsus, telotarsus and apotele; leg IV of the same species (fig. 119A) consists (apart from the coxal region) of trochanter, femur 1, femur 2, tibia, basitarsus, telotarsus and apotele. The promotor-remotor movements take place at the coxal region-trochanter articulation. The articulation between trochanter and

←

Fig. 118. Chthonius (Chthonius) tenuis L. Koch, mouthparts of adult; A, ventral view of rostrosoma; B, dorsal view of rostrosoma; C, lateral view; A–C, × 141.

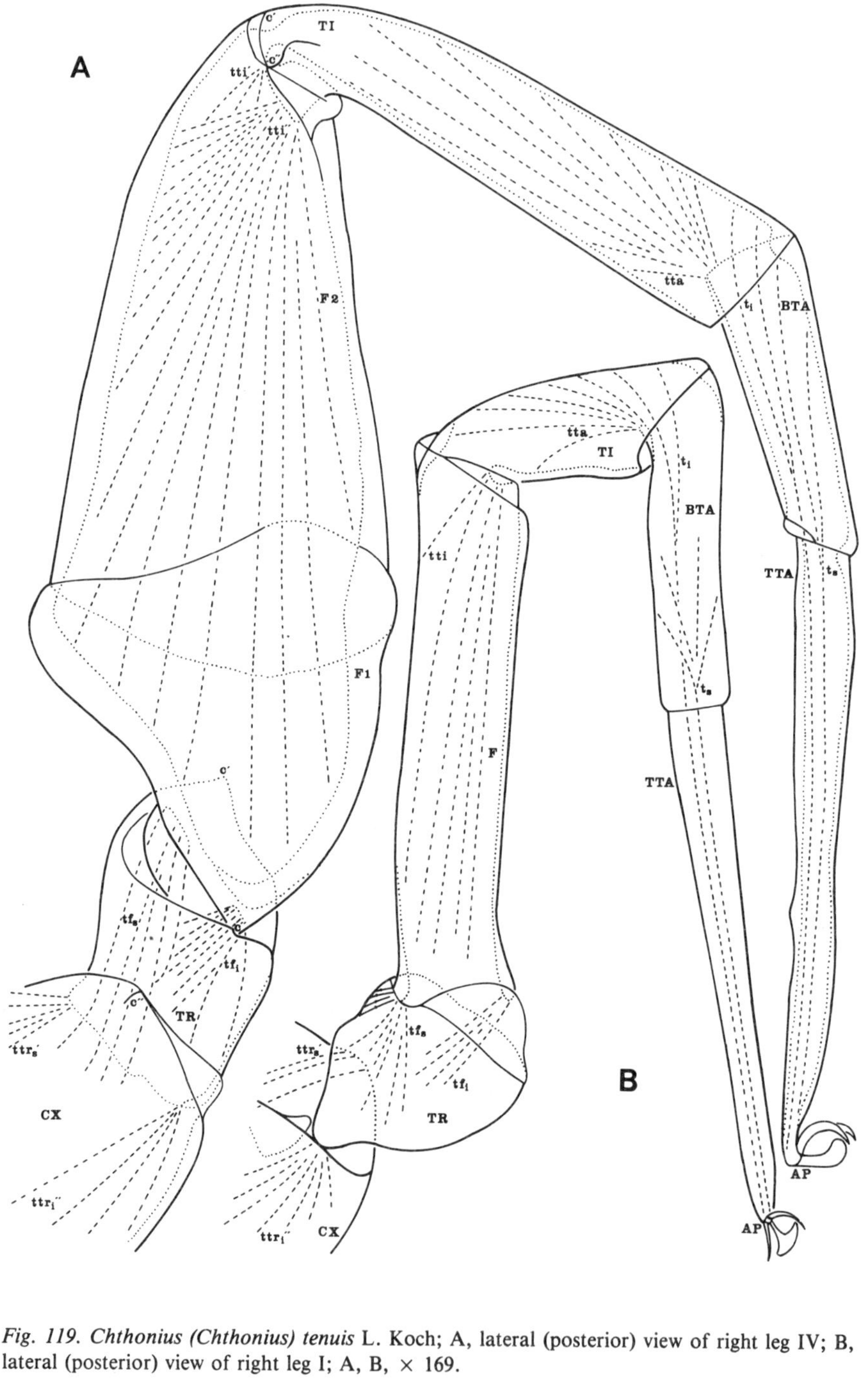

Fig. 119. Chthonius (Chthonius) tenuis L. Koch; A, lateral (posterior) view of right leg IV; B, lateral (posterior) view of right leg I; A, B, × 169.

femur (or femur 1) is constituted by a bicondylar, bidesmatic pivot joint. In leg IV of *Chthonius* the axis of movement of this joint is more or less longitudinal (as in Solifugae), for which reason it could be associated with leg rocking. The separation of femur 1 and femur 2 in leg IV of *Chthonius* is incomplete; the articulation is adesmatic and probably permits of little movement. The articulation between femur and tibia (or between femur 2 and tibia) is constituted by a hinge joint with two superior condyles and several inferior muscles. The tibia-basitarsus articulation is also constituted by a hinge joint; it is operated by inferior muscles, whilst the inferior tendon of the apotele passes the articulation (because it has its base in the tibia). The articulation between basi- and telotarsus is adesmatic; it is operated by the two tendons of the apotele. The telotarsus-apotele articulation (or the tarsus-apotele articulation) is bicondylar and bidesmatic. It may be remarked here that, in several species of Pseudoscorpionida, the femur 1-femur 2 articulation is constituted by a monodesmatic hinge joint, representing the ancestral condition of this joint (similar to that in Epimerata); the type of regression of the joint (disappearance of the muscles, followed by gradual disappearance of the separation of the segments) is also similar to that in Epimerata. Weygoldt & Paulus (1979: 95–96) homologized femur 2 with a patella. The position of the knee-joint, the above-mentioned similarities with Epimerata, and the general pattern of the evolutionary development of the Chelicerate appendages (see Van der Hammen, 1985: 50–51) do not support this hypothesis.

Internal anatomy. – Data with reference to Pseudoscorpionid internal anatomy were summarized by Vachon (1949: 449–464, figs. 209–225). The alimentary canal comprises the mouth, the pharynx (a transverse section presents four branches; the pharynx can function as a pharyngeal pump), the oesophagus, the midgut with numerous caeca, the hindgut and the anus. The excretory system comprises coxal glands and nephrocytes (Malpighian tubes are lacking, but the caeca of the midgut have excretory properties). The Pseudoscorpionid coxal gland belongs to segment V (the segment of leg III); as mentioned above, the orifice is at the posterior paraxial border of the coxisternal region of leg III (epimeron 3). According to Firstman (1973: 16, fig. 10) the endosternite is divided completely into separate anterior and posterior portions. The anterior endosternite is paired, lying on either side of the supra-oesophageal ganglionic mass; it receives the insertion of only a single pair of dorsoventral muscles. The posterior endosternite is a simple transverse, tendinous band; it is the place of insertion of a single pair of dorsoventral suspensor muscles, and at least three pairs of appendicular muscles are attached to it. The circulatory system is reduced to the heart and the anterior aorta. The respiratory system consists of tracheae and two pairs of stigmata (with lateroventral position), situated near the posterior border of sternites IX and X respectively; the tracheal tubes are highly branched. The central nervous system includes a supra-oesophageal syncerebrum and a sub-oesophageal ganglionic mass. The male reproductive system includes an unpaired testis with paired vasa deferentia and paired vesicula seminalis, a median prostatic reservoir, an ejaculatory canal atrium, an ejaculatory canal, and accessory glands; the atrium is associated with spermatophore production (see Legg, 1973). The female reproductive system includes an unpaired ovarium, paired oviducts, an unpaired oviduct and the genital chamber (with spermathecae and accessory glands). The glandular system includes prosomatic silk glands (orifice in the movable jaw of the chelicera) and venom glands (orifice in one or in both jaws of the chelicera).

Reproduction. – Pseudoscorpionid reproduction is bisexual. Sperm transfer is indirect; it is deposited on the substratum in the shape of spermatophores; it can be accompanied by courtship behaviour or not (in the case of courtship behaviour, spermatophores and the male genitalia are more complex (cf. Weygoldt, 1970a; Legg, 1973, 1975). Pseudoscorpionida are oviparous; oviposition is associated with nest-building, whilst the eggs are layed in a ventral brood pouch which remains attached to the female genital region.

Postembryonic development. – The ancestral number of Pseudoscorpionid instars is six: two larvae, three nymphs and an adult. The larvae are calyptostasic and remain inside the egg. The number of nymphs can be reduced to two, probably as a result of neoteny.

3. Additional remarks

The number of known species of Pseudoscorpionida is about 1100. The geographic distribution is world-wide, but species and specimens are particularly numerous in the subtropical and tropical parts of the world. The flattened shape of the body is especially adapted to life in fissures, under bark and stones, and in litter and moss; species are, however, also found in caves, in houses, in nests of mammals, on shrubs, etc. Phoresia is known from several species. Fossil species are known from the Oligocene and the Miocene; they were already fairly common at that time, and extinct representatives are distinctly related to extant species. Three main groups of Pseudoscorpionida are generally distinguished (in the hierarchy of the system introduced here, they represent orders): Chthoniida, Neobisiida and Cheliferida.

List of the notations of figs. 118–119

AP	apotele.
ap.ro	rostrosomatic apodeme.
at	line of attachment of cheliceral frame.
BTA	basitarsus.
c′, c″	condyles.
CE	cervix.
CH	chelicera.
CX	coxa.
CX.P	coxal region of palp.

F	femur.
F1	femur 1.
F2	femur 2.
LI	labium.
LL_i	inferior lobe of lateral lip.
LL_s	superior lobe of lateral lip.
LS	labrum.
OC	eye.
PH	pharynx.
t_i	inferior tendon of apotele.
t_s	superior tendon of apotele.
tf_i	inferior tendons and muscles of femur.
tf_s	superior tendons and muscles of femur.
TI	tibia.
TR	trochanter.
TR.P	trochanter of palp.
TTA	telotarsus.
tta	tendons and muscles of tarsus.
tti	tendons and muscles of tibia.
ttr_i	inferior tendons and muscles of trochanter.
ttr_s	superior tendons and muscles of trochanter.
′ (prime)	anterior (pertaining to the anterior face of an appendage).
″ (double prime)	posterior (pertaining to the posterior face of an appendage).

V. ANACTINOTRICHIDA

Important parts of my studies on Chelicerate morphology have been devoted to Anactinotrichid mites. In the course of time I studied the general morphology, the relationships, the evolution and the classification of the superorder Anactinotrichida (Van der Hammen, 1968a: 269–270, 273, 276–277; 1968c; 1970a, 1970b, 1970c, 1972, 1972a, 1973, 1975, 1977a, 1978, 1979, 1979a, 1980; 1985: 52–55, fig. 34; 1985a, 1986; 1986a: 46–47; 1986c: 224–228, fig. 3), as well as of the separate orders, viz., Opilioacarida (Van der Hammen, 1966, 1968b, 1969b, 1971, 1976, 1977; Coineau & Van der Hammen, 1979), Holothyrida (Van der Hammen, 1961, 1965; 1968a: 261–269, figs. 1–2; 1983b), Gamasida (Van der Hammen, 1964a; 1986c: 220–223, fig. 2) and Ixodida (Van der Hammen, 1964, 1983a). In the present chapter these studies are summarized and partly reconsidered. The chapter is subdivided into eight sections: an introduction, a general descriptive part, a discussion of the four orders, a discussion of the relationships of the four orders, and additional remarks.

1. Introduction

The name Anactinotrichida was suggested by Grandjean and introduced by me (Van der Hammen, 1961: 178–179) in order to replace the earlier name Anactinochitinosi, previously introduced by Grandjean (1935a: 123, 126; 1936: 85–89, 91, figs. 1–2; 1936b: 438–439); Grandjean (1962: 264) subsequently replaced the inappropriate term actinochitin by actinopilin. Grandjean restricted the names Anactinochitinosi and Anactinotrichida to a group including Holothyrida, Gamasida and Ixodida (see also Grandjean, 1970: 803–815). Opilioacarida were subsequently also classified by me with Anactinotrichida (Van der Hammen, 1968a: 276–277; 1968c: 403–404; 1972: 3, 34–36; 1972a: 274–278; 1973: 275–279). A discussion of the concept of the Acarida or mites (an artificial group comprising Actinotrichida and Anactinotrichida) can be found in the chapter on the Actinotrichida (chapter II of the systematic part of this book).

2. General descriptive part

The following survey of Anactinotrichid morphology is based on my previous studies, although corrections and supplementary details from literature are added. It includes also paragraphs on reproduction and postembryonic development. The present section constitutes a concise general introduction, adapted to the purpose of the present book; generally, little reference is made to the morphology of highly specialized (e.g., parasitic) groups.

Habitus. – Anactinotrichida constitute a group of rather small Chelicerata, although medium-sized representatives are found among Holothyrida and Ixodida. There are many different types of habitus, and the general shape varies from hemispherical to flat and from nearly round to elongate and narrow.

Divisions of the body, tagmata and segmentation. – The original subdivision of the body into prosoma and opisthosoma is still recognizable in Opilioacarida, but has disappeared in all other Anactinotrichida. A separate gnathosoma is present in all groups. An aspidosoma is recognizable in Opilioacarida. A podosoma is more or less clearly defined in all Anactinotrichida. Segmentation of the opisthosoma is distinctly present in Opilioacarida only. The subdivision of the dorsal shield of many Gamasida into two parts, by a notomeristic scissure (see Van der Hammen, 1970a: 8–9), is certainly not homologous with the original subdivision, by the disjugal furrow, into prosoma and opisthosoma. In comparison with Opilioacarida, the notomeristic scissure takes the place of the border of segments XII and XIII, and could be homologous with that intersegmental furrow.

Tegument. – In Opilioacarida the cuticle is leathery and little sclerotized (although remarkably tough when sectioned). In Holothyrida it is hard and leathery, and remarkably thick. In Gamasida and Ixodida, variously constituted sclerites are found, separated by soft skin.

Eyes. – Two or three pairs of eyes are present in Opilioacarida, one or two pairs in some Ixodida, and one pair in *Australothyrus ocellatus* Van der Hammen (Holothyrida). Eyes are not present in other Holothyrida, in Gamasida, and in part of the Ixodida.

Phanerotaxy. – In Opilioacarida, many types of phaneres are found, viz., papilliform or lobate setae, compound setae, mucronate setae, solenidia, coronidia, various types of ordinary setae (generally hollow), etc. Compound setae consist of an ornate base, with a striate distal part; they could represent a tubercle surmounted by a sensory phanere. Mucronate setae probably represent compound setae of which the distal part has become very small. On the Opilioacarid opisthosoma, regression in the number of papilliform setae can be associated with an increase in the number of lyrifissures. The number of types of phaneres is reduced in other groups of Anactinotrichida; compound and mucronate setae are apparently not known from Holothyrida, Gamasida and Ixodida. All Anactinotrichida are characterized by the absence of trichobothria, and by the absence of an actinopilinous setal core.

Dorsal aspect of idiosoma. – Apart from the presence of eyes (dealt with above), the dorsal surface of the idiosoma can be characterized by the presence of sclerites, setae, lyrifissures (or their homologues) and orifices of glands. Large sclerites, covering the entire dorsal surface of the idiosoma, are present in Holothyrida and many Gamasida; smaller sclerites are found in many other Gamasida and in ticks of the family Ixodidae. True dorsal sclerites are absent when the cuticle is more or less leathery.

Anal region. – The anus is terminal (the ancestral position) in Opilioacarida; in this group, the anal segment (segment XIX) is partly retractable. In other groups of Anactinotrichida, the anal opening is more or less ventral.

Ventral region of opisthosoma. – This region includes the anal and genital regions (dealt with separately). Anal and genital opening can be enclosed by a single ventral shield; in other cases, there can be one or more smaller ventral shields, or shields are absent (the cuticle being soft and leathery). The ventral region of the opisthosoma and the sternal region of the podosoma can be continuous or distinctly separated.

Genital region. – The transverse genital opening of Opilioacarida (a simple slit situated between coxae IV) is flanked by a pair of genital verrucae and evaginable papillae; they are considered homologous with exites and endites of the opisthosomatic appendages of segment VIII. In adult females of Opilioacarida a large ovipositor is present. It is an evaginable tube, the wall of which is a continuation of the vagina. The structure of the Opilioacarid ovipositor is very simple (no segmentation, no terminal lobes); it belongs to the most primitive ovipositors known from Chelicerata. In other groups of Anactinotrichida, the position of the genital opening can be more advanced. It can be closed by one or more sclerites, and present important differences between the two sexes. A primitive ovipositor is also known from Holothyrida and Ixodida.

Ventral aspect of podosoma. – A sternum (i.e., a neosternum) or sternal region is present in all Anactinotrichida. The basal segment of the legs is constituted by a coxa. Sternapophyses are known from Opilioacarida and some Holothyrida (a pair), and Gamasida (a fused pair). Ancestrally, the orifice of the coxal gland is posterior to coxa I, and the transport of its products is ancestrally associated with the sternapophyses and the subcapitular gutter (with ventral position in the infracapitulum).

Lateral aspect of idiosoma. – In Opilioacarida, four pairs of stigmata are present, situated in the laterodorsal parts of the segments IX–XII; the lateral region of the podosoma is characterized by the presence of lobes dorsal to the base of the anterior three pairs of legs. In Holothyrida, one pair of stigmata is present, surrounded by, or associated with, a peritreme (taenidium); in this group, orifices of one or two pairs of excretory glands can be present in the lateral region. The single paired lateral stigma of Gamasida is also associated with a peritreme. The single paired stigma of Ixodida is either posterior to coxa IV (Ixodidae) or lateral, above coxae III–IV (Argasidae), surrounded by a peritreme.

Gnathosoma. – The Anactinotrichid gnathosoma consists of a cheliceral frame and an infracapitulum. In most Gamasida, the dorsal part of the cheliceral frame (the tegulum) has developed a limbus, constituting a supracheliceral vault. The infracapitulum consists of a dorsal cervix, a ventral mentum, a pair of large lateral ridges (with the acetabula of the palps), and a pair of malapophyses (of which the ventral surface is called gena; the paired genae are a continuation of the mentum). The mouth is surrounded by three lips: the labrum or upper lip, and a pair of lateral lips; a labium (or underlip) is no more recognizable. The mouth presents three commissures; the inferior commissure occupies a more advanced position. The ventral surface of the infracapitulum presents (in Opilioacarida, Holothyrida and Gamasida) a subcapitular gutter which is, ancestrally, associated with the sternapophyses. The internal base of the gnathosoma is characterized by the presence of apodemes and the insertion or attachment of muscles. There can be a subcapitular apodeme (the internal continuation of the mentum) and a cervical apodeme (the internal continuation of the cervix). Extrinsic gnathosomatic muscles are inserted on the base of the tegulum and on the subcapitular apodeme (lateral muscles can also be present). Intrinsic muscles of the infracapitulum are attached to the cervical apodeme (the position of the cervical apodeme resembles that of the capitular apodeme of Actinotrichida; the last-mentioned apodeme, however, constitutes the place of insertion of extrinsic gnathosomatic muscles).

Chelicera. – The chelicera is three-segmented (trochanter, body of the chelicera, and apotele) in Opilioacarida, Holothyrida and Gamasida (ancestrally, it represents the prehensile type); it has been subject to considerable transformation in Ixodida. There are, ancestrally, a dorsal and an antiaxial lyrifissure (in the body of the chelicera). There are up to one or two setae on the trochanter (one in most Opilioacarida; two in the male of *Opilioacarus vanderhammeni* Juvara-Bals & Baltac) and up to four or five setae on the body of the chelicera (four in *Paracarus*, five in the male of *O. vanderhammeni*; in other Opilioacarida this number is three, whilst in other Anactinotrichida it is two or less).

Palp. – There are, originally, six free palpal segments: trochanter, femur, genu, tibia, tarsus and apotele. The apotele is terminal in Opilioacarida; in Holothyrida and Gamasida its position is paraxial (near the base of the tarsus). In Opilioacarida, Holothyrida and Gamasida, the border of tarsus and tibia is oblique (the paraxial surface is longer than the antiaxial). In Ixodida, the number of palpal segments is reduced to four (trochanter, femur, genu and tibiotarsus). The Anactinotrichid palp has no free coxa; the infracapitulum presents, however, a coxal region at the base of the palp. In Gamasida, this coxal region can present a supposed laterocoxal seta (*e*). The articulation between genu and tibia of the Anactinotrichid palp is represented in fig. 120. In Opilioacarida (fig. 120A) it is constituted by a bicondylar and bidesmatic joint, permitting of active flexion and extension. In Holothyrida, Gamasida and Ixodida (fig. 120B–E) it is also bicondylar, although it is apparently functioning as a hinge joint (permitting of active flexion). In Gamasida, however, it

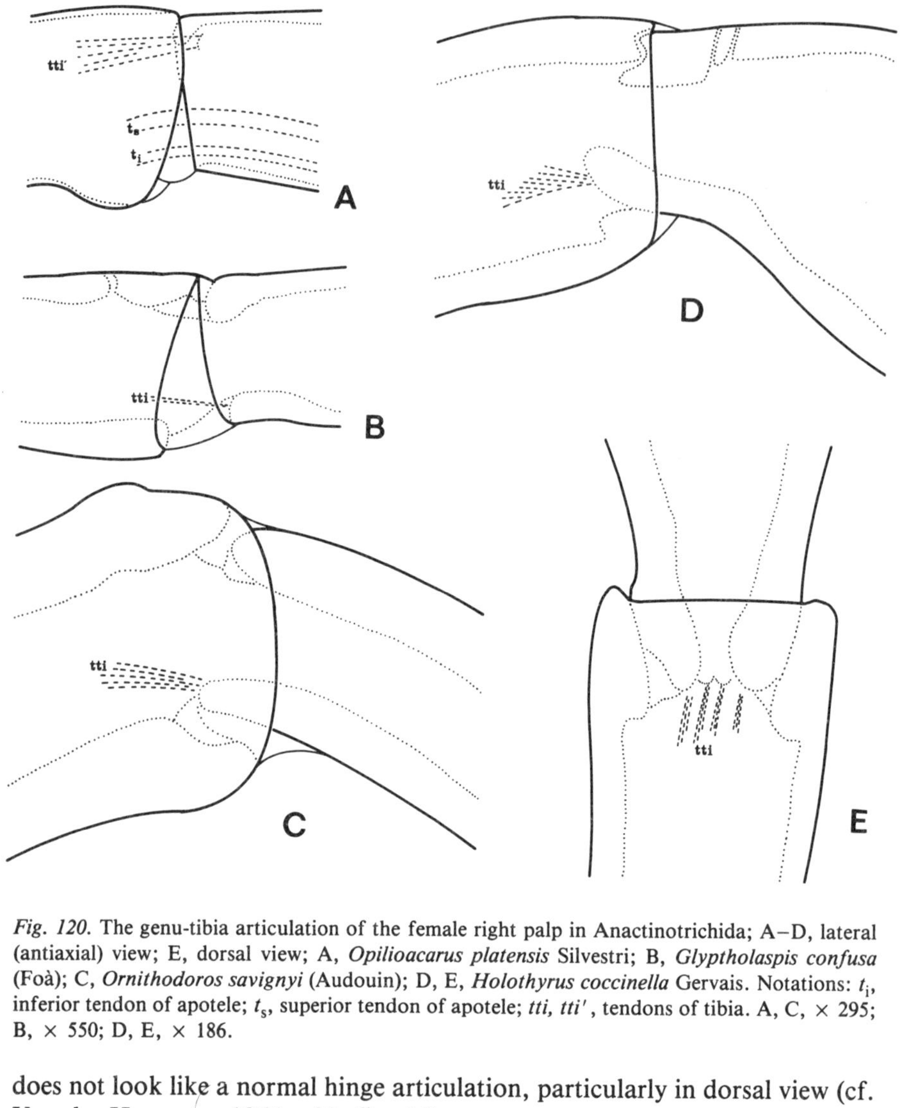

Fig. 120. The genu-tibia articulation of the female right palp in Anactinotrichida; A–D, lateral (antiaxial) view; E, dorsal view; A, *Opilioacarus platensis* Silvestri; B, *Glyptholaspis confusa* (Foà); C, *Ornithodoros savignyi* (Audouin); D, E, *Holothyrus coccinella* Gervais. Notations: t_i, inferior tendon of apotele; t_s, superior tendon of apotele; *tti, tti'*, tendons of tibia. A, C, × 295; B, × 550; D, E, × 186.

does not look like a normal hinge articulation, particularly in dorsal view (cf. Van der Hammen, 1964a: 25, fig. 14).

Legs. – The ancestral number of Anactinotrichid eudesmatic leg segments is seven or eight: coxa, trochanter (or trochanter 1 and 2 in legs III and IV of Opilioacarida), femur, genu, tibia, tarsus, and apotele. The apotele of leg I can be absent in Gamasida. The articulation between body and coxa is apparently acondylar and polydesmatic. The articulation between coxa and trochanter is bi- or monocondylar and bidesmatic (it is either a pivot or a rocking joint). The articulation between trochanter 1 and 2, and between trochanter or trochanter 2 and femur is bicondylar and bidesmatic (a pivot joint). The articulations between femur and genu, genu and tibia, and tibia and tarsus are constituted by mono- or bicondylar hinge joints. The articulation between tarsus and apotele is bicondylar and bidesmatic. The femur consists of two adesmatic segments, viz., basi- and telofemur, separated by the basifemoral ring. This articulation permits of little or no movement; it is generally found at the place of a bend at the base of the femur, and it is characterized by the presence of lyrifissures. The tarsus can comprise two or more adesmatic segments (besides the pretarsus), viz., basi- and telotarsus, separated by a basitarsal ring or scissure. In Opilioacarida, the telotarsus of legs II–IV is subdivided into two adesmatic segments. The Anactinotrichid tarsus I can present several types of sensory phaneres associated with pits or capsules, varying from the telotarsal organ in Opilioacarida to Haller's organ in Holothyrida and Ixodida.

Internal anatomy. – Recent summaries of the internal anatomy were prepared by T.E. Hughes (1959) and Legendre (1967, 1968). The endosternite and the circulatory system were studied by Firstman (1973), sperm morphology and spermiocytogenesis by Alberti (1980). The alimentary canal comprises mouth, pharynx, oesophagus, midgut and caeca, hindgut and anus. The excretory system comprises the Malpighian tubes and the coxal glands (the transport of the secretions is ancestrally associated with sternapophyses and subcapitular gutter). The endosternite is divided into a paired anterior and a posterior part. The anterior part receives the insertion of two pairs of dorsoventral muscles. The posterior part is associated with a single pair of dorsoventral muscles and one pair of transverse muscles. The circulatory system comprises a heart, an anterior aorta and a lacunar system. The respiratory system includes four (Opilioacarida) or one (other groups) pairs of stigmata, associated with tracheae. The central nervous system is concentrated to a single mass situated in the prosoma, and pierced by the oesophagus. The male reproductive system generally comprises paired testes, paired vasa deferentia, an unpaired vesicula seminalis and accessory glands. According to Alberti (1980), Anactinotrichid sperm morphology and spermiocytogenesis are fundamentally different from those of Actinotrichida. The female reproductive system can comprise paired or unpaired ovaria, paired oviducts, an unpaired uterus and a vagina which can develop into a primitive ovipositor (Opilioacarida, Holothyrida, Ixodida). In some Gamasida (Dermanyssina) a receptaculum seminis is present (at the base of coxa III or IV); its presence is associated with the absence of a vagina. The glandular system can include cheliceral glands, infracapitular glands and tegumental glands. The opisthosomatic muscular system of Opilioacarida comprises a great number (a network) of longitudinal and transverse subtegumental muscles, of which the sigilla are segmentally arranged. Dorsoventral opisthosomatic muscles are present in Holothyrida, Ixodida and Gamasida.

Reproduction. – Reproduction in Anactinotrichida is generally bisexual. Thelytokous parthenogenesis is sporadically found, in Ixodida, in species which normally reproduce bisexually. Sperm transfer is either indirect (spermatophory) or nearly direct (gonopody). In Ixodida and one group of Gamasida, the chelicerae (and, in Ixodida, also the gnathosoma) can function as a gonopod. Anactinotrichida are oviparous (in Ixodida, the number of eggs can be very great). In Opilioacarida, Holothyrida and Ixodida, oviposition takes place by means of a primitive ovipositor.

Postembryonic development. – The ancestral type of life-cycle (prelarva, larva, three nymphs and adult) is known from Opilioacarida; in this group, the prelarva is a calyptostase (nearly an elattostase). Three nymphs are known in the case of Holothyrida; there are, however, some indications that an octopod larva is also present. In Ixodida, the life-cycle comprises a larva, one nymph and an adult; in the family Argasidae, the nymph can pass through a series of growing moults. The life-cycle of Gamasida can comprise a larva, two nymphs and an adult.

3. Opilioacarida

At the beginning of the present chapter, I have listed the papers in which I have dealt with representatives of the Anactinotrichid order Opilioacarida. In these papers, detailed studies are made of Opilioacarid comparative morphology, postembryonic development, classification, relationships and geographical distribution. In the second volume of the Glossary of Acarological Terminology (Van der Hammen, 1976) I have, moreover, standardized Opilioacarid terminology.

The diagnostic characters of the order Opilioacarida are the following. The body is composed of precheliceral metameres and nineteen segments, among which thirteen distinctly recognizable opisthosomatic segments; owing to this, and because of the relatively very long legs, Opilioacarida have a superficial resemblance to small harvestmen. There are two or three pairs of eyes, and four pairs of stigmata (in segments IX–XII); stigma 1 is of tritonymphal origin, stigma 2 and 3 are protonymphal, and stigma 4 is deutonymphal. Opisthosomatic lyrifissures can be numerous. The opisthosomatic segments are separated by furrows, at the interior side of which numerous muscles are attached. The anal opening is situated in the terminal part of an anal tubercle which is retractable. The genital opening is represented by a transverse slit. A pair of evaginable genital papillae is present lateral to the genital opening; at rest (i.e., invaginated) they are covered by genital verrucae; papillae and verrucae are supposed to represent 'endites' and 'exites' of the opisthosomatic appendages of segment VIII. The paired orifices of the coxal glands debouch into paired sternal taenidia, which extend from the posterior border of coxae I to the subcapitular gutter. The paired sternapophyses are associated with the subcapitular gutter. The tegulum is anteriorly not produced (there is no supracheliceral limbus). The infracapitulum presents, close to the distal border, three pairs of paralabial setae which, as a rule, are specialized or hypertrophied: a small corniculus, With's organ (in *Paracarus* a forked seta), and a rutellum. Four pairs of circumbuccal setae and a rather great number of other infracapitular setae (among which at least one pair of laterodorsals) are situated posterior to the paralabials. The mouth is trifid, but a cross-section of the pharynx soon becomes hexagonal. The chelicera presents 0–2 setae on segment 1 (the trochanter) and 3–5 setae on segment 2 (the body of the chelicera). The palpal tarsus presents a paralaterodorsal area densely beset with hollow setae. The long legs present numerous setae more or less arranged in verticils and files (linear cosmiothrichy). A telotarsal organ, constituted by two solenidia, hidden in a small cavity, is present in leg I. Pretarsi II–IV present two pairs of setae. Legs III and IV present two trochanters (trochanter 2 is of tritonymphal origin). The postembryonic development includes; prelarva (a calyptostase, nearly an elattostase), larva, protonymph, deutonymph, tritonymph and adult; the tritonymphs are already sexed (male and female tritonymphs are distinctly differentiated).

The order Opilioacarida is very homogeneous and includes one family (Opilioacaridae) with six genera, viz., *Opilioacarus* (Southern Europe, India, North Africa, America; in all the American species, segment XVIII presents three setae, but it is doubtful whether this small character is sufficient to regard them as constituting a subgenus *Neocarus*), *Adenacarus* (Arabia), *Paracarus* (Kirgizya, Kazakhstan, Thailand), *Panchaetes* (West Africa), *Phalangiacarus* (Gabon, Ivory Coast), *Salfacarus* (Madagascar, South Africa, Tanzania); the localities in India and Thailand,[1] recorded here, refer to unpublished data (several undescribed species from the Americas are also known to me).

The present section is subdivided into two parts. In the first part a complete description is given of *Opilioacarus texanus* (Chamberlin & Mulaik); in the second part additional data with reference to other species are summarized.

3.1. *The morphology of* Opilioacarus texanus *(Chamberlain & Mulaik)*

The detailed description included in the present subsection is based on the study of five specimens (two females, one male and two nymphs), collected in the Big Bend National Park, Texas, U.S.A. The description was originally published nearly twelve years ago (Van der Hammen, 1966); the present subsection includes several corrections.

Measurements. – In the following records, the gnathosoma is supposed to be orientated as in fig. 123, i.e. curved downwards, and the anal tubercle to be extended (not retracted). The length is measured from the tip of the labrum

[1]Dr. Philippe Leclerc (Villejuif, France) kindly informed me that the material collected by him in Thailand represents a new genus (related to *Paracarus*). The description is in press and will be published in Acarologia. Some of the legs of the species are characterized by the presence of trichobothria (the first record of trichobothria for Anactinotrichida).

to that of the anal tubercle. Adult: length 1.60–1.70 mm; breadth 0.65–0.75. Tritonymph: length 1.55; breadth 0.62. Deutonymph: length 1.25; breadth 0.58. The legs of the tritonymph were carefully measured; data on separate segments are mentioned below. The total lengths (coxa not included) are the following: I 2.20 mm; II 1.09; III 1.04; IV 1.56. The sequence of the legs (according to decreasing length) is I, IV, II, III. Leg I is nearly one and a half times as long as the body, leg IV just as long, and leg II and III both about two thirds.

Habitus and colour. – The species has a superficial resemblance to a very small harvestman; in comparison with other mites it is, however, relatively large. When studied in a dry condition (on a carbon block), it appears that a prodorsal region is distinctly marked off by lateral and posterior furrows; this prodorsum presents a small, curved furrow in the posterior median part, and several deepened areas (e.g., in front of the eyes) at the places of muscle attachments. The opisthosoma presents dorsally a number of transverse furrows; the anterior three, especially the first one, are rather deep. Ventrally, there is a distinct furrow at the level of the greatest width of the idiosoma. The compound setae of the ventral surface (sternal and genital region), i.e. the setae consisting of a broad base and a narrow, pointed distal part (cf. the paragraphs in question), are also easily recognizable when a specimen is studied in a dry condition.

The ground-colour of the species is pale yellowish brown. The prodorsum bears two pairs of eyes, surrounded by blackish eye-patches; in the anterior and lateral parts, there are blue stripes, and there is a vague median blue band. With the exception of the anal valves, each segment of the opisthosoma presents dorsally and laterally a bluish band, separated form neighbouring bands by pale interspaces. The following parts of the legs are blue: leg I, nearly completely; leg II, femur, genu, tibia; leg III, femur, genu, tibia, basitarsus; leg IV, distal part of telofemur, genu, tibia, proximal part of basitarsus. As in other Opilioacarida, the colours are partly cuticular, partly caused by subcuticular pigment grains. The pigment grains are bluish or violet, and relatively widely separated from each other, except in the eye-patches where they are clustered closely together so that the patches appear to be black. The colour of the pigment grains, as well as the blue cuticular colour, disappear rather rapidly in lactic acid.

Food. – The gut of the specimens studied by me contained large masses of pollen grains; in some cases these belonged to one plant-species only, on one occasion distinctly to several. The pollen occurred as the only content, mixed with some brown balls of probably vegetable origin, or mixed with very few fragments of Arthropod appendages; only in one case (a female), the gut contained a mass of small, dark-brown fragments which could be pieces of chitin. It appears that the diet of the species, at least when the material was collected (April), especially consists of pollen grains. It is evident that Opilioacarida are not exclusively carnivorous (as has been supposed in literature); it may be remarked, however, that With (1903: 163, pl. 6 fig. 11) already discovered round bodies in the intestine, which he showed to botanists who regarded them as pollen grains or spores of fungi. The regular occurrence of Opilioacarids under rocks must be a diurnal habit; at night, several species will probably be found climbing plants.

Cerotegument. – This matter is difficult to recognize in *Opilioacarus*. On a carbon block, the cuticle has a slightly frosted appearance; it is probable that a very thin, finely granulated layer of cerotegument is indeed present.

Cuticle. – The cuticle is leathery and faintly sclerotized; it is easily deformed, but sectioning with a razor blade is thwarted by a remarkable resistance. The microsculpture consists of numerous small cones or warts. The chitinous tegument covers the idiosoma nearly completely; there are only interruptions of soft skin in the genital region (cf. the paragraph in question) and at the base of the anal tubercle. Although the paraxial side of the anal valves is nearly colourless and without microsculpture, the surface is rather smooth, not wrinkled and does apparently not represent soft skin. The presence of a nearly uninterrupted, faintly sclerotized cuticle, which is not subdivided into well-defined shields or sclerites, constitutes an interesting character of the Opilioacarida.

Idiosoma. – In dorsal view we can easily distinguish two main parts of the body: an anterior part with setae, but without lyrifissures; and a posterior part in which setae are only present in the anal region, and in which a great number of lyrifissures can be observed. These two parts, which are separated dorsally and laterally by a deep furrow, represent prosoma and opisthosoma; the deep furrow represents the disjugal furrow.

The segmentation of the opisthosoma can be reconstructed by a study of furrows, lyrifissures, and muscle attachments (sigilla). For that purpose, dorsal, ventral and lateral faces (figs. 121–123) must at first be investigated separately, and the results must thereupon be combined, in order to arrive at a clear three-dimensional image of the species. Muscle attachments (sigilla) are situated in the intersegmental furrows, in case these are present; otherwise, they are arranged in distinct, more or less transverse rows which represent the segmental borders. It appears that each opisthosomatic segment presents one or two rows of lyrifissures which are generally orientated parallel to the segmental borders. The anal tubercle represents one segment. An investigation of the lateral aspect reveals that segments have been pushed up in posterior direction, resulting in an obliteration of several pleural parts, whilst the dorsal part of one segment has apparently disappeared (it is pushed aside by the first and the 'second' opisthosomatic 'tergite'). As a result of the study of the three aspects (dorsal, ventral and lateral), we can now recognize a total number of thirteen opisthosomatic segments. Taking into account that the Chelicerate genital opening is situated in segment VIII, and that in *Opilioacarus* at least a dorsal and a ventral part of the pregenital segment is present, we arrive at the conclusion that the thirteen opisthosomatic segments represent the numbers VII–XIX. The prosoma includes the precheliceral metameres, the segments of chelicera and palp (I and II), and the segments bearing the legs (III–VI).

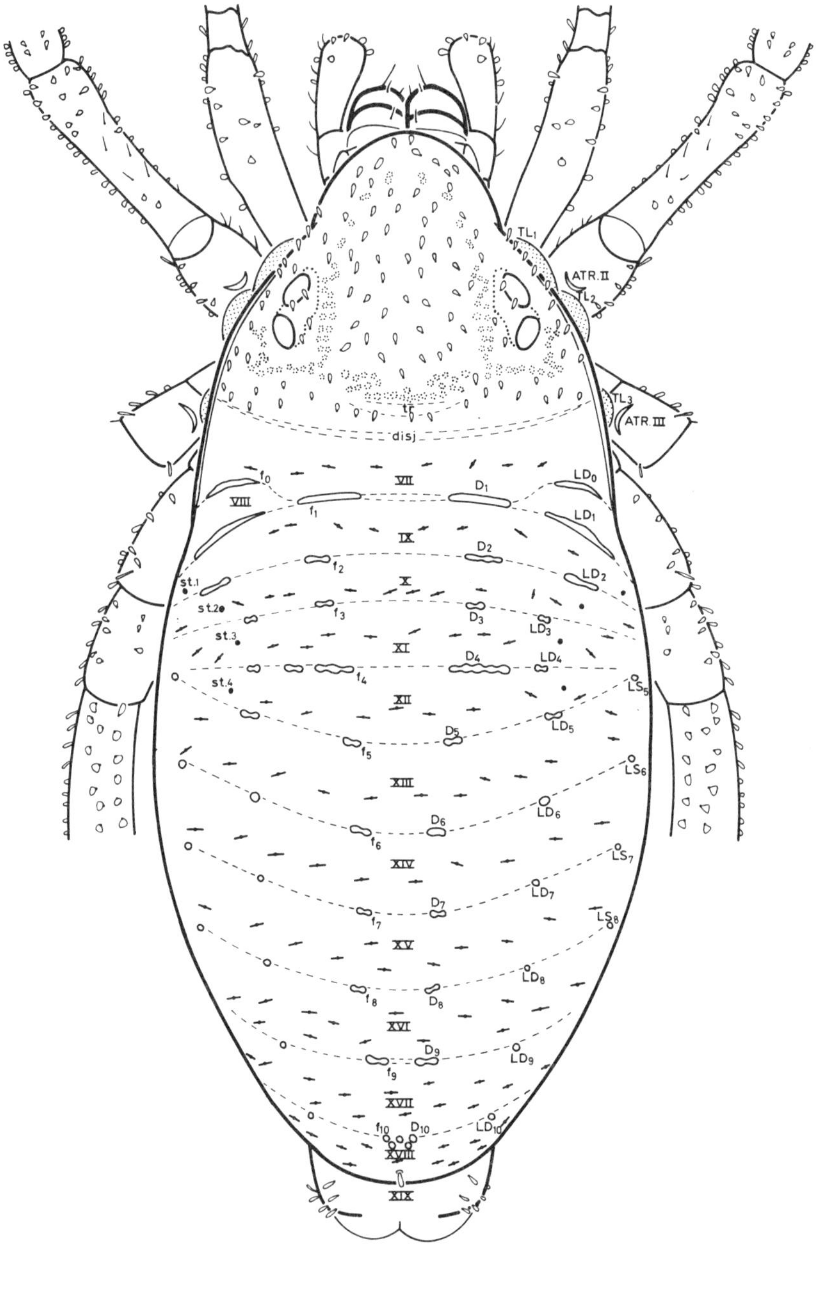

Fig. 121. Opilioacarus texanus (Chamberlin & Mulaik), dorsal view of female; × 92.

Prodorsum. – The anterior dorsal part of the idiosoma (fig. 121), bordered by distinct, deep, posterior and lateral furrows, is named prodorsum. As mentioned above, the posterior furrow represents the disjugal furrow (*disj, dj*), whilst the lateral is named abjugal furrow (*abj*). The last-mentioned furrow is best visible in lateral view (fig. 123). The outline of the prodorsum is rounded in the anterior part; it suddenly broadens in front of the first pair of legs, and reaches its greatest width at the posterior border. In lateral view (figs. 123, 129), the anterior part appears to be elevated, constituting a distinct rostral lobe. Although it is much broader, the lobe reminds one of the frontal protuberance of many, particularly primitive, Actinotrichida. The cheliceral frame is dorsally attached to the anterior (i.e. anteroventral) border of the lobe; the lobe could be of precheliceral origin.

In the posterior part of the prodorsum, there is a transverse furrow (*tr*) which is slightly curved forward (procurved). It is short, but distinct and rather broad; in the lateral part, it is obliterated. The narrow area bordered anteriorly by furrow *tr*, and posteriorly by the disjugal furrow *dj*, could represent the dorsal part of segments V and VI.

There are two pairs of eyes, with lateral position, situated in dark eye-spots constituted by violet pigment (figs. 121, 124A). A pair of circular spots (RE) is present on either side of the rostral lobe (figs. 127, 129); they are slightly concave, posteriorly bordered by a semi-circular recess of the prodorsum, and appear as white spots partly surrounded by dark pigment.

The prodorsum presents numerous setae, but no lyrifissures. The setae are papilliform; some setae of the posterior region are represented in fig. 124C. There are numerous muscle attachments, mainly anterior to furrow *tr*, and laterally at both sides of the eyes. The attachments are represented by small, round, smooth spots, easily recognizable in the microsculpture of the surface (fig. 124B). Numerous muscles are attached to a region extending from the eyes to the lateral recesses of the rostral lobe.

Opisthosoma. – In the present paragraph, dorsal, ventral and lateral aspects of the opisthosoma are described, with the exception of the anal tubercle and the genital region, which are dealt with below.

In dorsal view (fig. 121), the outline of the opisthosoma is more or less oval; it suddenly broadens near the border of segments VII and VIII, and the greatest breadth is between segments XIII and XIV. The disjugal furrow is slightly procurved (curved forward). The posterior borders of segments VII–X are recurved (curved backward). The borders of segments XI and XII are straight. All remaining opisthosomatic segments are procurved when studied in dorsal view. The dorsal part of segment VIII is pushed aside by segments VII and IX; its lateral parts, however, are still rather easily visible in dorsal view, because they are bordered by deep furrows. The intersegmental furrows have received, in figs. 121–123, the notations f_{0-10}. The deep furrows are f_0 and f_1; f_{2-5} and f_{10} are still rather distinct in the centro-dorsal part; the remaining dorsal furrows are indistinct, and particularly recognizable by means of the muscle attachments.

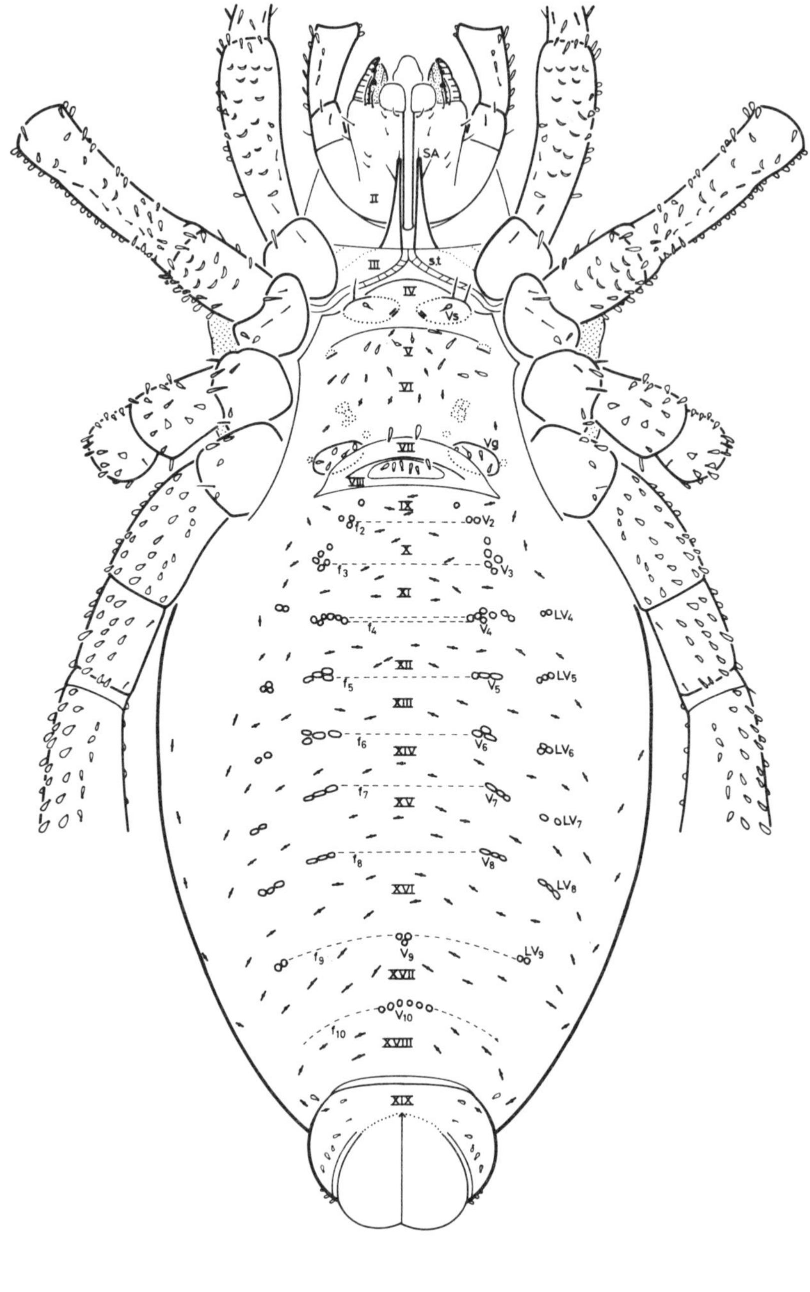

Fig. 122. Opilioacarus texanus (Chamberlin & Mulaik), ventral view of female; × 92.

In ventral view (fig. 122), the opisthosoma presents eleven postgenital segments, of which the posterior one is represented by the anal tubercle. The intersegmental furrow between segments XI and XII is distinct; the remaining ventral borders of the segments are particularly recognizable by a study of the muscle attachments.

The lateral parts of all opisthosomatic segments (fig. 123) have apparently been pushed up in posterior direction, as is already mentioned above. This is especially distinct in the case of segments XIII–XVII. The lateral parts of IX–XII have apparently fused (resulting in an obliteration of the borders), whilst it is difficult to establish the lateral borders of segments VII and VIII.

The microsculpture of the opisthosoma (fig. 124F) consists of small cones which appear as small warts when studied from above. They are arranged in rows that run more or less parallel to the borders of the segments (although slightly winding); at the same time, although in a less distinct way, the cones are also part of a more or less longitudinal system of rows.

The opisthosoma presents numerous muscle attachments (sigilla). Externally, these are recognizable because they appear as smooth, slightly concave spots in the microsculpture of the cuticle; the spots (at least the dorsals) are surrounded by granules which are much smaller than the cones (fig. 124D). The small granules are not present in the case of the sigilla of the prosoma. When a fresh specimen, in cold, diluted lactic acid, is studied in transmitted light, the muscles are easily visible; the greater part of the dorsals extend in lateroventral direction, but part of the posteriors extend in backward direction.

Apart from their intersegmental position, the muscle attachments (sigilla) are also arranged in distinct longitudinal rows. In dorsal view, three pairs of rows are visible: a pair of dorsal rows, a pair of laterodorsals, and a pair of laterals. In fig. 121 the sigilla are indicated as follows: D_{1-10} (dorsals), LD_{0-10} (laterodorsals) and LS_{5-8} (laterals). D_1, LD_0 and LD_1 are elongate, relatively large, and situated in rather deep furrows. D_2, LD_2 and D_4 are also relatively large; the remaining dorsal muscle attachments are smaller. The shape of the attachments varies according to the position: D_{2-9} and LD_{3-5} are more or less ∞-shaped; D_4 and sometimes also LD_2 are compound. The remaining muscle attachments visible in dorsal view are small and more or less circular; D_{10} is not a pair, but a median group of about five.

The ventral muscle attachments (fig. 122) are arranged in four rows: a pair of ventral rows (V_{2-8}) and a pair of lateroventral rows (LV_{4-9}). The posterior border of segment XVI presents a small unpaired group of median attachments (V_9) instead of a pair; the posterior border of XVII presents an unpaired transverse row (V_{10}). All ventral muscle attachments consist of groups of round or oval spots.

In lateral view (fig. 123), six rows of lateral muscle attachments are visible: LD_{0-10}, LS_{5-10}, $LL_{1,\,5-7}$, $LM_{1,\,5-8}$, $LI_{a-c,\,5-7}$, LV_{4-9}; three muscle attachments (L_{x-z}) have a separate position in VII or VIII. LS_{5-10} and $LI_{a-c,\,5-7}$ are

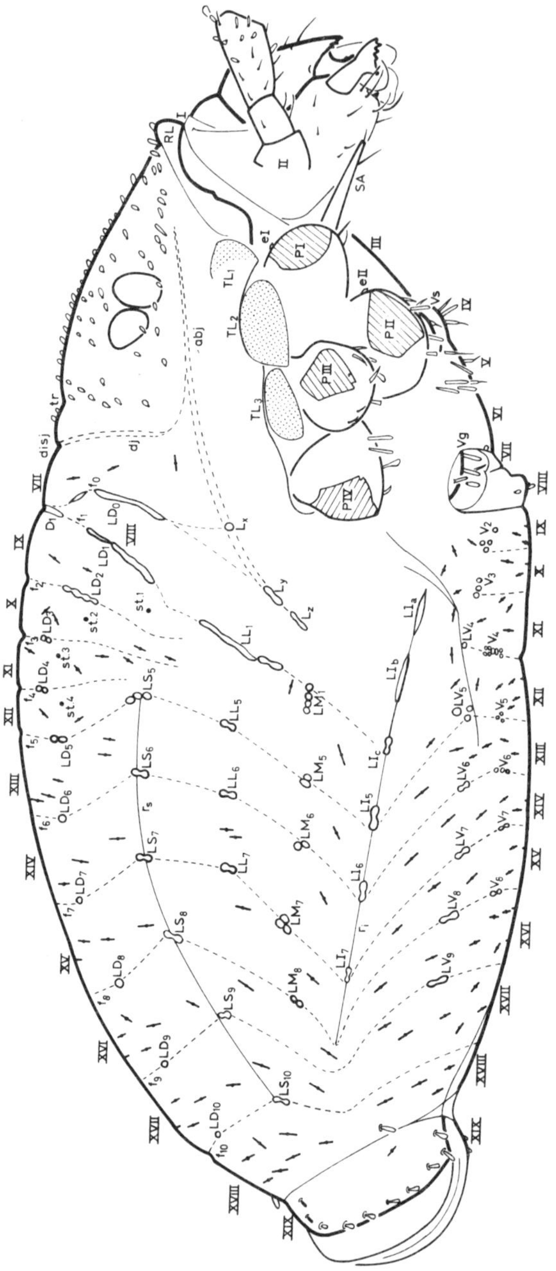

Fig. 123. Opilioacarus texanus (Chamberlin & Mulaik), lateral view of female; × 92.

situated on distinct ridges (r_s, respectively r_i). The number of groups of muscle attachments on the dorsal, ventral and lateral sides of the opisthosoma totals up to 119. The apparent shape of some of the lateral muscle attachments can slightly vary with changes in the orientation of the specimen in a slide.

Each segment presents a rather great number of lyrifissures. Dorsally, these are generally arranged in one transverse row to a segment; only in segments XVI, XVII and XVIII they are distinctly arranged in two dorsal rows. In the figured specimen, the numbers visible in dorsal view in segments VII, IX–XVIII are respectively: 8–8–12–12–8–13–10–11–14–14–12. The arrangement of the lyrifissures appears to be more or less symmetric. Two lyrifissures are represented separately in fig. 124G, H.

The ventral area, enclosed by the muscle attachments *V* and *LV* presents (in the figured specimen) the following disposition of the lyrifissures in segments IX–XVIII respectively: 3–4–8–10–8–7–8–9–8–6. The lyrifissures in segments IX–XI are apparently irregularly distributed; those in segments XIII and XIV are arranged in one row; in the remaining segments, the lyrifissures are more or less distinctly arranged in two rows.

Laterally, the rows of lyrifissures follow the backward translocation of the segments. Between muscle attachments *LD* and *LV*, there are at each side about 35 lateral lyrifissures, arranged in some six rows. Consequently, the total number of opisthosomatic lyrifissures in the specimen represented in figs. 121–123 amounts to 263 (the lyrifissures of the genital verrucae not included).

There are four pairs of stigmata (fig. 124F). Each pair of stigmata is situated in a separate segment: stigma 1 in IX, stigma 2 in X, stigma 3 in XI, and stigma 4 in XII. Part of the tracheal system is also represented in fig. 124F. The stigmata are unprotected, and the shape is slightly irregular; stigma 2 is represented separately in fig. 124E. Stigmata 1–4 are easily visible in lateral view; it appears that stigma 1 has a more ventral position than stigmata 2–4. It is interesting that the segments of the stigmata (IX–XII) are also the segments of which the pleural regions, ventral to the stigmata, have fused.

With the exception of the genital region, the anal tubercle, and segment XVIII, the opisthosoma is completely devoid of setae. Segment XVIII presents a small, unpaired dorsal seta and one pair of laterals; these setae are papilliform.

Anal tubercle. – The exact position of the anal tubercle is best visible in lateral view (fig. 123): it is directed backwards and slightly downwards. Ventrally, at the base of the tubercle, there is a transverse band of soft skin (fig. 122: *sk*). This area permits of further downward bending and partial retraction of the tubercle. The anal tubercle consists of two valves which can close the anal opening. In opened condition, each valve (figs. 122, 123) presents an antiaxial part with setae, a lyrifissure, and a distinct microsculpture, and a paraxial part which is smooth. When the valves are closed (and the tubercle partly retracted) the paraxial part is no more visible. The entire anal tubercle represents segment XIX.

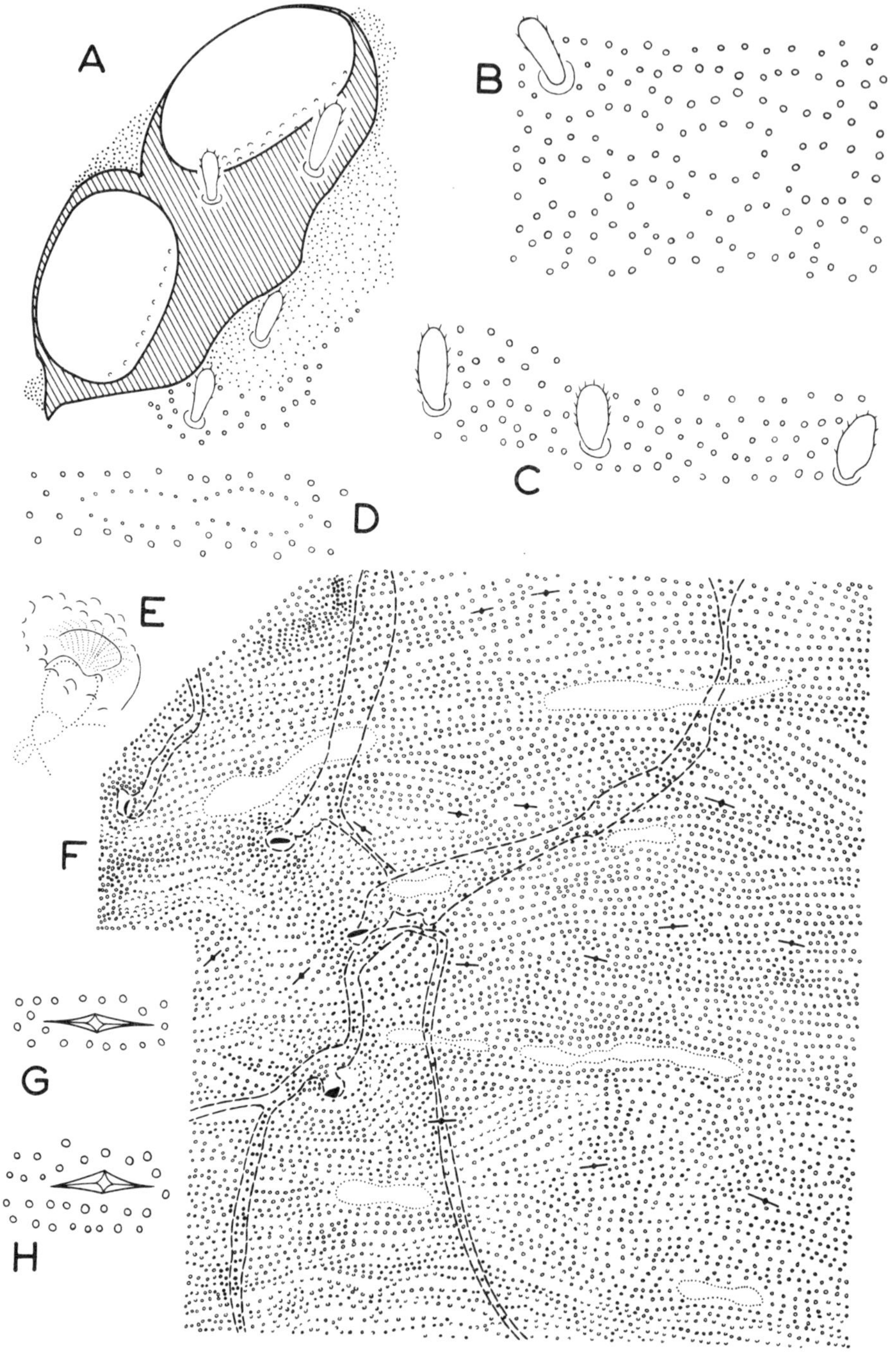
A
B
C
D
E
F
G
H

Genital region. – This region is regarded here as including the genital opening, the areas immediately anterior and posterior to it, and the genital verrucae. Because the material dealt with here includes females, a male and some nymphs, a few notes on juvenile and secondary sexual characters are added.

The adult genital opening (figs. 122, 123,125A, 126F: *og*) is a recurved slit, more or less situated in a fold; in my two nymphs, a small fold only is present at this place (fig. 125B). Posterior to the adult opening, a transverse ovate sclerite, separated by a small area of soft skin from the remaining sclerotized ventral surface, is distinctly recognizable. It presents a variable number of papilliform setae, viz., six or five setae arranged in one transverse row in my two females, and ten setae arranged in two rows (3 + 7) in the male. In my specimens, this genital sclerite is situated on a prominent fold (figs. 123, 126F); its surface presents the same microsculpture as the adjacent sclerotized areas. The genital sclerite is absent in my two nymphs (fig. 125B).

The area anterior to the genital opening can also be folded; this fold can even partly cover the genital verrucae (fig. 122). The pregenital area is different in the two sexes, as was already described for *Opilioacarus segmentatus* by Grandjean (1936b: 435). In the male specimen, it presents the same microsculpture as the remaining part of the idiosoma, whilst the female presents a distinctly bordered trapezoid smooth area (fig. 125A). The female area presents two pregenital papilliform setae (arranged in a pair), the male five (irregularly scattered). According to Grandjean (1936b: 435), the female pregenital area of *O. segmentatus* presents no setae and no trapezoid outline.

The genital verrucae are wart-like valves that can open by a posterior and paraxial slit. This slit gives entrance into a long internal sac (fig. 125A) which extends to the coxal region. The sac constitutes an evaginable papilla; it is present in both sexes. The function of the verruca and the papilla is unknown. The genital papillae remind us of the genital papillae of Actinotrichida. They are supposed to represent opisthosomatic appendages (the verruca is supposed to represent the 'exite', the papilla the 'endite' of an opisthosomatic appendage). In the paragraph on the sternal region, I return to these interesting structures. In the figures, verrucae and papillae are designated by the notations *Vg* (genital verruca) and *Pg* (genital papilla).

Each genital verruca presents one lyrifissure (*ivg*), one compound seta (*pvg*; it consists of a broad base and a long narrow tip), and a number of papilliform setae (three or four in the female, five or six in the male).

←

Fig. 124. Opilioacarus texanus (Chamberlin & Mulaik), female; A, left pair of eyes; B, posterior part of prosoma with seta, coniculate microsculpture and sigilla (external marks of muscle attachments); C, dorsal setae of the posterior part of the prosoma; D, sigilla between fourth and fifth opisthosomatic segments; E, lateral view of stigma 2; F, dorsal (slightly oblique) view of left region of stigmata 1–4; G, H, dorsal lyrifissures of the fifth opisthosomatic segment; A, × 550; B–D, G–H, × 720; E, × 565; F, × 295.

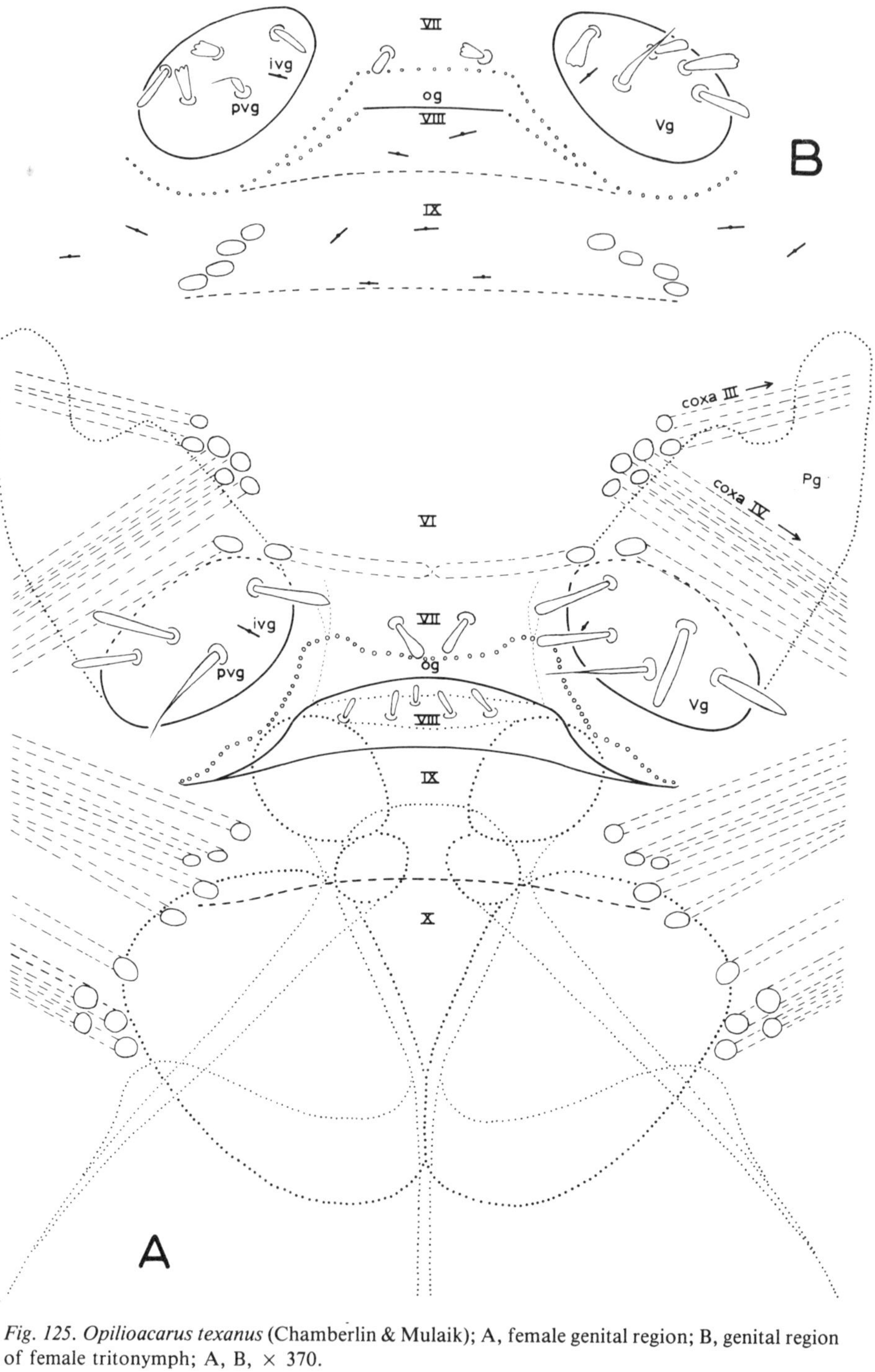

Fig. 125. Opilioacarus texanus (Chamberlin & Mulaik); A, female genital region; B, genital region of female tritonymph; A, B, × 370.

In figs. 122 and 125, an attempt is made to reconstruct the segmental origin of the genital region. The genital sclerite is supposed to belong to segment VIII, whilst the pregenital area presenting two or five setae (in female and male respectively) certainly represents segment VII. In fig. 125A I have represented part of the internal structures such as they were visible in a female in lactic acid. The muscles extending from the ventral muscle attachments in laterodorsal directions are also represented.

Eggs, removed from a female, are represented in fig. 126A–C. They have a remarkable shape which appeared to be constant in all eggs studied by me; this shape is probably not due to the action of lactic acid. All eggs presented two processes: a blunt process, and a narrow, elongate. The wall of these processes is slightly thicker than the wall of the remaining part of the egg; at the base of the processes, the wall of the egg is apparently partly porous (fig. 126B, C). In the largest egg (fig. 126A), the blunt process presents an interesting internal structure which reminds one of an orifice and a canal.

Sternal region of the podosoma. – The sclerotized part of this sternal region has a more or less pentagonal outline (fig. 122); the greatest breadth is posterior to coxae IV. Laterally, it is bordered by a faint ridge, with salient angles between coxae I and II, and between coxae II and III.

The sternal region of leg I consists for the greater part of soft skin. It bears a pair of sternapophyses which are associated with the subcapitular gutter. In my alcohol material, the sternapophyses (figs. 122, 123, 128, 129: *SA*) are orientated nearly parallel to the ventral surface of the infracapitulum; they present a lateroventral (antiaxial) seta, and a terminal one. The sternapophyses are associated with the subcapitular gutter. A pair of open grooves extend from between the bases of the sternapophyses up to the area between coxae I and II; they present a small tectum near coxa I (not figured). These open grooves represent sternal taenidia (*s.t*); they are associated with the orifice of the coxal gland. Anteriorly, the united taenidia are connected with the subcapitular gutter (the gutter also represents a taenidium).

The anterior part of the sternal region, which apparently belongs to the segment bearing leg II (segment IV), presents a pair of warts similar in shape to the genital verrucae. In ventral view, the anterior border of the warts is especially distinct; in various lateral orientations, the more or less hemispherical shape is easily visible (fig. 123). Just as the genital verrucae, they present a lyrifissure and setae; there are one or two spiniform setae at the anterior border, and one compound seta posterior to it. Because of the wart-like shape, the structures are named sternal verrucae (*Vs*). Sternal and genital verrucae are supposed to represent parts of homonomous structures (appendages), whilst the sternal verrucae could be homologous with Claparède's organ, known from prelarvae and larvae of Actinotrichida.

The sternal region posterior to the verrucae is apparently composed of segments V and VI. It presents concentrations of muscle attachments, which are represented in fig. 122 and, in detail, in fig. 125A. The last-mentioned figure

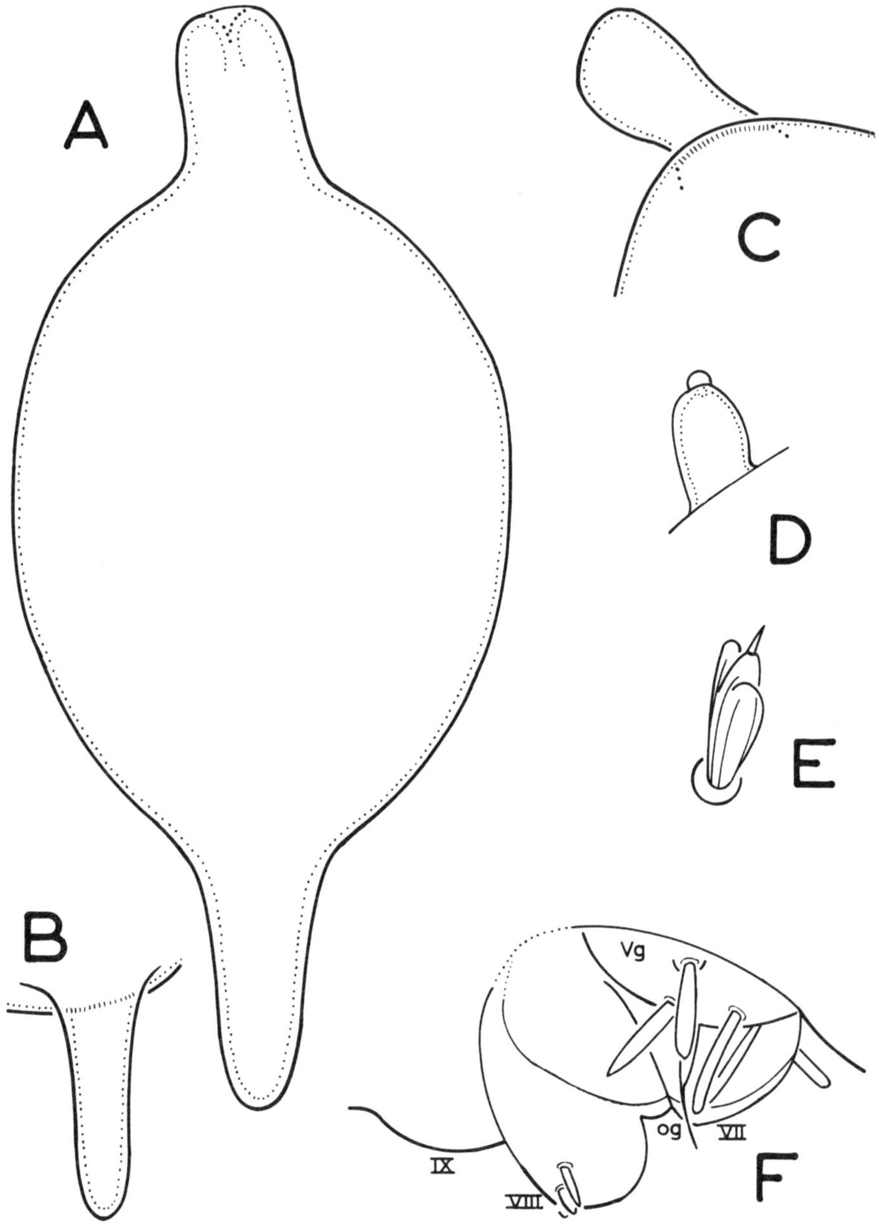

Fig. 126. Opilioacarus texanus (Chamberlin & Mulaik); A, egg; B, C, details (elongate, resp. blunt process) of a smaller egg, orientated obliquely; D, laterocoxal seta *eII*; E, distal seta *l'* of right basitarsus III; F, lateral view of female genital region; A–C, × 490; D–E, × 1750; F, × 490.

shows also the muscle bundles extending to coxae III and IV. A pair of transverse muscles anterior to the pregenital area appear to be attached to the endosternite. There are two irregular rows of lyrifissures in the sternal regions of segments V and VI: the anterior row (consisting of two large and two small lyrifissures) probably belongs to segment V, the posterior row (consisting of six small lyrifissures) probably belongs to segment VI. There are also a number of setae in this area, viz., four compound setae arranged in a trapezoid, and nine or ten (female) to thirteen (male) papilliform setae arranged in two irregular rows; in the nymphs the number of papilliform setae is smaller.

Lateral region of the podosoma. – In fig. 123, a female is represented of which the legs, with the exception of the coxae, have been removed. The coxae are free, movable segments, which are dealt with in the paragraph on the legs. The lateral region of the podosoma is characterized by the presence of lobes dorsal to the base of the anterior three pairs of legs. These lobes are partly also visible in dorsal and ventral view (figs. 121, 122). According to Chamberlin & Mulaik (1942: 128, pl. 2 fig. 12), the posterior two lobes (they did not observe the anterior) articulate with the dorsal processes (*ATR.II, ATR.III*) of trochanter II and trochanter III. In my opinion this is only part of the function of the lobes, the main function certainly being a protection of the basal parts of the legs. The lobes are named suprapedal lobes (TL_{1-3}). It may be remarked here that trochanter I does not present a dorsal process corresponding to the anterior suprapedal lobe (TL_1).

Gnathosoma. – The entire gnathosoma has more or less the shape of a cone, the broadest part being attached to the idiosoma. In my material, this buccal cone is, generally, inclined at an angle of about 45 degrees with the longitudinal axis of the idiosoma. According to Grandjean (1936b: 421), who observed a living specimen of *O. segmentatus*, the gnathosoma is, however, very mobile, being able of movements in the vertical plane as well as of movements from left to right (or the reverse). It is probable that the buccal cone can also be extended.

Just as in other mites, the gnathosoma consists of cheliceral frame and infracapitulum. It is described here in dorsal, ventral, lateral, frontal and posterior view. Sectioning is not necessary, because preparations in lactic acid of the entire gnathosoma are sufficiently clear for a study of internal structures. In the course of the investigations, the chelicerae must, however, be removed in order to study the extent of the cheliceral cavity (i.e., the posterodorsal cavity of the gnathosoma, covered by the tegulum), and the dorsal aspect of the infracapitulum.

When the entire gnathosoma is studied in dorsal orientation (fig. 127), it appears that the greater part is hidden from view by the relatively large chelicerae. The cheliceral sheaths pass proximally into a structure of which the extent can only be studied in detail after removal of the chelicerae, and consequently is described below. Parts of the rutella (*RU*), With's organs (*OW*) and the labrum (*LS*) are also visible in this orientation.

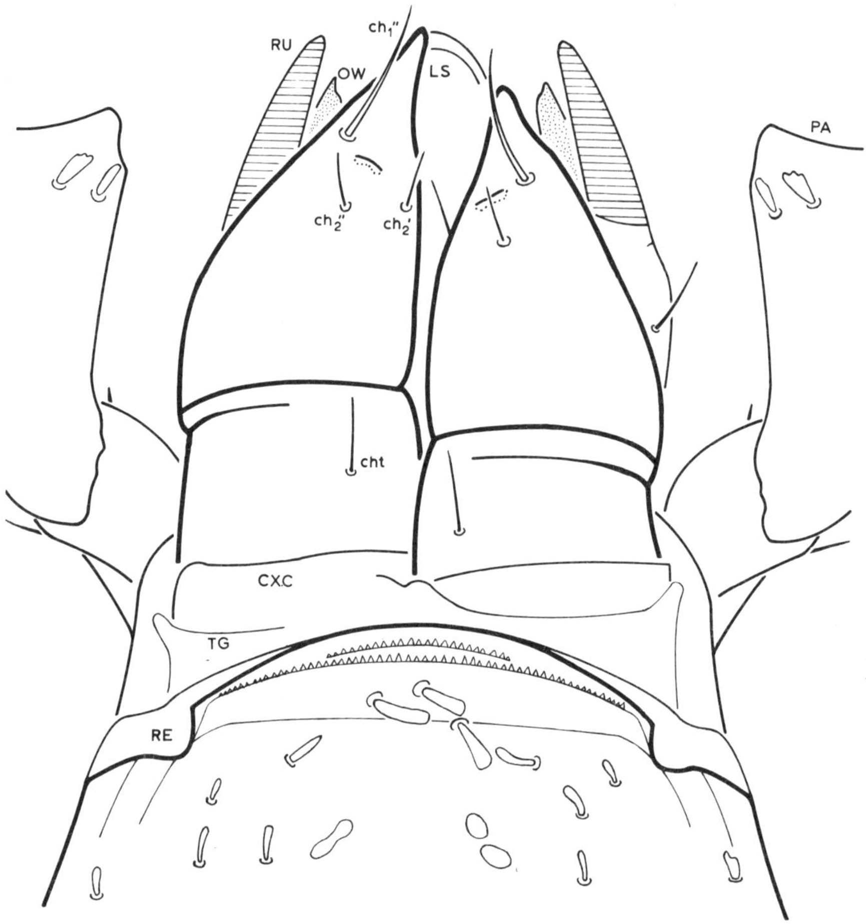

Fig. 127. Opilioacarus texanus (Chamberlin & Mulaik), female; dorsal view of gnathosoma and anterior part of idiosoma; × 370.

A ventral view of the gnathosoma (fig. 128) presents the following characters. Mentum (*H*) and genae (*G*) are not separated by a labio-genal articulation; the infracapitulum, consequently, is anarthric. There is a distinct subcapitular gutter extending from the lateral lips and the inferior oral commissure (*Ji*) to the posterior border of the infracapitulum; it continues in the soft skin (the arthrodial membrane) between gnathosoma and idiosoma, and joins the pair of sternal taenidia between the sternapophyses. In the paragraph on the sternal region of the podosoma, it was already mentioned that the subcapitular gutter also represents a taenidium. In both sexes, the genae have fused in the median part. In ventral orientation, the lateral lips have the appearance of thick lobes which join each other at the place of the inferior oral commissure.

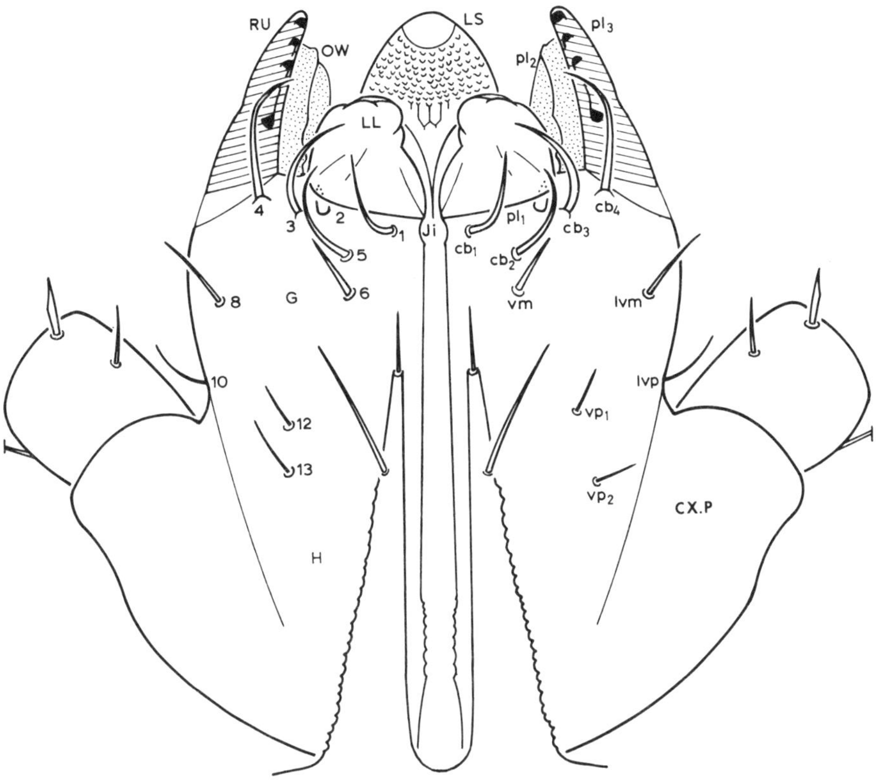

Fig. 128. Opilioacarus texanus (Chamberlin & Mulaik), female; ventral view of gnathosoma and sternapophyses; × 370.

I have made an attempt to classify the infracapitular setae. When we regard With's organ and the rutellum as hypertrophied setae, we can distinguish a row of three pairs of specialized setae, viz., (1) a small spine, resembling a small corniculus, close to the ventral border of the lateral lips (partly covered by them in ventral view), (2) With's organ, and (3) the rutellum. These are named here paralabial setae (pl_{1-3}). The small spine which is found in *O. segmentatus* close to the rutellum (cf. Grandjean, 1936b: fig. 1), is apparently absent in *O. texanus*; it is regarded here as a fourth paralabial seta (pl_4). The paralabials (possibly with the exception of the rutellum) are the only infracapitular setae which are not hollow.

Posterior to the paralabial setae, but very close to them, is a series of four pairs of curved setae, here named circumbuccal setae (cb_{1-4}). These probably have a special function, because each of them is in close relation to a special element of the gnathosoma: cb_1 and cb_2 are associated with the lateral lips; cb_3 with With's organ, and cb_4 with the rutellum. It is possible that they control certain movements of these organs.

LS
RU
LOW
LL
cb1
ogc
id
iα
CH
PA
cb4
cb3
pl1
cb2
ldm (=9)
lvm
vm
vp1
vp2
lvp
ogl
SY
p
dg
PH
SA
ap.ce
RL
RE

The remaining infracapitular setae can be classified according to their positions; ventrally, there are two pairs of median setae (*vm, lvm*), and three pairs of posteriors (vp_1, vp_2, *lvp*). In fig. 128 I have added Grandjean's notation (Grandjean, 1936b) at the left side of the figure; in this way the relations between his numbers and my notations are easy to establish. According to Grandjean, his numbers 10–11 (both *lvp*), and 12–13 (both *vp*) appeared sometimes to be replaced by one seta, whilst his seta 14 (vp_3) was mostly absent. In my material of *O. texanus*, 11 and 14 are not present. Grandjean's seta 7 (vm_2), constantly occurring in *O. segmentatus*, is also absent in *O. texanus*.

According to Grandjean's classification of rutella (Grandjean, 1957b: 252), the rutellum of *Opilioacarus* belongs to the category of the ordinary atelebasic rutella (occurring, for instance, in primitive Oribatid mites like *Parhypochthonius, Camisia* and *Hermannia*); this type of rutellum is not primitive and narrow (as in *Heterochthonius* or *Eniochthonius*), but nevertheless has no broad base extending to the sagittal plane of the infracapitulum. According to Grandjean (1957b: 273), With's organ probably represents a second rutellum; it can expand enormously under the influence of lactic acid, so that it can be difficult to represent its exact shape. The labrum is easily visible in ventral orientation; it presents numerous small teeth at the ventral surface.

Several parts of the gnathosoma can best be studied in lateral view. In fig. 129, a lateral membraneous part (*SY*) is represented, which was named synaptic tectum by Grandjean (1936b: 427). Dorsally, it is attached to the tegulum; a rounded prominence (*p*) is visible at its anterior laterodorsal border. Ventrally, the synaptic tectum is attached to the infracapitulum. The shape of labrum, With's organ, and rutellum is also easily recognizable in lateral view. With's organ appears as a more or less rounded disk with irregularly dentate or serrate anterior and dorsal border. The rutellum is large and presents five ventral teeth, of which the proximal one is the largest; it presents an internal core which extends into the malapophyses as a root. Apart from the ventral infracapitular setae, a laterodorsal pair is also present, just as in Holothyrida; it is indicated here as *ldm* (it corresponds with Grandjean's seta 9). In lateral view, the pharynx is also easily visible, together with the efferent duct (*dg*) of an infracapitular gland, which debouches ventral to the chelicerae.

There appear to be a pair of these infracapitular (probably salivary) glands: two ducts and two orifices (*ogi*) are present. These orifices are best visible when the chelicerae are removed (fig. 130). The position of *ogi* exactly corresponds with the position of foramen and orifice of infracapitular gland in Actinotrichida. The removal of the chelicerae should be executed with the utmost care; otherwise the small tegulum (fig. 130: *TG*) risks to get damaged or lost.

←

Fig. 129. Opilioacarus texanus (Chamberlin & Mulaik), female; lateral view of gnathosoma and anterior part of idiosoma; × 370.

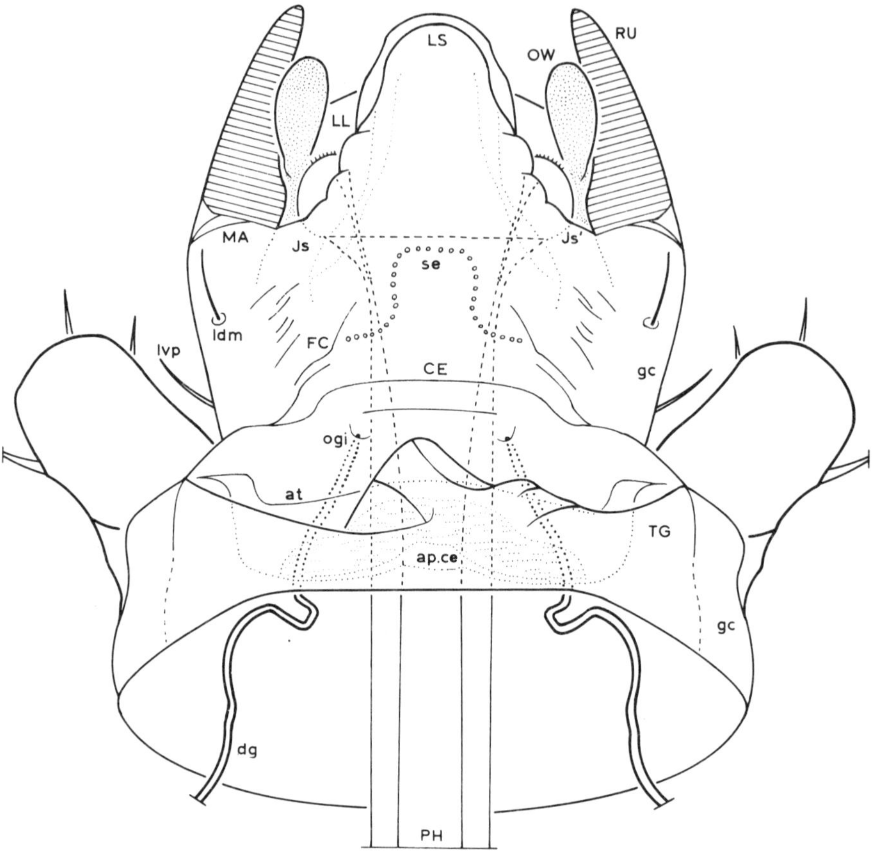

Fig. 130. Opilioacarus texanus (Chamberlin & Mulaik), female; dorsal (oblique) view of gnathosoma without chelicerae; × 370.

The Opilioacarid tegulum resembles the tegulum of Holothyrida, although it is smaller; in both groups, the tegulum is characterized by the absence of a supracheliceral limbus (an anterior dorsal extension).

Apart from the paired orifices *ogi*, the dorsal surface of the infracapitulum presents the following characters. The labrum has a very broad base. The superior oral commissures *Js* and *Js'* are easily visible. A pair of muscles (*tb*) extend from the labrum in posterior direction, and could be attached to the pharynx, at the place where the latter constitutes a pair of laterodorsal folds. With's organ, rutellum and the dorsal infracapitular seta *ldm* are also easily visible. The cervix is distinctly vaulted; the median elevated part (the capitular saddle *se*) presents a microsculpture. There is no laterocoxal seta of the palp. The ventral part of the cheliceral frame is attached to the cervix according to the line *at*. The distinctly thickened internal area, posterior to *at*, represents

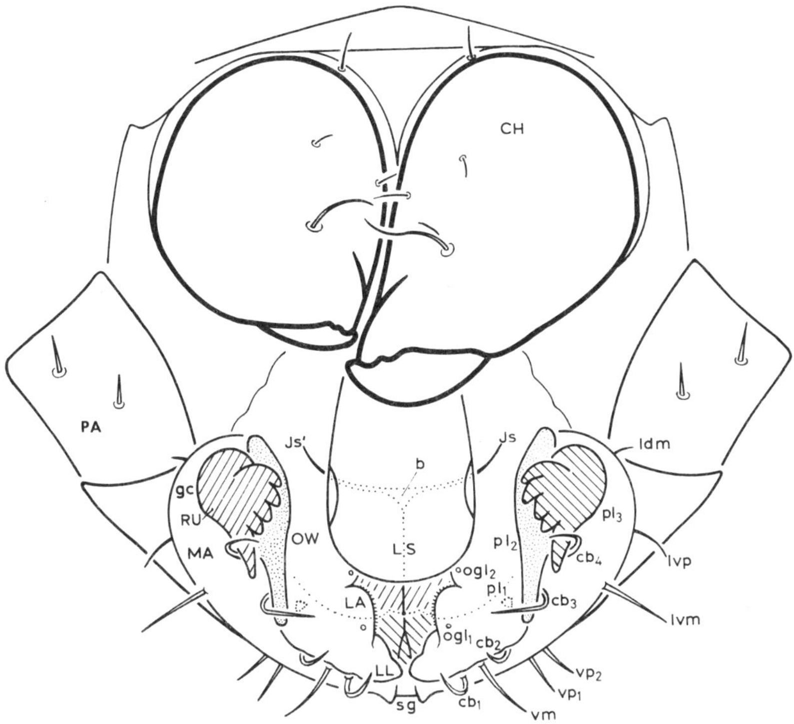

Fig. 131. Opilioacarus texanus (Chamberlin & Mulaik), female; frontal view of gnathosoma; × 370.

the cervical apodeme (*ap.ce*), which is also visible in lateral view (fig. 129).

The triangular shape of the mouth is easily visible when the gnathosoma is studied in frontal view. In fig. 131, I have indicated the superior commissures *Js* and *Js'*; the inferior commissure *Ji* is visible in the preoral cavity. The figure demonstrates that the lateral lips are remarkably thick lobes, ventrally and laterally surrounding the preoral cavity. The laterodorsal part of the lateral lips presents a small lobe, named labellum (*LA*); the paired labella are beset with small teeth which correspond with the teeth in the ventral part of the labrum. Two pairs of glands (labial glands) debouch in the laterodorsal part of the paired lateral lips; the orifices (ogl_1, ogl_2) as well as the ducts are easily visible. The curved shape of the circumbuccal setae (cb_{1-4}) is excellently visible in this orientation; as mentioned above, these setae are supposed to control the movements of lateral lips, With's organ and rutellum. Other interesting characters represented in fig. 131 are the laterodorsal position of seta *ldm*, and the profile of the subcapitular gutter (*sg*).

The shape of the pharynx can best be studied in posterior view (fig. 132). A transverse section of the pharynx, just behind the infracapitulum appears to

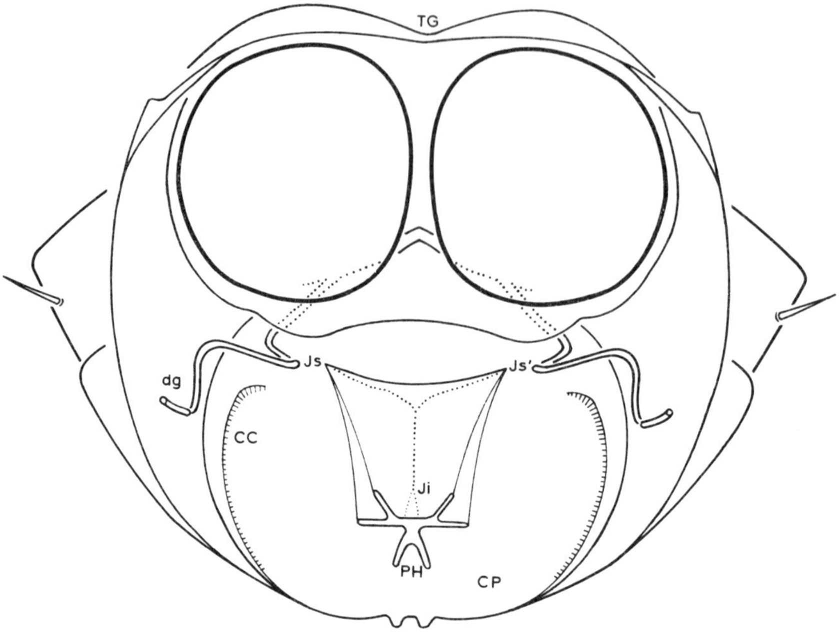

Fig. 132. Opilioacarus texanus (Chamberlin & Mulaik), female; posterior view of gnathosoma; × 370.

be hexagonal. The mouth, however, is triangular (or trifid; see fig. 131 and also fig. 132). This change of shape can partly be explained in the following way. As mentioned above, a pair of muscles (fig. 130: *tb*) are attached to the pharynx, just behind the mouth. At this place, the laterodorsal lobes of the pharynx (reference is made here to the 'closed' pharynx) are duplicated, giving rise to two pairs of folds. The single longitudinal section prepared by me was not appropriate for representation. According to this section, the muscles surrounding the pharynx continue as far as the superior oral commissures (*Js, Js'*); it appears that they do not extend to *Ji*. In fig. 132, the pharyngeal chamber (*CP*), i.e. the room enclosed by the infracapitulum, the entrance to the coxal cavities (*CC*), the ducts of the infracapitular glands (*dg*), and the tegulum (*TG*) are also represented.

The principal characters of the gnathosoma of *Opilioacarus texanus*, consequently, are the following: (1) the presence of a small tegulum without supracheliceral limbus; (2) the presence (at both sides) of rutellum, With's organ, a small horn-shaped paralabial spine or corniculus, and some ten hollow infracapitular setae (including four curved circumbuccals and a laterodorsal seta); (3) the lobe-shaped lateral lips with labellum and orifices of labial glands; (4) the absence of laterocoxal setae; (5) the presence of ducts and orifices of infracapitular glands; (6) the triangular (trifid) mouth; and (7) the hexagonal section of part of the pharynx.

Chelicera. – The chelicerae (fig. 133A–E) consist of three segments, viz., trochanter, body of the chelicera, and apotele. They are not very long, but large and rather high in comparison with the breadth. The antiaxial surface is distinctly curved. Trochanter and body of the chelicera articulate by means of a pair of lateral condyles (fig. 131A, B: *k'*, *k''*). The border of the segments is rather faintly sclerotized, so that the exact shape of the condyles is difficult to establish. Two small tendons are attached to the body of the chelicera, viz., a superior (tf_s) and an inferior (tf_i). In comparison with the legs (cf. the paragraph below), the articulation functions as a trochanter-femur articulation (a pivot joint). The articulation of the apotele (the movable jaw) with the body of the chelicera is also operated by two tendons (t_s, t_i). The superior tendon (t_s) is very strong, and extends into the trochanter. The trochanter bears one dorsal seta. The body of the chelicera presents three setae (one dorsal seta, two laterodorsals), and two lyrifissures, viz., a dorsal (*id*) and an antiaxial (iα). The chela is represented in opened condition in fig. 129, and in closed condition in fig. 133. Fixed and movable jaw both present about three irregularly shaped teeth. The movable jaw presents a rather distinct mark (fig. 129: *ogc*) at the base of the proximal tooth; this probably represents the orifice of a cheliceral gland. The arthrodial membrane of the articulation between the body of the chelicera and the apotele extends as a ventral oncophysis up to the distal third part of the movable jaw. This is especially visible in lateral view (fig. 133A, B: *opv*). It ends in a small ligule, best visible in ventral view (fig. 133E). The ligule is represented, more enlarged, in fig. 133C. It is difficult to establish the exact lateroventral border of the oncophysis.

Palp. – There are six palpal segments, viz., trochanter, femur, genu, tibia, tarsus and apotele. They are represented in fig. 134A–E. Just as in Holothyrida, tibia and tarsus have the shape of one large segment. Trochanter, femur and genu (fig. 134A) present ordinary setae; femur and genu present moreover a number of papilliform setae. The tibia and, especially, the tarsus present a variety of setae (fig. 134B–E). The tibia presents a great number of more or less lanceolate, ciliate setae, together with compound setae (i.e. setae composed of a relatively broad base and a narrow, solenidion-like terminal part) and phaneres which are strongly curved (probably solenidia). The tarsus presents five types of phaneres, arranged in the following way: (1) a dorsal area of fourteen hollow setae with more or less truncated tip; (2) a paraxial group of leaf-shaped setae; (3) three laterodorsal (slightly antiaxial) phaneres, probably representing solenidia; (4) compound setae (*sm*), especially in the terminal part; (5) ciliate setae in the ventral region. The tarsus presents, moreover, an antiaxial and a paraxial lyrifissure (*iα, iπ*). In dorsal and ventral view, the border of tarsus and tibia appears to be oblique: the paraxial side of the tarsus is distinctly longer than the antiaxial. The tendon of the tarsus has a paraxial position (fig. 134D, E: *tta*). The apotele is operated by two tendons, a superior and an inferior (fig. 134B–E: t_s, t_i). The distal paraxial part of the tibia presents a smooth area without microsculpture; the border of this area is represented in fig. 134C–E

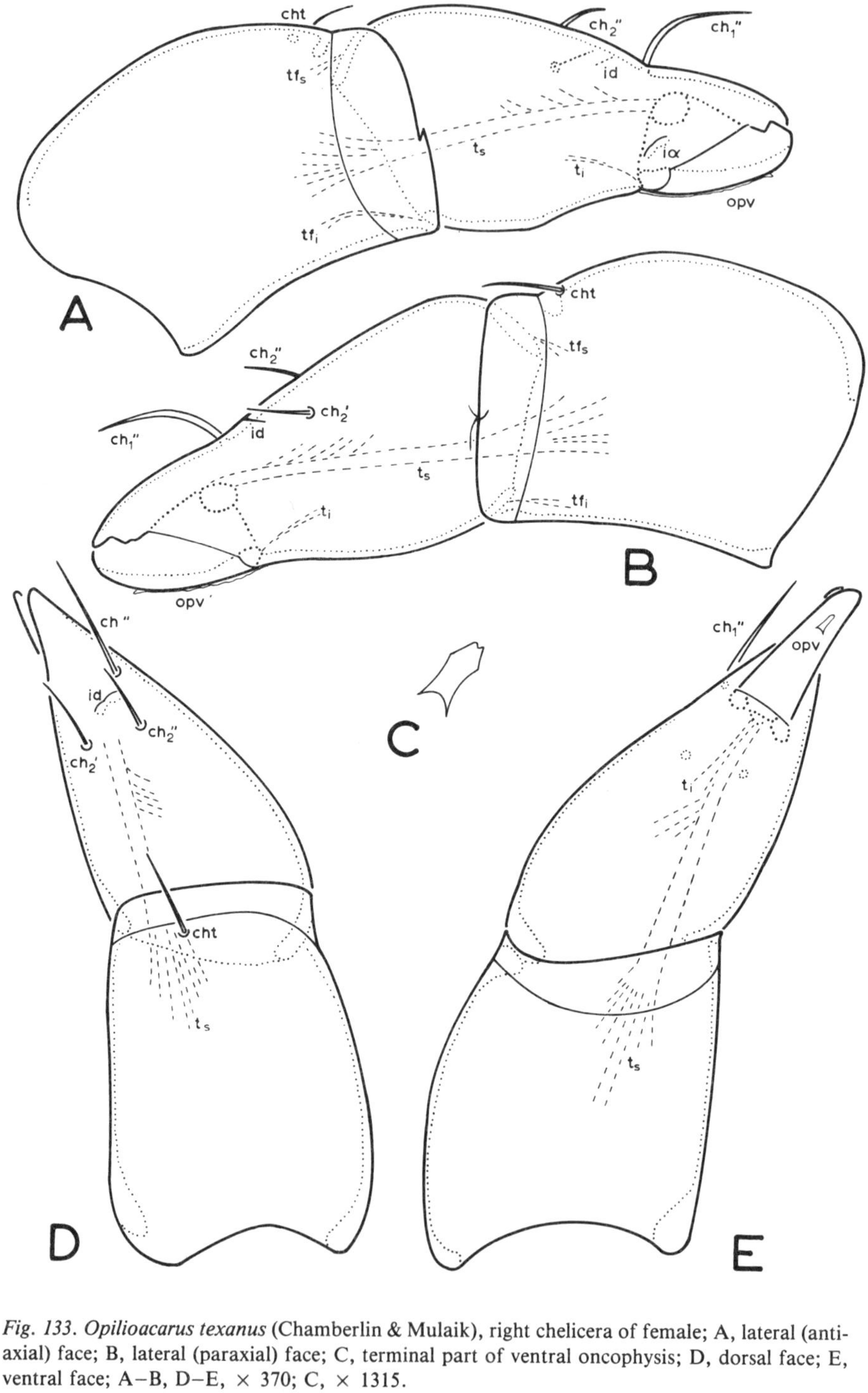

Fig. 133. Opilioacarus texanus (Chamberlin & Mulaik), right chelicera of female; A, lateral (antiaxial) face; B, lateral (paraxial) face; C, terminal part of ventral oncophysis; D, dorsal face; E, ventral face; A–B, D–E, × 370; C, × 1315.

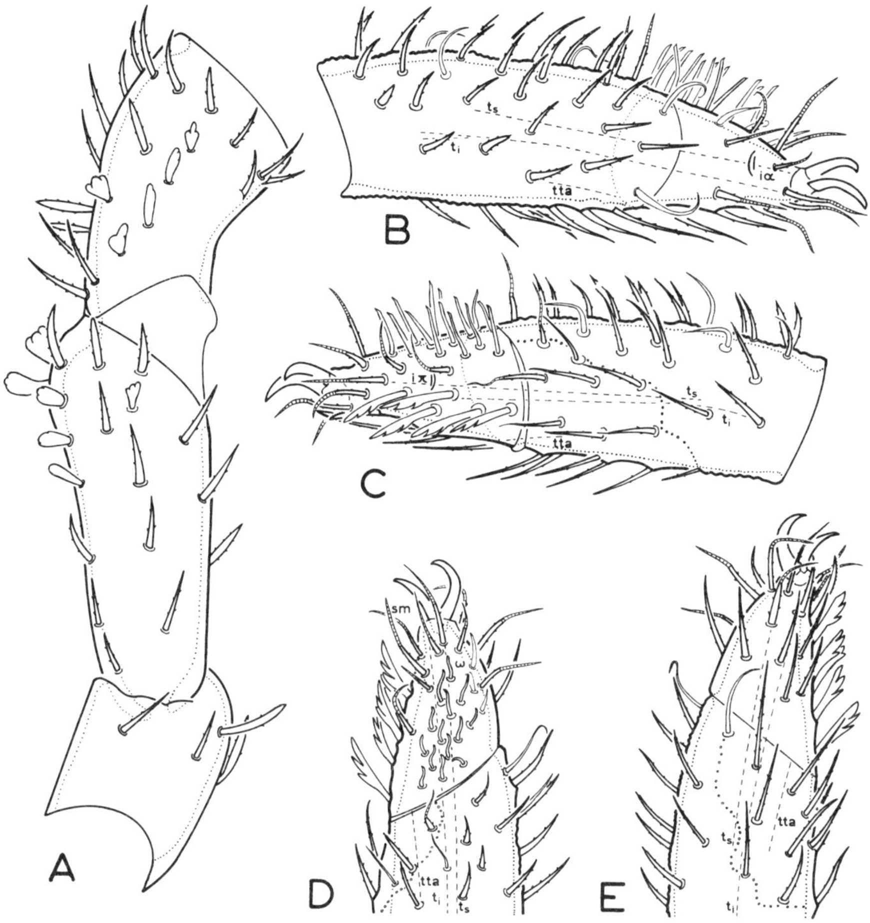

Fig. 134. Opilioacarus texanus (Chamberlin & Mulaik), right palp of male; A, lateral (antiaxial) view of trochanter, femur and genu; B, C, tibia, tarsus and apotele; B, lateral (antiaxial) view; C, lateral (paraxial) view; D, E, distal part of tibia, tarsus and apotele; D, dorsal view; E, ventral view; A–E, × 295.

by a dotted line. The tarsus is completely devoid of microsculpture.

Legs. – The legs are inserted laterally. The coxae of the first three pairs are protected dorsally by suprapedal lobes (fig. 123); ventrally, coxae II–IV are close to the lateral border of the sternal area (fig.122). Acetabula are rather indistinct when the coxae are not removed. In my material, legs I–II are directed forward, legs III–IV backward. Leg I is distinctly longer and more slender than the other legs; the ungues of its claw are much narrower, and the papilliform setae less numerous than in legs II–IV. According to the study by Grandjean (1936b: 420) of living specimens of *O. segmentatus*, leg I has only a restricted ambulatory function; its main function is supposed to be sensory. Part of the cuticle of telotarsus I is smooth (fig. 135A; distal to the dotted line).

As mentioned above, the sequence of the legs (according to decreasing length) is I, IV, II, III. The different legs consist of different numbers of segments, connected by different types of articulations. The segments are: coxa, trochanter (trochanter 1 and 2 in the case of legs III and IV), femur (basi- and telofemur are distinctly separated in the case of leg I), genu, tibia (basi- and telotibia in the case of leg I), basitarsus, telotarsus (telo- and acrotarsus, or telotarsus 1 and 2, in legs II–IV), pretarsus (only in legs II–IV) and apotele.

The legs of the tritonymph were carefully measured (with the exception of the coxae). The lengths of the segments (in mm) are the following. Leg I: trochanter 0.25, basifemur 0.05, telofemur 0.48, genu 0.36, basitibia 0.31, telotibia 0.25, basitarsus 0.24, telotarsus 0.26 (total length 2.20). Leg II: trochanter 0.12, femur 0.29, genu 0.13, tibia 0.14, basitarsus 0.16, telotarsus 0.17, acrotarsus 0.06, pretarsus + apotele 0.02 (total length 1.09). Leg III: trochanter 1 0.09, trochanter 2 0.06, femur 0.23, genu 0.11, tibia 0.14, basitarsus 0.16, telotarsus 0.17, acrotarsus 0.06, pretarsus + apotele 0.02 (total length 1.04). Leg IV: trochanter 1 0.14, trochanter 2 0.12, femur 0.38, genu 0.19, tibia 0.25, basitarsus 0.21, telotarsus 0.19, acrotarsus 0.06, pretarsus + apotele 0.02 (total length 1.56). It is interesting to compare the length of homonomous segments in legs II and III. These legs resemble each other, and have nearly the same total length; leg III presents, however, two trochanteres. It appears that the trochanter of leg II is smaller than trochanteres 1 + 2 of leg III. Femur and genu of leg II are, however, both longer than the homonomous segments in leg III. The exceptional length of leg I (when compared with leg IV) is particularly caused by the lengths of femur, genu and tibia.

The segments of the legs are movably connected by means of different types of articulations. The coxae have no condyles, but are movable in various directions by means of muscles. The articulation between coxa and trochanter presents one tendon (*ttr*); according to Grandjean (1936b: 430) it is associated with autotomy (appendotomy). A retrotectum protecting this articulation is visible in the figures of leg I (fig. 135B: *te*) and leg IV (fig. 139A: *te*); I do not know if this retrotectum is also present in legs II and III.

Trochanter 1 and 2, and trochanter and femur articulate by means of a pair of lateral condyles (k', k''), constituting a pivot joint; there are, in each case, a superior and an inferior tendon, with levator, respectively depressor muscles. These tendons are indicated as tr_s and tr_i, in the case of trochanter 2, and as tf_s and tf_i, in the case of the femur.

The articulation between basifemur and telofemur of leg I is an adesmatic articulation of very restricted movability. The suture separating both parts is complete and passes through an important number of lyrifissures (three antiaxials and some paraxials) of which the canals are distinct. In legs II–IV there are two large lyrifissures (but no suture) at the place of this articulation, an anterior (*if'*) and a posterior (*if''*). The articulations between femur and genu, and between genu and tibia, each present one tendon (*tg*, respectively *tti*), with flexor muscles; they represent hinge joints. The articulation between basitibia and telotibia of leg I is an adesmatic articulation of very restricted movability; it is situated halfway the tibia. The two parts are separated ventrally by a large lyrifissure, dorsally by a fold; laterally, the sclerotized cuticle is not interrupted. It may be remarked here that, according to Grandjean (1936b: 430), tibia I is completely divided in the adults of *O. segmentatus*. The articulation between tibia and basitarsus presents one tendon (*tta*) associated with flexor muscles; it represents a hinge joint.

The adesmatic articulation between basitarsus and telotarsus is different in leg I, on the one hand, and legs II–IV, on the other. Basi- and telotarsus of leg I (fig. 135E) are separated ventrally by two lyrifissures which have the shape of cupules. Dorsally and laterally, the segments are separated by soft skin. The ventral part of the articulation can probably function as a hinge. Basitarsus and telotarsus of legs II–IV (figs. 137, 138, 139E) are completely separated by soft skin; in the lateral region of the telotarsus the sclerotized cuticle is, however, produced and can certainly function as an articulation-point. This interesting articulation is represented in detail in fig. 140B–D. The basitarsus presents dorsally a pair of protuberances (*P'*, *P''*), corresponding with a pair of dorsal cavities (*S'*, *S''*) in the telotarsus. When the tarsus is elevated, the cavities rest against the protuberances. Anterior to each telotarsal cavity is a lyrifissure (*ite'*,*ite''*). Ventrally, just behind the articulation, the basitarsus presents a large lyrifissure (*ibv*). Telotarsus and acrotarsus (in legs II–IV) are dorsally and laterally separated from each other by soft skin (cf. figs. 137, 138); this articulation permits of flexion and extension. The apotele is movable by means of two tendons, a superior (t_s) and an inferior (t_i), associated with levator, respectively depressor muscles. The superior tendon (t_s) passes through pretarsus, acrotarsus (in the case of legs II–IV) and telotarsus, and continues in the basitarsus. The inferior tendon (t_i) passes through pretarsus, acrotarsus, telotarsus and basitarsus, and continues in the tibia. The pretarsus of legs II–IV (fig. 141A–C) presents an internal skeleton, consisting of two ventral bars which articulate distally with the apotele, proximally with the acrotarsus. The two tendons t_s and t_i are associated with the articulation of the apotele and the adesmatic tarsal segments. Contraction of the muscles associated with t_s will result in elevating the claw, elevating the pretarsus, extending the acrotarsus, and elevating the telotarsus. Contraction of the muscles associated with t_i will result in depressing claw, pretarsus and telotarsus.

The legs present an important number of lyrifissures, apart from those associated with articulations. I have discerned ventral lyrifissures in all segments of the legs, with the exception of coxa, acrotarsus, pretarsus and apotele; they are only partly represented in figs. 135, 137, 138, 139. There is a considerable

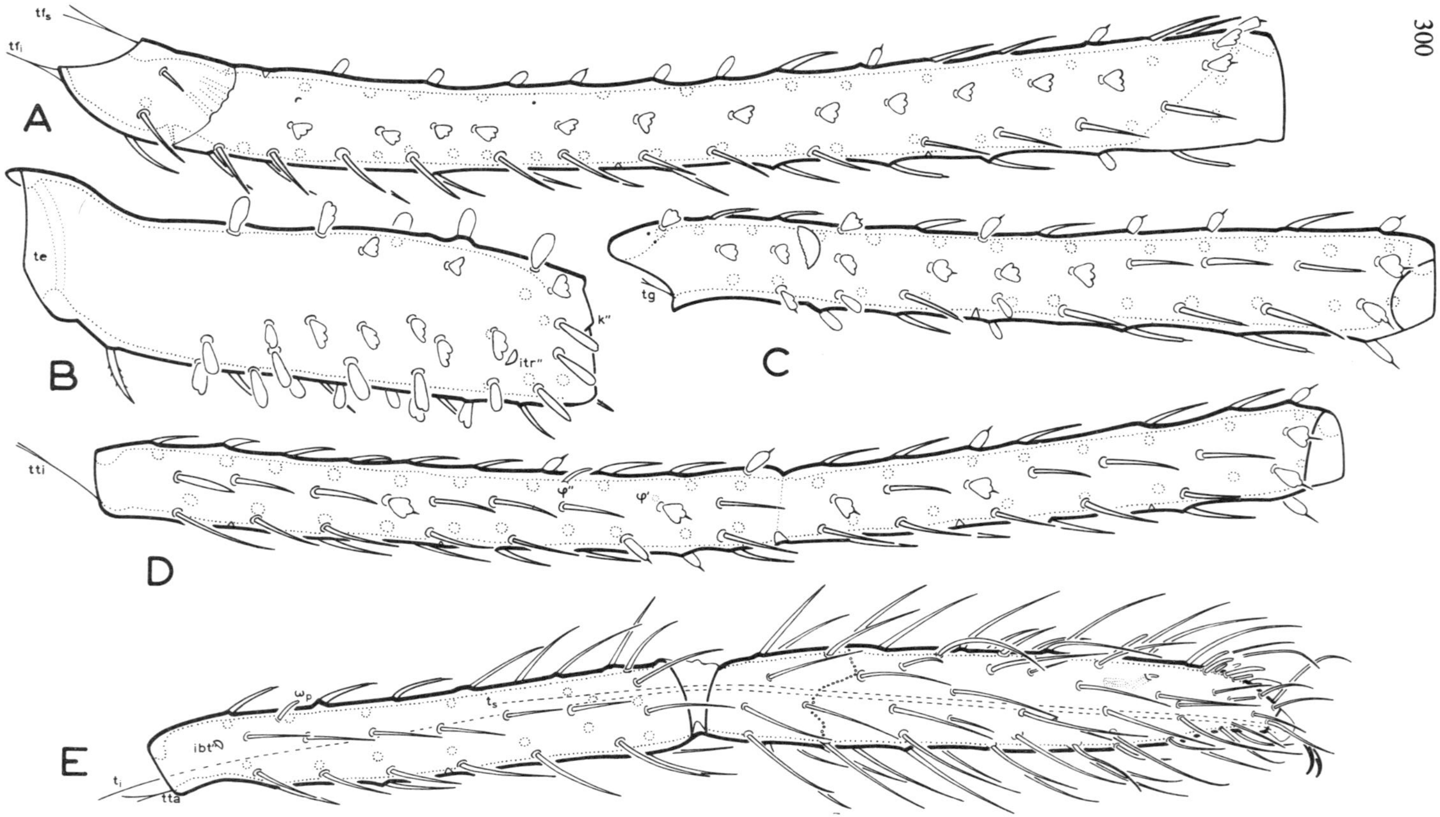

Fig. 135. Opilioacarus texanus (Chamberlin & Mulaik); lateral (posterior = antiaxial) view of right leg I of tritonymph; A, femur; B, trochanter; C, genu; D, tibia; E, tarsus and apotele; A–E, × 295.

variation in size of the ventral lyrifissures. A distinct posterior lyrifissure *ibt''* is present in the proximal part of the basitarsus; it occupies a laterodorsal position. Trochanter I presents an antiaxial (posterior) lyrifissure (*itr''*) in the distal part.

All legs present two ungues and a pulvillus. The ungues of leg I are very narrow, and the pulvillus is reduced (fig. 136A–C). The pulvilli of legs II–IV are distinctly recognizable, but not very large. Pulvillus III is represented in various orientations in fig. 141A–C; it has the shape of a little cushion under the ungues. The pretarsus of legs II–IV presents two pairs of setae: a pair of pectinate lateral setae (*l'*, *l''*), and a pair of curved laterodorsals (*ld'*, *ld''*).

All coxae (figs. 122, 123) bear a number of papilliform and ordinary setae. Coxae I and II present moreover a supposed laterocoxal seta (*eI*, respectively *eII*); these setae were first discovered by Grandjean (1936b: 430, fig. 4) in the case of *O. segmentatus*. They occupy the same position as those described by me in the case of *Glyptholaspis* (Gamasida), i.e. paraxial. The two laterocoxal setae are exactly similar in shape. Although they are reduced in several respects, they still distinctly belong to the mucronate type: there is relatively broad, hollow basal part, and a small, hollow, distinctly separated tip; both parts are apparently connected by a canal (fig. 126D).

There is an important difference in the chaetotaxy of leg I and that of legs II–IV, especially in the terminal part. This is certainly connected with the specialized sensory function of the first leg. Telotarsus I (fig. 135) presents a great number of long, smooth setae. An anterior dorsal area is conspicuous because of a number of smaller setae of different shapes. This area is represented, more enlarged and in different orientations, in fig. 136A, C. Ten hollow setae are easily visible; they are either broad and blunt, pointed, nearly rectangular in outline, or mucronate. Lateral to this area, two pairs of setae are present, with obliquely truncated tip. Several setae of the telotarsus adhere to each other, probably by means of some secretion (cf. fig. 136A, C). The distal, antiaxial (posterior) part of telotarsus I bears an interesting organ, already discovered by Grandjean (1936b: 433, fig. 5B, C) in *O. segmentatus*. It consists of a small cavity with two hollow phaneres (probably solenidia). One of these solenidia (*h*), which is clavate, is completely hidden; the other solenidion (*f*), which is setiform, is partly extending. The organ is represented in dorsal, ventral and lateral view in fig. 135A–C; it is named telotarsal organ.

The greater part of the setae of basitarsus I resemble those of the telotarsus. Some of the dorsal setae, however, are mucronate; they have the shape of an elongate ordinary seta with a distinctly separated point. The proximal part of basitarsus I presents a small laterodorsal (antiaxial = posterior) hollow phanere, probably a solenidion (ω_p).

Tibia I presents a great number of ordinary setae, mixed with some papilliform setae with distinct terminal point; the complete terminal verticil consists of papilliform setae. The basitibia presents in its distal part, antiaxially as well as paraxially, a small hollow phanere, probably a solenidion (fig. 135D: φ' φ'').

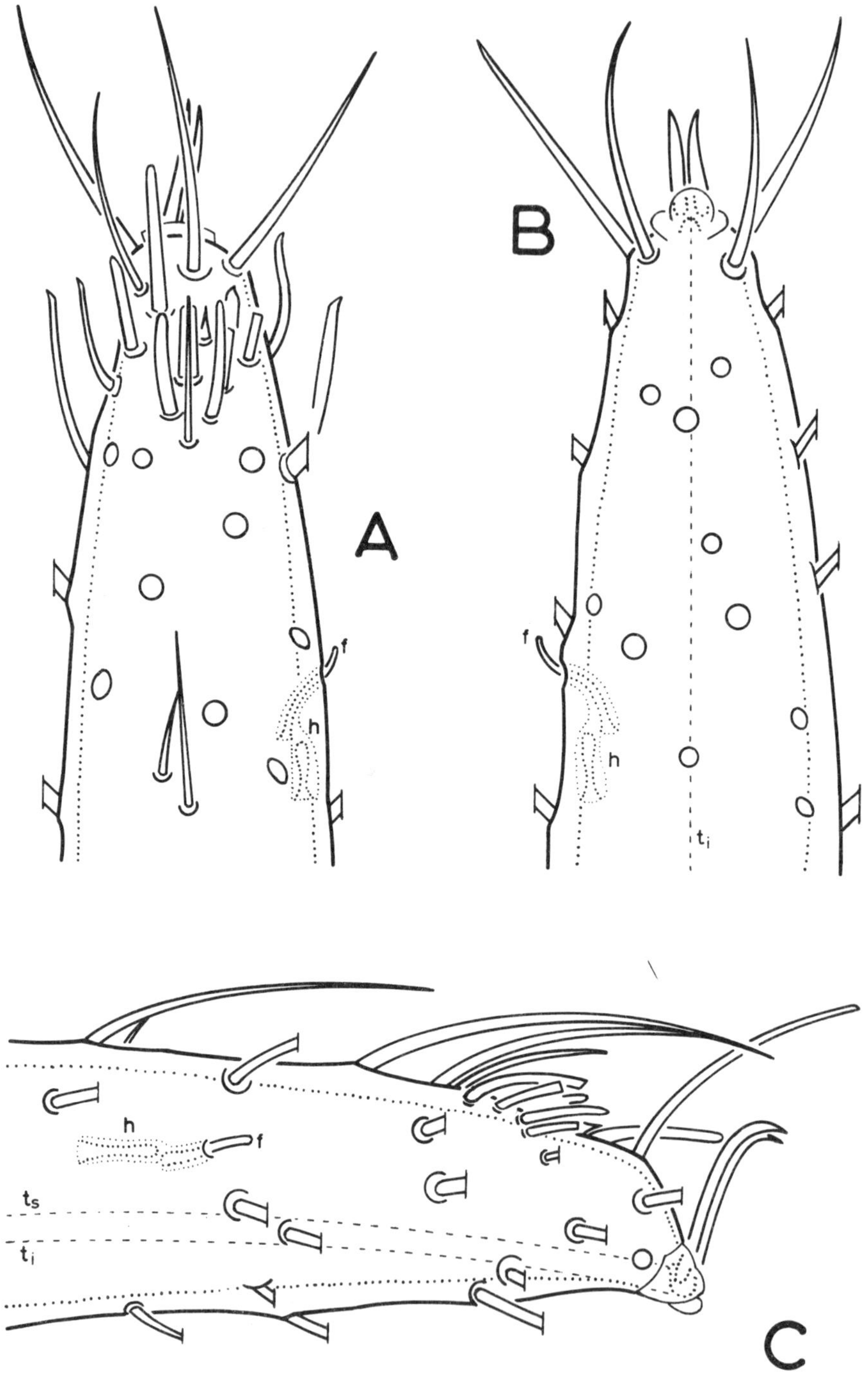
A
f
h
B
f
h
t_i
C
h
f
t_s
t_i

Fig. 137. Opilioacarus texanus (Chamberlin & Mulaik); lateral (posterior = antiaxial) view of right leg II of tritonymph; × 295.

←

Fig. 136. Opilioacarus texanus (Chamberlin & Mulaik); distal part of right tarsus I of female; A, dorsal view; B, ventral view; C, lateral (posterior = antiaxial) view; A–C, × 945.

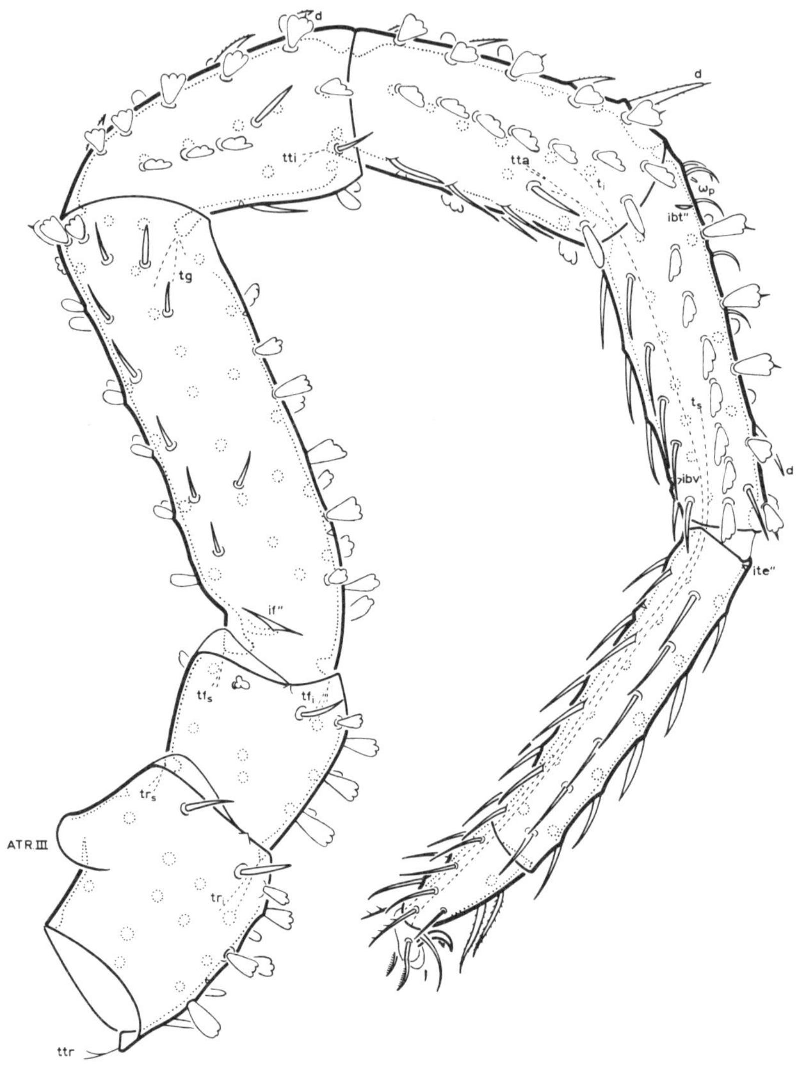

Fig. 138. Opilioacarus texanus (Chamberlin & Mulaik); lateral (posterior = paraxial) view of right leg III of tritonymph; × 295.

In genu I, papilliform setae are slightly more numerous than ordinary setae; all ordinary setae, part of the dorsal and lateral papilliform setae, and all papilliform setae of the terminal verticil are mucronate. Femur I presents also more papilliform than ordinary setae; part of the dorsal and terminal setae are mucronate. The setae of the trochanter are for the greater part papilliform; they mainly occupy ventral and antiaxial positions.

Trochanter II (fig. 137) presents a great number of papilliform setae, mainly with antiaxial and ventral positions. There is a large, dorsal, spoon-shaped process which, according to Chamberlin & Mulaik (1942: 128, pl. 2 fig. 12), articulates with suprapedal lobe II. Femur II presents a row of dorsal setae (*d*), two rows of laterodorsals (*ld*), two rows of laterals (*l*), and two rows of ventrals (*v*). The greater part of the setae are papilliform, but part of the *d* series, two *l'*, and all *v''* are ordinary setae. The ventral seta of the terminal verticil (probably *v''*, although it occupies a higher position) is papilliform; it is mucronate. Genu II presents a row of ordinary *d* setae, two rows of papilliform *ld* and *l* setae, and a number of irregularly inserted ventral setae of papilliform or ordinary shape; *l''* of the terminal verticil is mucronate. Nearly the same description applies to tibia II; the terminal *d* seta, however, is mucronate and strikingly large; one of the *ld''* setae, and the terminal *v''* are mucronate. Basitarsus II is characterized by the presence, in the proximal part, of a number of dorsal phaneres which have exactly the shape of a quarter of a circle; they are termed coronidia. There is also a small, straight, dorsal solenidion (ω_p) in the proximal part. Basitarsus II presents also some mucronate dorsal setae, ordinary and papilliform laterodorsals, papilliform laterals (the terminal one is mucronate), and ordinary ventrals. Telotarsus II presents six rows of ordinary setae (*ld, l, v*). Acrotarsus II is characterized by the presence of a laterodorsal (antiaxial) solenidion (ω_a), a bifid seta (probably *ld'*), a compound seta (probably *d*), a number of ordinary setae, and a pair of lateroventral terminal setae (*lv*) with some four ciliae of about equal length.

Leg III (fig. 138) resembles leg II in many respects. The number of posterior setae of trochanter 1 and 2 and femur, however, is small in comparison with the number of anteriors (the posterior face is paraxial). The dorsal spoon-shaped process belongs to trochanter 1. The paraxial *ld''* and *l''* of the femur are ordinary setae. Genu and tibia III present a strong resemblance to the corresponding segments of leg II; the terminal *d* of the tibia is also mucronate and strikingly large. Basitarsus III presents the same dorsal coronidia as basitarsus II, and the same solenidion ω_p. The terminal part of basitarsus III (fig. 140C) presents a small, rounded dorsal solenidion (ω_d). Seta *ld'* of the acrotarsus is not bifid; the terminal lateroventrals (*lv*) present about four ciliae.

The chaetotaxy of leg IV (fig. 139) resembles that of leg III in many respects. The number of posterior (paraxial) setae of the femur is strongly reduced; laterodorsally, just in the middle of the segment, a small paraxial solenidion (θ'') is, however, present. Trochanter 1 has no dorsal spoon-shaped process. The number of paraxial setae of the genu is reduced. The terminal seta *d* of the tibia is mucronate and longer than in leg II and III. The dorsal coronidia of the basitarsus, which have the shape of a quarter of a circle, occur in the greater part of the segment; they are absent only in the distal region; the solenidion ω_p is present. The greater part of the dorsal and terminal papilliform setae of genu, tibia and basitarsus are mucronate. All dorsal setae of the telotarsus are mucronate. Acrotarsus IV has the same number of setae as in leg III; the dorsal seta *d* is mucronate.

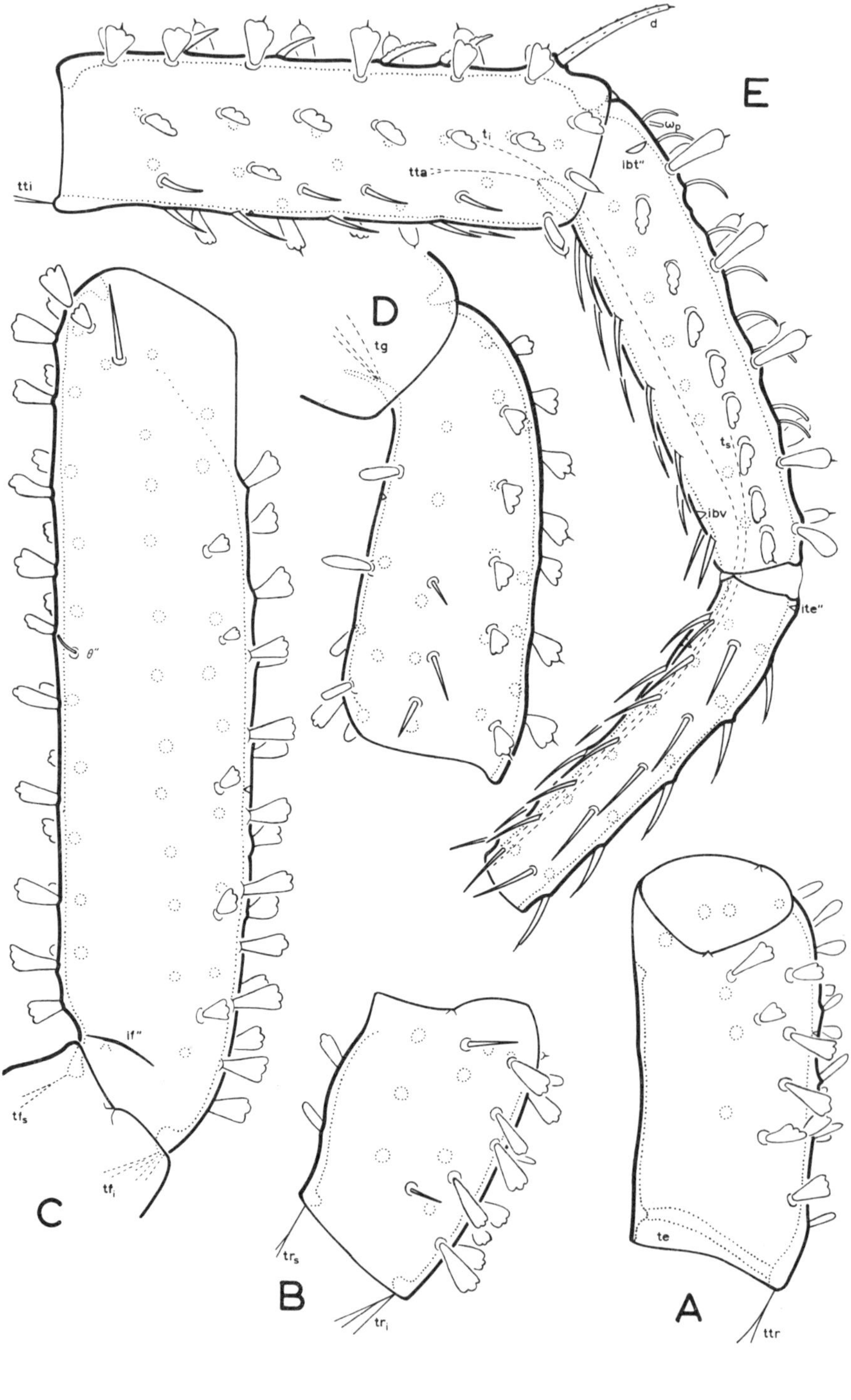
E
d
ω_p
ibt"
t_i
tta
tti
t_{s_i}
ibv
ite"
D
tg
C
θ"
lf"
tf_s
tf_i
B
tr_s
tr_i
A
te
ttr

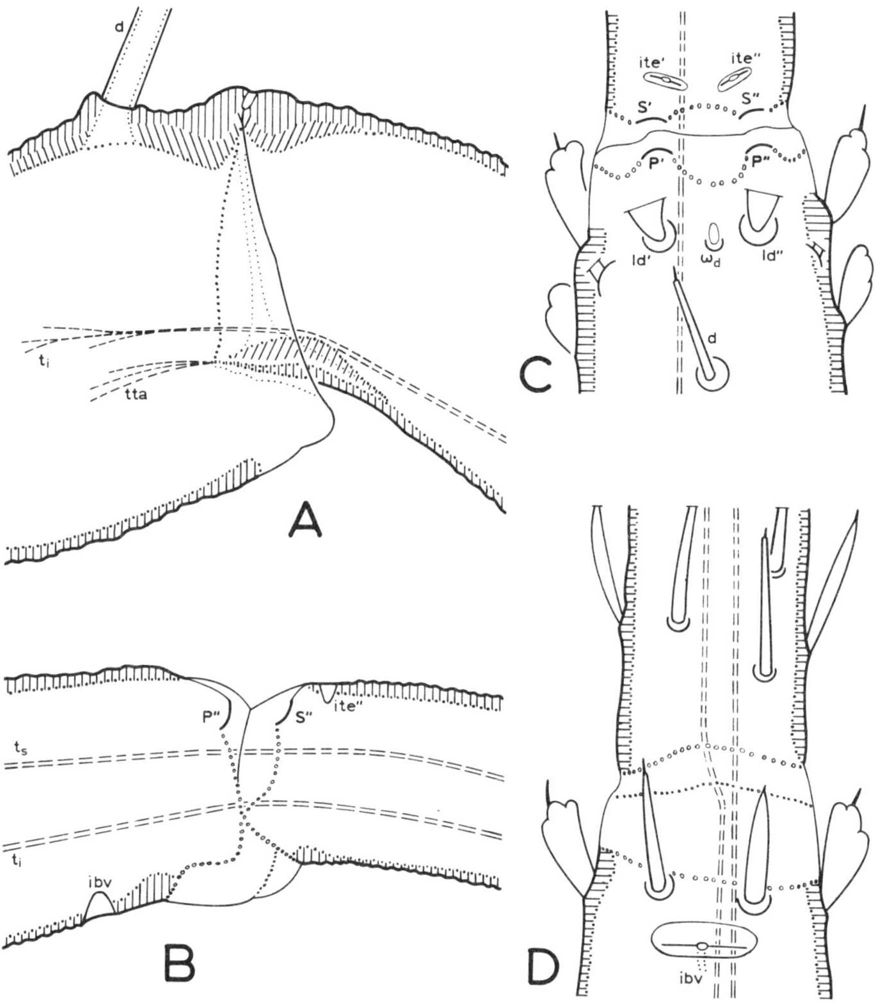

Fig. 140. Opilioacarus texanus (Chamberlin & Mulaik); A, lateral (posterior = paraxial) view of articulation between tibia and basitarsus of right leg IV of tritonymph (all setae omitted, except part of the large dorso-distal seta of the tibia); B–D, articulation between basi- and telotarsus of right leg III of female; B, lateral (posterior = paraxial) view (setae omitted); C, dorsal view; D, ventral view; A–D, × 710.

←

Fig. 139. Opilioacarus texanus (Chamberlin & Mulaik); lateral (posterior = paraxial) view of right leg IV of tritonymph; A, trochanter 1; B, trochanter 2; C, distal part of trochanter 2, and femur; D, distal part of femur, and genu; E, tibia, basitarsus and telotarsus; acrotarsus, pretarsus and apotele omitted; A–E, × 295.

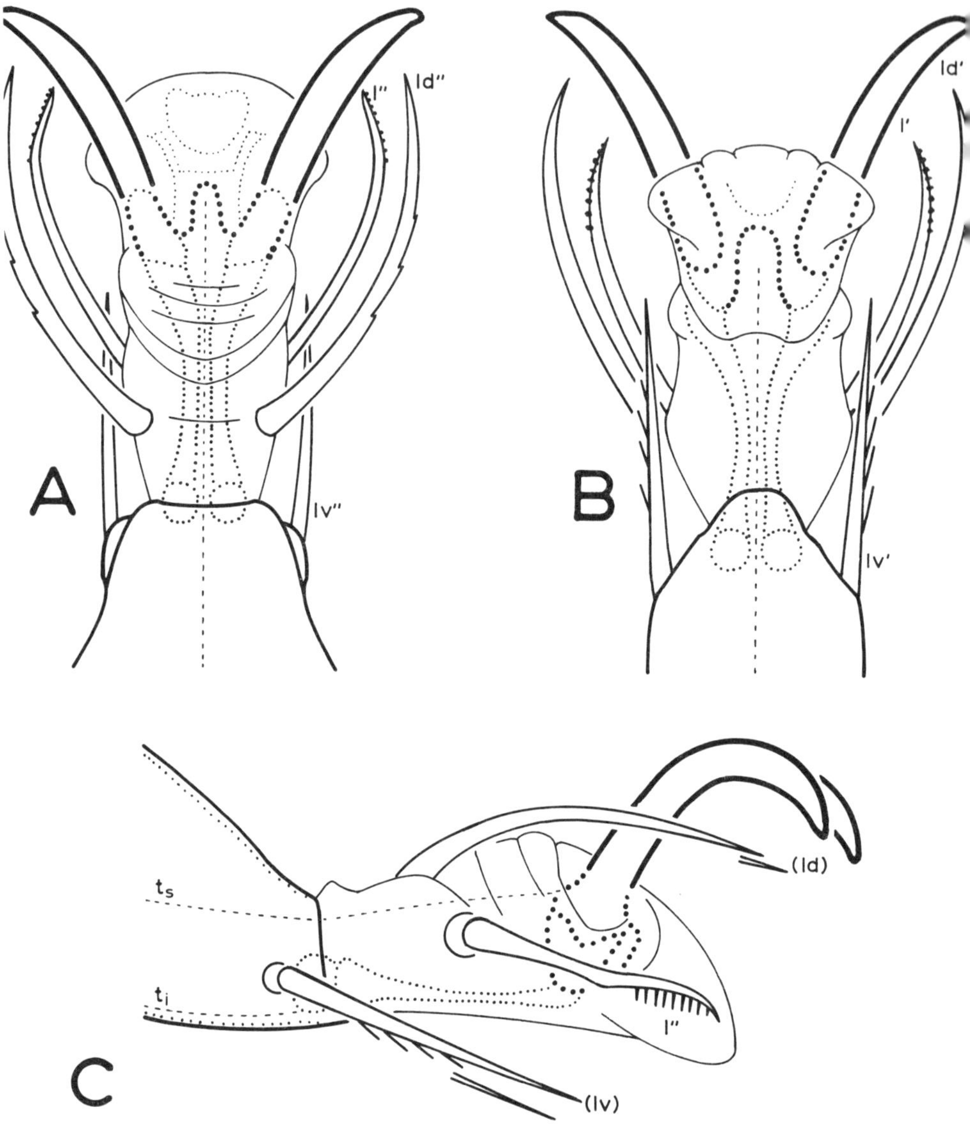

Fig. 141. Opilioacarus texanus (Chamberlin & Mulaik), distal part of acrotarsus, pretarsus and apotele of right leg III of female; A, dorsal view; B, ventral view; C, lateral (posterior = paraxial) view; A–C, × 1530.

Part of the terminal setae of the legs are birefringent. An investigation of leg III in polarized light demonstrated that birefringency is strong in the case of the ungues (including their basal part), but that the median process of the apotele (between the ungues) is isotropic. The four pretarsal setae (laterals and laterodorsals) are all strongly birefringent. The setae of the acrotarsus are birefringent to various degrees: distinctly birefringent in the ventrals, faintly in the laterals, and apparently not in the dorsal and laterodorsal setae. Birefringency is also exhibited by the distal *v'* of the telotarsus. It may be remarked here that the cuticle is also distinctly birefringent. A faint degree of birefringency can be observed in the tendons of the apotele: distinct in t_i, rather indistinct in t_s.

3.2. *Additional data*

Prodorsum. – Three pairs of eyes are present in species of the genus *Paracarus*. The additional eye is situated, on either side, between the ordinary pair and slightly higher (more paraxial); it is only slightly smaller than the other eyes (fig. 142A, C). An oculorostral groove, extending from the eyes to the semicircular recess on either side of the rostrum, is usually present in all Opilioacarida (fig. 142A).

Opisthosoma. – In the genera *Panchaetes* and *Salfacarus*, opisthosomatic setae are not only present in the genital region and on segments XVIII and XIX, but also on segments VII–XVII (the post-embryonic development of the opisthosomatic setae is dealt with below). The presence of these setae is associated with a decrease in the number of lyrifissures (opisthosomatic lyrifissures in these genera are probably idionymous). The number of setae of segment XVIII is characteristic for several genera and species groups: 0 in *Paracarus* and *Adenacarus*, 3 in the American *Opilioacarus*, 7 in the Old World species of *Opilioacarus* and in *Phalangiacarus* (in the last-mentioned genus, one of these setae can be vertitionally absent); the setae of segment XVIII are numerous and variable in number in the genera *Panchaetes* and *Salfacarus*.

Anal tubercle. – A comparative study of the anal tubercle in numerous specimens of *Opilioacarus platensis* Silvestri, distinctly demonstrated that it consists of one segment; there is only one series of muscles. The anal opening cannot only be closed by the paired valves, but the segment can also be retracted, resulting in an apparent disappearance of the separate tubercle (fig. 143).

Genital region. – A primitive ovipositor is now known from several species. In evaginated condition it is represented by a short unsegmented tube with two indistinct terminal lobes (fig. 144A, C). The genital papillae can be evaginated (associated with opening of the verrucae) simultaneously with the ovipositor (fig. 144B, C).

Sternapophyses. – In *Adenacarus arabicus* (With) (fig. 145A, B) the sternapophyses present four or five setae instead of two, viz., one apical seta and three or four lateral setae.

Gnathosoma. – In *Paracarus hexophthalmus*, With's organ is replaced by a bifurcate seta (figs. 142B, 146). Two laterodorsal infracapitular setae, instead of one, are present in *Paracarus* and *Adenacarus* (figs. 142B, 145C).

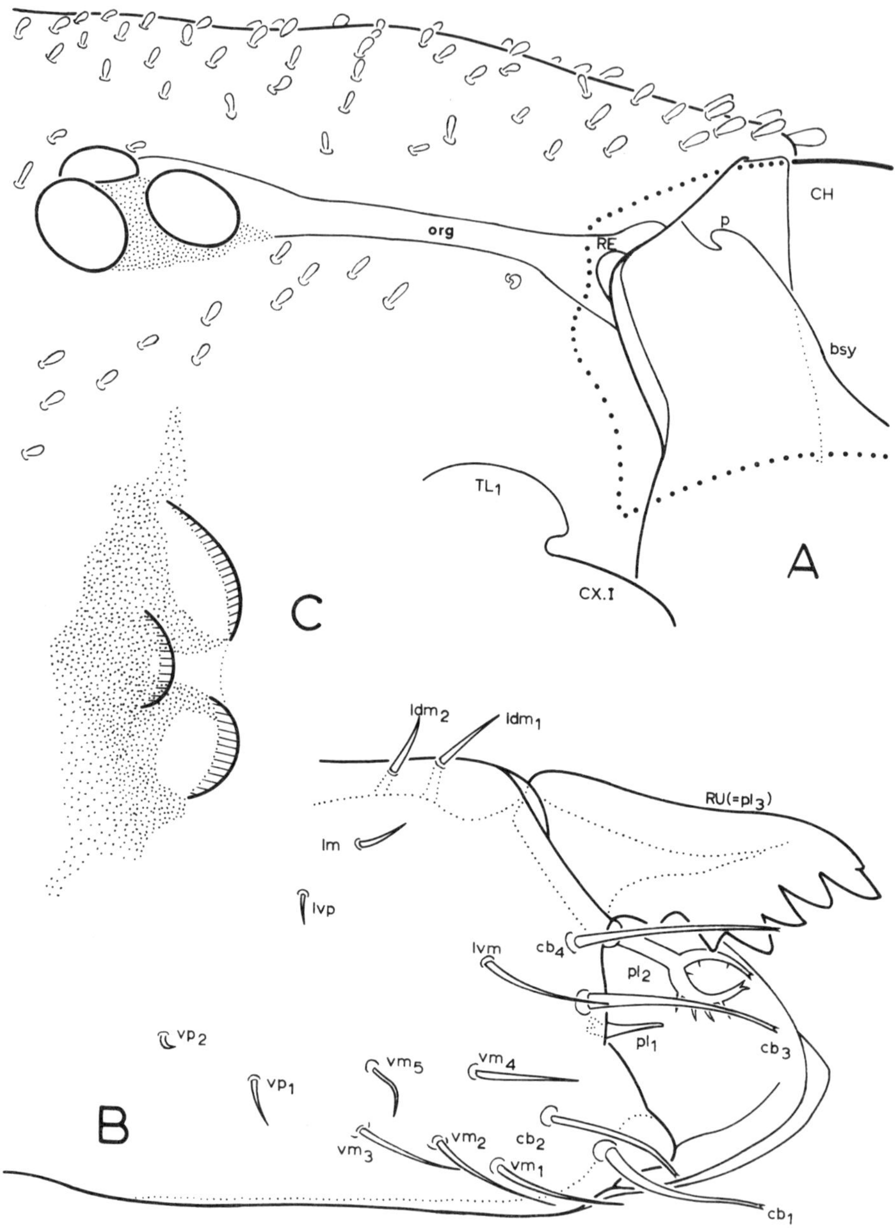

Fig. 142. Paracarus hexophthalmus (Redikorzev), female; A, lateral view of frontal part of idiosoma and basal part of gnathosoma; B, lateral view of infracapitulum (labrum omitted); C, right group of eyes, viewed obliquely from above; A, × 230; B, C, × 370.

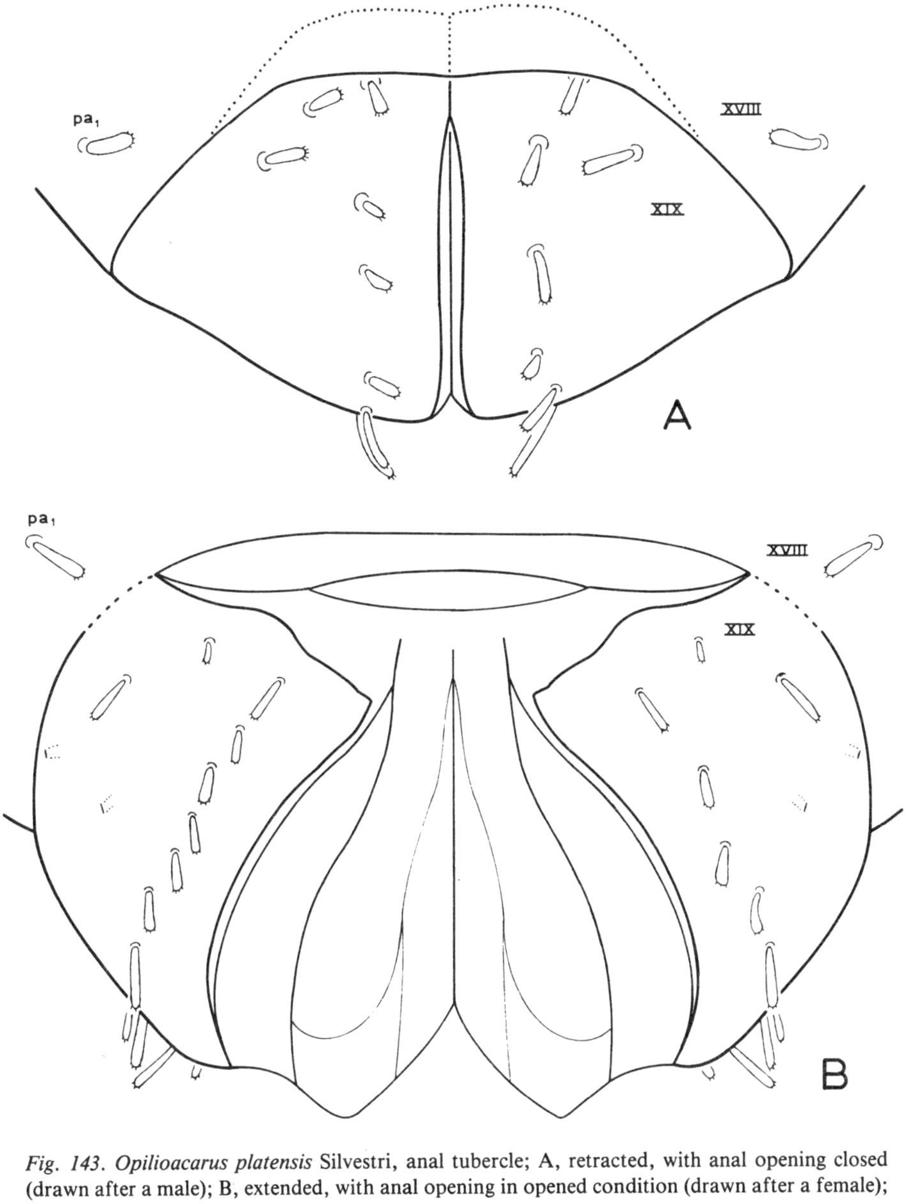

Fig. 143. Opilioacarus platensis Silvestri, anal tubercle; A, retracted, with anal opening closed (drawn after a male); B, extended, with anal opening in opened condition (drawn after a female); A, B, × 370.

Chelicera. – In the male of *Opilioacarus vanderhammeni* Juvara-Bals & Baltac, the chelicera presents additional setae: there are two setae on the trochanter (instead of the normal single one) and five setae on the body of the chelicera (instead of the normal three of the female). This is apparently the

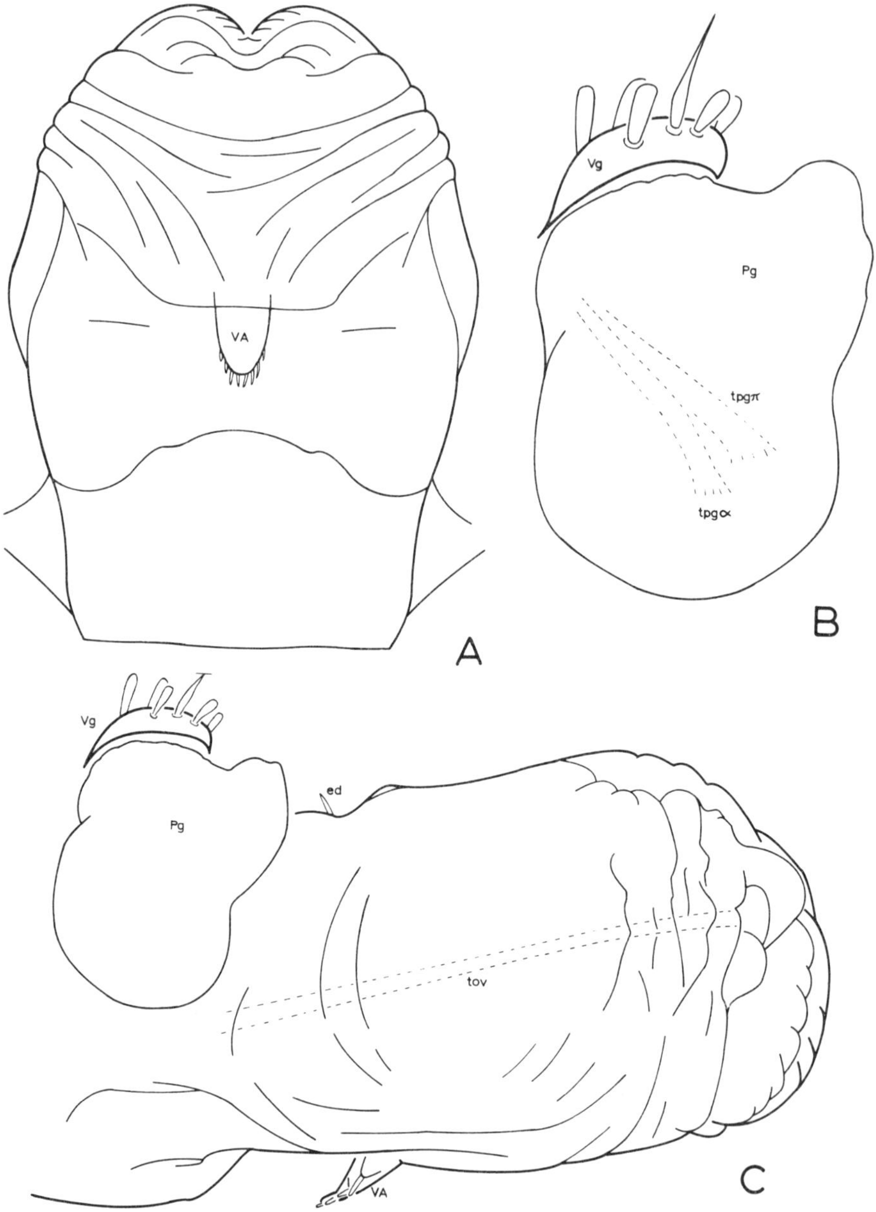

Fig. 144. Paracarus hexophthalmus (Redikorzev), female; A, ventral view of extended ovipositor; B, lateral view of right genital verruca with evaginated genital papilla; C, lateral view of extended ovipositor and evaginated genital papilla; A, × 140; B, × 295; C, × 185.

Fig. 145. Adenacarus arabicus (With), male; A, lateral (antiaxial) view of right sternapophysis; B, anteroventral view of sternapophyses; C, lateral view of infracapitulum; A–C, × 490.

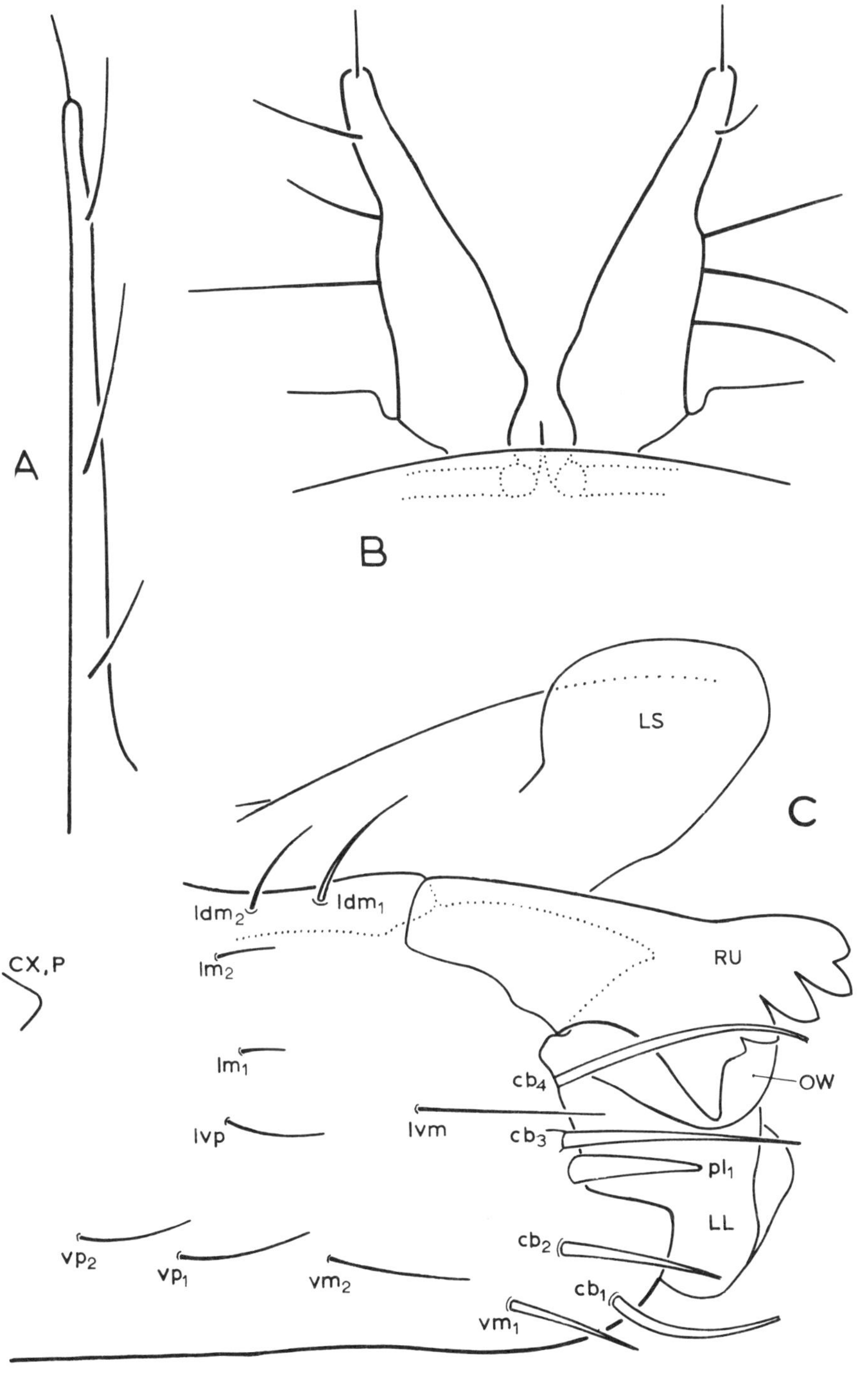
A
B
LS
C
ldm2
ldm1
lm2
CX, P
RU
lm1
cb4
OW
lvp
lvm
cb3
pl1
LL
cb2
vp2
vp1
vm2
cb1
vm1

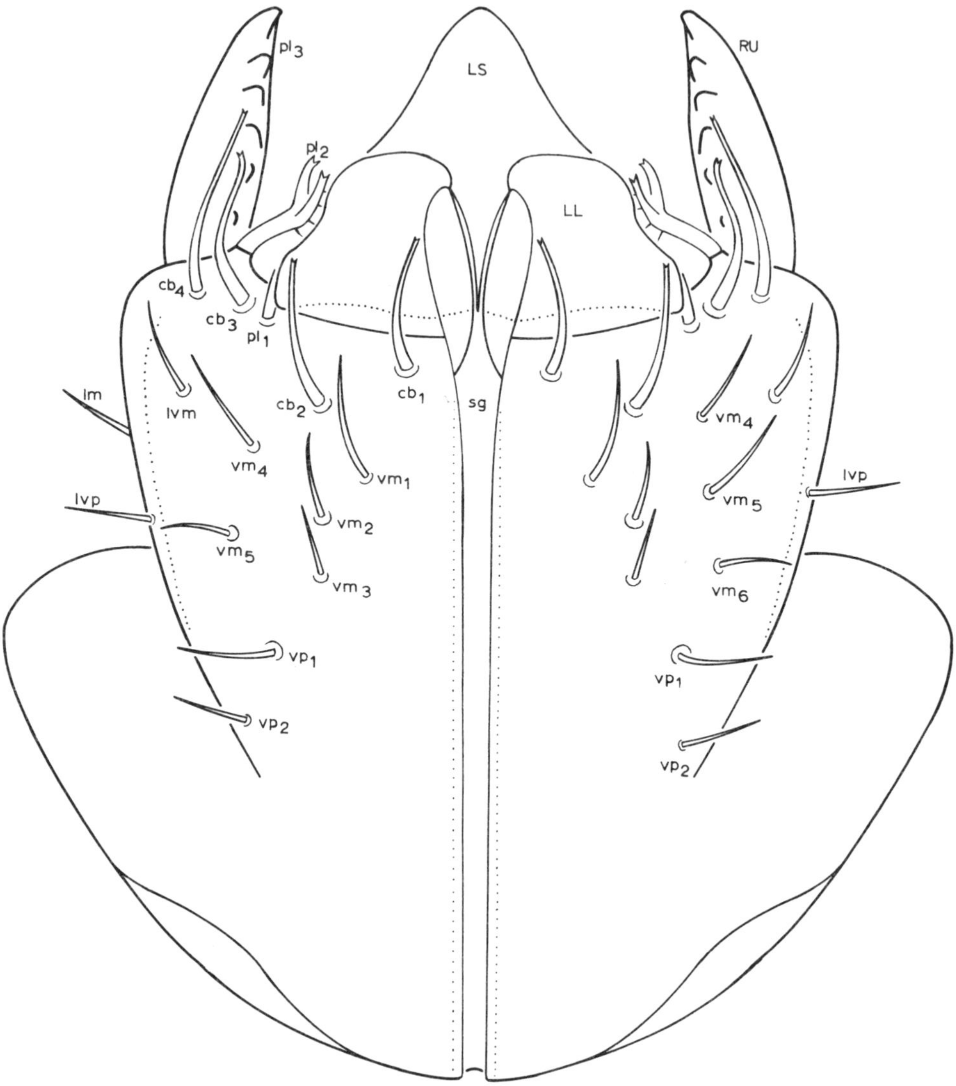

Fig. 146. Paracarus hexophthalmus (Redikorzev), female; ventral (slightly oblique) view of infracapitulum (labrum schematized); × 370.

greatest number of cheliceral setae known from mites. In the single specimen of *Paracarus hexophthalmus* studied by me (a female), the body of the chelicera presents four setae, whilst no seta is present on the trochanter.

Legs. – In the genus *Panchaetes*, coronidia are not only present on the basitarsus of legs II–IV, but also on tibiae II–IV (fig. 147). In the related genus *Salfacarus* (fig. 151A, B), the occurrence of coronidia is restricted to the basitarsus.

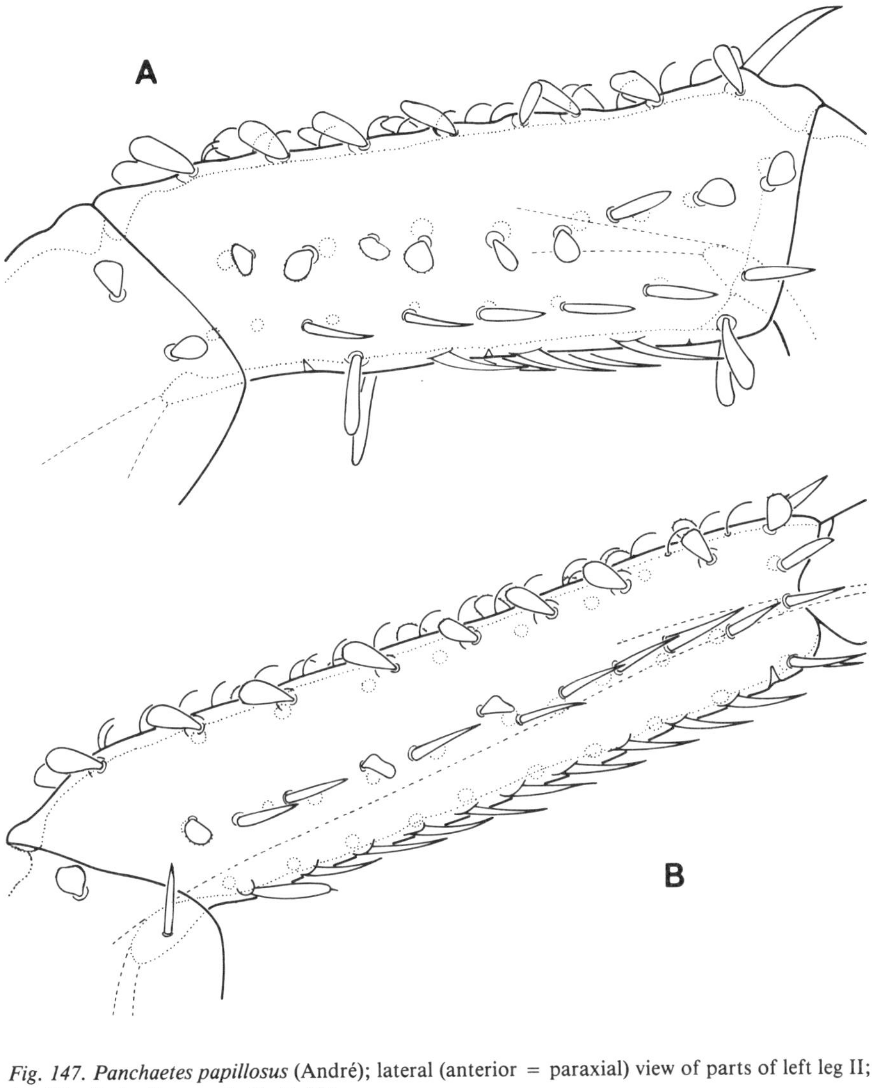

Fig. 147. Panchaetes papillosus (André); lateral (anterior = paraxial) view of parts of left leg II; A, tibia; B, basitarsus; A, B, × 295.

Postembryonic development. – The life-cycle is completely known in the case of *Phalangiacarus brosseti* Coineau & Van der Hammen. Incomplete series of data are known from other species. The following data refer to *Phalangiacarus* (figs. 148–149). There are six stases: prelarva, larva, protonymph, deutonymph, tritonymph and adult.

The prelarva is a calyptostase (nearly an elattostase). It presents three pairs of long legs without setae, but with distinct segmentation. The idiosoma presents no setae. The gnathosoma is only slightly regressive.

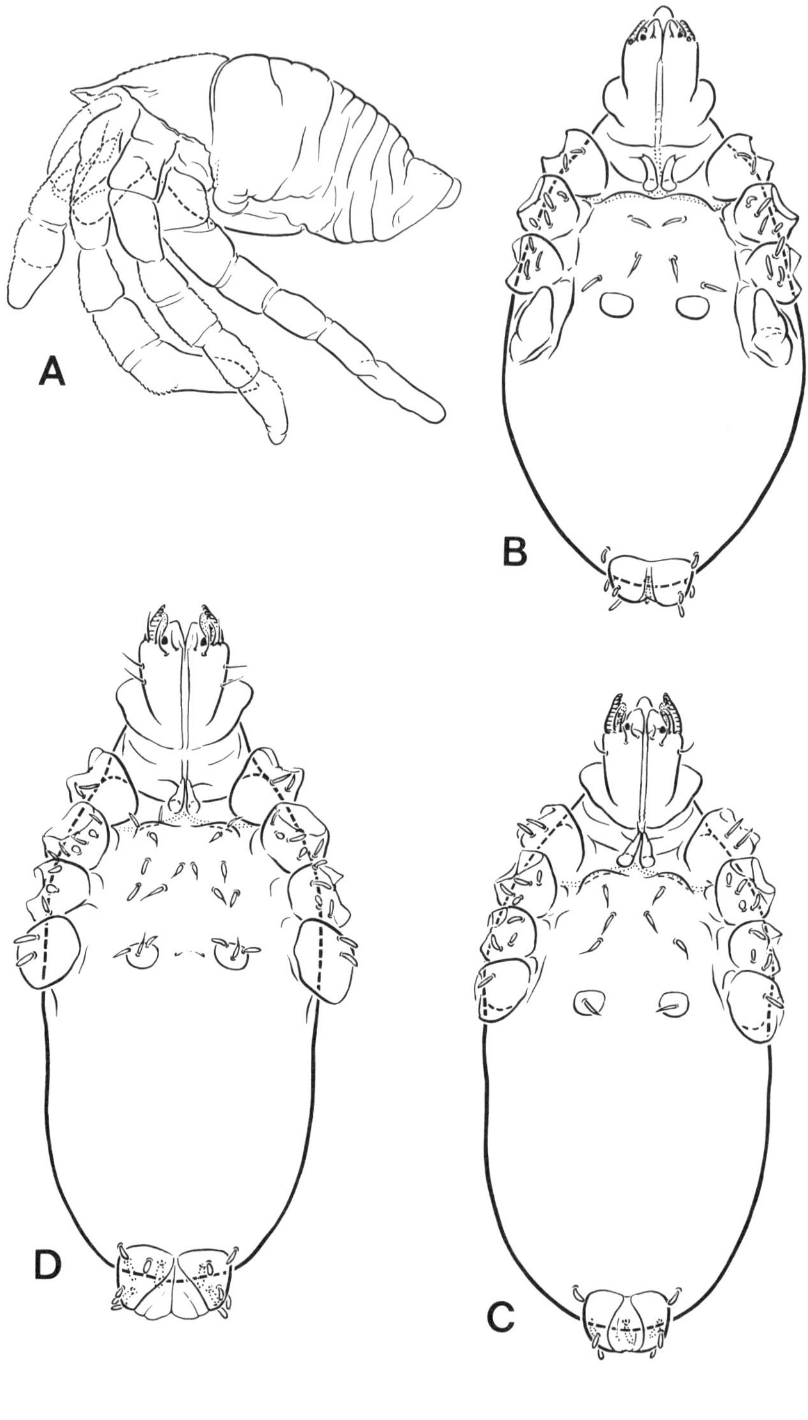
A
B
D
C

The larva has three pairs of normal legs, and a rather large vestige of leg IV. It is further characterized by the absence of stigmata, and the absence of setae on sternal and genital verrucae. The mouthparts present a number of regressive characters: the chela (of the chelicera) is slightly reduced, just as the rutellum and With's organ (apparently the beginning of an elattostasic regression).

The protonymph presents two pairs of stigmata (*st.2, st.3*). It has no acrotarsus at legs II–IV. Legs III and IV have no trochanter 2. The cheliceral trochanter bears no setae.

The deutonymph presents three pairs of stigmata (*st.2, st.3, st. 4*). Leg IV presents an acrotarsus. Legs III and IV have no separate trochanter 2, and the cheliceral trochanter is glabrous.

The tritonymph presents four pairs of stigmata (*st. 1–4*). Legs II–IV present an acrotarsus. Legs III and IV have two trochanters. Male and female tritonymph are distinctly differentiated.

The adult is characterized by the complete development of the genital organs and by an increased number of setae.

The number of setae of the opisthosoma, the sternal region and the infracapitulum gradually increases in the course of postembryonic development. The postembryonic development of the genital region, from deutonymph to adult, in *Opilioacarus platensis* Silvestri, is represented in fig. 150. The base level of the seta of the cheliceral trochanter (when present) is tritonymphal or adult.

In *Salfacarus robustipes* Van der Hammen, the postembryonic development of trochanter 2 was studied in detail. Although the base level of the segment is tritonymphal, the joint manifests itself incompletely already in the deutonymph (fig. 151C): the basifemoral ring presents characters of the trochanter 2-femur articulation (the posterior articulation-point and the superior tendon) as well as characters of the true basifemoral ring (the anterior lyrifissures and the associated suture, and the absence of the inferior tendon).

List of the notations of figs. 121–151

abj,	abjugal furrow.
ap.ce,	cervical apodeme.
at,	line of attachment of cheliceral frame to infracapitulum.
ATR.II–III,	dorsal apophyses of trochanters II–III.
b,	mouth.
bsy,	border of synaptic tectum.

←

Fig. 148. The postembryonic development of *Phalangiacarus brosseti* Coineau & Van der Hammen (1); A, prelarva; B, larva; C, protonymph; D, deutonymph; A–D, × 956. (After a drawing by Coineau, in Van der Hammen, 1976).

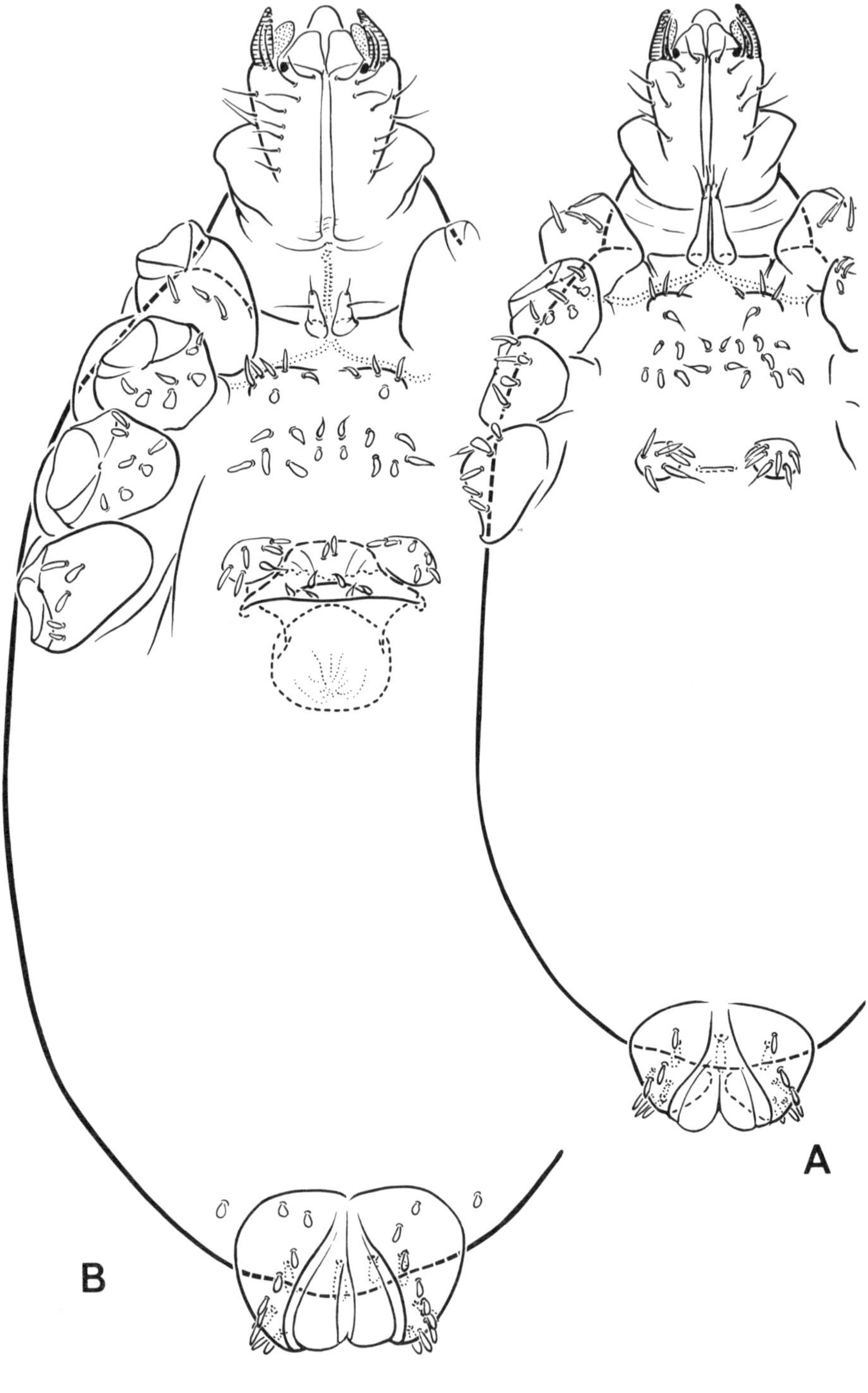
A
B

cb_{1-4}	circumbuccal setae.
CC	coxal cavity of infracapitulum.
CE	cervix.
ch_1'', ch_2', ch_2''	setae of the body of the chelicera.
CH	chelicera.
cht	seta of cheliceral trochanter.
CP	pharyngeal chamber.
CX.C	cheliceral sheath.
CX.P	coxal region of palp.
CX.I–IV	coxae I–IV.
d	dorsal seta of the appendages.
D_{1-10}	dorsal sigilla of opisthosoma.
dg	duct of infracapitular gland.
disj, dj	disjugal furrow.
ed	dorsal spine of ovipositor.
eI, eII	laterocoxal setae of legs I–II.
f	projecting solenidion of telotarsal organ.
f_{0-10}	intersegmental furrows of opisthosoma.
FC	cheliceral groove.
G	gena.
gc	large lateral ridge of infracapitulum.
h	hidden solenidion of telotarsal organ.
H	mentum.
ibt''	laterodorsal lyrifissure of basitarsus.
ibv	ventrodistal lyrifissure of basitarsus.
id	dorsal lyrifissure of chelicera.
if', *if''*	lyrifissures of basifemoral ring.
ite', *ite''*	basidorsal lyrifissures of telotarsus.
itr''	posterior lyrifissure of trochanter 1.
ivg	lyrifissure of genital verruca.
$i\alpha$	antiaxial lyrifissure (of body of chelicera and palpal tarsus).
$i\pi$	paraxial lyrifissure of palpal tarsus.
Ji	inferior commissure of mouth.
Js, Js'	superior commissures of mouth.
k', *k''*	condyles.
l', *l''*	lateral setae of legs.
L_{x-z}	lateral sigilla of opisthosoma.
LA	labellum.

←

Fig. 149. The postembryonic development of *Phalangiacarus brosseti* Coineau & Van der Hammen (2); A, female tritonymph; B, adult female; A–B, × 956. (After a drawing by Coineau, in Van der Hammen, 1976).

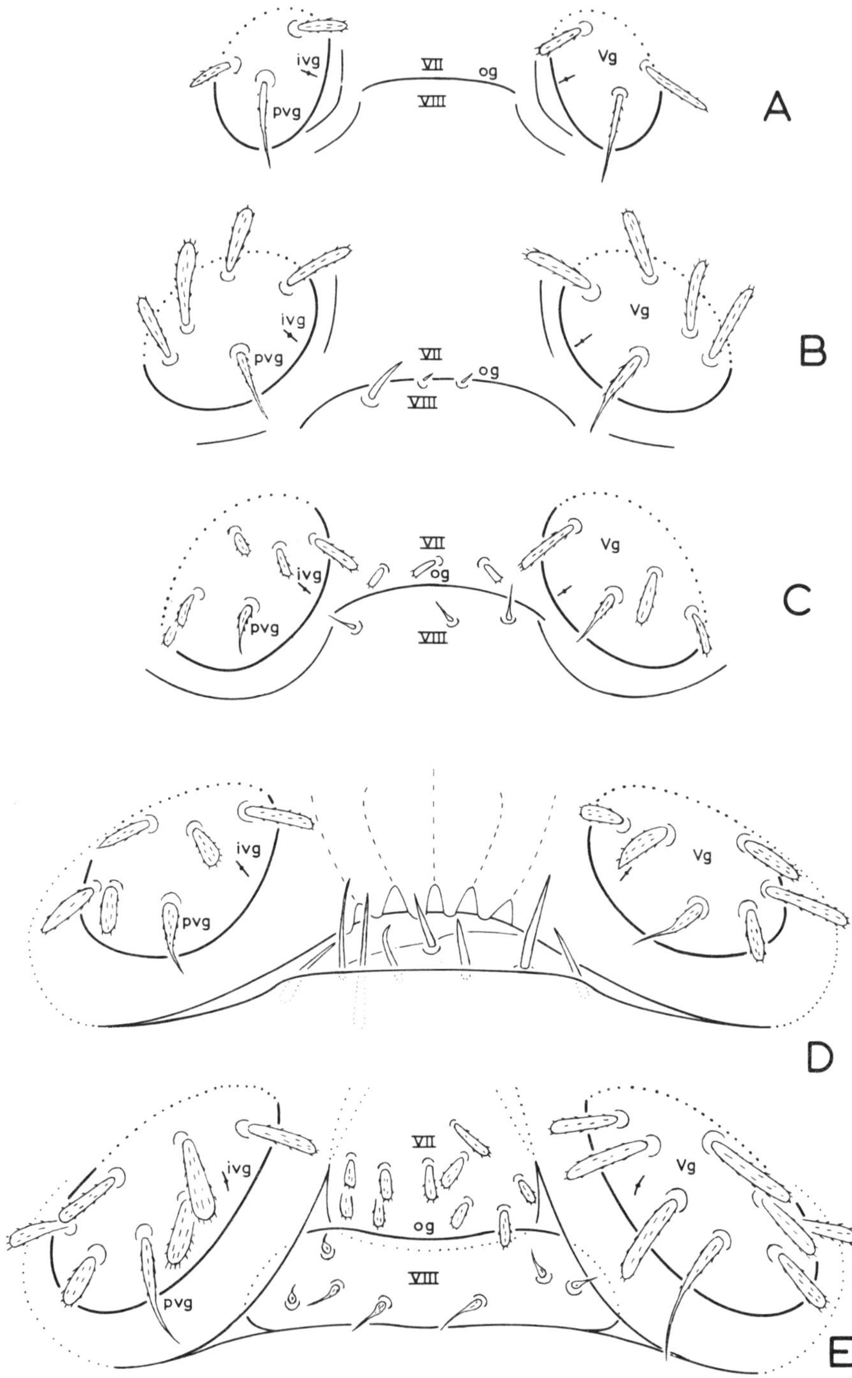

ivg
pvg
VII
og
VIII
vg
A
B
C
D
E

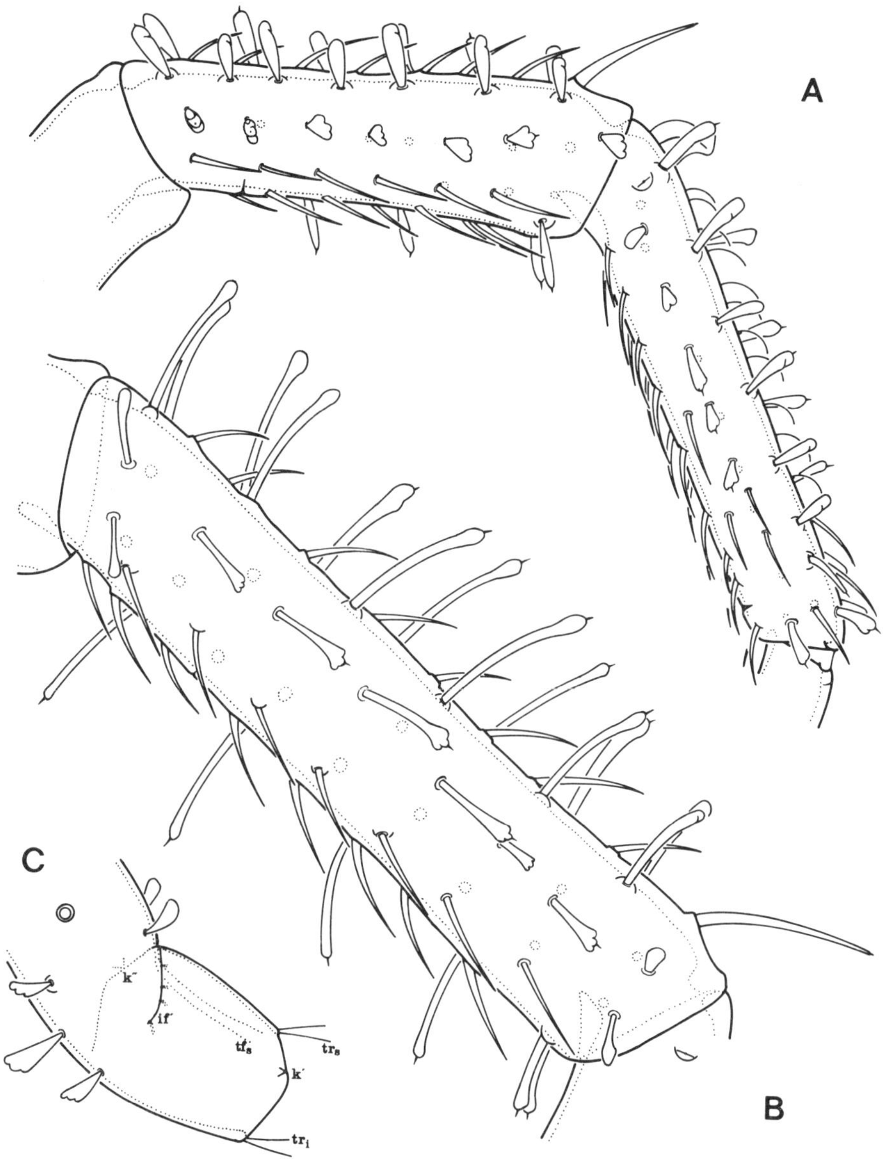

Fig. 151. Salfacarus robustipes Van der Hammen; A, tibia and basitarsus of right leg II of adult female, lateral (antiaxial) view; B, tibia of right leg IV of adult female, lateral (paraxial) view; C, lateral (anterior = paraxial) view of basal part of deutonymphal leg III, with 'basifemoral ring' presenting some characters of the trochanter 2-femur articulation; A, B, × 185; C, × 295.

←

Fig. 150. Opilioacarus platensis Silvestri, ventral view of genital region; A, deutonymph; B, female tritonymph; C, male tritonymph; D, female; E, male; A–E, × 370.

ld', *ld''*, (*ld*)	laterodorsal setae of legs.
LD_{0-10}	laterodorsal sigilla of opisthosoma.
ldm, ldm_{1-2}	laterodorsal infracapitular setae.
$LI_{a-c,\ 5-7}$	ventrolateral sigilla of opisthosoma
LL	lateral lips.
$LL_{1,\ 5-7}$	lateral sigilla of opisthosoma.
lm_{1-3}	mediolateral infracapitular setae.
$LM_{1,\ 5-8}$	mediolateral sigilla of opisthosoma.
LS	labrum.
LS_{5-10}	laterodorsal sigilla of opisthosoma.
lv', *lv''*, (*lv*)	lateroventral setae of legs.
LV_{4-9}	lateroventral sigilla of opisthosoma.
lvm	lateroventral median infracapitular seta.
lvp	lateroventral posterior infracapitular seta.
MA	malapophysis.
og	genital opening.
ogc	orifice of cheliceral gland.
ogi	orifice of infracapitular gland.
ogl_{1-2}	orifices of labial glands.
opv	ventral oncophysis of chelicera.
org	oculorostral groove.
OW	With's organ.
p	laterodorsal prominence of synaptic tectum.
P', *P''*	laterodorsal distal protuberances of basitarsi II–IV.
pa_1	lateroventral seta of segment XVIII.
PA	palp.
Pg	genital papilla.
PH	pharynx.
pl_{1-3}	paralabial setae.
pvg	compound seta of genital verruca.
PI–IV	legs I–IV.
RE	circular concavity on either side of the rostral lobe.
RL	rostral lobe.
RU	rutellum.
S', *S''*	laterodorsal basal cavities of telotarsi II–IV.
SA	sternapophysis.
se	capitular saddle.
sg	subcapitular gutter.
sm	compound seta of palpal tarsus.
s.t	sternal taenidium.
st.1–4	stigmata 1–4.
SY	synaptic tectum.
t_i	inferior tendon of apotele.
t_s	superior tendon of apotele.
te	retrotectum of trochanter.
tf_i	inferior tendon of femur.
tf_s	superior tendon of femur.
tg	tendon of genu.

TG	tegulum.
TL_{1-3}	suprapedal lobes.
tov	tendons of ovipositor.
$tpg\alpha$	antiaxial tendon of genital papilla.
$tpg\pi$	paraxial tendon of genital papilla.
tr	transverse prodorsal furrow.
tr_i	inferior tendon of trochanter 2.
tr_s	superior tendon of trochanter 2.
tta	tendon of tarsus.
tti	tendon of tibia.
ttr	tendon of trochanter (or trochanter 1).
V_{2-10}	ventral sigilla of opisthosoma.
VA	ventral apophysis of ovipositor.
Vg	genital verruca.
vm, vm_{1-5}	medioventral infracapitular setae.
vp_{1-2}	posteroventral infracapitular setae.
Vs	sternal verruca.
1–13	infracapitular setae (Grandjean's notation).
I–XIX	segments of the body.
θ (thêta)	solenidion of femur.
φ (phi)	solenidion of tibia.
ω (omega)	solenidion of tarsus.
ω_a	basidorsal solenidion of acrotarsus II.
ω_d	dorsodistal solenidion of basitarsus III.
ω_p	basidorsal solenidion of basitarsi I–IV.
′ (prime)	anterior face of an appendage; the right one of two symmetrical structures.
″ (double prime)	posterior face of an appendage.

4. Holothyrida

At the beginning of chapter V of the systematic part, I have listed the papers in which I have dealt with representatives of the Anactinotrichid order Holothyrida. In these papers, detailed studies are made of Holothyrid comparative morphology, postembryonic development, classification, relationships and geographical distribution. The third part of the Glossary of Acarological Terminology, devoted to Holothyrid terminology, is in course of preparation.

The diagnostic characters of the order Holothyrida are the following. The idiosoma presents a single dorsal shield covering the complete dorsal surface, and overlapping the ventral shield. One pair of eyes is present in a species of the genus *Australothyrus*; eyes are not present in any other species now known. There is one pair of stigmata, with lateral position, associated with a peritreme (a taenidium); one or two pairs of orifices of supposed excretory glands are also present in the lateral region. The anal opening (with ventral position) is closed by two shields, each with several setae (more than two). The genital opening of the female is closed by four shields (pregenital, postgenital and the paired laterogenitals). The genital opening of the male is closed by two shields; it can be situated in a distinctly differentiated region (the epiandrium, present in Holothyridae). There is a transverse sternal groove in the anterior part of the sternum (as mentioned in the general part, the sternum is really a neosternum); sternal setae, when present, are not idionymous. Paired sternapophyses are present in the genus *Allothyrus*. The gnathosoma is covered dorsally by the rostral tectum. The tegulum is not produced (there is no supracheliceral limbus). The infracapitulum presents a large subcapitular apodeme. There are numerous infracapitular setae, among which one or two pairs of laterodorsals and one pair of corniculi. A cross section of the pharynx is quadrangular. The chelicerae are long; they consist of three segments (trochanter, body of the chelicera, and apotele), and present two lyrifissures and one or two setae. The palp presents paraxially, at the base of the tarsus, a transformed bi- or tripartite apotelic claw. Palpal tarsus with a paralaterodorsal area of hollow setae. Legs with numerous setae. Tarsi II–IV subdivided by a basitarsal ring into basi- and telotarsus; femora I–IV with basifemoral ring. The terminal part of tarsus I is characterized by the presence of Haller's organ; in Allothyridae, tarsus I is subdivided into a bathrotarsus and a very small apicotarsus. Pretarsi II–IV bear one pair of setae. The legs (including the ambulacrum) of Holothyridae can present secondary sexual characters. Postembryonic development incompletely known: there are three nymphs (the largest nymphs are sexed); there are some indications that the larva is octopod.

The order Holothyrida comprises three families: Allothyridae (Australia and New Zealand), Neothyridae (South America) and Holothyridae (islands in the Indian Ocean, New Guinea, Ceylon).

The present section is subdivided into three subsections. In the first subsection, a detailed description is given of a species of Allothyridae; the second subsection deals with the morphology and the postembryonic development of a species of Holothyridae; in the third subsection, additional data with reference to other species are mentioned.

4.1. *The morphology of* Australothyrus ocellatus *Van der Hammen*

The present subsection is based on a study of material from Northeast Queensland, Australia (Mt. Bellendon Ker, Mt. Bartle Frere). Numerous specimens of the species were collected from rainforest litter, at altitudes of 1560 and 1620 m.

Measurements. – Male and female are of equal size; the length is about 2.4 mm, the breadth 1.3 mm. The lengths of the legs (of the female represented in figs. 162–163, 165–166; there are no striking differences, in this respect, between male and female) are the following: leg I 2.3 mm, leg II 2.0 mm, leg III 2.1 mm, leg IV 2.9 mm.

Habitus and colour. – The adult specimens of the species are relatively robust mites, with legs about as long as the body. The cuticle is glossy, and presents numerous small tubercles (associated with setae) and large, smooth sigilla. The colour of the adult specimens is reddish brown.

Tegument. – The tuberculate areas (not the sigilla) of the dorsal shield are covered with cerotegument and adhering dirt; ventrally, the cerotegument layer can be particularly well-developed in the sternal concavities and at the posterolateral border of the ventral shield. Each tubercle of the idiosomatic microsculpture is associated with a spiniform seta, and contains a cupuloid (a rounded sensory organ, resembling a cupule, and probably homologous with a lyrifissure). Numerous pores of tegumental glands are present in the areas between the tubercles.

Dorsal shield (figs. 152, 153A). – The dorsal shield is oval in shape, with the greatest breadth about the middle; it is narrowed in the rostral region. The large sigilla of the dorsal shield are arranged in a characteristic pattern. There is an elongate unpaired median sigillum which is relatively very long. It is flanked, in the median and posterior part, by pairs of oval sigilla. There are some three pairs of oblique elongate laterodorsal sigilla, and five pairs of rounded posteromarginals. There are, moreover, some scattered smaller sigilla. The areas between the sigilla present tubercles, setae and pores of tegumental glands; there are, however, no tubercles in the rostral part of the idiosoma. Although the sigillotaxy presents elements of a segmental arrangement, it is strongly different from the ancestral segmented pattern found in Opilioacarida.

There is one pair of eyes (figs. 152, 153A, 157A, B) in the anterolateral region of the dorsal shield. Each eye consists of a cornea (a transparent, rounded, convex part of the cuticle, functioning as a lens), situated in a rounded light spot of the cuticle. The lateral stigma is surrounded by a peritreme (fig. 153A). There are, on either side, two orifices of large glands: one posterior to the eye, and one posterior to the peritreme (fig. 153A). The orifice behind the eye is situated in a light spot (its duct is even visible through the cuticle); it probably constitutes the orifice of an excretory gland. The orifice behind the peritreme (the so-called stigma 2) is connected with a colourless tube (similar to that of the anterior orifice), and not with a chitinous atrium (as in Holothyridae) previously supposed to represent a respiratory organ. This confirms the conclusion, recently drawn by Travé (1982: 205–206), that the organ represents an excretory gland. According to observations by Dr. David Lee (personal communication), a drop of liquid secretion can be present, in living Allothyrids, in this area.

Ventral shield (fig. 154A, B). – The ventral shield is contiguous with the sternal shield; it surrounds the anal region and the posterior part of the genital region. It presents tubercles with cupuloids and setae, which are arranged in oblique rows; some twelve rows can be distinguished, and these could represent traces of the original segmentation (in this case, the anal opening would be in

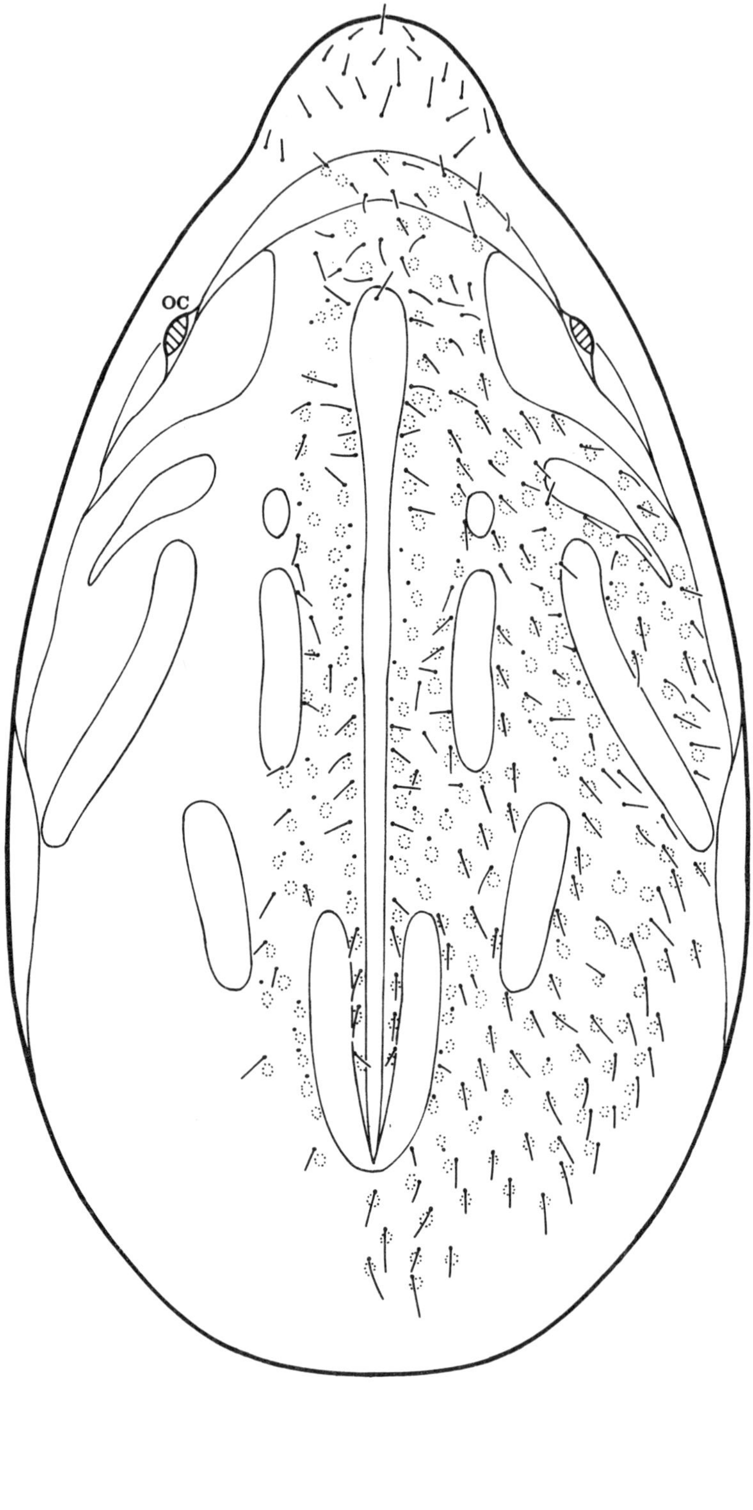
OC

segment XIX as in Opilioacarida). A marginal row of long setae is present in the lateral and posterior part of the shield. Numerous pores of tegumental glands are present, scattered among the tubercles. The peridium has the same position as in other species of Allothyridae (behind coxa IV). A layer of glandular cells, lining part of the internal wall of the peridium, appears to be present; each of these cells passes into a papilla (fig. 157C, D).

Anal region. – The anal opening is covered by a pair of anal shields; these can be closed by strong muscles. Each of the shields or covers presents a tubercle, some six setae, and pores of tegumental glands (fig. 154).

Genital region. – The female genital region is closed and covered by four shields (figs. 154, 155): a pregenital shield, a pair of laterogenitals, and a postgenital. The postgenital shield is the largest; it is not markedly constricted in the anterior part, and slightly wider than long. The pregenital shield is much wider than long; the laterogenital shields are narrow. The shape of the female genital area is strongly different from that in *Allothyrus constrictus* (Domrow), in which the pregenital shield is triangular, the postgenital shield markedly constricted behind anterior horns, and the laterogenital shields rather broad; it is more related to that in *Allothyrus australasiae* (Womersley), in which the postgenital shield is, however, much longer than wide. The four shields present tubercles (with cupuloids), setae and scattered pores of tegumental glands. A primitive ovipositor is present; it can be evaginated (fig. 153B), and in that condition it is relatively short.

The male genital region (fig. 156) is covered by the usual two shields; these present tubercles (with cupuloids), setae and pores of tegumental glands. There is no concave epiandrium.

Sternal region (figs. 155, 156). – The sternal region presents a transverse sternal groove (with tubercles, cupuloids, setae and pores) in the anterior part of the sternal region, a transverse intersegmental area (with thin cuticle, and without tubercles) at the border of segments IV and V (between acetabula II and III) and a pair of sternal cavities (with tubercles, cupuloids, setae and pores) between coxae III. The last-mentioned cavities are particularly concave in the lateral part, where they extend in posterior direction. There are no internal sternal apodemes, and no sternapophyses.

Gnathosoma. – In rest, the gnathosoma is situated in the camerostome (figs. 153C, 159B). It can be completely extended, as well as bent in ventral direction (perpendicular to the idiosoma) (fig. 157A); these particular movements of the gnathosoma (which are also found in ticks) are associated with the presence of a subcapitular apodeme and its tendons (and of muscles inserted on the base of the tegulum). There is an internal continuation of the cervix (a cervical apodeme) which supports the invaginated chelicerae; no extrinsic muscles are inserted on this apodeme.

←

Fig. 152. Australothyrus ocellatus Van der Hammen, dorsal view of adult female, × 75.

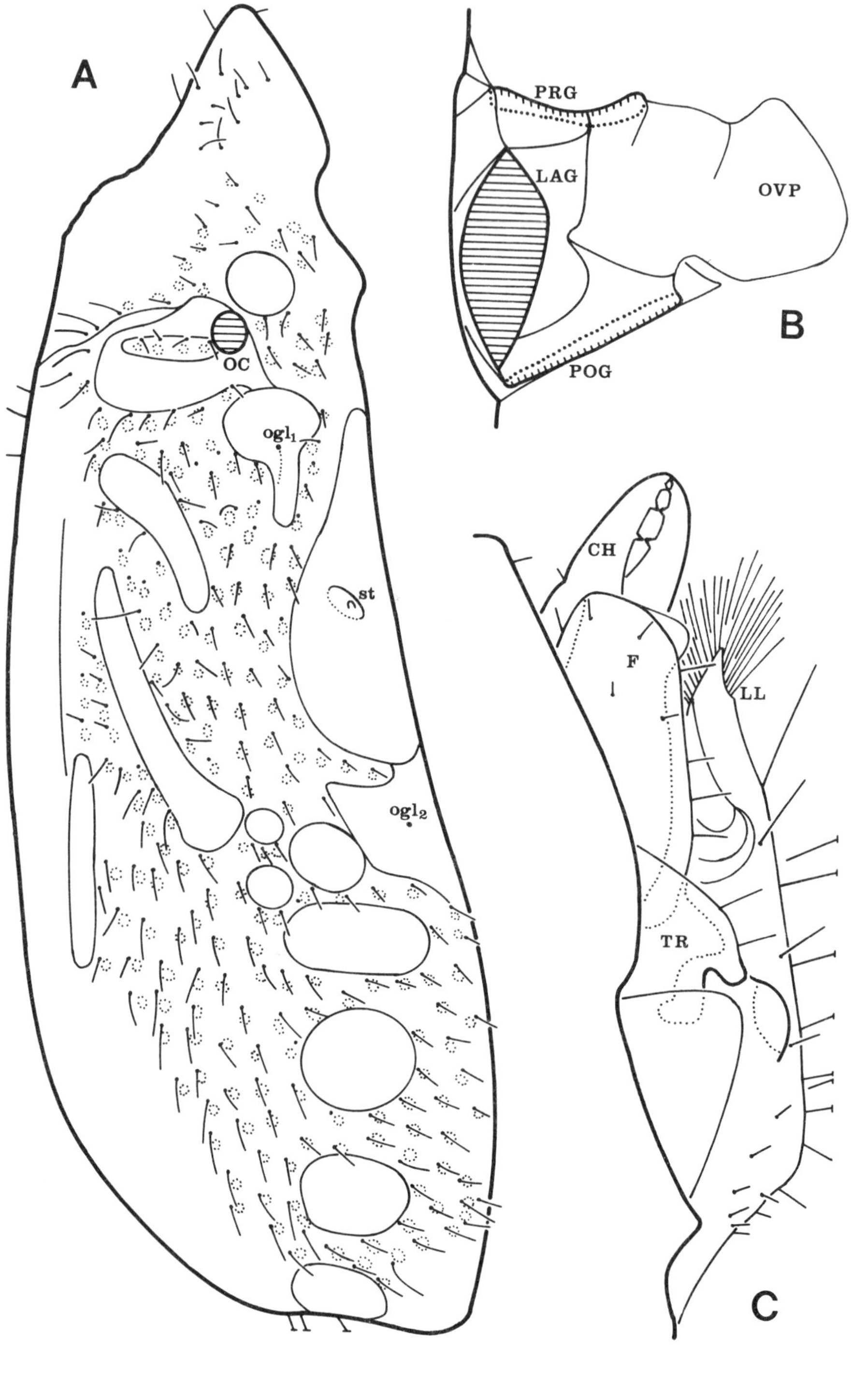
A
PRG
LAG
OVP
B
POG
OC
ogl_1
CH
st
F
LL
ogl_2
TR
C

There are two pairs of dorsal infracapitular setae (ldm_1, ldm_2), as in the Opilioacarids *Paracarus hexophthalmus* (Redikorzev) and *Adenacarus arabicus* (With); the posterior of the two setae (ldm_2) is small. The corniculus is more or less cone-shaped, without lateral tooth. The paired orifices of the infracapitular glands are in the posterior part of the cervix, close to the line of attachment of the cheliceral frame (as in Opilioacarida, Ixodida and Oribatida; see fig. 158A). The ventral infracapitular setae (fig. 158B) are numerous (eighteen, on either side, in the figured specimen). The subcapitular gutter is narrow. The labrum is long and tapering, and presents papillae (particularly in the terminal part). The lateral lips are also tapering; they present long papillae in the dorsal, lateral and ventral parts (fig. 159A, C). As in other Allothyridae, there is a large, lateral apophysis, associated with an apophysis of the palpal trochanter.

Chelicera (fig. 160A, B). – The chelicera is long, and consists of three segments (trochanter, body of the chelicera, and apotele) as in other Holothyrida. There are two cheliceral setae, as in *Allothyrus constrictus* (Domrow); ch_1 (with dorsal position) and ch_2' (with paralaterodorsal position); they are apparently homologous with ch_1'' and ch_2' in Opilioacarida. The dorsal lyrifissure *id* is situated between the two setae (as in Opilioacarida); the antiaxial lyrifissure $i\alpha$ is situated near the articulation point of the apotele and the body of the chelicera. Both jaws of the chela present three large teeth and rows of very small teeth.

Palp (fig. 161). – The palp consists of the usual six segments: trochanter, femur, genu, tibia, tarsus and apotele. There are no fringed setae on the genu, and there is no paraxial brush (a group or comb of stiff setae) on the tibia. The relatively small tarsus presents, among other phaneres, a dorsal area with hollow setae, and a large curved dorsodistal sensory seta with broad, lanceolate terminal part. The paraxial part of the tarsus presents a membraneous lobe. The apotelic claw (with paraxial position) is tripartite: the median and the more antiaxial part are normally developed, but the more paraxial part is small, rounded and membraneous.

Legs (figs. 162–166). – Leg IV is slightly longer than the body, leg I about as long as the body, and legs II and III slightly shorter. There is no sexual dimorphism in shape and chaetotaxy of the legs. All legs consist of coxa, trochanter, basi- and telofemur (separated by a basifemoral ring), genu, tibia, tarsus and apotele. In legs II–IV, the tarsus is subdivided, by a basitarsal ring, into basi- and telotarsus. Leg I is subdivided, as in other Allothyridae, into a bathrotarsus and a very small apicotarsus. Haller's organ (fig. 160C), situated in the dorsoterminal part of the bathrotarsus, contains a number of sensory

---

←

Fig. 153. Australothyrus ocellatus Van der Hammen, adult female; A, lateral view of dorsal shield; B, lateral view of evaginated ovipositor; C, lateral view of gnathosoma in camerostome; A, B, × 75; C, × 176.

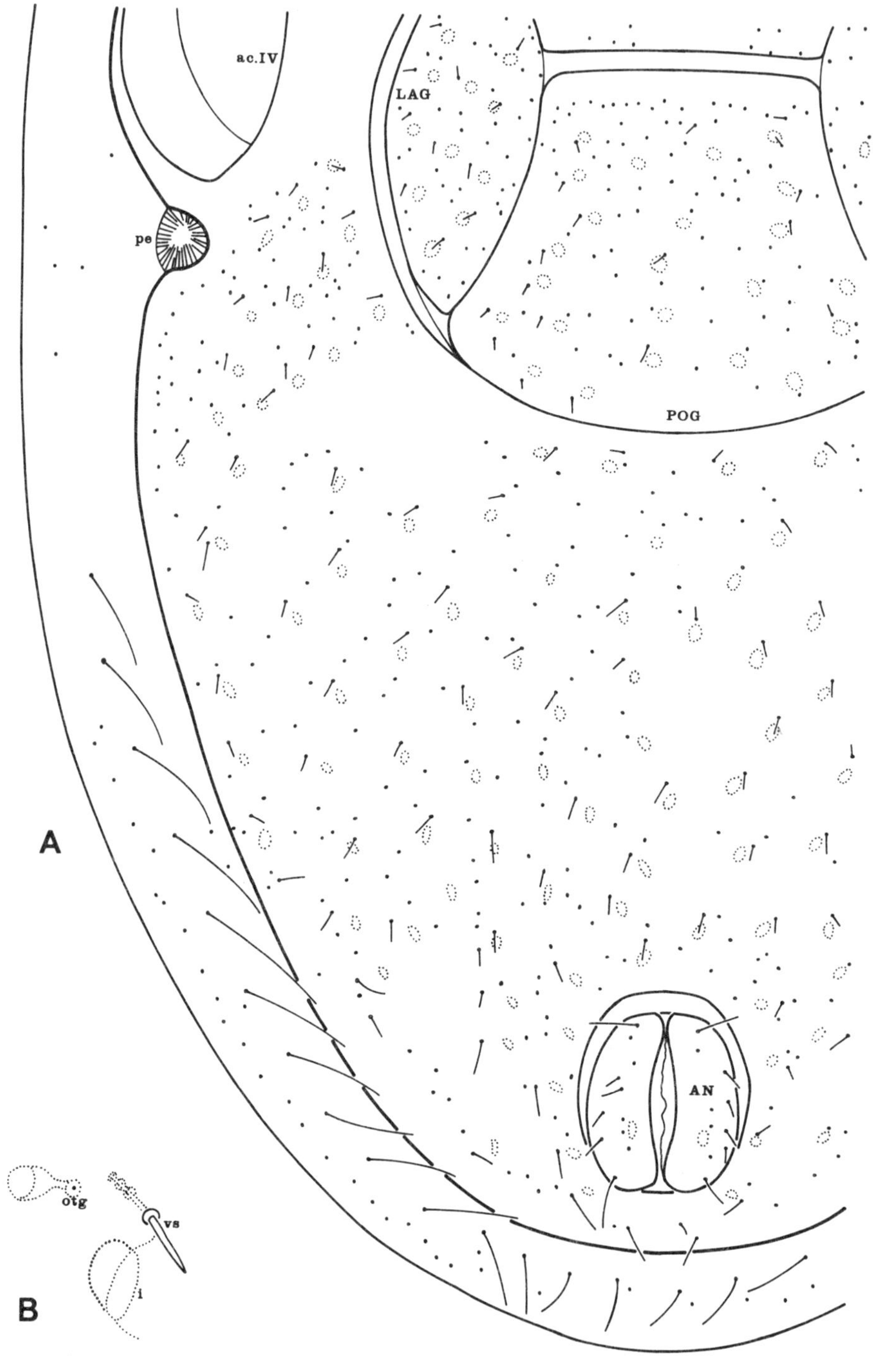
ac.IV
LAG
pe
POG
AN
A
otg
vs
i
B

phaneres. The area dorsal to it presents also several sensory phaneres (among them one phanere with a broad, lanceolate terminal part). All segments present many setae (particularly numerous in the case of the tarsus) and pores (orifices of tegumental glands). The ambulacrum presents a bidactyl claw and, in legs II–IV, a pulvillus and a pretarsus with one pair of laterodorsal setae.

4.2. Morphology and postembryonic development of Holothyrus coccinella *Gervais*

The data with reference to the present species, previously published in two papers (Van der Hammen, 1965; 1983b: 36–40, figs. 23–25), are summarized in the present subsection. The two papers were based on material from Mauritius, which included males, females and numerous immature specimens.

Measurements. – The length of the adult specimens (males and females) varies from 4.3 to 5.5 mm; the breadth of the smallest specimen is 3.1 mm, the height 1.5 mm. The lengths of the legs of an adult male of 5.0 mm are the following: leg I 8.0 mm, leg II 6.0 mm, leg III 6.3 mm, leg IV 8.0 mm.

Habitus and colour. – The species presents the typical habitus of a Holothyrid: relatively large, with highly arched, very shining idiosoma which is broadly elliptic in outline; in some respects, the habitus reminds one of a very large Gamasid mite. The colour is mainly reddish brown; there are darker spots in the lateral and marginal regions, and darker lines in the median and anterior marginal part.

Cerotegument. – The idiosoma presents dorsally a very thin, whitish layer of cerotegument; in dry condition (when observed on a carbon block), it reminds one of a thin layer of white frost. This cerotegument is also found on the ventral surface of the idiosoma, on the legs, and in larger masses under the extending border of the dorsal shield. The layer is much thicker and presents a different structure (it could be a different secretion) in taenidium 1 (cf. fig. 167D, 169E, 176A). There is a brownish, more or less tube-shaped mass of secretion near the office of the excretory gland; this certainly is no cerotegument.

Cuticle. – The cuticle is solid, just as in other species of Holothyridae; the surface is shining. The remarkable thickness is easily visible in fig. 167A (a transverse section through the idiosoma). As a result of this solidity and thickness, the idiosoma is difficult to section.

There are two shields: a dorsal shield and a holoventral shield, connected by a chitinous plicature band. Consequently, the idiosoma is completely covered by the exoskeleton; the surface presents no areas of soft skin. The rostral

←

Fig. 154. Australothyrus ocellatus Van der Hammen, adult female; A, ventral shield with anal and part of genital opening; B, seta, cupuloid, and orifice of tegumental gland of ventral shield; A, × 141; B, × 550.

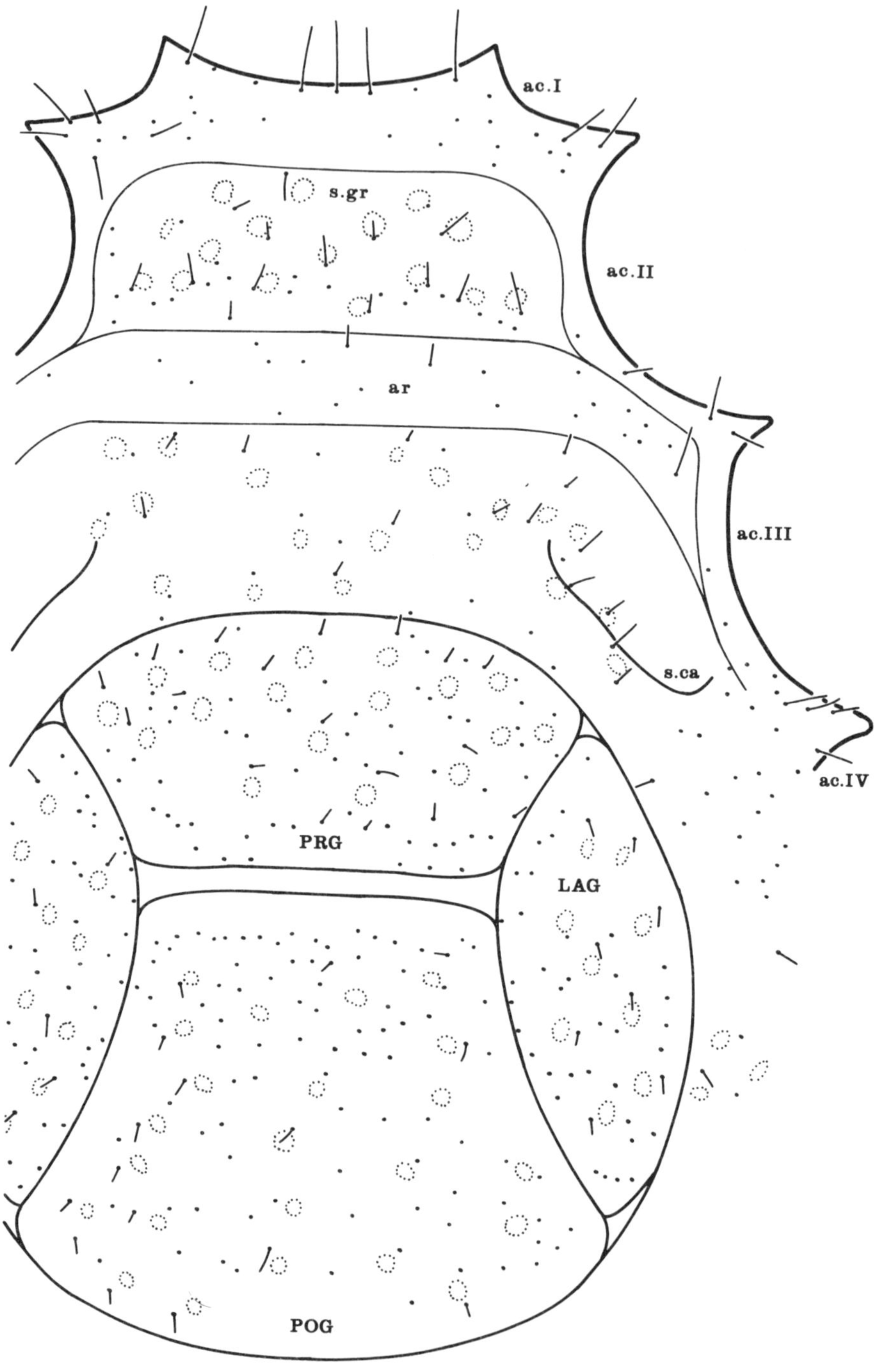

Fig. 155. Australothyrus ocellatus Van der Hammen, ventral view of sternal and genital region of adult female; × 176.

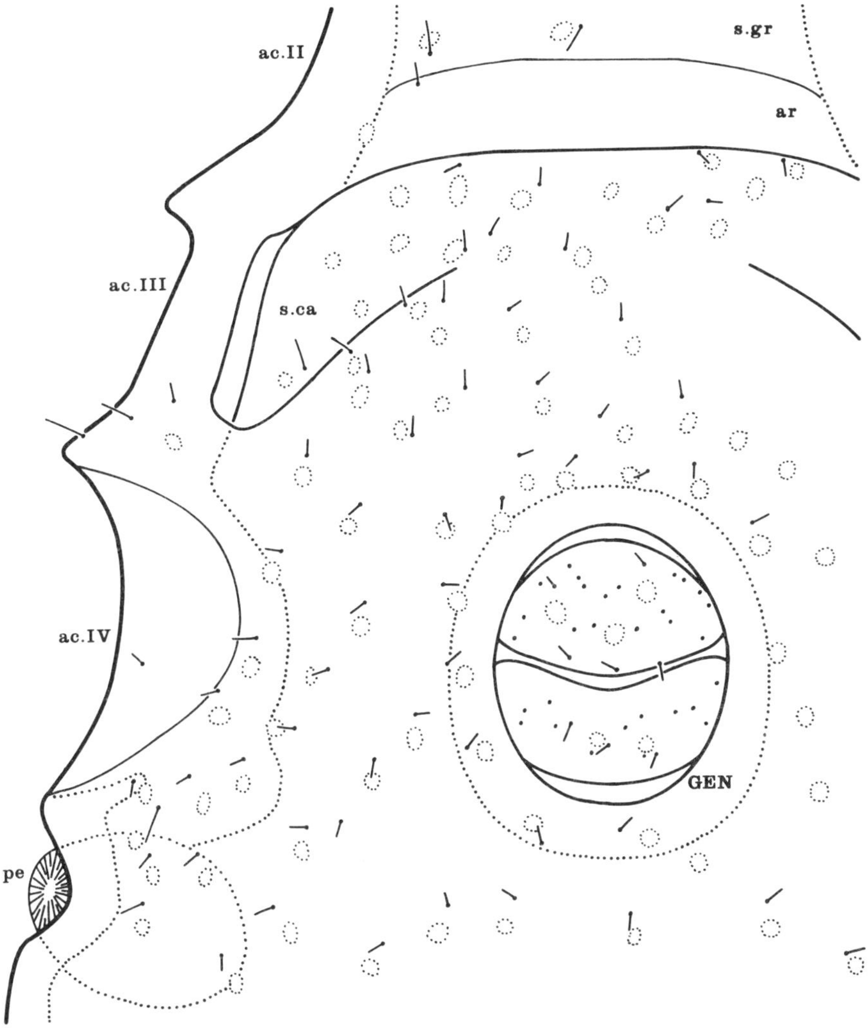

Fig. 156. Australothyrus ocellatus Van der Hammen, ventral view of genital and part of sternal region of adult male; × 176.

tectum (fig. 169A) is rather thin. Sections demonstrate that the reddish brown colour of the specimens is indeed the colour of the cuticle; there are no sub-cuticular pigment grains.

Pores and lyrifissures. – The idiosoma presents a great number of small, simple pores; their distribution is apparently a disorderly one. According to my observations, the idiosoma presents only one pair of normal lyrifissures (fig. 169A, E: i_1); a more detailed study of the cuticle could perhaps demonstrate the occurrence of other lyrifissures or their homologues.

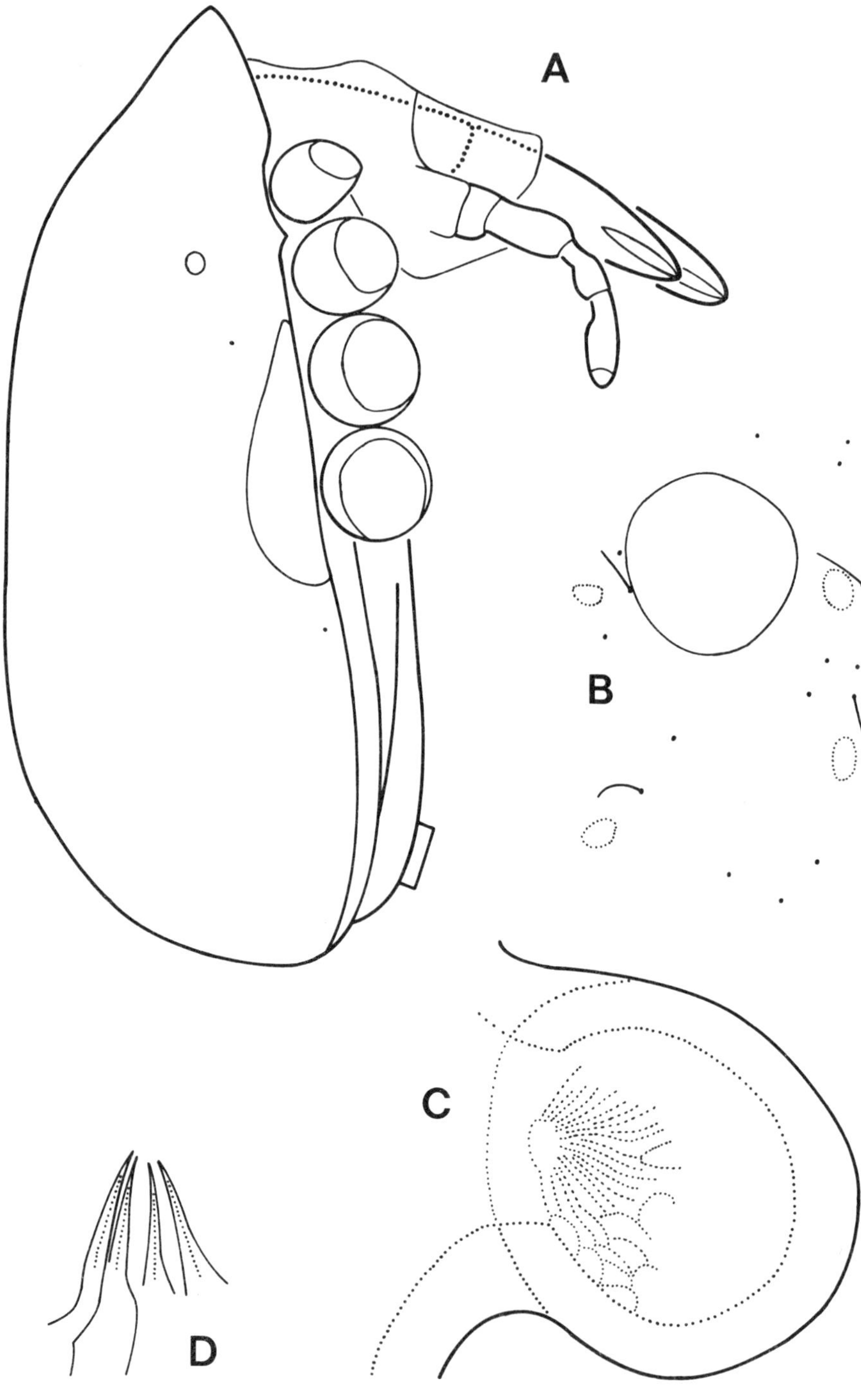
A
B
C
D

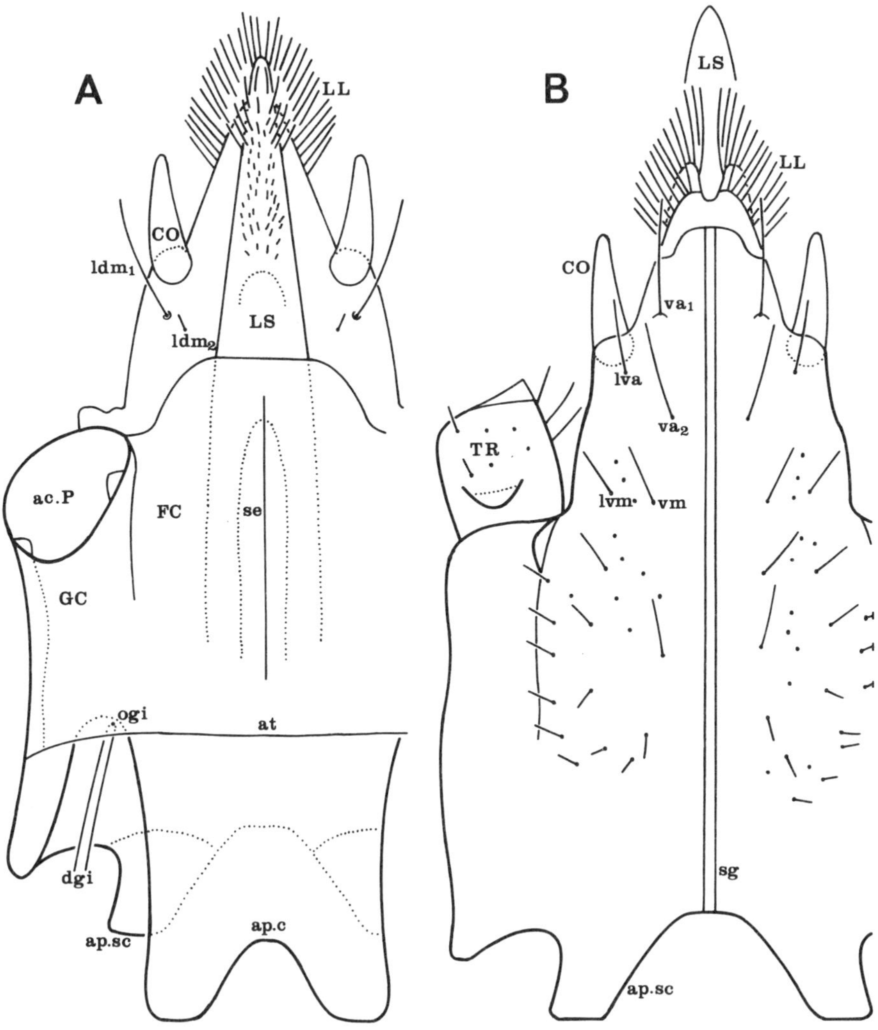

Fig. 158. Australothyrus ocellatus Van der Hammen, infracapitulum of adult female; A, dorsal view; B, ventral view; A, B, × 176. (N.B. for *ap.c* read *ap.ce*).

←

Fig. 157. Australothyrus ocellatus Van der Hammen; A, lateral view of adult male with extended gnathosoma; B, lateral view of ocular region of adult female (same orientation as figs. 153A, 157A); C, internal (= dorsal) view of left peridium of adult female; D, detail of peridium with supposed glandular cells and papillae; A, × 55; B, × 330; C, × 368; D, × 736.

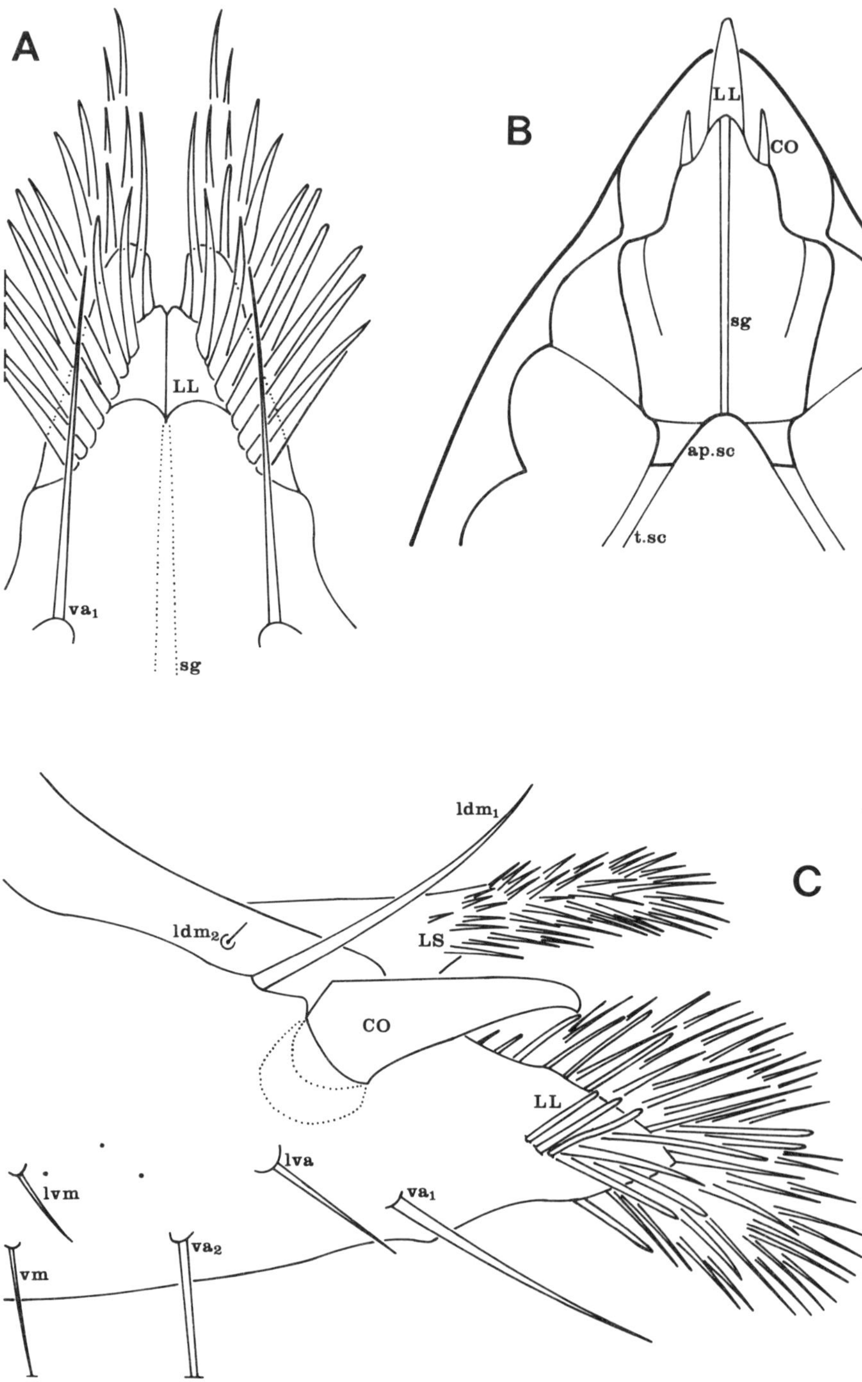
A
LL
va1
sg
B
LL
CO
sg
ap.sc
t.sc
C
ldm1
ldm2
LS
CO
LL
lva
lvm
va1
va2
vm

Idiosoma. – The dorsal shield, the ventral face, and the lateral region of the idiosoma are described in separate paragraphs. As mentioned above, dorsal and holoventral shield are connected by a chitinous plicature band; the dorsal shield is distinctly extending laterally beyond the holoventral shield. The greatest width of the idiosoma is just behind acetabulum IV.

Dorsal shield. – There are numerous very small, irregularly distributed dorsal setae and pores. As mentioned above, there is only one pair of normal lyrifissures (fig. 169A, E: i_1). The thin, transparent rostral tectum presents a pair of lateral lobes. The dorsal shield presents at both sides the orifice of a supposed excretory gland; the orifice is situated in a small depression, from which a groove in the dorsal shield (apparently a taenidium) runs closely parallel to the lateral border, as far as the rostrum (figs. 167D, 169A, E, 176A). In fig. 176B, I have represented a sagittal section through the chitinous atrium of the supposed excretory gland. A tube runs from the orifice through the cuticle and debouches into a system of saccules; these saccules are partly in the cuticle. According to Thon (1905) and Hughes (1959), the system extends from the orifice to the rectum. In the present species it is smaller and apparently more or less globular. Judging from my transverse section, the chitinous atrium presents internal longitudinal ridges. As mentioned above, a mass of a brownish secretion is usually found on the dorsal shield near the orifice. It will be interesting to reinvestigate the system in various species, because it is not impossible that in some species it indeed constitutes a transition to a respiratory system.

Ventral face. – The ventral face presents a holoventral shield which encloses a genital and an anal region. The anterior part of the holoventral shield, between the acetabula of the legs, constitutes the sternal region which presents numerous small setae and pores, but no normal lyrifissures and no specialized sternal setae. The sternal region consists of a posterior elevated part and an anterior depression (the sternal groove) which is mainly situated between the second pair of legs. In fig. 167E, the line of greatest depth of the sternal groove is indicated by *sb*, the anterior border of the elevated part by *sr*. The anterior border of the sternal region constitutes the border of the camerostome.

The female genital region presents four shields: a large postgenital shield, a narrow anterior pregenital, and a pair of narrow laterogenitals. They are represented in fig. 167B, C; these figures also demonstrate the way of closing of the four shields. The male genital region, or epiandrium (fig. 168A), is shallow (in contradistinction to the sternal groove, which is deep); it is more or less pentagonal, although the posterior border is rather indistinct. There is no epiandrial tooth. The male genital opening (closed by two shields) is wider than long. The area which anteriorly borders the epiandrium, is slightly convex in

←

Fig. 159. Australothyrus ocellatus Van der Hammen, infracapitulum of adult female; A, ventral view of distal part; B, ventral view of gnathosoma in camerostome (sternal shield removed); C, lateral view of terminal part (labrum raised); A, × 470; B, × 75; C, × 368.

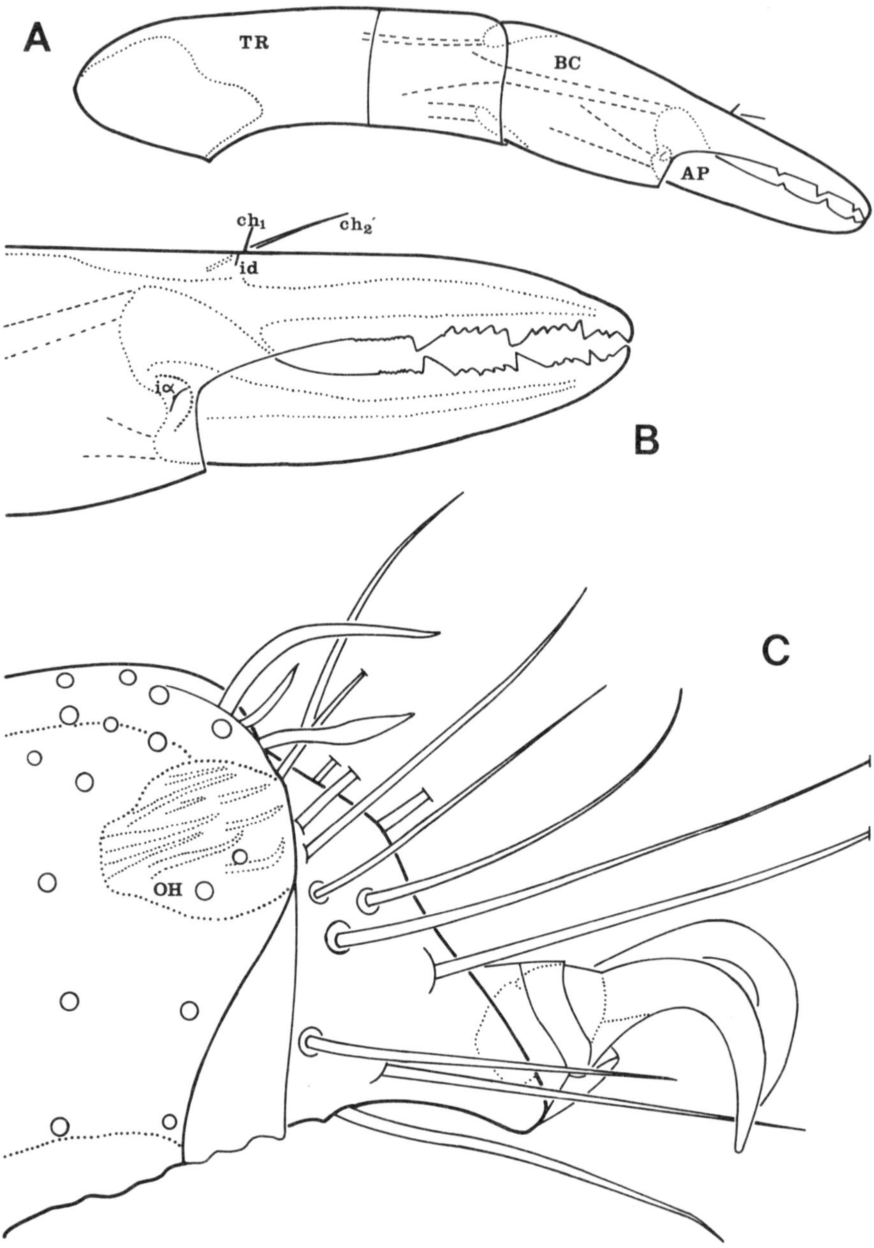

Fig. 160. Australothyrus ocellatus Van der Hammen, adult female; A, B, lateral (antiaxial) view of right chelicera; A, trochanter, body of the chelicera, and apotele; B, chela; C, lateral (posterior) view of terminal part of right leg I; A, × 75; B, × 176; C, × 736.

the median part. Each anal valve presents a row of nine setae, and numerous pores. There is an anterior and a posterior anal locking-piece (fig. 169D).

Lateral region. – In lateral view, the stigma, the orifice of the excretory gland, and taenidia 1 and 2 are visible (fig. 167D). The stigma is rather large and oval, and resembles the stigma of Gamasida; it is situated in the posterior part of taenidium 1 (the peritreme) which is relatively broad, and extends from a region ventral to the orifice of the excretory gland to the rostral region. Taenidium 1 is situated in a special area ventral to the extending lateral border of the dorsal shield; consequently, taenidia 1 and 2 run exactly parallel separated only by the extending border of the shield (see fig. 167A). A continuous, well-defined strip of a granular cerotegument or other secretion covers the bottom of taenidium 1. It may be repeated here, that the complete lateral region of the idiosoma is projecting; it consists of the lateral part of the dorsal shield (with taenidium 2), the area of taenidium 1 and the plicature band (fig. 167A). The plicature band consists of thin, flexible chitin; lateral muscles are apparently inserted on and attached to ventral and dorsal shield (fig. 167F: *ml*).

Gnathosoma. – In the present description of the gnathosoma, dorsal, ventral, lateral, anterior and posterior faces, and transverse as well as longitudinal sections are dealt with. It may be remarked here that there appears to be a slight difference between a gnathosoma which is studied in retracted position in the camerostome (fig. 169B), and a gnathosoma which is separated from the idiosoma (fig. 171). This is due to the flexibility of the large lateral ridges (with the coxal vaults) which are more curved inwards in retracted condition. The degree of sclerotization of the gnathosoma is different in the different regions. The anterior and median parts, particularly of the mentum, are heavily sclerotized. The posterior regions of tegulum, cervix and mentum are, however, rather soft, pale in colour, and not well-defined. As a result of this, sectioning is difficult, and dissection must be carried out with the utmost care. The complete gnathosoma should preferably be separated from the idiosoma by dissection of the cuticle surrounding the camerostome. Even after these precautions the posterior part, although undamaged, remains difficult to study because of the membraneous parts attached to it. The complete gnathosoma, which has more or less the shape of a cone (hence the name buccal cone), consists of a relatively large posterior part, and a smaller narrower part anterior to mouth and palpal acetabula; the anterior, preoral part is named coniculus.

A dorsal view of a gnathosoma in which the chelicerae are still in the original position, is represented in fig. 170. This figure clearly demonstrates the position of the relatively small tegulum, and the way in which the cheliceral sheaths are attached to its anterior border *bt*: at the line *bt*, the simple tegulum passes into the two separate sheaths. The duct of the infracapitular gland is, on either side, in the pharyngeal chamber; the orifice is close to the line of attachment of the cheliceral frame to the infracapitulum (visible after removal of the chelicerae). A taenidium appears to extend from the orifice in anterior direction (it constitutes, more or less, the border between cervix and large lateral ridges).

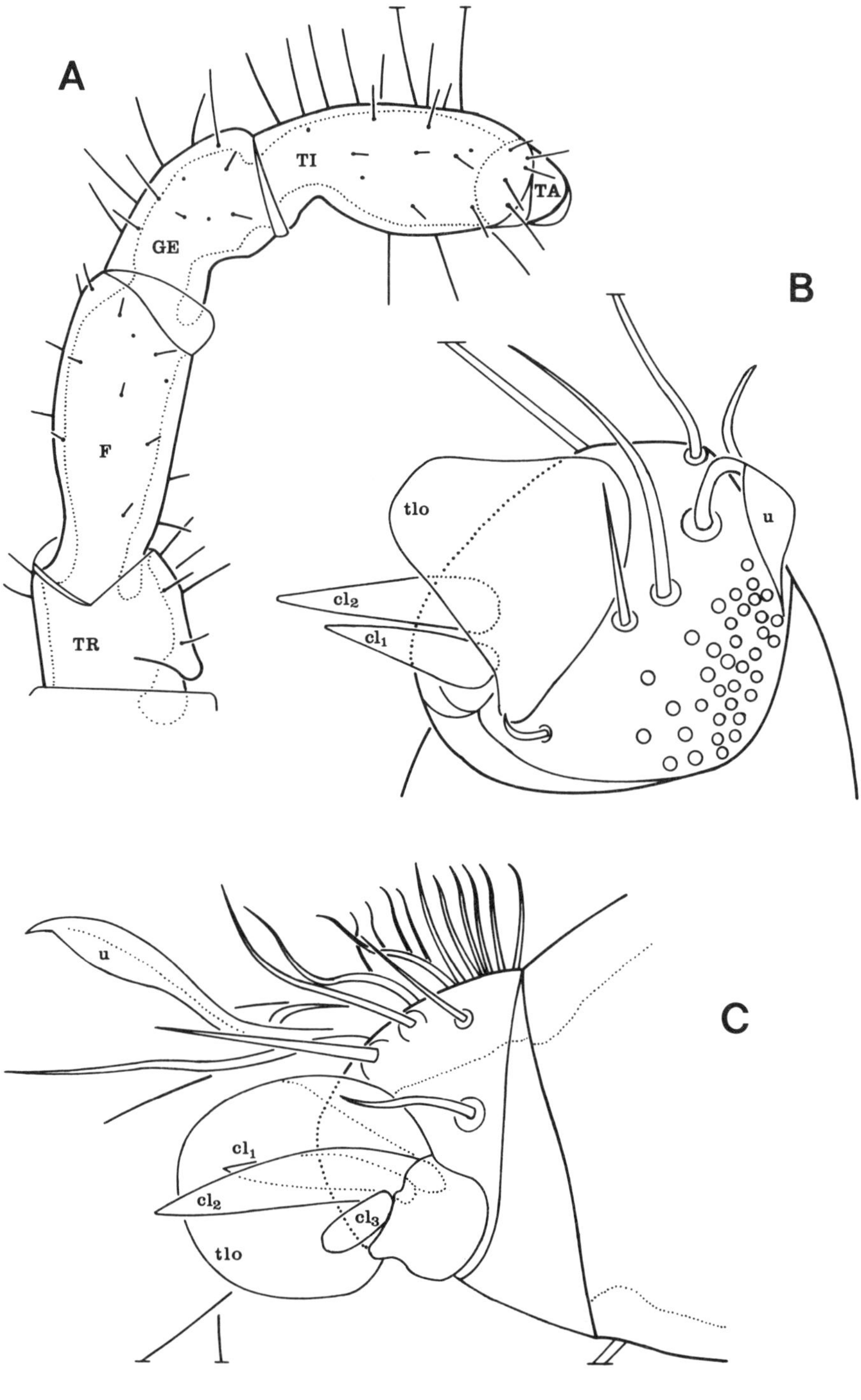
A
TI
TA
GE
F
TR
B
tlo
u
cl2
cl1
C
u
cl1
cl2
cl3
tlo

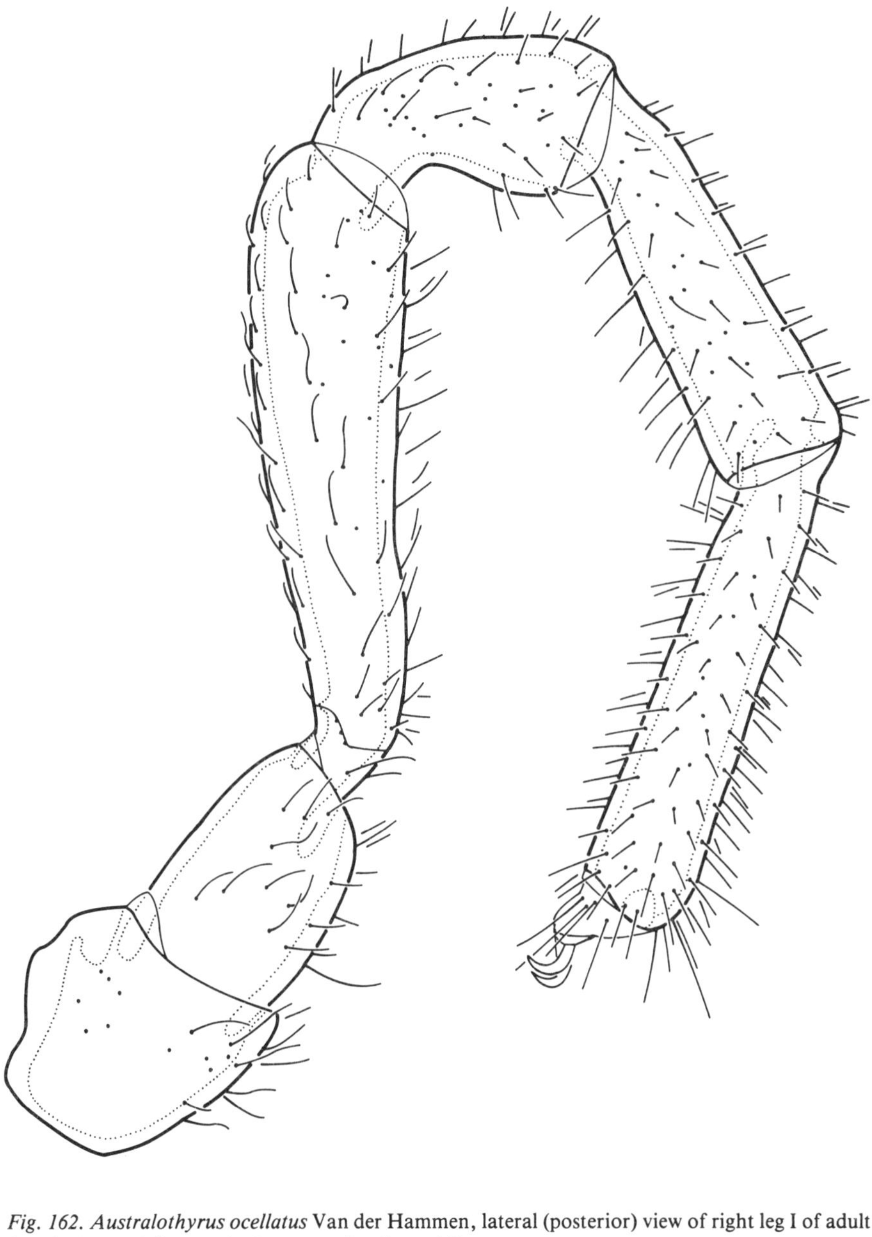

Fig. 162. Australothyrus ocellatus Van der Hammen, lateral (posterior) view of right leg I of adult female (part of the terminal setae omitted); × 141.

←

Fig. 161. Australothyrus ocellatus Van der Hammen, right palp of adult female; A, lateral (antiaxial) view; B, C, tarsus and apotele; B, dorsal view; C, lateral (paraxial) view; A, × 176; B, C, × 736.

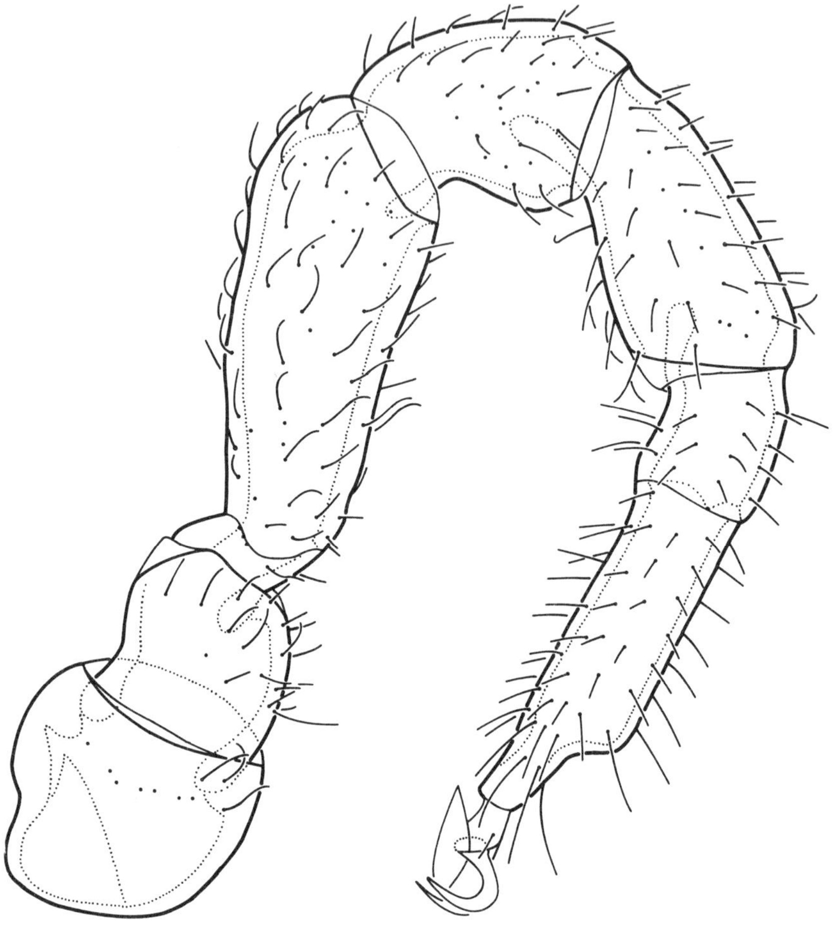

Fig. 163. Australothyrus ocellatus Van der Hammen, lateral (posterior) view of right leg II of adult female; × 141.

A ventral view of the gnathosoma (fig. 171) presents the same infracapitular setae as in other species of the family Holothyridae (*va, vm, vp, lva, lvm, lvp*). Characters of the ventral face, also present in other species of Holothyrida, include: (a) the absence of a separation between mentum and genae; (b) the presence of a pair of laciniae (flaps with a papillose fringe anteriorly bordering the genae); and (c) the presence of a subcapitular apodeme at the posterior border (in many Holothyridae, it is more heavily sclerotized than the remaining posterior part of the gnathosoma, and can be easily separated as a whole). The subcapitular apodeme constitutes the internal continuation of the mentum, posterior to the line of attachment *mb* of the arthrodial membrane of the

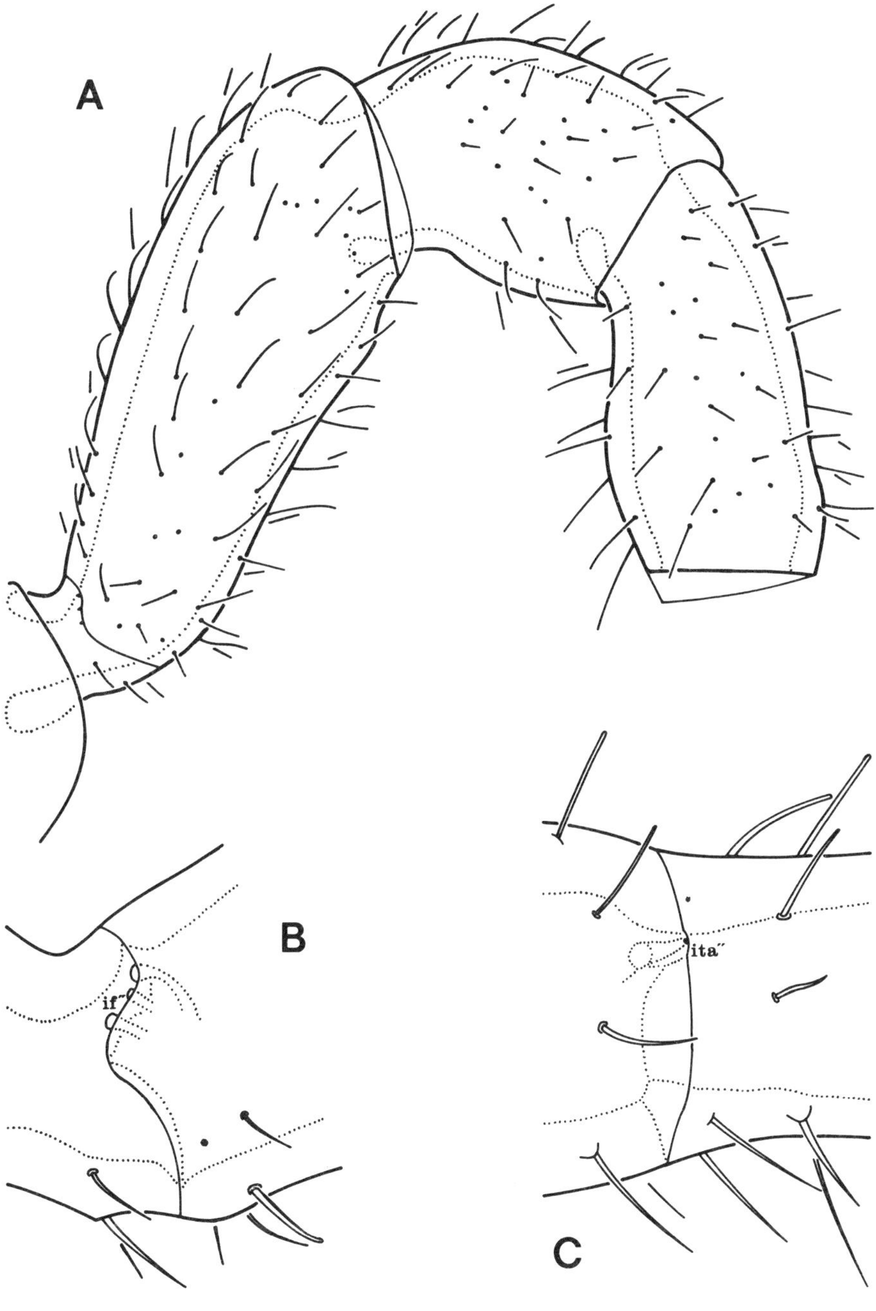

Fig. 164. Australothyrus ocellatus Van der Hammen; A, lateral (posterior) view of femur, genu and tibia of right leg II of adult male; B, C, lateral (posterior) view of details of right leg III of adult female; B, basifemoral ring; C, basitarsal ring; A, × 176; B, C, × 368.

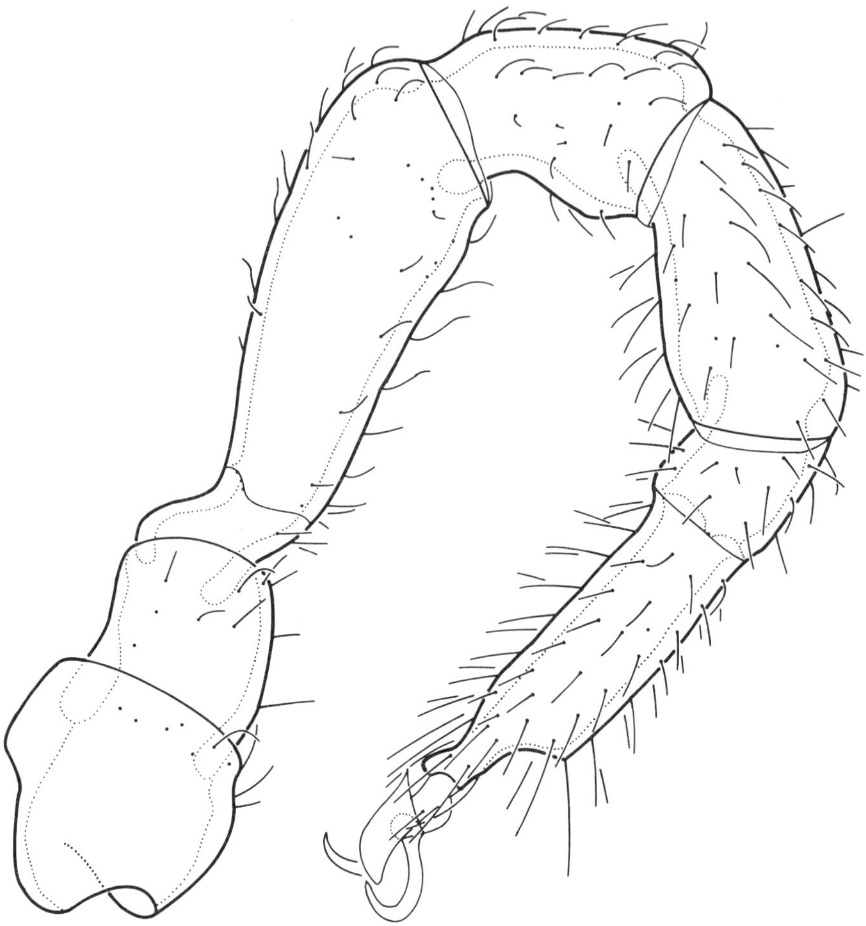

Fig. 165. Australothyrus ocellatus Van der Hammen, lateral (posterior) view of right leg III of adult female; × 141.

gnathosoma. The ventral surface of the infracapitulum of *H. coccinella* is particularly characterized by the presence of a distinct apophysis in the coxal region of the palp.

Fig. 172, the lateral aspect, presents a number of characters also visible in dorsal and ventral views, such as the infracapitular setae, the subcapitular apodeme (of which the ventral position is distinctly visible), and the coxal apophysis. The figure clearly demonstrates the attachment of the cheliceral sheaths to the tegulum. The lateral mark *gl* in the trochanter of the palp probably represents the orifice of a gland. I have represented also the border *mb* of the arthrodial membrane connecting the gnathosoma with the idiosoma.

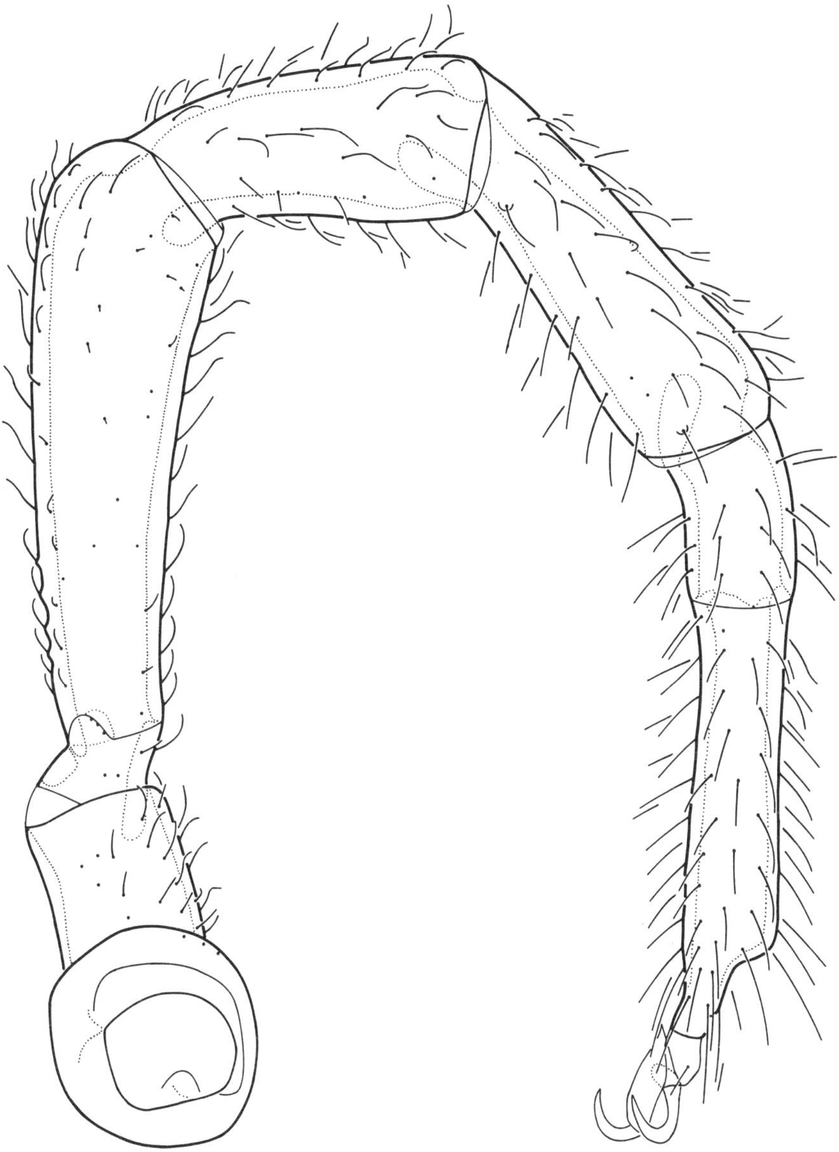

Fig. 166. Australothyrus ocellatus Van der Hammen, lateral (posterior) view of right leg IV of adult female; × 118.

I have used my specimens of *H. coccinella* also to prepare longitudinal and transverse sections of the gnathosoma. The longitudinal section, represented in fig. 173, is in a plane parallel to the median plane: the pharynx is just left entire, and both paralabral teeth and the labrum are present (the labrum of the

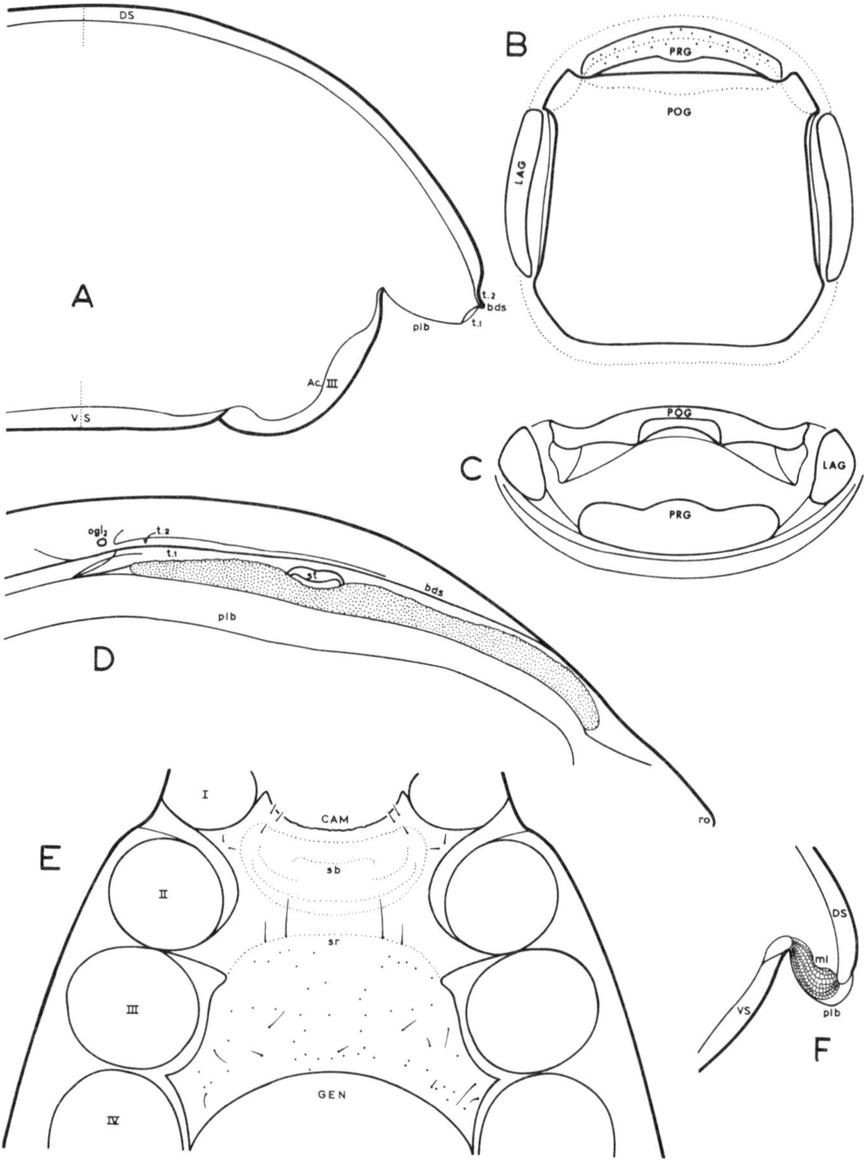

Fig. 167. Holothyrus coccinella Gervais, female; A, transverse section through idiosoma between stigma and genital region; B, ventral view of genital region; C, anterior view of genital region; D, lateral part of idiosoma with taenidia, stigma and supposed orifice of excretory gland, viewed obliquely from below; E, sternal region; F, transverse section through lateral part of idiosoma, posterior to genital region; A–F, × 36.5.

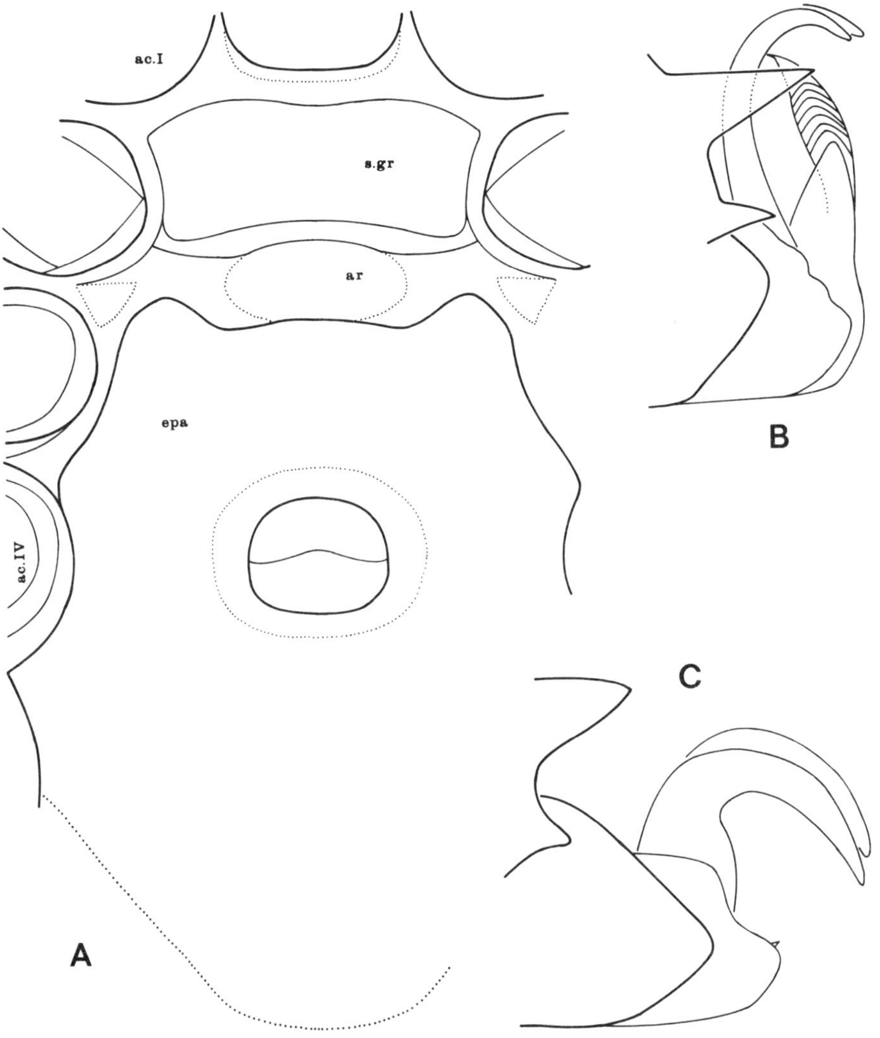

Fig. 168. Holothyrus coccinella Gervais; A, ventral view of sternal and genital region of adult male; B, C, lateral (posterior) view of terminal part of tarsus and ambulacrum of right leg I; B, adult male; C, adult female; A, × 44; B, C, × 295.

specimen, however, was damaged and had to be schematized). The figure demonstrates a number of important details. The labrum is represented in retracted position. The labral tendon *tls* is inserted on a chitinous piece. The dorsal wall of the pharynx continues as the ventral wall of the labrum; the ventral wall of the pharynx continues as the ventral wall of the preoral cavity. One of the laciniae (with papillose fringe) is visible, dorsally partly covered by a thin

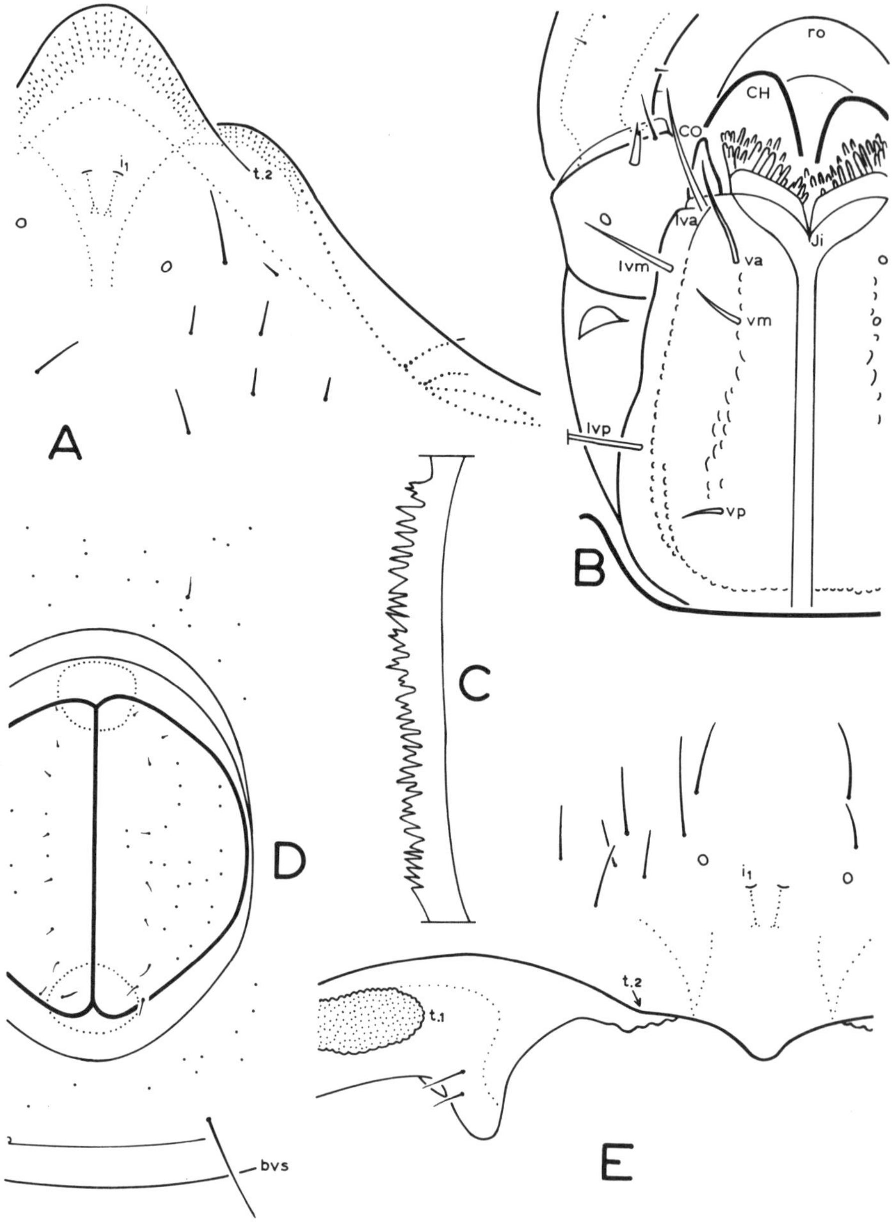

Fig. 169. Holothyrus coccinella Gervais, female; A, dorsal view of rostral region; B, ventral view of gnathosoma in retracted position in the camerostome; C, row of teeth on the left wall of the labial gutter, paraxial face; D, anal region; E, frontal view of rostral region; A, × 115; B, × 110; C, × 710; D, × 55; E, × 115.

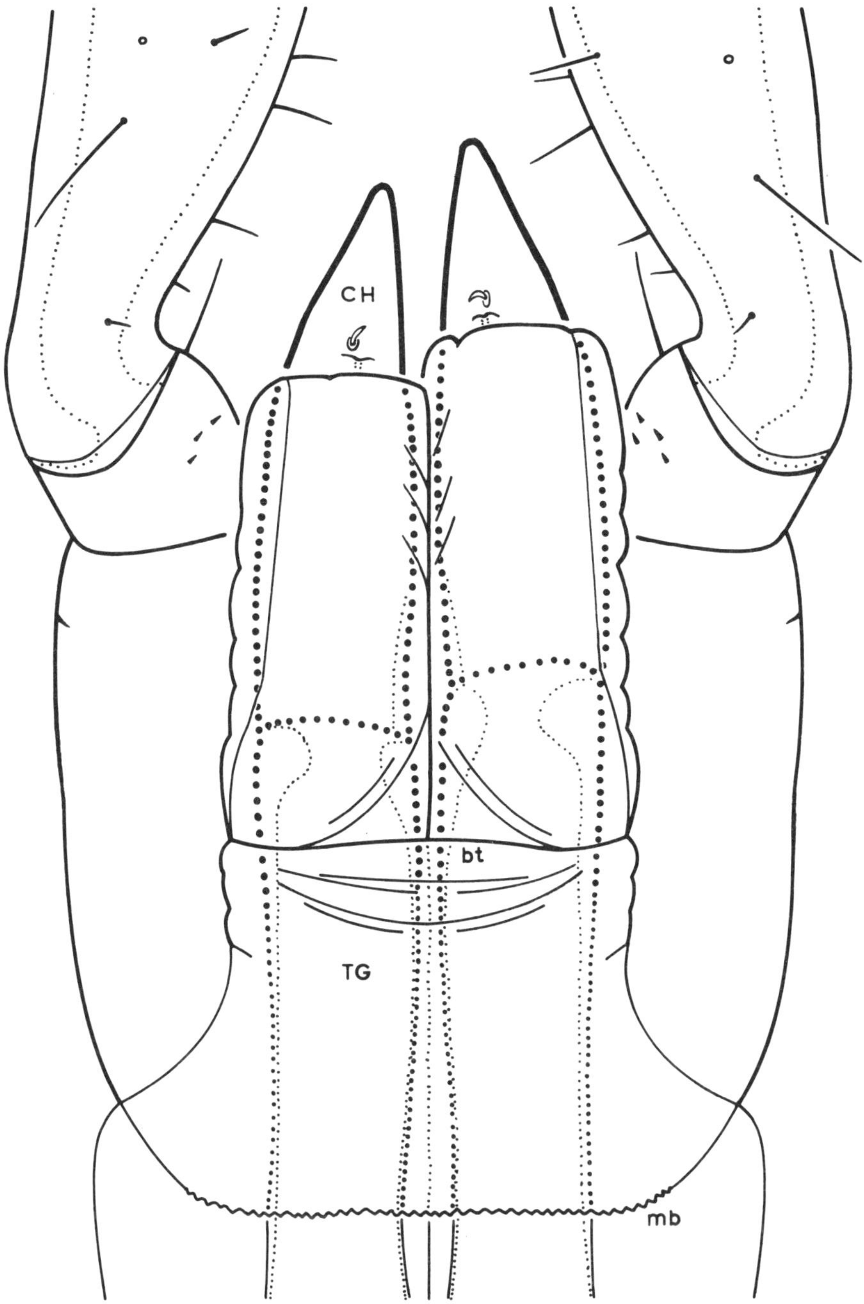

Fig. 170. Holothyrus coccinella Gervais, female, dorsal view of gnathosoma, × 145.

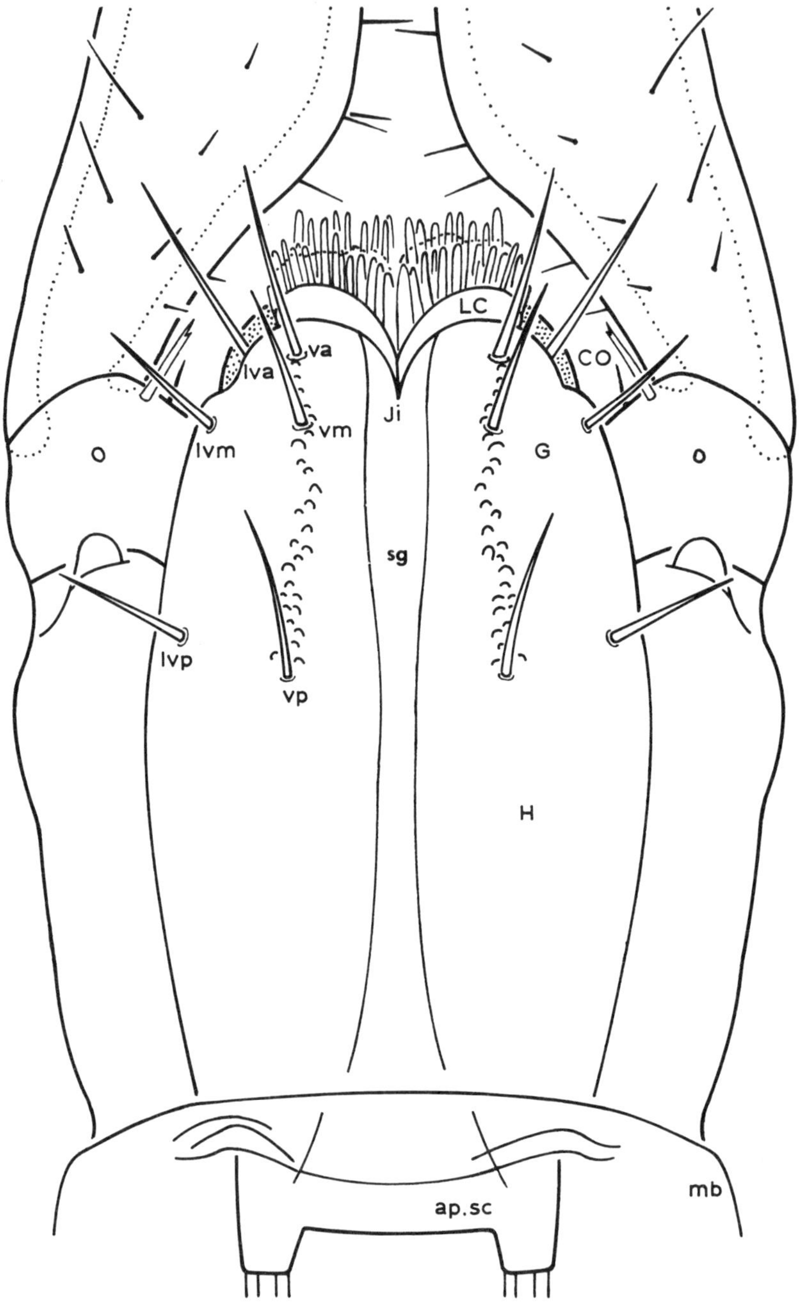

Fig. 171. Holothyrus coccinella Gervais, female, ventral view of gnathosoma, × 145.

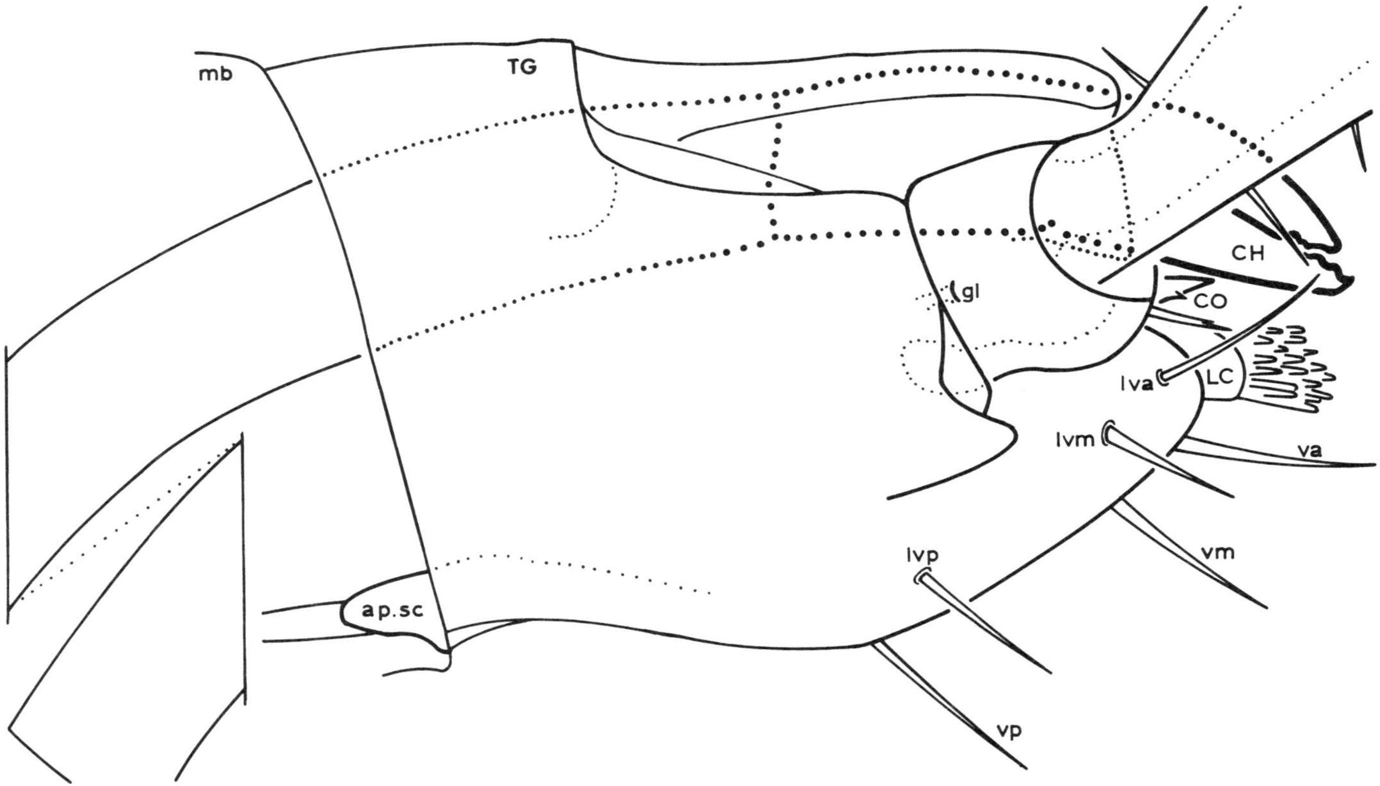

Fig. 172. Holothyrus coccinella Gervais, female, lateral view of gnathosoma, × 145.

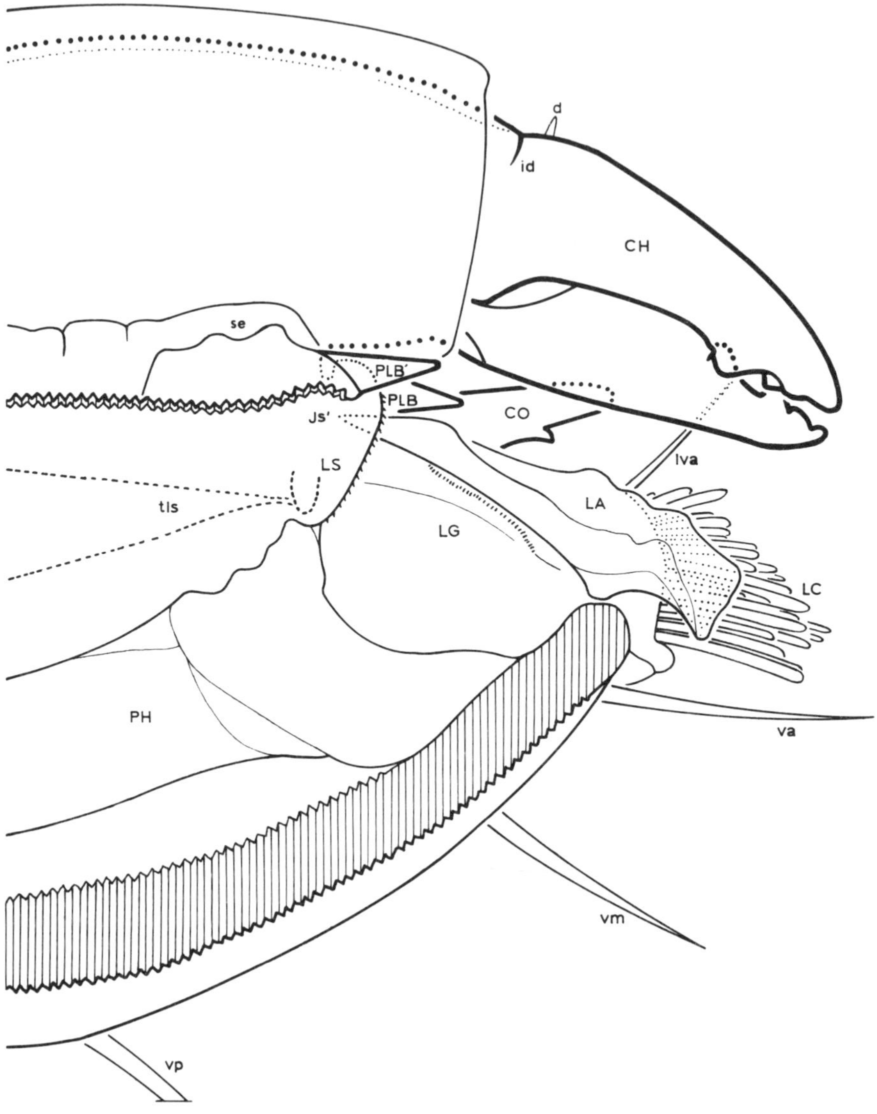

Fig. 173. Holothyrus coccinella Gervais; longitudinal section through anterior part of gnathosoma (labrum schematized because of damage and retracted position); × 30.

flap which certainly represents the labellum. The labial gutter has a longitudinal row of teeth; this row is represented separately in fig. 169C. One of the superior commissures of the mouth (*Js'*) is visible, lateral to the paralabral tooth *PLB'*. A dorsal ridge (the capitular saddle *se*) separates the chelicerae.

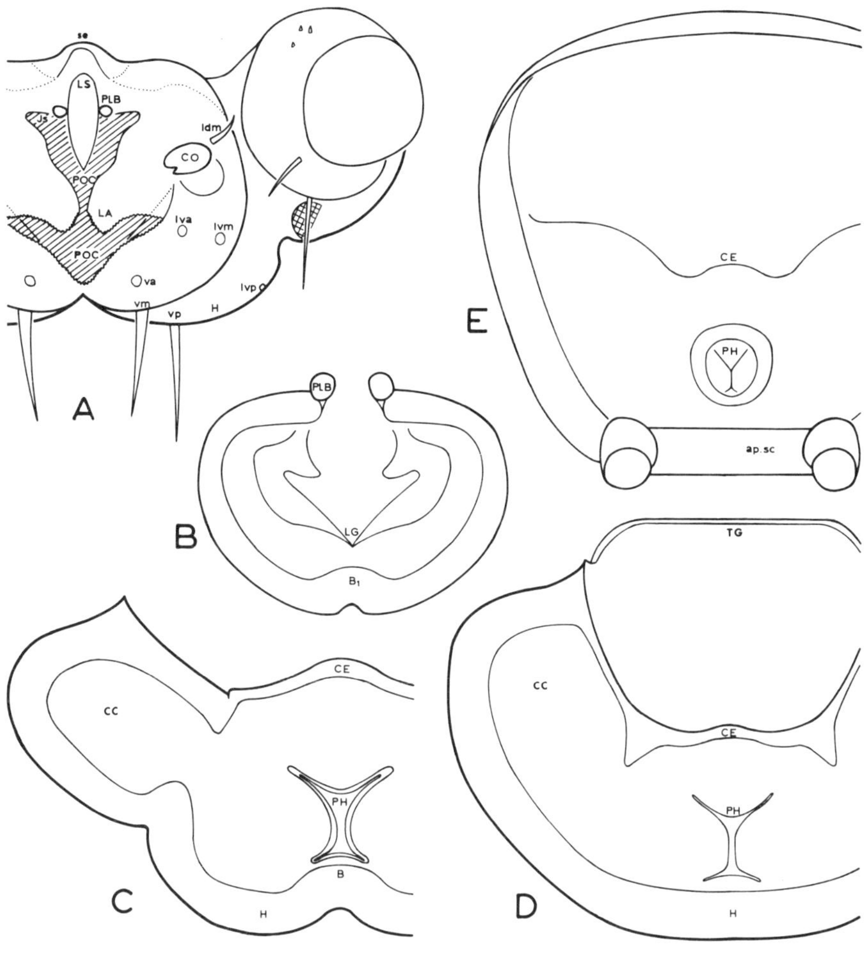

Fig. 174. Holothyrus coccinella Gervais, female; A, frontal view of gnathosoma (distal part of labella and laciniae removed); B, transverse section through coniculus, just anterior to the base of the labrum (the section is viewed in distal direction); C, transverse section through gnathosoma, just posterior to the acetabulum of the palp; D, transverse section through gnathosoma, in the region of the tegulum; E, posterior view of gnathosoma; in all cases the chelicerae have been removed; A–E, × 120.

In fig. 174A, I have represented a frontal view of the gnathosoma, in which the distal part of labella and laciniae is removed. The preoral cavity is visible in this figure between the labella (cf. Van der Hammen, 1961: fig. 2; this figure demonstrates the normal position of the labella). The dorsal position of the infracapitular seta *ldm*, posterior to the corniculus, is distinctly visible in frontal view.

A posterior view of the gnathosoma (fig. 174E) demonstrates that the posterior part of the cervix is indeed very thin; it does apparently not present an apodeme. The distinctly sclerotized subcapitular apodeme is visible at the posterior border of the mentum. Coxal vaults are not present in the posterior part of the gnathosoma. A posterior section of the pharynx is quadrangular; it is surrounded by a muscular layer.

Transverse sections, respectively anterior to the mouth, just behind the mouth, and in the region of the tegulum, are represented in fig. 174B, C, E. It may be remarked here that sectioning of the gnathosoma of *Holothyrus* is difficult because of the considerable differences in the degree of sclerotization between dorsal and ventral parts. Consequently, fig. 174B–E must be regarded as more or less schematized.

Fig. 174B represents a transverse section through the coniculus. The paralabral teeth are visible; the labrum is removed. The lateral border of the labial gutter can be recognized in the distal part of the section (the proximal part was damaged). A low boss (B_1) is visible at the bottom of the preoral cavity (see below).

A transverse section posterior to the mouth (fig. 174C) demonstrates that the coxal cavity is not separated from the pharyngeal chamber: there is a lateral (paraxial) connection. The pharynx is quadrangular. A distinct ventral boss (*B*) is present at the bottom of the infracapitular chamber.

A transverse section in the region of the tegulum finally shows the space occupied by the chelicerae, the pharyngeal chamber and the quadrangular pharynx. A ventral boss at the bottom of the pharyngeal chamber is not present at the level of this section.

The transverse sections, together with frontal and posterior views, demonstrate a number of important characters. It appears that the coxal cavity and the space included in the large lateral ridge, on either side of the infracapitulum, are lengthwise connected with the pharyngeal chamber. The pharynx appears to be quandrangular in the sectioned parts. The ventral bosses *B* and B_1 are parts of a median ridge which starts posterior to the mouth and continues in the preoral cavity (it is not visible in fig. 173, because that section is not a sagittal one). This ridge could represent the incorporated labium (the presence of a labium, besides the three other lips, is generally associated with a quadrangular section of the pharynx).

Chelicera. – The chelicerae are three-segmented and very long. The entire chelicera is represented in fig. 172, the distal part (in different views) in fig. 175A–D. There is a dorsal lyrifissure *id*, a dorsal seta *d*, and an antiaxial lyrifissure *iα*. The antiaxial face of the fixed jaw presents a proximal mark

→

Fig. 175. Holothyrus coccinella Gervais, female; A–D, distal part of right chelicera; A, lateral (antiaxial) face; B, lateral (paraxial) face; C, dorsal face; D, ventral face; E–F, lateral (paraxial) views of right palp; E, group of stiff setae on the paraxial side of the tibia; F, distal part of the genu, with two fringed setae; A–F, × 185.

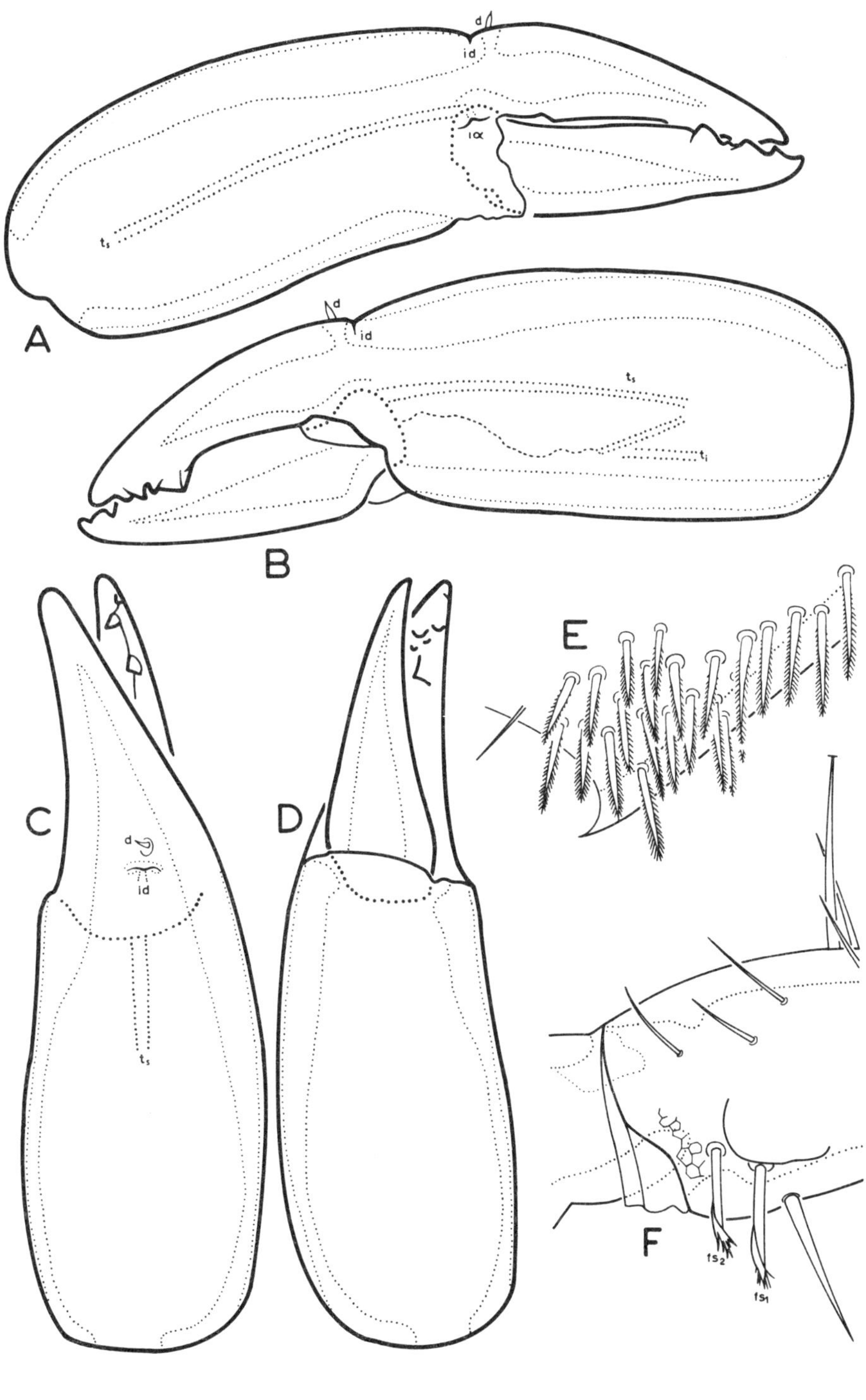
d
id
iα
t_s
A
d
id
t_s
t_i
B
C
d
id
t_s
D
E
F
fs_2
fs_1

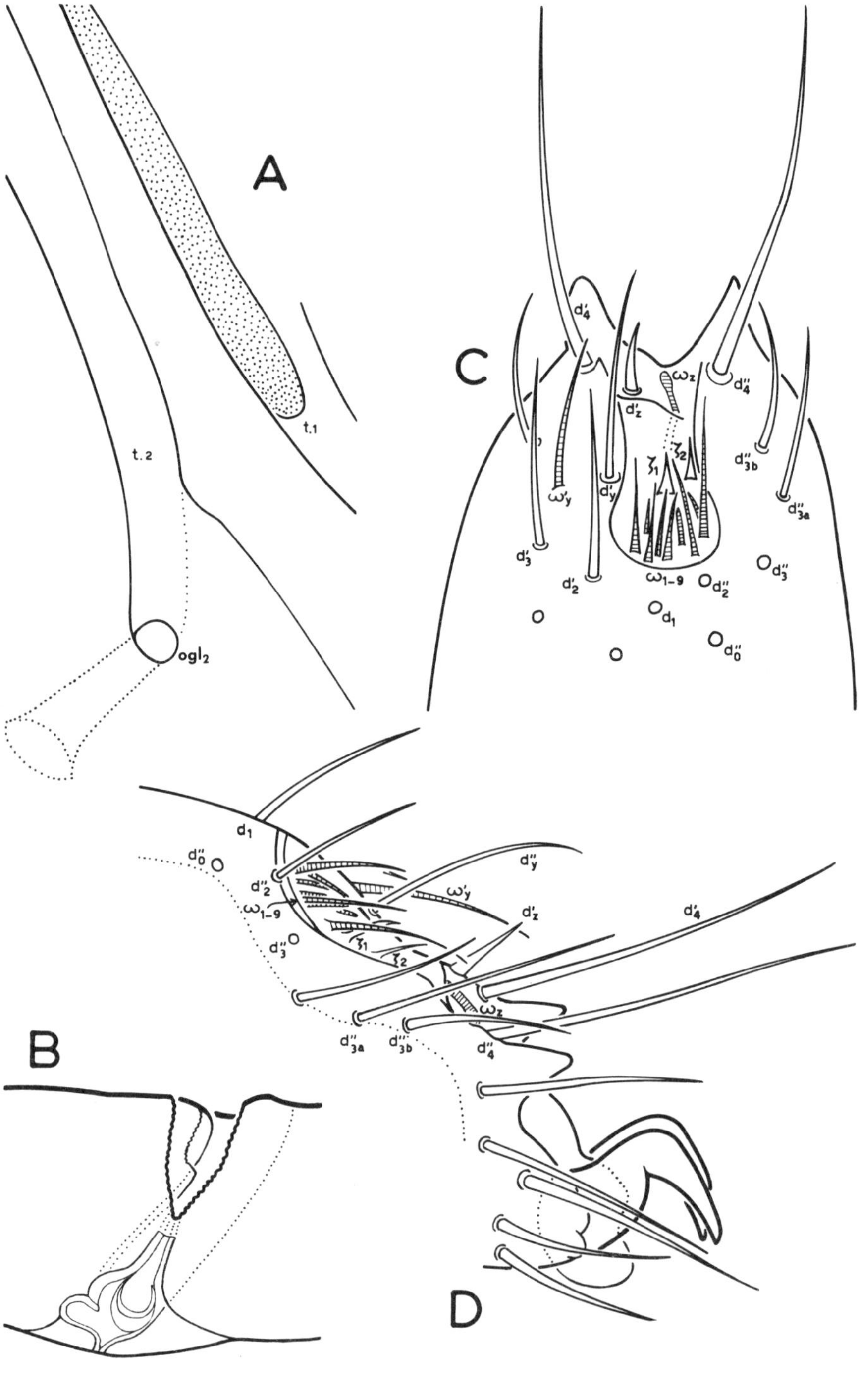
A
t.1
t.2
ogl2
C
d'4
d''4
ω2
d'z
ω'y
d'y
ζ1
ζ2
d''3b
d''3a
d'3
d'2
ω1-9
d''2
d''3
d1
d''0
B
d1
d''0
d''2
ω1-9
d''3
ω'y
d''y
d'z
d'4
ζ1
ζ2
ωz
d''3a
d''3b
d''4
D

which probably constitutes the orifice of a gland. The fixed jaw is distinctly shorter than the movable jaw. Teeth are found in the distal part of both jaws of the chela.

Palp. – The palp consists of six segments. Details of genu and tibia are represented in fig. 175E, F. Tarsus and tibia together have the shape of one large segment. Lyrifissures could not be discerned. There is a strong resemblance to the palp of *Hammenius (Thonius) grandjeani* (Van der Hammen): apotele, dorsal paraxial group of hollow setae, and paraxial and terminal setae are nearly similar in the two species (cf. Van der Hammen, 1961: fig. 6). In most segments, a great number of setae is present. The palp mainly differs from that of *Hammenius (Thonius) grandjeani* by the presence on the genu of a ventral paraxial group of 25–33 stiff ciliate setae (fig. 175E) instead of a comb, and by the presence on the tibia of two paraxial fringed setae of moderate length (fig. 175F) instead of one large seta.

Legs. – It is impossible to prepare formulae representing the chaetotaxy of the legs, because of the considerable number of setae. In the case of the setae surrounding Haller's organ (dorsally, in the distal part of tarsus I) a notation of comparative value is, however, possible; in this case the setae (and other phaneres) appear to be idionymous.

The legs present adesmatic joints in femora I–IV and tarsi II–IV. Lyrifissures are associated with the basifemoral and basitarsal rings; one of the lyrifissures of the femur is a compound one.

Coxa I does not present a laterocoxal seta. The femora present numerous small pores. Tibiae and tarsi have no ventral teeth. Tarsi II–IV bear a pair of bifurcate terminal setae as in *Hammenius (Thonius) grandjeani*; in the case of tarsi II and III the outer branch of the fork is the largest. Leg I has no pretarsus; the other legs present a pretarsus bearing a pair of ciliate setae. Pulvilli II–IV have nearly the same shape as in *Hammenius (Thonius) grandjeani*. As in other species of Holothyridae, leg I of the male (fig. 168B) presents a long pulvillus, whilst pulvillus I is reduced to a vestige in the female (figs. 168C, 176D). In the female of *H. coccinella*, the ungues of leg I are, moreover, much heavier than in immature and male specimens (this secondary sexual character of the female is not known from species of *Hammenius*). Haller's organ closely resembles that of *Hammenius (Thonius) grandjeani*; in fig. 176C, D, I have used nearly the same notation (cf. Van der Hammen, 1961: fig. 8). I observed the following differences: at the base of the paraxial lobe a clavate solenidion (ω_z) is present; ω_y'' is apparently absent, and an ordinary seta d_{3a}'' is found at this place; the two sensory setae in the groove (ζ_1, ζ_2) resemble eupathidia. The

←

Fig. 176. Holothyrus coccinella Gervais, female; A, orifice of supposed excretory gland and taenidia 1 and 2, viewed obliquely from above; B, transverse section through chitinous atrium of supposed excretory gland; C–D, distal part of right tarsus I with Haller's organ; C, dorsal view; D, lateral (antiaxial) view; A, × 230; B, × 370; C–D, × 230.

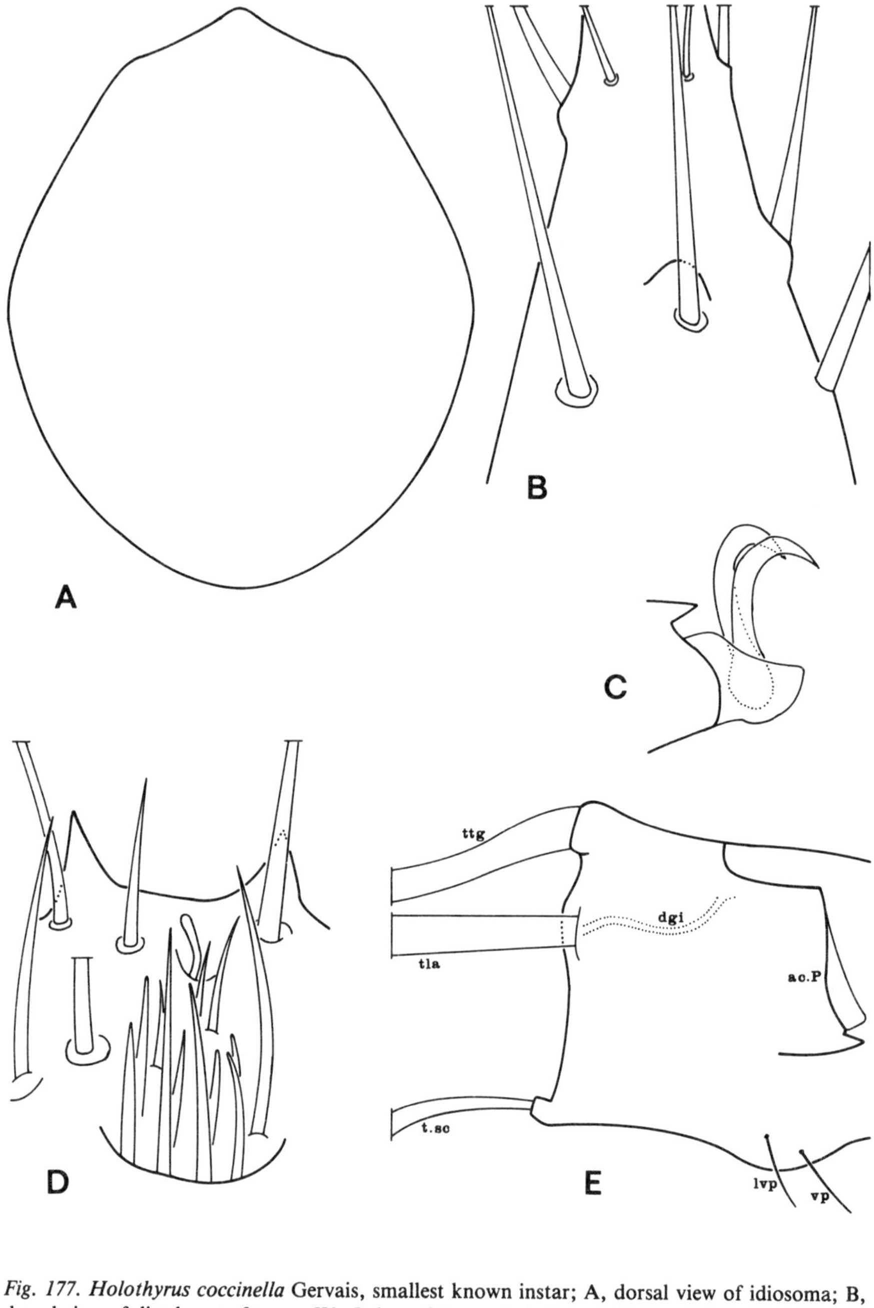

Fig. 177. Holothyrus coccinella Gervais, smallest known instar; A, dorsal view of idiosoma; B, dorsal view of distal part of tarsus IV; C, lateral (posterior) view of distal part of tarsus and ambulacrum of leg I; D, dorsal view of Haller's organ; E, lateral view of basal part of gnathosoma; A, × 42; B, D, × 550; C, × 295; E, × 141.

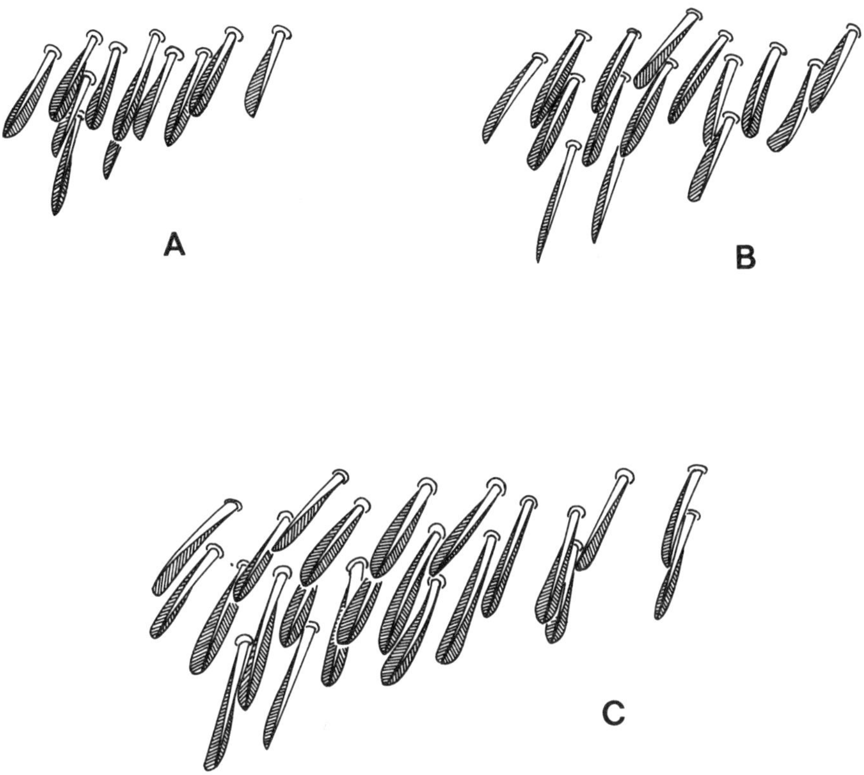

Fig. 178. Holothyrus coccinella Gervais, lateral (paraxial) view of brush of right palpal tibia; A, protonymph; B, deutonymph; C, tritonymph; A–C, × 295.

dorsal seta d_1 has a median position; d_0'' is strongly developed.

Postembryonic development. – The 41 immature specimens of the material studied by me, belong to four groups of distinctly different size. The idiosomatic lengths of these groups range from 1.7–1.8, 2.1–2.3, 2.6–2.8 and 3.4–4.4 respectively. Three groups can be distinguished by the number of setae in the paraxial brush of the palpal tibia (fig. 178). In the first and second size class, it presents 10–11 setae, in the third 15–16, and in the fourth 19–23 (in the adult, the number ranges from 25–33). I could find no morphological differences (besides size) between representatives of the first and second size class (I did, however, not make a detailed comparative study of leg chaetotaxy). The smallest instar could represent an octopod larva; the other instars certainly represent proto-, deuto- and tritonymph. The length of most of the tritonymphs ranges from 3.4–3.6 mm. The idiosoma of the smallest instar (fig. 177A) presents already the characteristic broad shape of the adult. Some characters of the gnathosoma can easily be studied in this instar: the duct of the infracapitular gland, and the presence of three pairs of gnathosomatic

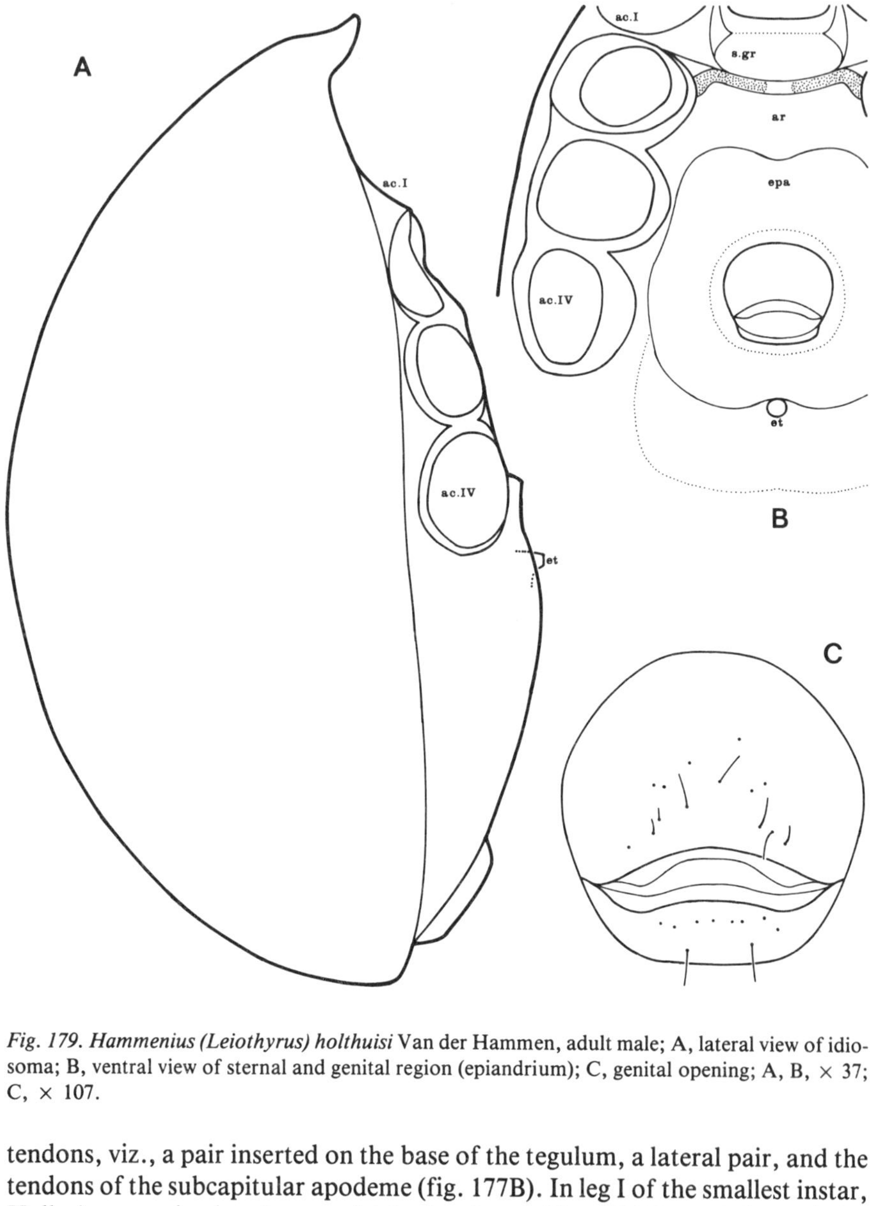

Fig. 179. Hammenius (Leiothyrus) holthuisi Van der Hammen, adult male; A, lateral view of idiosoma; B, ventral view of sternal and genital region (epiandrium); C, genital opening; A, B, × 37; C, × 107.

tendons, viz., a pair inserted on the base of the tegulum, a lateral pair, and the tendons of the subcapitular apodeme (fig. 177B). In leg I of the smallest instar, Haller's organ is already completely developed (fig. 177D), and the ambulacrum presents a reduced pulvillus and slender ungues (fig. 177C). In leg IV of the smallest instar, the number of dorsodistal teeth of the tarsus is reduced (fig. 177B): the lateral teeth are absent, and the median tooth is reduced to a rounded tubercle.

4.3. *Additional data*

Tegument. – In species of the genus *Hammenius* (fam. Holothyridae), the tegument is characterized by the presence of dark purple or violet subcuticular pigment grains (absent at the places of muscle attachments). Lyrifissures of normal shape, in the tegument of the dorsal shield, were described by Travé (1983: 334, fig. 1G); it will be interesting to reinvestigate the occurrence of idiosomatic lyrifissures in other species of Holothyridae.

Dorsal shield. – In *Neothyrus*, the position of the orifice of the posterior excretory gland is laterodorsal (Lehtinen, 1981: 4, 5, fig. 1; Lehtinen still indicated it as stigma 2); it is characterized by the presence of a protruding membraneous funnel. The stigma of *Neothyrus* is associated with an elongate peritreme (as in Holothyridae; the peritreme is much broader in Allothyridae).

Ventral shield. – The peridium, characteristic for the family Allothyridae, was also described and figured by me for *Allothyrus constrictus* (Domrow) (see Van der Hammen, 1968a: 263–265, fig. 1).

Genital region. – In species of the family Holothyridae, the shape of the male epiandrium is characteristic for the different species; in some species, it is characterized by the presence, in the posterior part, of a so-called epiandrial tooth (fig. 179A, B). There can also be differences in the shape of the male genital covers (fig. 179B, C).

Sternal region. – Paired sternapophyses are now known from *Allothyrus constrictus* (Domrow) only (see Van der Hammen, 1968a: 263, fig. 2A; and fig. 180A); they are not present in *Australothyrus*.

Gnathosoma. – The infracapitulum is holotrichous in Holothyridae only. The radula-shaped labrum, a characteristic attribute of species of the family Holothyridae, was represented by Van der Hammen (1961: 187–188, figs. 2, 4); it is associated with a pair of paralabral teeth (fig. 181).

Chelicera. – Two cheliceral setae are present in Allothyridae, one cheliceral seta in Holothyridae and Neothyridae. In Allothyridae, the jaws of the chela present two or three large teeth (and interjacent smaller teeth). In Holothyridae, the shape and distribution of the teeth of the chela is irregular (fig. 182B). In Neothyridae, the jaws present a distal and a median tooth, and interjacent smaller teeth.

Palp. – In *Hammenius (Leiothyrus) holthuisi* Van der Hammen, the coxal region of the palp and the palpal trochanter present associated apophyses (fig. 182A). In Holothyridae, the palpal genu is characterized by the presence of one or two fringed setae: one fringed seta in *Hammenius* (figs. 182D, 183D), two fringed setae in *Holothyrus*. In this family, the palpal tibia is characterized by the presence of a tibial brush; in *Holothyrus* this brush has the shape of a group of stiff setae, in *Hammenius* it has the shape of a comb (fig. 183A). In Allothyridae, the tibial brush is absent; in Neothyridae, it consists of a few setae only. The phanerotaxy of the palpal tarsus (there are, e.g., distal sensory setae and dorsal hollow phaneres arranged in a well-defined area) presents essential

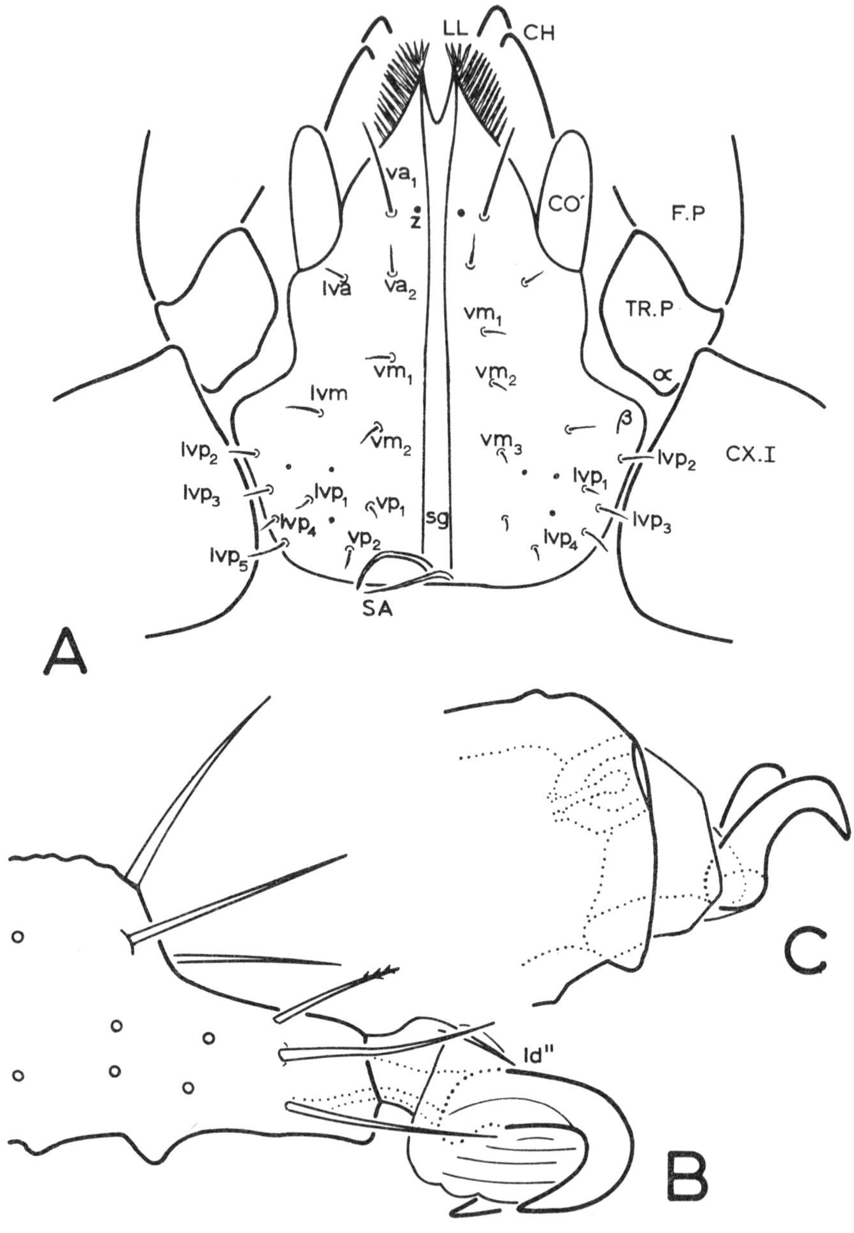

Fig. 180. Allothyrus constrictus (Domrow), female; A, ventral view of gnathosoma in camerostome; B, distal part of tarsus and apotele of one of the posterior legs, lateral view; C, lateral (antiaxial) view of distal part of right leg I (setae omitted); A, × 235; B, C, × 490.

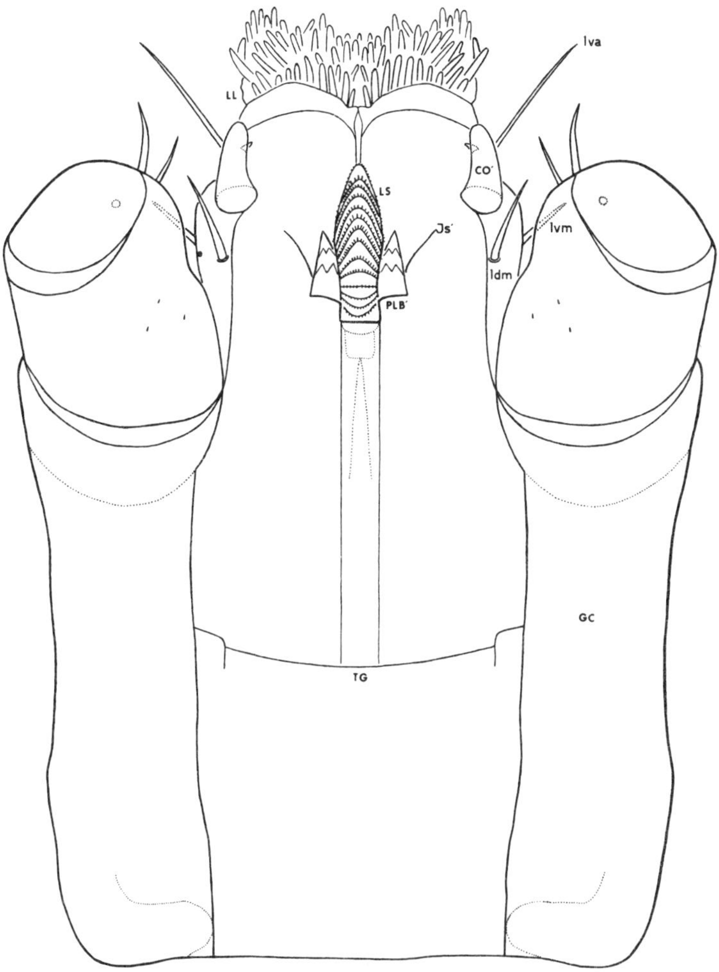

Fig. 181. Hammenius (Thonius) grandjeani (Van der Hammen), adult male; dorsal view of gnathosoma (chelicerae removed), × 125.

similarities (and differences) in the three families (cf. fig. 183A–C).

Legs. – The basifemoral ring, the basitarsal ring and the ambulacrum of *Hammenius (Thonius) grandjeani* were described and figured by Van der Hammen (1961: 191–193, figs. 5b, 7, 9). The basitarsal ring and the ambulacrum of this species are also represented in figs. 184–185. The terminal part of tarsus I of *Allothyrus constrictus* (Domrow) was described and figured by Van der Hammen (1968a: 267–268, fig. 2B, C; see also fig. 180B, C). Secondary sexual characters in the legs of species of Holothyridae (presence of teeth, differences in shape and size of ungues and pulvillus) were also studied by Van der Hammen (1983b: 35, 44, figs. 20D, 21B–D, 22, 27C, E).

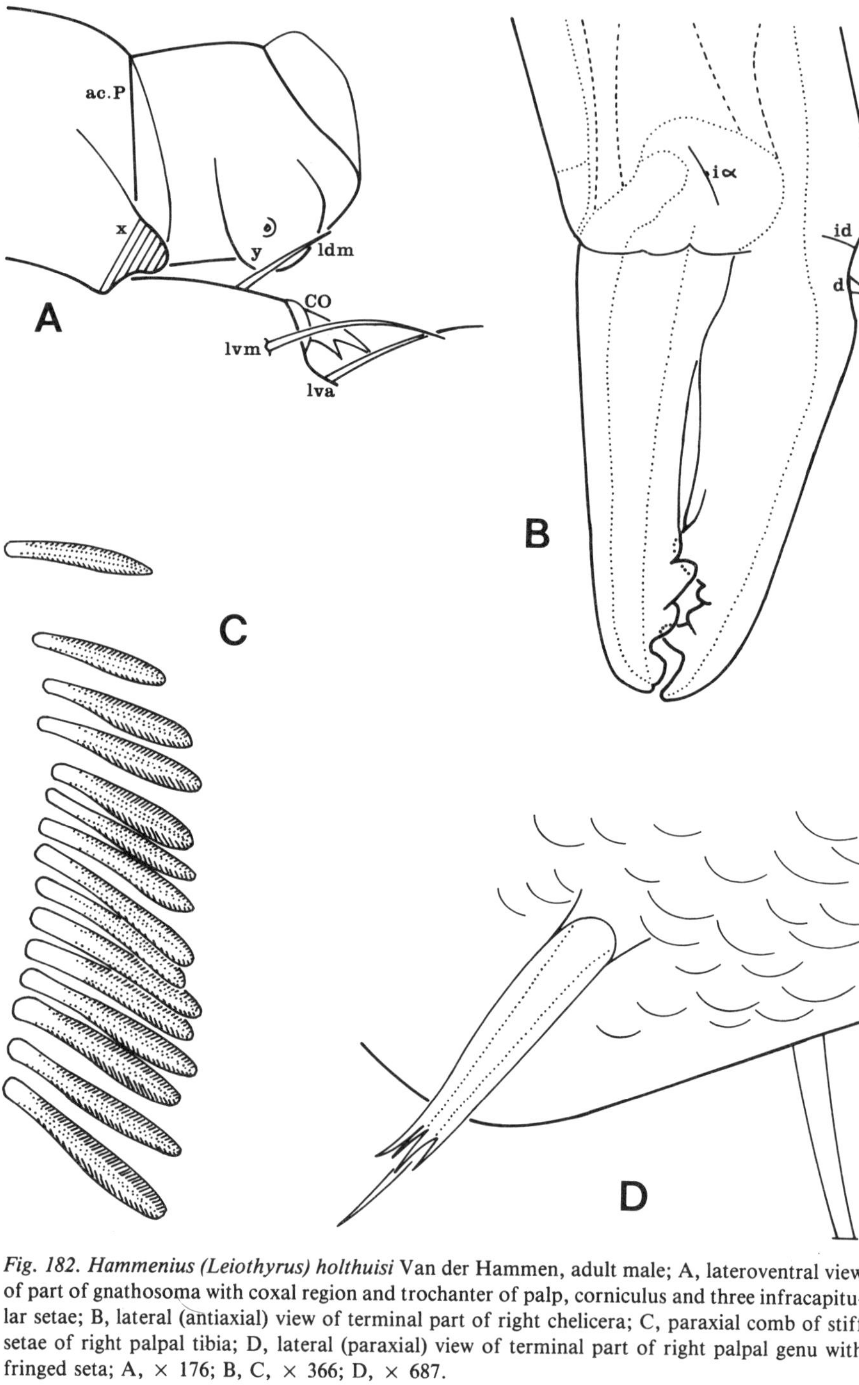

Fig. 182. Hammenius (Leiothyrus) holthuisi Van der Hammen, adult male; A, lateroventral view of part of gnathosoma with coxal region and trochanter of palp, corniculus and three infracapitular setae; B, lateral (antiaxial) view of terminal part of right chelicera; C, paraxial comb of stiff setae of right palpal tibia; D, lateral (paraxial) view of terminal part of right palpal genu with fringed seta; A, × 176; B, C, × 366; D, × 687.

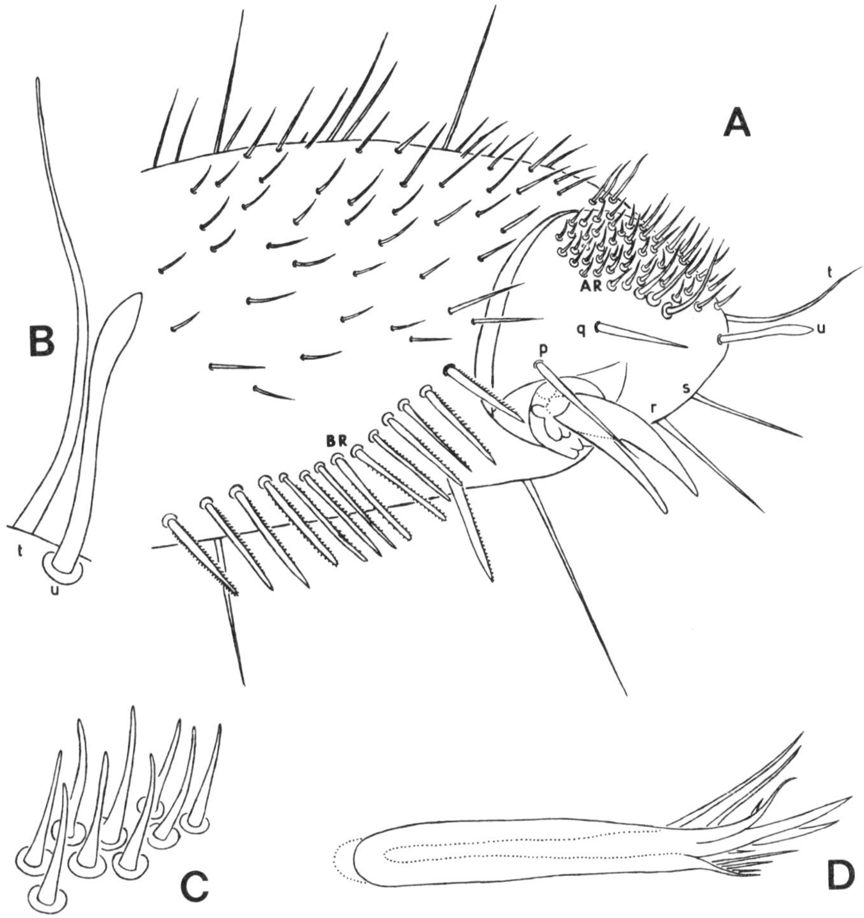

Fig. 183. Hammenius (Thonius) grandjeani (Van der Hammen), adult male; A, distal part of tibia, tarsus and apotele of left palp, lateral (paraxial) view; B, distal sensory setae of palpal tarsus; C, part of the paralaterodorsal area with hollow setae of the palpal tarsus; D, fringed seta of palpal genu; A, × 190; B–D, × 450.

List of the notations of figs. 152–185

ac.P	acetabulum of palp.
ac.I	acetabulum of leg I.
ac.II	acetabulum of leg II.
ac.III, Ac.III	acetabulum of leg III.
ac.IV	acetabulum of leg IV.
AN	anal covers.
AP	apotele.

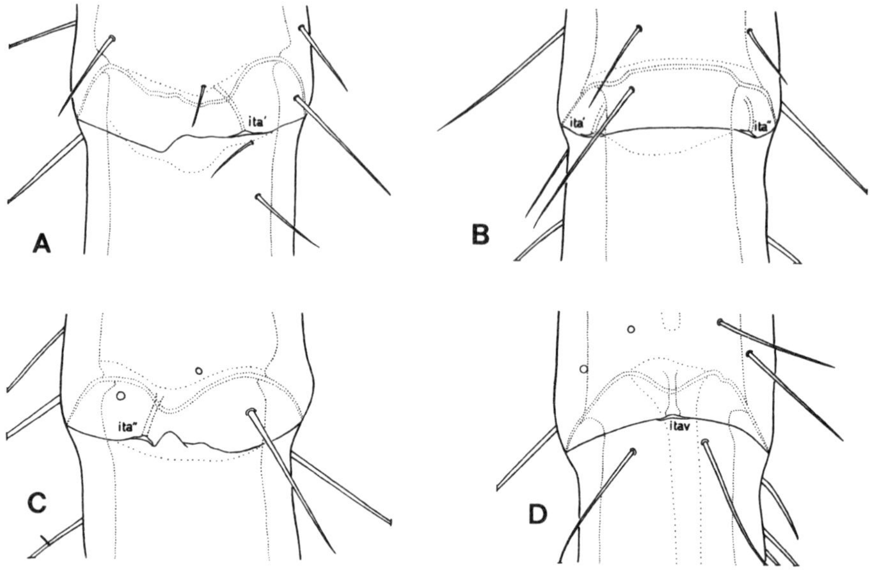

Fig. 184. Hammenius (Thonius) grandjeani (Van der Hammen), adult male, part of left tarsus III with adesmatic articulation (basitarsal ring) and associated lyrifissures; A, lateral (antiaxial) view; B, dorsal view; C, lateral (paraxial) view; D, ventral view; A–D, × 125.

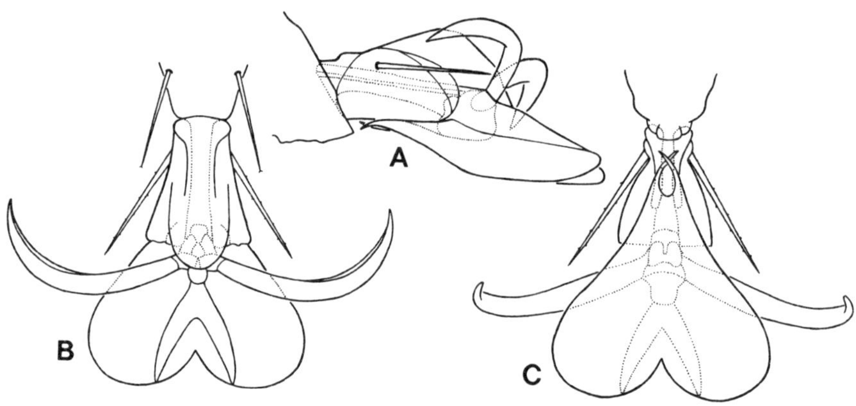

Fig. 185. Hammenius (Thonius) grandjeani (Van der Hammen), ambulacrum of adult male; A, lateral view; B, dorsal view; C, ventral view; A–C, × 125.

ap.ce	cervical apodeme.
ap.sc	subcapitular apodeme.
ar	sternal area posterior to transverse sternal groove.
AR	paralaterodorsal area with hollow setae of palpal tarsus.
at	line of attachment of cheliceral frame to infracapitulum.

B, B_1	proximal, resp. distal section of an internal median ridge of the ventral infracapitular wall.
BC	body of chelicera.
bds	border of dorsal shield.
BR	paraxial brush of palpal tibia.
bt	anterior border of tegulum.
bvs	border of ventral shield.
CAM	camerostome.
CC	coxal cavity of infracapitulum.
CE	cervix.
CH	chelicera.
ch_1, ch_2'	cheliceral setae.
cl_{1-3}	ungues of palpal apotele.
CO, CO'	corniculi.
d	dorsal seta.
d_{0-4}, d_z	setae surrounding Haller's organ.
dgi	duct of infracapitular gland.
DS	dorsal shield.
epa	epiandrium.
et	epiandrial tooth.
F	femur.
FC	cheliceral groove of infracapitulum.
F.P	femur of palp.
G	gena.
GC	large lateral ridge of infracapitulum.
GE	genu.
GEN	genital covers.
gl	supposed orifice of gland in palpal trochanter.
H	mentum
i	lyrifissure, cupuloid.
i_1	anterior lyrifissure of dorsal shield.
id	dorsal lyrifissure of chelicera.
if''	posterior compound lyrifissure associated with basifemoral ring.
ita', *ita''*	lateral lyrifissures associated with basitarsal ring.
itav	ventral lyrifissure associated with basitarsal ring.
$i\alpha$	antiaxial lyrifissure of chelicera.
Ji	inferior commissure of mouth.
Js, Js'	superior commissures of mouth.
LA	labellum.
LAG	female laterogenital shield.
LC	lacinia.
ld''	posterior laterodorsal seta (of pretarsus).
ldm, ldm_{1-2}	laterodorsal infracapitular setae.
LG	labial gutter.
LL,	lateral lips.

LS	labrum.
lva	anterior lateroventral infracapitular seta.
lvm	median lateroventral infracapitular seta.
lvp, lvp_{1-5}	posterior lateroventral infracapitular setae.
mb	arthrodial membrane of gnathosoma.
OC	eye.
ogi	orifice of infracapitular gland.
ogl_{1-2}	orifices of supposed excretory glands.
OH	Haller's organ.
otg	orifice of tegumental gland (pore).
OVP	ovipositor.
p	seta of palpal tarsus.
pe	peridium.
PH	pharynx.
pl.b	plicature band.
PLB, PLB'	paralabral teeth.
POG	postgenital shield of female.
PRG	pregenital shield of female.
q	seta of palpal tarsus.
r	seta of palpal tarsus.
ro	rostrum.
s	seta of palpal tarsus.
SA	sternapophyses.
sb	line of greatest depth of sternal groove.
s.ca	sternal cavity.
se	capitular saddle.
sg	subcapitular gutter.
s.gr	transverse sternal groove.
sr	anterior border of elevated sternal area.
st	stigma.
t	seta of palpal tarsus.
TA	tarsus.
TG	tegulum.
TI	tibia.
tla	lateral tendon of gnathosoma.
tlo	lobe of palpal tarsus.
tls	tendon of labrum.
TR	trochanter.
TR.P	trochanter of palp.
t.sc	tendon of subcapitular apodeme.
ttg	tendon inserted on base of tegulum.
t.1	taenidium 1.
t.2	taenidium 2.
u	distal lanceolate sensory seta of palpal tarsus.
va_{1-2}	anterior ventral infracapitular setae.

vs	ventral idiosomatic seta.
vm, vm_{1-2}	median ventral infracapitular setae.
vp, vp_{1-2}	posterior ventral infracapitular setae.
VS	ventral shield.
x	apophysis of coxal region of palp.
y	apophysis of palpal trochanter.
z	supposed vestige of infracapitular seta.
α (alpha)	apophysis of palpal trochanter.
β (beta)	apophysis of infracapitulum.
ζ_{1-2} (zeta)	hollow sensory setae (supposed eupathidia) of Haller's organ.
ω_{1-9}, ω_y', ω_z (omega)	hollow sensory phaneres (supposed solenidia) of Haller's organ.
I–IV	acetabula of legs I–IV.
′ (prime)	anterior face of an appendage; the right one of two symmetrical structures.
″ (double prime)	posterior face of an appendage.

5. Gamasida

The few papers in which I have dealt with Gamasid morphology are listed at the beginning of chapter V. In these papers, detailed studies are made of one species of Gamasida, viz., *Glyptholaspis confusa* (Foà). They are summarized below as an example of the morphology of the order.

The diagnostic characters of the order Gamasida are the following. The subdivision of the body into prosoma and opisthosoma is generally completely lost (a dorsodisjugal furrow, with advanced position, is perhaps still present in *Rhodacarus*). The dorsal surface of the idiosoma presents generally one or more shields. There are no eyes. There is one pair of stigmata, with lateral position, surrounded by a peritreme. The paired anal valves present each one seta at the most. The genital opening of the female is closed by one, three or four shields, that of the male by two shields. The sternal area (a neosternum) can present idionymous setae and lyrifissures. As a rule, an unpaired sternapophysis is present (absent in highly specialized parasitic species). There are up to four pairs of infracapitular setae (among which no laterodorsals) and one pair of corniculi (hypertrophied infracapitular setae). Generally, the tegulum is produced into a well-developed supracheliceral limbus. The cross-section of the pharynx is triangular or quadrangular. As a rule, the palp consists of five free segments (fusion of segments is sometimes found) and an apotele with paraxial position at the base of the palpal tarsus (the apotelic claw is mostly bi- or tripartite). The legs generally consist of seven segments (including the apotele). Ambulacrum ancestrally with two ungues and a pulvillus; the ambulacrum of leg I can have been subject to regression. Generally, the postembryonic development comprises four stases: larva, protonymph, deutonymph and adult.

The order Gamasida includes numerous species, presenting a wide variety of habitus. The geographic distribution is world-wide. Many species are parasitic. The present section includes only one subsection: the detailed description of the female of a species of Macrochelidae.

5.1. *The morphology of the female of* Glyptholaspis confusa *(Foà)*

The present description is based on material from manure, collected in Oregon, U.S.A.. It deals with general structure, and specific characters which have been described in full detail by other authors (such as the microsculpture of the shields) are not especially emphasized. I refer to the description published by Evans & Browning (1956: 36, sub *Macrocheles plumiventris*; only the description of the female refers to the present species), and particularly to the paper by Filipponi & Pegazzano (1960: 154). Special attention is paid to those characters which are generally omitted in the description of species.

Measurements. – According to my material, the length of the female idiosoma is 1.175–1.485 mm, the breadth 0.740–1.050 mm. Detailed data with reference to measurements are published by Filipponi & Pegazzano (1960: 159); the material of these authors, however, presents less variety than my specimens.

Habitus and colour. – The species is a large Gamasid mite. The greatest width of the idiosoma is in the posterior part, at about 2/3 of its total length. The dorsal shield is entire. The large shields of the idiosoma present a conspicuous reticulate microsculpture. The colour of the sclerotized parts is brown; the soft skin is whitish. The greater part of the setae are more or less clavate.

The legs are relatively long. Legs I and II are directed more or less forward, leg III sideward, and leg IV backward. Leg IV is the longest. Leg I, inserted in a region of soft skin, is more slender than the other legs; it bears no ambulacrum.

Cerotegument. – When studied in a dry condition (on a carbon block), the cerotegument appears to be granular and whitish. It occurs in irregular masses. Concentrations are found on the tegulum, the ventral surface, and between the coxae of the legs. Much less, if anything, is found on the dorsal shield and the legs.

Cuticle. – The idiosoma of the adult female presents the following sclerites (cf. figs. 186, 187, 188A): dorsal, sternal, metasternals, genital, paragenitals (the so-called accessory sclerites), ventri-anal, peritrematals (fused anteriorly with the dorsal shield), holopodals (a fusion of exo- and endopodal shields), and metapodals. In lateral view (fig. 188A) two other small shields are visible: posterior to the peritrematal shields, and posterior to the metapodal shield. The borders of some of the large shields are covered by soft skin. Several of the smaller shields are completely covered by soft skin; the two pairs of laterals,

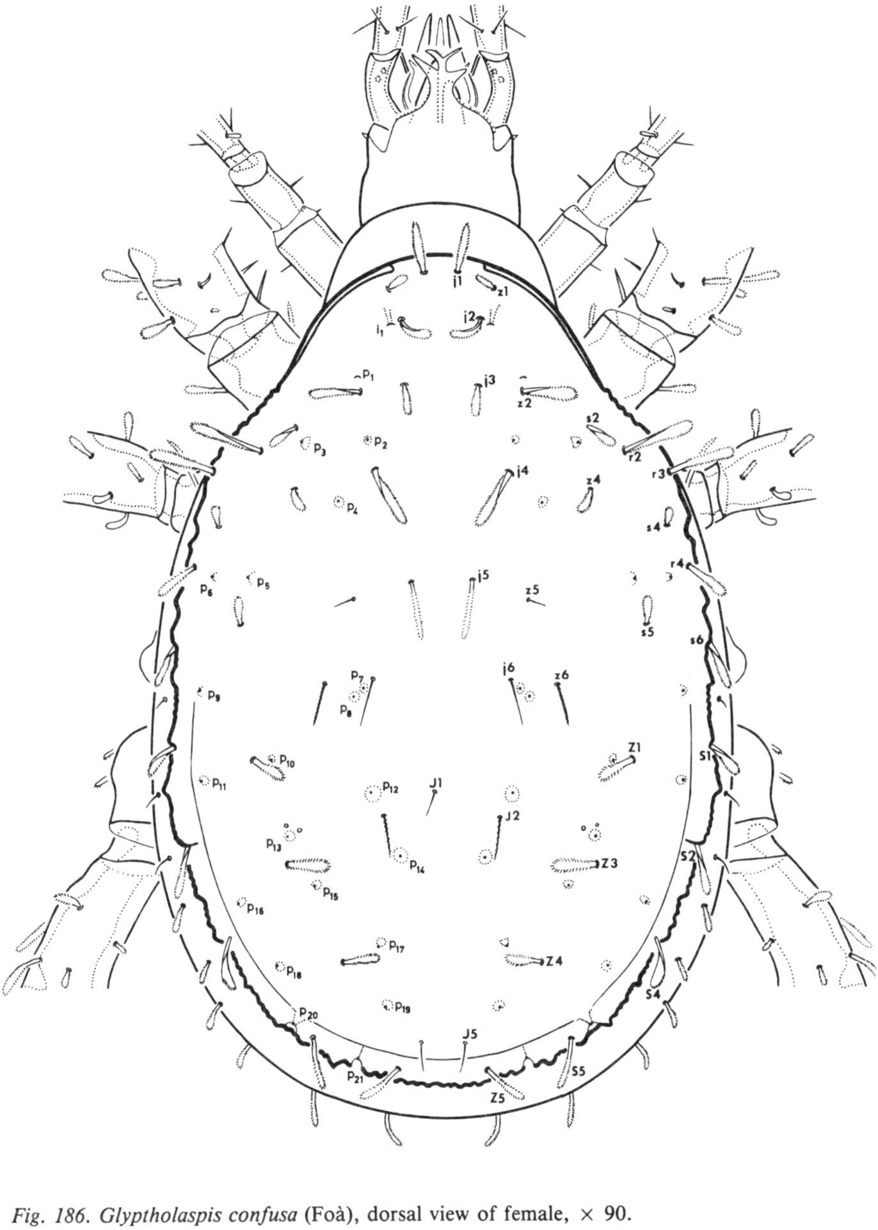

Fig. 186. Glyptholaspis confusa (Foà), dorsal view of female, × 90.

the metasternals, and the metapodals represent these covered shields.

The microsculpture of the shields is described and figured (particularly by photographs) by various authors, e.g., by Filipponi & Pegazzano (1960). Details of the microsculpture of the dorsal shield are represented here in fig.

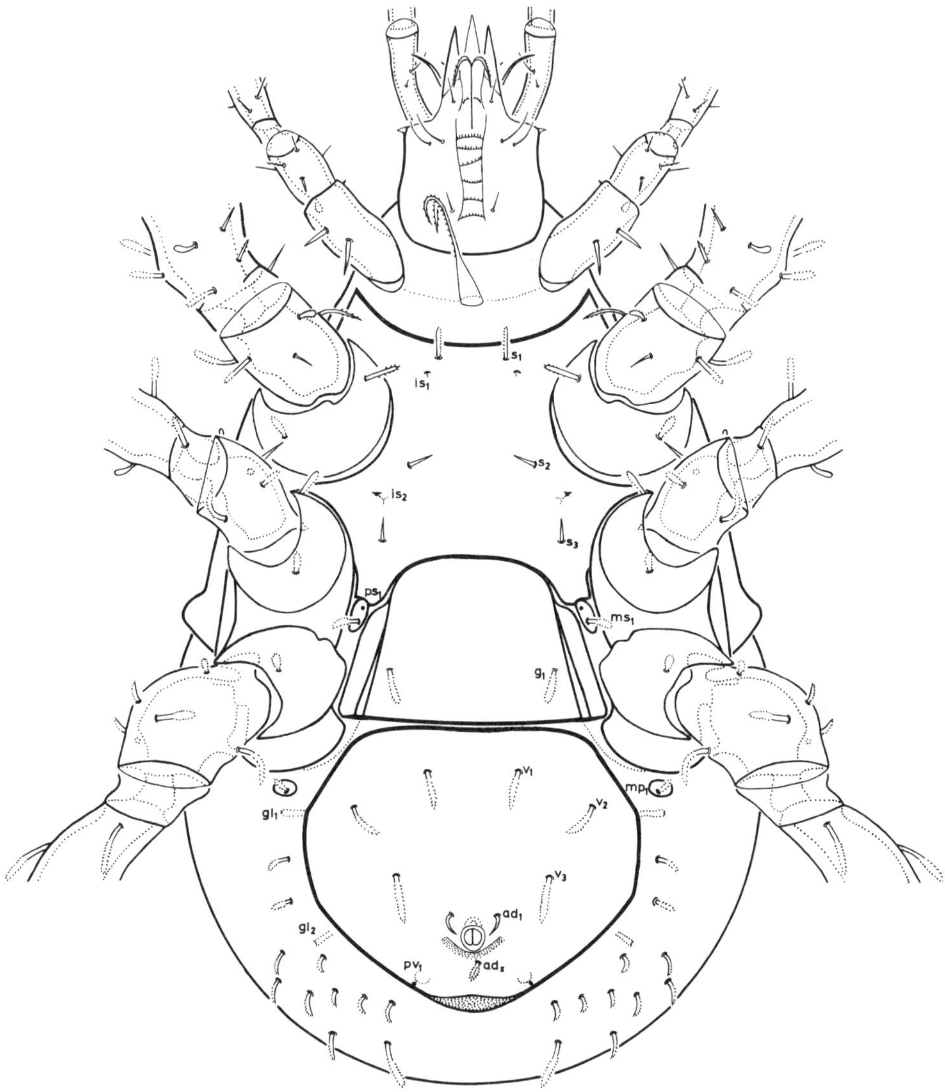

Fig. 187. Glyptholaspis confusa (Foà), ventral view of female; × 90.

189A, D. The reticulations (and smaller elements) of the microsculpture are positive, i.e., the mesh-work and the granules are slightly projecting.

In fig. 189C, I have represented a transverse section through a part of the dorsal shield. The figure shows that the shield continues laterally under a wrinkled soft skin. It appears that the uncovered sclerotized part consists of two layers; the upper layer is not present in the covered part.

Lyrifissures and orifices of glands. – The cuticle of idiosoma and appendages presents a rather great number of lyrifissures (or their homologues) and orifices of glands. Typical lyrifissures are present in the cuticle of the appendages, e.g., associated with basifemoral and basitarsal ring; they are dealt with below, in the paragraphs on chelicera, palp and legs. A transverse section through a basitarsal ring, showing three lyrifissures, is represented in fig. 189B. The dorsal shield presents an anterior lyrifissure near seta *j2* (fig. 189E); a transverse section through this region (fig. 189F) demonstrates that this organ indeed represents a lyrifissure. The sternal shield presents two pairs of lyrifissures. The dorsal shield, and the lateral and ventral parts of the idiosoma further show a rather great number of poroid structures. Transverse sections through two poroids of the idiosoma are represented in fig. 189G, H. They have an interesting structure. A tube runs from the epiostracum (often from a small depression in the surface of the shield) through the upper layer of the ectostracum, and debouches into a more or less bell-shaped court, a space in the deeper layer of the cuticle. The greater part of these poroid structures are homologues of lyrifissures (they can be replaced by lyrifissures in other species of Gamasida). There are, however, also a number of orifices of tegumental glands. In literature (including my 1964 paper) poroid structures and orifices of tegumental glands are, generally, not clearly distinguished (differences in structure and arrangement are dealt with by Athias-Henriot, 1969, 1969a). In my 1964 paper, it was already mentioned that, in other species of Gamasida, lyrifissures can take the place of the poroid structures of *Glyptholaspis*. In the present subsection, my 1964 observations are repeated.

Idiosoma. – The idiosoma is rather high and oval in outline; the greatest width is at the level of coxa IV. It is represented in different orientations in figs. 186, 187, 188A. The dorsal shield occupies the greater part of the dorsal surfaces. In dorsal view, the peritrematal shield appears to project at the level of the stigma. In lateral and ventral views, it appears that important lateral, posteroventral and posterior regions are not sclerotized. A part of the idiosoma, anterior to the dorsal shield, is also not sclerotized; the coxa of leg I is inserted in the soft skin of this area (not in the holopodal shield).

Dorsal shield. – There are 28 pairs of setae and 1 (according to literature sometimes 2) unpaired on the dorsal shield; they have received the current notations (see, e.g., those applied to *Macrocheles* by Norton, 1973). The dorsal shield presents one pair of typical lyrifissures (i_1) and 21 pairs of poroid structures (p_{1-21}). Lyrifissure i_1 is situated lateral to *j2*. The disposition of setae and poroid structures is represented in fig. 186. The setae belong to three types: they are (1) simple, pointed and smooth, or (2) simple, pointed and slightly ciliate, or (3) plumose. The exact shape of the plumose setae is represented in fig. 189A. Setae *z5, j6, J1* (unpaired) and *J5* are simple; *z6* and *J2* are slightly ciliate; the remaining setae of the dorsal shield are plumose. The position (direction) of the dorsal setae is easily visible in lateral view (fig. 188A). The external lateral border of the dorsal shield is more or less crenulate. Under the

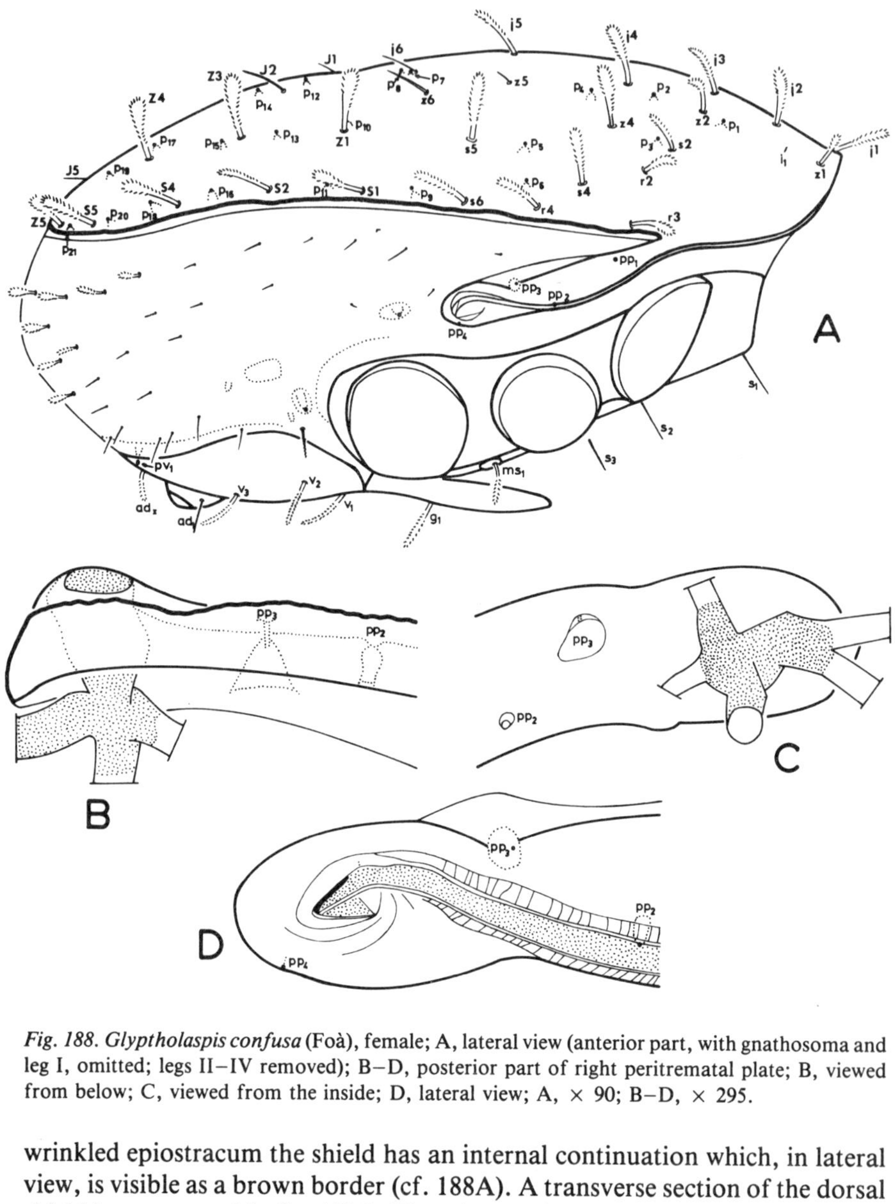

Fig. 188. Glyptholaspis confusa (Foà), female; A, lateral view (anterior part, with gnathosoma and leg I, omitted; legs II–IV removed); B–D, posterior part of right peritrematal plate; B, viewed from below; C, viewed from the inside; D, lateral view; A, × 90; B–D, × 295.

wrinkled epiostracum the shield has an internal continuation which, in lateral view, is visible as a brown border (cf. 188A). A transverse section of the dorsal shield is represented in fig. 189C. The microsculpture of the shield is reticulate; granulations are present inside the mesh-work (fig. 189A, D).

Ventral surface (fig. 187). – The ventral shields have partly fused. Exo- and endopodal shields are integrated into a holopodal shield. The holopodal shield has partly fused with the sternal shield, viz., anterior to acetabulum II (*x*) and anterior to acetabulum III (*y*). There is also a pair of paragenital shields (generally called accessory sclerites). The remaining shields of the ventral surface are free (not fused), viz., the metasternals, the genital, the metapodals, and the ventrianal. The sternal shield presents two pairs of lyrifissures (is_1, is_2), and three pairs of setae (s_{1-3}). The metasternal shield is covered by wrinkled soft skin; it presents one poroid structure (ps_1) and one seta (ms_1). The genital shield presents one pair of setae (g_1). The metapodal shield is covered by soft skin; it presents one seta (mp_1). The ventrianal shield presents four pairs of setae and one unpaired seta; three pairs are regarded here as ventrals (v_{1-3}), the remaining setae as adanals (ad_1, ad_x). There is one pair of poroid structures (pv_1) in the ventrianal shield, near the posterior border. The anal opening is situated on a projecting part of the ventrianal shield (cf. fig. 188A). The punctate parts of the shield (on the declivity behind the anal opening, and at the posterior border of the shield) are probably caused by cerotegument (possibly together with a general porosity of the cuticle). The soft skin between legs I bears, in the median part, the unpaired sternapophysis, which is bifurcate; it is associated with the subcapitular gutter of the gnathosoma.

Lateral aspect. – In lateral view (fig. 188A) the peritrematal shield, some small lateral shields, the holopodal shield, and the greater part of the regions of soft skin can be studied. The peritrematal shield has fused, in its anterior part, with the dorsal shield. It contains the peritreme, four poroid structures (probably tegumental glands and their orifices), and the stigma. The peritreme (fig. 183D) is an open groove (a taenidium). It extends as far as the posterior border of the stigma; at this point, a slit in the bottom of the groove appears to be connected with the stigma; the bottom of the peritreme as well as the entrance to the stigma are both punctate. The peritrematal shield is distinctly projecting laterally, at the place of the stigma. As mentioned above, there are four poroid structures (pp_{1-4}). Poroid structure pp_3 has a relatively very large court with irregular outline; pp_2 is at the border of the peritreme; pp_4 has a narrow canal. Poroid structures pp_2 and pp_3 are represented in various positions in fig. 188B–D. When the peritrematal shield is separated from the idiosoma, the tracheal trunk can be studied; it subdivides into four branches, the posterior of which subdivides again into two branches. The surface of the tracheal trunk and the bases of the primary subdivisions are brown and punctate (fig. 188B, C). As mentioned above, exo- and endopodal shields have fused and constitute a holopodal shield. This can easily be observed after separation of the legs. In fig. 188A, all legs have been removed. Posterior to the peritrematal shield, a small covered shield with one poroid structure is found. Posterior to the holopodal shield, two other small covered lateral shields are found; one of these presents a relatively large poroid structure. The soft skin, particularly visible in lateral orientation, bears many setae, mostly more or less plumose or ciliate. There are two orifices (gl_1, gl_2) lateral to the ventrianal shield; they could represent orifices of glands.

Gnathosoma (figs. 190–198). – The gnathosoma of *Glyptholaspis confusa* is tube-shaped; a transverse section is nearly circular. In order to understand

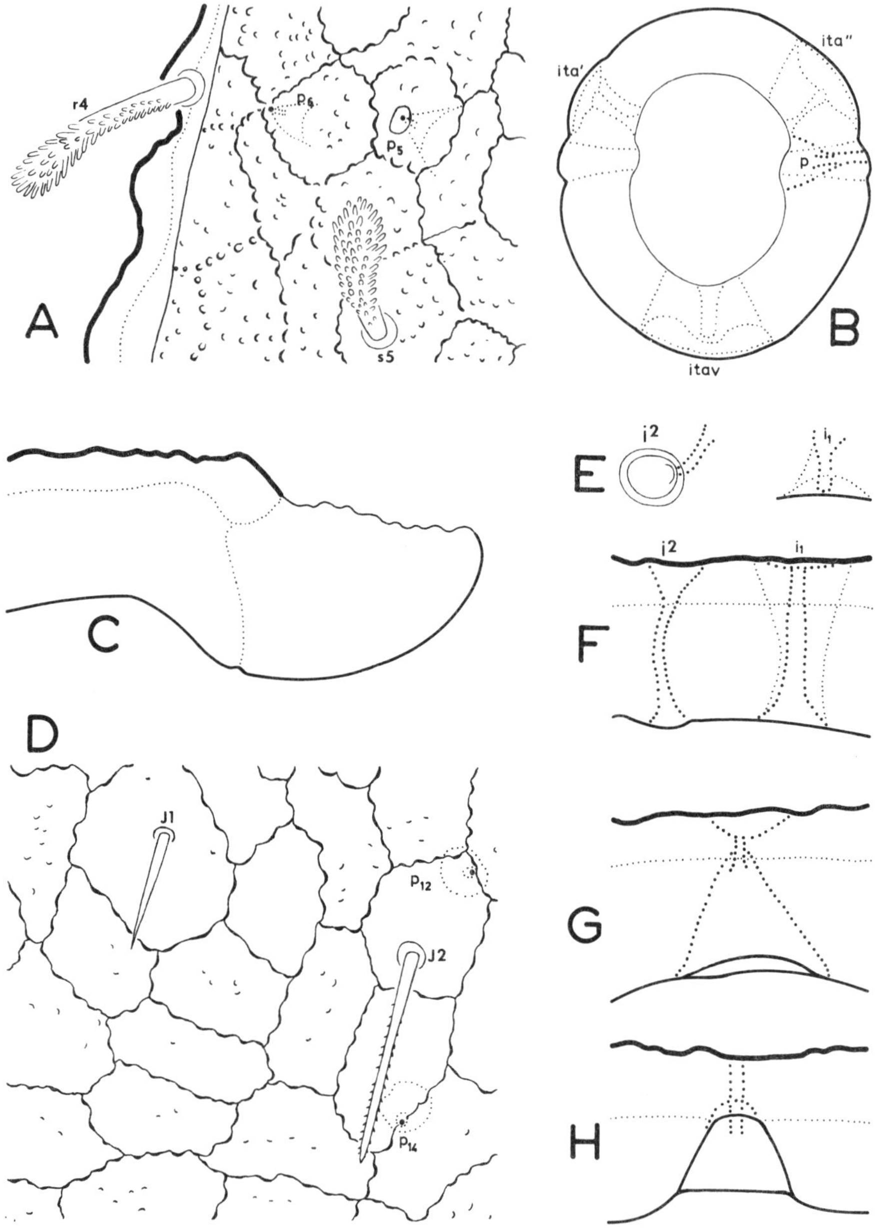

Fig. 189. Glyptholaspis confusa (Foà), female; A, fragment of lateral part of dorsal shield; B, transverse section through basitarsal ring of left leg III, showing three lyrifissures and a pore; C, transverse section through dorsal shield, showing the lateral continuation of the shield; D, central part of dorsal shield; E, lyrifissure i_1 and place of insertion of seta j_2 of the dorsal shield; F, transverse section through region of fig. E; G, transverse section through dorsal shield, just beside one of the large central pores; H, section through a posterior pore (p_{19}) of the dorsal shield; A, D, × 370; B, C, E–H, × 710.

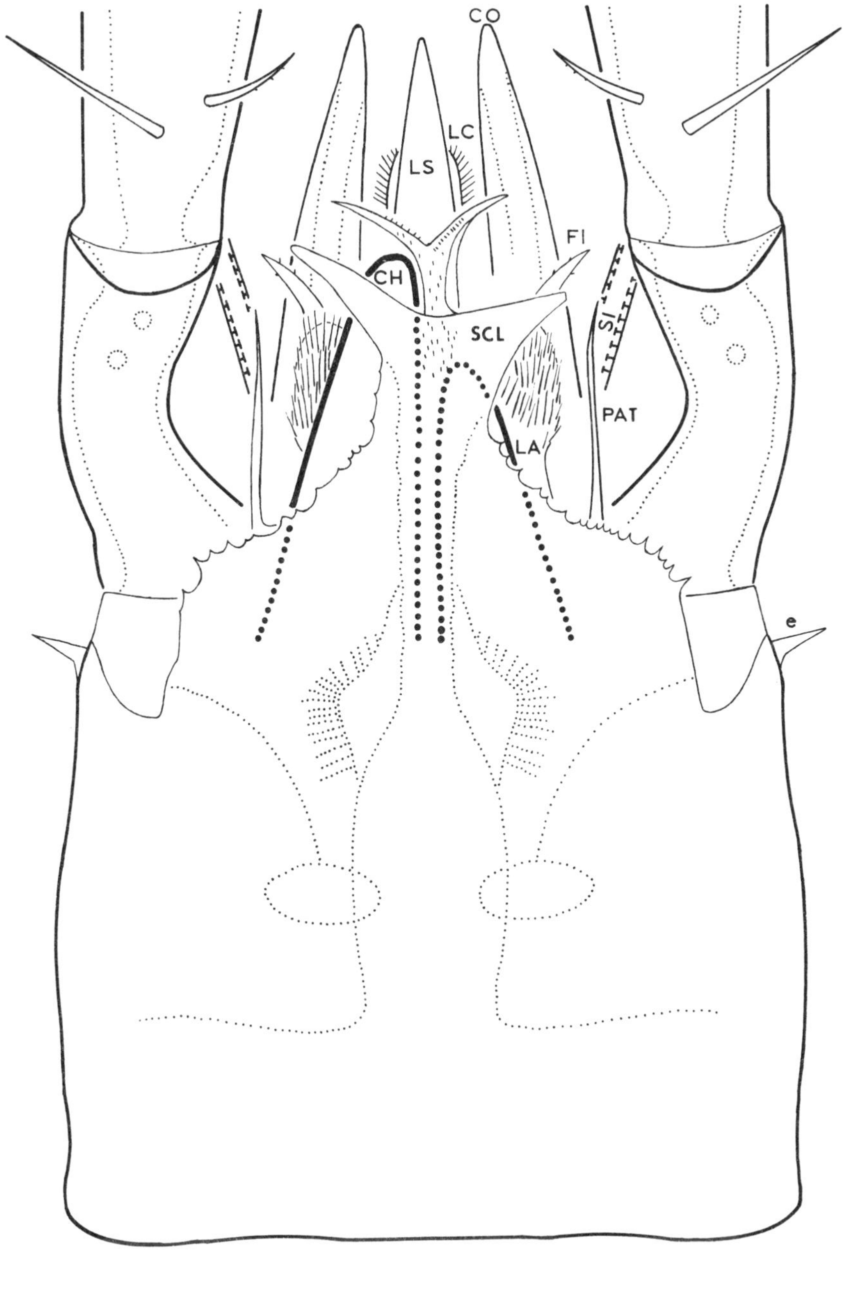

Fig. 190. Glyptholaspis confusa (Foà), female; dorsal view of gnathosoma; × 440.

its complicated structure, it is necessary to study it in dorsal, ventral and lateral view, as well as in frontal and posterior view. After these observations, it is also necessary to remove several parts, and to study transverse as well as longitudinal sections.

A dorsal view of the entire gnathosoma (fig. 190) presents the following characters. The supracheliceral limbus (*SCL*) which is usually characterized as tripartite or trifid, has an irregularly dentate anterior border which, in the median part, is more or less asymmetrically produced into a tripartite extension, of which the median part is bifurcate at the end. A ligament is attached to one of the terminal branches. Very small hairs can be observed in the anterior median part. Under the surface the structure of the limbus is more or less striate. The supracheliceral limbus covers the chelicerae (*CH*) when these are retracted.

At the lateral border of the infracapitulum, near the place of attachment of the supracheliceral limbus, and not far from the acetabulum of the palp, a seta with reduced core is present. It is whitish and pointed, and presents a broadened base. It is supposed to represent a laterocoxal seta (*e*). A similar seta is present in the case of coxa I; this seta has the same position as the mucronate seta with reduced core of coxae I and II of *Opilioacarus*, homologized by Grandjean (1936b: 430, fig. 4) with laterocoxal setae.

The dorsal face of the infracapitulum, which is more or less hidden from view by the supracheliceral limbus and the chelicerae, presents an extremely complicated structure; the elements of this structure can be studied after removal of chelicerae and limbus. In the anterior part, the corniculi (*CO*) (sometimes called external malae) are inserted; they are long and conspicuous. Between the corniculi, a long, tapering labrum (*LS*) is also easily visible (this structure is often called epipharynx). At a lower level, partly under the labrum, a pair of pilose processes is present; these are called here laciniae (*LC*) (in literature they are also indicated as internal malae, hypopharyngeal or hypostomal processes, etc.). Lateral to the corniculi, two paired structures are visible that are generally called salivary styli and duplicationes stylorum. The salivary styli represent the free parts of efferent ducts of the paired infracapitular glands which are situated in the idiosoma. The free parts of the chitinous ducts are enveloped in soft skin. In 1964, I used the term siphunculus in reference to the salivary styli (Van der Hammen, 1964a: 14). According to Vitzthum (1940), the duplicationes stylorum are the result of a basal bifurcation of the salivary styli. They can, in fact, be characterized as protective elements that arise just above and slightly lateral to the base of the styli, apparently from the soft skin. Because of their position, I introduced the name patrocinium (*PA*) for these structures (Van der Hammen, 1964a: 14).

At the same level as the patrocinia, but paraxial to these, a pair of flaps beset with hairs is partly visible. These are apparently the structures called paralabra interna by Vitzthum (1940), paralabra externa by T.E. Hughes (1949), and hypopharyngeal styli by Gorirossi (1955). In 1964, I introduced for these structures the name labellum (*LA*), which is more appropriate as will appear from

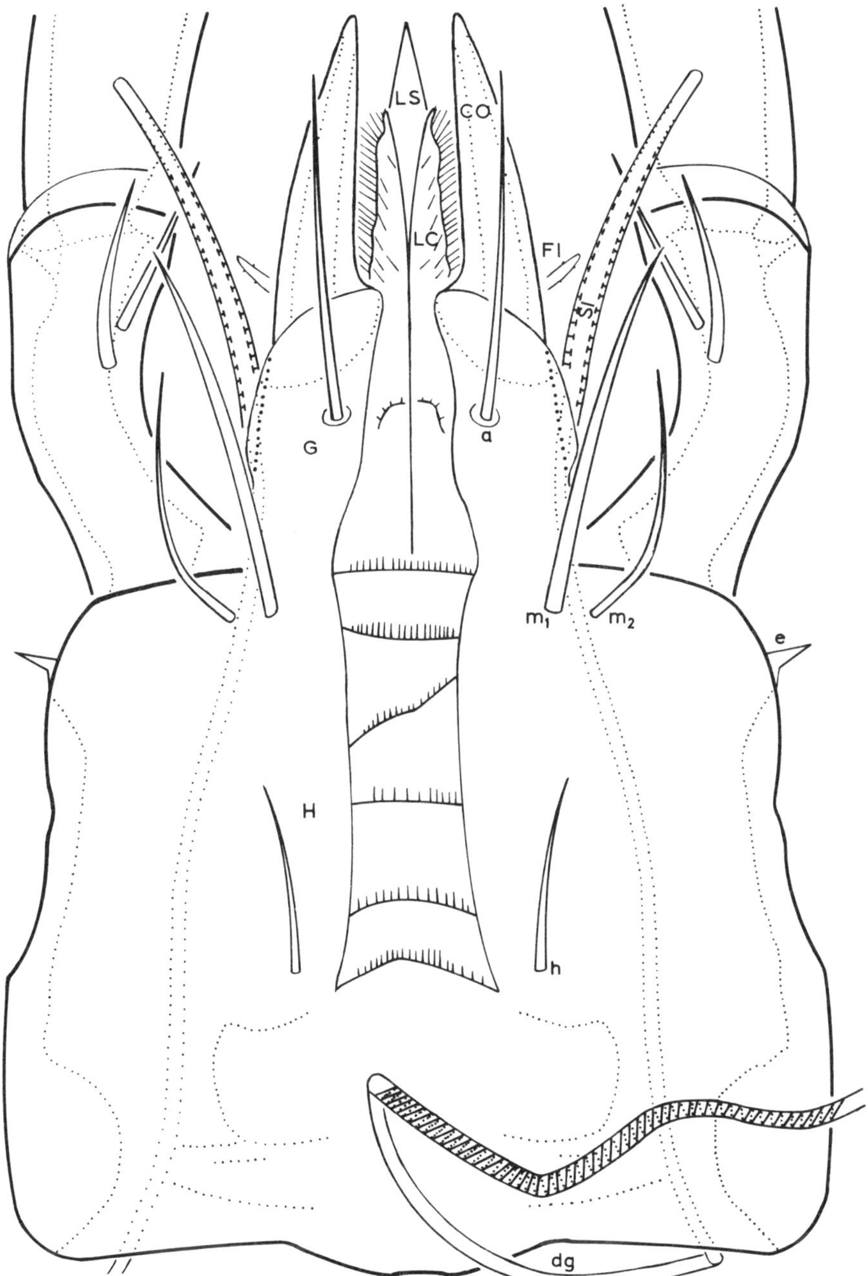

Fig. 191. Glyptholaspis confusa (Foà), female; ventral view of gnathosoma; × 440.

the study of sections (Van der Hammen, 1964a: 14). Anterior to the labellum, a thin, pointed structure rises from below, which was termed fimbrilla (*Fl*) (Van der Hammen, 1964a: 14). I do not know if this structure is identical with that called hypopharyngeal process by Gorirossi & Wharton (1953). The exact topography of all these structures will gradually become clear from a study of the various aspects of the gnathosoma.

A ventral view of the gnathosoma (fig. 191) presents the following characters. The infracapitulum is anarthric: mentum (*H*) and genae (*G*) are not separated by a labiogenal articulation. The usual four pairs of ventral infracapitular setae are present: h, m_1, m_2 and a. These setae were called respectively: capitular seta (c.s.), external and posterior rostrals (ext. p.r., int. p.r.) and rostral seta (r.) by Evans (1957), and C_4, C_3, C_2 and C_1 by Hirschmann (1959).

The median part of the ventral infracapitular surface presents a rather large subcapitular gutter (named protosternum + deutosternum by Gorirossi & Wharton, 1953; capitular groove by Evans, 1957; and Hypostom by Hirschmann, 1959). The gutter extends from the posterior part of the mentum (behind the infracapitular setae *h*) to the anterior border of the genae. In literature, the gutter is often regarded as a less sclerotized sternal element separating the enlarged coxae of the palps. The subcapitular gutter represents, in fact, a highly specialized structure (a taenidium), associated with the sternapophyses. In the case of *Glyptholaspis*, the gutter presents six transverse ridges (one of them is oblique), beset with small teeth, and a pair of smaller curved ridges between setae *a*. The anterior part of the subcapitular gutter presents a longitudinal suture which extends up to the point where the laciniae come together. It appears that, as a result of ventral fusion of the lateral lips, the inferior oral commissure has moved to a more distal position (Ji_2).

Anteriorly, the bottom of the subcapitular gutter passes into a pair of long pilose processes which are generally curved downwards in the distal part. These structures are called here laciniae (Vitzthum: laciniae; Evans: internal malae; Gorirossi: hypopharyngeal or hypostomal processes). In ventral view, the labrum is easily visible between the laciniae. The corniculi (hypertrophied setae) are inserted in the apical part of the so-called malapophyses (the paired anterior regions of the infracapitulum, of which the ventral surface is called gena). Their position is nearly terminal; they are long and pointed, and are directed to the front. In ventral view, the salivary styli (or siphunculi) are also conspicuous. The chitinous efferent ducts (*dg*) can easily be discerned inside, and posterior to, the infracapitulum. The posterior part of each duct is constituted by a spiral tube; in fig. 191 it is represented at one side (removed in distal direction, for the sake of space). The tips of the fimbrillae project lateral to the corniculi. The laterocoxal seta *e* is also conspicuous in ventral view.

In lateral view (fig. 192), the position of the chelicerae, under the suprachelіceral limbus, is easily visible. The figure represents also the coxal region of the palp with the palpal acetabulum, and the laterocoxal seta *e*. It is clear that the corniculi have a terminal position. Salivary stylus (siphunculus), labrum, label-

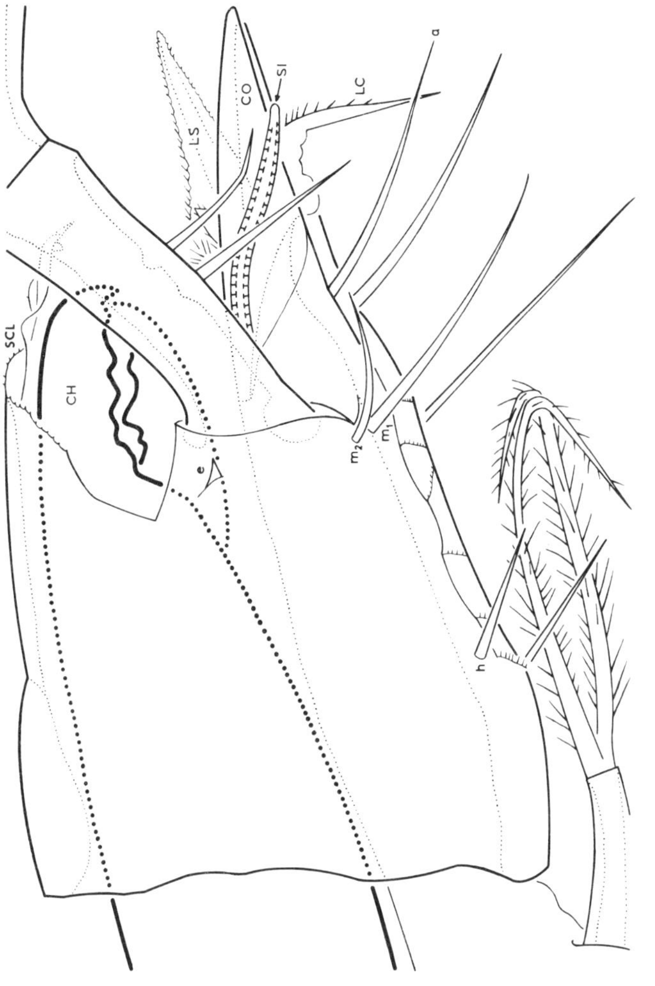

Fig. 192. Glyptholaspis confusa (Foà), female; lateral view of gnathosoma; × 440.

lum, fimbrilla and lacinia (curved downwards in the distal part) are also partly represented in fig. 192.

Fig. 193 represents a frontal view of the gnathosoma. It is now evident that one of the terminal lobes of the suprachelicerал limbus is attached by means of a solid, asymmetrical ligament (*li*) to a ridge of the dorsal surface of the infracapitulum (the capitular saddle *se*). When the supracheliceral limbus is removed, part of the ligament remains attached to the capitular saddle; it is not impossible that a similar fragment has been regarded as a labrum by several authors. Fig. 193 represents all the elements which surround the mouth. The labrum has a broad base and a narrow tip. It is flanked by the pilose labella. The fimbrillae are visible in front of the labella. The position of patrocynium

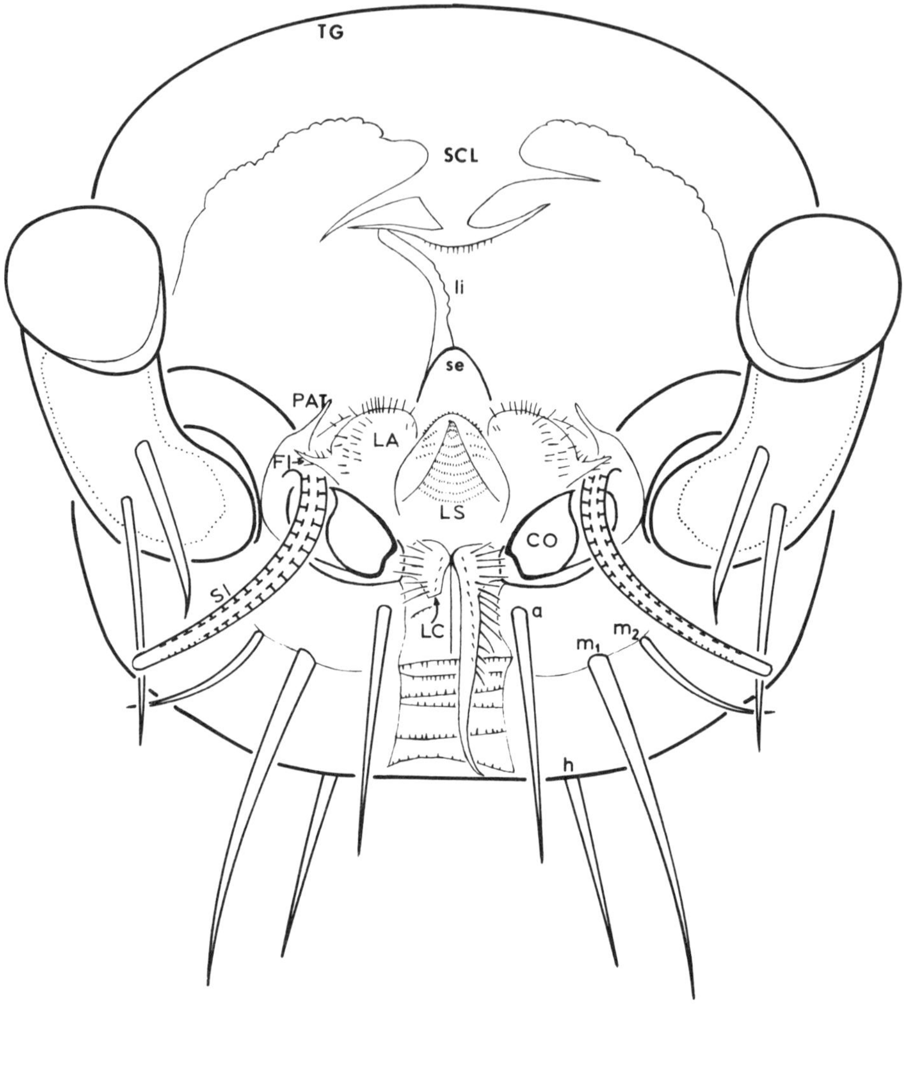

Fig. 193. Glyptholaspis confusa (Foà), female; frontal view of gnathosoma; × 440.

and salivary stylus (siphunculus), lateral to the labella, is also clear. Fig. 194B represents a posterior view of the gnathosoma; it shows the cheliceral sheats, three posterior bars (apodemes), the left efferent duct entering the coxal cavity, and the pharynx.

For a clear understanding of the exact position of several structures it is necessary to dissect the gnathosoma. In fig. 194C, the greater part of the palp is removed; this figure shows the connection of the suprachelicer al limbus with the capitular saddle by a ligament. After removal of suprachelicer al limbus and

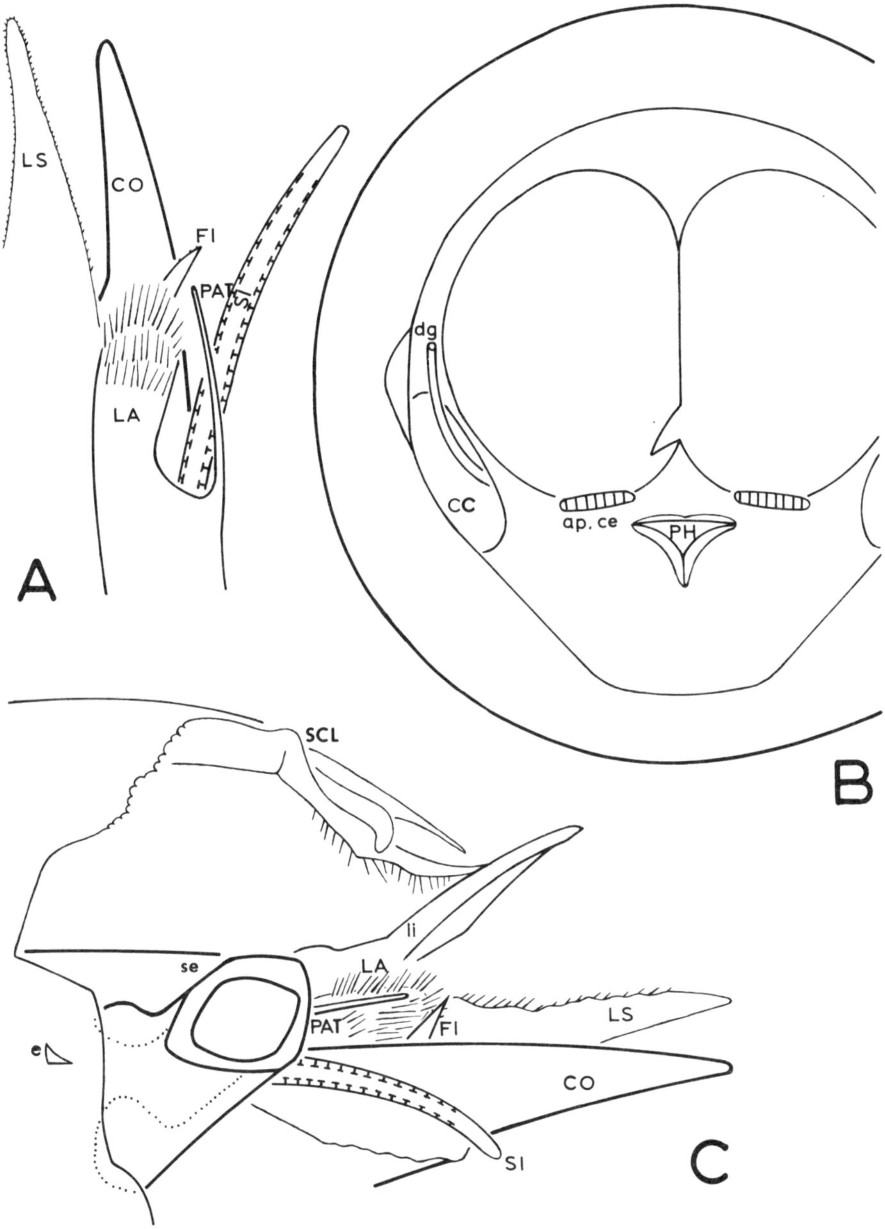

Fig. 194. Glyptholaspis confusa (Foà), gnathosoma of female; A, anterior part viewed obliquely from above (supracheliceral limbus removed); B, posterior view (the chelicerae are extended and, in this view, only the cheliceral sheaths are visible); C, lateral view of anterior part (chelicerae and part of palpal trochanter removed); A–C, × 440.

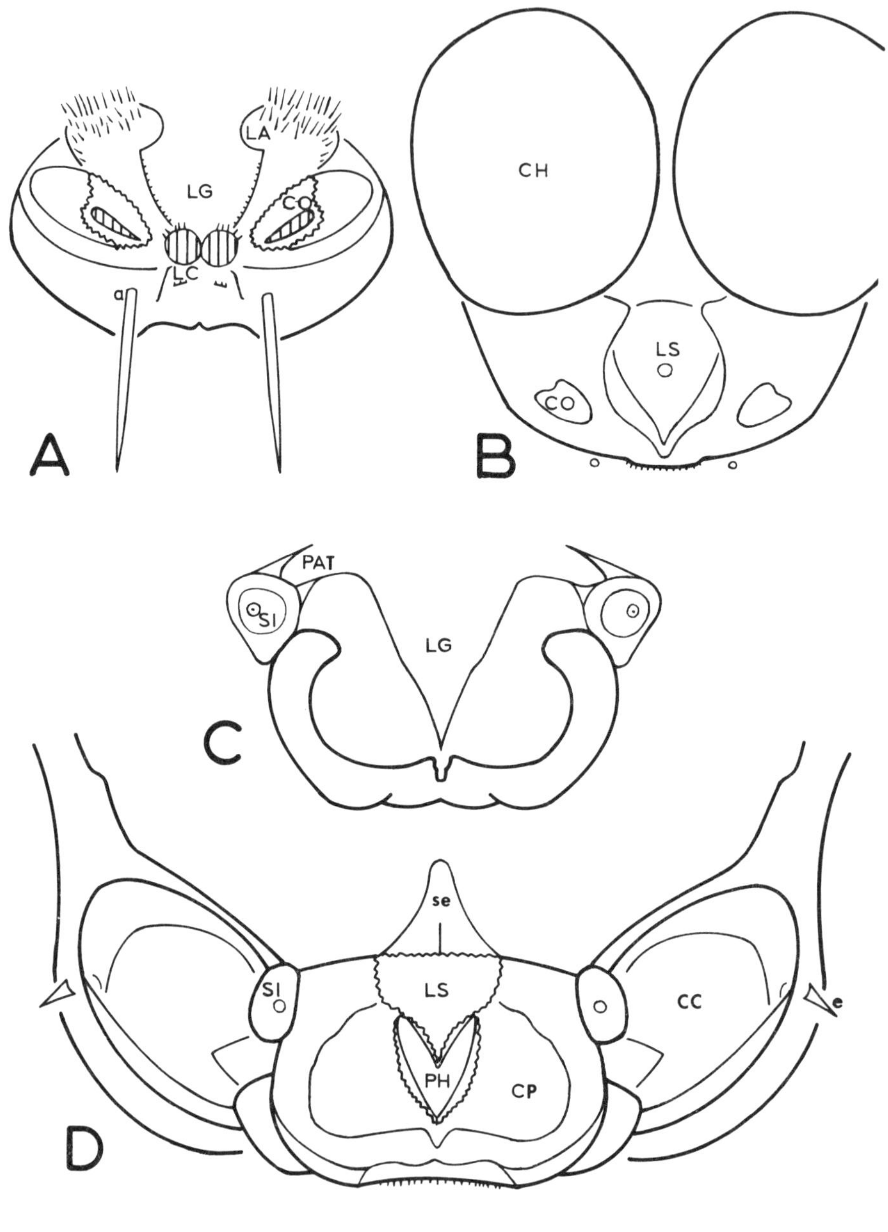

Fig. 195. Glyptholaspis confusa (Foà), female, transverse sections through gnathosoma; A, section at the level of the base of corniculi and laciniae (labrum and chelicerae removed); B, section anterior to infracapitular setae *a*; C, section anterior to the mouth, viewed in distal direction (labrum and chelicerae removed); D, section anterior to the mouth, viewed in proximal direction (chelicerae, suprachelicera l limbus and tegulum removed); A–D, × 440.

chelicerae, several other details become visible. Fig. 194A demonstrates that labellum and patrocinium have a basal connection, and that the salivary stylus (siphunculus) passes under the patrocinium. Fig. 196B represents a dorsal view of the posterior part of the infracapitulum. It shows the cervix (*CE*) with the capitular saddle (*se*) which separates the paired cheliceral grooves (*FC*). Posterior to the line of attachment of the cheliceral frame, the (external) cervix passes into the (internal) cervical apodeme (*ap.ce*) which is represented by three bars.

Further details of the gnathosoma are demonstrated by sections. In fig. 195, a number of transverse sections are represented. It appears that, in rest, the labrum is situated in a labial gutter (fig. 195B: *LG*), and that the walls of the gutter are partly beset with teeth (fig. 195A). The labella constitute a dorsal continuation of the lateral walls of the gutter; they are overhanging in the anterior part. The laciniae are a continuation of the ventral wall of the gutter. A section anterior to the mouth (fig. 195D), viewed in posterior direction, shows that the mouth and the entrance to the pharynx are more or less heart-shaped. The pharynx is contained in the central cavity of the infracapitulum, which is called pharyngeal chamber (*CP*). The pharyngeal chamber is connected on both sides with the coxal cavities (*CC*). The duct of the salivary stylus (or siphunculus) is in the coxal cavity, close to the lateral wall of the pharyngeal chamber.

The dorsal wall of the pharynx continues as the ventral wall of the labrum. The lateral walls of the pharynx continue as the paraxial walls of the lateral lips which, in Gamasida, are very complicated and consist of several parts. Lateroventrally, the lateral pharyngeal walls pass into the walls of the labial gutter (fig. 195C). As mentioned above, the dorsal part of the gutter passes into the labella; these are posteriorly also connected with the cervix. Labial gutter as well as labella must be regarded as parts of the lateral lips. It may be remarked here that, in literature, the labial gutter has received a confusing terminology; it was named hypopharynx (although it is situated anterior to the mouth), as well as paralabrum internum.

The anterior part of the infracapitulum, cut off by a transverse section anterior to the mouth, is represented in posterior view in fig. 185C. The posterior part of the labial gutter is visible here, as well as its dorsal continuation (the smooth basal part of the labellum). The figure demonstrates also its connection with the patrocinium and the soft skin enveloping the salivary styli (siphunculi).

The final solution of the remaining problems can be achieved by means of a study of longitudinal sections. Fig. 197 represents a sagittal section (exactly through the plane of symmetry); the pharynx (*PH*) is slightly removed, in order to show the underlying structures. The figure represents the gnathosoma as a tube, subdivided by the cervix into a dorsal part which contains the chelicerae, and a ventral part which contains the pharynx. The dorsal 'wall' of the tube consists, in fact, of the tegulum (which is part of the body wall) and the project-

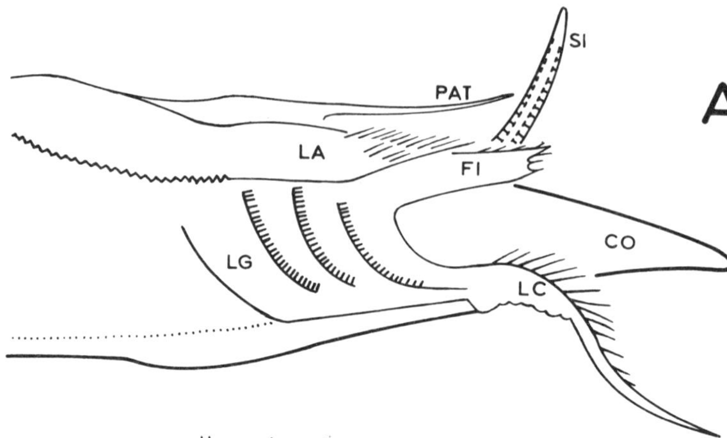

A

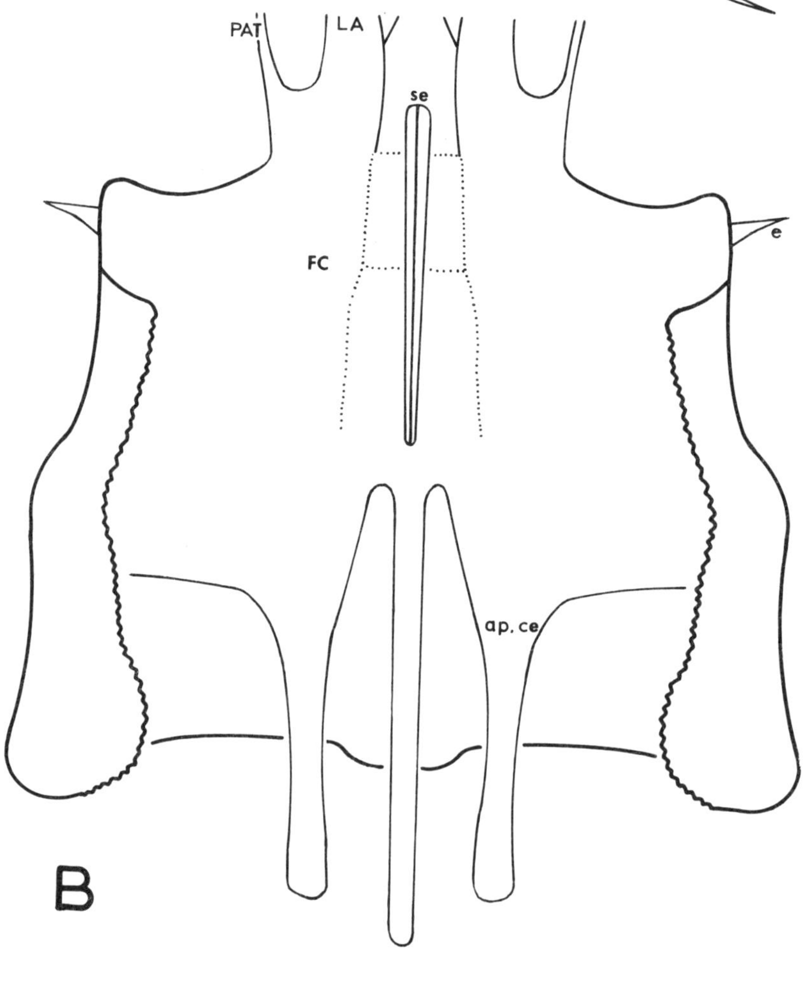

B

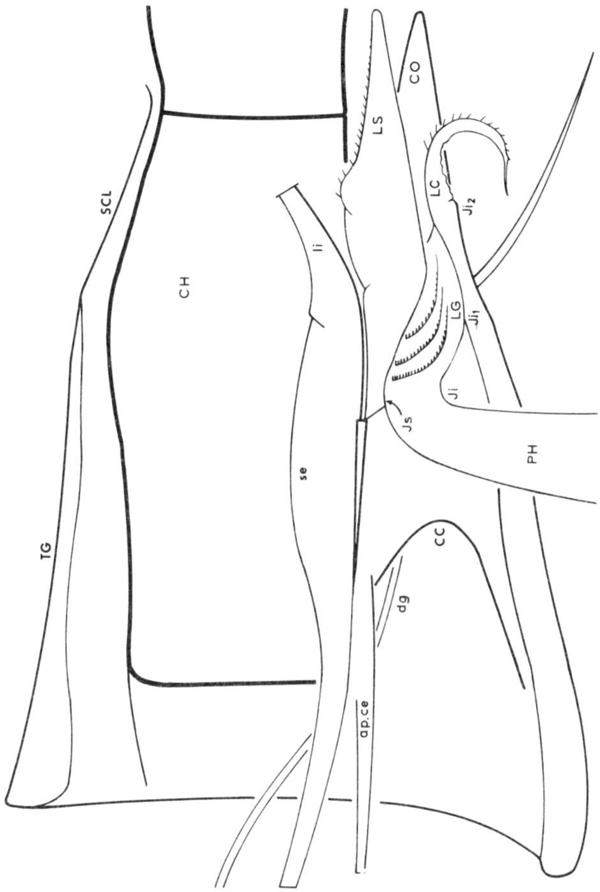

Fig. 197. Glyptholaspis confusa (Foà), female, longitudinal section through gnathosoma (pharynx moved to new position, in order to show the entrance to the coxal cavity; × 312.

ing supracheliceral limbus. The membraneous cheliceral frame, which constitutes the body wall between the anterior border of the tegulum and the attachment of the frame to the dorsal wall of the infracapitulum (anterior to the cervical apodeme), is not represented in this section. The central cavity of the infracapitulum (the pharyngeal chamber) is laterally connected with the coxal cavity; in fig. 197, it is demonstrated that the chitinous duct of the infracapitular gland enters into the coxal cavity. Tegulum and lateral and ventral parts of the infracapitulum constitute the ring-shaped base of the gnathosoma. Fig. 197 represents also the median ridge of the cervix (the capitular saddle), the ante-

←

Fig. 196. Glyptholaspis confusa (Foà), female; A, longitudinal section through anterior part of gnathosoma (labrum removed), showing the labial gutter; B, dorsal view of posterior part of gnathosoma (tegulum and supracheliceral limbus removed); A, B, × 440.

rior ligament, and the posterior bars (the cervical apodeme). Interesting details of the longitudinal section are constituted by the anterior continuation of the walls of the pharynx. The dorsal wall of the pharynx passes into the ventral wall of the labrum. Ventral and lateral to the labrum, the left half of the labial gutter is visible. It presents three rows of teeth. It appears that the anterior part of the labial gutter has fused with the genae (which is normal in the case of lateral lips); the ventral part projects anteriorly as the laciniae. The oral commissures are also indicated in fig. 197. One of the superior commissures (*Js*) is represented at the base of the labrum. It is difficult to determine the exact position of the unpaired inferior commissure (*Ji*). I have indicated a point ventral to *Js* as *Ji*; it constitutes the real corner of the mouth. The point where the labial gutter has fused with the bottom of the subcapitular groove is indicated as Ji_1. The point where the bottom of the gutter subdivides into the laciniae is indicated as Ji_2. In fig. 196A, a longitudinal section is represented, in which the labrum was removed. This section demonstrates that the fimbrilla is the anterior dorsal process of the labial gutter. Fimbrillae as well as laciniae must also be regarded as parts of the lateral lips.

The details from the transverse and longitudinal sections complete the description of the extremely complicated gnathosoma. The exact position of all elements is demonstrated, and figs. 190–197 together represent the complete topography of the gnathosoma of *Glyptholaspis*. The muscles by which the gnathosoma as a whole is operated are described separately in the following paragraphs (see fig. 198).

In *Glyptholaspis confusa*, the muscles operating the movements of the gnathosoma are inserted on the posterior (internal) border, i.e., on the basal part of the so-called basis capituli (which represents the fused bases of tegulum and mentum) (fig. 198A). There is a series of dorsal muscles (fig. 198B: *mtg*) inserted on the base of the tegulum, and a series of ventral muscles (fig. 198C: *msc*) inserted on the base of the mentum. There are, however, no extrinsic muscles inserted on the cervical apodeme (the internal continuation of the cervix; in Actinotrichida the capitular apodeme, which has a similar position, constitutes the main place of insertion of the muscles operating gnathosoma movements). The function of the cervical apodeme in Anactinotrichida is not investigated here; its shape resembles that of the Uropygid median and lateral apodemes and its function could be more or less similar (place of attachment of intrinsic muscles of the gnathosoma).

In a transverse section, it can easily be observed that the dorsal muscles are attached to the dorsal shield of the idiosoma, anterior to the place of attachment of the muscles inserted on the bases of the chelicerae (fig. 198D). This condition is very similar to that found in Holothyrida (see Van der Hammen,

---

→

Fig. 198. Glyptholaspis confusa (Foà), female; A–C, gnathosoma separated from idiosoma; A, lateral view; B, dorsal view of base; C, ventral view of base; D, longitudinal (nearly sagittal) section of gnathosoma and anterior part of idiosoma; A–C, × 221; D, × 106.

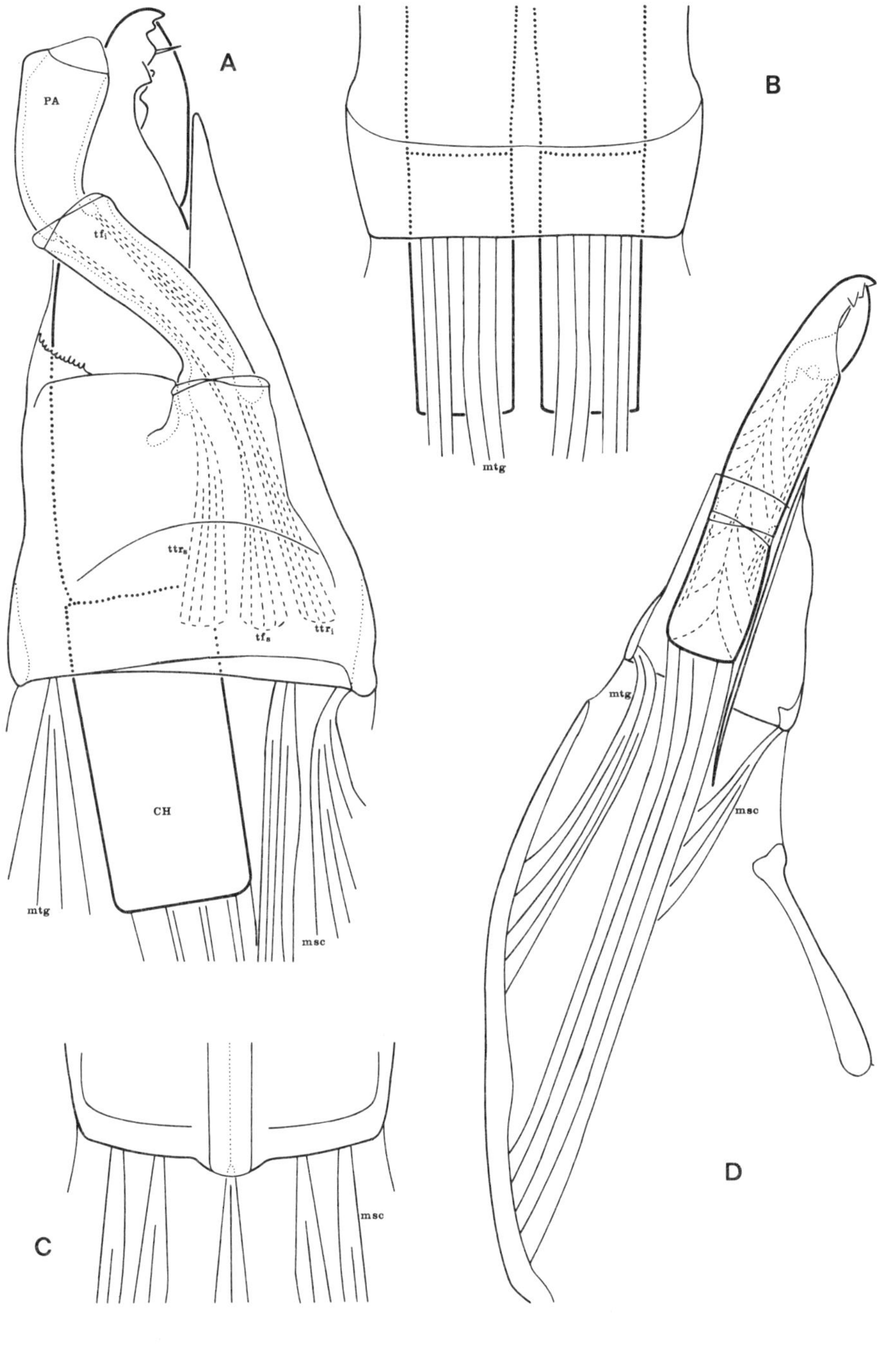
A
PA
tf$_i$
ttr$_s$
tf$_s$
ttr$_i$
CH
mtg
msc
B
mtg
C
msc
D
mtg
msc

1983b: 11, 39–40, figs. 7B, 8B, 24E; and section 4 of this chapter) and Ixodida (see Van der Hammen, 1983a: 218, fig. 11B; and section 6 of this chapter), and probably also to that in Opilioacarida.

As mentioned already above (in chapter V section 2 of the systematic part), the condition in Actinotrichida is quite different. In this group, the movements of the gnathosoma are operated by muscles inserted on the capitular apodeme. A detailed description of these muscles can be found in my paper on the gnathosoma of *Hermannia convexa* (C.L. Koch), an Oribatid mite (Van der Hammen, 1968: 18, 23–24, figs. 7, 8B, 9B, 10A, 12; see also chapter II subsection 4.5 of the systematic part). Subcapitular muscles, besides the muscles inserted on the capitular apodeme, are mentioned for Caeculidae (Coineau, 1971: 130, fig. 29; 1974: 78, fig. 28). Evidently the insertion, on a capitular apodeme, of muscles operating the movements of the gnathosoma, constitutes a fundamental character of the Actinotrichida, by which this group is radically different from Anactinotrichida and Ricinulei (i.e., Cryptognomae). In Ricinulei, the muscles operating the movements of the gnathosoma are inserted, ventrally and laterally, on the basal part of the infracapitulum (see Van der Hammen, 1979a: 9, figs. 6A, D, 7A, E, in which paper the lateral muscles are erroneously referred to as inserted on a capitular apodeme; see also chapter VI of the systematic part).

As mentioned already in previous chapters (see also Van der Hammen, 1971: 472; 1972a: 274–275), the evolution of the Actinotrichid gnathosoma is associated with the evolution of a podocephalic canal (with laterodorsal position), whilst the evolution of the Anactinotrichid and Ricinuleid (i.e., Cryptognomic) gnathosoma is associated with the evolution of sternal taenidia, subcapitular gutter and sternapophyses (with ventral position). For this reason, the process of gnathosomatization in Actinotrichida must have been independent from that in Cryptognomae; this means that it did not start in a common ancestor (it must have been the result of an evolution which could be characterized as partially parallel).

The Cryptognomic gnathosoma is, however, also different from that of the Actinotrichida by the considerable anterior extension of the mentum (see also Van der Hammen, 1979a: 24), and the advanced position of the inferior oral commissure; in Ricinulei, it can still be observed that the labium is completely incorporated in the mentum (Van der Hammen, 1979a: 9, fig. 6B; see also chapter VI of the systematic part).

Previous students of the gnathosoma were particularly struck by the similarities in its structure, and have paid little attention to the fundamental differences. The similarities are, in fact, attributable to general similarities in the structure of the Chelicerate mouthparts (see also Van der Hammen, 1986a), particularly those found in Apatellata and Arachida s. str.

→

Fig. 199. Glyptholaspis confusa (Foà), distal part of right chelicera of female; A, lateral (antiaxial) face; B, lateral (paraxial) face; C, dorsal view; D, ventral view; E, ventral oncophyses *opv'* and *opv''*; A–D, × 370; E, × 710.

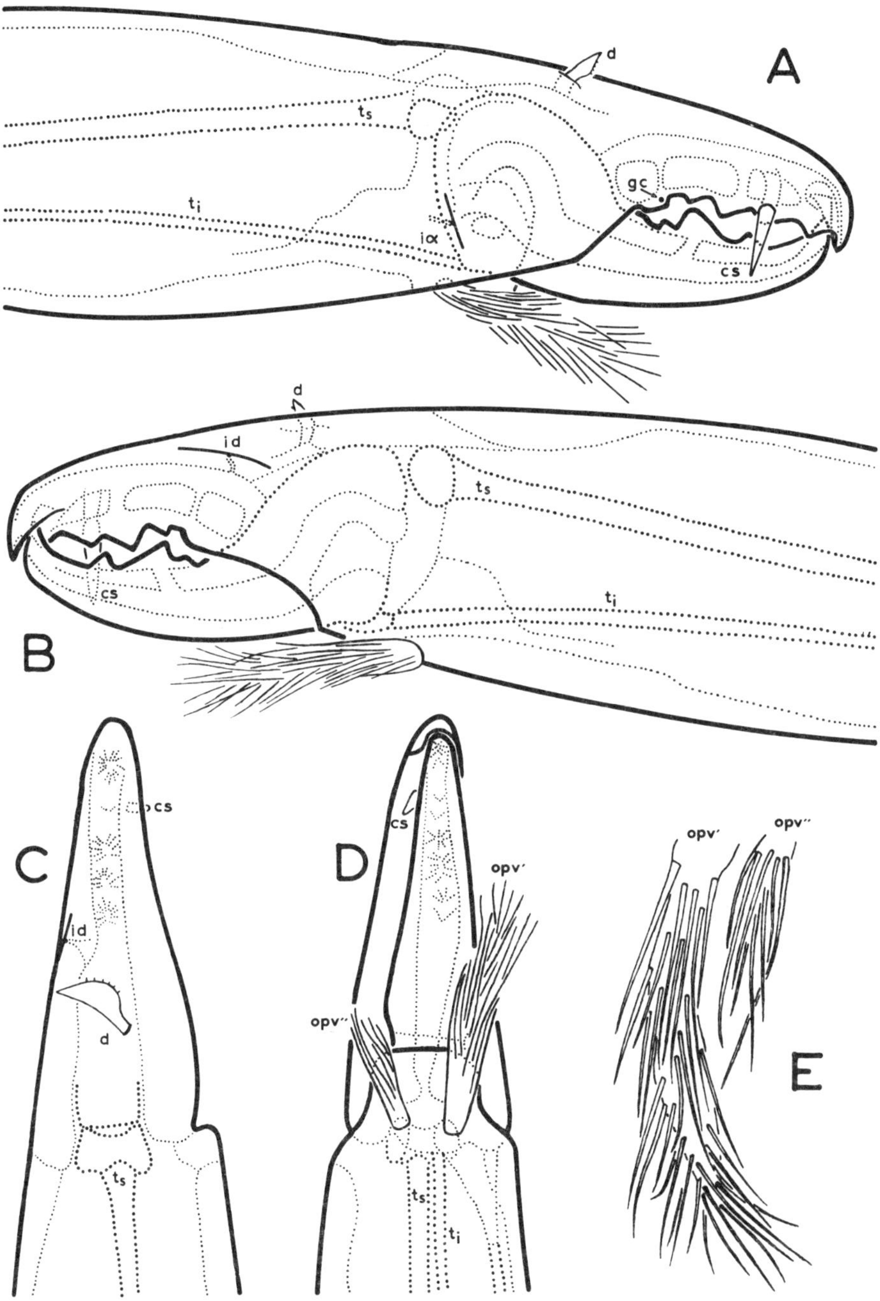
A
d
t_s
t_i
gc
iα
cs
B
d
id
t_s
t_i
cs
C
cs
id
d
t_s
D
cs
opv′
opv″
t_s
t_i
E
opv′
opv″

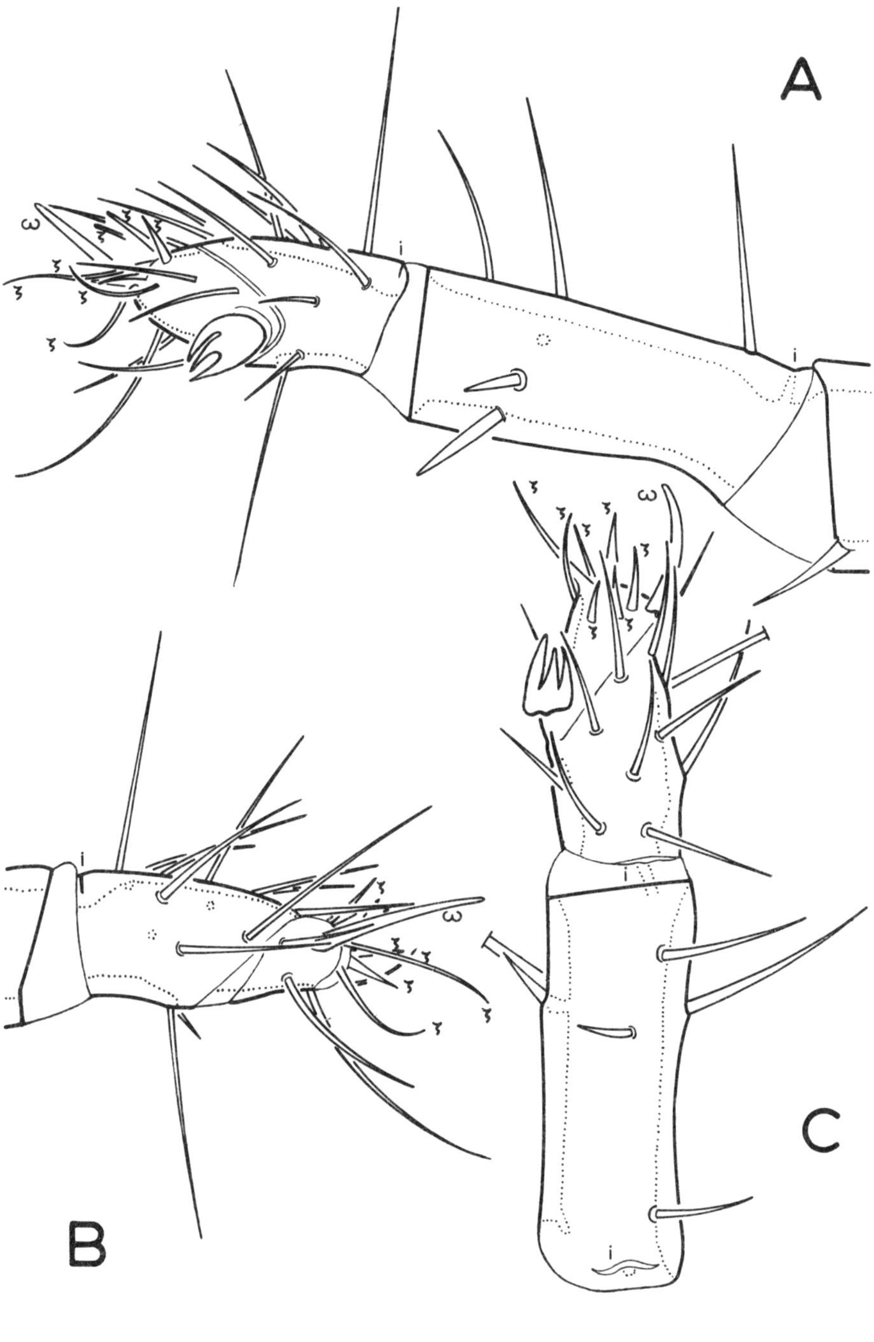
A
B
C
ω
ω
ω
i
i
i
i

Chelicera. – The chelicerae are three-segmented (trochanter, body of the chelicera, and apotele). The trochanter is represented in fig. 197. Fig. 199 depicts the distal part of the right chelicera. The movable jaw represents the apotele, the fixed jaw is the terminal part of the body of the chelicera. The various faces of the chelicera are distinctly different. The antiaxial face (fig. 199A) presents a lateral lyrifissure *iα*, a dorsal seta d, the small orifice of a gland (*gc*) at the base of the fixed jaw, and a specialized process generally indicated as pilus dentilis (called cheliseta *cs* by me in 1964). The antiaxial ventral brush *opv''* is the smaller one of the pair of plumose brush-like elements which constitute the ventral oncophysis. The paraxial face (fig. 199B) presents a dorsal lyrifissure *id*, situated in the fixed jaw. Part of seta *d* is also visible in this orientation. The paraxial ventral brush *opv'* is the larger one of the two elements which constitute the ventral oncophysis. A dorsal view (fig. 199C) distinctly shows the position of the dorsal seta *d* and lyrifissure *id*. The pair of brushes which constitute the ventral oncophysis is best visible in ventral view (fig. 199D, E). In dorsal as well as in ventral view, lines on the teeth of the jaws are visible through the cuticle. The apotele of the chelicera is movable by means of the usual two tendons (t_s and t_i).

Palp. – The palp consists of six segments (trochanter, femur, genu, tibia, tarsus and apotele). The proximal segments are partly represented in figs. 190–192, the distal segments in fig. 200. Tarsus and tibia together have the shape of one segment. The apotele (fig. 200A, C) is represented by a trifid element with paraxial lateroventral position at the base of the tarsus. Genu and tibia present a rather large dorsal lyrifissure at the posterior border (fig. 200C). The tarsus presents a dorsal group of three hollow setae, and a more or less terminal group of five; these probably represent eupathidia (cf. fig. 200A–C: ζ). The large antiaxial hollow phenere ω probably represents a solenidion (fig. 200B, C). The phanerotactic formula of the palp (from trochanter up to the tarsus) is about 2–5–6–13–15.

Legs. – Leg I is inserted in soft skin, legs II–IV in the acetabula of the holopodal shield. Leg I, which lacks an apotele, and presents numerous hollow setae at the tip, certainly has a sensory function. Legs II–IV present pretarsus and apotele. Legs I and II are directed forward, leg IV backward, and leg III more or less sideward. The sequence of the legs, arranged according to decreasing length (pretarsus and apotele not included in the measurements), is: IV, I, II, III. In comparison with II–IV, leg I is slender, with rather differently shaped segments. The phanerotactic formulae of the legs, from the coxa up to and including the apotele, are the following (sensory phaneres included): I (2+1–5–13–11–12–18+∞–0); II (2–5–11–11–10–18–2); III (2–5–6–7–7–18–2); IV (1–5–6–6–7–18–2); the supposed laterocoxal seta of coxa I,

---

←*Fig. 200. Glyptholaspis confusa* (Foà), right palp of female; A, lateral (paraxial) face of femur (distal part), genu, tibia, tarsus and apotele; B, lateral (antiaxial) face of genu (distal part), tibia and tarsus; C, dorsal view of genu, tibia, tarsus and apotele; A–C, × 440.

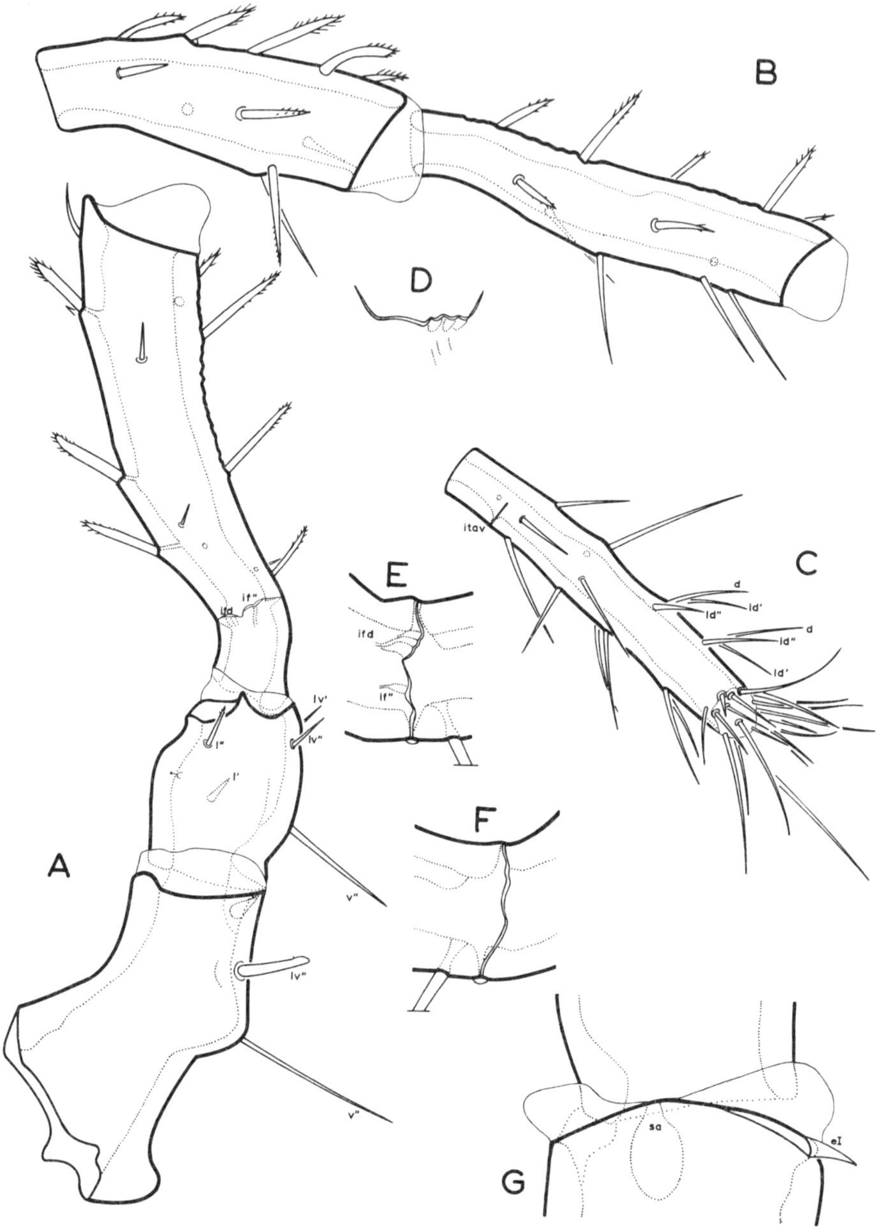

Fig. 201. Glyptholaspis confusa (Foà), right leg I of female; A–C, lateral (antiaxial) face; A, coxa, trochanter and femur; B, genu and tibia; C, tarsus, D–F, basifemoral ring; D, obliquely from above, showing compound lyrifissure *ifd*; E, lateral (antiaxial) view, showing lyrifissures *ifd* and *if''*; F, lateral (paraxial) view; G, ventral view of distal part of coxa and proximal part of trochanter, showing coxal 'sacculus' and supposed laterocoxal seta *eI*; A–C, × 245; D–G, × 470.

and the sensory phaneres of the distal part of the tarsus are mentioned separately.

Leg I is represented in fig. 201. Coxa I bears two ordinary setae with a more or less ventral position (*lv''* and *v''*), of which *v''* is long and pointed, and *lv''* rather blunt. At the paraxial face, near the anterior border, a specialized seta with reduced core is present, which is rather small, whitish, and pointed, and presents exactly the same shape as the supposed laterocoxal seta at the base of the palp. It is supposed to represent the laterocoxal seta *eI* of leg I; this seta is represented in fig. 201G. A sacculus (*sa*) debouches at the anterior ventral border of coxa I (fig. 201A, G); its function is unknown, but it probably represents a gland. Trochanter I has no dorsal setae, two laterals (*l'*, *l''*), two lateroventrals (*lv'*, *lv''*), and one ventral (*v''*); they are rather short, except *v''* which is long and pointed as the corresponding seta of the coxa. The antiaxial face presents a very small laterodorsal pore (probably the orifice of a tegumental gland). The basifemoral ring is complete (fig. 201A, D–F). It presents two lyrifissures, both at the antiaxial face: one dorsal (*ifd*) and one lateral (*if''*); *ifd* is probably composed of three lyrifissures, as represented in fig. 201D, E. The basifemur bears no setae; the telofemur presents thirteen setae, some of which are relatively large and pilose, whilst others are small, pointed and smooth. The setae of the telofemur are apparently arranged in three verticils. The eleven setae of the genu are mostly strong and pilose. The same applies to the twelve setae of the tibia; some of the ventral tibial setae are long and pointed. Tarsus I (fig. 201C) has no basitarsal ring; at the place of the ring, a ventral lyrifissure (*itav*) is present. It is the only lyrifissure of the segment. Pretarsus and apotele are lacking, but tarsus I presents at the tip an area with numerous hollow sensory phaneres; this area is not sclerotized. The hollow phaneres remind one of eupathidia, with the exception of a paraxial small and hollow phanere (not figured), inserted among the supposed eupathidia; this phanere is shorter and less pointed, and possibly represents a solenidion. The setae of tarsus I are arranged in at least four verticils. Two anterior dorsal groups of three setae (*ld''*, *d*, *ld'*) probably also represent eupathidia. The setae of the tarsus are relatively long and pointed.

Leg II is represented in fig. 202. Coxa II presents two plumose lateroventral setae. Trochanter II has no dorsal setae, four plumose laterals (*ld'*, *lv'*, *ld''*, *lv''*) and one simple ventral seta that probably represents an antiaxial one (*v''*). The basifemoral ring is incomplete; it is interrupted in the ventral antiaxial part. There is an isolated antiaxial lyrifissure (*if''*), and a dorsal paraxial one (*if'*) that is part of the ring. The setae of femur II are apparently arranged in three verticils. There are apparently two dorsal setae (a dorsal seta is not present in the case of the basifemoral verticil). The paraxial setae are more or less simple, the dorsals small and only slightly ciliate, the antiaxials more or less plumose. The setae of genu II are arranged in three verticils. Three plumose setae have a dorsal position. Besides these setae, genu II bears two paraxial laterals, four antiaxial laterals, and two ventrals. The setae of tibia II are ar-

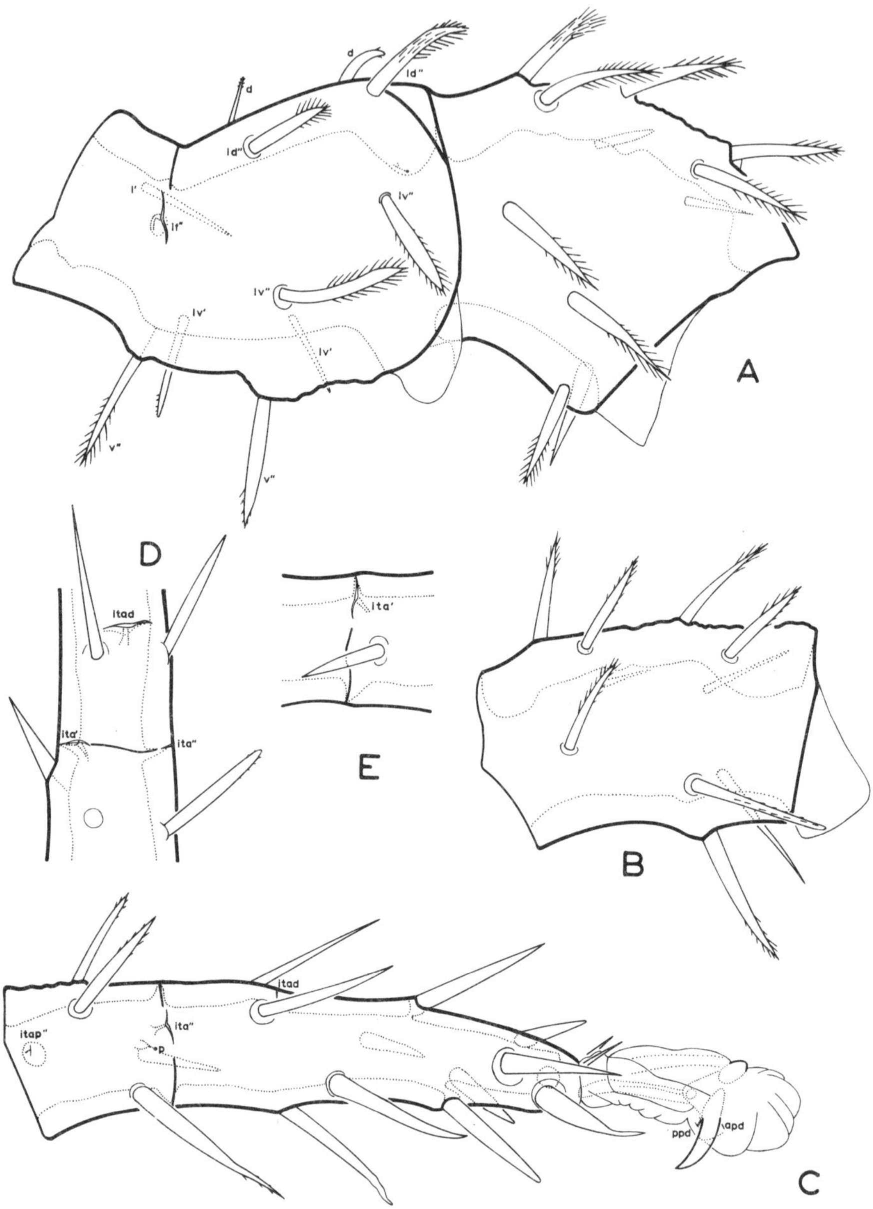

Fig. 202. Glyptholaspis confusa (Foà), right leg II of female; A–C, lateral (antiaxial) view; A, femur and genu; B, tibia; C, tarsus and apotele; D, dorsal view of part of tarsus; E, lateral (paraxial) view of basitarsal ring; A–E, × 245.

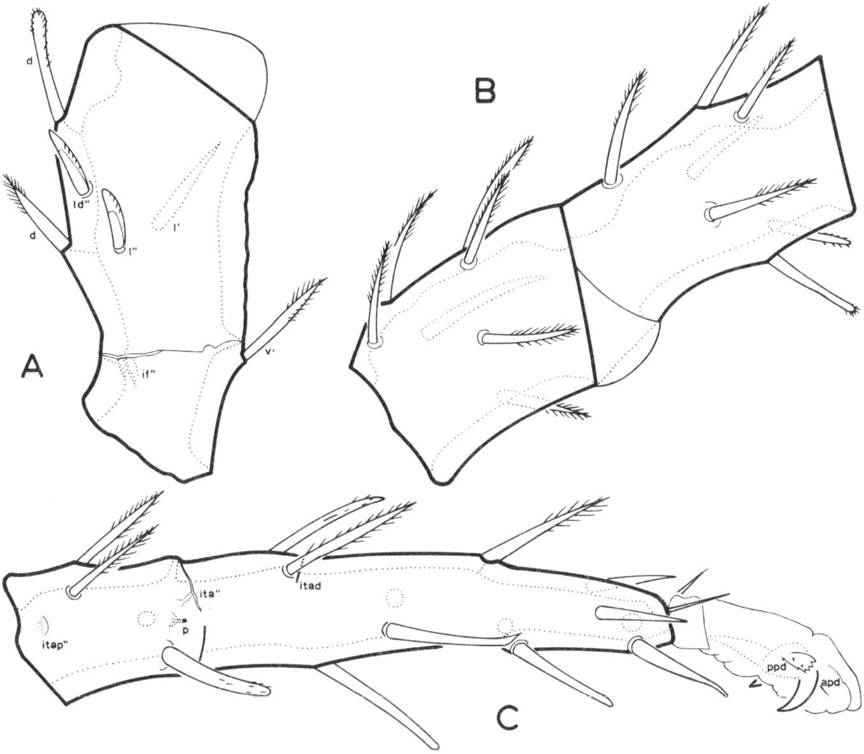

Fig. 203. Glyptholaspis confusa (Foà), lateral (posterior) view of right leg III of female; A, femur; B, genu and tibia; C, tarsus and apotele; A–C, × 245.

ranged in two verticils. There are apparently two dorsal setae which are plumose. The paraxial tibial setae are more or less simple, the antiaxials more or less plumose. The basitarsal ring presents small interruptions in the antiaxial and paraxial parts. The adesmatic articulation presents three lyrifissures: an antiaxial (*ita''*), a ventral (*itav*), and a dorsal paraxial (*ita'*). Proximally of the antiaxial interruption of the ring is a pore (probably the orifice of a tegumental gland). The tarsus presents also a dorsal lyrifissure (*itad*), in the middle of the segment. A very small antiaxial fissure at the base of the tarsus possibly represents the proximal lyrifissure *itap''*. Pretarsus and apotele are similar to those of leg III. They are dealt with below.

Leg III is represented in fig. 203; pretarsus and apotele moreover in fig. 205. Coxa III bears two plumose ventral setae. Trochanter III presents five setae. The basifemoral ring is complete; there are two lyrifissures (*if''* and *if'*). The setae of femur III are apparently arranged in three verticils. There are two dorsal setae. The basifemoral verticil is reduced to *v'*. Femoral setae *l''* and *ld''* have a characteristic shape (fig. 203A). The setae of genu III are arranged in

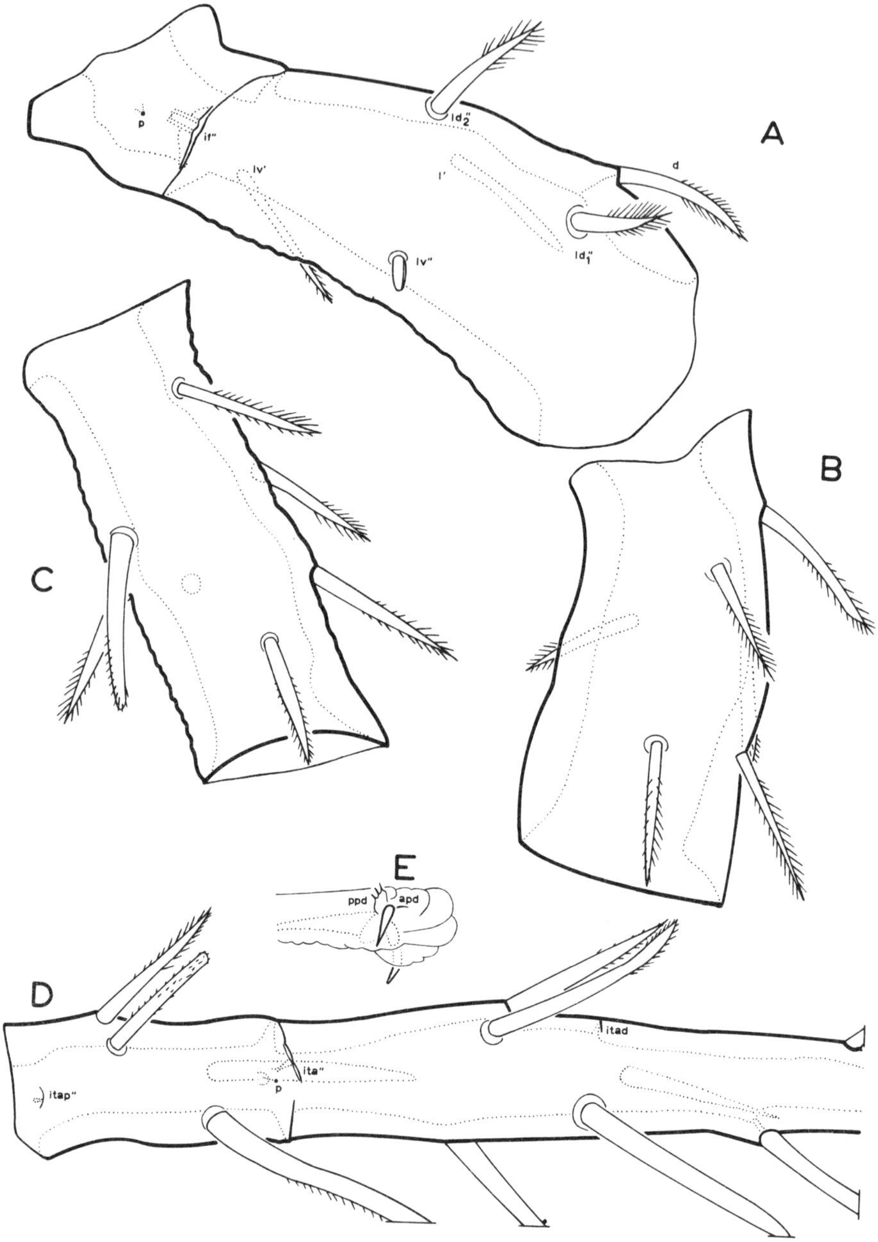

Fig. 204. Glyptholaspis confusa (Foà), lateral (paraxial) view of right leg IV of female; A, femur; B, genu; C, tibia; D, part of tarsus; E, part of pretarsus and apotele (oblique view); A–E, × 245.

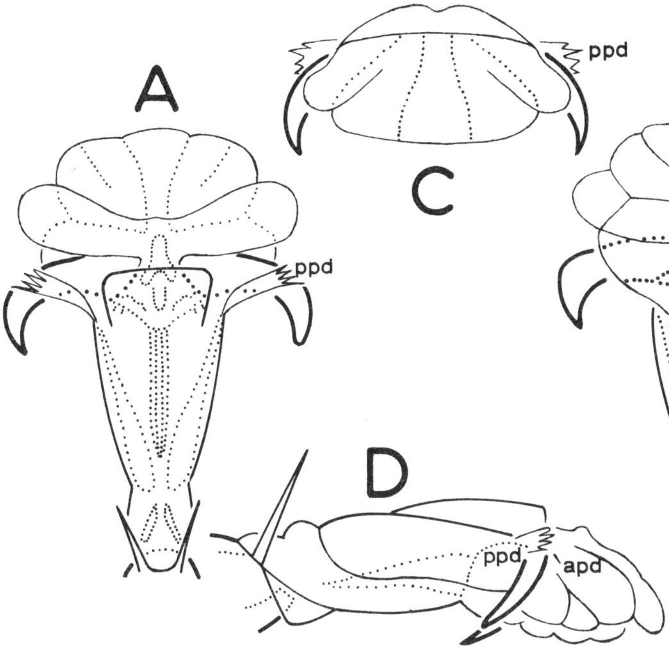

Fig. 205. Glyptholaspis confusa (Foà), ambulacrum (pretarsus and apotele) of right leg III of female; A, dorsal view; B, ventral view; C, frontal view; D, lateral (posterior) view; A–D, × 490.

two verticils which include altogether seven plumose setae. Tibia III bears also seven more or less plumose setae, apparently arranged in two verticils. Tarsus III presents eighteen setae. This segment nearly exactly corresponds with tarsus II; it presents lyrifissures and a pore at similar places. A section through the basitarsal ring is represented in fig. 189B; this figure clearly demonstrates that the tarsal wall is thickened internally as well as externally at the places where the ring is interrupted. Pretarsus and apotele III are represented separately, in various projections, in fig. 205. The pretarsus has a pair of lateral internal chitinous pieces; the usual two tendons of the apotele are also present. The pulvillus is apparently a continuation of the soft skin of the pretarsus; it is more or less lobate. There are two ungues which have a lateral position. Posterior and slightly dorsal to the ungues, there is a pair of 'laterodistal pretarsal elements' (Krantz, 1962). They have a dentate distal border; when observed obliquely (fig. 203C), the surface presents some teeth. There is also a pair of anterior 'pretarsal elements'; these have the shape of a simple seta. The structures were called anterior and posterior paradactyli (*apd*, *ppd*) in my 1964 paper.

The greater part of leg IV is represented in fig. 204. Coxa IV (cf. fig. 187) presents one plumose ventral seta. Trochanter IV has five plumose setae. The basifemoral ring has a ventral paraxial interruption; there are a paraxial lyrifissure (*if''*) and a dorsal antiaxial (*if'*) which is not compound. The paraxial face of the basifemur presents a pore (probably the orifice of a tegumental gland). There are no basifemoral setae. The telofemoral setae have received the notations *d*, ld_1'', ld_2, *l'*, *lv''*, *lv'*; they are more or less plumose, with the exception of *lv''* which is conical in shape. The setae of genu IV are arranged in two verticils; there are six plumose setae. Tibia IV bears seven plumose setae, probably representing two verticils. Tarsus IV is similar to tarsus II and III, except for its being more elongate. Pretarsus and apotele IV are represented in oblique view in fig. 204E; an anterior setiform, and a posterior dentate paradactylus are visible in this orientation.

List of the notations of figs. 186–205

a	anterior infracapitular seta.
ad_1, ad_x	adoral setae.
ap.ce	cervical apodeme.
apd	anterior paradactylus.
CC	coxal cavity of infracapitulum.
CH	chelicera.
CO	corniculus.
CP	pharyngeal chamber.
cs	pilus dentilis (cheliseta).
d	dorsal seta (of the appendages).
dg	duct of gland.
e	supposed laterocoxal seta of palp.
eI	supposed laterocoxal seta of leg I.
FC	cheliceral groove.
FI	fimbrilla.
g_1	seta of genital shield.
G	gena.
gc	orifice of gland in the fixed jaw of the chelicerae.
gl_{1-2}	supposed orifices of glands, lateral to ventrianal shield.
h	posterior infracapitular seta.
H	mentum.
i_1	anterior lyrifissure of dorsal shield.
id	dorsal lyrifissure of chelicera.
if', *if''*	lateral (anterior, resp. posterior) lyrifissures of femur.
ifd	dorsal lyrifissure of femur.
is_{1-2}	lyrifissures of sternal shield.
ita', *ita''*	lateral (anterior, resp. posterior) lyrifissures of tarsus.
itad	dorsal lyrifissure of tarsus.

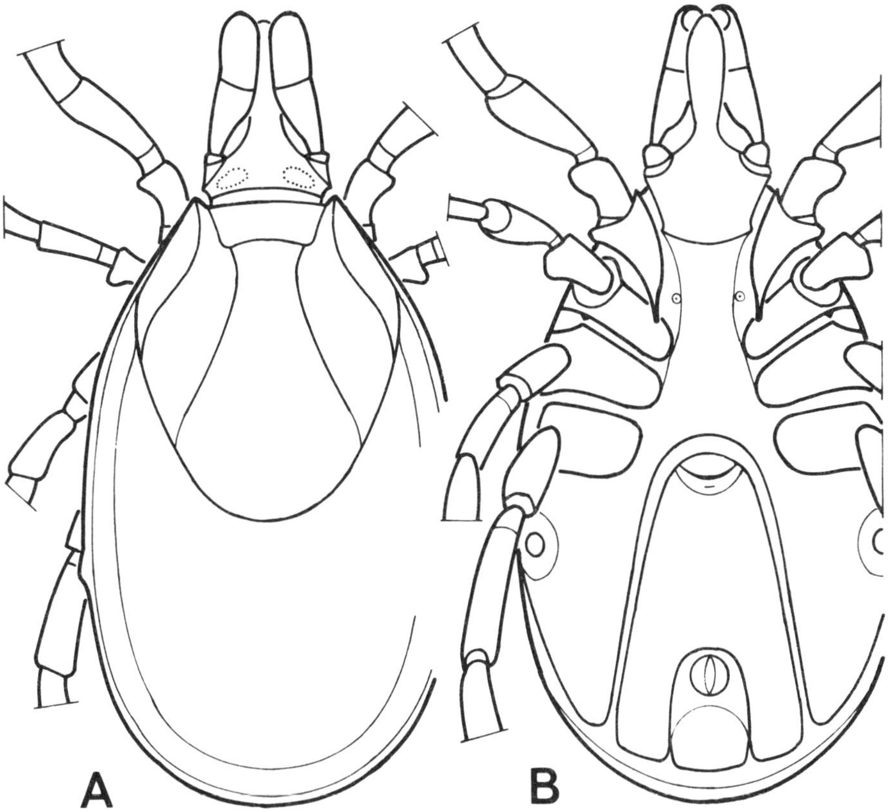

Fig. 206. Ixodes ricinus (Linnaeus), adult female; A, dorsal view; B, ventral view (slightly schematized); A, B, × 32.

itap''	proximal lyrifissure of tarsus.
itav	ventral lyrifissure of tarsus.
iα	antiaxial lyrifissure of chelicera.
Ji, Ji_1, Ji_2	inferior oral commissure.
Js	superior oral commissure.
j1–6, J1–5	idiosomatic setae of dorsocentral series.
l', l''	lateral setae (of the appendages).
LA	labellum.
LC	lacinia.
ld', ld''	laterodorsal setae (of the appendages).
li	ligament attaching supracheliceral limbus to capitular saddle.
LG	labial gutter.
LS	labrum.
lv', lv''	lateroventral setae (of the appendages).

m_{1-2}	median infracapitular setae.
mp_1	seta of metapodal shield.
ms_1	seta of metasternal shield.
msc	inferior muscles of gnathosomatic base (subcapitular muscles).
mtg	superior muscles of gnathosomatic base (muscles of tegulum).
p_{1-21}	'pores' of the dorsal shield.
PA	palp.
PAT	patrocinium.
PH	pharynx.
pp_{1-4}	'pores' of peritrematal shield.
ppd	posterior paradactylus.
ps_1	'pore' of metasternal shield.
pv_1	'pore' of ventrianal shield.
r_{2-4}	idiosomatic setae of marginal series.
s_{1-3}	sternal setae.
sa	sacculus of coxa I.
SCL	supracheliceral limbus.
se	capitular saddle.
SI	salivary stylus (siphunculus).
s_{2-6}, S_{1-5}	idiosomatic setae of lateral series.
t_i	inferior tendon of apotele.
t_s	superior tendon of apotele.
tf_i	inferior tendon and muscles of femur.
tf_s	superior tendon and muscles of femur.
ttr_i	inferior tendon and muscles of trochanter.
ttr_s	superior tendon and muscles of trochanter.
v_{1-3}	ventral setae of ventrianal shield.
v', v''	ventral setae (of the appendages).
x	area of fusion of sternal and holopodal shield (anterior to acetabulum II).
y	area of fusion of sternal and holopodal shield (anterior to acetabulum III).
z_{1-6}, Z_{1-5}	idiosomatic setae of marginal series.
ζ (zeta)	supposed eupathidium.
ω (omega)	supposed solenidion.
' (prime)	pertaining to the anterior face of an appendage.
'' (double prime)	pertaining to the posterior face of an appendage.

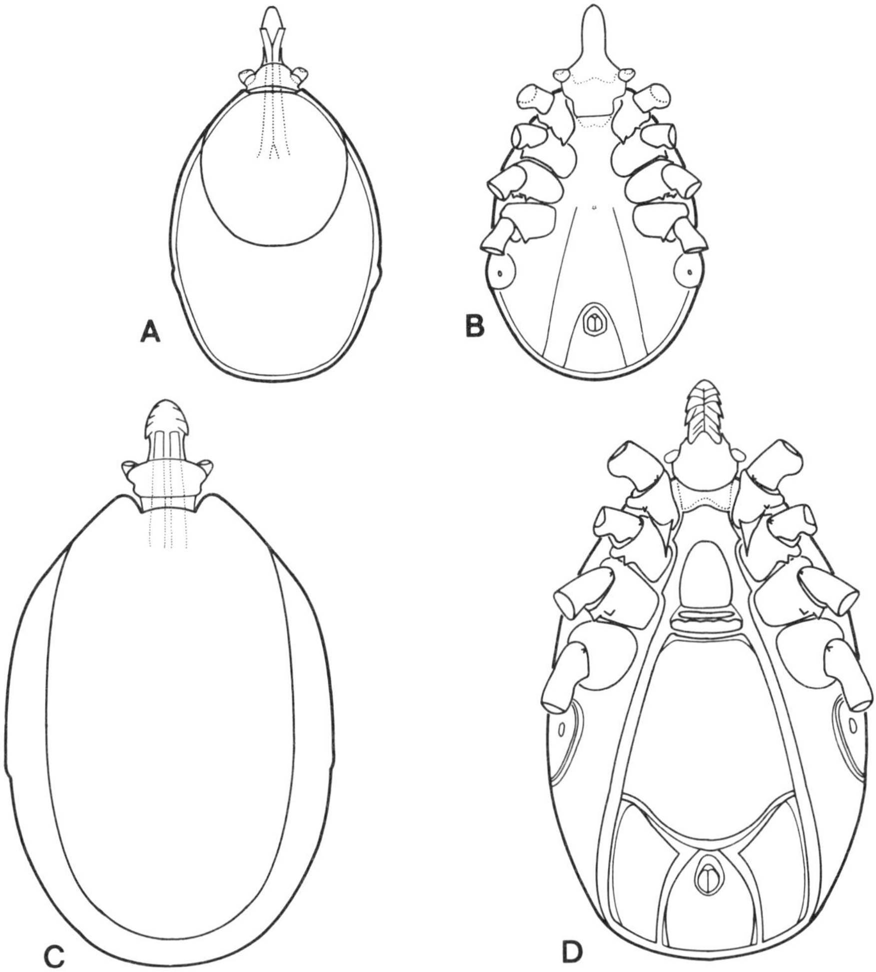

Fig. 207. Ixodes ricinus (Linnaeus); A, B, nymph; A, dorsal view; B, ventral view; C, D, adult male; C, dorsal view; D, ventral view (slightly schematized); A–D, × 32.

6. Ixodida

At the beginning of chapter V, I have mentioned the two papers in which I have dealt with representatives of the Anactinotrichid order Ixodida. Although the morphology of ticks had been treated at great length in literature (e.g., André, 1949; Arthur, 1946, 1951, 1953, 1953a, 1956, 1956a, 1957, 1962; Babos, 1964; Bertram, 1939; T.E. Hughes, 1959; Nuttal, Cooper & Robinson, 1908, 1908a, 1908b; Pagenstecher, 1861; Robinson & Davidson, 1913, 1913a,

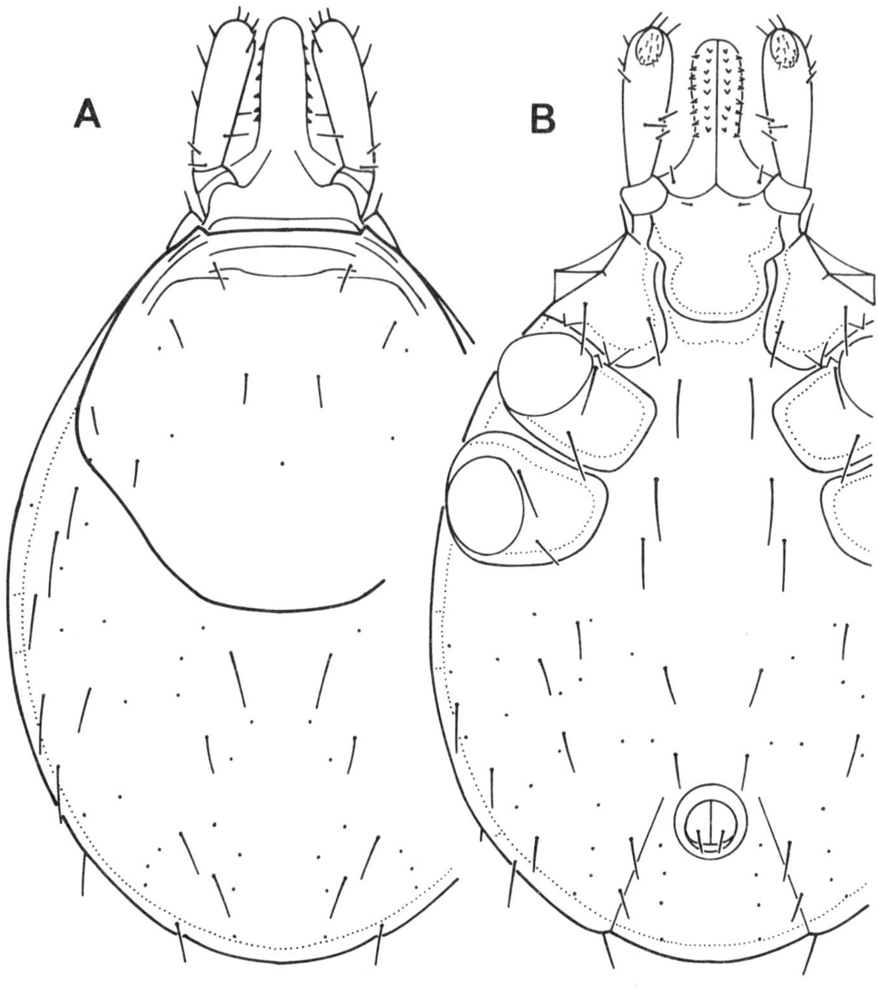

Fig. 208. Ixodes ricinus (Linnaeus), larva; A, dorsal view; B, ventral view; A, B, × 161.

1914; Ruser, 1933; Schulze, 1932, 1935, 1941, 1942; Snodgrass, 1948; Vitzthum, 1931, 1940, 1941, 1942, 1943; Zebrowsky, 1926), the morphological data could not easily be included in Chelicerate comparative morphology until the publication of my Notes on the Comparative Morphology of Ticks (Van der Hammen, 1983a).

The diagnostic characters of Ixodida are the following. A distinct subdivision of the body, into prosoma and opisthosoma, is lost. The idiosoma can present a dorsal shield (Ixodina) or a leathery cuticle (Argasina). One or two pairs of eyes can be present, or eyes are absent. There is one pair of stigmata (absent in the larva), situated posterior to coxa IV, or laterally, above coxae

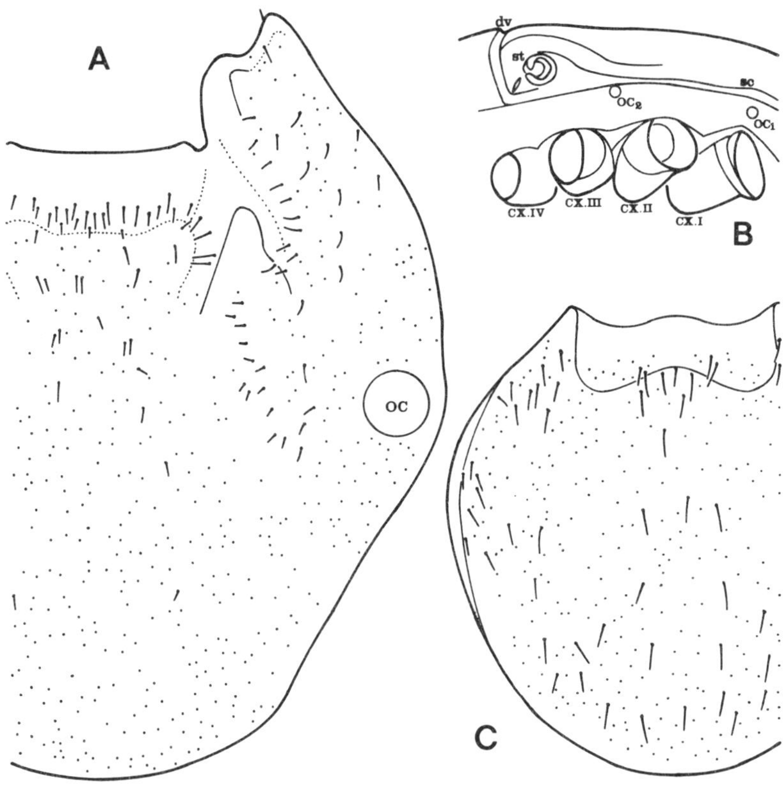

Fig. 209. A, *Hyalomma dromedarii* C.L. Koch, female, dorsal view of scutum; B, *Ornithodoros savignyi* (Audouin), female (not engorged), lateral view of 'pleural' region; C, *Ixodes ricinus* (Linnaeus), female, dorsal view of scutum; A, C, × 44; B, × 8.

II–III. The paired anal covers each bear one or more setae. There are no sternapophyses. The lateral lips are elongate; they are armed with recurved teeth. Anterior to the mouth, the preoral cavity has a V-shaped section. A transverse section of the pharynx is hexagonal. Tibia and tarsus of the palp have coalesced; the palpal apotele is lacking. The legs generally present seven eudesmatic segments, among which the apotele; generally, the coxae are little or not movable. The tarsus of leg I is characterized by the presence of Haller's organ. The postembryonic development generally includes three stases: larva, nymph and adult; Argasid nymphs can pass through a number of growing moults.

Ticks are, with few exceptions, parasitic on Vertebrates (mainly on Mammals, Birds and Reptiles). The total number of described species was recently estimated at 800 (Krantz, 1978).

According to Camicas & Morel (1977), the order Ixodida is subdivided into two suborders: Ixodina and Argasina. The suborder Ixodina includes two superfamilies: Ixodoidea (with the families Ixodidae and Amblyommidae) and

Nuttallielloidea (with the family Nuttalliellidae, based on one incompletely known species). The suborder Argasina includes the superfamily Argasoidea (with the single family Argasidae).

Ixodina (hard ticks) are generally characterized by the presence of a dorsal shield, whilst the gnathosoma is visible from above; in nymphs and adults, a pair of stigmata is present, posterior to coxae IV.

Argasina (soft ticks) are characterized by the absence of a dorsal shield (the cuticle is leathery or wrinckled), whilst, in nymphs and adults, the gnathosoma is not visible from above; the paired stigmata are situated, in nymphs and adults, above coxae III–IV.

The comparative morphology and the postembryonic development of both suborders (Ixodina and Argasina) is dealt with in the following subsection.

6.1. *Comparative morphology and postembryonic development of Ixodida*

The present subsection is based on the study of three species of ticks, viz., *Ixodes ricinus* (Linnaeus) (family Ixodidae), *Hyalomma dromedarii* C.L. Koch (family Amblyommidae), and *Ornithodoros savignyi* (Audouin) (family Argasidae). The material of *Ixodes ricinus* originates from various localities in The Netherlands, the material of *Hyalomma dromedarii* and *Ornithodoros savignyi* from localities in Egypt (cf. Van der Hammen, 1983a: 213).

Constitution of the body. – The Ixodid body is divided into two pseudotagmata: gnathosoma and idiosoma. The original subdivision into prosoma and opisthosoma is no more recognizable. The podosoma is restricted to a small sternal (neosternal) area and the circumpedal regions.

Tegument. – According to data in literature (Pagenstecher, 1861; Nordenskiöld, 1908; Arthur, 1962; Babos, 1964; Nathanson, 1967), the skin consists of a living layer (hypodermis) and several non-living layers which are secreted by the hypodermis. Generally, three non-living layers can be distinguished: chitonostracum or procuticle (the thickest layer) and epiostracum or epicuticle (together constituting the cuticle; the epiostracum represents the superficial cuticular layer) and the cerotegument (consisting of a lipid layer with wax blooms and, in Argasidae, an outer cement layer, both developed by exudation through pores in the cuticle). The chitonostracum generally consists of three layers: ectostracum, endostracum and Schmidt's layer. The cuticle can be soft or sclerotized; it contains numerous pore canals extending from the hypodermis to the epiostracum. The soft skin has the ability to stretch enormously during engorgement. The tegument contains tegumental glands and various types of tegumental sense organs. Setae and other sensory phaneres are inserted on the tegument.

Chaetotaxy. – Larval chaetotaxy (fig. 208) appears to be idionymous (setae being capable of receiving an individual designation). Numbers of setae generally increase progressively in nymphs and adults (fig. 209A, C), and chaetotaxy becomes adelonymous (setae being no more capable of receiving an individual designation).

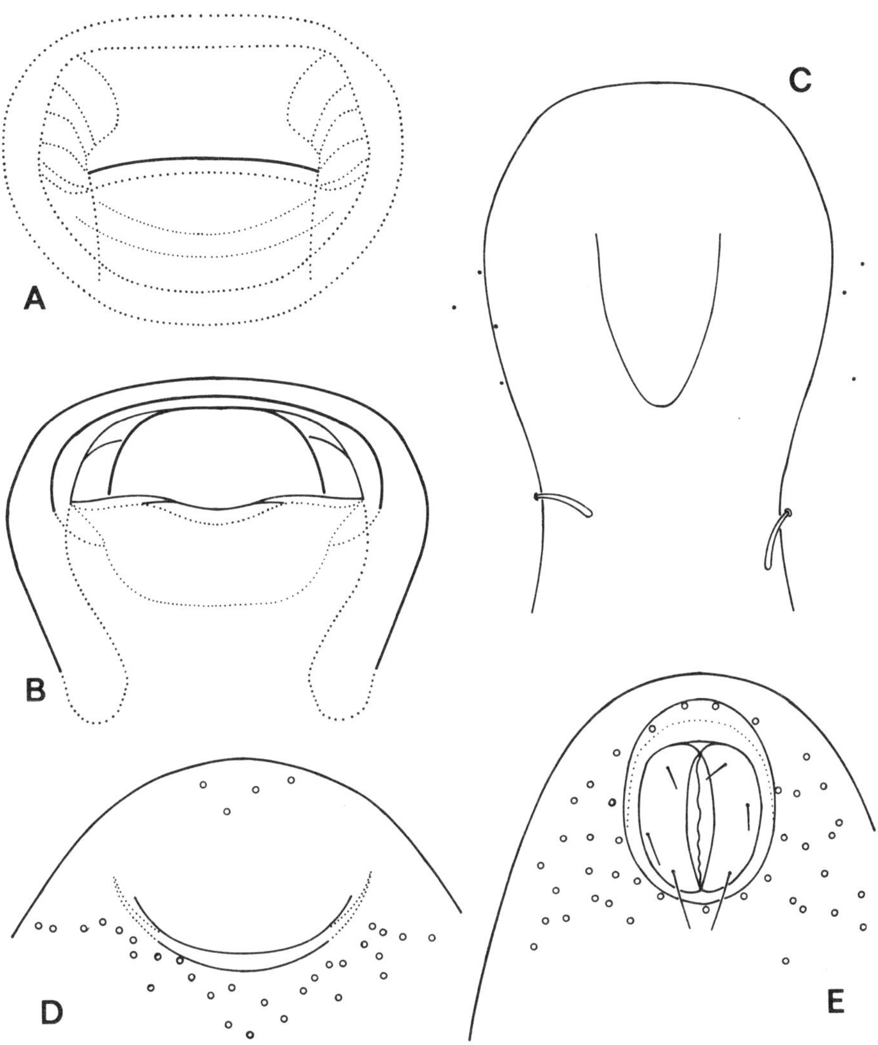

Fig. 210. A, B, *Ornithodoros savignyi* (Audouin), ventral view of genital region; A, female; B, male; C, *Hyalomma dromedarii* C.L. Koch, female, ventral view of genital region; D, E, *Ixodes ricinus* (Linnaeus), female; D, ventral view of genital region; E, ventral view of anal region; A, × 44; B, × 88; C, × 186; D, E, × 141.

Eyes. – Eyes are either absent (fig. 209C), or there are on both sides one or two eyes. In Ixodoidea the eyes are situated, in the female, close to the lateral border of the scutum (fig. 209A) and, in the male, in a corresponding position. In Argasidae the eyes, in case they are present, have a lateral, supracoxal position (fig. 209B). Eyes, in ticks, are simple ocelli that consist of a rounded, transparent part of the cuticle, functioning as a lens, and sensory cells.

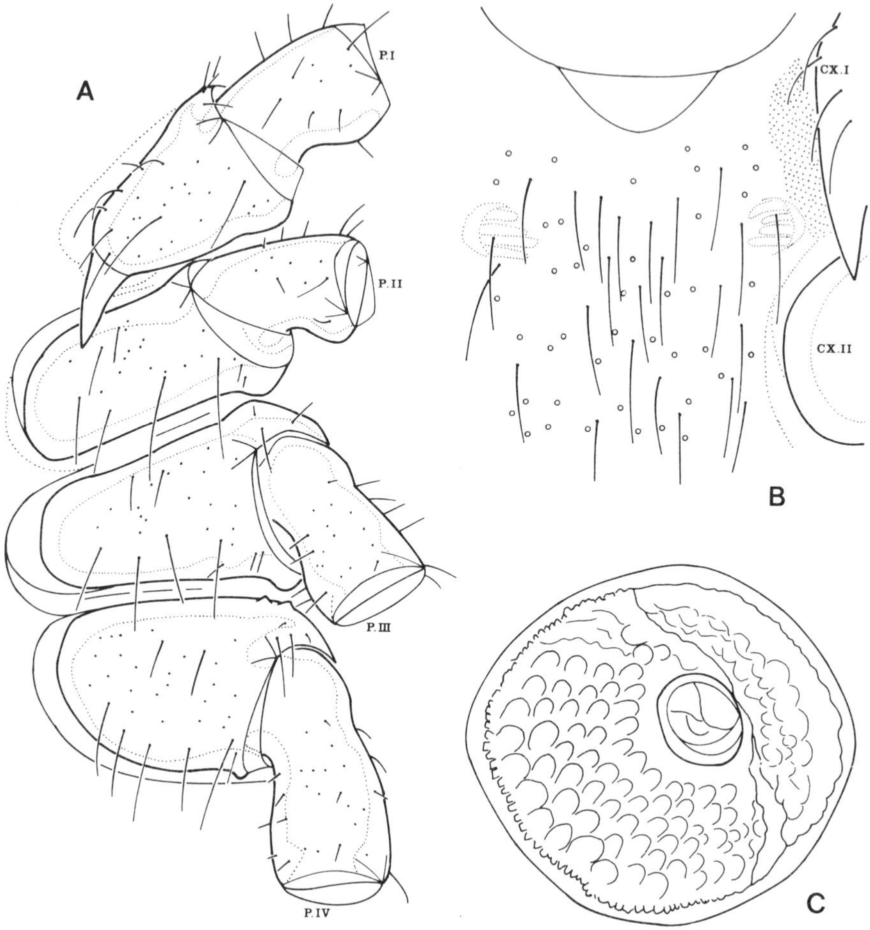

Fig. 211. Ixodes ricinus (Linnaeus), adult female; A, coxae and trochanteres of left legs I–IV, ventral view; B, anterior part of sternal region, ventral view; C, right peritreme and stigma, ventral (slightly oblique) view; A, × 77; B, × 118; C, × 155.

Dorsal aspect of idiosoma. – As mentioned above, the dorsal side of the idiosoma of Ixodoidea presents a scutum (large, occupying the entire surface in the male; small, restricted to the anterior part, in the female and the immature stases); in Argasidae the dorsal side is leathery or wrinkled. Grooves (cervical, lateral, posteromedian, paramedian, accessory, marginal, etc.) apparently correspond with muscle insertions. The festoons along the posterior border of Amblyommidae, and the paired foveae dorsales of the same group, constitute sensory organs.

Ventral aspect of idiosoma. – Apart from the podosoma, the respiratory organs, the genital region and the anal region, which are dealt with in separate paragraphs, the ventral side of the idiosoma presents several grooves (postanal median, anal, genital, etc.). The ventral side of males of the genus *Ixodes* presents a number of ventral sclerites (fig. 207D).

Anal region. – The anus is constituted by a median longitudinal opening in the posterior part of the ventral region. It is bordered by the anal valves (or paraprocts) which can be surrounded by a sclerotized ring (fig. 210E). The valves bear one or more setae; the number of anal setae can increase in the course of postembryonic development. In several taxa the anal chaetotaxy could be neotrichous.

Genital region. – The genital opening appears, in the course of postembryonic development, in the adult stase. It has an advanced position at the level of coxae III and IV, and is constituted by a transverse, more or less rounded slit. In Argasidae, the male genital region has the shape of a horseshoe; in females it is more or less oval (fig. 210A, B). In Ixodina, the genital region is more or less oval and can be surrounded, anteriorly and laterally, by the genital groove (figs. 206B, 207D, 210C, D).

Respiratory organs. – The paired stigmata, which are of postlarval (nymphal) origin, are situated posterior to coxa IV (Ixodina; figs. 206A, 207B, D) or laterally, above coxae III–IV (Argasina; fig. 209B). The split-shaped stigmata are surrounded by a so-called peritreme (fig. 211C), a taenidium with a characteristic microsculpture (this microsculpture is apparently mainly attributable to the microsculpture of the cerotegument).

Lateral aspect of idiosoma. – Unengorged ticks are flat, and the lateral part of the idiosoma is narrow. The lateral region includes the stigmata (described above) and the lateral part of the podosoma (described below). In Argasina, the eyes have a lateral position above the coxae, close to a lateral (supracoxal) fold (fig. 209B).

Podosoma. – The podosoma consists of a sternal (i.e., neosternal) and a circumcoxal region. The sternal region extends from the base of the gnathosoma to the genital region, and laterally to the circumcoxal region (fig. 206B). In males of the genus *Ixodes*, there is a small sternal shield (fig. 207D). The sternal region presents numerous setae and pores (fig. 211B). The circumcoxal region (fig. 211A) is not sclerotized. Legs II–IV do not possess a coxa-body articulation, although muscles are inserted on the base of the coxae. The coxae are probably movable because of the flexibility of the circumcoxal tegument.

Gnathosoma (figs. 212–216). – In all Anactinotrichida, the gnathosoma consists (as in Actinotrichida) of two parts: cheliceral frame and infracapitulum. The cheliceral frame constitutes the body wall between the rostral region of the idiosoma and the dorsal part of the infracapitulum; it includes the tegulum (the dorsal part of the so-called basis capituli; in ticks, muscles are apparently inserted on its base). The cheliceral frame is attached to the dorsal part of the infracapitulum according to the line *at* (fig. 215A, C, D). The cheliceral sheaths are movably attached to the cheliceral frame. In adult females, the tegulum presents a pair of areae porosae (fig. 212).

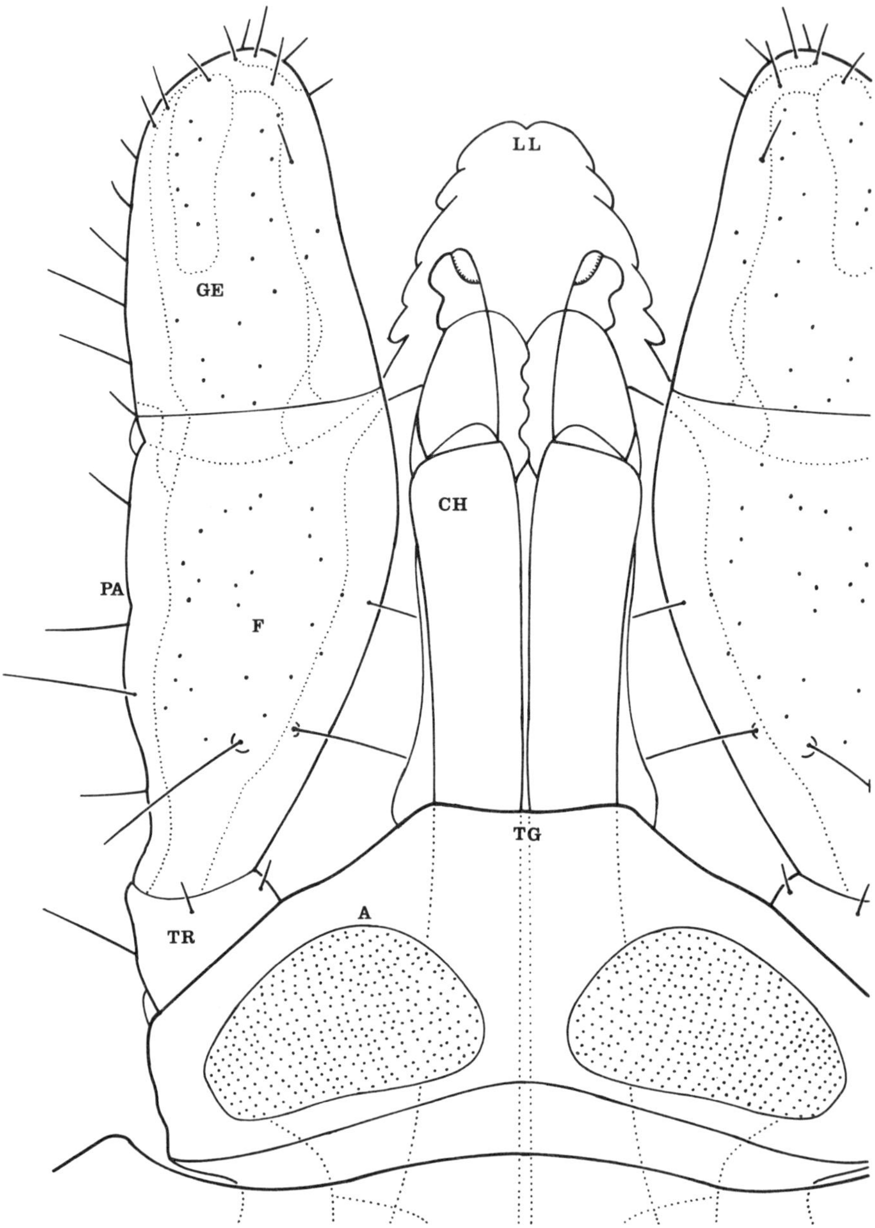

Fig. 212. Ixodes ricinus (Linnaeus), adult female, dorsal view of gnathosoma; × 176.

→

Fig. 213. Ixodes ricinus (Linnaeus), adult female, ventral view of gnathosoma; × 176.

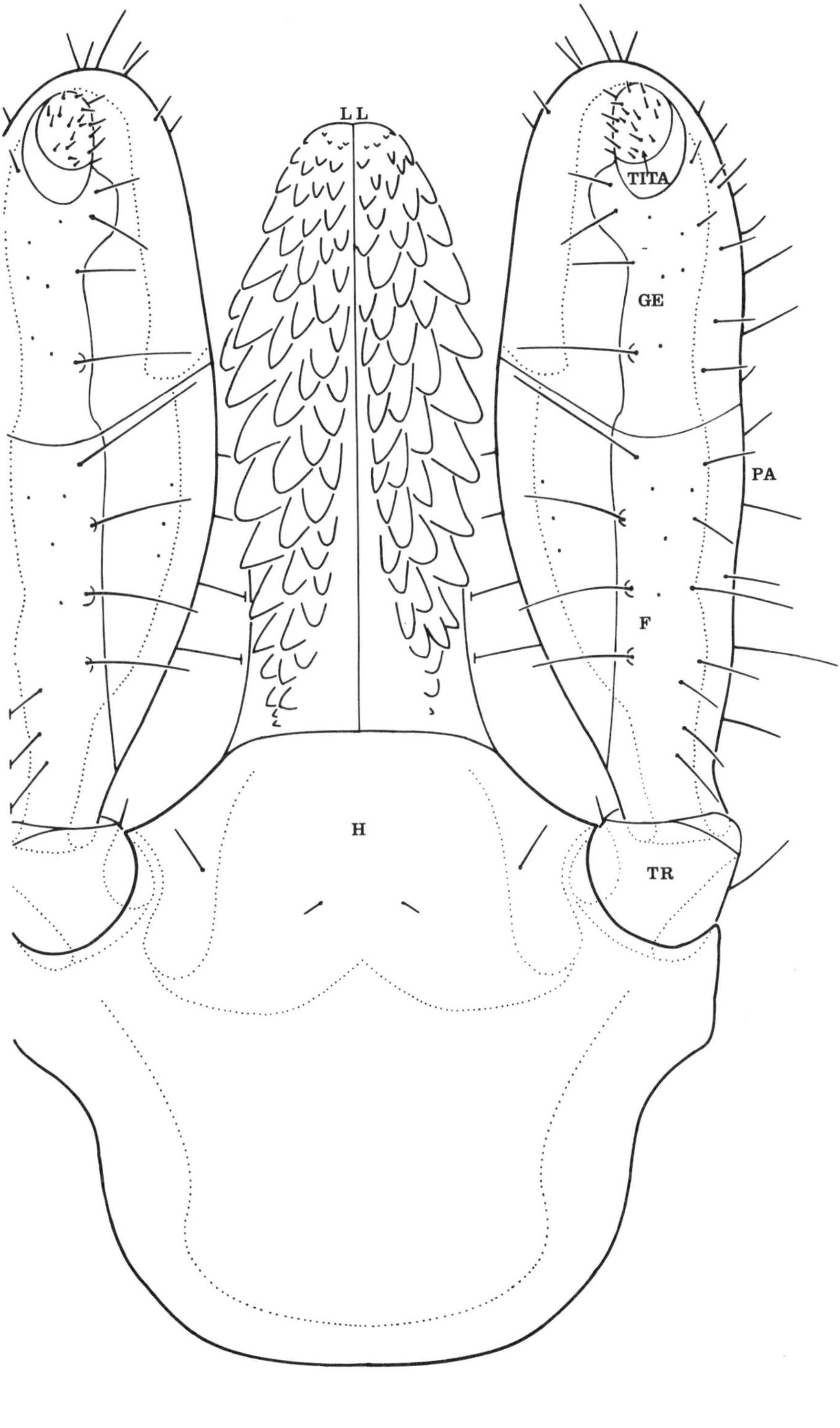
LL
TITA
GE
PA
F
H
TR

The infracapitulum consists of a ventral region (mentum), a small dorsal region (cervix), and the large lateral ridges (including the coxal cavities, and the acetabula of the palps). It bears the lips and the palps, and contains mouth and pharynx. The mentum (figs. 213, 214A, B) presents several infracapitular setae; muscles are inserted on its internal prolongation (the subcapitular apodeme; see fig. 216B). There are three lips (fig. 215): a labrum (upper lip) and the paired lateral lips. The lateral lips present a ventral and a dorsal part. The ventral part constitutes the so-called hypostome (with a denticulate ventral surface, and a membraneous dorsal wall; see figs. 212, 213, 214A, B, D). In *Ornithodoros* (Argasina; see fig. 215C, D), the paired labella (which constitute the dorsal part of the lateral lips) have the shape of a pair of laterodorsal lobes. In *Hyalomma* (Ixodina; see fig. 215A, B), the paired labella have fused (in literature they have erroneously been regarded as part of the cheliceral sheaths), and constitute an anterior prolongation of the cervix, completely covering the sunken labrum (in literature the labrum has been referred to as tongue-like process or flap); there is a distinct longitudinal ridge in the median part. The more or less triangular opening between the lateral lips constitutes the entrance to the preoral cavity. The labral sclerite is associated with the lateral walls of the pharynx, probably constituting part of a suctorial apparatus (figs. 215A, 216A). It may be remarked here that the architecture of this part of the infracapitulum can only be studied after removal of the tegulum and the chelicerae by dissection. The posterior border of the cervix is indicated by the line *at*; anteriorly, the cervix passes into the labella. Laterally, the cervix is bordered by the lateral walls of the gnathosoma, and the large lateral ridges. A transverse section of the pharynx is trifid, each of the branches being split again near the end (fig. 214C). The ducts of the paired infracapitular glands enter the infracapitulum (fig. 214C) and have an orifice in the cheliceral frame close to the line *at* (*Ornithodoros*; see fig. 215C, D), or in the cervix (*Hyalomma*; see fig. 215A, B).

Chelicera. – The paired chelicerae consist of three segments: trochanter, body of the chelicera and apotele (fig. 215D); the exact limits of these are not easily distinguishable; muscles are inserted on the base of the trochanter (fig. 216B: ttr_s, ttr_i). Trochanter and body of the chelicera are apparently incompletely separated by an area of soft tegument with more or less dorsal position (fig. 217D). The shape of the chela is considerably different from that in other groups of Anactinotrichida (figs. 212, 213, 217A–C). It consists of a so-called hood, which represents the fixed jaw of other Anactinotrichida, and a so-called digit representing the apotele or movable jaw. The apotele is operated by the usual two tendons (t_s and t_i). The digit consists of a so-called internal article and a so-called external article. The tendons are inserted on the internal article which articulates with the body of the chelicera. The external article articulates with the internal article. A so-called dorsal process, or third apophysis, is found in the distal part of the internal article.

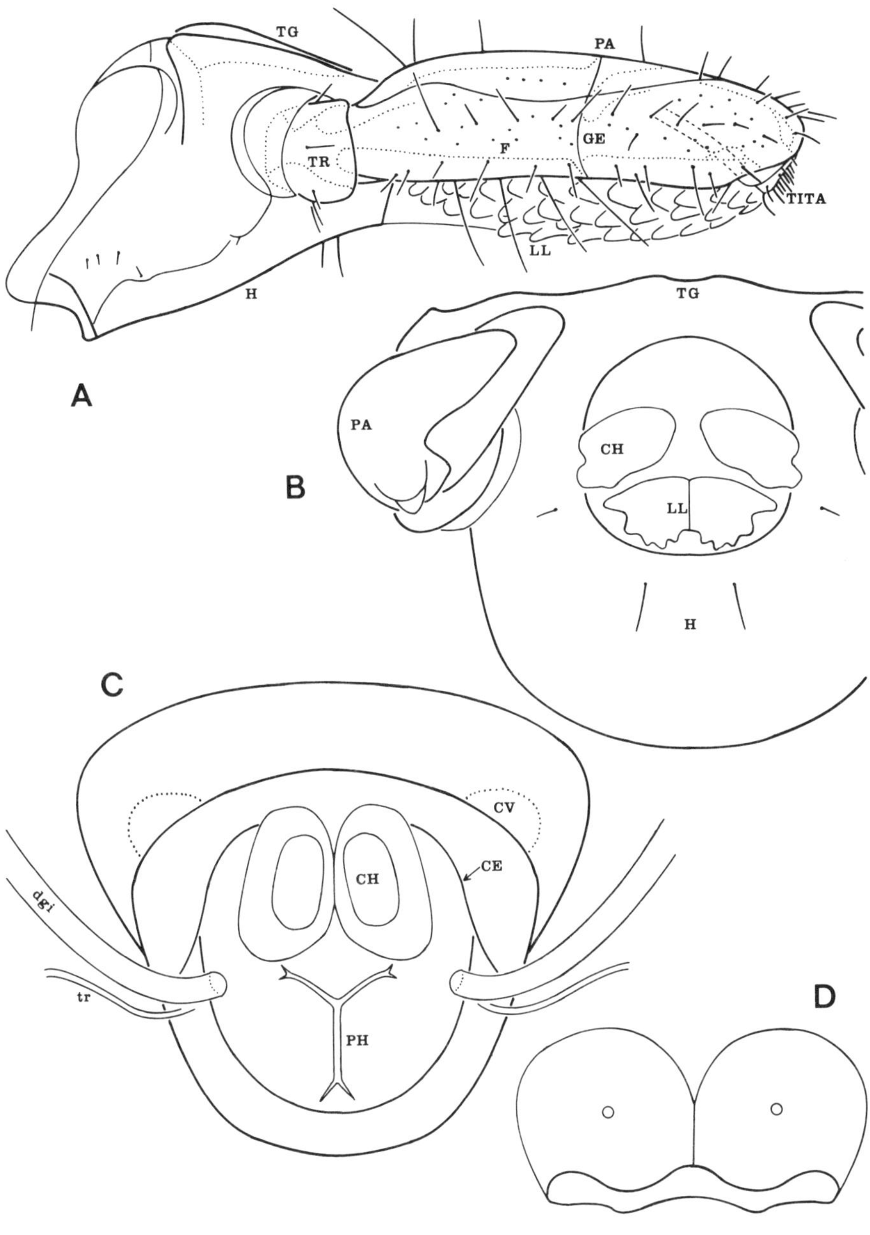

Fig. 214. Ixodes ricinus (Linnaeus), adult female; A, lateral view of gnathosoma; B, frontal view of gnathosoma; C, posterior (oblique) view of gnathosoma; D, transverse section (near the base) of the lateral lips; A, × 92; B, × 141; C, × 108; D, × 224.

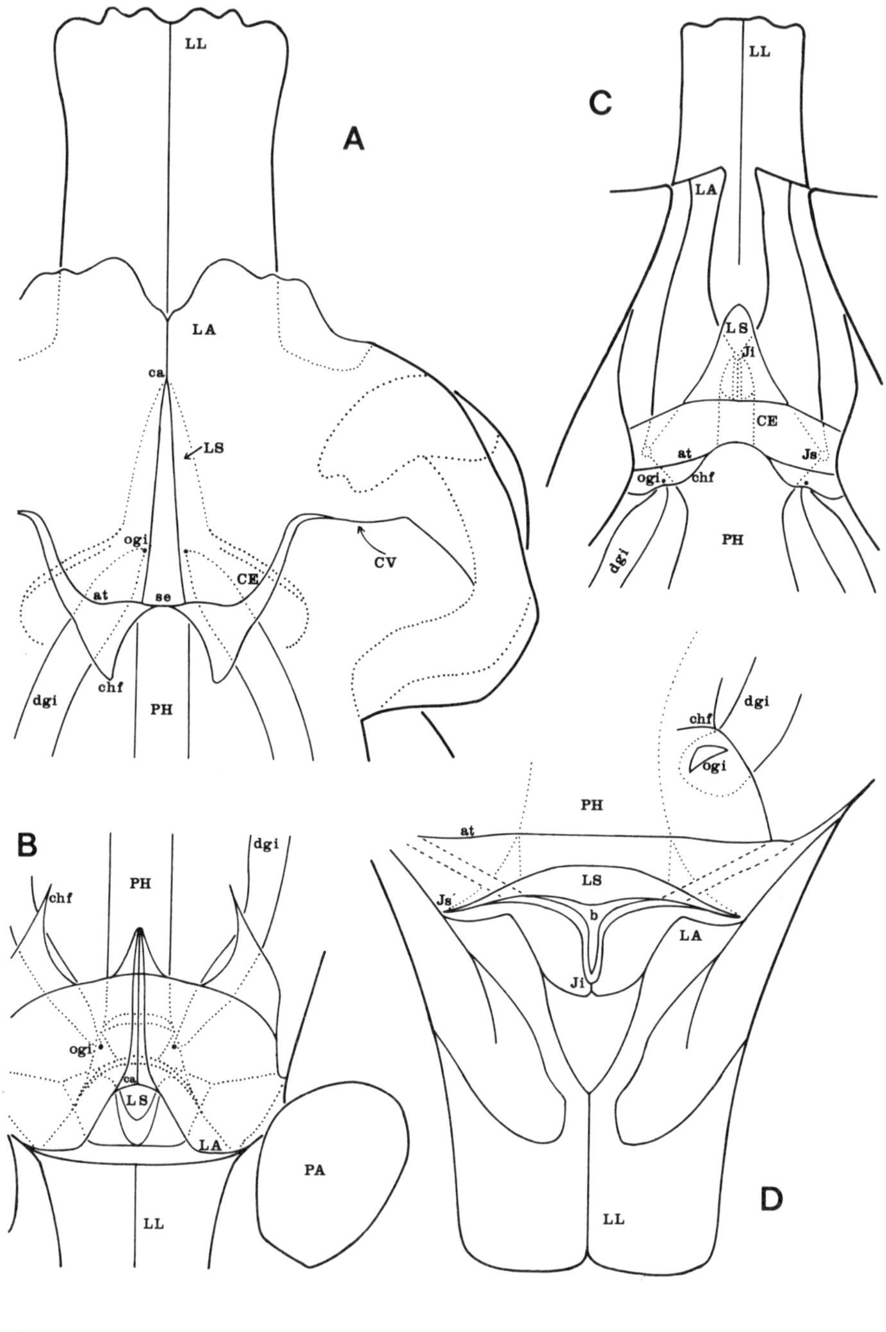

Fig. 215. A, B, *Hyalomma dromedarii* C.L. Koch, gnathosoma of adult female (tegulum and chelicerae removed); A, dorsal view; B, dorsofrontal view; C, D, *Ornithodoros savignyi* (Audouin), gnathosoma of adult female (tegulum and chelicerae removed); C, dorsal view; D, frontal view; A–C, × 88; D, × 141.

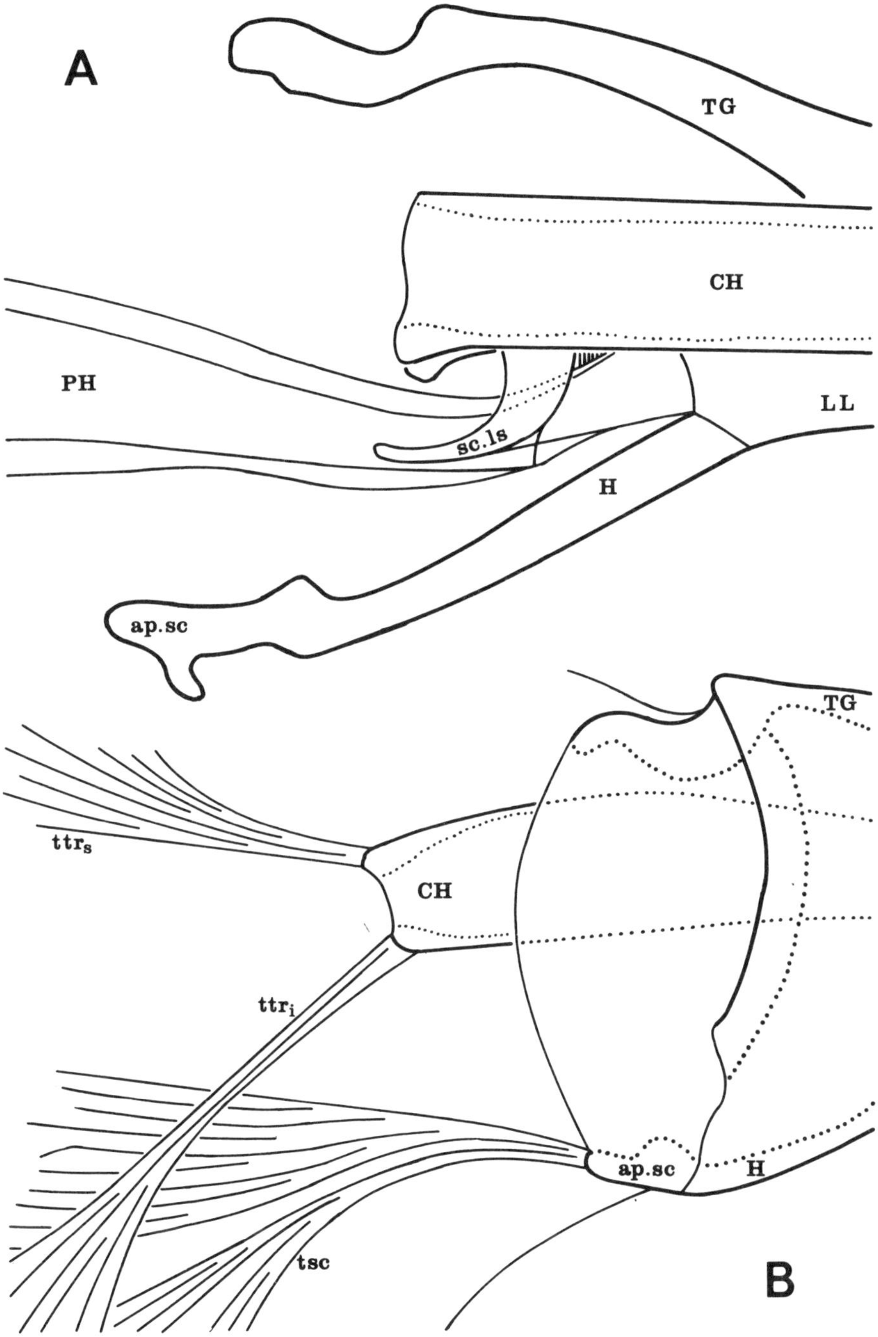

Fig. 216. For legend see page 416.

Palp. – The Ixodid palp (figs. 212, 213, 214, 218, 219) consists of four segments: trochanter, femur, genu and tibiotarsus. In Opilioacarida, Holothyrida and Gamasida, the tarsus of the palp is relatively small, whilst tibia and tarsus together have the shape of a single segment; in Ixodida tibia and tarsus have apparently been integrated into a single tibiotarsus. There is no palpal apotele. In Argasidae (fig. 219) the shape of the palp resembles that of a leg. In Ixodoidea (fig. 218) the palp has a specialized shape: it is flattened and presents a paraxial concavity, adapted to the shape of the lateral lips (see fig. 214B). In Argasidae, the tibiotarsus has the shape of a normal terminal segment. In Ixodoidea, the tibiotarsus is relatively very small and has shifted to a ventrodistal position. The articulation between gnathosoma and trochanter is constituted by a rocking joint, the trochanter-femur articulation by a pivot joint. The articulation between femur and genu allows of little mobility (the segments have been integrated in the case of the larva). The articulation between genu and tibiotarsus is constituted by a hinge joint with flexor muscles. The palp presents setae and other phaneres (eupathidia-like phaneres are found in the distal part of the tibiotarsus). In Ixodoidea, setae of the paraxial side can constitute a comb extending over femur and genu; the comb is apparently associated with the lateral lips. Pores can be found in all segments.

Legs (figs. 220–228). – The legs of Ixodina and Argasina consist of seven eudesmatic segments: coxa, trochanter, femur, genu, tibia, tarsus and apotele (figs. 220A, 222A, 223A, 224A, B, 226, 227). The femur of legs I–IV (figs. 221A, 222B, 223B, 224C) is subdivided by a basifemoral ring into two adesmatic segments: basifemur and telofemur (the adesmatic articulation allows of very little mobility). The tarsus of legs II–IV (figs. 222F, 223E, 224F) is subdivided by a basitarsal ring into two adesmatic segments: basitarsus and telotarsus (the adesmatic articulation allows of very little mobility).

The articulation between coxa I and body allows of some mobility. There is no articulation between the body and coxae II–IV, but (as mentioned in the paragraph on the podosoma) the elasticity of the circumcoxal tegument probably allows of some coxal movability by the muscles inserted on the base of the coxa. The eudesmatic articulation between coxa and trochanter is apparently a bidesmatic rocking joint. The eudesmatic trochanter-femur articulation is constituted by a bidesmatic bicondylar pivot joint with levator and depressor muscles. The femur-genu, genu-tibia, and tibia-tarsus articulations are constituted by hinge joints with broadened and flattened superior hinge; they present inferior flexor muscles. The study of leg musculature is thwarted by the thick and dark cuticle of many ticks. In the case of *Hyalomma dromedarii*, in which species the cuticle of the appendages is less dark, I could establish the

←

Fig. 216. Hyalomma dromedarii C.L. Koch, basal part of gnathosoma of adult female; A, longitudinal section, close to pharynx (basal part of chelicera removed); B, lateral view of entire base; A, × 132; B, × 90.

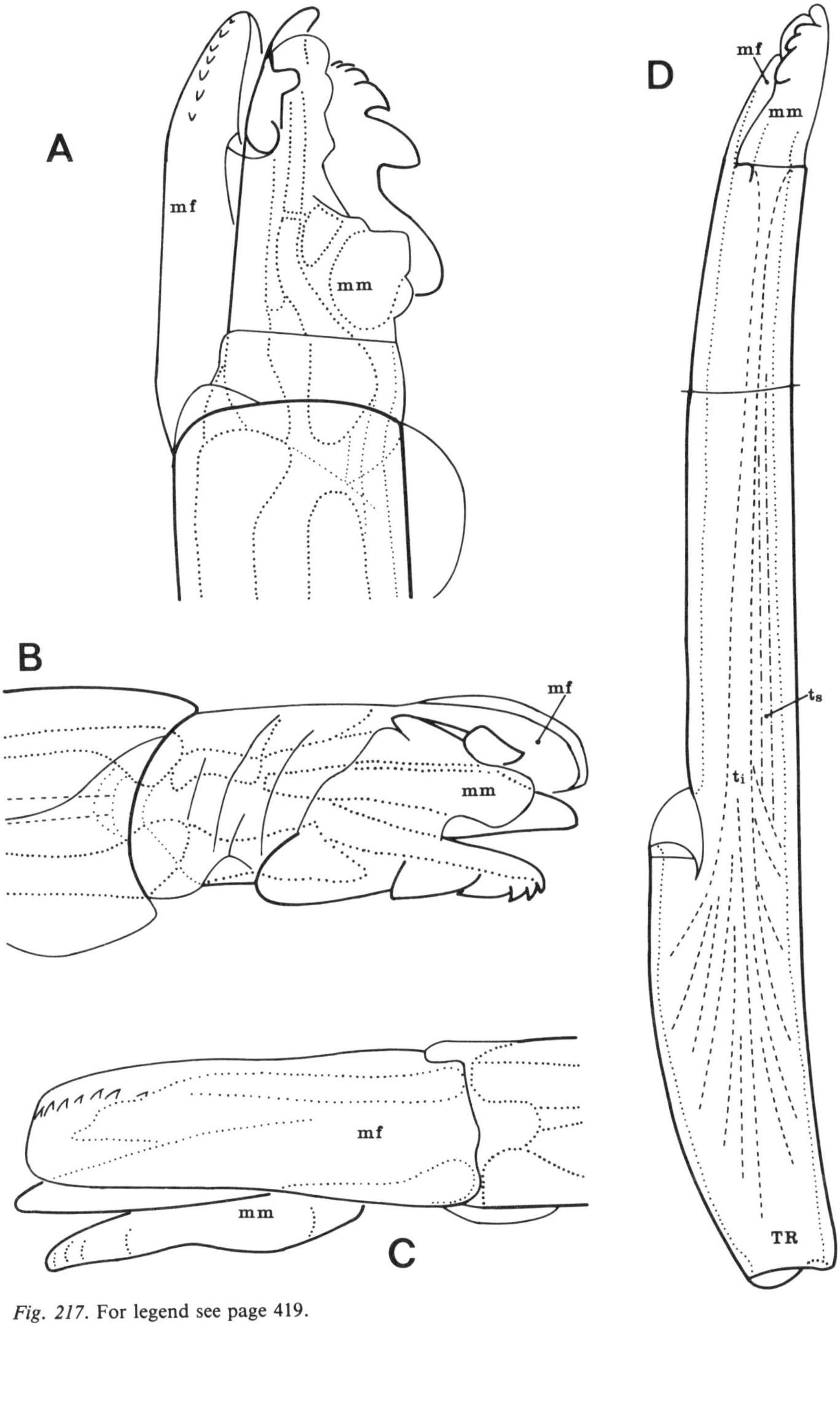

Fig. 217. For legend see page 419.

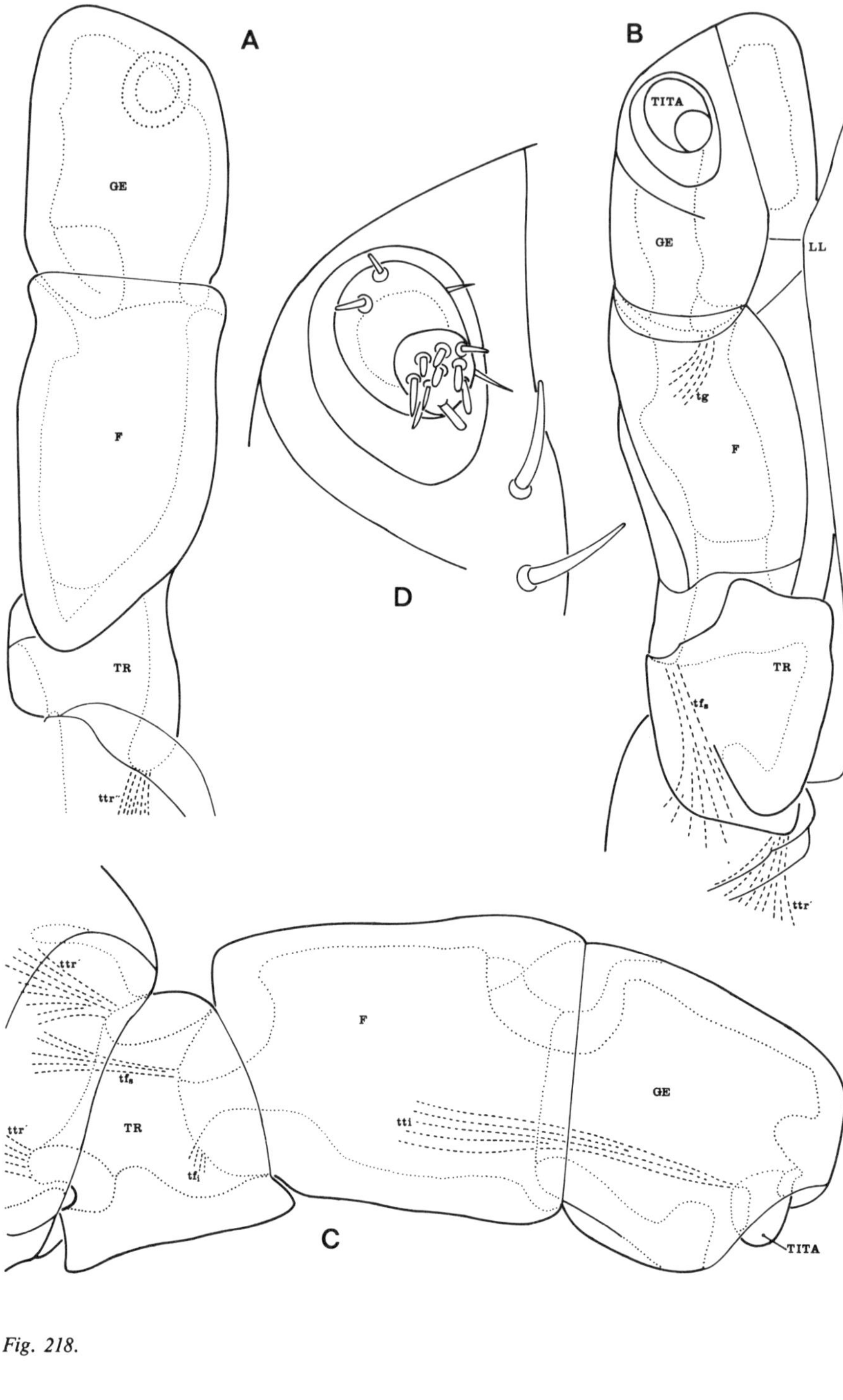

Fig. 218.

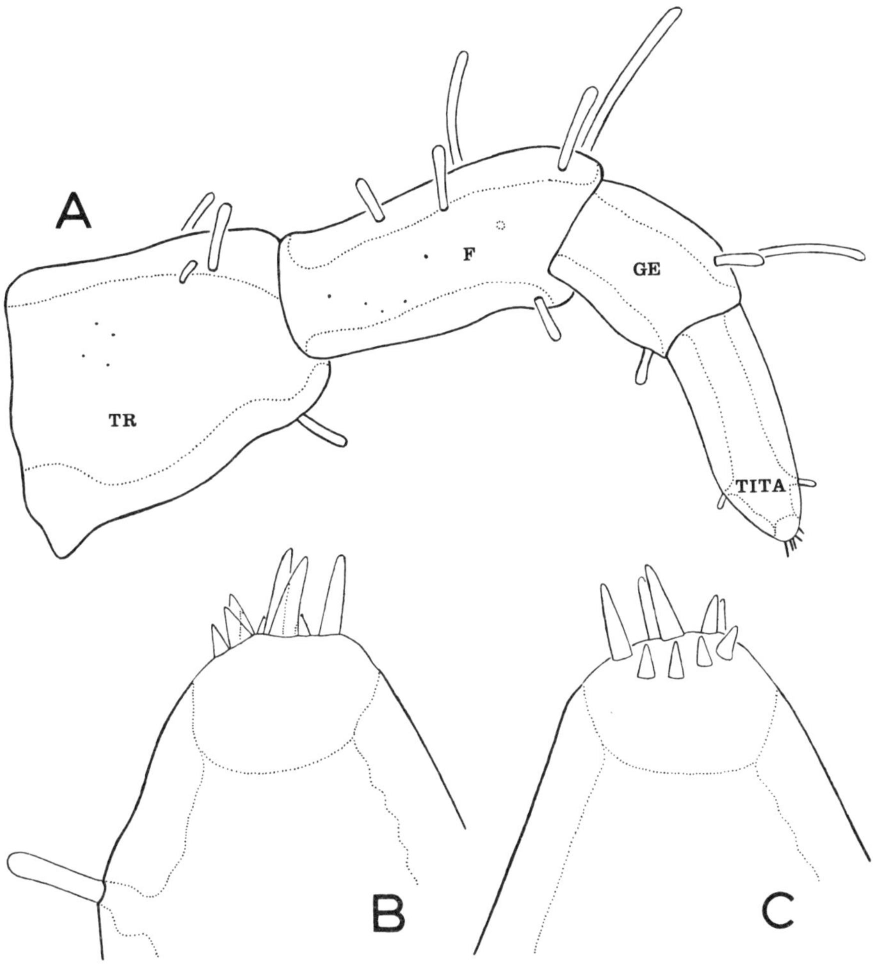

Fig. 219. Ornithodoros savignyi (Audouin), right palp of adult female; A, lateral (antiaxial) view; B, C, distal part of tibiotarsus; B, lateral (antiaxial) view; C, lateral (paraxial) view; A, × 120; B, C, × 830.

← ←

Fig. 217. A–C, *Ixodes ricinus* (Linnaeus), distal part of righ chelicera of adult female; A, dorsal (slightly oblique) view of chelicera in situ; B, lateral (antiaxial) view; C, lateral (paraxial) view; D, *Hyalomma dromedarii* C.L. Koch, right chelicera of adult female, lateral (antiaxial) view; A–C, × 368; D, × 110.

←

Fig. 218. Hyalomma dromedarii C.L. Koch, right palp of adult female; A, dorsal view; B, ventral view; C, lateral (antiaxial) view (in figs. A–C, all phaneres are omitted); D, ventral view of terminal part; A–C, × 118; D, × 246.

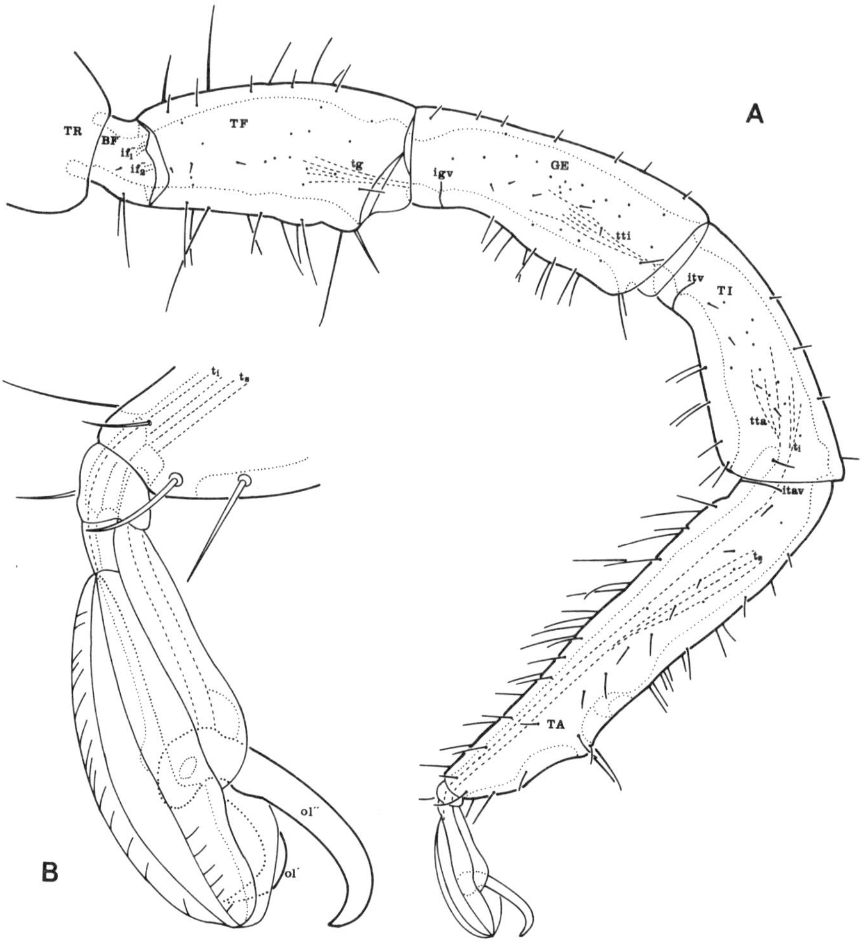

Fig. 220. Ixodes ricinus (Linnaeus), adult female, lateral (antiaxial = posterior) view of right leg I; A, distal part of trochanter, femur, genu, tibia, tarsus and apotele; B, distal part of tarsus, and apotele; A, × 77; B, × 246.

occurrence of extensor muscles associated with the femur-genu articulation (fig. 226); this extensor was already described by Ruser (1933; in her opinion, part of the flexor muscles could also function as extensors, which does not seem very probable). The tarsus-apotele articulation is constituted by a bidesmatic bicondylar pivot joint with levator and depressor muscles (there is a superior tendon t_s and an inferior tendon t_i).

The basifemoral ring is constituted by an external suture (corresponding with an internal cuticular fissure) associated with several (sometimes compound) lyrifissures. The basitarsal ring of legs II–IV is constituted by a suture

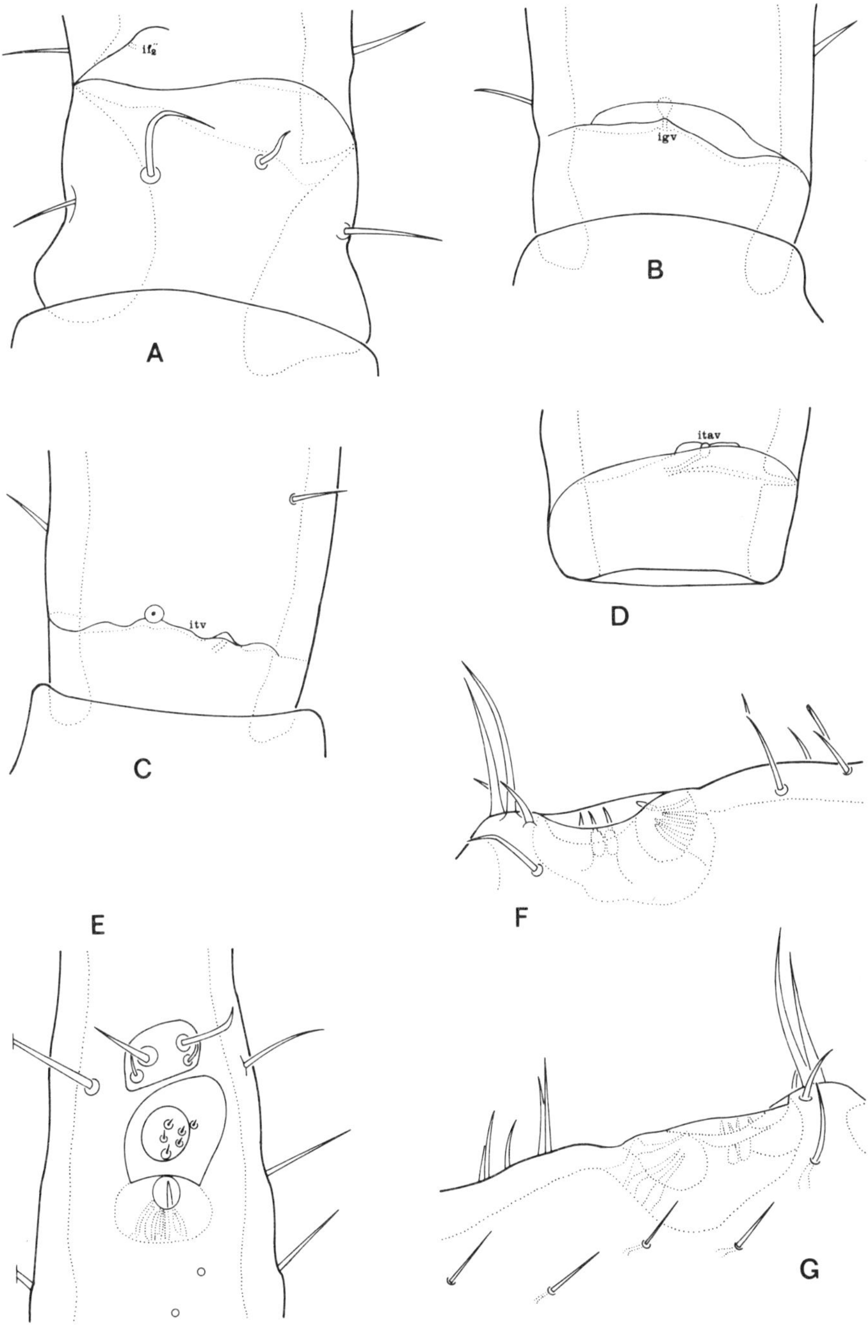

Fig. 221. For legend see page 423.

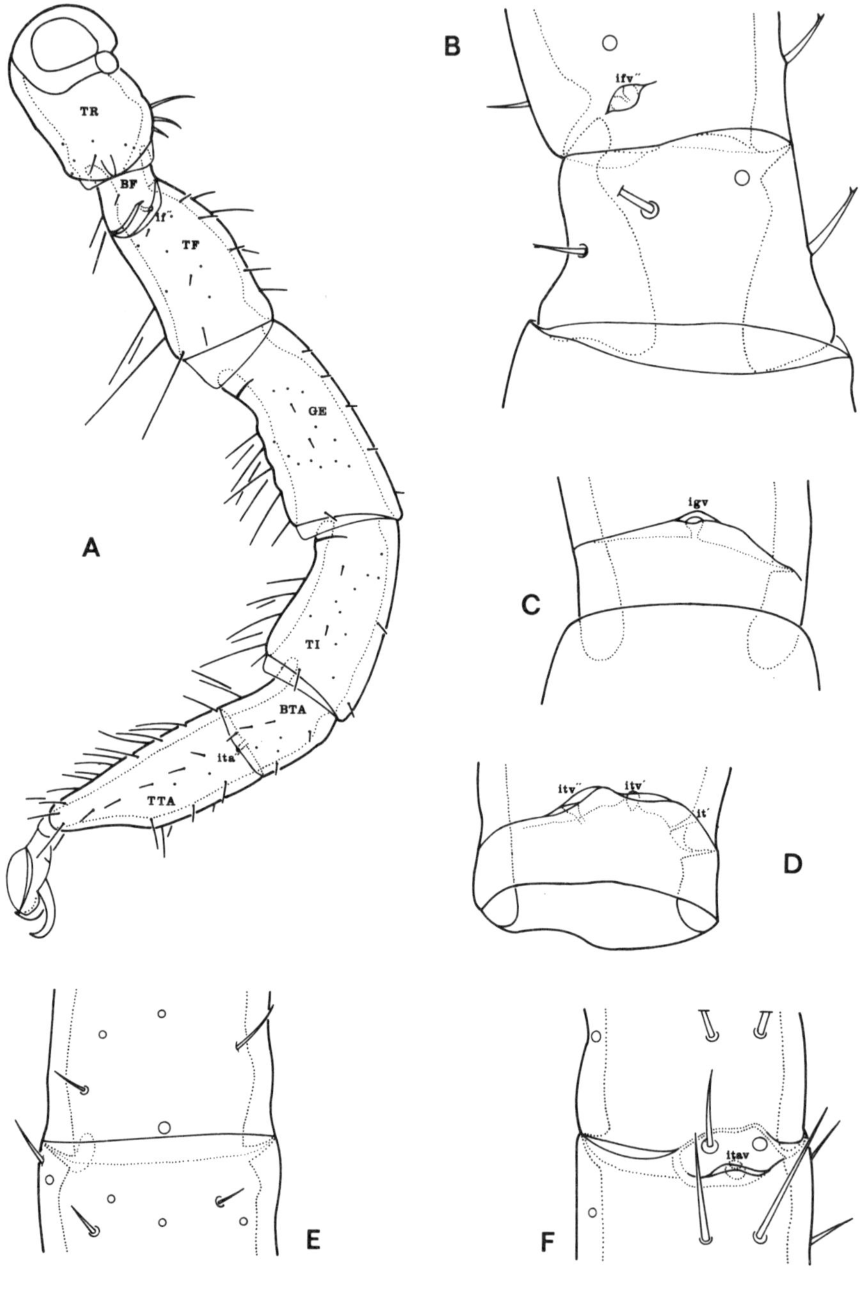

Fig. 222. For legend see page 423.

corresponding with an internal cuticular fissure and associated with lateral and ventral lyrifissures. In Ixodoidea (not in Argasidae) the ventral lyrifissure *itav* is associated with a 'sclerite' (with two setae) which is bordered by a suture (corresponding with an internal cuticular fissure). Externally, the cuticle of the 'sclerite' is continuous with that of the surrounding part of the tarsus; consequently, it does not constitute an 'intercalary sclerite', as supposed by Edwards & Evans (1967: 598, 601, fig. 2B). The function of the 'sclerite' is probably associated with the sensory function of the ventral lyrifissure *itav* (apparently the detection of minute deformations in the exoskeleton). Genu and tibia I–IV (figs. 221B, C, 222C, D, 223C, D, 224D, E) and tarsus I (fig. 221D) present one or more basiventral lyrifissures: one ventral lyrifissure (*igv*) in the case of the genu, three lyrifissures (*it''* or *itv''*, *itv*, and *it'* or *itv'*) in the case of the tibia, and one ventral lyrifissure (*itav*) in the case of tarsus I.

The legs present setae and other phaneres. Several types of sensory phaneres are concentrated in Haller's organ (in leg I); this organ consists of an anterior trough and a capsule (figs. 221E–G, 228B, C). Various types of pores are present in the cuticle of the legs.

The ambulacrum consists of a small pretarsus, an apotele with two ungues, and a pulvillus (figs. 220B, 225). The pulvillus is reduced in the family Argasidae (fig. 227C).

Internal anatomy. – Summaries of data on the internal anatomy were published by T.E. Hughes (1959), Arthur (1962) and Babos (1964). Recently (Van der Hammen, 1979a), I prepared a concise summary of our knowledge on Cryptognomic and Anactinotrichid internal anatomy. The following summary is entirely based on data from literature. The alimentary canal comprises pharynx, oesophagus, midgut and caeca, and hindgut. The excretory system comprises the paired Malpighian tubes (debouching into the terminal part of the hindgut) and the paired coxal glands (in fully engorged or ovigerous fe-

← ←

Fig. 221. Ixodes ricinus (Linnaeus), details of right leg I of adult female; A, basal part of femur, ventral view; B, basal part of genu, ventral view; C, basal part of tibia, ventral view; D, basal part of tarsus, ventral view; E–G, Haller's organ; E, dorsal view; F, lateral (paraxial = anterior) view; G, lateral (posterior) view; A–G, × 246.

←

Fig. 222. Ixodes ricinus (Linnaeus), right leg II of adult female; A, lateral (posterior) view of trochanter, femur, genu, tibia, tarsus and apotele; B, basal part of femur, ventral view; C, basal part of genu, ventral view; D, basal part of tibia, ventral view; E, F, basitarsal ring; E, dorsal view; F, ventral view; A, × 77; B–F, × 246.

→

Fig. 223. Ixodes ricinus (Linnaeus), right leg III of adult female; A, lateral (posterior) view of trochanter, femur, genu, tibia, tarsus and apotele; B, basal part of femur, ventral view; C, basal part of genu, ventral view; D, basal part of tibia, ventral view; E, basitarsal ring, ventral view (in figs. B–E, several setae are represented by their alveolus only); A, × 77; B–E, × 246.

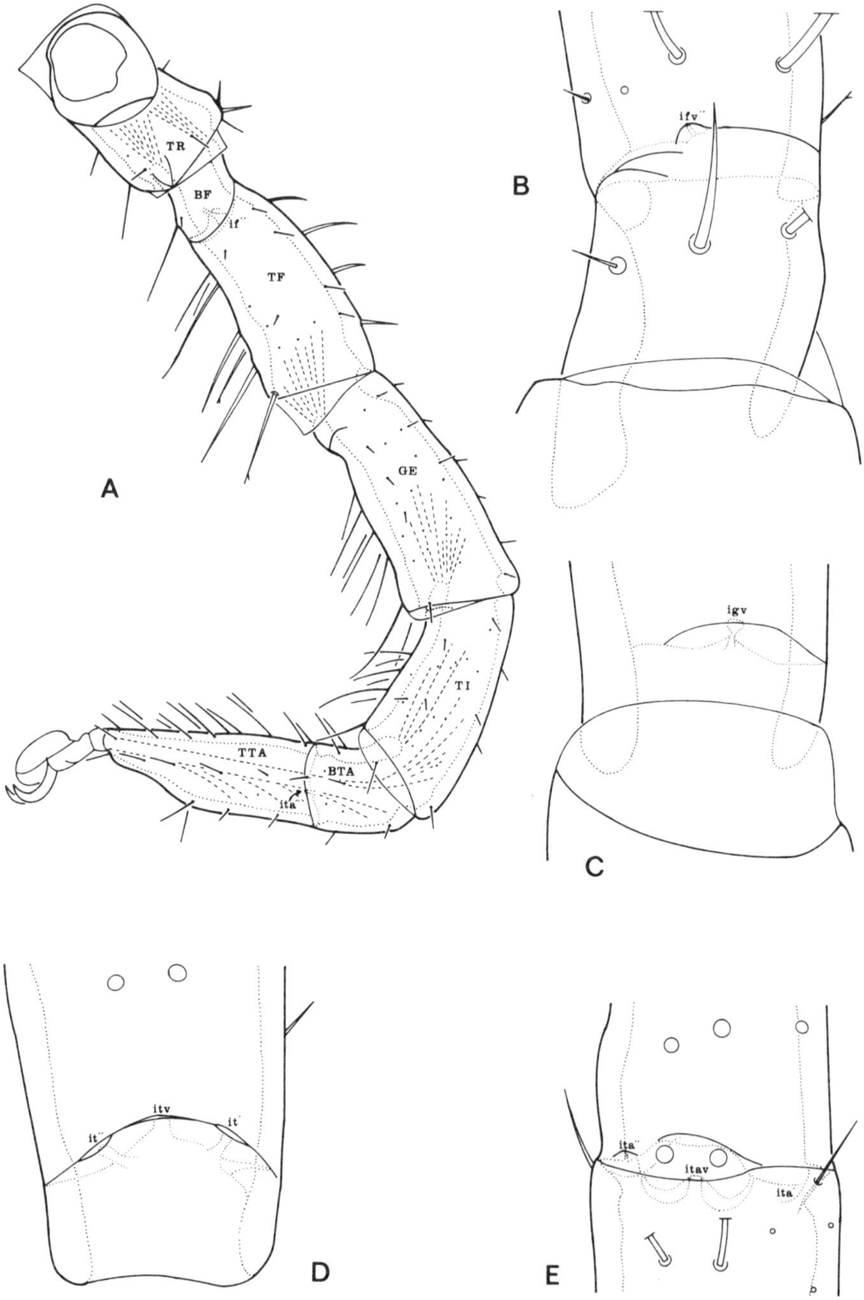

Fig. 223. For legend see page 423.

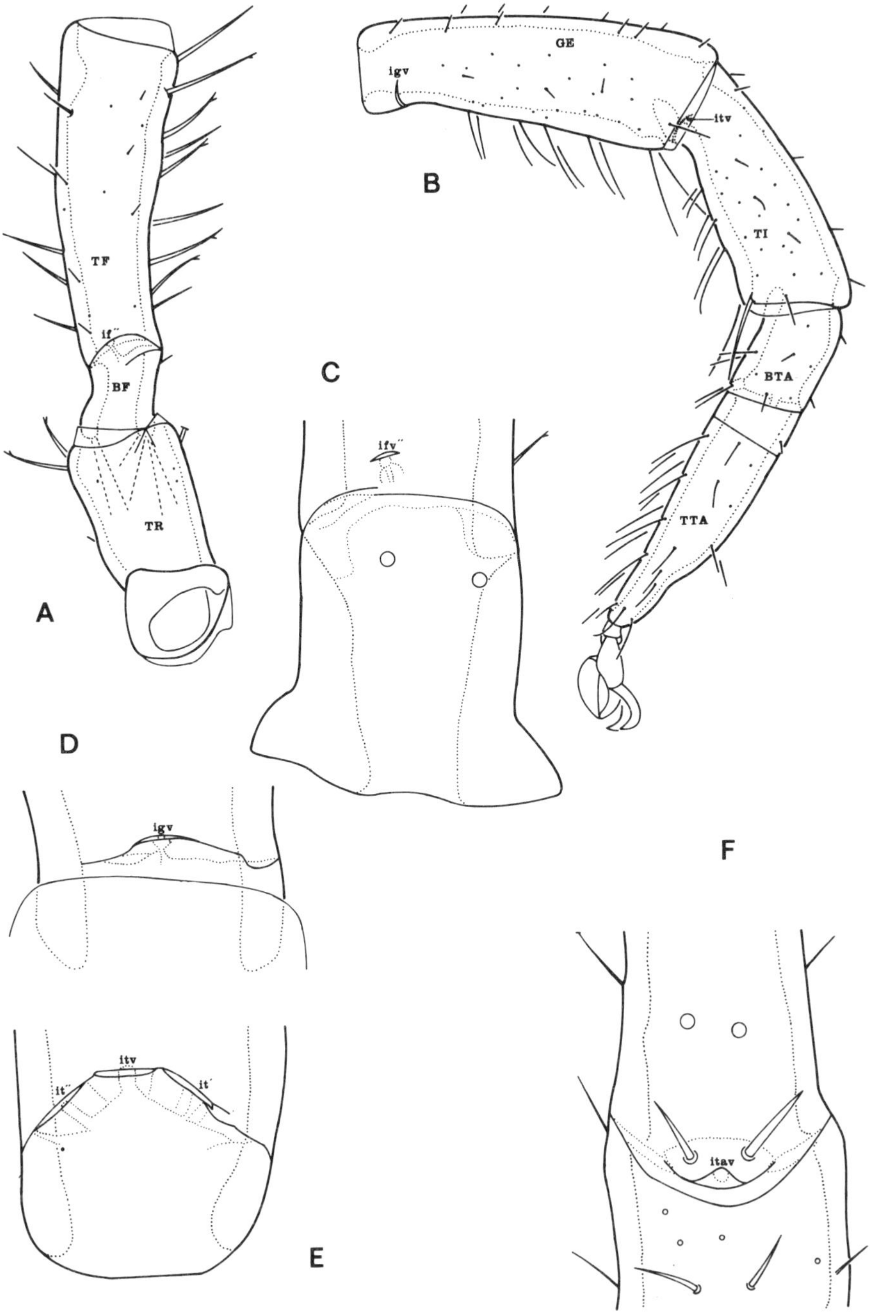

Fig. 224. For legend see page 426.

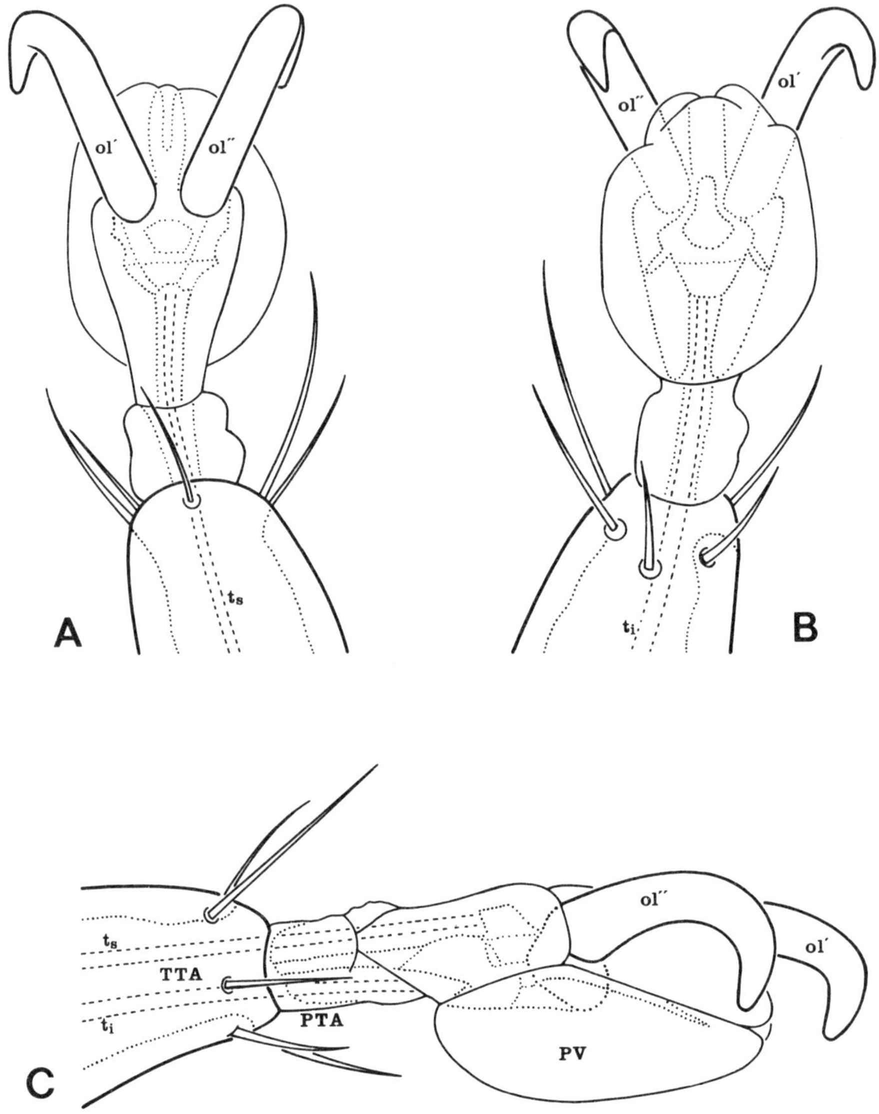

Fig. 225. Ixodes ricinus (Linnaeus), adult female, ambulacrum of right leg II; A, dorsal view; B, ventral view; C, lateral (posterior) view; A–C, × 368.

←

Fig. 224. Ixodes ricinus (Linnaeus), right leg of adult female; A, lateral (posterior) view of trochanter and femur; B, lateral (posterior) view of genu, tibia, tarsus and apotele; C, basal part of femur, ventral view; D, basal part of genu, ventral view; E, basal part of tibia, ventral view; F, basitarsal ring, ventral view; A–B, × 77; C–F, × 246.

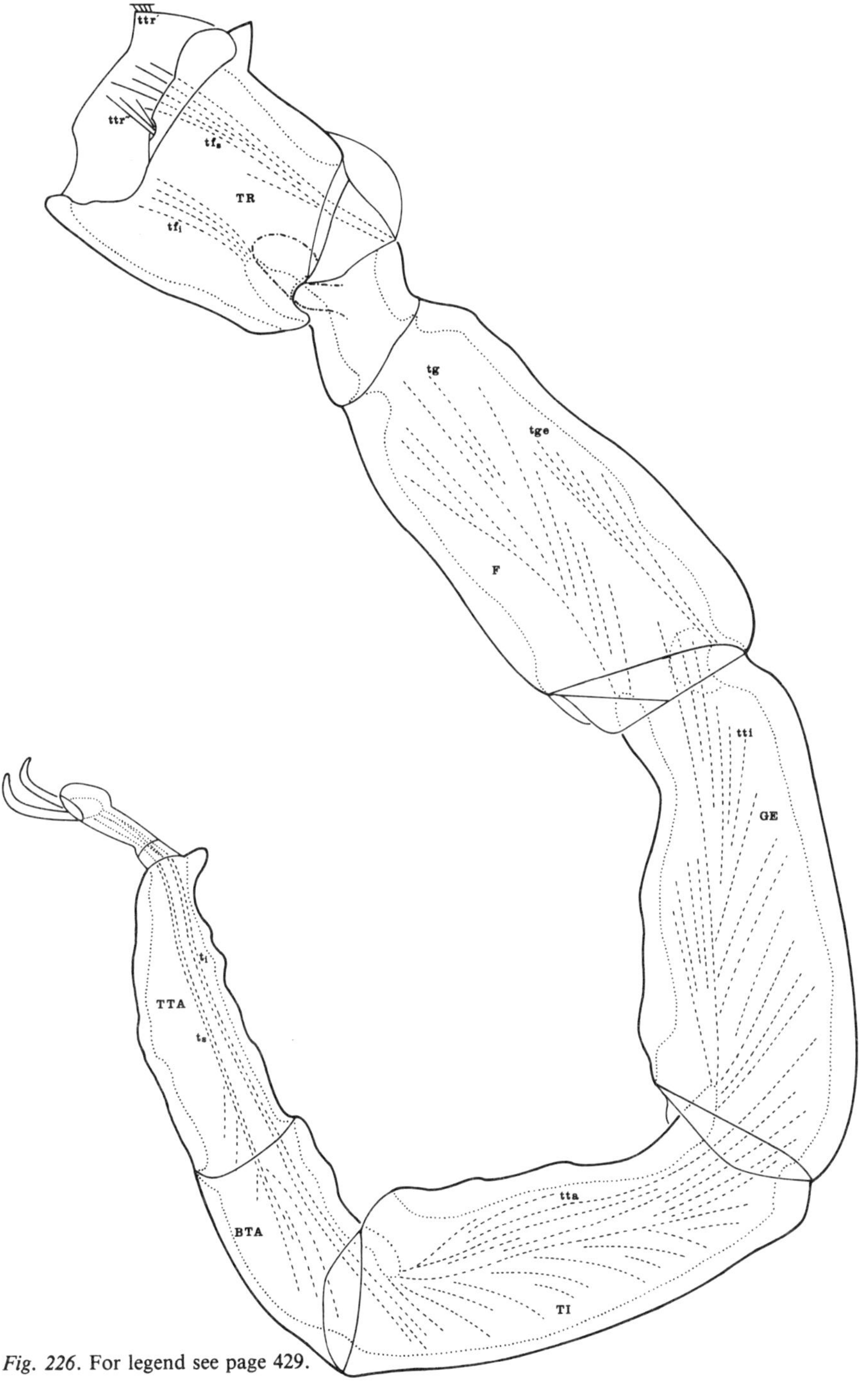

Fig. 226. For legend see page 429.

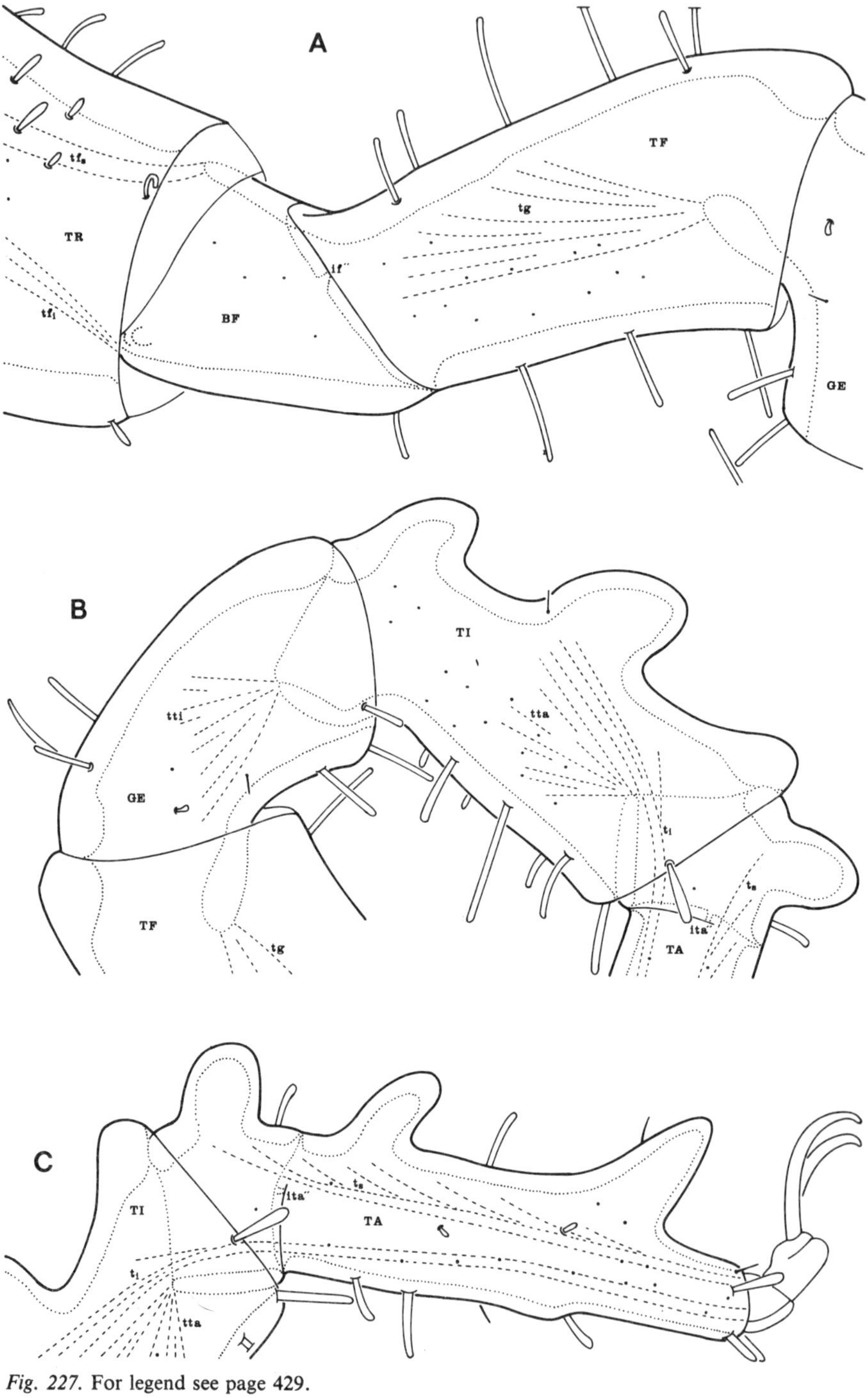

Fig. 227. For legend see page 429.

males they fulfil the function of the Malpighian tubes; the orifice is posterior to coxa I). According to Firstman (1973), the endosternite has a single pair of persistent transverse suspensors and three pairs of dorsoventral suspensors; it can be characterized as rather primitive. The circulatory system consists of a heart, a pair of ostia, an aorta, a periganglionic sinus, and arteries. The respiratory system consists, at both sides, of a stigma, a subostial space, an atrium and tracheae. The central nervous system is concentrated to a single mass, pierced by the oesophagus. The supraoesophageal mass consists of the cerebral and cheliceral ganglia. The pedipalpal ganglia lie level with the entry of the oesophagus in the brain (or somewhat lower). The suboesophageal nerve mass includes the four pairs of ganglia associated with the podosomatic segments. Posterior to the fourth pair of pedal ganglia, lie several pairs of opisthosomatic ganglia. The male testes are elongate structures; narrow vasa deferentia arise from their frontal end, and join to form a common ejaculatory duct (accessory glands open, near the base of the ejaculatory duct, into the lumen). The paired female ovaria are tubular; the oviducts join to form a common oviduct which passes into the vagina (in some cases the oviducts pass into a bilobate uterus; the vagina can function as a primitive ovipositor). A paired infracapitular gland is situated in the anterior part of the idiosoma; as mentioned above, the orifices are in the gnathosoma.

Reproduction.—Reproduction in ticks is generally bisexual. Thelytokous parthenogenesis is sporadically found, in Argasina and Ixodina, in species which normally reproduce bisexually (Oliver, 1971: 291–294). Sperm transfer is nearly direct; the male implants a spermatophore in the female system by means of chelicerae and gnathosoma (this is a case of gonopody). Oviposition takes place by means of a primitive ovipositior.

Postembryonic development (figs. 206–208).—As far as known, the Ixodid life-cycle comprises three stases: larva, nymph and adult. A prelarva has not been found, although a calyptostasic prelarva could remain enclosed in the egg (a prelarva is known from Opilioacarida). The larva is hexapod. The single nymph could be homologous with the Opilioacarid protonymph (in that case the Ixodid adult would be neotenous). In Argasidae, the nymph can pass through growing- or repetition-moults; the forms which are the result of this type of moulting are not different from the preceding form (except in size), and are called isophena of the nymph.

← ←

Fig. 226. Hyalomma dromedarii C.L. Koch, segmentation and articulation of right leg II of adult female, lateral (posterior) view (coxa not included; setae omitted); × 73.

←

Fig. 227. Ornithodoros savignyi (Audouin), right leg III of adult female, lateral (posterior) view; A, distal part of trochanter, femur and basal part of genu; B, distal part of femur, genu, tibia and basal part of tarsus; C, distal part of tibia, tarsus and apotele; A–C, × 77.

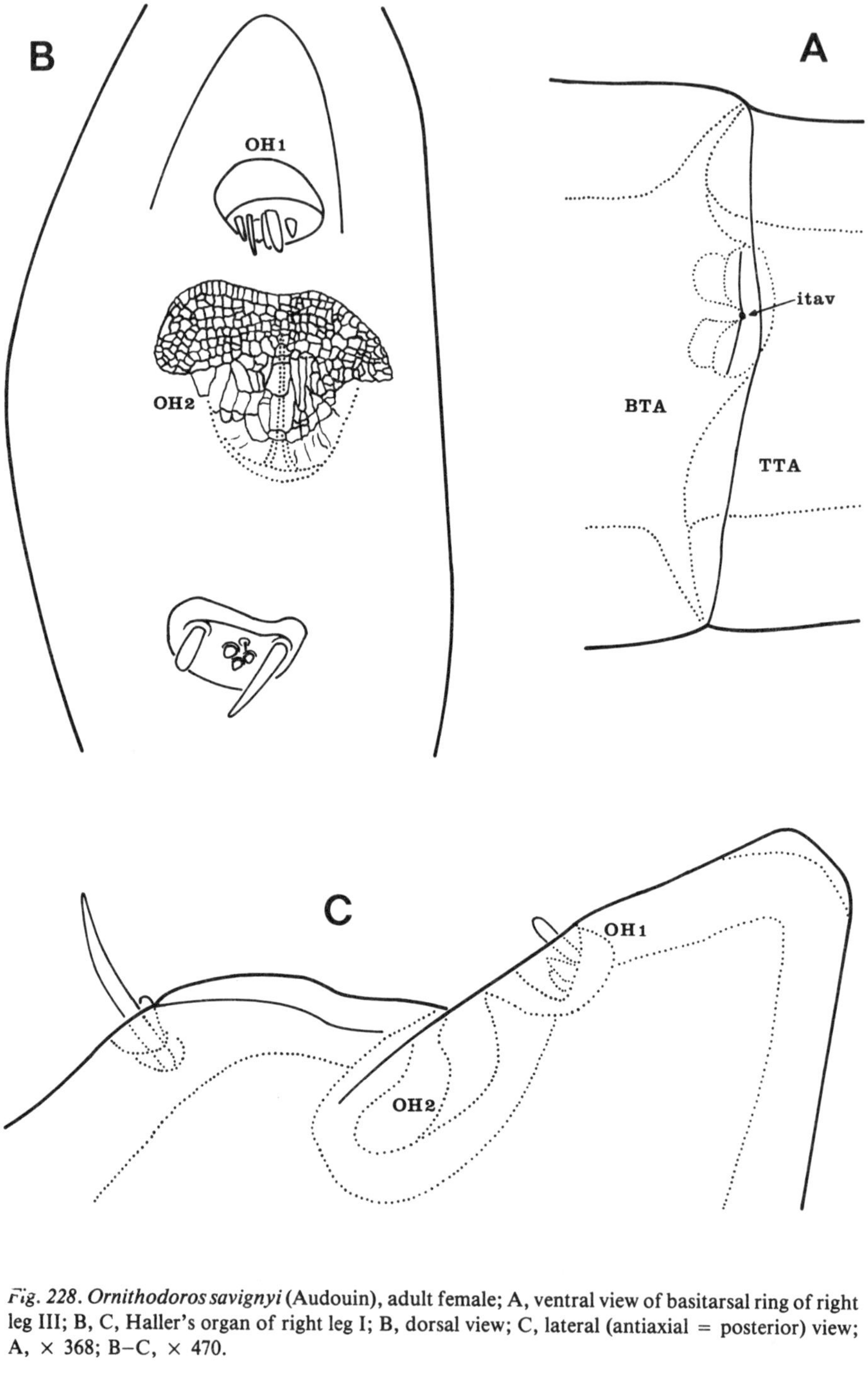

Fig. 228. Ornithodoros savignyi (Audouin), adult female; A, ventral view of basitarsal ring of right leg III; B, C, Haller's organ of right leg I; B, dorsal view; C, lateral (antiaxial = posterior) view; A, × 368; B–C, × 470.

List of the notations of figs. 206–228

A	area porosa.
ap.sc	subcapitular apodeme.
at	line of attachment of cheliceral frame to infracapitulum.
b	mouth.
BF	basifemur.
BTA	basitarsus.
ca	carina.
CE	cervix.
CH	chelicera.
chf	cheliceral frame.
CV	coxal cavity of the infracapitulum.
CX I–II	coxae I–II.
dgi	duct of intracapitular gland.
dv	dorsoventral groove.
F	femur.
GE	genu.
H	mentum.
if'', if_1'', if_2''	posterolateral lyrifissures of basifemoral ring.
ifv''	posteroventral lyrifissure of basifemoral ring.
igv	ventral lyrifissure of genu.
it'	anterolateral lyrifissure of tibia.
it''	posterolateral lyrifissure of tibia.
ita'	anterolateral lyrifissure of basitarsal ring.
ita''	posterolateral lyrifissure of basitarsal ring.
itav	ventral lyrifissure of basitarsal ring.
itv	ventral lyrifissure of tibia.
itv''	posteroventral lyrifissure of tibia.
Ji	inferior oral commissure.
Js, Js'	superior oral commissures.
LA	labellum.
LL	lateral lips.
LS	labrum.
mf	fixed jaw of chelicera.
mm	movable jaw of chelicera.
OC, OC_1, OC_2	eyes.
ogi	orifice of infracapitular gland.
OH_1	anterior trough of Haller's organ.
OH_2	capsule of Haller's organ.
ol'	anterior lateral ungues.

ol''	posterior lateral unguis.
PA	palp.
PH	pharynx.
PTA	pretarsus.
PV	pulvillus.
PI–IV	legs I–IV.
sc	supracoxal fold.
sc.ls	labral sclerity.
se	capitular saddle.
st	stigma.
t_i	inferior tendon of apotele.
t_s	superior tendon of apotele.
TA	tarsus.
TF	telofemur.
tf_i	inferior tendon of femur.
tf_s	superior tendon of femur.
tg	tendon (flexor) of genu.
TG	tegulum.
tge	tendon (extensor) of genu.
TI	tibia.
TITA	tibiotarsus.
tr	trachea.
TR	trochanter.
tsc	tendon of subcapitular apodeme.
tta	tendon of tarsus.
TTA	telotarsus.
tti	tendon of tibia.
ttr_i	inferior tendon of trochanter.
ttr_s	superior tendon of trochanter.
ttr'	anterior tendon of trochanter.
ttr''	posterior tendon of trochanter.

7. Relationships of the Anactinotrichid orders

Anactinotrichida constitute a homogenous group characterized by a great number of shared derived characters. Investigations, by Alberti (1980), of sperm morphology and spermiocytogenesis have confirmed the monophyletic character of the superorder. It is more difficult to demonstrate special affinities (i.e. closer relationships) between some of the orders. Some characters important for an analysis of these special relationships are the following.

It is not impossible that the four pairs of opisthosomatic respiratory organs of Opilioacarida originated from excretory glands, and that three of them are homologous with the single pair of respiratory organs and the two pairs of excretory glands of Holothyrida (two excretory glands are described for Allothyridae). The single pair of respiratory organs of Holothyrida is certainly homologous with the single pair found in Gamasida and Argasina (the homology of the respiratory organs in Ixodina is not certain); this pair could be homologous with stigma 3 (in segment XI) of Opilioacarida.

The Holothyrid, Opilioacarid and Ixodid tegulum is not produced into a supracheliceral limbus. The last-mentioned structure is present in most (but not all) Gamasida (cf. Van der Hammen, 1972a: 279).

In Holothyrida and Ixodida, the gnathosoma can be completely extended, as well as bent in ventral direction (perpendicular to the idiosoma). The functional morphology of the gnathosoma (and its movements) in Anactinotrichida is, however, still insufficiently studied, and I do not know if this character is also found in other orders.

The paired orifice of the infracapitular glands is ancestrally close to the line of attachment of the cheliceral frame to the infracapitulum. In Gamasida, this orifice is considerably advanced (salivary styli); in this case, the paired ducts of the glands are found in the coxal cavities of the infracapitulum.

The dorsal parts of the lateral lips (the labella) are faintly or moderately developed in Opilioacarida and Gamasida. They present a considerable secondary extension in Holothyrida and Ixodida.

Ancestrally, subcapitular gutter, sternapophyses and orifices of coxal glands are associated structures (as in Opilioacarida). In Gamasida, sternapophysis and subcapitular gutter are still associated with each other. In most Holothyrida (except *Allothyrus*) and in Ixodida the sternapophyses have disappeared.

The apotelic claw of the palp is terminal in Opilioacarida, lateral (paraxial) in Holothyrida and Gamasida, absent in Ixodida. The tibial brush of the palp in Holothyridae, and the brush of the palpal trochanter and femur in Ixodidae, constitute interesting analogies.

Sensory areas in the distal part of leg I are found in all groups of Anactinotrichida, but Haller's organ in ticks (consisting of capsule and trough) has a distinct resemblance to Haller's organ in the family Holothyridae (consisting of a dorsal cavity and an elongated groove). The supposed ancestral condition (telotarsal organ or antiaxial capsule, and dorsal group of sensory setae) is still present in Opilioacarida and Allothyridae.

There are some indications that the Holothyrid larva is octopod. It will be very important to reinvestigate the postembryonic development of this group.

It is evident that Holothyrida share a number of ancestral characters with Opilioacarida. Holothyrida share some interesting derived characters (pertaining to Haller's organ, to the regression of sternapophyses and the secondary extension of labella, and to gnathosoma movements) with Ixodida. It will be interesting to investigate whether the two groups of ticks present special affinities with different groups of Holothyrida.

8. Additional remarks

Fossil species of Anactinotrichida (Gamasida, Ixodida), not very different from extant species, are known from the Tertiary. Nothing is known about the earliest Anactinotrichida, although the related Ricinulei presented the characteristic Ricinuleid habitus already in the Carboniferous.

Opilioacarida and Holothyrida, which present many ancestral characters, could be regarded as living fossils. These groups have a restricted geographical distribution. The number of known species is rather small, and representatives are often regarded as rare.

Gamasida and Ixodida constitute much more successful groups of Anactinotrichida: the diversity and numerousness of extant species is great. The greatest degree of diversity, in morphology as well as biology, is found in Gamasida; species of this group can be free-living or parasitic, epigeous or hypogeous, cavernicolous, epizoic, halophilous, etc.

VI. RICINULEI

A detailed study of Ricinuleid morphology and postembryonic development was included in the first part of my Comparative Studies in Chelicerata (Van der Hammen, 1979a). The results of this study, reconsidered and corrected, are now included in the present chapter. It is subdivided into three parts: an introduction, a descriptive part, and additional remarks.

1. Introduction

Comparative Ricinuleid morphology was founded by Hansen & Sørensen (1904), when these authors published their classic work on two orders of Arachnida. Little had changed by 1932 when Kaestner (1932b) published his chapter on Ricinulei in Kükenthal's Handbuch. Segmentation, supposed primitiveness, and internal anatomy were subsequently studied by Millot (1945, 1946, 1947), and at the time when the chapter on Ricinulei in Grassé's Traité de Zoologie was published (Millot, 1949e) important progress was made.

From about 1967 onwards, our knowledge of Ricinulei has again considerably increased. Cooke (1967) studied the biology of Ricinulei (among which sperm transfer); Kennaugh (1968) made a study of the cuticle; Brignoli (1973) studied the female genital apparatus. Detailed comparative morphological papers were published by Pittard & Mitchell (1972), Dumitresco & Juvara-Bals (1973, 1976, 1977, 1977a) and Legg (1976). Tuxen (1974) studied the systematics of the African Ricinulei. Several authors published descriptions of new species of American Ricinulei.

The present discussion of Ricinuleid morphology and postembryonic development is based on original studies, supplemented with data from literature.

2. Descriptive part

The material on which the original part of the present descriptions is based consisted of the following four species: *Ricinoides karschii* (Hansen & Sørensen, 1904), from Gabon; *Ricinoides sjoestedtii* (Hansen & Sørensen, 1904), from

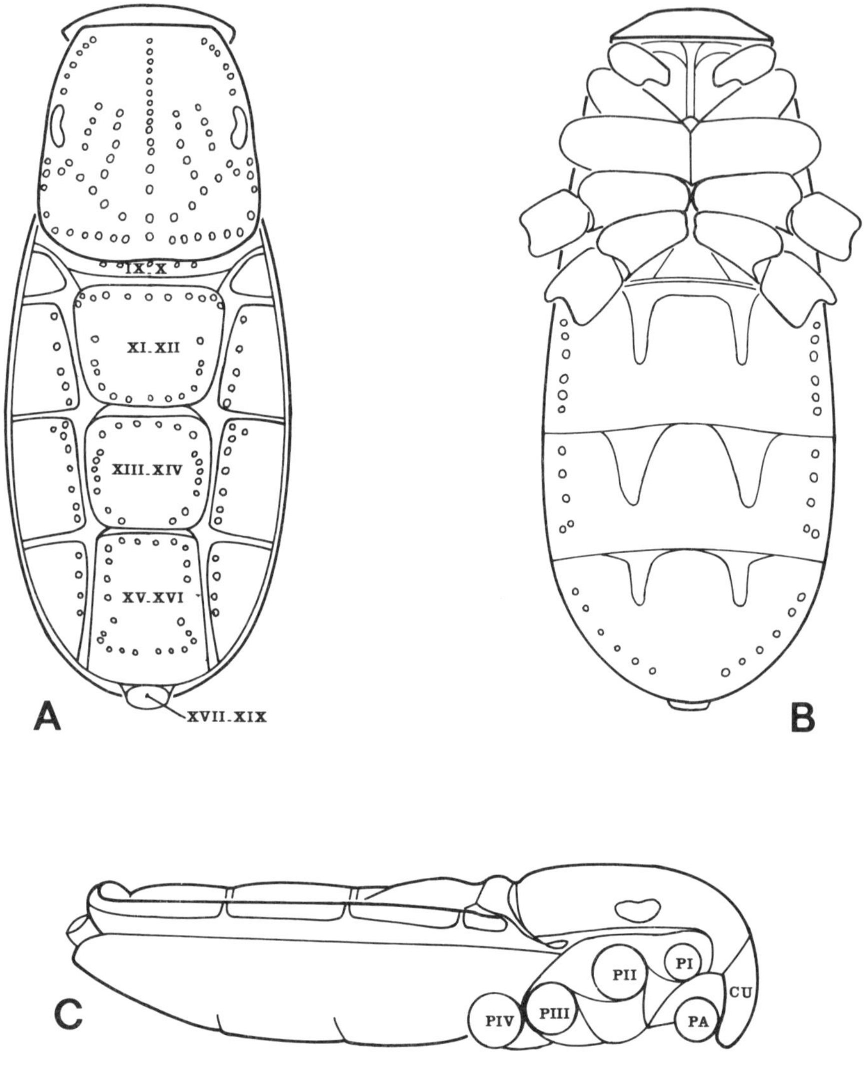

Fig. 229. Cryptocellus magnus Ewing, male; A, dorsal view; B, ventral view; C, lateral view (podosoma schematized); A–C, × 12.5.

South Nigeria; *Cryptocellus simonis* Hansen & Sørensen, 1904, from Belém, Brazil; and *Cryptocellus magnus* Ewing, 1929, from N. Colombia. In my 1979a paper, the *Cryptocellus simonis* material was recorded as *C. foedus* (I had received identified material from Belém; according to Platnick & Shadab, 1977, material from this locality belongs to *C. simonis*); the *C. magnus* material was classified, in my 1979a paper, with the genus *Heteroricinoides* which, ac-

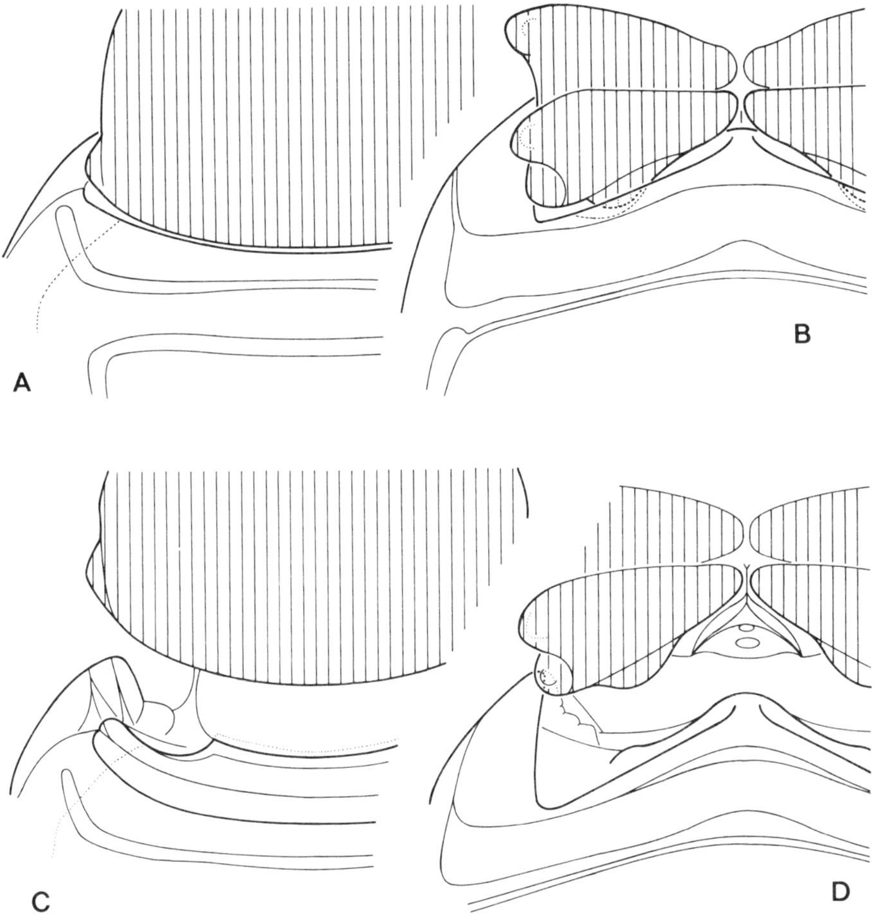

Fig. 230. Ricinoides karschii (Hansen & Sørensen), deutonymph; pedicel and associated regions of prosoma and opisthosoma (the prosoma indicated by ruling); A, B, locked position (pedicel concealed); C, D, uncoupled; A, C, dorsal view; B, D, ventral view; A–D, × 37.

cording to Platnick & Paz (1979), has the status of a species group.

Habitus. – Ricinulei are clumsy, rather slow-moving animals, of medium size (adult not exceeding 10 mm); they are flattened, with a movable flap (cucullus) attached to the front edge of the body. All species rapidly sham death when alarmed.

Divisions of the body (fig. 229). – The body is divided into prosoma and opisthosoma. The prosoma is constituted by the precheliceral metameres and the segments I–VI. The opisthosoma is here regarded as constituted by the segments VII–XIX (partly fused), the segments VII and VIII constituting a 'pedicel' which enables coupling of prosoma and opisthosoma. Part of the an-

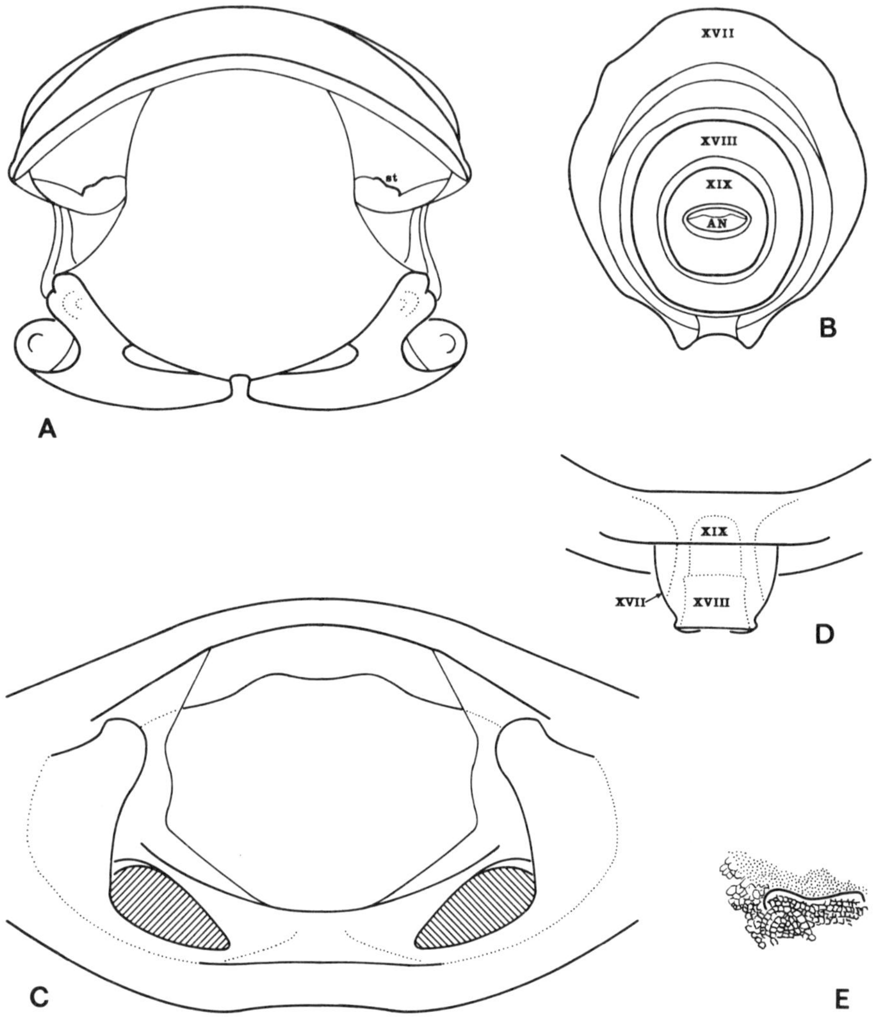

Fig. 231. Ricinoides karschii (Hansen & Sørensen), deutonymph; A, posterior view of prosoma; B, posterior view of pygidium (segments XVIII and XIX retracted); C, anterior view of opisthosoma (without pedicel); D, dorsal view of pygidium (segments XVIII and XIX retracted); E, posterior view of left stigma; A, C, D, × 37; B, E, × 118.

terior prosomatic segments constitute a movable gnathosoma (the mouthparts) which is located in a camerostome. An abjugal furrow separates a dorsal region of the prosoma (the aspidosoma) from the podosoma. Details with reference to the present interpretation are given below.

Cuticle. – The cuticle is extraordinary thick and more or less leathery; it is very hard, and difficult to section. It presents lyrifissures, several types of

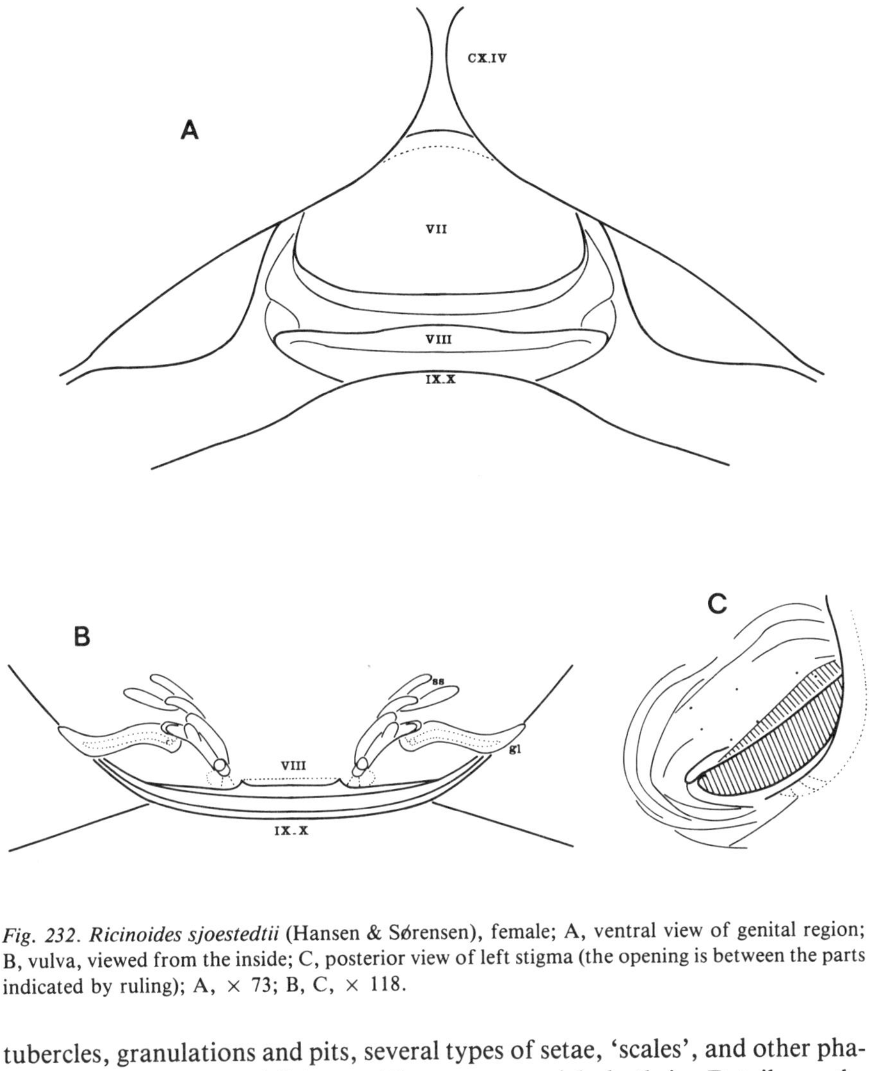

Fig. 232. Ricinoides sjoestedtii (Hansen & Sørensen), female; A, ventral view of genital region; B, vulva, viewed from the inside; C, posterior view of left stigma (the opening is between the parts indicated by ruling); A, × 73; B, C, × 118.

tubercles, granulations and pits, several types of setae, 'scales', and other phaneres (solenidia, coronidia), etc. There are no trichobothria. Details on the cuticular organs are discussed below, in the paragraphs in question.

Aspidosoma (fig. 229A). – The anterior part of the aspidosoma constitutes the cucullus (fig. 233A), a movable hood which can be raised to a horizontal plane, and lowered to a vertical position (covering the camerostome). Although the cucullus has been observed to assist in several activities (capture of prey, egg carrying), the original function certainly pertains to the concealment of the mouthparts and to the protection of the camerostome. Its presence constitutes a specialized (derived) character of the Ricinulei.

The surface of the aspidosoma presents various types of tubercles, various phaneres (setae), and pits (with tubercles); the pits probably represent sigilla. In many species of Ricinulei, the aspidosoma presents, on both sides, a light area, above coxa II. In my specimen of *Cryptocellus magnus* it is removed from the lateral border of the aspidosoma, and slightly bilobate at the paraxial side. It occupies the same position as the eyes of Opilioacarida, and it is here regarded as a vestige of two eyes. Because the lenses and the dark pigment are lacking, light sensitivity will certainly be reduced or absent. Functional eyes, with similar position, were present in fossil Ricinulei (Selden, 1983).

Pedicel and associated regions (figs. 230, 231A, C). – The anterior, narrowed (but not very narrow) part of the opisthosoma constitutes the pedicel. It comprises the segments VII and VIII. The genital region of the pedicel (segment VIII) is dealt with below; the sclerite situated in females anterior to the genital opening, is attributed here to segment VII.

Apart from the small sclerites, the cuticle of the pedicel is soft. It can be folded when prosoma and opisthosoma are coupled (it reminds one of a concertina-passage in a corridor train). In a locked position, the pedicel is completely concealed. The locking device (fig. 231A, C) involves a posterodorsal ridge of the aspidosoma corresponding with an anterodorsal groove of the opisthosoma, and a pair of anteroventral opisthosomatic sockets corresponding with coxae IV and its posterior keels.

The pedicel is strongly different from that in other groups of Chelicerata, not only because it comprises segment VIII, but also because of its function. Its main function is to enable coupling of prosoma and opisthosoma, in which condition genital opening and stigmata are concealed and protected. The position of the single pair of stigmata is apparently prosomatic; it is dealt with below.

Opisthosoma (figs. 229, 231B, D). – The opisthosoma is divided into the pedicel (dealt with above), a main body, and the pygidium (dealt with below). The pedicel is constituted by two segments (VII and VIII), the pygidium by three. The 'segments' of the main body are relatively very long; they are regarded here as fusions of respectively the following segments: IX + X, XI + XII, XIII + XIV, and XV + XVI (they present double rows or sigilla). The pygidium is constituted by three segments, viz., the numbers XVII, XVIII and XIX. According to this interpretation, the number of segments is the same as in Opilioacarida. The cuticle of the opisthosomatic body presents tubercles, setae and sigilla.

Pygidium (fig. 231B, D). – The narrow pygidium is constituted by the segments XVII–XIX. The anal opening is situated in segment XIX. The segments can be withdrawn, each into the preceding segment, like a telescope, and then retracted into the opisthosomatic body. It may be remarked here that in Opilioacarida segment XIX is retractable.

Genital region (fig. 232A, B). – The female genital region is represented in fig. 232A, B. It is constituted by a genital shield (segment VIII), situated posterior to the pregenital shield (segment VII). When studied from the inside,

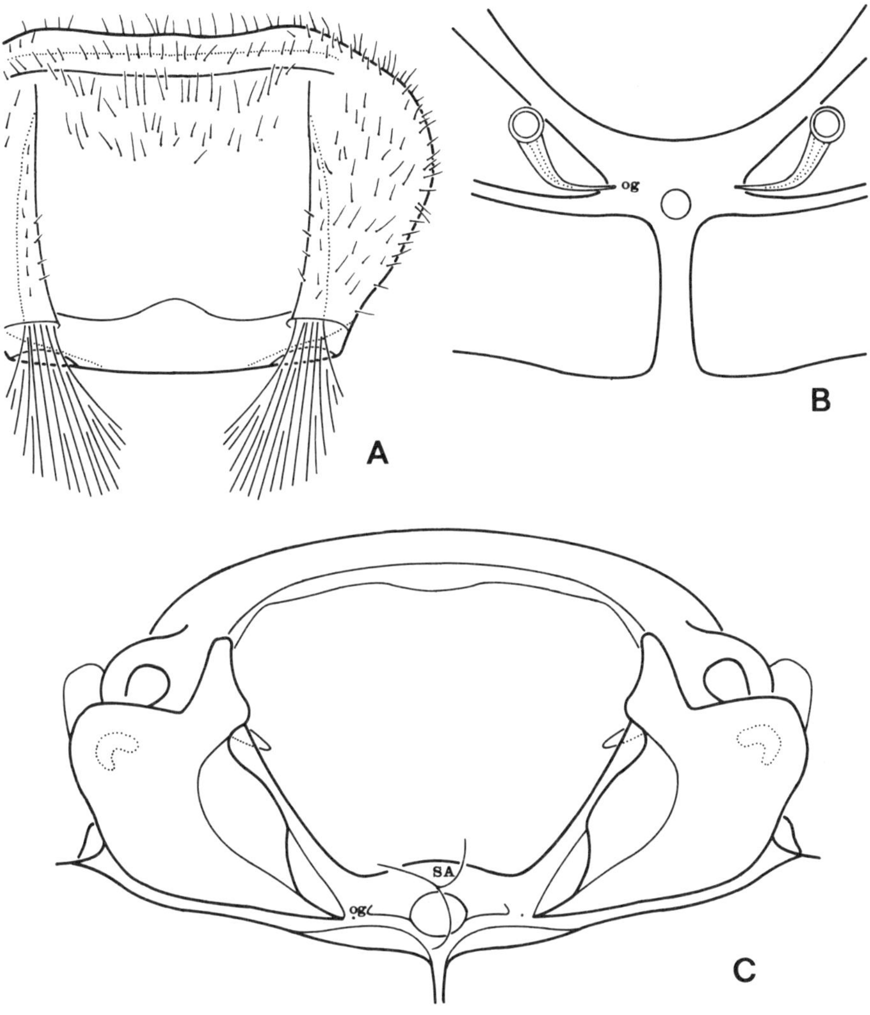

Fig. 233. Ricinoides karschii (Hansen & Sørensen), tritonymph; A, ventral view of cucullus; B, coxal and sternal regions of segments III and IV (with pair of excretory ducts of coxal glands), viewed from the interior; C, ventral view of camerostome and coxal and sternal regions of segment III (gnathosoma and cucullus removed); A–C, × 44.

the structure of the vulva becomes visible. It has been studied by Brignoli (1973: 154–155, fig. 1A–D), Legg (1976: 37–39, fig. 28A), Platnick & Shadab (1976: 2, figs. 14, 16, 17), and especially by Dumitresco & Juvara-Bals (1976: 160–165, figs. 6–8; 1977: 259–261, fig. 1). In all genera there are a pair of supposed spermathecae (simple or with a diverticulum); in the genus *Ricinoides* a number of sac-like structures are also present (in my specimen of *Ricinoides sjoestedtii*, the number is six). Millot (1947) interpreted the so-called spermathecae as glands; because of their thick wall and narrow lumen, I am inclined to support his view. The function of the sac-like structures is unknown.

The male genital region was described in detail by Pittard & Mitchell (1972: 13, figs. 42, 46–48) and Dumitresco & Juvara-Bals (1973: 268, fig. 7B). It is much smaller than that of the female. It is characterized by the presence of a conical, sclerotized, projecting piece bearing the genital opening at the postero-ventral side. The structure is apparently not a penis or a spermatopositor. Its composition of two segments (as supposed by Pittard & Mitchell) is not evident. The male presents no separate pregenital sclerite.

Podosoma (figs. 229, 233B, C). – The podosoma extends from the circumcapitular furrow (the border of the gnathosoma) to the disjugal furrow (the anterior border of the pedicel). Laterally, it is bordered by the abjugal furrow (at the lateral border of the aspidosoma). The sternal region of the podosoma is extremely small; the ventral surface of the podosoma is nearly entirely occupied by the large coxae of the legs. Coxae I–III are fixed; only coxae IV are movable. The small sternal sclerite of segment III (the segment bearing legs I) presents two or three sternapophyses (fig. 233C), similar in shape to those of the genus *Allothyrus* (Holothyrida), and homologous with the sternapophyses of Anactinotrichida.

Paraxial to coxa I, or between coxae I and II (fig. 233C), the orifice of the coxal glands is present (just as in Opilioacarida). It is supposed here that the products of these glands are transported to the subcapitular gutter (as in Opilioacarida). Just as in Anactinotrichida, the Ricinuleid sternapophyses are associated with the subcapitular gutter.

The stigma is generally supposed to belong to the prosoma; it could, originally, have been opisthosomatic (as in Anactinotrichida). It is situated above coxae III or IV, close to the border of the pedicel.

Gnathosoma (figs. 233C, 234, 235, 237D). – The gnathosoma is situated in a camerostome (fig. 233C); it is movable with respect to the idiosoma. Just as in Anactinotrichida, two parts can be distinguished: a cheliceral frame (the dorsal part) and an infracapitulum. The membraneous cheliceral frame constitutes the body wall between the infracapitulum and the aspidosoma; the chelicerae are movably attached to it. In comparison with Anactinotrichida, the Ricinuleid infracapitulum is rather primitive (the cervix is small; there are no lateral lips; the mouth presents two commissures only, and is consequently crescent-shaped), although it presents a specialization particular to Ricinulei only, viz., the enlarged coxal regions of the palp. The labrum is large; its dorsal surface presents long setae (fig. 234A). After removal of the labrum, the labi-

---

→

Fig. 234. Ricinoides karschii (Hansen & Sørensen), infracapitulum of tritonymph; A, dorsal view; B, dorsal view after removal of the labrum, showing the labium; C, fragment of radula-like surface of labium; D, ventral view; E, supposed laterocoxal setae of the right side; A, B, D, × 55; C, × 368; D, × 687.

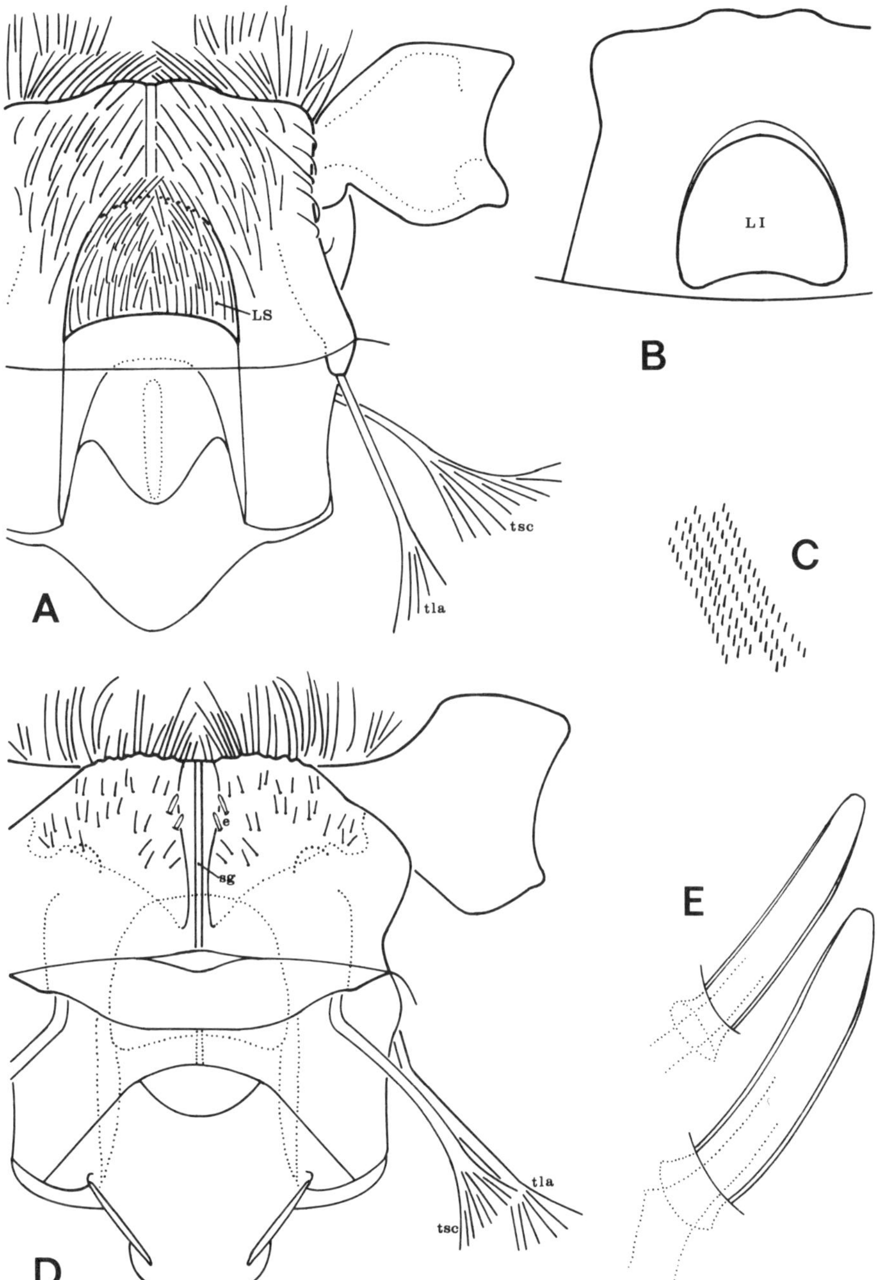
A
LS
tsc
tla
B
LI
C
D
e
sg
tla
tsc
E

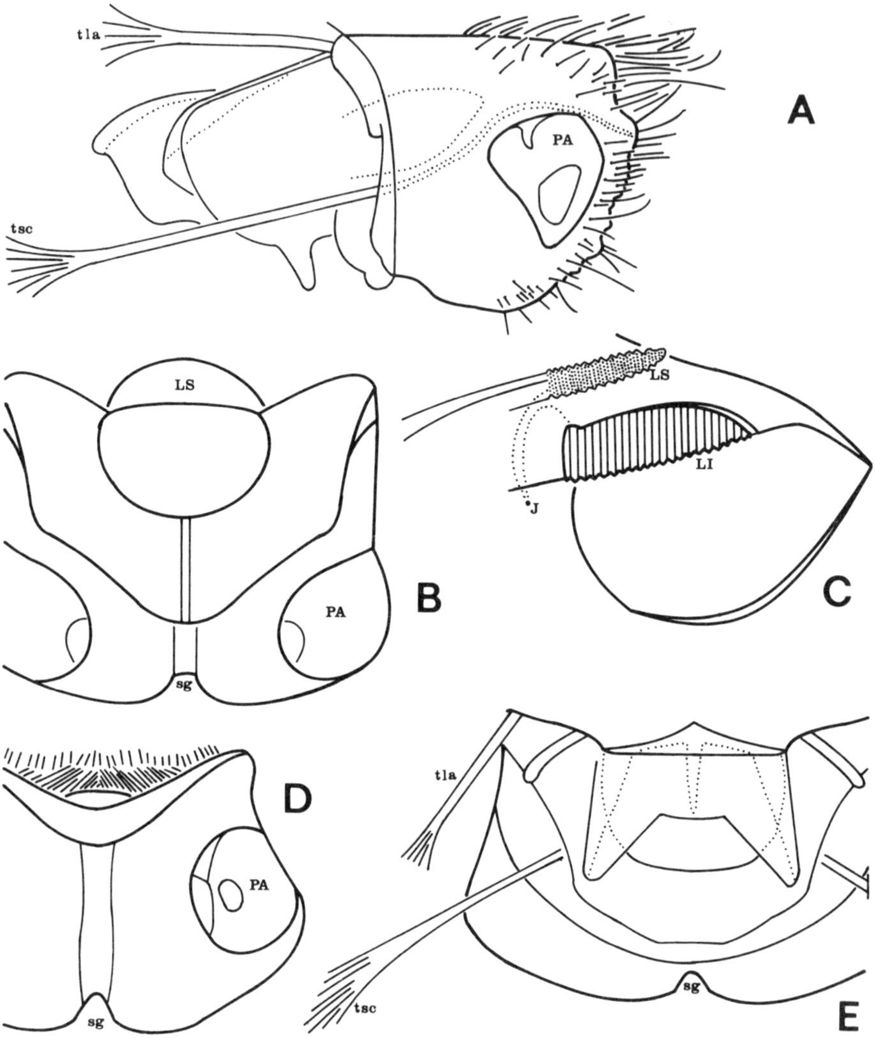

Fig. 235. Ricinoides karschii (Hansen & Sørensen), infracapitulum of tritonymph; A, lateral view; B, frontal view (obliquely from above); C, sagittal section through anterior part, viewed from the paraxial side; D, frontal view; E, posterior view; A–E, × 55.

→

Fig. 236. Ricinoides karschii (Hansen & Sørensen), right chelicera of tritonymph; A, dorsal view; B, ventral view; C, lateral (paraxial) view (setae omitted); D, distal part of seta of the ventral surface; E, lateral (antiaxial) view (setae omitted); A–C, E, × 92; D, × 550.

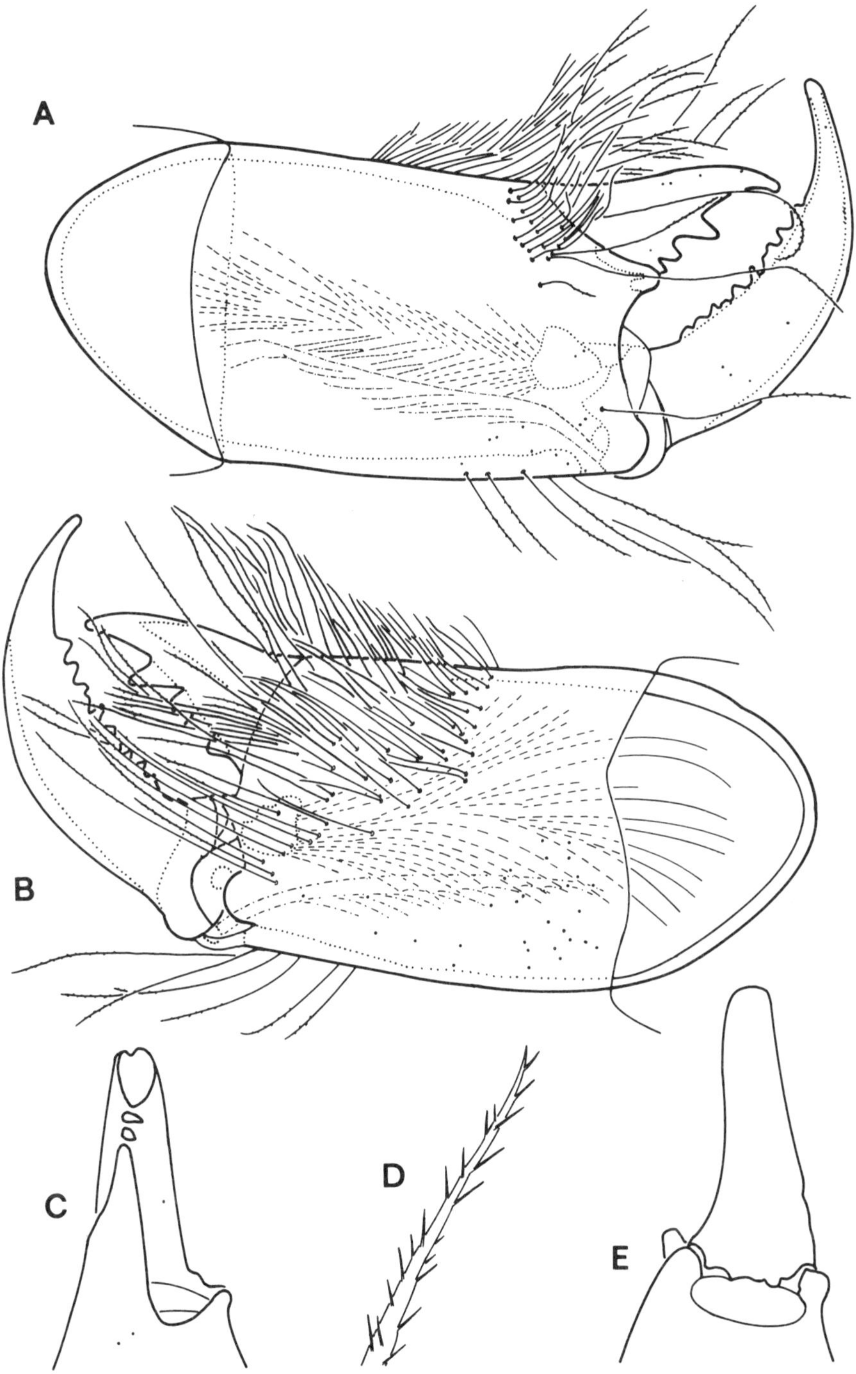
A
B
C
D
E

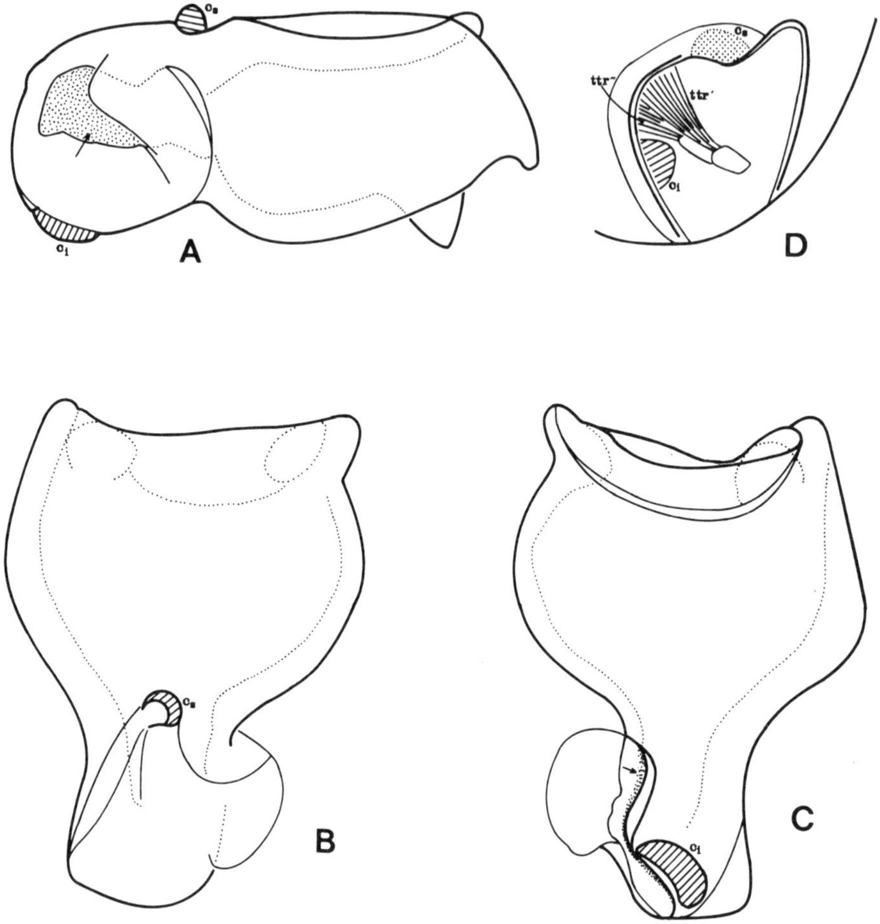

Fig. 237. Ricinoides karschii (Hansen & Sørensen), tritonymph; A–C, trochanter of right palp (the arrows in figs. A and C indicate the proximal internal opening); D, acetabulum of right palp; A, lateral (posterior) view; B, dorsal view; C, ventral view; D, laterofrontal view; A–D, × 88.

um becomes visible (fig. 234B); its surface is covered with radula-like teeth. The labium is completely incorporated in the mentum (which, evidently, must have extended in distal direction).

The mouth (fig. 235C) has two commissures (*J* and *J'*). Gnathosoma movements are operated by tendons and muscles: lateral tendons (*tla*) are inserted on the internal lateral border of the infracapitulum, subcapitular tendons (*tsc*) on the subcapitular apodeme (figs. 234A, D, 235A, E).

The coxal regions of the palp (there is no indication that there has been a free palpal coxa in the ancestors of Ricinulei) are much enlarged in paraxial direction (figs. 234D, 235A, C), although the subcapitular gutter has remained open

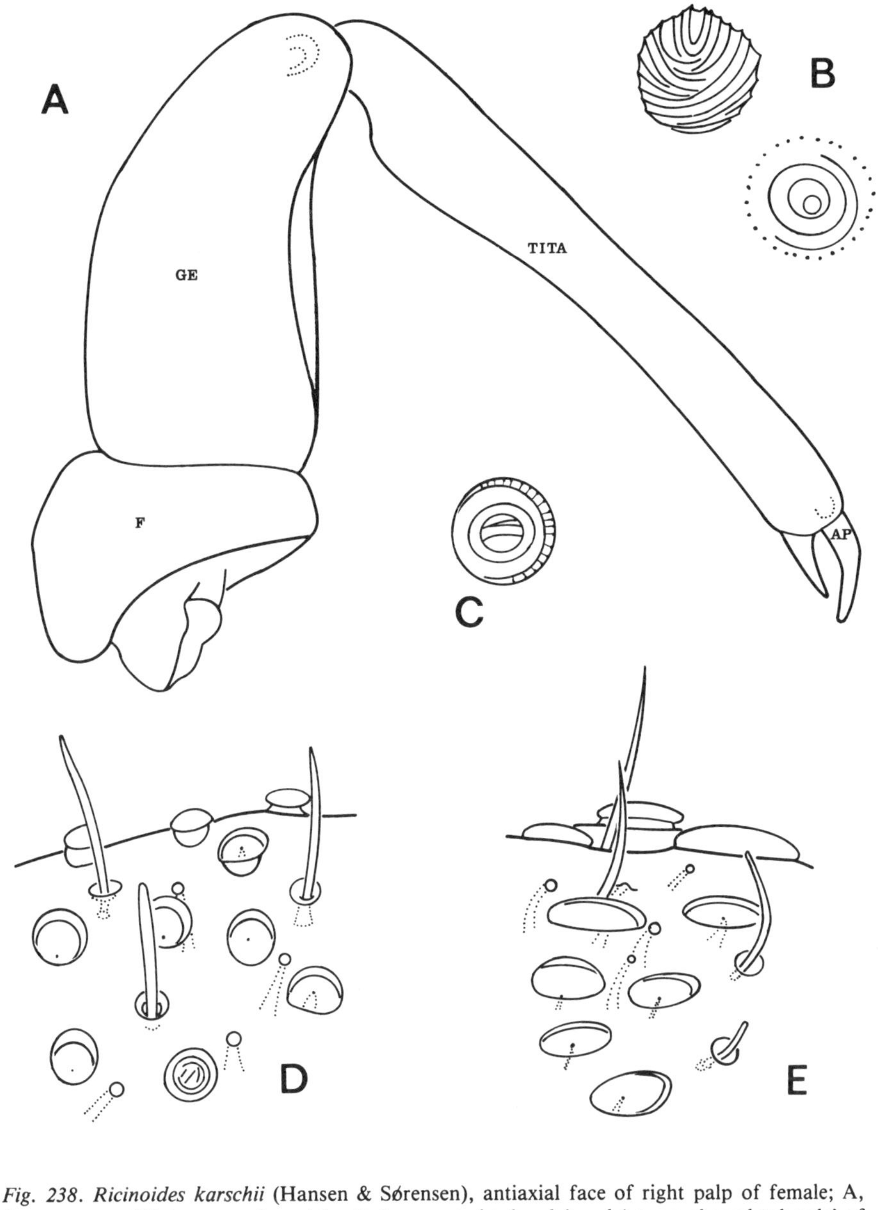

Fig. 238. Ricinoides karschii (Hansen & Sørensen), antiaxial face of right palp of female; A, femur, genu, tibiotarsus and apotele; B, 'corrugated tubercle' and 'saucer-shaped tubercle' of femur; C, 'saucer-shaped tubercle' of the base of the tibiotarsus; D, basal part of tibiotarsus with phaneres, tubercles and orifices of tegumental glands; E, median part of tibiotarsus with phaneres, tubercles and orifices of tegumental glands; A, × 55; B, C, × 687; D, E, × 368.

(fig. 234D). The characteristic two pairs of setae, situated on either side of the subcapitular gutter (fig. 235D, E), are here homologized with laterocoxal setae

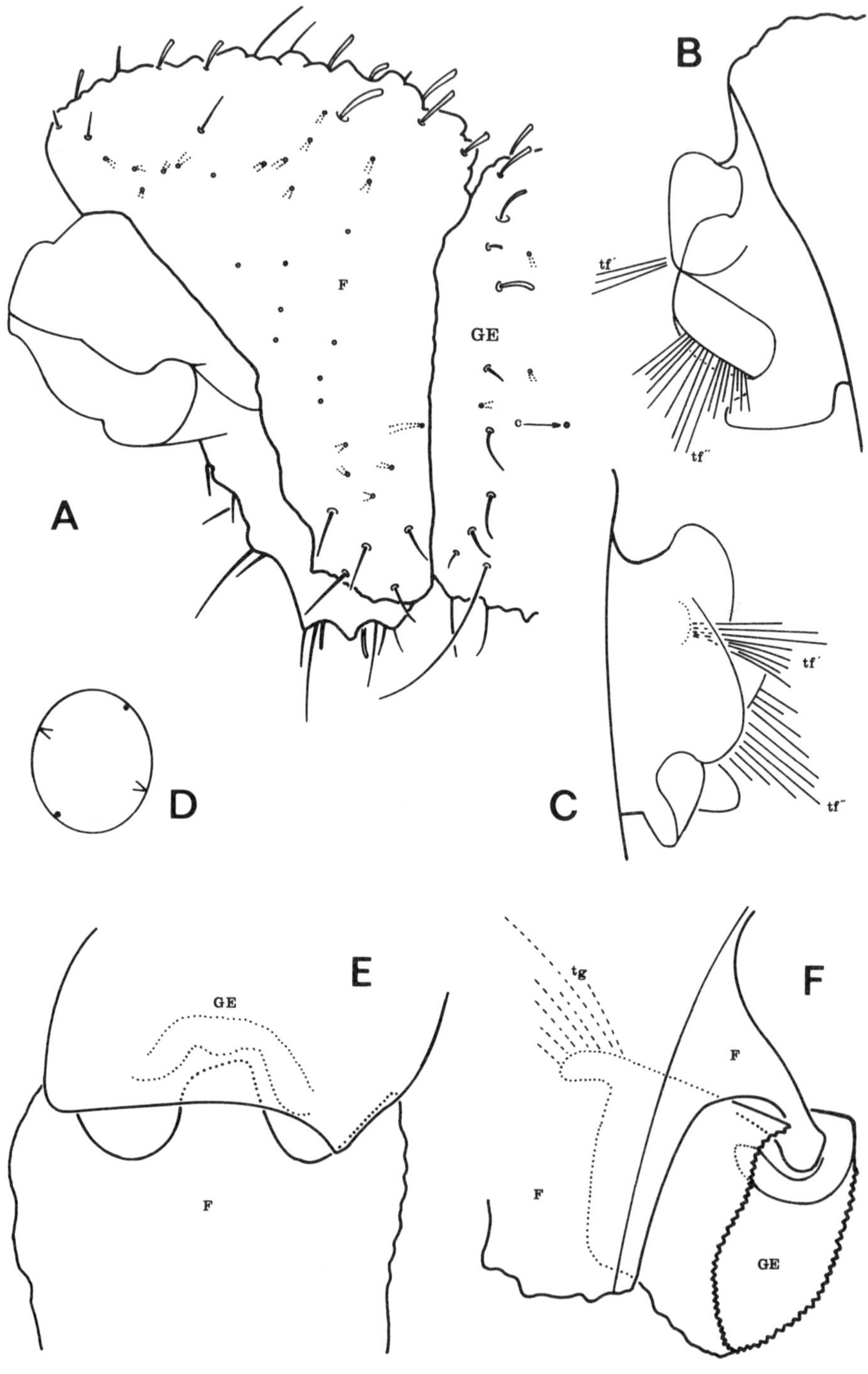
A
B
C
D
E
F
F
GE
c
tf´
tf˝
GE
F
tg
F
F
GE

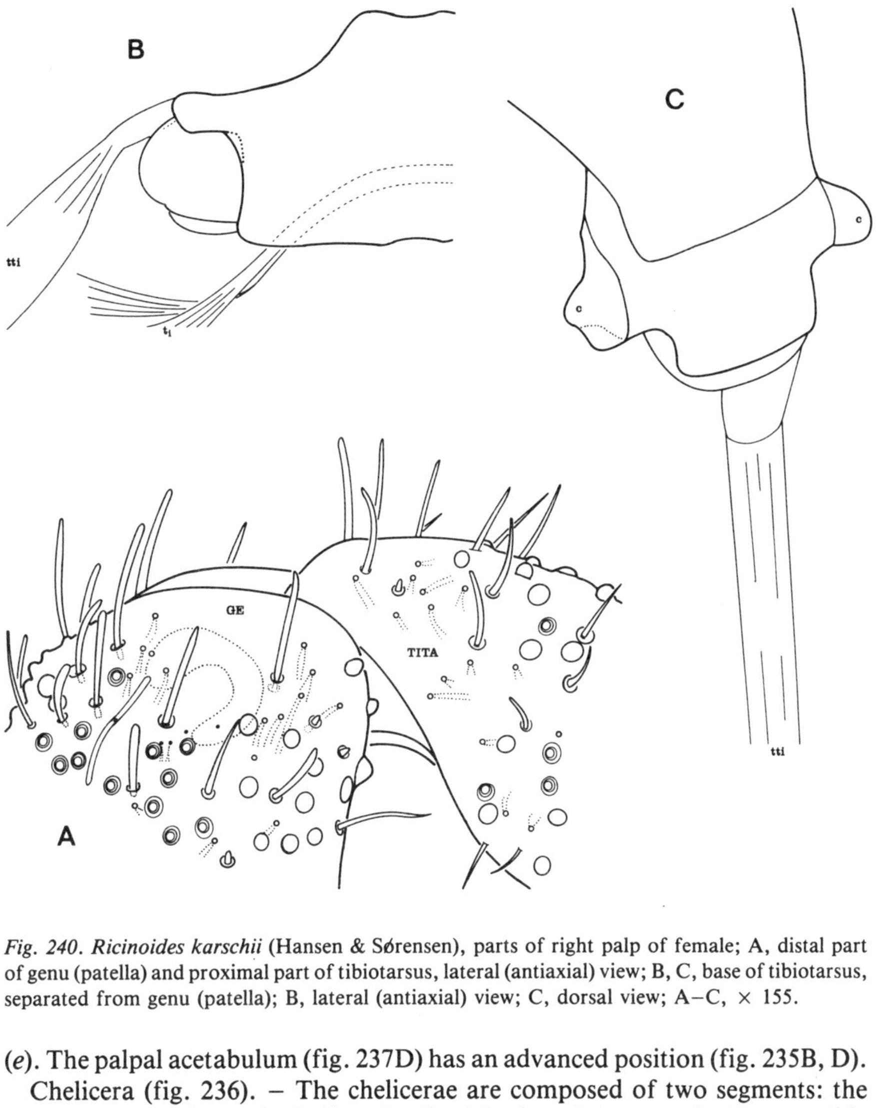

Fig. 240. Ricinoides karschii (Hansen & Sørensen), parts of right palp of female; A, distal part of genu (patella) and proximal part of tibiotarsus, lateral (antiaxial) view; B, C, base of tibiotarsus, separated from genu (patella); B, lateral (antiaxial) view; C, dorsal view; A–C, × 155.

(*e*). The palpal acetabulum (fig. 237D) has an advanced position (fig. 235B, D).

Chelicera (fig. 236). – The chelicerae are composed of two segments: the body of the chelicera (including the fixed jaw) and the apotele (the movable jaw); there is no separate trochanter. The fixed jaw is much smaller than the movable jaw. The base of the movable jaw presents the usual two tendons (t_s

←

Fig. 239. Ricinoides karschii (Hansen & Sørensen), parts of right palp of female; A, femur and proximal part of genu (patella), lateral (antiaxial) view; B, C, basal part of femur; B, lateral (antiaxial) view; C, lateral (paraxial) view; D, schematic representation of the articulation between trochanter and femur; E, F, distal part of femur and proximal part of genu (patella); E, dorsal view; F, lateral (antiaxial) view; A–C, E, F, × 110.

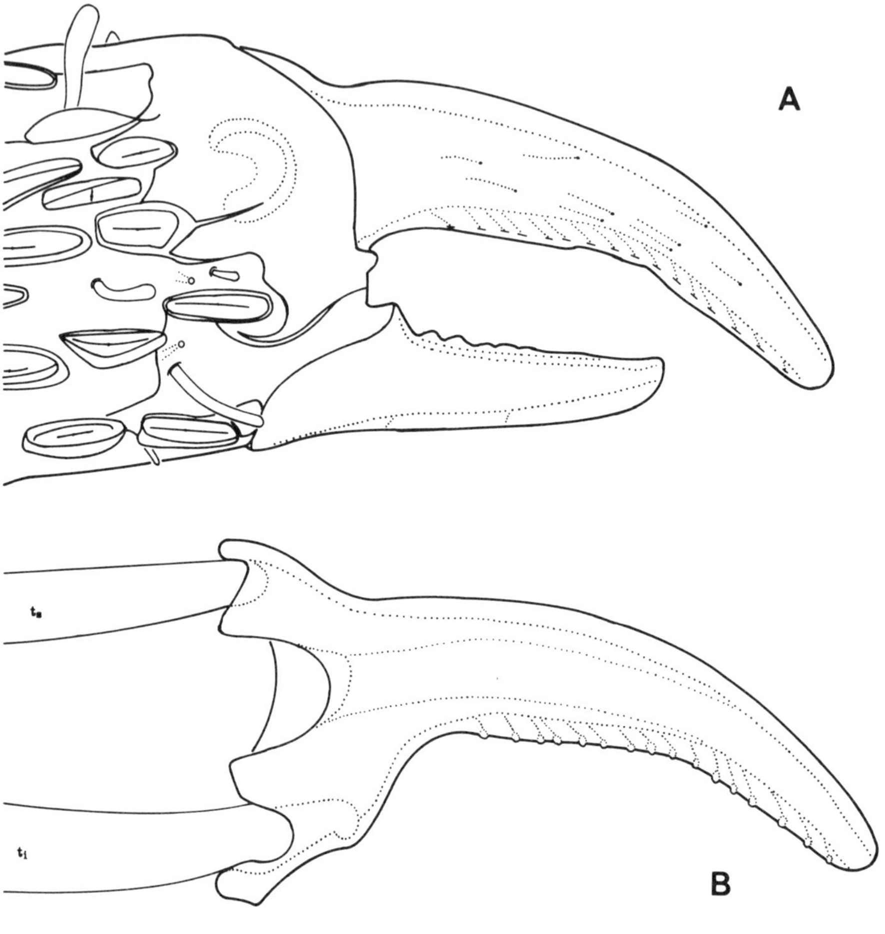

Fig. 241. Ricinoides karschii (Hansen & Sørensen), antiaxial face of part of right palp of female; A, distal part of tibiotarsus, and apotele; B, apotele (separated from tibiotarsus) with superior tendon t_s and inferior tendon t_i; A, B, × 295.

and t_i), inserted on it. The position of the chelicerae has changed; the jaws operate in a plane parallel to the infracapitulum (not perpendicular to it). The chelicerae are beset with numerous long setae; they present many orifices of tegumental glands.

→

Fig. 242. Ricinoides karschii (Hansen & Sørensen), right leg I of female; A, antiaxial face of trochanter, femur, genu (patella), tibia, basitarsus, telotarsus and apotele; B, tree-like setae (supposed famuli) of telotarsus, antiaxial view; C, laterodorsal phaneres of telotarsus, antiaxial view; D, lateral phaneres of telotarsus, antiaxial view; A, × 44; B–D, × 550.

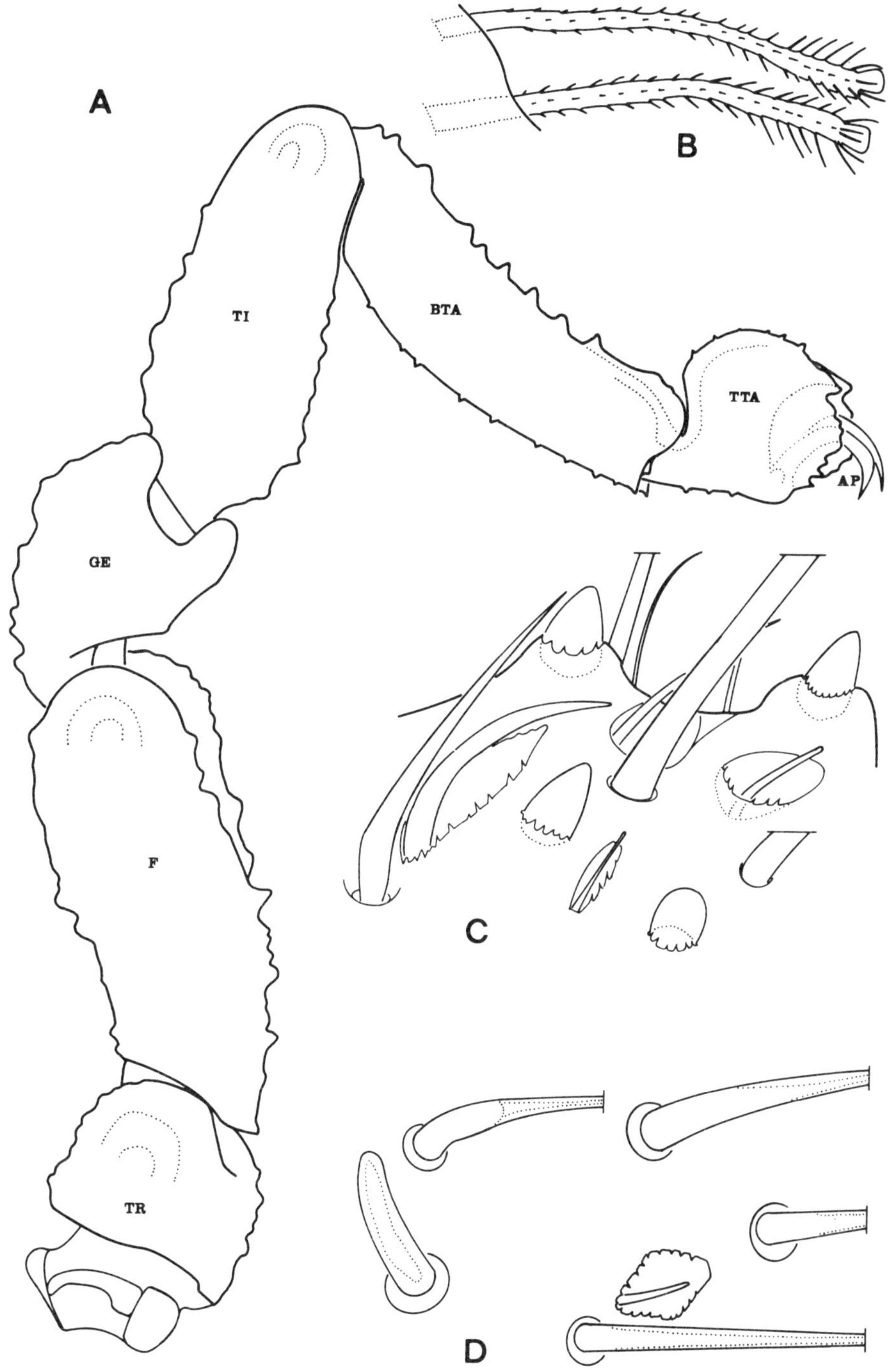
A
TI
BTA
TTA
AP
GE
F
TR
B
C
D

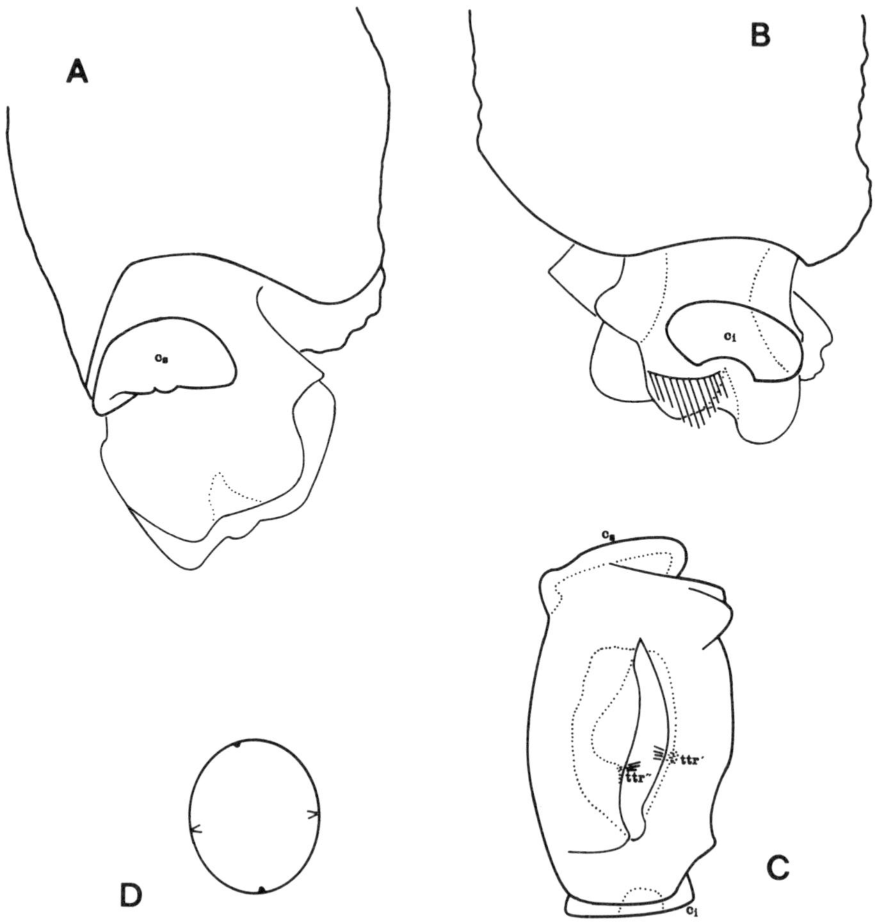

Fig. 243. Ricinoides karschii (Hansen & Sørensen); A–C, base of trochanter of right leg I of female; A, dorsal view; B, ventral view; C, posterior view; D, schematic representation of the articulation between coxa and trochanter; A–C, × 88.

Palp (figs. 237–241). – The palp is composed of the following segments (figs. 237, 238A): trochanter, femur, genu, tibiotarsus and apotele (the claw-segment); as will be demonstrated below, the second segment is not a

→

Fig. 244. Ricinoides karschii (Hansen & Sørensen), parts of right leg I of female; A, distal part of trochanter and proximal part of femur, lateral (antiaxial) view; B, proximal part of femur and fragment of distal part of trochanter (paraxial half removed), dorsal view; C, D, trochanter-femur articulation; C, dorsal view; D, ventral view; E, F, base of femur; E, lateral (antiaxial) view; F, lateral (paraxial) view; A, × 92; B–F, × 88.

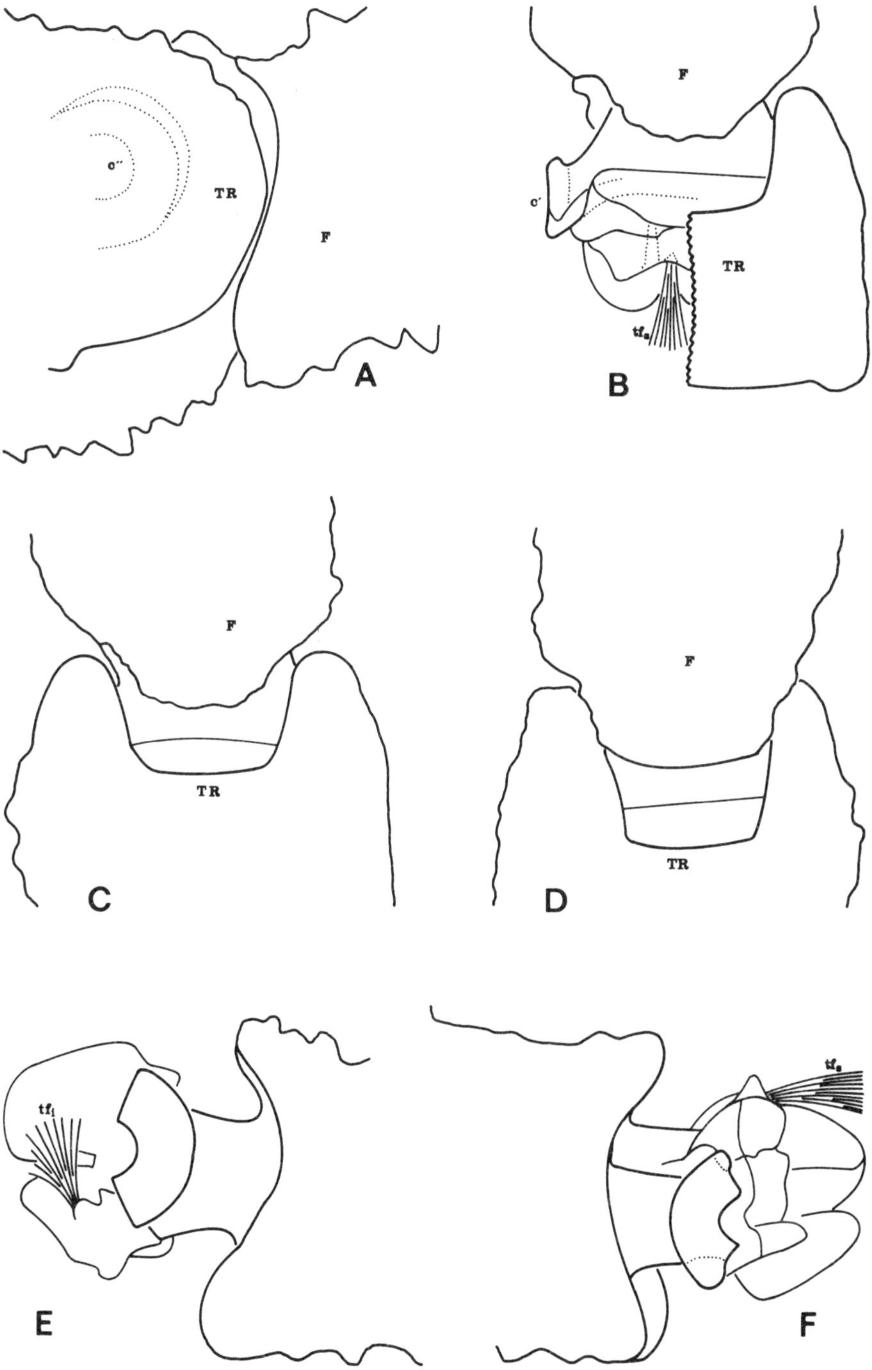
c''
TR
F
A
F
c'
TR
tf₈
B
F
TR
C
F
TR
D
tf₁
E
tf₈
F

trochanter 2. The terminal part of the tibiotarsus (the fixed jaw) and the apotele (the movable jaw) constitute a chela. The articulation of the palpal trochanter with the infracapitulum (fig. 237) is bidesmatic and bicondylar. The articulation between trochanter and femur (fig. 239A–D) is also bidesmatic and bicondylar (the femur is relatively very short). The articulation between femur and genu (fig. 239A, E–F) is monodesmatic and monocondylar; the tendon insertion has developed a superior position. Because of the monodesmatic distal articulation, the segment distinctly represents a femur and not a trochanter 2. The articulation between genu and tibiotarsus (fig. 240) is bicondylar; a superior tendon is inserted on the base of the tibiotarsus, whilst one of the tendons of the apotele is also involved in the articulation. The genu-tibiotarsus joint permits of active flexion and extension. The articulation between tibiotarsus and apotele (fig. 241) is bicondylar and bidesmatic (as usual); the position of the apotele is superior instead of inferior. The cuticle of the palp (fig. 238B–D) presents various types of tubercles ('corrugated', 'saucer-shaped', and 'mushroom-shaped'), lyrifissures, orifices of tegumental glands, and various types of phaneres (some of which are associated with pits). The phaneres associated with pits are probably solenidia. The long and narrow tubercles in the distal part of the palp of *Ricinoides karschii* (fig. 241A) are characterized by the presence of a line (lengthwise) and a central canal. They could represent transformed lyrifissures. There is a gradual transition from the long and narrow tubercles to the saucer-shaped. For this reason, a comparative investigation of the ultrastructure of lyrifissures and tubercles could be interesting.

Legs (figs. 242–251). – Legs I and II are composed of the following eudesmatic segments (fig. 242A): coxa, trochanter, femur, genu, tibia, tarsus and apotele; legs III and IV present moreover a trochanter 2. Coxae I–III are fixed, whilst coxa IV is movable. All tarsi of the legs are divided into a basi- and a telotarsus (an ancestral, adesmatic division), whilst telotarsi II–IV present a number of adesmatic divisions (a derived character of the Ricinulei). The number of adesmatic telotarsal segments of the adult is five in the case of legs II and IV (telotarsi 1–5), four in the case of leg III (telotarsi 1–4). The number of immature telotarsal segments will be dealt with below.

The articulation between coxa IV and the body is bidesmatic (figs. 253E, 255A) and apparently monocondylar. The articulation between coxa and trochanter (fig. 243) is bidesmatic and bicondylar, just as the articulation between trochanter 1 and trochanter 2 of legs III and IV (fig. 249C, E–G). The articulation between trochanter and femur of legs I and II (fig. 244) and between trochanter 2 and femur of legs III and IV is also bicondylar and bidesmatic.

The articulation femur-genu (patella) is a complicated hinge joint (fig. 245). It presents (in the case of leg I, investigated here) two superior condyles (which take the place of the single superior condyle of ancestral Cryptognomae) and several tendons (viz., a tendon inserted, distal to the articulation points, on the

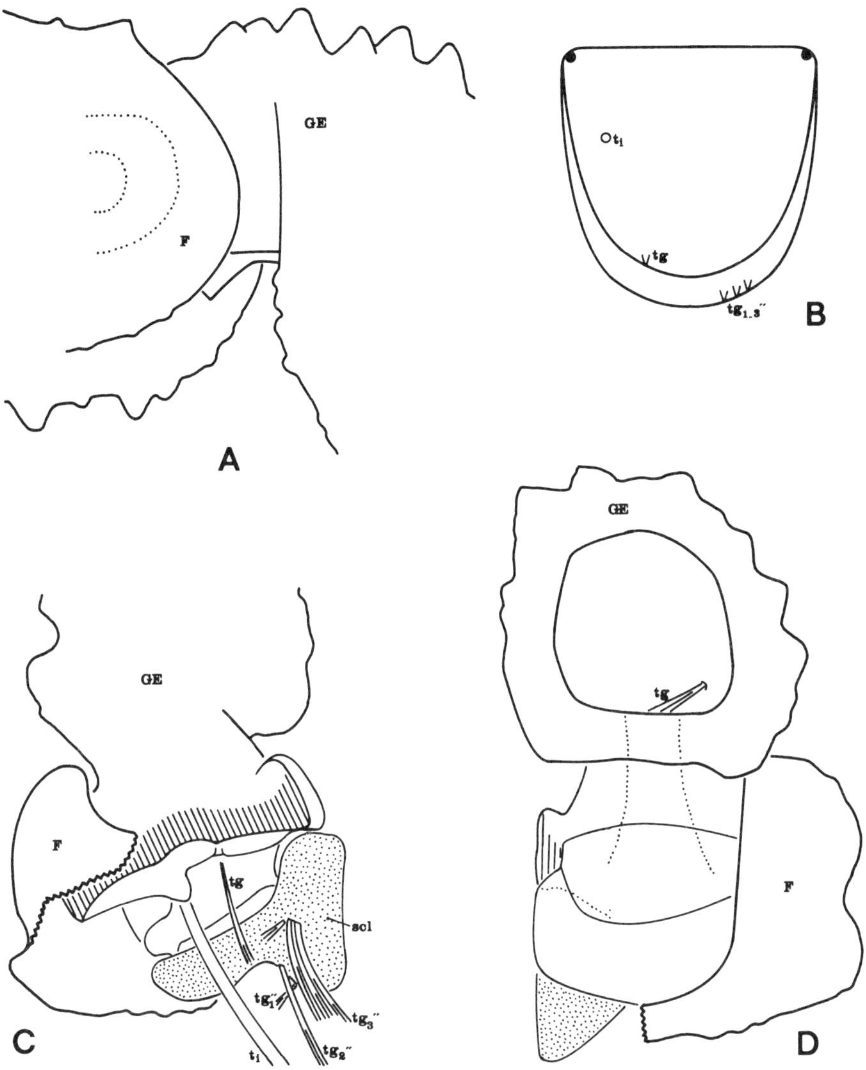

Fig. 245. Ricinoides karschii (Hansen & Sørensen), part of right leg I of female; A, distal part of femur and proximal part of genu (patella), lateral (antiaxial) view; B, schematic representation of the articulation between femur and genu (patella); C, D, proximal part of genu (patella), and fragment of distal part of femur (antiaxial half removed); C, dorsal view; D, ventral view; A, × 88; C, D, × 92.

genu, and some three tendons inserted on a 'sclerite' at the base of the genu). A large tendon (apparently t_1, one of the tendons of the apotele) continues into the femur.

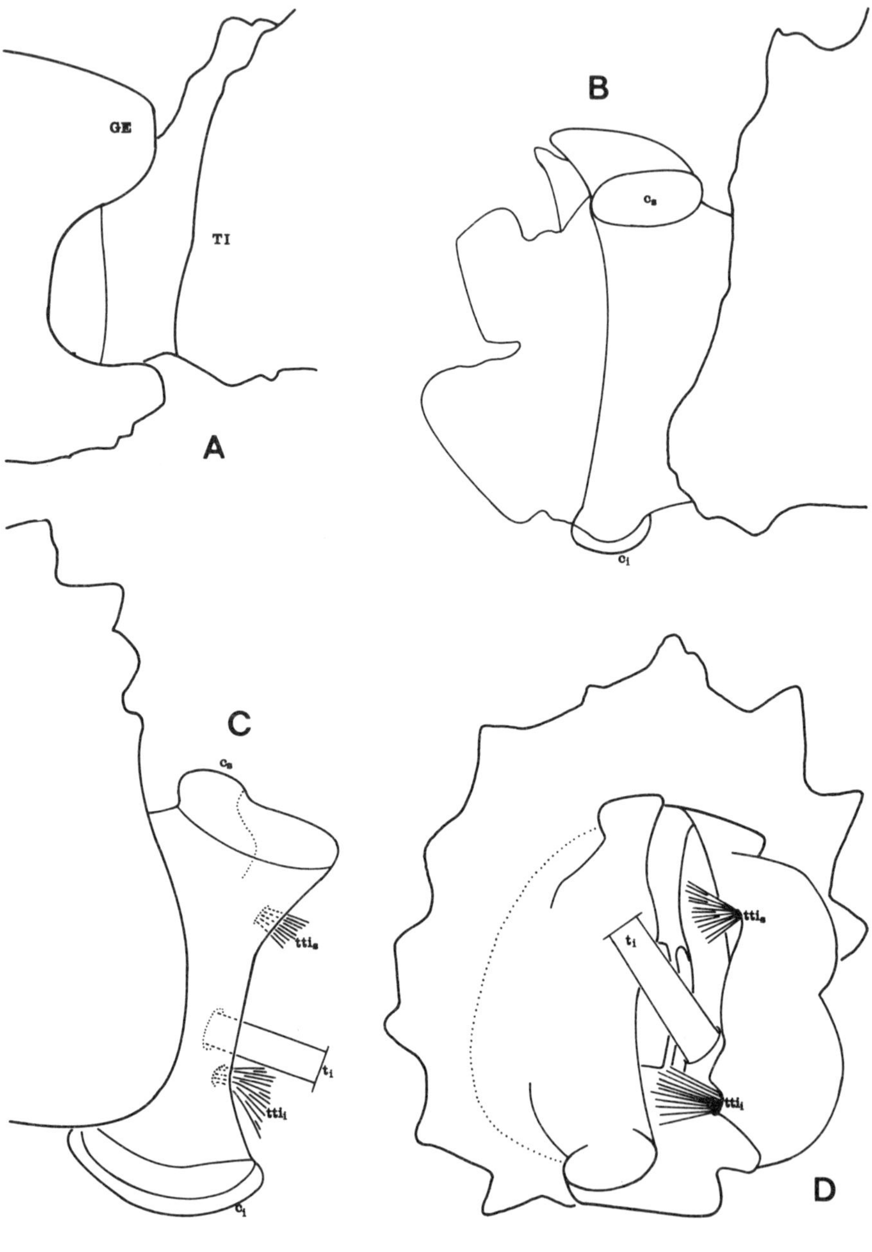

Fig. 246. Ricinoides karschii (Hansen & Sørensen), parts of right leg I of female; A, articulation between genu (patella) and tibia, lateral (antiaxial) view; B–D, base of tibia (separated from genu); B, lateral (antiaxial) view; C, lateral (paraxial) view; D, posterior view; A, × 92; B–D, × 141.

The articulation genu-tibia (fig. 246) is a hinge joint of which the axis of rotation is nearly vertical instead of horizontal. It is bicondylar (the two condyles, which are antiaxial in leg I, take the place of the single condyle of the ancestral hinge joint) and bidesmatic (the two tendons, which are paraxial in leg I, take the place of the single tendon of the ancestral hinge joint). One of the tendons of the apotele apparently continues into the genu (and the femur).

The articulation tibia-basitarsus (fig. 247) is a hinge joint. It is bicondylar (the two superior condyles take the place of the ancestral single condyle). In the case of leg I, there is apparently one paraxial tendon (*tta*), inserted on a basal 'sclerite', whilst one of the tendons of the apotele (apparently t_i) continues into the tibia (and even as far as the femur).

The adesmatic articulation between basi- and telotarsus (fig. 248A–C), and between the various telotarsi, are operated by two tendons of the apotele, which are guided by openings situated superior, resp. inferior to the axis of articulation of the joints in question.

The articulation between tarsus and apotele (fig. 248E–F) is a pivot joint (bicondylar, bidesmatic). In the case of leg I, investigated here, one of the tendons continues as far as the femur.

The cuticle of the legs presents: lyrifissures (fig. 249D: the lyrifissures ibt_{1-2}'' of the basitarsus occupy the same position as ibt'' in Opilioacarida); several types of tubercles (fig. 242C), just as on body and palp; compound phaneres (fig. 242C, 249A, B), i.e., ordinary tubercles surmounted by a corrugated tubercle; and various other types of phaneres, such as solenidia (often associated with pits) (fig. 242C, D), coronidia (dorsally, on tibia and basitarsus, just as in Opilioacarida) (fig. 249A), 'treelike' setae (fig. 242B), and 'ordinary' setae (fig. 242C, D). The compound phaneres of Ricinulei, and the compound and mucronate setae of Opilioacarida could be homologous organs. Coronidia and solenidia have also been described from Opilioacarida. The 'treelike' setae could be homologous with the bifid seta of Opilioacarida (on the acrotarsus of leg II); they could represent famuli. Solenidia associated with pits are known also from Anactinotrichida.

Leg III of the adult male (figs. 250, 251) is a gonopod. The segments involved in sperm transfer are basitarsus, telotarsus 1 and telotarsus 2. The male basitarsus presents a relatively large apophysis (the basitarsal process) and a depression in which telotarsus 1 recedes. The male telotarsus 1 presents two or three processes, viz., a fixed process (here called falcicula), the so-called accessory piece (here called rastellum), and a small process (here called docidium); all three processes are present in the male specimen of *Cryptocellus magnus* investigated by me, whilst the docidium is lacking in my male specimen of *Cryptocellus simonis*. The male telotarsus 2 presents a so-called lamina cyathiformis (here called spathula), in which the telotarsal processes fit. In *Cryptocellus simonis*, trochanter 2, tibia and basitarsus of the male gonopod (fig. 251A) each present a distal apophysis; in the male leg I of this species, apophyses are present on femur and tibia (fig. 251B), whilst trochanter 1 of the male leg IV also presents an apophysis (fig. 251C).

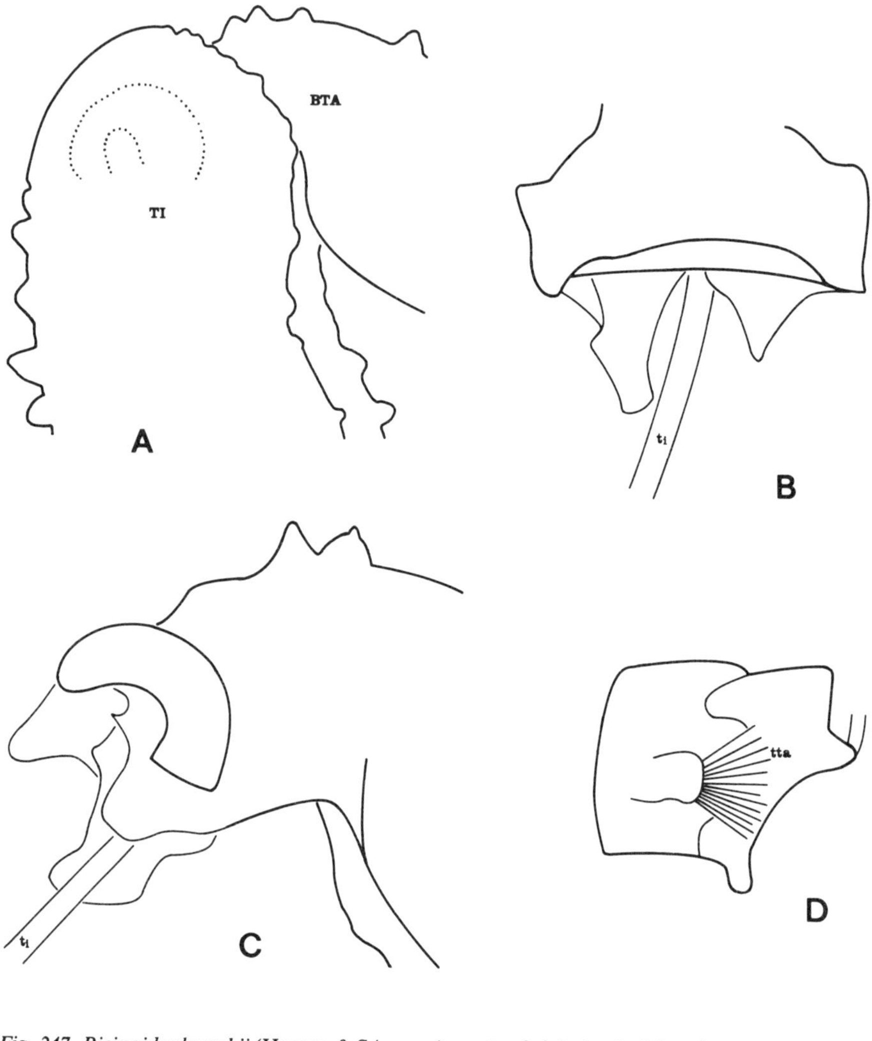

Fig. 247. Ricinoides karschii (Hansen & Sørensen), parts of right leg I of female; A, articulation between tibia and basitarsus, lateral (antiaxial) view; B, C, base of basitarsus, separated from tibia (without basal 'sclerite'); B, dorsal view; C, lateral (antiaxial) view; D, basal 'sclerite' of basitarsus, posterior view; A, × 92; B–D, × 141.

→

Fig. 248. Ricinoides karschii (Hansen & Sørensen), parts of right leg I of female; A, articulation between basitarsus and telotarsus, lateral (antiaxial) view; B, basal part of telotarsus, lateral (antiaxial) view; C, transverse section through telotarsus, with the two guided tendons, viewed in the direction of the basitarsus; D, distal part of telotarsus, and apotele, lateral (antiaxial) view; E, apotele, separated from telotarsus, ventral view; F, distal part of telotarsus and apotele, frontal view; A, D, × 92; B, C, F, × 141; E, × 295.

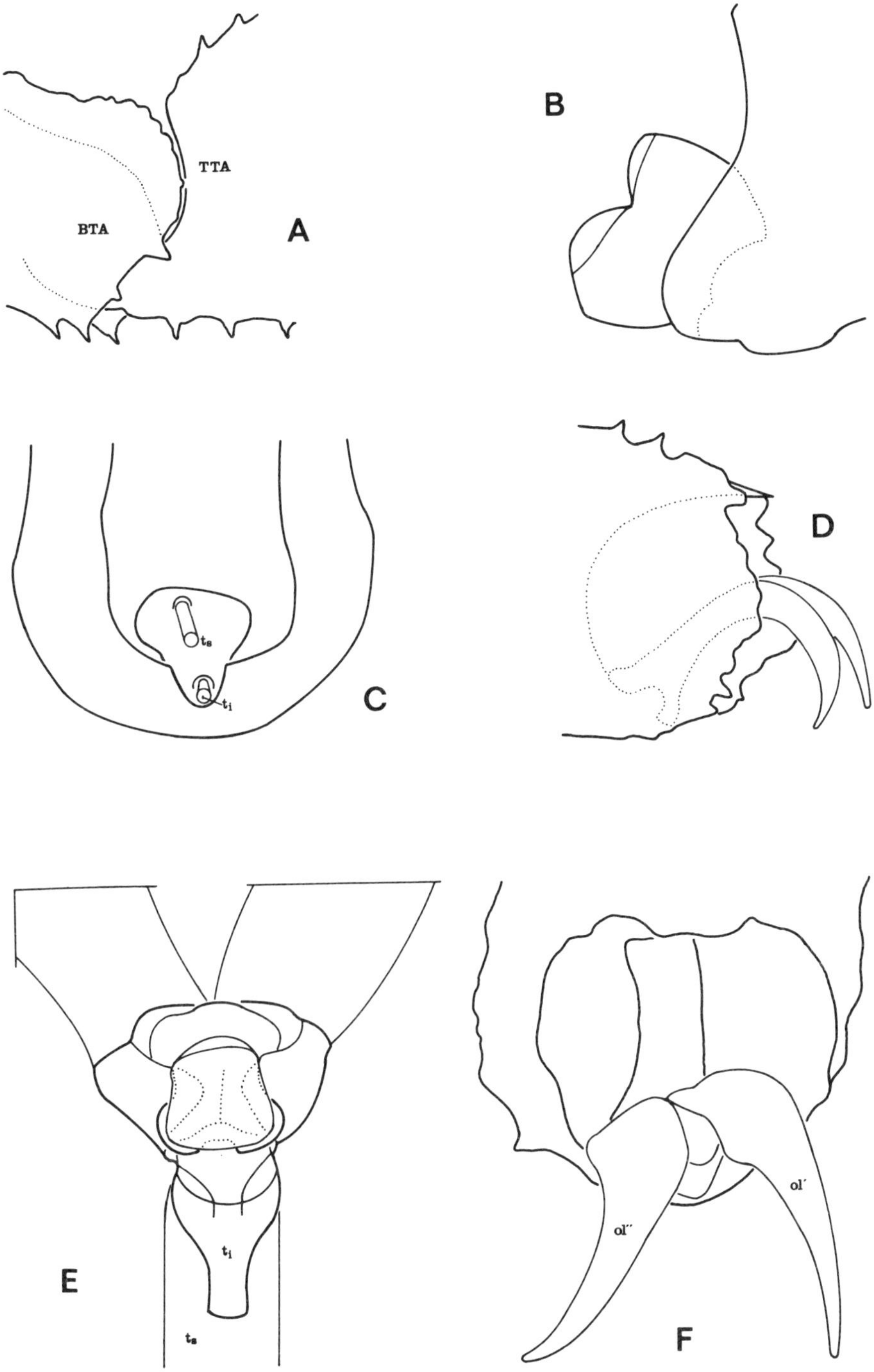
TTA
BTA
A
B
t_s
t_i
C
D
E
t_i
t_s
ol´´
ol´
F

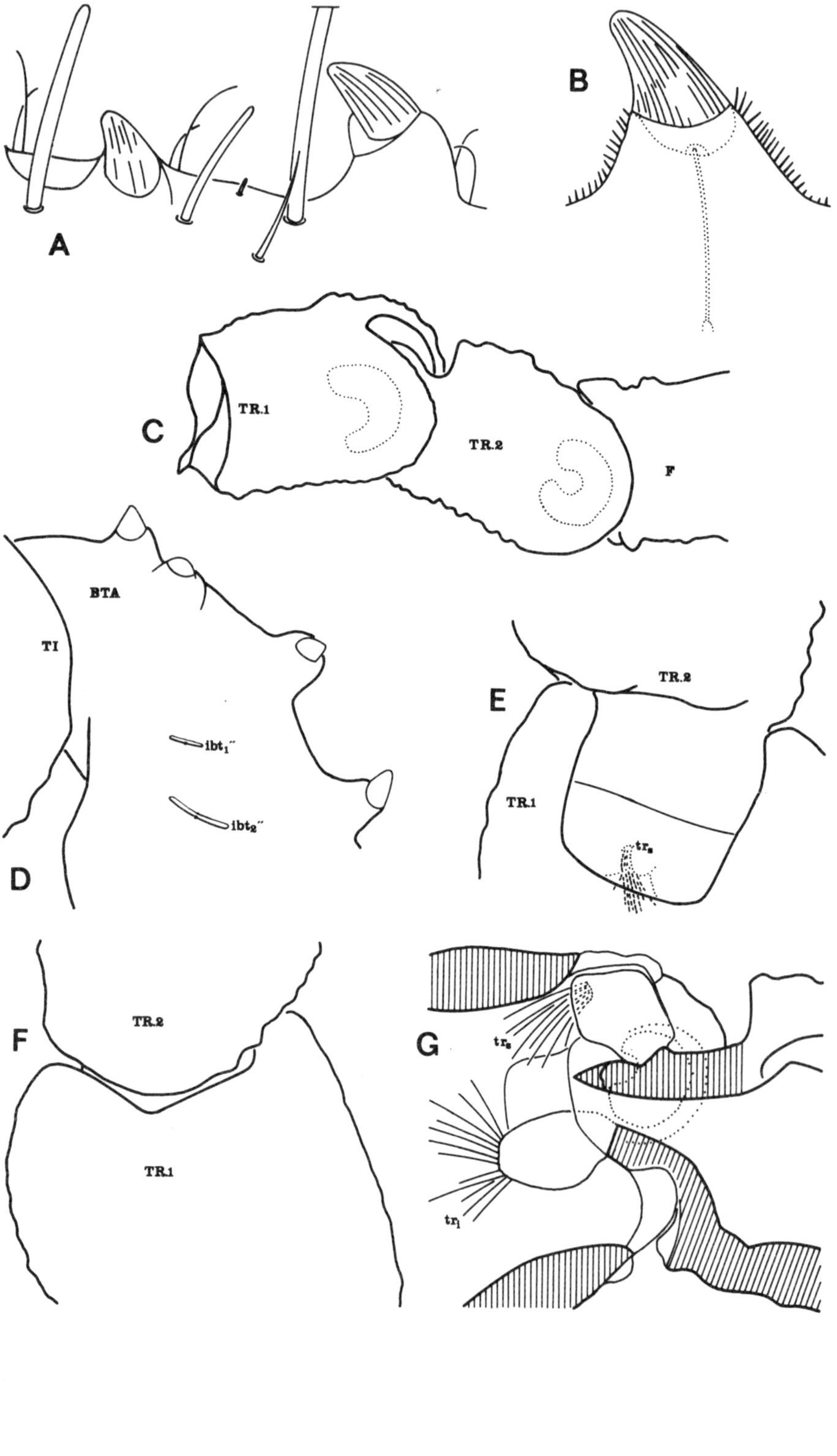
A
B
C
TR.1
TR.2
F
D
BTA
TI
ibt_1''
ibt_2''
E
TR.2
TR.1
tr_s
F
TR.2
TR.1
G
tr_s
tr_l

Internal anatomy. – The internal anatomy of Ricinulei was studied by Millot (1947). He made observations with reference to the central nervous system (all ganglia fused to a solid mass pierced by the oesophagus), the respiratory system (each stigma gives entrance into an atrium from which numerous tracheal tubules are starting), the circulatory system (reduced), the alimentary canal (mouth, pharynx, oesophagus, midgut and caeca, hindgut and anus), the excretory system (Malpighian tubes, coxal glands), the prosomatic glands, the musculature, and the genital organs. Firstman (1973: 16–18, fig. 11) subsequently studied the arterial system (characterized as typically apulmonate) and the endosternite (vestigial).

Reproduction. – Cooke (1967) has made a study of sperm transfer. After an initial phase of courtship behaviour, the male mounts the female, collects a spermatophore (a globule with an apparently hardened surface), from his genital opening, in one of the gonopods (leg III) and applies it to the genital opening of the female; it takes some time to introduce the spermatophore (or its contents) into the vulva.

Postembryonic development (figs. 252–255). – The immature stases differ from the adults by several characters, among which the wider separation of opisthosomatic tergites and sternites, the incomplete development of genital region (fig. 255A) and gonopods, and a smaller number of telotarsal segments (except in the tritonymph). The telotarsal formulae, from larva to adult, or (in the case of leg IV) from protonymph to adult, are the following: leg I, 1–1–1–1–1; leg II, 2–4–5–5–5; leg III, 2–3–4–4–4; leg IV, 2–4–5–5. A case of supposed regeneration is shown in fig. 255E: in a tritonymph of *Ricinoides karschii*, the right leg III presented the abnormal number of two telotarsi, whilst the corresponding left leg (fig. 255D) presented the normal number of four; it is supposed that the right telotarsus was broken off in one of the preceding stases. Leg IV of the larva is vestigial (fig. 252D, E). In the case of *Ricinoides karschii* it is present as a pair of stumps. A qualification as leg buds is not correct (it is a case of regression). A prelarva is not yet known from Ricinulei, but could be present as a calyptostase or an elattostase (cf. Van der Hammen, 1978).

←

Fig. 249. Ricinoides karschii (Hansen & Sørensen), right leg IV of female; A, dorsal phaneres (hollow setae, mucronate tubercles, and coronidia) of the basitarsus, lateral (antiaxial) view; B, dorsal mucronate tubercle of the basitarsus, lateral (antiaxial) view; C, trochanter 1, trochanter 2, and base of femur, lateral (antiaxial) view; D, base of basitarsus with lyrifissures ibt_1'' and ibt_2'', lateral (antiaxial) view; E, F, trochanter 1-trochanter 2 articulation; E, dorsal view; F, ventral view; G, sagittal section through trochanter 1-trochanter 2 articulation (the paraxial half is viewed from the inside); A, × 295; B, × 550; C, × 44; D, × 141; E–G, × 88.

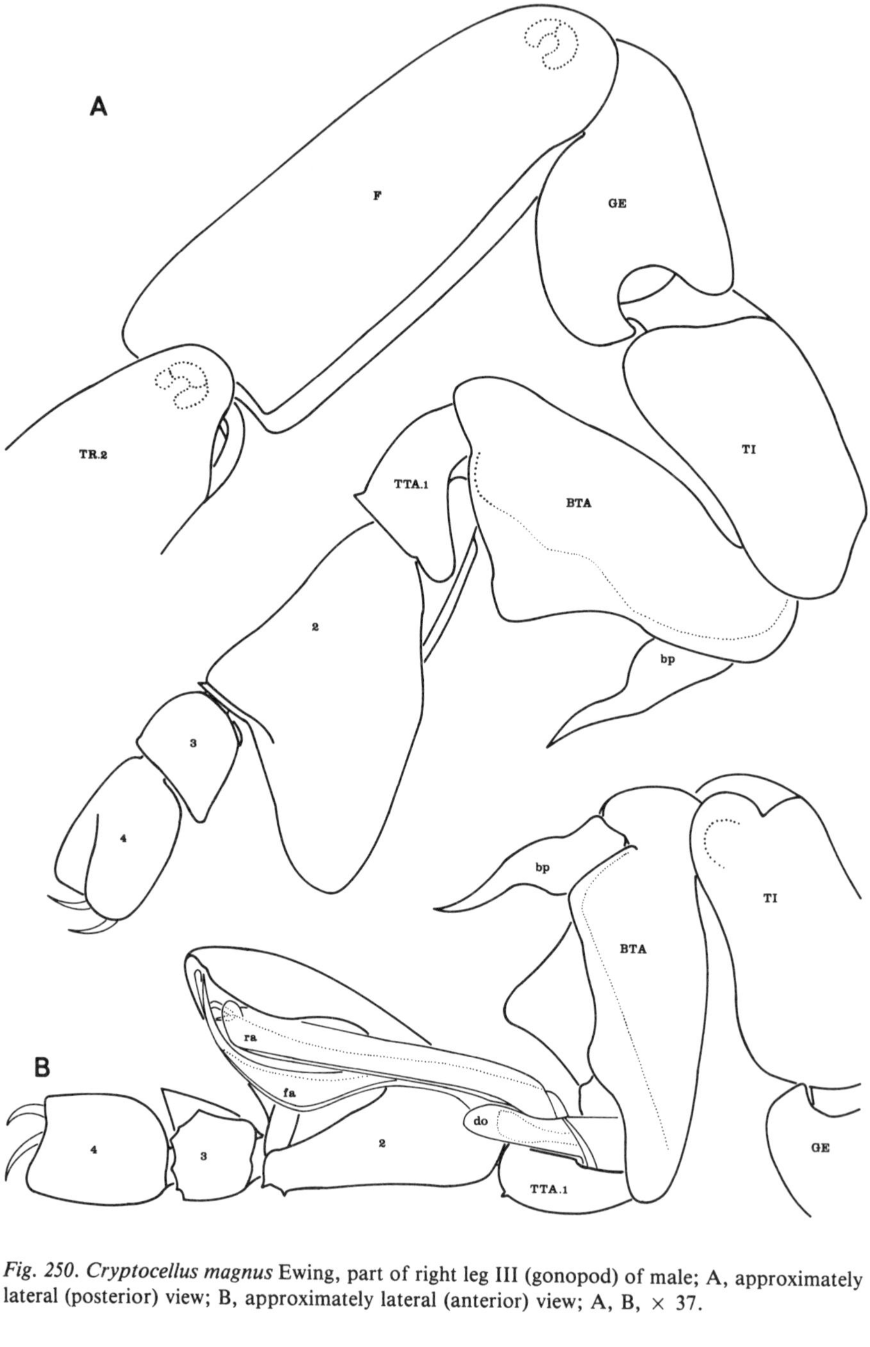

Fig. 250. Cryptocellus magnus Ewing, part of right leg III (gonopod) of male; A, approximately lateral (posterior) view; B, approximately lateral (anterior) view; A, B, × 37.

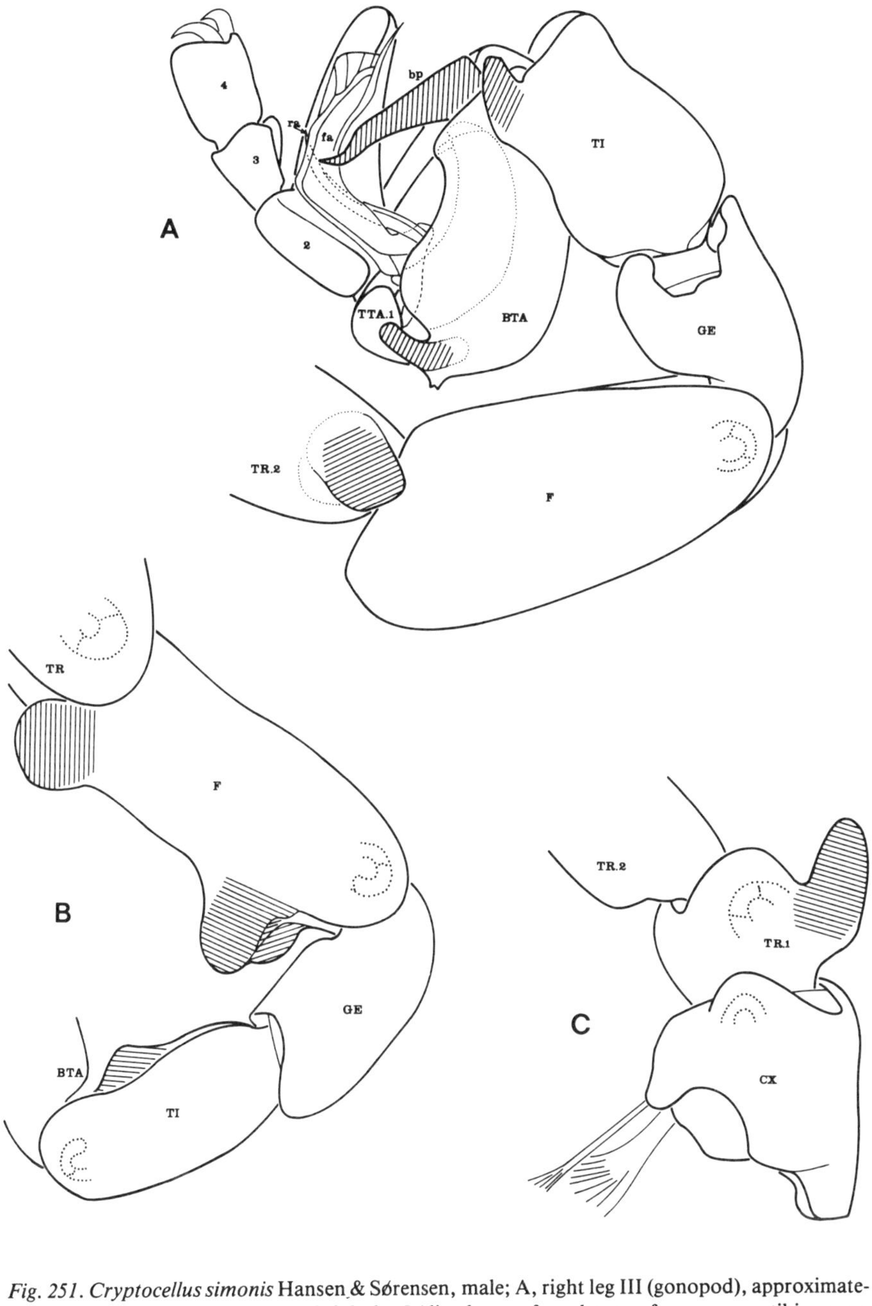

Fig. 251. Cryptocellus simonis Hansen & Sørensen, male; A, right leg III (gonopod), approximately lateral (anterior) view; B, part of right leg I (distal part of trochanter, femur, genu, tibia, proximal part of basitarsus), approximately lateral (posterior) view; C, coxa, trochanter 1 and part of trochanter 2 of right leg IV, approximately lateral (posterior) view; A–C, × 37.

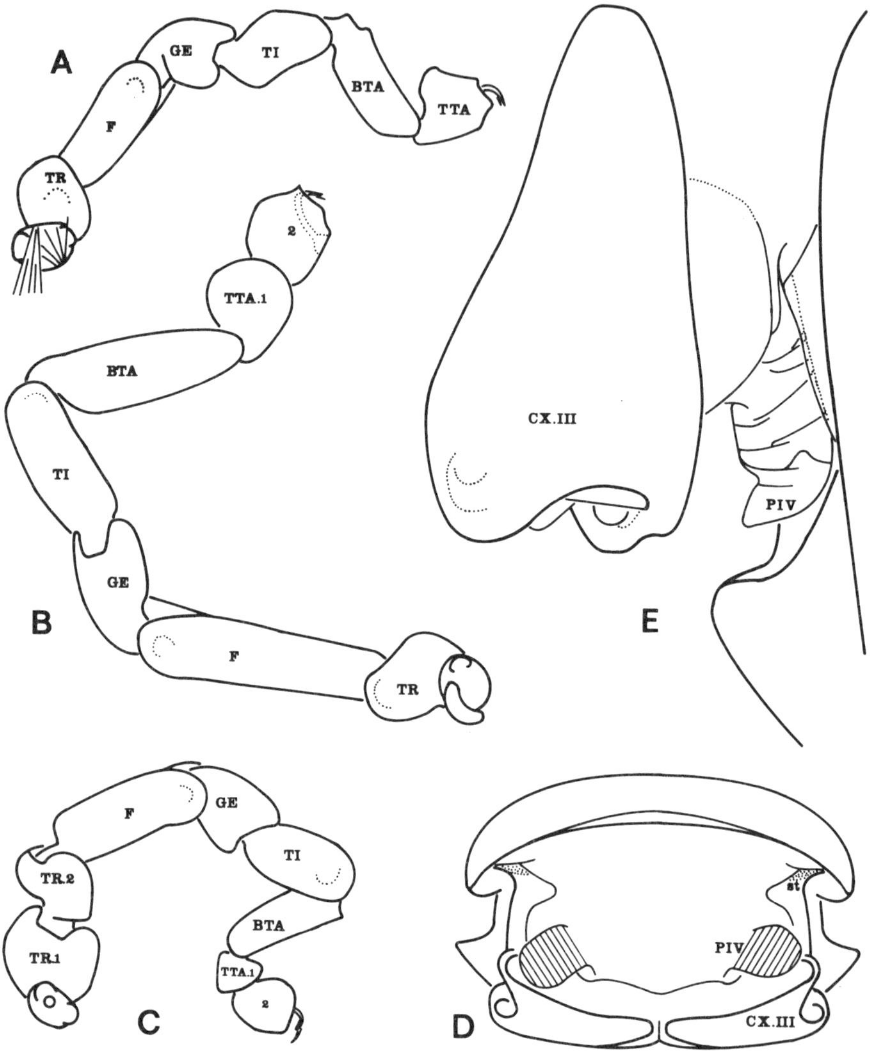

Fig. 252. Ricinoides karschii (Hansen & Sørensen), larva; A, right leg I, lateral (posterior) view; B, right leg II, lateral (posterior) view; C, right leg III, lateral (posterior) view; D, posterior view of prosoma with vestiges of leg IV (indicated by ruling); E, ventral view of posterior prosomatic region (with coxa III and vestige of leg IV), and anterior region of opisthosoma (rigth side); A–D, × 44; E, × 141.

3. Additional remarks

The distribution of extant Ricinulei is essentially Gondwanian. Species are now known from West Africa (from former Portuguese Guinea to Gabon) and

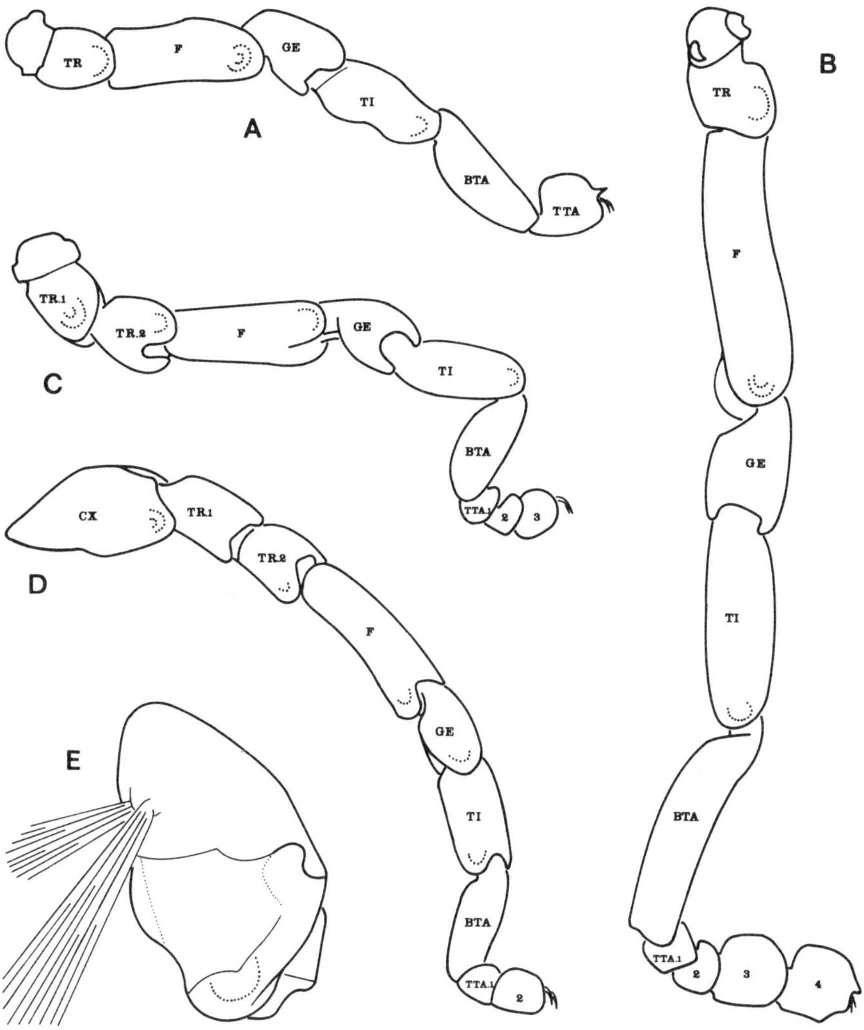

Fig. 253. Ricinoides karschii (Hansen & Sørensen), protonymph; A–D, right legs in lateral (posterior) view; A, leg I; B, leg II; C, leg III; D, leg IV (including the movable coxa); E, dorsal view of right coxa IV (separated from the prosoma); A–D, × 37; E, × 73.

from part of the Americas (from southern Texas to Brazil). Two extinct families of the group (Poliocheridae and Curculioididae) are known from the Carboniferous of Europe and North America (cf. Petrunkevitch, 1955). Evidently, the group at that time had a different and much larger distribution.

Of the two fossil families, the family Poliocheridae is closest to extant Ricinulei, whilst the family Curculioididae is especially different by the

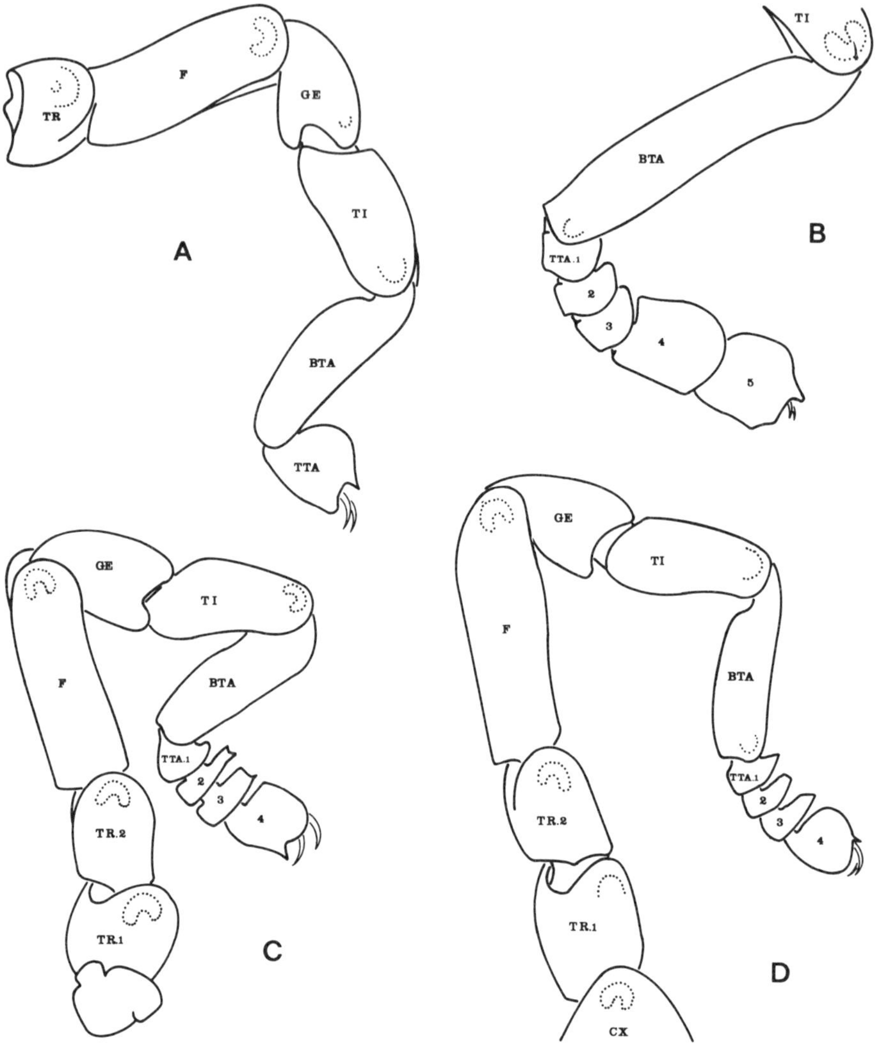

Fig. 254. Ricinoides karschii (Hansen & Sørensen), deutonymph; A–D, right legs in approximately lateral (posterior) view; A, leg I; B, terminal part of leg II; C, leg III; D, leg IV; A–D, × 37.

presence of a single dorsal opisthosomatic shield (longitudinally divided) and the presence of divided opisthosomatic sternites.

Fossil and extant Ricinulei must be characterized as highly specialized Chelicerata. Specialization seems to be linked with a particular kind of protection: all openings (mouth, stigmata, genital opening, anus) can be concealed; the arthodial membranes of the appendicular articulations are protected by complicated joints; the cuticle is extraordinary thick and hard. In spite of the

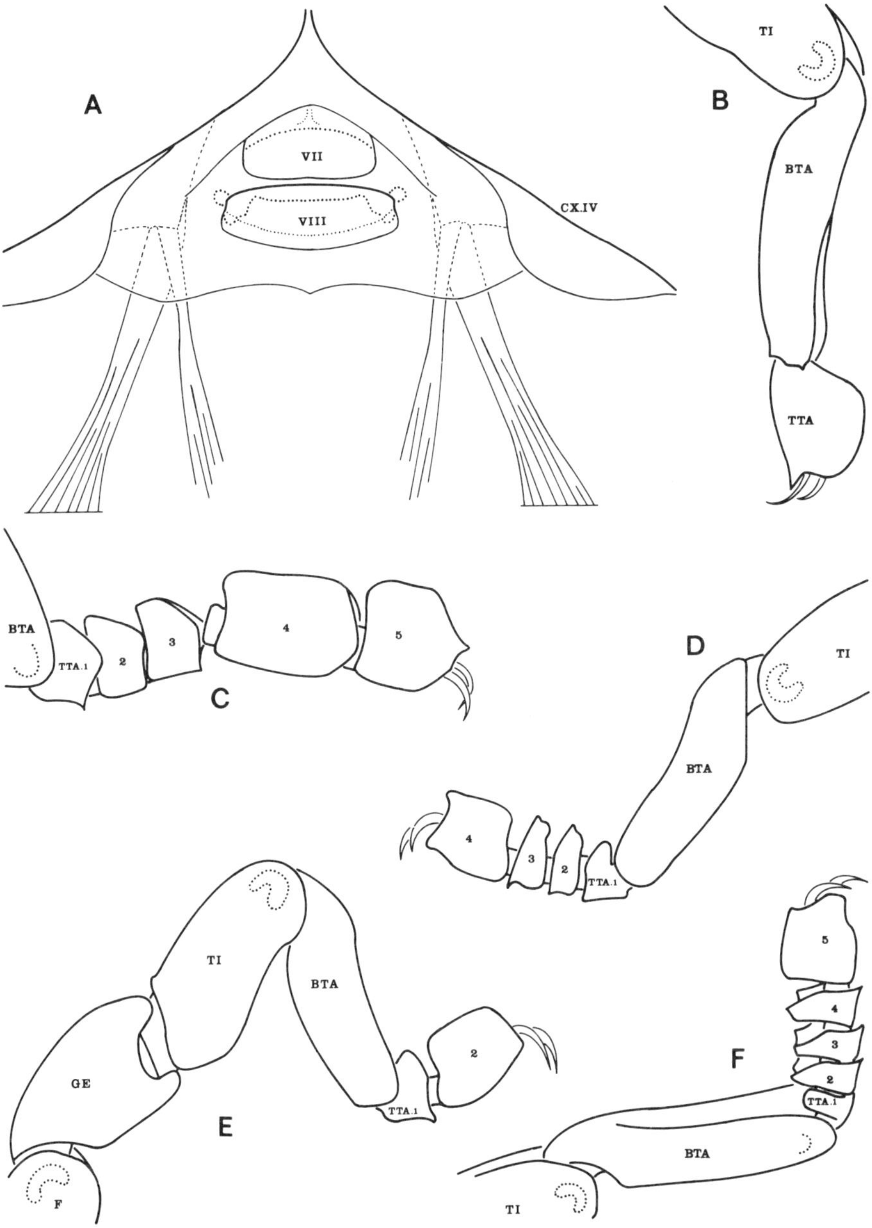

Fig. 255. Ricinoides karschii (Hansen & Sørensen), female tritonymph; A, ventral view of genital region (and posterior border of coxae IV); B–D, legs in approximately lateral (posterior) view; B, terminal part of right leg I; C, terminal part of right leg II (joint between telotarsi 3 and 4 dislocated); D, terminal part of left leg III; E, terminal part of abnormal right leg III (probably a case of regeneration; the corresponding normal left leg is represented in fig. D); F, terminal part of right leg IV; A, × 73; B–F, × 37.

highly specialized joints of the legs, locomotion seems to be slow. Extant species are known to sham death in case of danger. Nothing seems to be known about the enemies of Ricinulei. Protective specialization seems to be excessive, and other hypotheses with reference to its adaptive value could perhaps be introduced. The colonization by Ricinulei of Mexican and Central American caves seems to be a recent favourable development in the evolution of the group.

List of the notations of figs. 229–255

AN	anal opening.
AP	apotele.
BF	basifemur.
bp	basitarsal process.
BTA	basitarsus.
c, c', c''	condyle, articulation point.
c_i	inferior condyle.
c_s	superior condyle.
CU	cucullus.
CX	coxa.
CX.III, CX.IV	coxae III–IV.
disj	disjugal furrow.
do	docidium.
e	laterocoxal setae of palp.
F	femur.
fa	falcicula.
GE	genu (patella).
gl	gland.
ibt_{1-2}''	proximal, posterolaterodorsal lyrifissures of basitarsus.
J	commissure of mouth.
LI	labium.
LS	labrum.
og	orifice of coxal gland.
ol'', ol''	ungues.
PA	palp.
PI–IV	legs I–IV.
ra	rastellum.

SA	sternapophyses.
scl	'sclerite'.
sg	subcapitular gutter.
ss	sac-like structures of female genital region.
st	stigma.
t_i	inferior tendon of apotele.
t_s	superior tendon of apotele.
tf_i	inferior tendon of femur.
tf_s	superior tendon of femur.
TF	telofemur.
tf', tf''	tendons of femur.
tg, tg_{1-3}''	tendons of genu.
TI	tibia.
TITA	tibiotarsus.
tla	lateral tendon of infracapitulum.
tr_i	inferior tendon of trochanter 2.
tr_s	superior tendon of trochanter 2.
TR	trochanter.
TR.1	trochanter 1.
TR.2	trochanter 2.
tsc	tendon of subcapitular apodeme.
tta	tendon of tarsus.
TTA	telotarsus.
TTA.1	telotarsus 1.
tti_i	inferior tendon of tibia.
tti_s	superior tendon of tibia.
ttr', ttr''	tendons of trochanter (or trochanter 1).
2, 3, 4, 5	telotarsi 2–5.
VII–XIX	opisthosomatic segments VII–XIX.
' (prime)	anterior face of an appendage.
'' (double prime)	posterior face of an appendage.

VII. UROPYGI

In the fourth part of my Comparative Studies in Chelicerata (Van der Hammen, 1986a: 8–12, 14, 18, 23, 32–38, figs. 2A, B, 5A, B, 7A, B, 10, 11A, 17, 18) I have dealt with some parts of Uropygid morphology, viz., segmentation of the body, respiratory system, orifices of coxal glands, mouthparts, and segmentation and articulation of the legs. Previously, I had mentioned Uropygi only in passing, in some general papers dealing with Chelicerate morphology, postembryonic development and classification (Van der Hammen, 1974: 386, 387; 1977a: 310, 312, 313, 317, tab. 1–3; 1978: 45, 49; 1985: 55, fig. 34; 1985a: 137, 139, 142, 143, 144, 145, figs. 3, 7); Uropygi are also mentioned in my final survey of Chelicerate relationships (Van der Hammen, 1986c: 225, 227, fig. 3). The results of these studies are summarized in the present chapter, supplemented with data from literature. The chapter is subdivided into three parts: an introduction, a descriptive part, and additional remarks.

1. Introduction

Data with reference to the comparative morphology of Uropygi have been summarized by Millot (1949b) and Kaestner (1932; 1956: 546–550). Snodgrass (1948: 26–31, fig. 9) studied the mouthparts, Firstman (1973: 22–24, figs. 15–17) the endosternite and the circulatory system, Weygoldt (1970b, 1971) reproductive behaviour and postembryonic development, and Alberti & Palacios-Vargas (1987) sperm morphology and spermatogenesis.

In the past, Uropygi have been classified together with Amblypygi as Pedipalpi; it is now generally admitted that Amblypygi are more closely related to Araneida. Several modern authors still regard Schizomida and Holopeltida as separate groups; the two groups are, however, closely related and are regarded by me as orders of the superorder Uropygi.

The original part of the present chapter is based on a study of two species: the Schizomid *Trithyreus* spec. (material from former Netherlands New Guinea); and the Holopeltid *Tetrabalius seticauda* (Doleschall) (material from the former Netherlands East Indies).

2. Descriptive part

Habitus. – Uropygi are Chelicerata with segmented opisthosoma, and two types of habitus can be distinguished. Schizomida are small, and present a subdivided prodorsum and a short flagellum, whilst the palp is not strikingly robust. Holopeltida are medium-sized or large, the prodorsum is not subdivided, the flagellum is long and multi-jointed, and the palp is robust (the terminal part constitutes a sort of chela).

Divisions of the body, tagmata and segmentation. – Data with reference to Uropygid segmentation have been summarized by Kaestner (1932: 4–7, figs. 4, 5, 9) and Millot (1949b: 534–536, figs. 288–291). The two groups of Uropygi (Schizomida and Holopeltida) are treated here separately, particularly because of differences in the structure of prodorsum and flagellum; a species of Schizomida is represented, in lateral view, in fig. 3A, a species of Holopeltida in fig. 3B.

The Schizomid prodorsum is generally regarded as being subdivided into a relatively large propeltidium, a mesopeltidium consisting of two small sclerites, and a metapeltidium. The two sclerites constituting the so-called mesopeltidium are here regarded as arthrodial sclerites (see also fig. 260E), and the other two sclerites are now called pro- and postpeltidium. The ventral surface of the prosoma is for the greater part occupied by the coxae of the legs; there is a triangular sternite between the coxae of legs I and II (the sternite between the coxae of legs IV is supposed to represent the sternite of segment VII). The opisthosoma consists of the segments VII–XVIII and a 'telson' (the flagellum). Segment VII is distinctly narrowed, and consists of a small tergite and probably also of a narrowed sternite (the triangular sternite between coxae IV). Segments VIII–XV each consist of a tergite and a sternite, connected by pleural regions of soft skin. The genital opening is constituted by a slit at the posterior border of sternite VIII. The segments XVI–XVIII constitute a distinctly narrowed metasoma, in which the tergites and sternites are not differentiated. The telson is represented by a short flagellum which is subdivided into a small number of 'segments' that can present additional annulations. The anal opening is terminal in segment XVIII, ventral to the base of the flagellum.

The Holopeltid prodorsum is represented by a large, undivided sclerite; the ventral surface of the prosoma resembles that of the Schizomida (it is for the greater part occupied by the coxae of the appendages; there is a triangular sternite between the coxae of legs I–II). The Holopeltid opisthosoma consists of the segments VII–XVIII; segment VII is narrowed, the genital opening is constituted by a transverse slit at the posterior border of sternite VIII, segments VIII–XV each consist of a tergite and a sternite separated by a pleural region of soft skin, and the segments XVI–XVIII constitute a narrow metasoma in which the tergites and sternites are not differentiated. The Holopeltid 'telson' (the flagellum) is long and narrow, and multi-jointed. The anal opening is terminal in segment XVIII, ventral to the base of the flagellum.

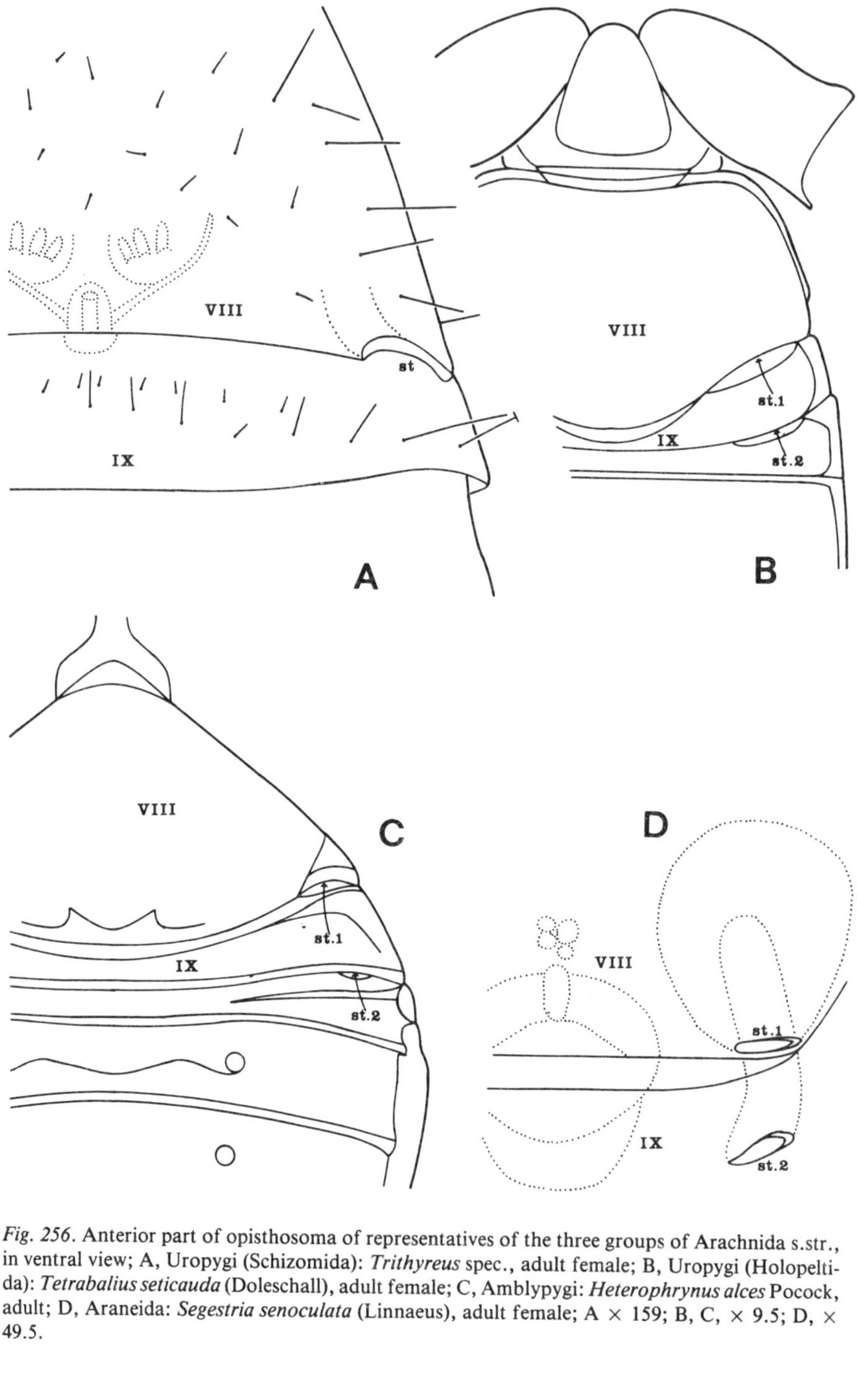

Fig. 256. Anterior part of opisthosoma of representatives of the three groups of Arachnida s.str., in ventral view; A, Uropygi (Schizomida): *Trithyreus* spec., adult female; B, Uropygi (Holopeltida): *Tetrabalius seticauda* (Doleschall), adult female; C, Amblypygi: *Heterophrynus alces* Pocock, adult; D, Araneida: *Segestria senoculata* (Linnaeus), adult female; A × 159; B, C, × 9.5; D, × 49.5.

Tegument. – The sclerotized parts of the cuticle are relatively thin in Schizomida, rather heavy and hard in Holopeltida. Spines, ordinary setae, trichobothria and lyrifissures are present.

Aspidosoma. – As mentioned above, the aspidosoma (or prodorsum) comprises one sclerite (Holopeltida), or two sclerites connected by a pair of arthrodial sclerites (Schizomida; the postpeltidium can be divided by a median suture). Holopeltida present a pair of median eyes and, on both sides, a group of three lateral eyes (two vestigial lateral eyes can also be present). Schizomida can present regressive or vestigial lateral eyes.

Dorsal and laterodorsal region of opisthosoma. – This region comprises the opisthosomatic sclerites, and the intersegmental areas and pleural regions of soft skin. Segment VII is distinctly narrowed. Segments XVI–XVIII constitute a narrow metasoma, without pleural regions of soft skin.

Flagellum. – In Holopeltida, the flagellum is long, narrow and multijoined. In Schizomida, it is short; it can, in extant species, present 1–4 'segments' (1–4 in the female, 1–2 in the male), whilst the 'segments' can also present annulations. The male flagellum can be broadened and present characteristic specific shapes.

Ventral and lateroventral region of opisthosoma. – The sternite of segment VII is supposed to be represented by the triangular sclerite between coxae IV. The genital opening is represented by a transverse slit at the posterior border of segment VIII. The Uropygid respiratory organs are represented by booklungs and stigmata with ventral position: one pair (stigmata in segment VIII) in Schizomida (fig. 256A), two pairs (stigmata in segments VIII and IX) in Holopeltida (fig. 256B). Lateral regions of soft skin are not present in the narrowed metasoma (segments XVI–XVIII). The anus is in segment XVIII, ventral to the base of the flagellum.

Ventral aspect of podosoma. – The greater part of the ventral surface of the podosoma is occupied by the coxae of the legs. A triangular sclerite (the sternum, or neosternum) is present between coxae I and II; a very small sclerite can be present between coxae III. The orifice of the coxal gland is paraxial, at the base of coxa I (figs. 257A, B, 258B). In Schizomida as well as Holopeltida, it is associated with a taenidium and a setiferous sternapophysis. The taenidia of both sides are connected with an unpaired intercoxal gutter (also a taenidium), with ventral position, which belongs to the mouthparts, and is situated between the palpal coxae. The long setae of the sternapophysis are normally lying in the intercoxal gutter, exactly as in the case of sternapophyses and subcapitular gutter in the two groups of Cryptognomae (Anactinotrichida and Ricinulei).

Lateral aspect of podosoma. – The lateral region of the podosoma is occupied by the antiaxial part of the coxae of the legs and by soft skin.

Mouthparts. – The Schizomid mouthparts comprise (besides the chelicerae): the labrum, the cervix (posteriorly bordered by the line of attachment of the cheliceral frame), median and lateral apodemes, the coxapophyses of the

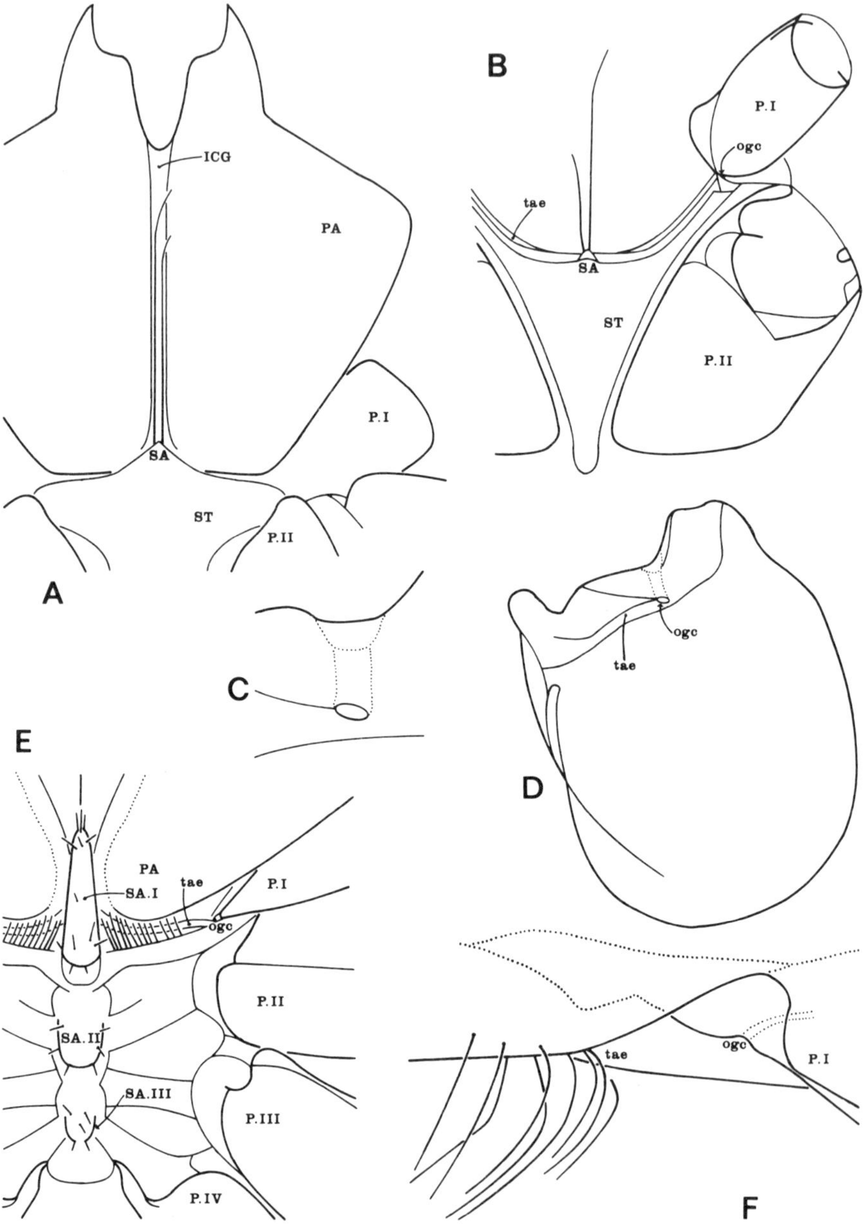

Fig. 257. Coxal and sternal region of Arachnida s.str.; A, B, *Tetrabalius seticauda* (Doleschall), adult male; A, ventral view of anterior part of coxal and sternal region; B, anteroventral view of coxal and sternal region of legs I and II; C, D, *Segestria senoculata* (Linnaeus), adult female; C, orifice of coxal gland (detail of D); D, basal view of right coxa I (the upper part of the figure is dorsal, the left part anterior); E, F, *Heterophrynus alces* Pocock, adult; E, ventral view of coxal and sternal region with sternapophyses I–III; F, posterior view of coxal and sternal region of right leg I, with orifice of coxal gland and taenidium (the upper part of the figure is dorsal); A, B, E, × 12.8; C, × 250.8; D, F, × 83.6.

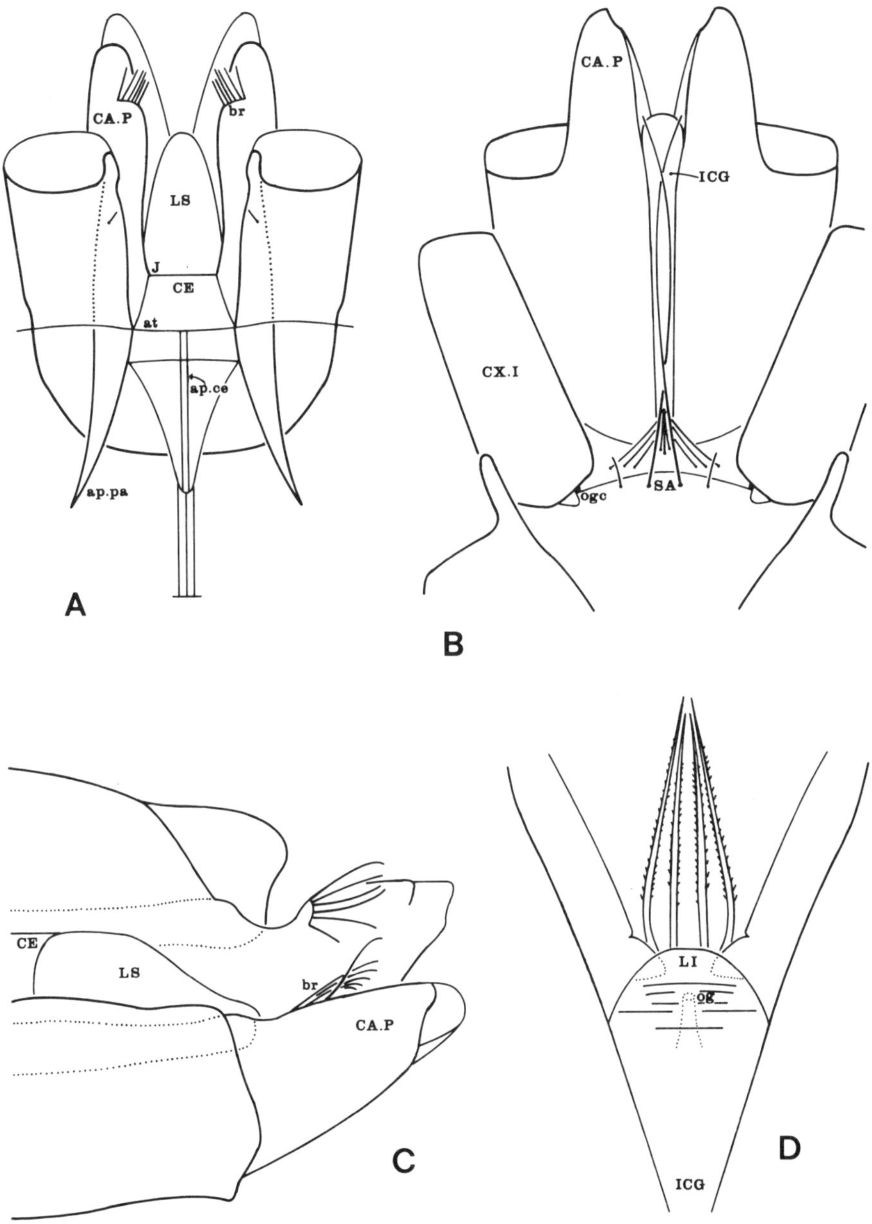

Fig. 258. Trithyreus spec., mouthparts of adult female; A, dorsal view; B, ventral view; C, oblique laterodorsal view; D, ventral view of labium and adjacent parts of coxapophyses and intercoxal gutter; A–B, × 88; C, × 141; D, × 295.

palps, an intercoxal gutter (with ventral position, between the fused palpal coxae) and a labium; there are no lateral lips. A dorsal view of the mouthparts (chelicerae removed) is represented in fig. 258A, a ventral in fig. 258B, and a lateral in fig. 258C. The unpaired median apodeme is long and narrow, and muscles are inserted on it. The paired lateral apodemes are a continuation of the internal apodeme of the palpal coxa (homologous with the anterior coxal apodeme of all Arachnida s.str.), and palpal muscles are attached to it. The coxapophyses of the palp present a dorsal brush. As mentioned above, the intercoxal gutter is associated with the sternapophyses and the orifices of the coxal glands (as in Cryptognomae). The labium is an anterior extension of the bottom of the intercoxal gutter, and is laterally fused with the palpal coxapophyses; an orifice of a labial gland is visible in ventral view (fig. 258D).

The Holopeltid mouthparts are very similar to those of the Schizomida. They are represented, in dorsal view (after removal of the chelicerae), in fig. 259A; in this figure, the attachment of palpal muscles to the lateral apodeme (the anterior apodeme of the palpal coxa) is also represented. The similarity of the mouthparts in Schizomida and Holopeltida demonstrates again the close relationship of two groups of Uropygi.

Chelicera. – The chelicerae are two-segmented; they consist of the body of the chelicera and the apotele. They resemble a pocket-knife (although the body of the chelicera presents a distal projection which resembles a fixed jaw of a chela), and are attached to the rostral part of the prosoma.

Palp. – The Uropygid palp consists of six segments: coxa (or coxal region; it is not a free segment), trochanter, femur, patellotibia, tarsus and apotele. The paraxial parts of the coxae are connected to each other by a narrow 'sternal' area which constitutes the intercoxal gutter; the distal part of the coxa presents a paraxial coxapophysis. The apotele is relatively large; in Holopeltida, the apotele and a projection of the tarsus constitute a sort of chela.

Legs. – The Uropygid legs are studied here in both groups (Schizomida and Holopeltida). Leg IV of a *Trithyreus* species (Schizomida) is represented in fig. 260. The Schizomid legs consist of coxa, trochanter, femur, patella, tibia (patella and tibia have apparently fused in leg I, as in the case of the palp), basitarsus, telotarsus (subdivided into several adesmatic segments) and apotele (absent in leg I). Leg I consists (in *Trithyreus*) of coxa, trochanter, femur, patellotibia, basitarsus, and telotarsi 1–7. Legs II–IV consist (in *Trithyreus*) of coxa, trochanter, femur, patella, tibia, basitarsus, telotarsi 1–3, and apotele. The coxae of the legs occupy the ventral surface of the prosoma; they are regarded here as coxae at an advanced stage of evolutionary development (not as coxae which secondarily have become fixed). The coxa-trochanter articulation (fig. 260E) presents an anterior condyle and a posterior arthrodial sclerite; the arthrodial sclerite is movably connected with the coxa by a relatively broad hinge, and with the trochanter by an articulation point (as appears from studies in Amblypygi and Araneida, the arthrodial sclerite evolved by desclerotization of the surrounding cuticle). This complicated joint (similar

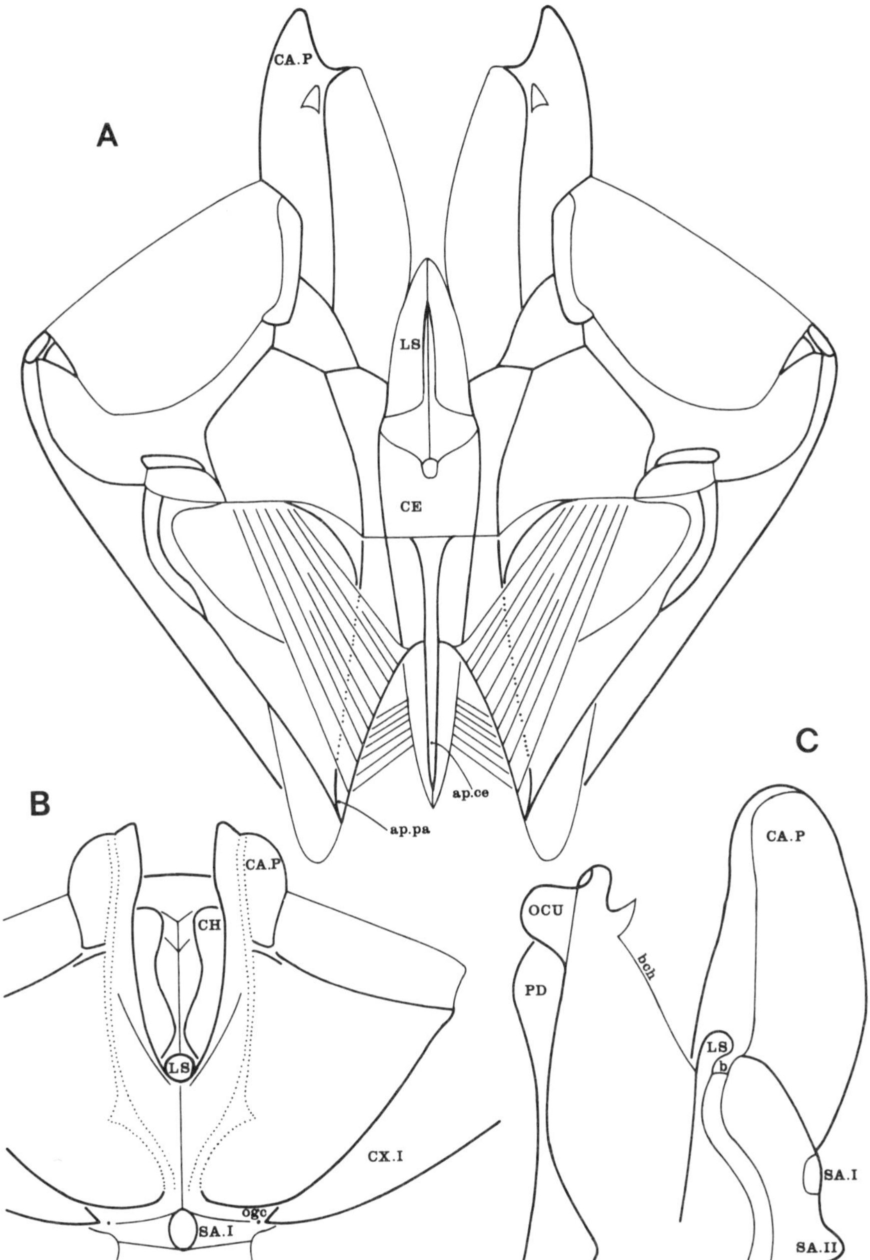

Fig. 259. Mouthparts of Arachnida s.str.; A, Uropygi (Holopeltida): *Tetrabalius seticauda* (Doleschall), adult male, dorsal view of mouthparts (chelicerae removed); B, C, Amblypygi: *Heterophrynus alces* Pocock, adult; B, anteroventral (oblique) view of mouthparts (sternapophysis I cut off); C, longitudinal section of mouthparts (chelicerae removed); A, × 20.3; B, C, × 4.8.

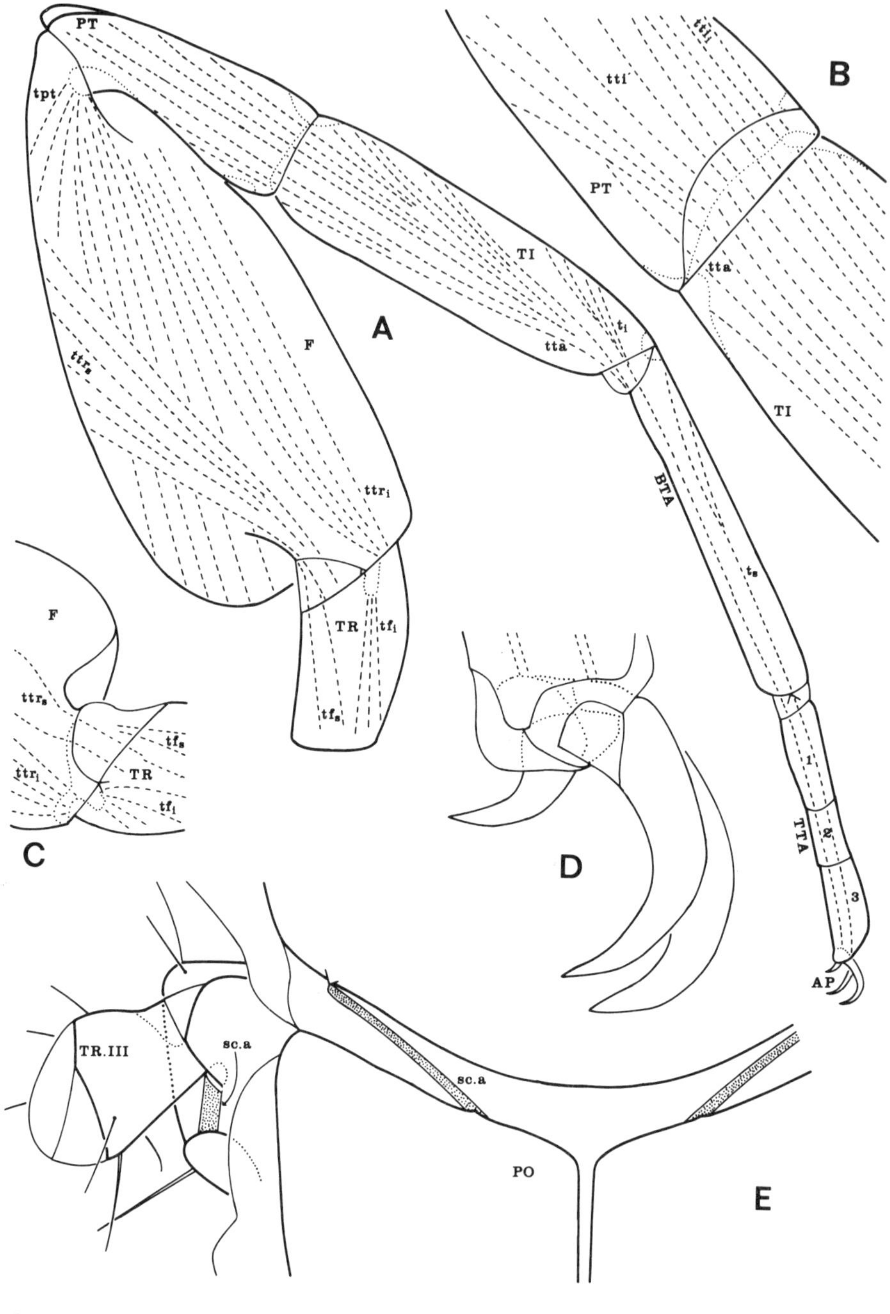

Fig. 260. Trithyreus spec.; A–D, right leg IV of adult female; A, lateral (posterior) view of trochanter, femur, patella, tibia, basitarsus, telotarsi 1–3 and apotele; B, lateral (anterior) view →

joints are present in other Arachnida s.str.) permits of promotor-remotor movements associated with leg rocking. The trochanter-femur articulation is constituted by a bicondylar pivot joint with superior and inferior muscles. The muscles of the trochanter are in the femur (as in other Arachnida s.str.). The femur-patella articulation is constituted by a bicondylar hinge joint with inferior muscles. The patella-tibia articulation (fig. 260A, B) is constituted by a particular type of hinge joint (similar types are found in other groups of Arachnida s.str.); it is characterized by the presence of a superior condyle, whilst the entire posterior part of the joint can function as a hinge (flexion is in anterior direction); there are anterolateral and antero-inferior muscles, whilst the flexor muscles of the tarsus extend into the patella. The tibia-basitarsus articulation is represented by a hinge joint which is operated by the flexor muscles of the tarsus (the inferior tendon of the apotele extends, moreover, into the tibia). The adesmatic basitarsus-telotarsus articulation and the adesmatic articulations in the telotarsus are operated by the tendons of the apotele. The tarsus-apotele articulation is bicondylar; it is operated by the usual two tendons (superior and inferior). The apotele of legs II–IV (fig. 260D) is characterized by the presence of two ungues and a so-called pseudonychium (an apophysis of the apotele, not a true unguis).

Leg IV of a *Tetrabalius* species (Holopeltida) is represented in fig. 261. Holopeltid legs generally consist of coxa, trochanter, femur, patella, tibia (patella and tibia have apparently fused in the case of leg I, as in the palp), basitarsus, telotarsus (subdivided into several adesmatic segments) and apotele (absent in the case of leg I). In *Tetrabalius*, leg I consists of coxa, trochanter, femur, patellotibia, basitarsus, and telotarsi 1–9; legs II–IV consist of coxa, trochanter, femur, patella, tibia, basitarsus, telotarsi 1–3, and apotele. The coxae of the Holopeltid legs occupy the ventral surface of the prosoma and are regarded (as in the case of Schizomida) as coxae at an advanced stage of evolutionary development. An internal apodeme (fig. 261C) is present at the anterior side of the coxae (as in other groups of Arachnida s.str.); it is probably homologous with the apodeme that is usually associated with an epimeron. The coxa-trochanter articulation (fig. 261C–E) presents an anterior condyle, distal to the coxal apodeme, and a posterior arthrodial sclerite (broad hinge with the trochanter, articulation point with the coxa). This joint permits of promotor-remotor movements associated with leg rocking. The other joints in the legs of *Tetrabalius* (fig. 261A) resemble those of Schizomida. The patella-tibia joint (fig. 261A, B) presents a superior condyle, whilst the posterior part of the joint can function as a broad hinge; the joint permits also of leg rocking.

---

of patella (distal part) and tibia (proximal part) (the lower part of the figure is dorsal); C, lateral (anterior) view of trochanter (distal part) and femur (proximal part); D, lateral (posterior) view of distal part of telotarsus 3 and apotele; E, dorsal view of part of prosoma (with arthrodial sclerites between pro- and postpeltidium) and coxa and trochanter of right leg III; A, C, × 88; B, × 186; D, × 550; E, × 141.2.

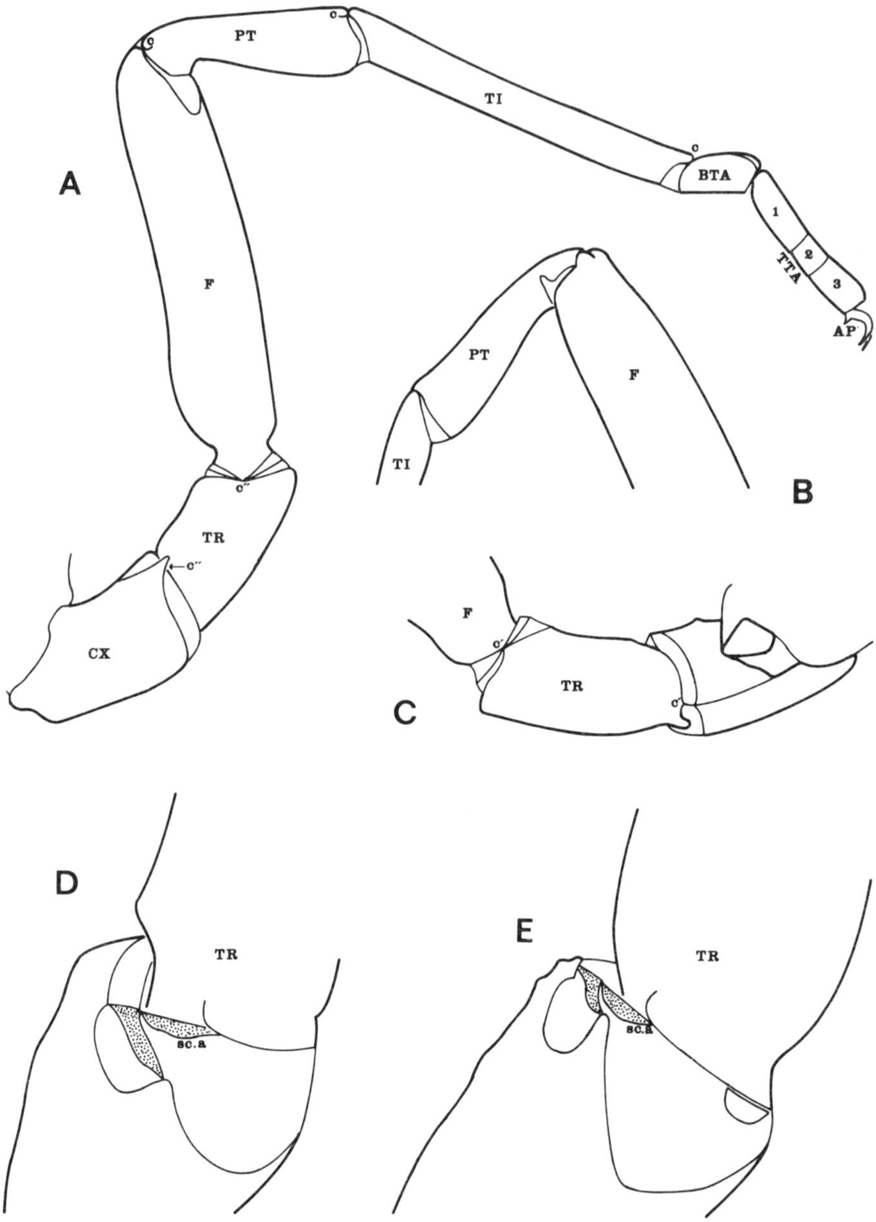

Fig. 261. Tetrabalius seticauda (Doleschall), right leg IV of adult male; A, lateral (posterior) view; B, lateral (anterior) view of distal part of femur, patella, and proximal part of tibia; C, lateral (anterior) view of coxa, trochanter and proximal part of femur; D, E, articulation between coxa and trochanter, lateral (posterior) face; D, oblique view in the direction of the base of the leg (laterodorsal orientation); E, oblique view in the direction of the base of the leg (lateral orientation); A–C, × 8.3; D–E, × 20.3.

Internal anatomy. – The alimentary canal (digestion is external: food is rendered fluid outside the alimentary canal and ingested in this state) comprises: mouth, pharynx, oesophagus, midgut and caeca (up to four pairs of prosomatic caeca, typically eight pairs of opisthosomatic caeca), hindgut and anus. The excretory system comprises Malpighian tubes, nephrocytes and coxal glands. The Uropygid coxal gland (Kaestner, 1932: 37–38, figs. 52, 54; Millot, 1949b: 550–551, fig. 310), with its two sacculi, labyrinth and duct, extends from segment IV to segment VI; it is probably the result of a fusion of two or more glands. As mentioned above, the orifice of the gland is at the base of coxa I. The endosternite consists of two bands connected by two (Schizomida) or three (Holopeltida) transverse bridges, which consequently enclose one or two fenestra; the endosternite receives the insertions of four pairs of dorsal suspensors, three (Holopeltida) or four (Schizomida) pairs of ventral suspensors, and one (Schizomida) or two (Holopeltida) pairs of transverse suspensors. The circulatory system is well-developed; it consists of a heart with five (Schizomida) or nine (Holopeltida) pairs of ostia, and a well-developed arterial system (among which an anterior and a posterior aorta and a prosomatic sinus). The respiratory system is represented by booklungs and stigmata with ventral position: one pair (stigmata in segment VIII) in Schizomida (fig. 256A), two pairs (stigmata in segments VIII and IX) in Holopeltida (fig. 256B). The central nervous system includes a supraoesophageal syncerebrum, a suboesophageal ganglionic mass and an opisthosomatic mass. The male reproductive system comprises paired testes, paired vasa deferentia, a genital atrium and accessory glands. A study of sperm morphology and spermatogenesis has confirmed the close relationship of Schizomida and Holopeltida. The female reproductive system comprises paired ovaria, paired oviducts, a uterus and receptacula seminis. The glandular system of Holopeltida includes paired anal glands. The opisthosomatic musculature comprises dorsal and ventral longitudinal muscles, lateral muscles, and eight (Holopeltida) or seven (Schizomida) pairs of tergo-sternal muscles.

Reproduction. – Reproduction is bisexual. Sperm transfer is indirect (spermatophory) and includes courtship behaviour; spermatophores are deposited by the male, and are subsequently taken up by the female. Uropygi are oviparous; eggs are laid during a stay in a subterranean chamber, and remain attached to the venter.

Postembryonic development. – The postembryonic development comprises six stases: an elattostasic larva (carried on the back of the mother), four nymphs and an adult.

3. Additional remarks

The superorder Uropygi includes over a hundred species which are found in tropical and subtropical regions (Holopeltida are not found in Europe, Africa and Australia). Extinct Schizomida are known from the Upper Tertiary; extinct Holopeltida (presenting already the same general habitus as extant species) are known from the Carboniferous.

List of the notations of figs. 256–261

AP	apotele.
ap.ce	apodeme of cervix (median apodeme).
ap.pa	apodeme of palp (lateral apodeme).
at	line of attachment of cheliceral frame.
b	mouth.
bch	base of chelicera.
br	brush of palpal coxapophysis.
BTA	basitarsus.
c, c', c''	condyles.
CA.P	coxapophysis of palp.
CE	cervix.
CH	chelicera.
CX	coxa.
CX.I	coxa of leg I.
F	femur.
ICG	intercoxal gutter.
J	oral commissure.
LI	labium.
LS	labrum.
OCU	ocularium.
og	orifice of labial gland.
ogc	orifice of coxal gland.
PA	palp.
PD	prodorsum.
PO	postpeltidium.
PT	patella.
P.I–IV	legs I–IV.
SA	sternapophysis.
SA.I–III	sternapophyses of the segments bearing legs I–III.
sc.a	arthrodial sclerite.
ST	sternum.
st, st.1–2	stigmata.
t_i	inferior tendon of apotele.
t_s	superior tendon of apotele.
tae	taenidium.

tf_i	inferior tendons and muscles of femur.
tf_s	superior tendons and muscles of femur.
TI	tibia.
tpt	tendons and muscles of patella.
TR	trochanter.
TR.III	trochanter of leg III.
TTA	telotarsus.
TTA1–3	telotarsi 1–3.
tta	tendons and muscles of tarsus.
tti	tendons and muscles of tibia.
tti_i	inferior tendons and muscles of tibia.
ttr_i	inferior tendons and muscles of trochanter.
ttr_s	superior tendons and muscles of trochanter.
VIII, IX	segments VIII, IX.
′ (prime)	anterior face of an appendage.
″ (double prime)	posterior face of an appendage.

VIII. AMBLYPYGI

The two orders of the superorder Araneomorpha are dealt with in separate chapters, because our knowledge of comparative morphology is not yet sufficient for a general discussion: chapter VIII deals with the order Amblypygi, chapter IX with the order Araneida.

In the fourth part of my Comparative Studies in Chelicerata (Van der Hammen, 1986a: 10–12, 14–16, 18, 23, 38, figs. 2C, D, 5C, 7E, F, 11B, C, 19) I have dealt with some parts of Amblypygid morphology, viz., segmentation of the body, respiratory system, orifices of coxal glands, mouthparts, and segmentation and articulation of the legs. Previously, I had mentioned Amblypygi only in passing, in some general papers dealing with Chelicerate morphology, postembryonic development and classification (Van der Hammen, 1974: 385, 386, 387; 1977a: 310, 312, 313, 317, tab. 1, 3; 1978: 45, 49, 52; 1985: 55, fig. 34; 1985a: 137, 138, 139, 142, 143, 144, 145, figs. 3, 7). Amblypygi are also mentioned in my final survey of Chelicerate relationships (Van der Hammen, 1986c: 225, 227, fig. 3). The results of these studies are summarized in the present chapter, supplemented with data from literature. The chapter is subdivided into three parts: an introduction, a descriptive part, and some additional remarks.

1. Introduction

Data with reference to the comparative morphology of Amblypygi have been summarized by Millot (1949c) and Kaestner (1932; 1956: 550–551). Snodgrass (1948: 26–27, fig. 8) studied the mouthparts, Firstman (1973: 24, fig. 18) the endosternite and the circulatory system, Weygoldt, Weisemann & Weisemann (1972) the reproductive system, and Weygoldt (1969, 1970) reproduction and postembryonic development.

The original part of the present chapter is based on a study of adult specimens of *Heterophrynus alces* from Surinam.

2. Descriptive part

Habitus. – Amblypygi are relatively small or medium-sized to relatively large, dark-coloured Chelicerata with flattened body, distinctly separated prosoma and opisthosoma, distinctly segmented opisthosoma, large raptorial palps (armed with long spines, and with well-developed palpal apotele), and long legs.

Divisions of the body, tagmata and segmentation. – Data with reference to Amblypygid segmentation have been summarized by Kaestner (1932: 4–7, figs. 7–8) and Millot (1949c: 564–565, figs. 318–320). A species of Amblypygi is represented, in lateral view, in fig. 3C, D. The Amblypygid prodorsum consists of a single, large prodorsal shield. The ventral surface of the prosoma is occupied by a sternum (a neosternum), surrounded by the coxae of the appendages. The opisthosoma consists of the segments VII–XVIII. Segment VII is narrowed; its tergite is hidden by the prodorsum (compare figs. 3C and 3D) and its sternite is represented by a triangular sclerite between coxae IV. The tergite of segment VIII is small and hidden by the prodorsal shield; its sternite is much larger, whilst the genital opening is a transverse slit at its posterior border. Segments IX–XVII each consist of a tergite and a sternite, separated by a pleural region of soft skin; the sternites of segments IX and X are distinctly shortened. In segment XVIII, tergite and sternite are not distinctly differentiated; the anal opening is terminal in this segment. In Amblypygi, segments XVI–XVIII do not constitute a narrowed metasoma, whilst a 'telson' is not present.

Tegument. – The sclerotized parts of the cuticle are hard and relatively well-developed. The pleural skin is relatively soft. Ordinary setae, spines, various types of sensory phaneres (among which trichobothria) and lyrifissures are present.

Aspidosoma. – The aspidosoma (or prodorsum) is constituted by a single large sclerite. Generally, it presents two median eyes and two groups of three lateral eyes.

Dorsal and laterodorsal region of opisthosoma. – This region comprises the opisthosomatic sclerites and the intersegmental and pleural regions of soft skin. Tergite VII is narrowed, and tergite VIII is small; both are hidden by the prodorsal shield.

Ventral and lateroventral region of opisthosoma. – The sternite of segment VII is represented by a triangular sclerite between coxae IV. The genital opening is represented by a transverse slit at the posterior border of sternite VIII. The Amblypygid respiratory system (fig. 256C) is represented by booklungs and paired stigmata with ventral position (in segments VIII and IX); a pair of small evaginable ventral sacs, with unknown function, can also be present in the Amblypygid segment IX (see Millot, 1949c: 572–573, fig. 332). The anal opening is constituted by a terminal slit in segment XVIII. There is no flagellum.

Ventral aspect of prosoma. – The ventral surface of the prosoma is occupied by a relatively small sternum (a neosternum), surrounded by the large coxae of palp and legs. Amblypygi are characterized by the presence of one or two pairs of coxal glands: a pair of large ones in segment III, and a pair of smaller ones (present in the family Charontidae only) in segment V. The orifice of the last-mentioned gland, if present, is paraxial, at the posterior border of coxae III. The orifice of the first-mentioned gland (fig. 257E, F) is paraxial, at the base of coxa I; it is associated with a taenidium and a sternapophysis. The Amblypygid sternum generally presents three sternapophyses, of which those associated with coxae II and III are more or less vestigial (they could originally have been associated with the orifices of the coxal glands of the segments IV and V).

Lateral aspect of prosoma. – The lateral region of the prosoma is occupied by the antiaxial part of the coxae of palp and legs, and by soft skin.

Mouthparts. – The Amblypygid mouthparts are very simple, and comprise (besides the chelicerae): a small labrum (cervix not distinctly differentiated), and the coxae and coxapophyses of the palp. There is no labium and there are no lateral lips. A ventral view of the mouthparts is represented in fig. 259B, a longitudinal (nearly sagittal) section in fig. 259C. The position of the mouth is rostral. The mouthparts are associated with the anterior sternapophyses and the orifices of the coxal glands.

Chelicera. – The chelicerae are two-segmented: they consist of the body of the chelicera and the apotele. They are not chelate, but resemble a pocket-knife. A stridulating organ is known from the chelicera of *Musicodamon* (cf. Millot, 1949c: 565, fig. 348).

Palp. – The large, raptorial palp consists of six segments: coxa, trochanter, femur, patellotibia, tarsus and apotele. The coxa presents a coxapophysis; palpal coxa and coxapophysis are associated with the mouthparts. The palp is armed with strong spines, and the apotele is well-developed.

Legs. – The legs of *Heterophrynus alces* are studied here as an example of the Amblypygid legs; leg IV of this species is represented in fig. 262. In the specimen investigated by me, leg I consists of coxa, trochanter, femur, patella, tibia 1–31, tarsus 1–71 (basitarsus and telotarsus are indistinguishable) and a reduced apotele (of which the ungues have fused at the base); the numbers of tibial and tarsal segments are certainly variable in this species. Legs II–IV consist of coxa, trochanter, femur, patella, tibia (tibia 1–3 in the case of leg IV), basitarsus, telotarsus 1–4, and apotele. The coxae of palp and legs are arranged around a sternum; they are boat-shaped, and present an anterior apodeme (fig. 262D). The coxa-trochanter joint presents an anterior condyle, distal to the coxal apodeme, and a posterior arthrodial sclerite (broad hinge-articulation with the coxa, articulation point with the trochanter; see fig. 262C); the muscles of the trochanter are in the femur. The coxa-trochanter joint permits of promotor-remotor movements and is associated with leg rocking. The Amblypygid legs demonstrate that the arthrodial sclerite of the coxa-trochanter joint arose by desclerotization of the posterior distal part of the coxal cuticle (the desclerotized parts constitute 'fenestrae' and the transition from sclerotized to desclerotized parts can be gradual). The trochanter-femur articulation is constituted by a bicondylar pivot joint, associated with levator

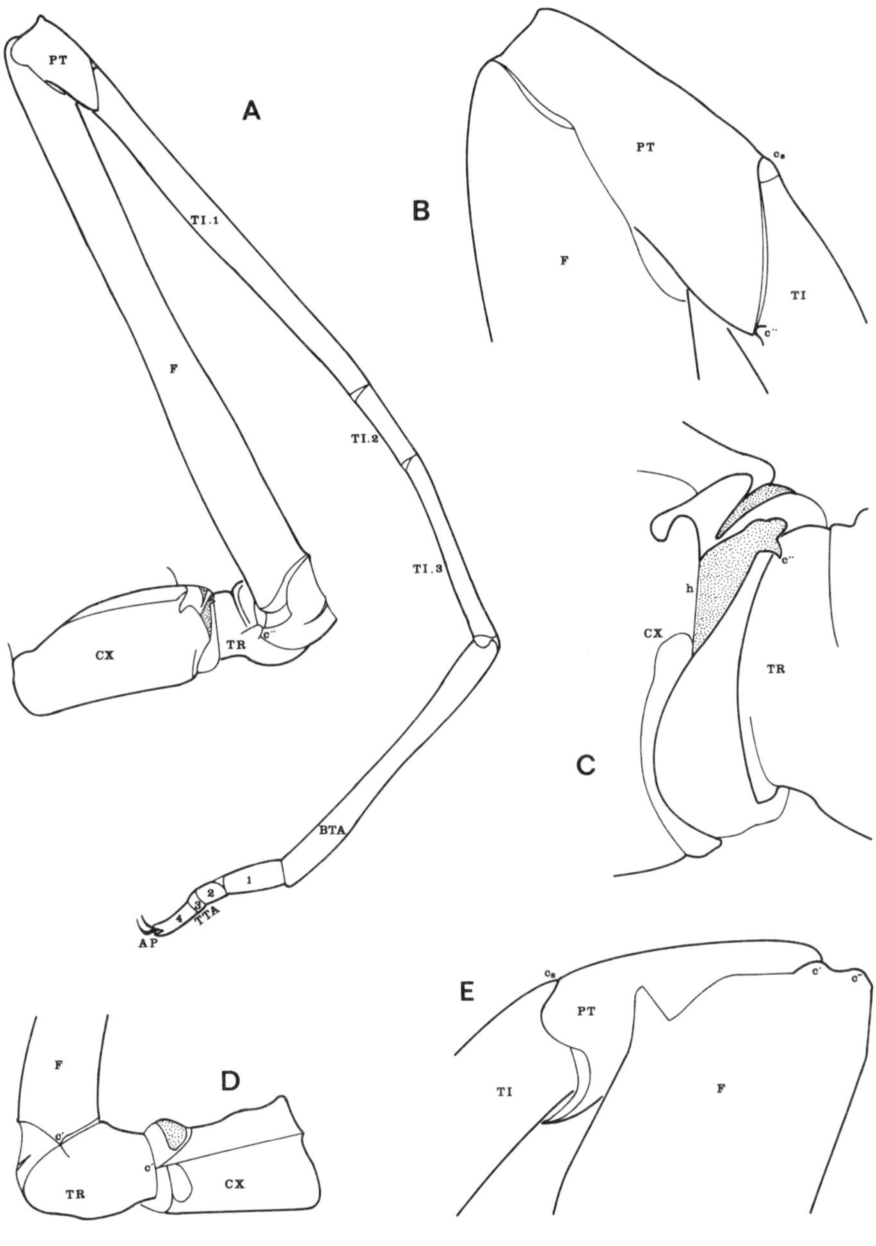

Fig. 262. Heterophrynus alces Pocock, right leg IV of adult; A, lateral (posterior) view; B, lateral (posterior) view of distal part of femur, patella and proximal part of tibia; C, lateral (posterior) view of articulation between coxa and trochanter; D, lateral (anterior) view of coxa, trochanter and proximal part of femur; E, lateral (anterior) view of distal part of femur, patella and proximal part of tibia; A, D, × 4.2; B, C, E, × 16.9.

and depressor movements. The femur-patella articulation is constituted by a hinge joint which permits of considerable flexion. The patella-tibia articulation (fig. 262B, F) is constituted, in legs II–IV, by a bicondylar hinge joint of which the hinge is posterolateral (there is a superior and a posterolateral condyle); in leg I there is only a superior condyle, but anterior flexion can occur because of the presence of a ventral sclerotized structure, by which ventral flexion is restricted. Adesmatic joints are present in tibia IV. The tibia-basitarsus articulation is constituted by a hinge joint. The basitarsus-telotarsus articulation and the adesmatic articulations in the telotarsus are operated by the tendons of the apotele. The telotarsus-apotele articulation is constituted by the usual bicondylar pivot joint with superior and inferior tendons. The apotele bears, in legs II–IV, two ungues; often (Charontidae) the ambulacrum presents also a pulvillus.

Internal anatomy. – The alimentary canal (digestion is external: food is rendered fluid outside the alimentary canal and ingested in this state) comprises mouth, pharynx, pumping gizzard, midgut with four pairs of simple prosomatic caeca and four pairs of complex opisthosomatic caeca, hindgut and anus. The excretory system includes coxal glands, Malpighian tubes and nephrocytes. As mentioned above, there are one or two pairs of coxal glands: a large one in segment III, and a smaller one (present in the family Charontidae only) in segment V. The endosternite (it has no fenestrations) presents four pairs of dorsal suspensors, three pairs of transverse suspensors and homologues of two pairs of ventral suspensors (of which at least the posterior has developed into a pair of non-contractile tendinous processes); the endosternite extends in posterior direction to the second opisthosomatic segment. The circulatory system includes a heart (limited to the opisthosoma) with six pairs of ostia, and a well-developed arterial system (among which an anterior and a posterior aorta and a prosomatic sinus). The respiratory system is represented by two pairs of booklungs and stigmata (in segments VIII and IX). The central nervous system comprises a supraoesophageal syncerebrum and a suboesophageal ganglionic mass (which includes the opisthosomatic ganglia). The male reproductive system consists of paired testes, paired vasa deferentia, paired vesiculae seminales, paired ventral and lateral glands, and a genital atrium with a cone-shaped spermatophore organ (which originates from a pair of 'gonopods'). Sperm morphology and spermatogenesis confirm the close relationship with Araneida. The female reproductive system consists of paired ovaria, paired oviducts, an unpaired uterus, a genital atrium with 'gonopods' and receptacula seminis. There are apparently no poison glands and no repugnatorial glands. The opisthosomatic musculature includes dorsal and ventral longitudinal muscles, lateral muscles and eight pairs of tergosternal muscles.

Reproduction. – Reproduction is bisexual. Sperm transfer is indirect (spermatophory) and includes courtship behaviour; spermatophores are deposited by the male, and are subsequently taken up by the female. Amblypygi are oviparous; eggs remain attached to the female venter, surrounded by a layer of secretion.

Postembryonic development. – The Amblypygid postembryonic development comprises an elattostasic larva (which remains attached to the brood pouch or to the venter of the mother), a relatively great but variable number of nymphoids, an adult and isophena of the adult (the isophena are the result of growing or repetition moults).

3. Additional remarks

The order Amblypygi includes some 70 species; the geographic distribution is restricted to the tropical and subtropical parts of Africa, Asia and America. The life-habitats are nocturnal and several species are cavernicolous (either troglobiotic or troglophilous). Extinct species, presenting already the characteristic habitus of extant species, are known from the Carboniferous.

List of the notations of fig. 262

AP	apotele.
BTA	basitarsus.
c', c''	anterior, resp. posterior condyle.
c_s	superior condyle.
CX	coxa.
F	femur.
h	hinge.
PT	patella.
TI.1–3	tibiae 1–3.
TR	trochanter.
TTA.1–4	telotarsi 1–4.

IX. ARANEIDA

In the fourth part of my Comparative Studies in Chelicerata (Van der Hammen, 1986a: 10–12, 16, 18, 23–26, 38–43, figs. 2E, 5D, 7C, D, 12, 20, 21) I have also dealt with Araneid morphology, viz., segmentation of the legs, respiratory system, orifices of coxal glands, mouthparts, and segmentation and articulation of the legs. Previously, I had mentioned Araneida only in passing, in some general papers dealing with Chelicerate morphology, postembryonic development and classification (Van der Hammen, 1974: 385, 387; 1977a: 310, 312, 313, 314, 316, 317, tab. 1–3; 1978: 45, 49, 52, 54; 1985: 52, 55, fig. 34; 1985a: 137, 138, 139, 141, 142, 143, 144, 145, figs. 3, 7). Araneida are also mentioned in my final survey of Chelicerate relationships (Van der Hammen, 1986c: 225, 227, fig. 3). The results of these studies are summarized in the present chapter, and supplemented with data from literature. The chapter is subdivided into three parts: an introduction, a descriptive part, and some additional remarks.

1. Introduction

Data with reference to the comparative morphology of Araneida have, among others, been summarized by Gerhardt & Kaestner (1937–1938) and Millot (1949d). Numerous papers with reference to Araneid morphology and postembryonic development have since been published, but only some of these are mentioned below.

The original part of the present chapter is based on material of *Liphistius desultor* Schiödte (from Malaya), and of *Segestria senoculata* (Linnaeus) and *Araneus quadratus* Clerck (both from The Netherlands).

2. Descriptive part

Habitus. – Araneida are small to relatively large Chelicerata with a distinct constriction at the border of prosoma and opisthosoma, due to the modification of the first opisthosomatic segment into a narrow pedicel. Prosoma with a single prodorsal shield. Opisthosoma with two to four pairs of spinnerets.

Divisions of the body, tagmata and segmentation. – Data with reference to the constitution of the Araneid body have been summarized by Gerhardt & Kaestner (1937: 398–408, figs. 488–505) and Millot (1949d: 590–600, figs. 350–365). Although vestiges of segmentation can be found in various groups of spiders (and particularly in the first instars), true segmentation is found in the spider family Liphistiidae only. A species of *Liphistius* is represented, in lateral view, in fig. 3E. The Araneid prosoma is constituted by precheliceral metameres and the segments I–VI. The dorsal surface of the Araneid prosoma is represented by a single prodorsal shield. The ventral surface of the prosoma is constituted by a sternum (i.e., a neosternum); part of the ventral prosomatic surface can also be occupied by the coxae of the appendages (as in *Liphistius*). The Araneid opisthosoma is constituted by the segments VII–XVIII. Segment VII constitutes a narrow pedicel. In *Liphistius*, the dorsal parts of the segments VIII–XVIII are represented by isolated sclerites. The genital opening is represented by a transverse slit at the posterior border of segment VIII. The four pairs of Liphistiid spinnerets (this number is reduced in others spiders) represent the opisthosomatic appendages of segments X and XI. Segment XVIII is constituted by the anal tubercle.

Tegument. – The prosoma is covered by a single prodorsal sclerite. In Liphistiidae, the dorsal surface of the opisthosoma presents segmentally arranged sclerites (tergites). In other Araneida, the skin of the opisthosoma is generally soft, although there are some groups in which the opisthosoma presents sclerites (in some cases covering the entire opisthosoma). Setae, spines, various types of sensory phaneres (among which trichobothria) and lyrifissures are present.

Aspidosoma. – The single prodorsal shield generally presents a V-shaped furrow. There are typically eight eyes, but this number can be reduced to six, four, two or zero.

Dorsal and laterodorsal region of opisthosoma. – The pedicel (segment VII) is always very narrow; it is generally (but not always) short and hidden by the aspidosoma. In dorsal view, the heart can often be seen through the soft cuticle; muscle attachments (sigilla) can also be distinctly visible. The terminal part of the opisthosoma is constituted by the anal tubercle (segment XVIII).

Ventral and lateroventral region of opisthosoma. – The narrow pedicel is generally visible in ventral view. The genital orifice is represented by a transverse slit (the epigastric fold) at the posterior border of segment VIII; in the females of the 'higher' Araneida, the genital region is characterized by the presence of a sclerite (epigyne). Paired stigmata with ventral position are present in the segments VIII and IX (fig. 256D); they are associated with either booklungs or tracheae. The ancestral position of the spinnerets (opisthosomatic appendages of the segments X and XI) is still present in Liphistiidae; in other groups, they are found in the posterior part of the opisthosoma. In Liphistiidae, four pairs of spinnerets ('endites' and 'exites' of the opisthosomatic appendages) are present; in other spiders this number is reduced to three pairs, two pairs or one pair. The anteromedian pair can be transformed into a so-called calamistrum (a chitinous plate with orifices of silk glands), or be vestigially represented by a small conical appendage (colulus); in *Heptathela* (Liphistiidae) the anteromedian pair is reduced to a single spinneret.

Ventral aspect of prosoma. – The ventral surface of the prosoma is occupied by the sternum (a neosternum); it is surrounded by the coxae of palp and legs. In Liphistiidae, this sternum is narrow, whilst the large, boat-shaped coxae occupy a ventral position; in other Araneida, the sternum is generally less narrow and the coxae have a more lateral position. One or two pairs of coxal glands are present. The orifice of the anterior gland (present in all Araneida) is paraxial, at the base of coxa I (fig. 275C, D). The orifice of the posterior gland (present in Liphistiidae and Mygalomorphae) is at the base of coxa III. The orifice of the anterior gland can be associated with a taenidium. Sternapophyses are not present in Araneida.

Lateral aspect of prosoma. – The lateral region of the prosoma is occupied by the antiaxial part of the coxae (in case the coxae occupy a ventral position) or by the bases of the coxae (in case the coxae occupy a lateral position). The coxae are surrounded by soft skin.

Mouthparts. – The rather simple Araneid mouthparts comprise (besides the chelicerae): the labrum and the cervix, the coxapophyses of the palp (not present in Liphistiidae and many Mygalomorphae), the palpal coxae, and the labium; there are no lateral lips. The mouthparts of *Segestria senoculata* are represented in dorsal and ventral view, and in a longitudinal section, in fig. 263A–C. A longitudinal section of the mouthparts of *Araneus quadratus* is represented in fig. 263D, a ventral view of the mouthparts of *Liphistius desultor* in fig. 263E. The position of the mouth is more or less rostral (fig. 263C); in many higher Araneida (fig. 263D), however, the chelicerae, the labrum and the labium (including the preoral cavity) are curved downwards, although the mouth has preserved its original position. The Araneid labium is generally regarded as representing the sternite of segment II. The absence of the palpal coxapophyses (fig. 263E) in more primitive groups of Araneida could either be an indication of the secondary origin of these apophyses in other Araneida, or of a suppression or an unrealized potentiality (coxapophyses are present in Uropygi and Amblypygi).

Chelicera. – The chelicerae are two-segmented: they consist of the body of the chelicera and the apotele. They are not chelate, but resemble a pocket-knife. They are either directed forwards or downwards. The orifice of a poison gland is in the terminal part of the apotele. A stridulating organ (or part of the organ) can be present on the body of the chelicera.

Palp. – The palp consists of coxa, trochanter, femur, patella, tibia and tarsus; an apotele is either present or absent. A coxapophysis, associated with ingestion, is present in most higher Araneida. In male spiders, the palp can function as a gonopod; this is associated with important morphological transformations.

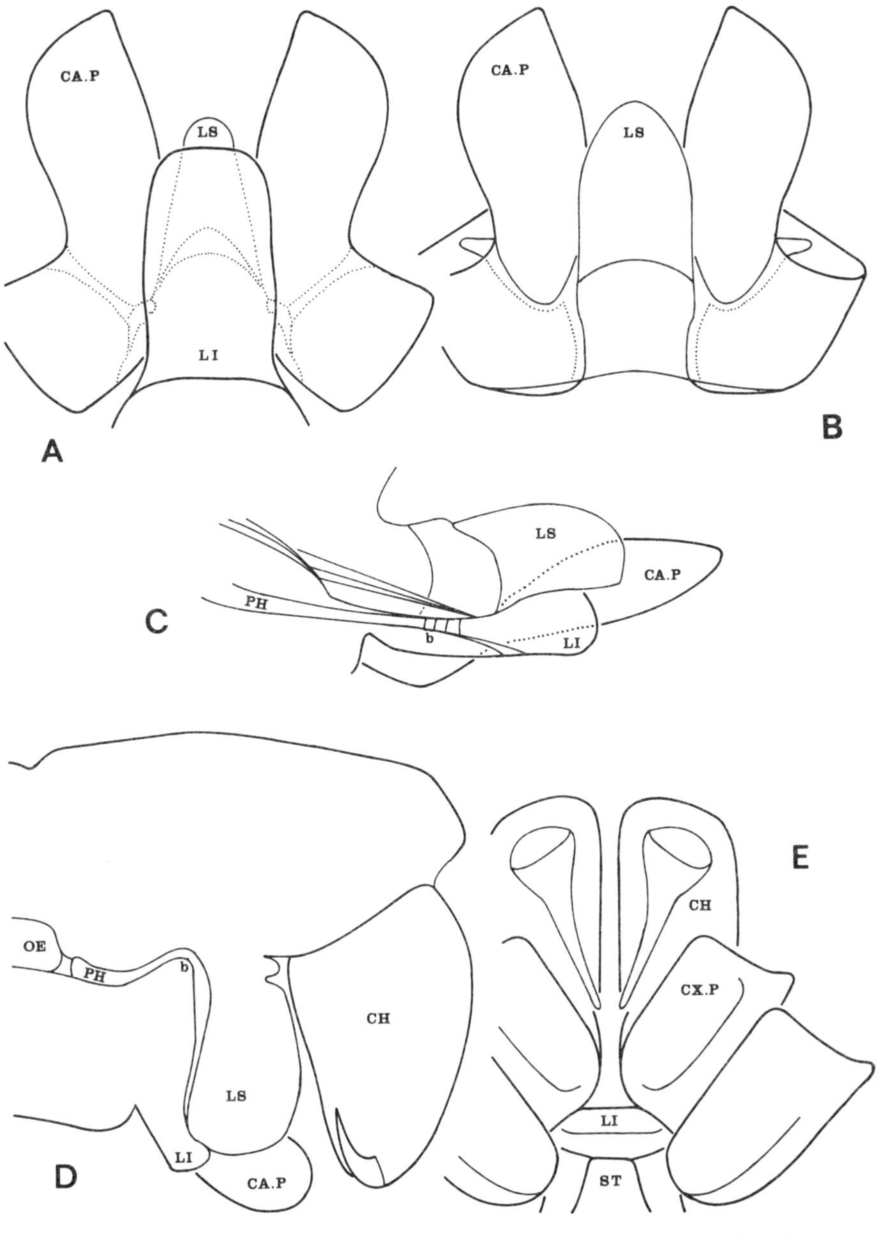

Fig. 263. Mouthparts of Araneida; A–C, *Segestria senoculata* (Linnaeus), adult female; A, ventral view; B, dorsal view; C, longitudinal section; D, *Araneus quadratus* Clerck, adult female, longitudinal section; E, *Liphistius desultor* Schiödte, adult female, ventral view; A–C, × 44; D, × 20.3; E, × 5.1

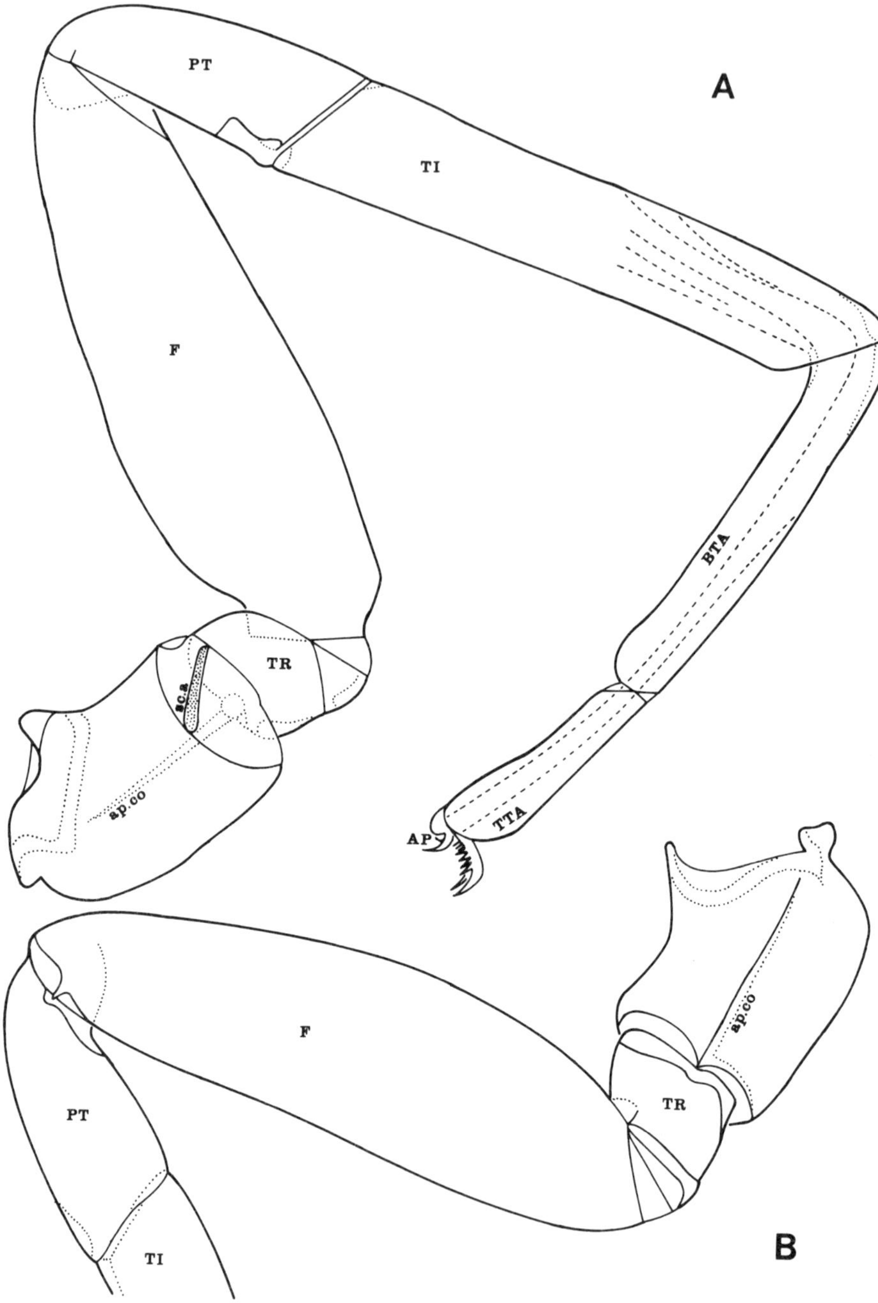

Fig. 264. Segestria senoculata (Linnaeus), right leg IV of adult female; A, lateral (posterior) view; B, lateral (anterior) view of coxa, trochanter, femur, patella and proximal part of tibia; A, B, × 46.8.

Legs. – The Araneid legs are studied here in two species: *Liphistius desultor* (fig. 265) and *Segestria senoculata* (fig. 264). An important and detailed study of the legs of *Heteropoda venatoria* was recently published by Clarke (1984); in this paper, a discussion of previous literature can be found. The segments present in the Araneid legs are: coxa, trochanter, femur, patella, tibia, basitarsus, telotarsus and apotele. The coxae of *Liphistius* are large and boat-shaped (as in Amblypygi), and are arranged on both sides of a narrow sternum; they are surrounded by soft skin, by which condition they are movable. The coxae of *Segestria* (as in most other species of Araneida) are short and cylindrical; they have a much more lateral position, and have a loose articulation with the sternum. All coxae present an anterior apodeme (figs. 264B, 265C). The coxae in *Liphistius* are regarded here as representing an advanced stage in the evolutionary development of free coxae from epimera, the coxae in *Segestria* (and most other species of Araneida) as the final stage. The possibilities of movement at the body-coxa joint are restricted, particularly in *Liphistius*. The coxa-trochanter joint presents an anterior condyle, distal to the coxal apodeme, and a posterior arthrodial sclerite (figs. 264A, 265B). The last-mentioned sclerite (often hidden from view by a fold of the arthrodial membrane) takes the place of the triangular sclerite of Uropygi and Amblypygi, but does not present the broad hinge at one of the ends; it is supposed here that movements at the coxa-trochanter joint are guided and restricted by the presence of this sclerite. A study of the sclerite in other groups of Arachnida s.str., particularly Amblypygi, demonstrates that the arthrodial sclerite arose by desclerotization of the posterior distal cuticle of the coxa (in this way, the particular type of joint evolved from a pivot joint). The smaller sclerites, which are now associated with appendotomy (see Roth & Roth, 1984) represent the ancestral distal border of the coxa. The coxa-trochanter joint is associated with promotor-remotor movements, and with some dorso-ventral movement and leg-rocking. As in other Arachnida s.str., the muscles of the trochanter are in the femur. The trochanter-femur articulation is constituted by a bicondylar pivot joint associated with levator and depressor movements. The femur-patella articulation is constituted by a hinge joint (with two superior condyles) and inferior flexor muscles; this joint allows of considerable flexion. The Araneid patella-tibia articulation is very interesting from an evolutionary point of view. In *Liphistius* (fig. 265A, C), this joint is similar in all four legs (and in the palp); it is constituted by a main superior articulation point, and a very small inferior one. The main movements are antero-posterior, whilst dorso-ventral movements are restricted by the presence of the ventral articulation point (which functions when the joint is at its maximum flexion). In *Segestria* (fig. 264A, B), the patella-tibia joint of legs I–III is more or less similar to that in *Liphistius*, whilst in leg IV patella and tibia are in close permanent contact at the posterior side (constituting a posterior hinge, allowing of antero-posterior movements only). According to the data published by Clarke (1984: 184), legs I and II of *Heteropoda* are similar to legs I–IV of *Liphistius* and legs I–III

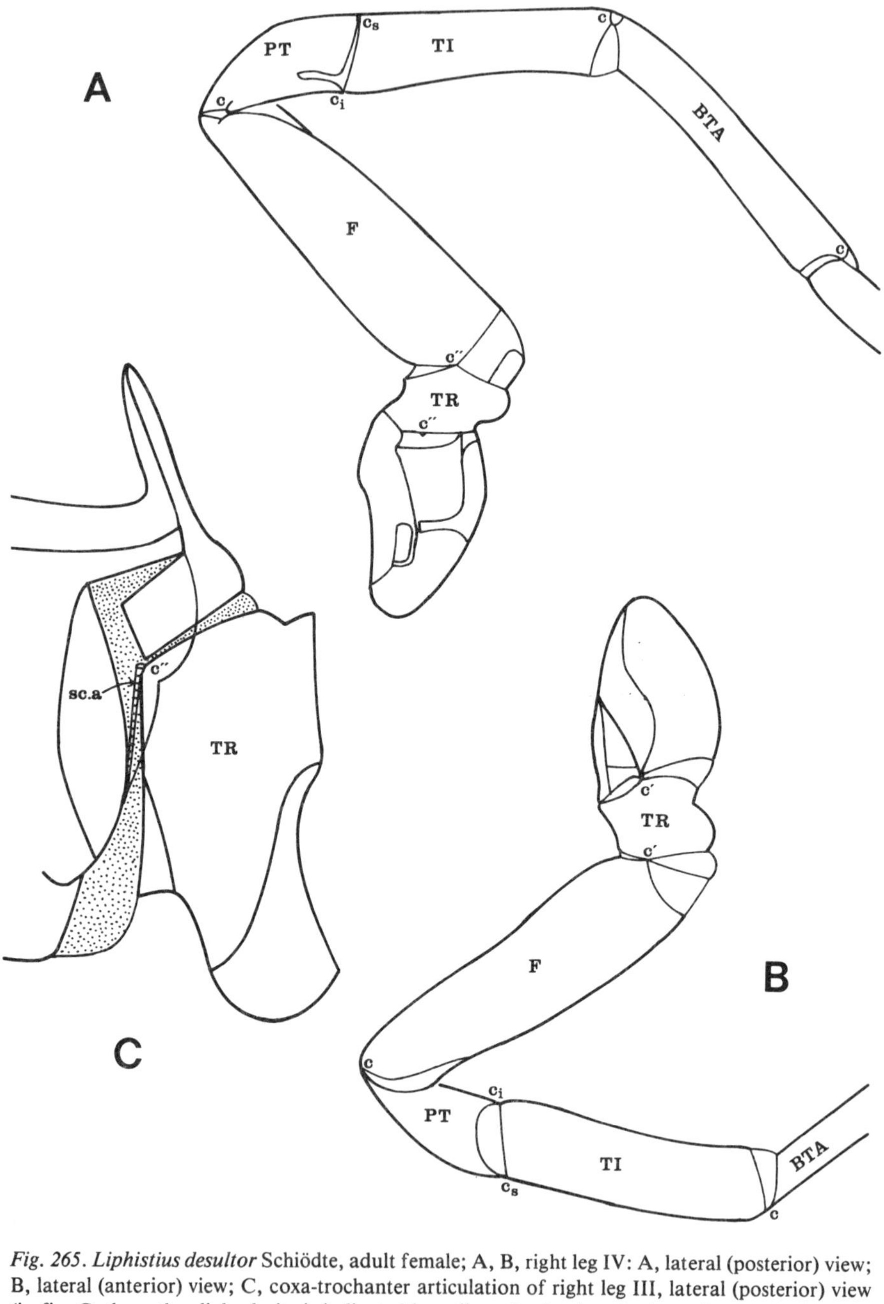

Fig. 265. Liphistius desultor Schiödte, adult female; A, B, right leg IV: A, lateral (posterior) view; B, lateral (anterior) view; C, coxa-trochanter articulation of right leg III, lateral (posterior) view (in fig. C, the arthrodial sclerite is indicated by ruling, the desclerotized coxal tegument by stippling); A, B, × 4.3; C, × 16.9.

of *Segestria*, whilst legs III and IV are similar to leg IV of *Segestria*. Evidently, the evolution of the patella-tibia joint, in the direction of a structure allowing of antero-posterior movements only, has progressively manifested itself from primitive to higher Araneida, starting with leg IV; because the evolution has manifested itself in the three groups of Arachnida s.str., it must represent a case of parallelism attributable to similarities in the evolutionary potentialities of these groups. The Araneid tibia-basitarsus articulation is constituted by a hinge joint with superior articulation; it is associated with flexion. The adesmatic basitarsus-telotarsus articulation is operated by the tendons of the apotele. The telotarsus-apotele articulation is constituted by a bicondylar pivot joint with the usual two tendons (superior and inferior). The apotele bears two ungues, whilst the pseudonychium is present in many groups of Araneida.

Internal anatomy. – The alimentary canal comprises mouth, pharynx, oesophagus, gizzard (pumping organ), midgut with prosomatic and opisthosomatic caeca and stercoral pouch, short hindgut and anus. The excretory system includes Malpighian tubes, nephrocytes and coxal glands. One or two pairs of coxal glands are present. The anterior pair, present in all Araneida, belongs to segment III; its orifice is at the base of coxa I. The posterior gland, present in Liphistiidae and Mygalomorphae, belongs to segment V; its orifice is at the base of coxa III. The prosomatic endosternite (Firstman, 1973: 24–29, figs. 19–21) is constituted by a single unfenestrated structure which receives the origins of various muscles; there are four pairs of dorsal suspensors and one to three pairs of transverse suspensors (in Mygalomorphae, each dorsal suspensor is continuous with a noncontractile tendon; together they represent the entire dorsoventral muscles). The circulatory system includes a heart with two to five pairs of ostia (highest number in pulmonate spiders, lowest number in tracheate spiders), an arterial system (among which an anterior aorta which divides into two branches, a prosomatic sinus, and a posterior aorta), and ventral lacunae; blood is pumped forward and backward. The respiratory system is represented by paired stigmata in segments VIII and IX, associated with either booklungs or tracheae. It is generally accepted that booklungs have arisen, in the course of evolution, from gills, and are homologous with opisthosomatic appendages; it is supposed that the Araneid tracheae have replaced the booklungs in the course of evolution. The central nervous system comprises a supraoesophageal syncerebrum and a suboesophageal ganglionic mass (incorporating also all opisthosomatic neuromeres). The male reproductive system includes paired testes, paired vasa deferentia and an unpaired atrium. Sperm morphology and spermatogenesis confirm the close relationship with Amblypygi. The female reproductive system comprises paired ovaria, paired oviducts, a uterus, a vagina and a variable number of spermathecae (receptacula seminis). The glandular system includes poison glands (in the prosoma; orifice in the cheliceral apotele), salivary glands and silk glands. The opisthosomatic musculature includes a regressive dorsal musculature, ventral longitudinal and transverse muscles, and a variable number (4–0) of paired tergosternal (dorsoventral) muscles.

Reproduction. – Araneid reproduction is generally bisexual (parthenogenesis has been described from a few species), and various degrees of sexual dimorphism can occur. Various types of courtship behaviour are known. Sperm transfer is nearly direct: the male palp can function as a gonopod (gonopody), sperm is transported to the female receptacula seminis, and fertilization is internal. Araneida are oviparous; eggs are nearly always deposited in a silk cocoon which is guarded (or even carried) by the female.

Postembryonic development. – The Araneid postembryonic development can include 1–2 calyptostasic prelarvae, 1–2 elattostasic larvae, 5–10 nymphoids, the adult (males can be neotenous) and, in some groups, isophena of the adult (in the case of growing or repetition moults).

3. Additional remarks

Araneida present a great diversity of external shape, associated with a wide variety of ecological adaptation and a world-wide geographical distribution. Some 20.000 species are known. Araneida represent the highest stage of Chelicerate evolutionary development, particularly because of visual faculties and spinning behaviour. Extinct species are recorded from the Carboniferous (there are, moreover, uncertain or probable records with reference to Devonian species).

List of the notations of figs. 263–265

AP	apotele.
ap.co	coxal apodeme.
b	mouth.
BTA	basitarsus.
c, c', c''	condyles.
c_i	inferior condyle.
c_s	superior condyle.
CA.P	coxapophysis of palp.
CH	chelicera.
F	femur.
LI	labium.
LS	labrum.
OE	oesophagus.
PH	pharynx.
PT	patella.

sc.a	arthrodial sclerite.
ST	sternum.
TI	tibia.
TR	trochanter.
TTA	telotarsus.
′ (prime)	anterior.
″ (double prime)	posterior.

X. OPILIONIDA

A detailed study of Opilionid morphology, including notes on the postembryonic development, was published as the third part of my Comparative Studies in Chelicerata (Van der Hammen, 1985). Previously (and also shortly afterwards), I had mentioned Opilionida in some general papers dealing with Chelicerate morphology, postembryonic development, evolution and classification (Van der Hammen, 1974: 386, 387; 1977a: 310, 312, 313, 314, 315, 316, tabs. 1–3; 1978: 45, 49, 55; 1985a: 137, 139, 141, 142, 143, 144, 145, figs. 3, 7), whilst Opilionida are also mentioned in my final survey of Chelicerate relationships (Van der Hammen, 1986c: 225, fig. 3). The results of these studies are now included in the present chapter, which is subdivided into three parts: an introduction, a descriptive part, and some additional remarks.

1. Introduction

The original part of the present chapter is based on a detailed comparative study of representatives of the three orders (Cyphophthalmi, Laniatores, Palpatores) into which Opilionida are currently subdivided, viz., *Siro rubens* Latreille (Cyphophthalmi), *Scotolemon lespesi* Lucas (Laniatores), and *Paroligolophus agrestis* (Meade) (Palpatores). The material of *Siro rubens* originated from Dordogne (France), the material of *Scotolemon lespesi* from Arriège (France), and the material of *Paroligolophus agrestis* from The Netherlands. During the study of this material, special attention was paid to mouthparts and appendages.

2. Descriptive part

Habitus. – Opilionida are small to medium-sized Chelicerata, with rounded or oval body, more or less distinct segmentation, and inconspicuous separation of prosoma and opisthosoma. The condition of the cuticle varies from soft and thin to heavily sclerotized; in many cases the cuticle of body, palp and legs is characterized by the presence of more or less heavy spines. The legs can be comparatively short, but are more often very, sometimes even extremely, long (the characteristic habitus of a 'harvestman' being that of an Opilionid with small, rounded body and very long legs).

Divisions of the body and segmentation. – The Opilionid body consists, as in other Chelicerata, of prosoma and opisthosoma; these two tagmata are broadly joined (the body is latigastric). Opilionida generally present a more or less distinct external segmentation of the body. As a result of the study of the embryonic and postembryonic development of the body in the family Phalangiidae, by Winkler (1957), the numbering of the segments is now known with more or less certainty. Winkler discovered that, in Phalangiidae, segment VII (the first opisthosomatic segment, reduced in several other groups of Chelicerata) is normally developed in the case of the dorsal part (the tergite), reduced in the case of the ventral part (the sternite). The prosoma consists of an unknown number of precheliceral metameres and the segments I–VI. The tergites of the prosoma constitute the prodorsum; most of these tergites have fused, but tergites V and VI are often more or less distinctly separated. The labium (if present) represents, according to Winkler, the sternite of segment III. As mentioned below (in the paragraph on the mouthparts), chelicerae and palps, and the coxal and sternal regions of the segments bearing the anterior legs, constitute a separate myliosoma (a pseudotagma) which is movably attached to the main part of the body.

The opisthosoma consists of the segments VII–XVI. Ventrally, the genital region occupies an advanced position between the coxae of the legs. The operculum genitale of the Phalangiidae consists, according to Winkler, of part of segment VIII and probably sternite VI (sternite VII being reduced). The posterior segments of the opisthosoma have been subject to regression. In Cyphophthalmi (fig. 266) the position of the anal opening is more ventral than in other groups of Opilionida; this certainly represents a secondary translocation (the ancestral position of the Chelicerate anal opening is terminal). The Cyphophthalmic anal cover (operculum anale) represents segment XVI; the surrounding plate represents a fusion of tergite XV and sternites XIV and XV (according to Juberthie, 1970, sternite XIV has a separate position in some Cyphophthalmi; in all Cyphophthalmi, tergite XIV has a separate dorsal position). In Laniatores (figs. 4, 275E), tergite XIV has a separate dorsal position, whilst sternites XIV and XV have fused; the anal cover articulates with the tergite of segment XIV, and could represent a fusion of tergite XV and segment XVI (this notation is added to fig. 275E). In Palpatores (figs. 267, 268), the posterior segments have fused in various measures. In *Paroligolophus agrestis*, tergites XIV and XV have apparently fused; sternite XIV has fused with sternite XIII, and sternite XV is no more recognizable, whilst the anal cover represents segment XVI. It is interesting that in all Opilionida segments XIV and XV (and not XVI) have been subject to a regression which is associated with fusion. The opisthosomatic segments, particularly the tergites, can have fused in various ways (with each other, as well as with the prodorsum), constituting various types of shields (see: Martens, 1978: 16–17, figs. 2–5). In *Scotolemon lespesi*

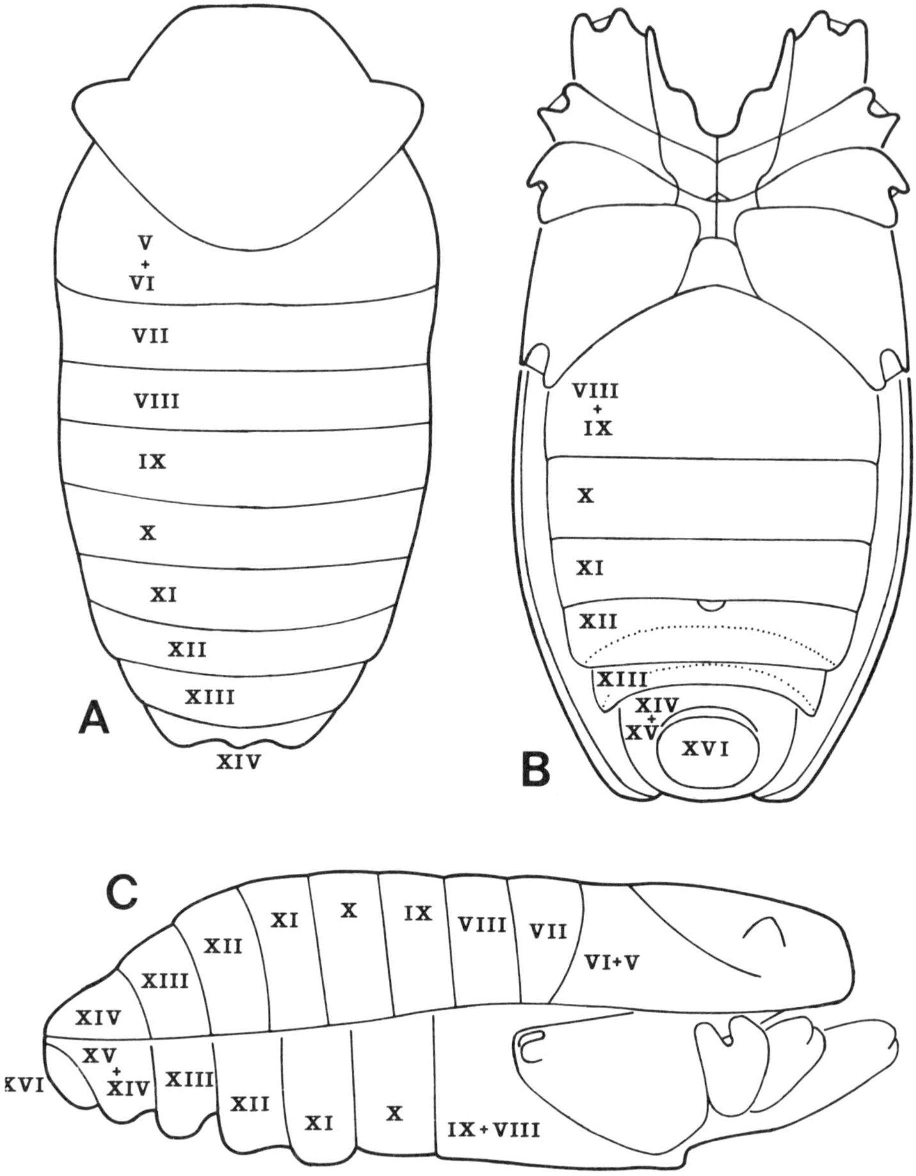

Fig. 266. Siro rubens Latreille, adult (chelicerae and palps removed); A, B, male; A, dorsal view; B, ventral view; C, lateral view of female (posterior segments extended); A–C, × 63.

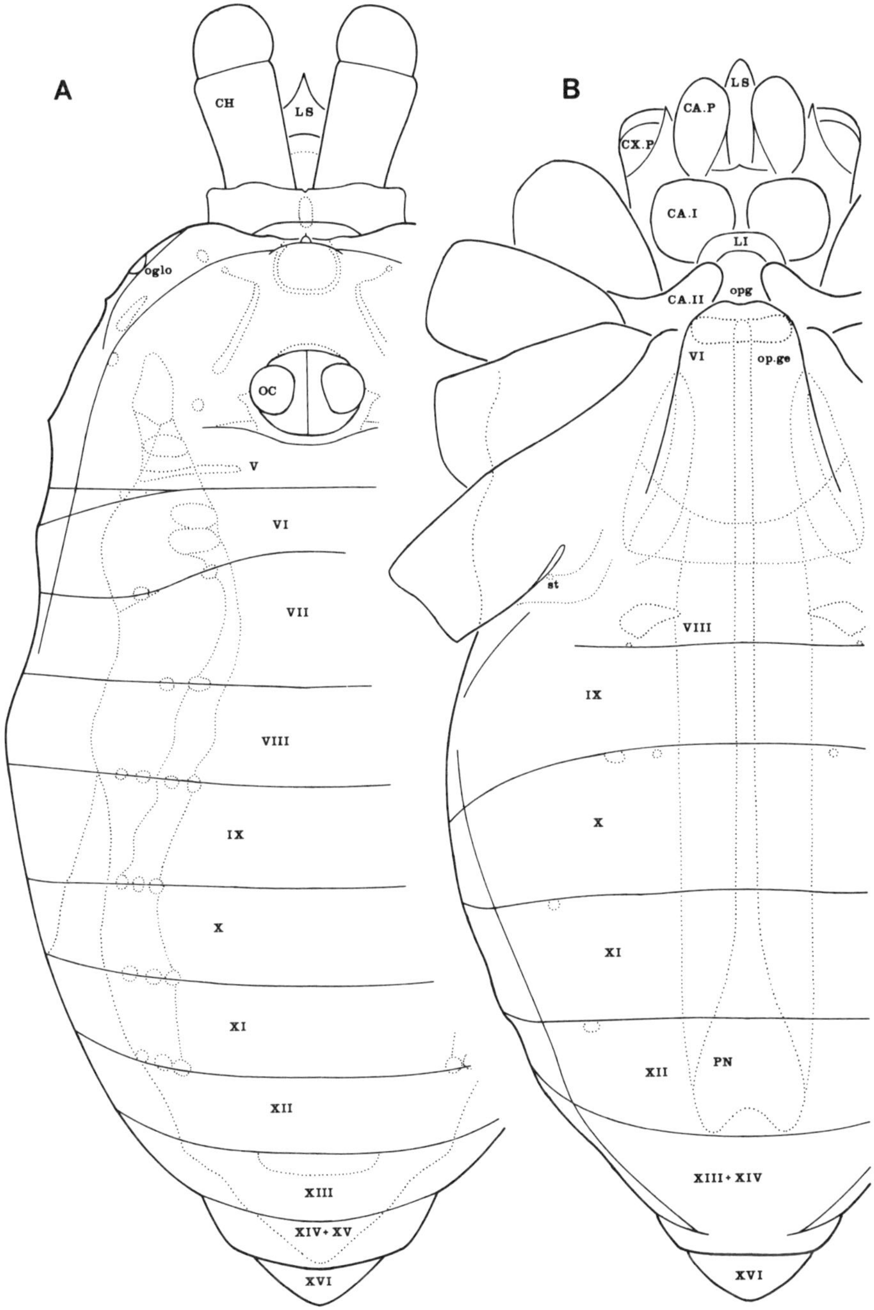

Fig. 267. Paroligolophus agrestis (Meade), adult male (anal cover extended); A, dorsal view; B, ventral view; A, B, × 37.

(fig. 4), tergites VII–XI have fused with the prodorsum; tergites XII–XIV and sternites X–XIV are partly covered by soft skin.

Tegument. – The cuticle varies from more or less heavily sclerotized (Cyphophthalmi, Laniatores, Trogulidae, Nemastomatidea) to soft (Phalangiidae). It presents a variety of phaneres, such as spines, tubercles (often with apical spine or seta; see fig. 287B), papillae, various types of setae, and solenidia (fig. 292A, F). The cuticle can present various types of microsculpture, among which an aciculate ornamentation which resembles a dense arrangement of small setae (figs. 292E, 294B, 297E); the aciculate ornamentation can be associated with a reticulate microsculpture. Lyrifissures can occur as single slits, or arranged in groups, rings or lyriform organs; they are regularly found on the appendages (chelicera, palp, legs), often associated with the articulations, but they can also occur on the coxapophyses (see figs. 276A, B, 282, 285C, 287B, 288, 294A, C, 295A–C, 296A, C–E, G, 297F).

Prodorsum (figs. 4A, C, 266A, C, 267A, 268, 277A). – The prodorsum, or dorsal shield of the prosoma, includes the tergites of the segments I–VI; it includes also material of precheliceral origin (such as the eyes). All or most of the tergites can have fused, but tergites V and VI often constitute separate sclearites. The prodorsum can also be integrated in a larger dorsal shield which includes a number of opisthosomatic tergites (up to eight). The lateral border of the prodorsum is above the coxae of the legs. Generally, there are two eyes, situated on an ocular tubercule or ocularium; the ocularium can be lacking, or the eyes can be situated on separate tubercles. In some cases (e.g., *Siro*) there are no eyes. The orifices of the odoriferous glands (one on either side) are situated in the lateral part of the prodorsum, or just below its lateral border (figs. 275A, 276A, 277E); they can each be associated with a taenidium (see: Van der Hammen, 1980: 154; Juberthie, 1961), an external, narrow gutter or canal, of a ribbon-like shape, running at the surface of the tegument. In Cyphophthalmi, the orifices are near the apex of a lateral tubercle. Various more or less distinct sigilla are also present in the prodorsal region, among which the sigilla of the dorsal suspensor muscles of the endosternite, and the sigilla of the large depressor muscles of the chelicerae.

Dorsal and laterodorsal region of opisthosoma. – This region (which includes the tergites VII–XV) consists of a variable number of separate sclerites, ranging from a single dorsal shield (fused with the prodorsum) to eight free sclerites. Tergite XIV of the male of *Siro* (fig. 277C, D) contains the orifices of the male postero-opisthosomatic glands (see: Juberthie, 1967, 1967a; Rafalski, 1958, regarded them as respiratory organs).

Ventral and lateroventral region of opisthosoma. – In the discussion of this region, anal and genital regions are not included (they are discussed in separate paragraphs). The ventral region of the opisthosoma is influenced by a posterior shortening of the body, and a forward translocation of sternites. Sternites VIII and IX have fused; in the case of the Phalangiidae, a forward projection of sternite VIII has probably fused with the prosomatic sternite VI, constitut-

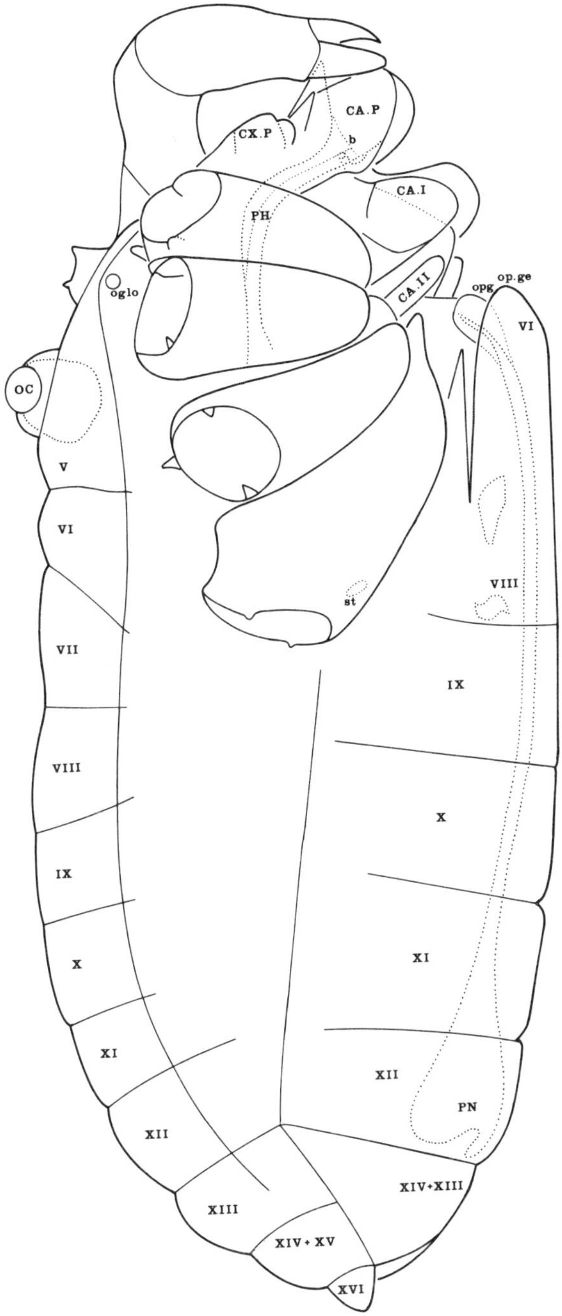

Fig. 268. Paroligolophus agrestis (Meade), adult male (anal cover extended), lateral view; × 37.

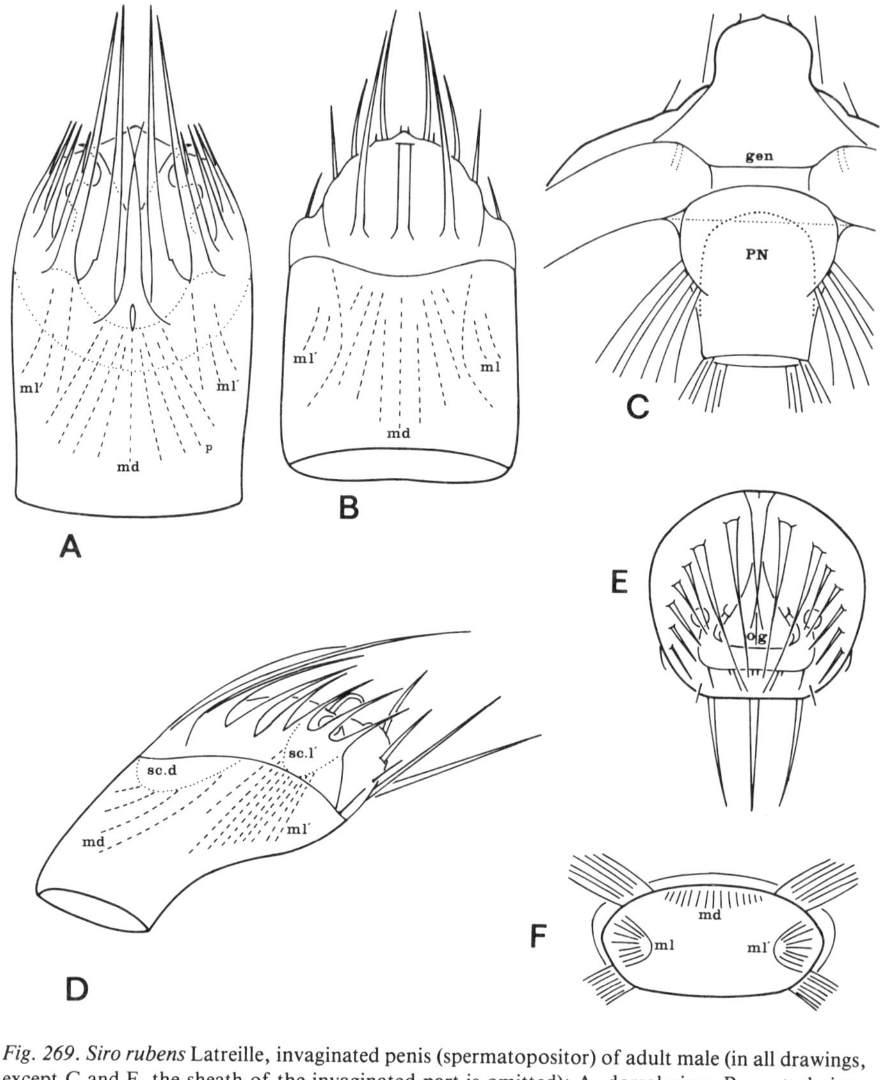

Fig. 269. Siro rubens Latreille, invaginated penis (spermatopositor) of adult male (in all drawings, except C and F, the sheath of the invaginated part is omitted); A, dorsal view; B, ventral view; C, ventral view of invaginated penis (spermatopositor), after removal of opisthosomatic sternites, with sheath, retractor muscles, and genital opening; D, lateral view (the left upper part of the figure is dorsal); E, apical view (the upper part of the figure is dorsal); F, basal view, with proximal retractor muscles and internal muscles *md, ml*, and *ml'* (the upper part of the figure is dorsal); A–B, D–F, × 295; C, × 141.

ing the genital cover (operculum genitale). The stigma (figs. 275B, 276B, 277A) is situated in sternite VIII, either free on the surface of the sternite (Sironidae, many Laniatores), or hidden in a fold posterior to coxa IV (Phalangiidae, Trogulidae). In *Siro rubens* and *Scotolemon lespesi*, sternites

XIII and XIV + XV are movable, and can be retracted and extended.

Anal region (figs. 4B, C, 266B, C, 267B, 268). – The anal cover (operculum anale) presents a dorsal hinge; it is opened by pressure and closed by muscles inserted ventral to (and close to) the hinge (fig. 275C). The anal cover is generally regarded as representing segment XVI. In several cases, particularly when the anal cover is movably attached to the dorsal shield, it could perhaps also include the tergite of segment XV; this hypothesis is introduced in one of the figures of *Scotolemon* (fig. 275C). In Cyphophthalmi (fig. 277F), the anal opening has a less terminal and more ventral position than in other groups of Opilionida; it is surrounded by a sclerite representing a fusion of tergite XV and sternites XIV and XV (as mentioned above, sternite XV has, in some cases, a separate position). Fig. 266C demonstrates that, in *Siro*, tergite XV separates the anal cover and the dorsal shield; a comparison of figs. 266C and 4C supports the hypothesis of an inclusion of tergite XV in the anal cover of *Scotolemon*.

Genital region (figs. 4B, 266B, 267B). – The Opilionid genital opening is at the anterior border of segment VIII (Winkler, 1957: 365–367). It has an advanced position, between the coxae of the legs, probably because of the special type of Opilionid sperm transfer, which is associated with copulation (not known from Cyphophthalmi). Female Opilionida are characterized by the presence of an ovipositor, male Opilionida by the presence of a penis (or, in Cyphophthalmi, a spermatopositor; it is probably associated with the transfer of a spermatophore). Ovipositor and penis (and spermatopositor) originate from extensions (prolongations) of the internal reproductive system; consequently, the eugenital opening (the ancestral opening), which has the shape of a fissure, is now in the distal part of ovipositor and penis (its current name is secondary genital opening). The genital opening (it is currently named primary genital opening), i.e., the opening at the anterior border of segment VIII, through which ovipositor and penis are evaginated (fig. 275D), is unprotected (Cyphophthalmi) or protected by a genital cover (operculum genitale) which originates from a forward projection (a fold) of segment VIII. In Phalangiidae, it can be relatively long, and probably includes the sternite of segment VI (Winkler, 1957: 370). The opening between genital cover and prosomatic venter is here named progenital opening (it is currently named tertiary genital opening). Originally (in the ancestors), eugenital and genital opening must have coincided. Penis and ovipositor can be very long; in invaginated position they can extend as far as the posterior border of segment XII (fig. 268).

There are various types of penis (spermatopositor) and ovipositor (see: Dumitrescu, 1970; Martens, 1976, 1978; Martens, Hoheisel & Götze, 1981). The spermatopositor of *Siro* (fig. 269; see also: Janczyck, 1956: 516–518, figs. 25–26; Rafalski, 1958: 539–550, figs. 15–18; Juberthie, 1967: 166, fig. 8A) is very short and flattened, and not distinctly subdivided into a so-called truncus and a so-called glans. The apical part is surrounded by a verticil of some 22 setae. The eugenital opening (a fissure) is bordered, on either side, by

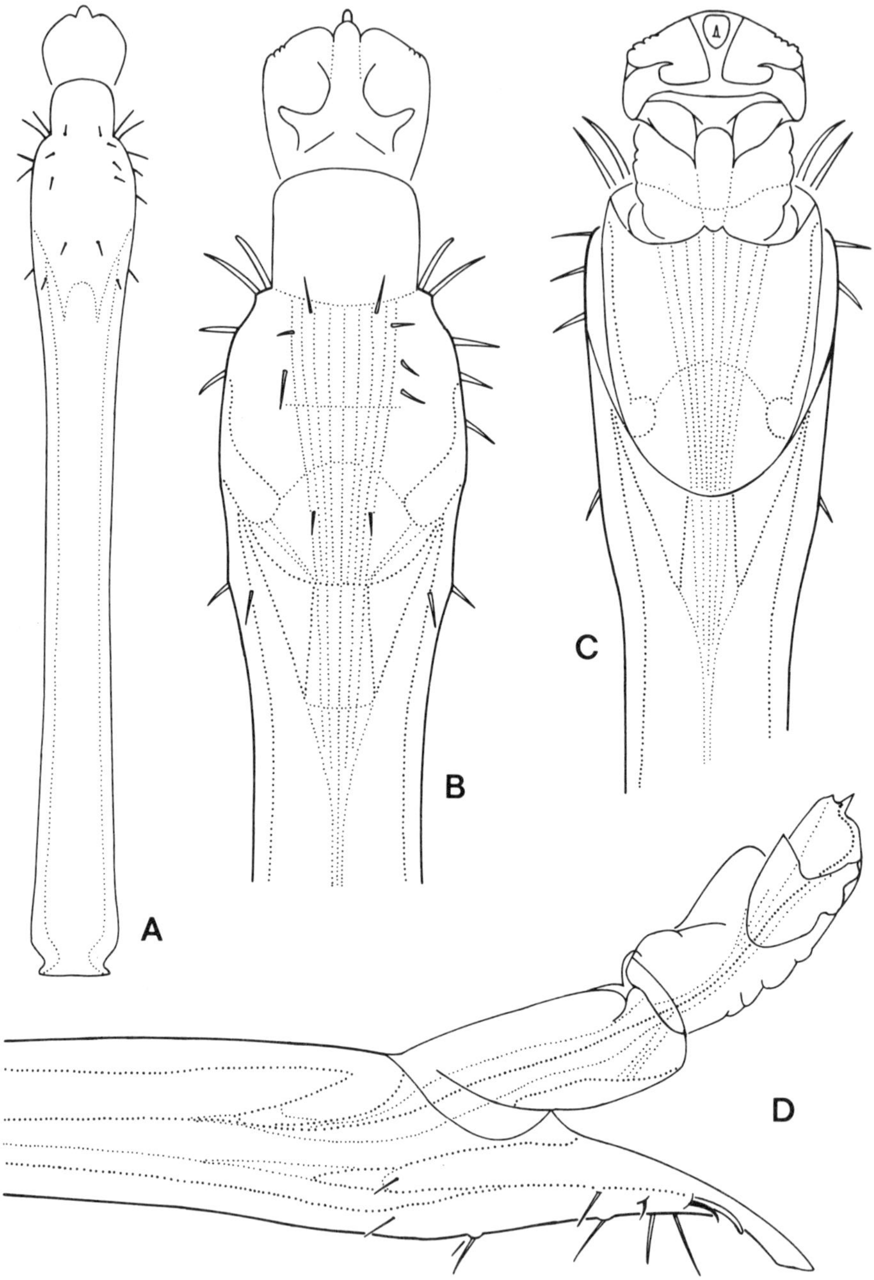

Fig. 270. Scotolemon lespesi Lucas, penis of adult male; A, ventral view; B–D, distal part; B, ventral view; C, dorsal view; D, lateral view; A, × 141; B–D, × 295.

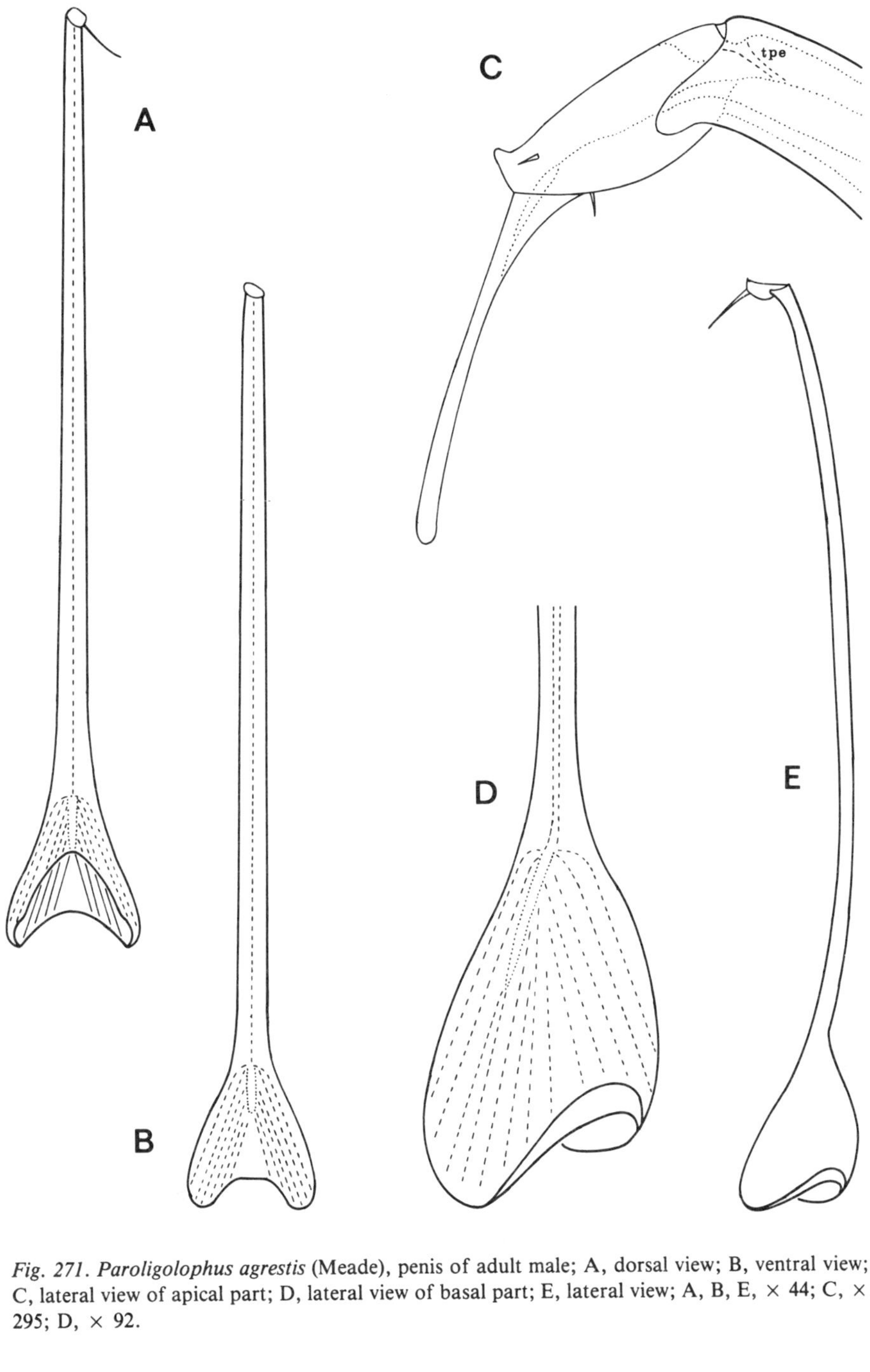

Fig. 271. Paroligolophus agrestis (Meade), penis of adult male; A, dorsal view; B, ventral view; C, lateral view of apical part; D, lateral view of basal part; E, lateral view; A, B, E, × 44; C, × 295; D, × 92.

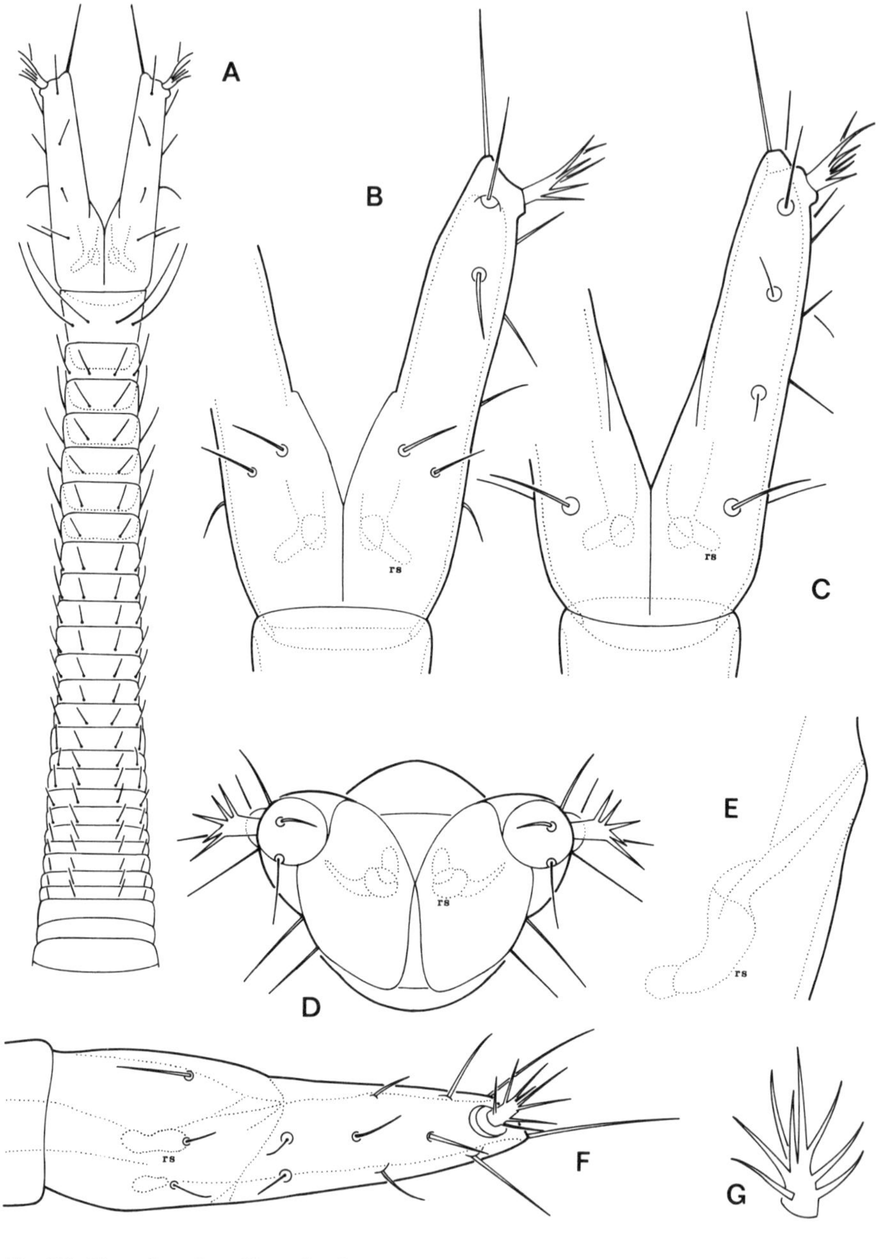

Fig. 272. Siro rubens Latreille, ovipositor of adult female; A, ventral view of invaginated ovipositor (after removal of opisthosomatic sternites); B, C, apical part; B, dorsal view; C, ventral view; D, apical view; E, paralateral (oblique) view of part of one of the apical lobes (with the receptacula seminis); F, antilateral view of apical part; G, antilateral view of ramified apical seta; A, × 118; B–D, F, × 246; E, G, × 458.

a lobe bearing a claw-shaped clutch and a rounded tubercle. There are three muscle bundles (a dorsal one and two laterals), inserted on internal sclerites.

The penis of *Scotolemon lespesi* (fig. 270) is large and presents two apical lobes, a dorsal and a ventral, which are not operated by muscle bundles, but by hydrostatic pressure. Setae are present in the ventral and lateral regions of the ventral apical lobe; the sperm duct is in the dorsal lobe. The ventral lobe is flattened, and apically truncate; the dorsal lobe is membraneous, and consists of various movable parts.

The penis of *Paroligolophus agrestis* (fig. 271) is large; it distinctly consists of a truncus and a distal glans. A long tendon is inserted on the base of the glans. In the genus *Paroligolophus*, the penis is characterized by a very narrow truncus and a very large base; the muscle bundle, operating the tendon of the glans, is concentrated here in the large base. In rest the glans is at an oblique angle with the truncus; it terminates in a long stylus.

The ovipositor of *Siro rubens* (fig. 272; see also: Janczyck, 1956: 518–520, fig. 28; Rafalski, 1958: 550–552, figs. 19–20; Juberthie, 1967: 166, fig. 8B) is long and annulated. There are some 27 rings and a distal furca. With the exception of the basal ones, the rings present rows of setae. The furca consists of a pair of externally unsegmented lobes, each characterized by the presence of simple setae and a subapical ramified seta (on a low antilateral process). Each of the two lobes presents moreover, near its base, an internal receptaculum seminis.

The ovipositor of *Scotolemon lespesi* (fig. 273) is short, not annulated, with two terminal lobes. The eugenital opening (a fissure) is between the bases of the lobes; it is surrounded by a verticil of some seven pairs of setae. There is a ventral muslce bundle, inserted on the base of the lobes, by which these are probably slightly retractable. There are four receptacula seminis, radially arranged around the distal part of the oviduct (see also: Martens, Hoheisel & Götze, 1981: 47–49, figs. 30B, 33, 43B).

The ovipositor of *Paroligolophus agrestis* (fig. 274) is long and annulated. There are some thirty rings (the three rings of the elements of the furca included). The paired receptacula seminis extend from segment 3 (the basal segment of the furca) to segment (ring) 7. The furca consists of two three-segmented lobes. The terminal part of each of the two lobes presents a rounded protuberance with numerous short setae.

In all of these ovipositors, the eugenital opening is represented by a fissure between the base of the lobes. The lobes probably represent the opisthosomatic appendages of segment VIII. In the penis of *Siro rubens*, the apical part presents also two lobes, one on either side of the eugenital opening; these could also represent the opisthosomatic appendages of segment VIII. In the penis of *Scotolemon lespesi* and *Paroligolophus agrestis*, the ancestral paired lobes (i.e., the supposed opisthosomatic appendages) have apparently fused.

Ventral aspect of prosoma. – The coxal and sternal regions associated with palp and legs (segments II–VI) are represented in figs. 4B, 266B, 267D and,

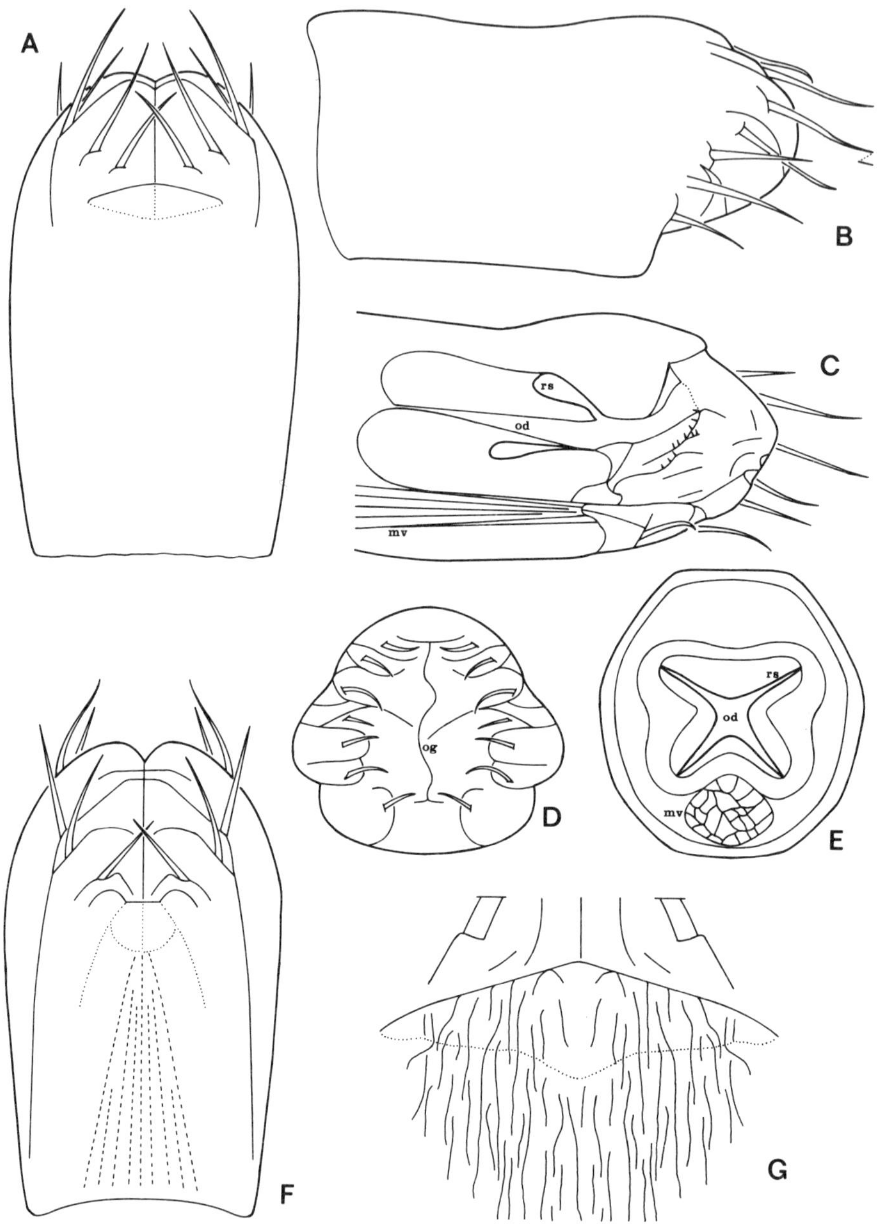

Fig. 273. Scotolemon lespesi Lucas, ovipositor of adult female; A, dorsal view; B, lateral view; C, longitudinal (sagittal) section, lateral view; D, apical view; E, transverse section; F, ventral view; G, dorsal view of microsculpture posterior to the two lobes; A–F, × 155; G, × 458.

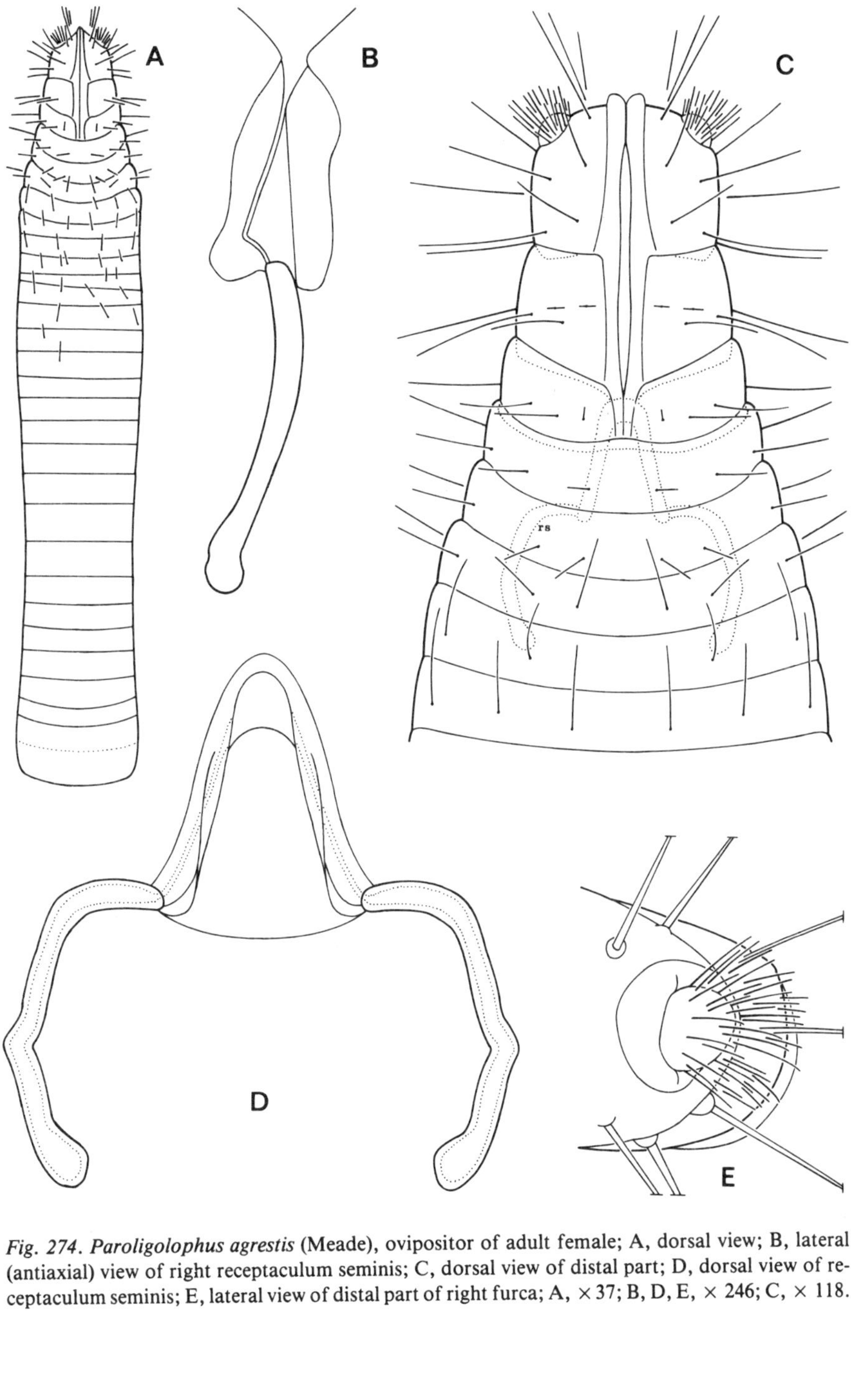

Fig. 274. Paroligolophus agrestis (Meade), ovipositor of adult female; A, dorsal view; B, lateral (antiaxial) view of right receptaculum seminis; C, dorsal view of distal part; D, dorsal view of receptaculum seminis; E, lateral view of distal part of right furca; A, × 37; B, D, E, × 246; C, × 118.

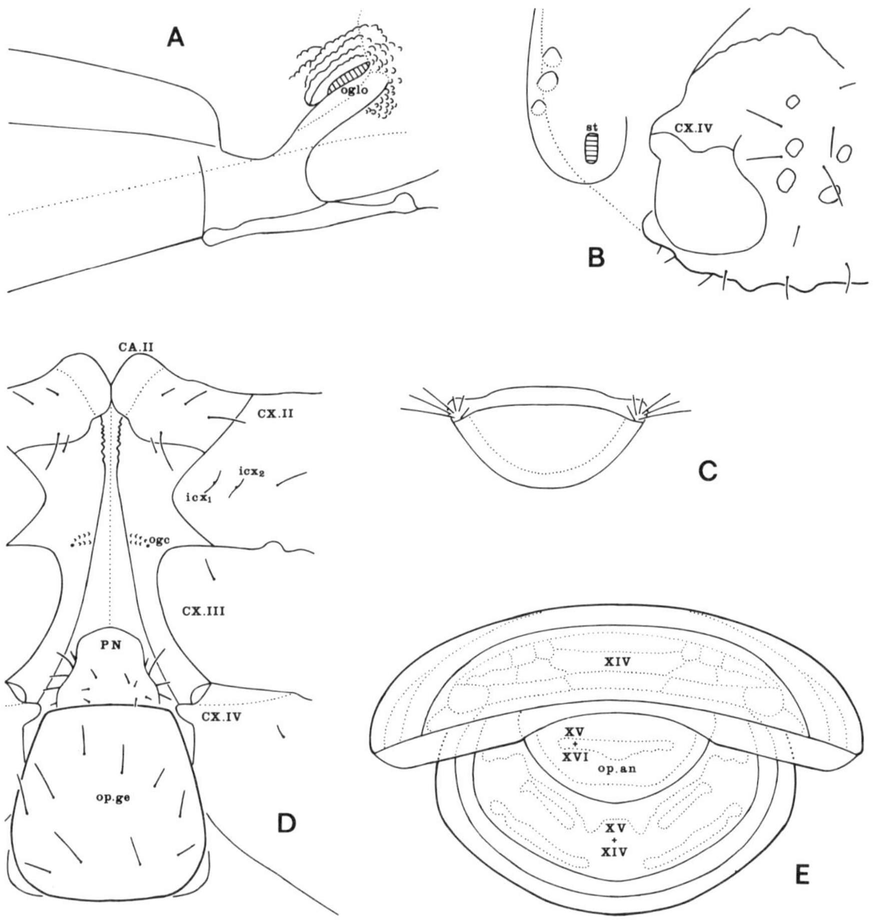

Fig. 275. Scotolemon lespesi Lucas, adult male; A, part of lateral region of prosoma, dorsal to coxa II, with orifice of odoriferous gland, lateral view; B, lateral view of right coxa IV and stigma; C, anal cover, separated from opisthosoma, viewed from inside; D, sternal and genital region (penis partly evaginated), ventral view; E, posterior part of opisthosoma, posterior view; A, × 147; B, D, × 118; C, E, × 37.

in more detail, in figs. 275D, 276C, 277A, B. The coxae of the legs are large and immovably integrated in the ventral surface of the prosoma, with the exception of the anterior coxae which, when they participate in feeding, are able to perform adductor-abductor movements (see the paragraph on the mouthparts). The coxae of palp and several legs are associated with coxapophyses, or coxal lobes, which are regarded as 'endites' (paraxial branches of the appendages); they are supposed to be homologous with the sternapophyses of other groups of Chelicerata (see: Van der Hammen, 1980: 149–150). The occurrence of large coxae is associated with the small size of the sternal area. The

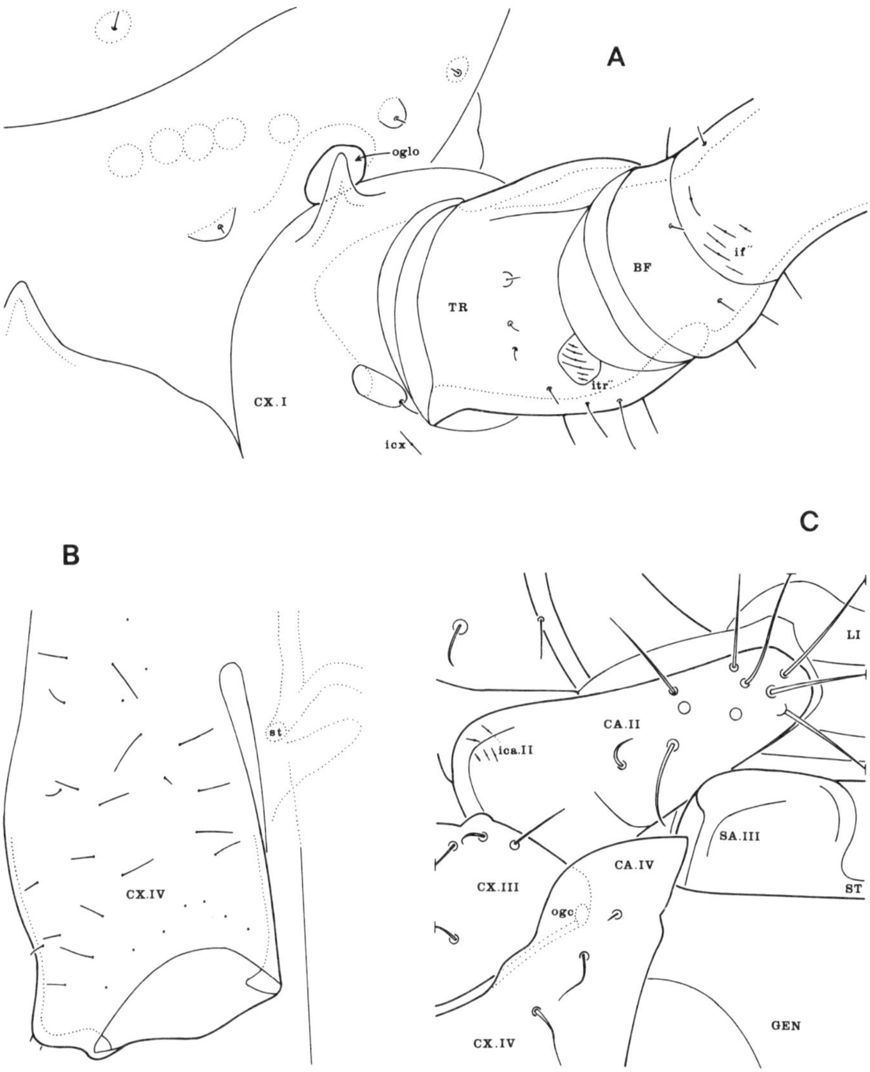

Fig. 276. Paroligolophus agrestis (Meade), adult female; A, lateral view of part of prosoma, with orifice of odoriferous gland, and base of leg I; B, ventral view of coxa IV and stigma; C, ventral view of coxal and sternal region, with orifice of coxal gland; A, C, × 118; B, × 73.

→

Fig. 277. Siro rubens Latreille, adult male; A, B, ventral view of coxisternal region (chelicerae and palps removed); A, coxae contiguous; B, coxae I uncoupled and separated from coxae II by soft skin; C, dorsal view of posterior border of opisthosoma; D, posterior view of opisthosoma; E, dorsal view of rostral region; F, ventral view of anal region; A–F, × 118.

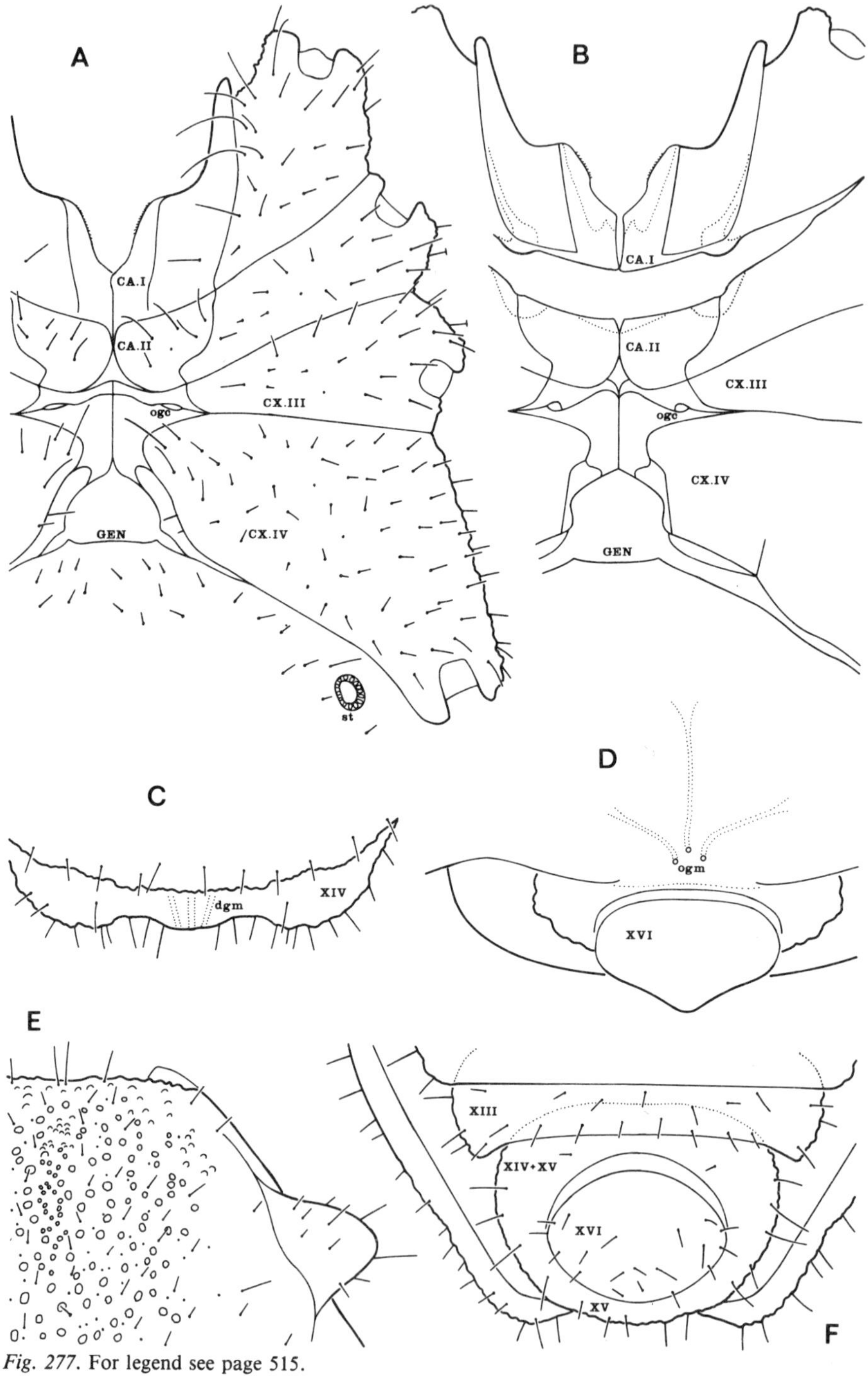

Fig. 277. For legend see page 515.

last-mentioned area is also reduced in size because of the advanced position of the genital opening. The orifice of the coxal gland is in the sternal area, close to the paraxial border of coxa III.

In *Siro rubens* (figs. 266, 277A, B), the coxa of the palp, and coxae I and II present large coxapophyses which have completely fused with the coxae. They are separated from each other, and from coxae III, by an area of soft skin, which can be extended (fig. 277B). There is no labium. It is difficult to decide whether the area between coxae III and IV consists of the sternites of segments V and VI, or of the coxapophyses III and IV. In the last-mentioned case, the coxapophyses must be completely integrated in the sternal area, because the orifice of the coxal gland is at the surface, at the border of segments V and VI (fig. 277A).

In *Scotolemon lespesi* (figs. 4B, 275D), the coxae of palp and legs I and II present large coxapophyses which have fused with the coxae. The presence of the lyrifissures icx_1 and icx_2 is certainly associated with the adductor-abductor movements of coxa II. Although a labium is sometimes supposed to be absent in Laniatores, it is here supposed to be represented by the membraneous projection between coxapophyses I and II (in this case it could be homologous with the sternite of segment III). The remaining part of the sternal area is constituted by a narrow plate, fused with the coxae (and coxapophyses II). The orifice of the coxal gland is near the paraxial border of coxa III (more advanced than in *Siro* or *Paroligolophus*); it is connected with the median part of the sternum by two rows of very small tubercles (fig. 275D).

In *Paroligolophus agrestis* (figs. 267B, 276C), the coxae of palp and legs I, II and IV are associated with coxapophyses. These apophyses are more or less flexibly attached to the coxae. The presence of lyrifissures at the base of coxapophyses II (fig. 276C: *ica.II*) is evidently associated with this movability. The sternite of segment III is partly represented by the labium. The sternite of segment V is probably partly represented by the transverse sclerite between legs III; its presents, on either side, a rounded prominence which could represent the sternapophyses of segment V (there are no coxapophyses in this segment). According to Winkler (1957: 370), the anterior part of the genital cover, which overlaps the greater part of the sternal area, could represent the sternite of segment VI (i.e., part of it). The remaining part of the sternal area, which partly consists of soft skin, certainly includes sternal parts of other prosomatic segments. The orifices of the coxal glands are at the paraxial border of coxa III, covered by coxapophyses IV (fig. 276C).

Lateral aspect of prosoma. – A study of the prosoma in lateral view (figs. 4C, 266C, 268) demonstrates the more or less upward direction of the coxae. Different aspects of shape and size of the coxapophyses are visible in this orientation. The orifice of the odoriferous gland is best visible in lateral view. In *Siro rubens* (fig. 266C), the orifice is near the apex of a cone-shaped tubercle, in the lateral part of the prodorsum. In *Scotolemon lespesi* (figs. 4B, 275A),

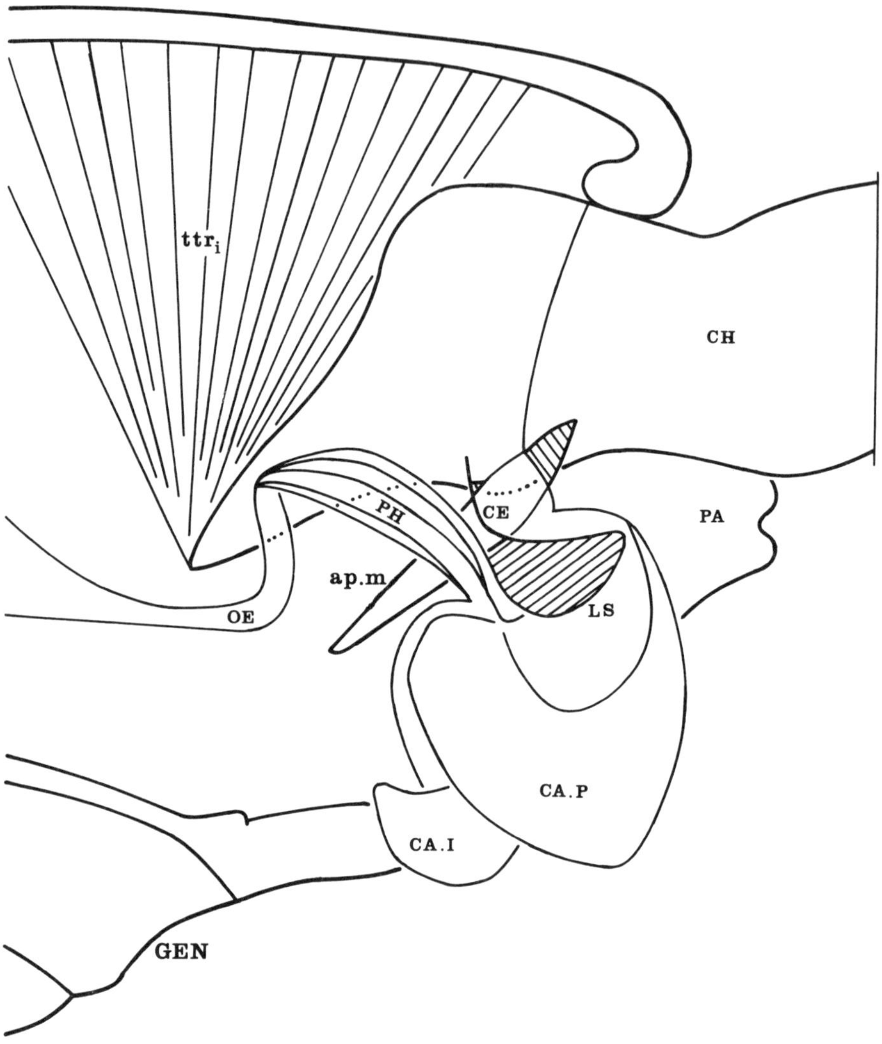

Fig. 278. Siro rubens Latreille, adult female, longitudinal (sagittal) section of anterior region of body with mouthparts; × 112.

it is situated at the border of the prodorsum; it is associated with a taenidium (see: Juberthie, 1961). In *Paroligolophus agrestis* (figs. 268, 276A), it is ventral to the lateral border of the prodorsum.

Mouthparts. – The Opilionid mouthparts (see: Snodgrass, 1948: 43–51, figs. 15–17) include: the chelicerae (see following paragraph) and the cheliceral frame (i.e., the part of the tegument to which the chelicerae are attached); labrum, cervix (generally called epistome or clypeus; see Van der Hammen,

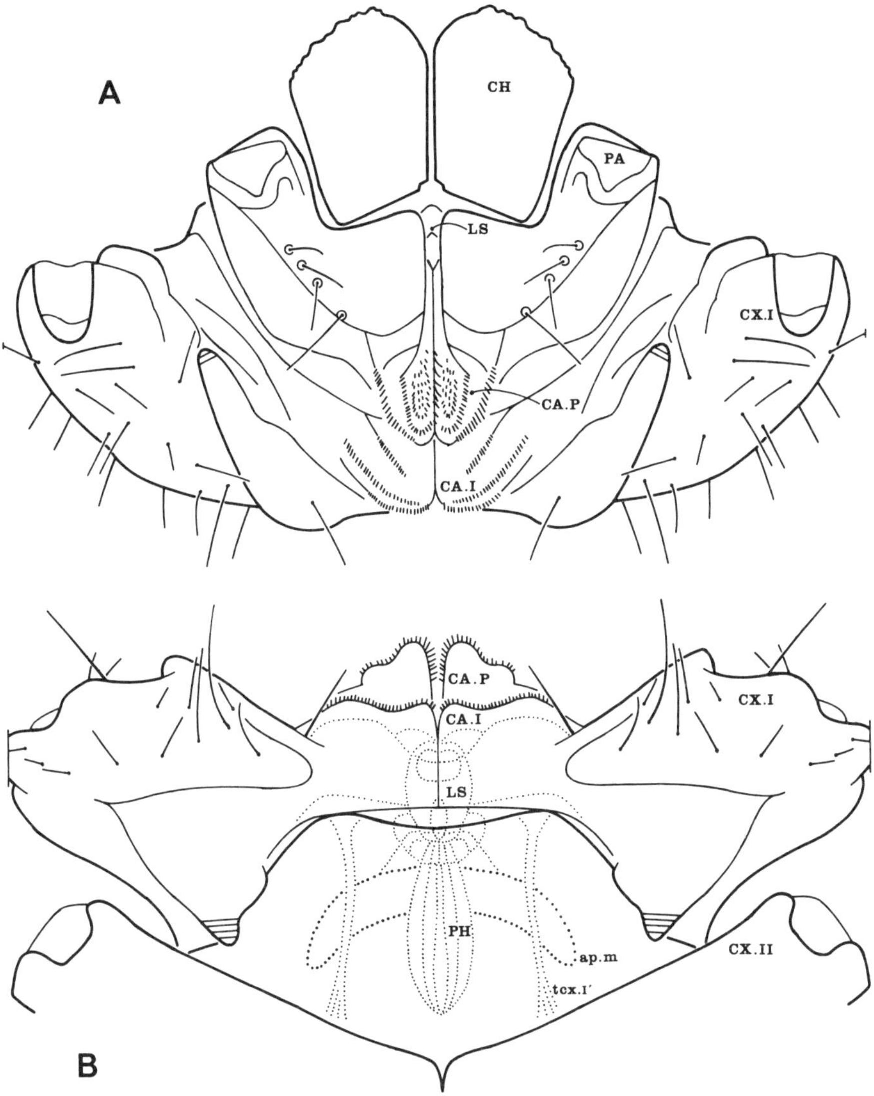

Fig. 279. Siro rubens Latreille, mouthparts of adult female; A, frontal view (chelicerae partly cut off, and represented by a transverse section of the trochanter); B, ventral view (coxae I and II widely separated by extension of the area of soft skin); A, B, × 141.

1980: 28–29) and apodeme (i.e., myliosomatic apodeme); labium, mouth and pharynx; palp (dealt with in a separate paragraph), and the coxapophyses of palp and anterior legs. The mouthparts are particularly characterized by the more or less ventral position of the mouth, the curved pharynx and the par-

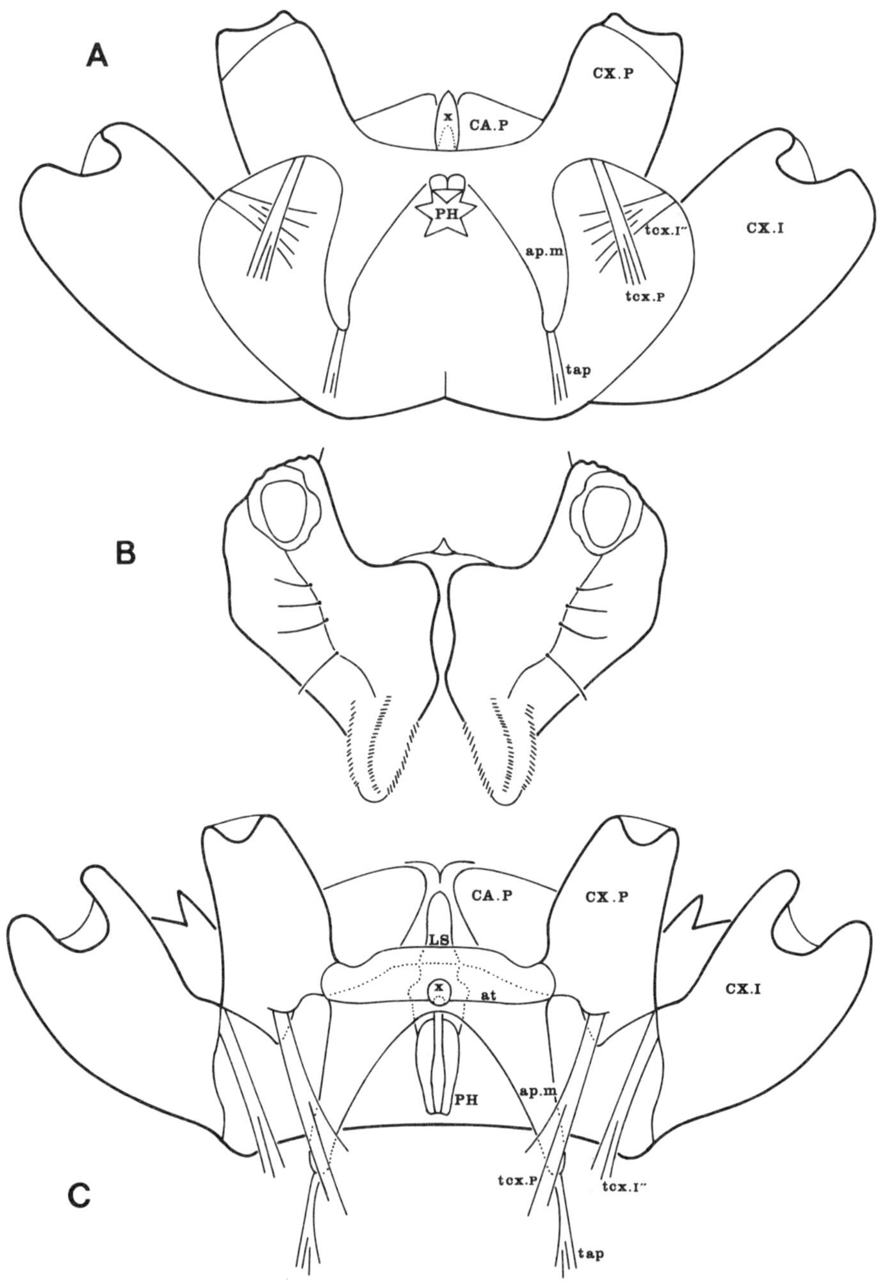

Fig. 280. Siro rubens Latreille, mouthparts of adult (A, C, female; B, male); A, posterior view; B, frontal view of (widely separated) palpal coxae; C, dorsal view; A–C, × 141.

ticipation of the coxapophyses of palp and anterior legs in food-intake (see fig. 268). The type of ingestion associated with these mouthparts (it is also found in Xiphosura and Scorpionida) is called coxisternal feeding. The compound structure made up of the elements involved in coxisternal food-intake is called myliosoma. Among the groups with coxisternal feeding, Opilionida and Xiphosura are characterized by the ingestion of solid food and internal digestion, Scorpionida by the ingestion of fluid food and external digestion.

A myliosoma is a more or less cone-shaped subdivision of the body. It has a complicated structure, which is studied here in various orientations, and in sections. Dorsal view generally refers here to the dorsofrontal view, frontal view to the ventrofrontal view, posterior view to the dorsoposterior view (the ancestral position of the mouthparts has considerably changed, in connection with the ventral translocation of the mouth, and its association with the coxapophyses of the legs).

The cheliceral frame, which is relatively small, extends from the cervix (to which it is attached according to the line *at*, see fig. 280C) to the rostral border of the prodorsum. Part of the articulation point of a chelicera is constituted by a condyle near the line *at* (figs. 278, 282); the greater part of the arthrodial membrane is dorsal. The chelicerae are operated by a large inferior depressor muscle (ttr_i), attached to the prodorsum, and a smaller superior levator muscle (ttr_s). The cervix (*CE*) extends from the base of the labrum to the myliosomatic apodeme (*ap.m*) (see: figs. 278, 281A, C, 282, 283A, F); generally, it is rather heavily sclerotized. Near the labral base, the cervix presents a distinct median convexity. The myliosomatic apodeme constitutes the internal continuation of the cervix; a strong median tendon (*tm*) and paired lateral tendons can be inserted on it (figs. 280A, C, 281A–C, 283A, B, E, F). The labrum (*LS*) constitutes the anterior prolongation of the cervix (figs. 278, 279, 280C, 281A, C, 282, 283A, C, E, F). It is partly attached to the coxapophyses of the palp, and can present a pointed terminal part, which is directed downwards. The labium (*LI*), if present, is situated between coxapophyses I and II (it overlaps coxapophyses I); as mentioned above, it is homologous with the sternal region of segment III. It is distinctly sclerotized in Palpatores (fig. 267B). The mouth (*b*) has a ventral position, and is bordered (figs. 278, 282, 283F) by labrum, coxapophyses of palp and leg I, and labium (when present). The pharynx (*PH*) has a hexagonal transverse section (fig. 280A); it is curved downwards in the anterior part (figs. 268, 278, 282, 283F). The coxapophyses of palp and legs I and II (figs. 278–283) are constituted by paraxial lobes of the appendages; generally, these lobes are regarded as 'endites'. According to Dumitrescu (1975), salivary glands are present in the coxapophyses of palp and leg I. The palpal coxae participate in the general movements of the palp; the coxapophyses are operated by the adductor and abductor muscles *tcx'*, *tcx''* and tcx_i (figs. 286, 287A, 288A). The coxae of the legs do not participate in the walking movements of these appendages; adduction of the coxae (and the coxapophyses) of legs I and II, during ingestion, is operated by the adductors *tcx*

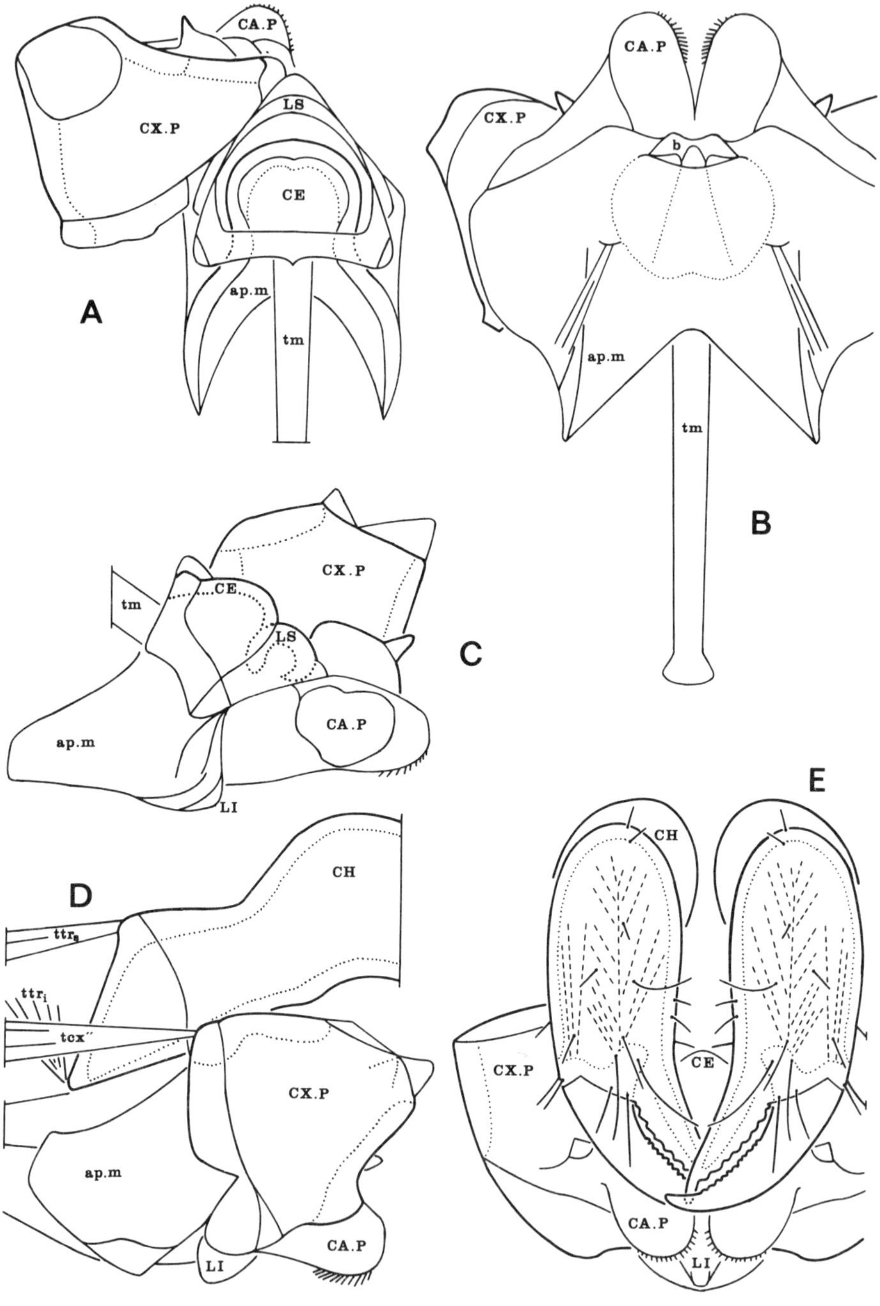

Fig. 281. Scotolemon lespesi Lucas, mouthparts of adult female; A, dorsal view (chelicerae and right palpal coxa removed); B, ventral view; C, lateral view (chelicerae and right palpal coxa removed); D, lateral view; E, frontal view; A–E, × 88.

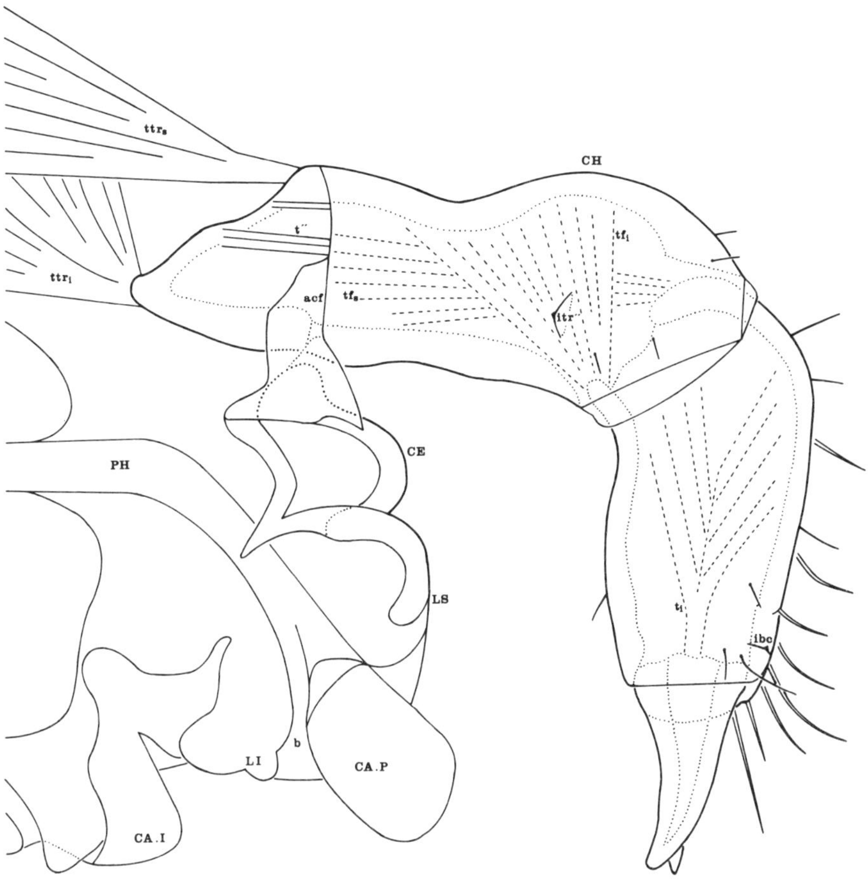

Fig. 282. Scotolemon lespesi Lucas, adult male; longitudinal section of anterior prosomatic region (with mouthparts), antiaxial to right chelicera; × 118.

(figs. 293B, C, 294A, 296F); abduction probably takes place by hydrostatic pressure. As mentioned above, in the paragraph on the ventral aspect of the prosoma, the coxae of palp and legs I and II can be separated by soft skin. In these cases, the soft skin can be extended (fig. 277B) by hydrostatic pressure, whilst the myliosoma turns around an articulation point at the laterofrontal border of the prodorsum. Retraction takes place by the retractor muscles tcx_r (fig. 293C).

The mouthparts of *Siro rubens* (see figs. 277A, B, 278–280, 286A, 293B, C) are particularly characterized by the absence of a labium, and by the possibility of coxisternal extension and retraction (associated with myliosomatic articulation). The cuticle of the palpal coxapophyses is rather soft, particularly in the distal part. The concavity, laterally bordered by the coxae of leg I, which contains the coxapophyses of palp and leg I, is generally called stomotheca; its

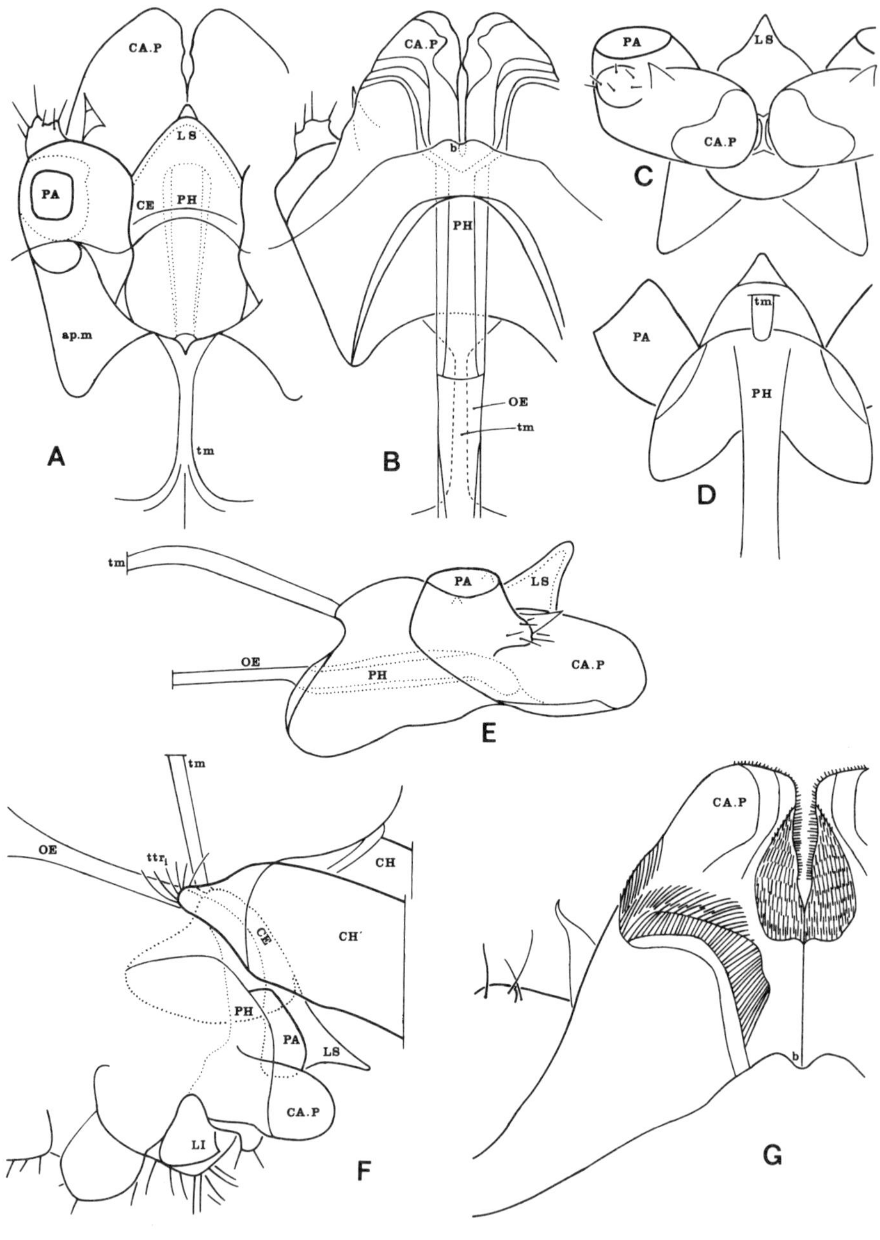

Fig. 283. Paroligolophus agrestis (Meade), mouthparts of adult male (chelicerae removed, except in the case of F); A, dorsal view; B, ventral view; C, frontal view; D, posterior view; E, lateral view; F, longitudinal section (antiaxial to right chelicera); G, ventral view of mouth and palpal coxapophyses; A–F, × 118; G, × 236.

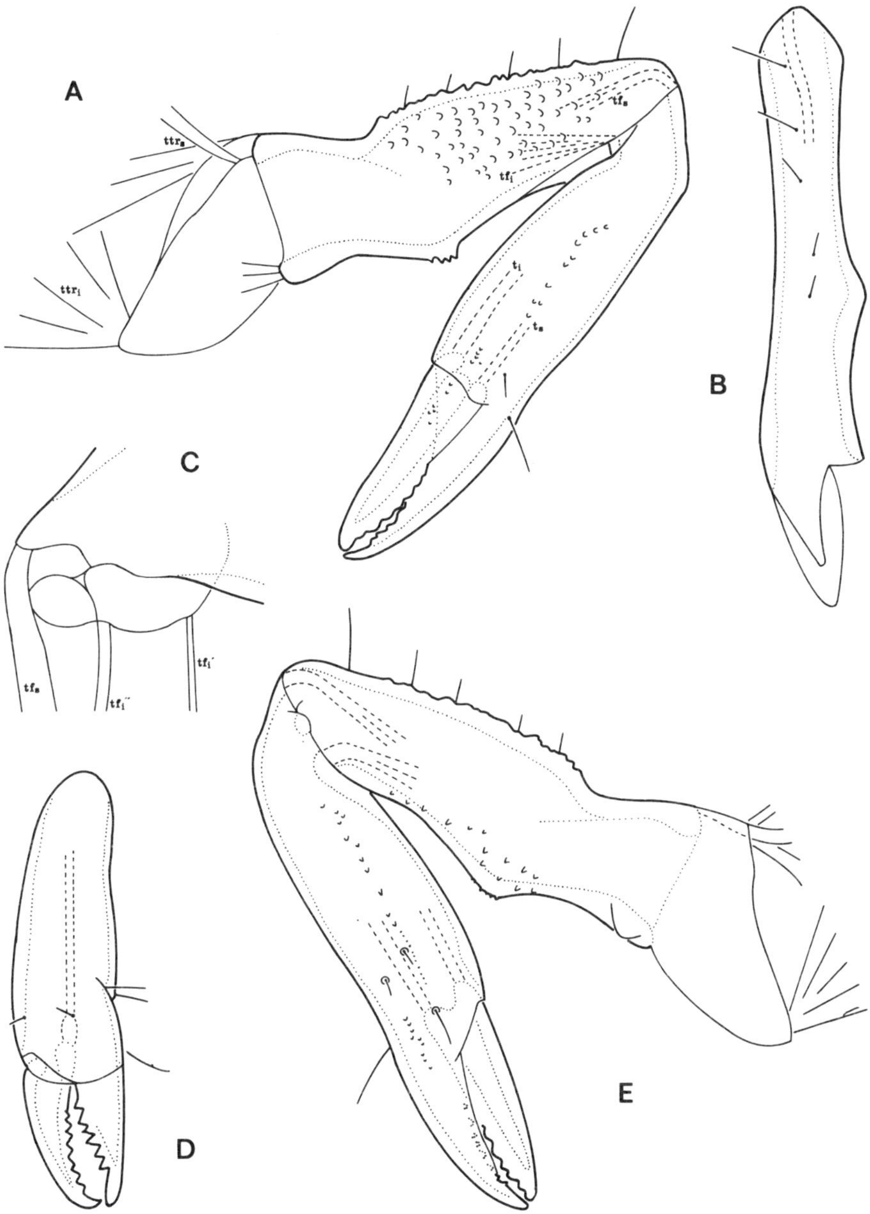

Fig. 284. Siro rubens Latreille, right chelicera of adult male; A, lateral (antiaxial) view; B, dorsal view of trochanter; C, lateral (antiaxial) view of proximal region of body of chelicera (separated from trochanter); D, ventral (oblique) view of body of chelicera and apotele; E, lateral (paraxial) view; A–B, D–E, × 118; C, × 246.

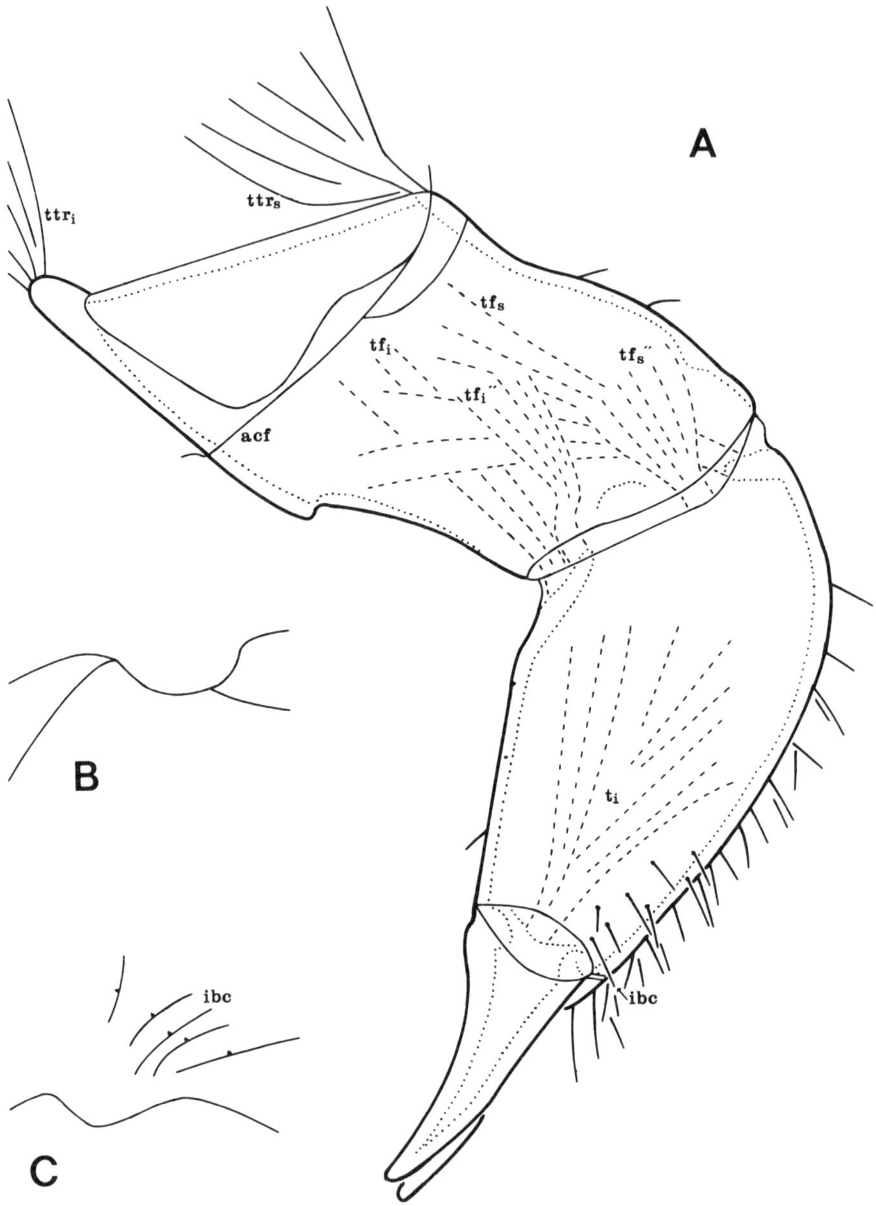

Fig. 285. Paroligolophus agrestis (Meade), right chelicera of adult male; A, lateral (antiaxial) view; B, C, articulation between body of chelicera and apotele; B, ventral view; C, dorsal view; A, × 110; B, C, × 368.

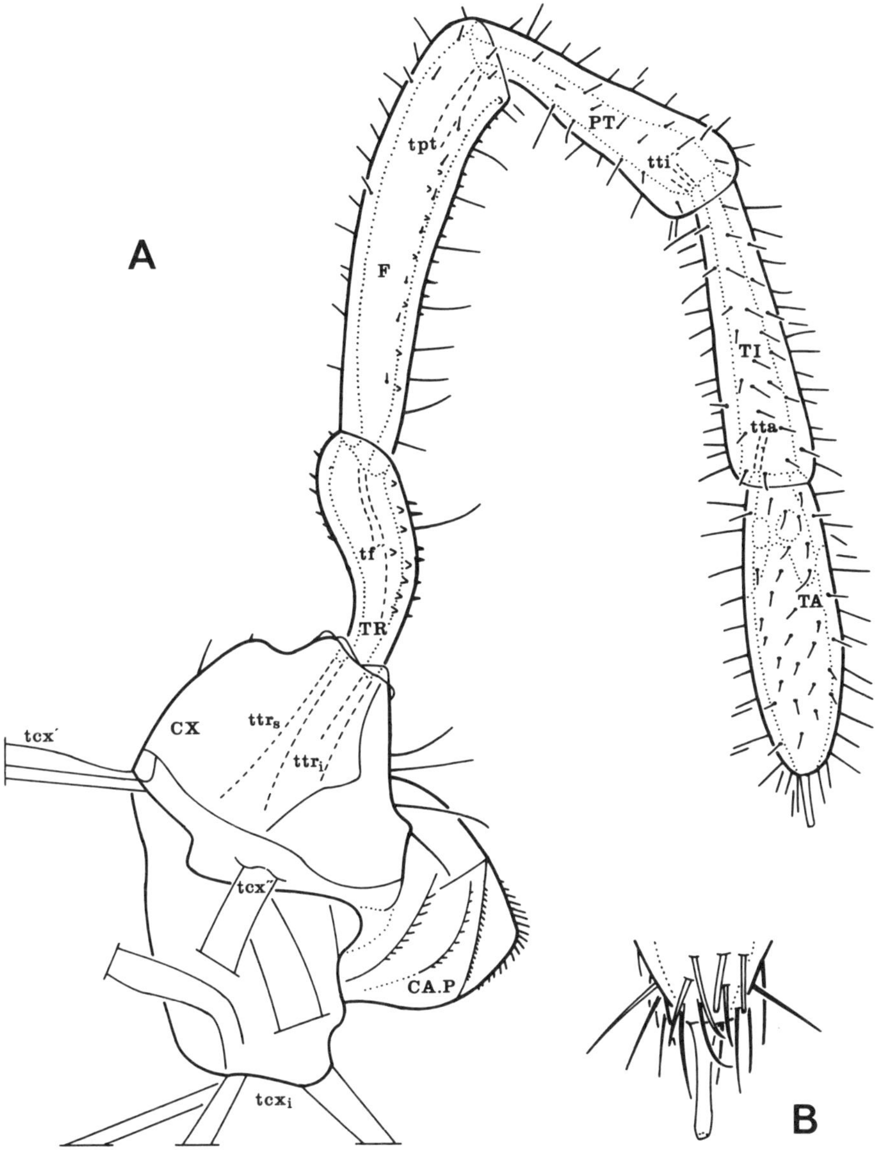

Fig. 286. Siro rubens Latreille, lateral (antiaxial) view of right palp of adult male; A, coxa, trochanter, femur, patella, tibia, tarsus and apotele (phanerotaxy of terminal part schematized); B, distal part of tarsus and apotele; A, × 176; B, × 368.

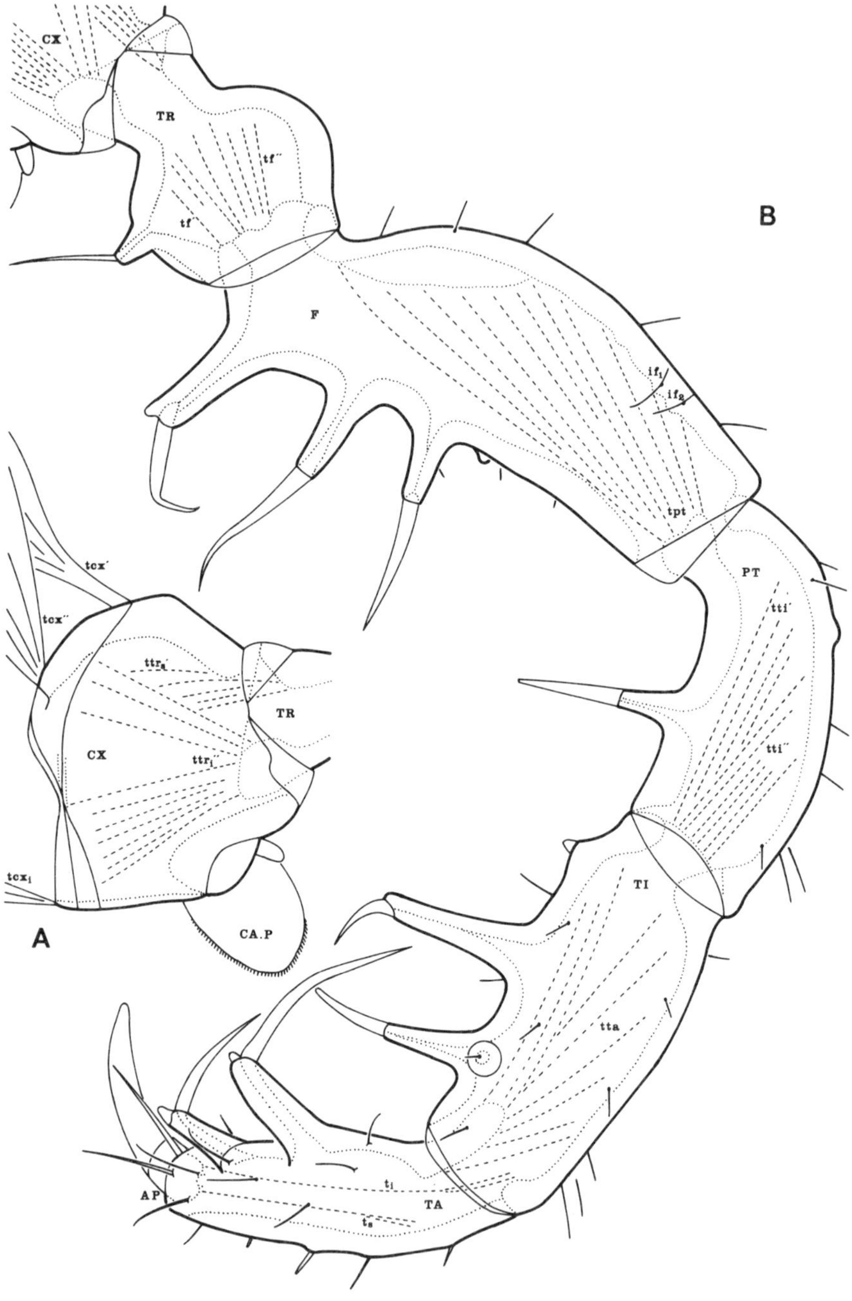

Fig. 287. Scotolemon lespesi Lucas, right palp of adult male, lateral (antiaxial) view; A, coxa and proximal part of trochanter; B, distal part of coxa, trochanter, femur, patella, tibia, tarsus and apotele; A, B, × 118.

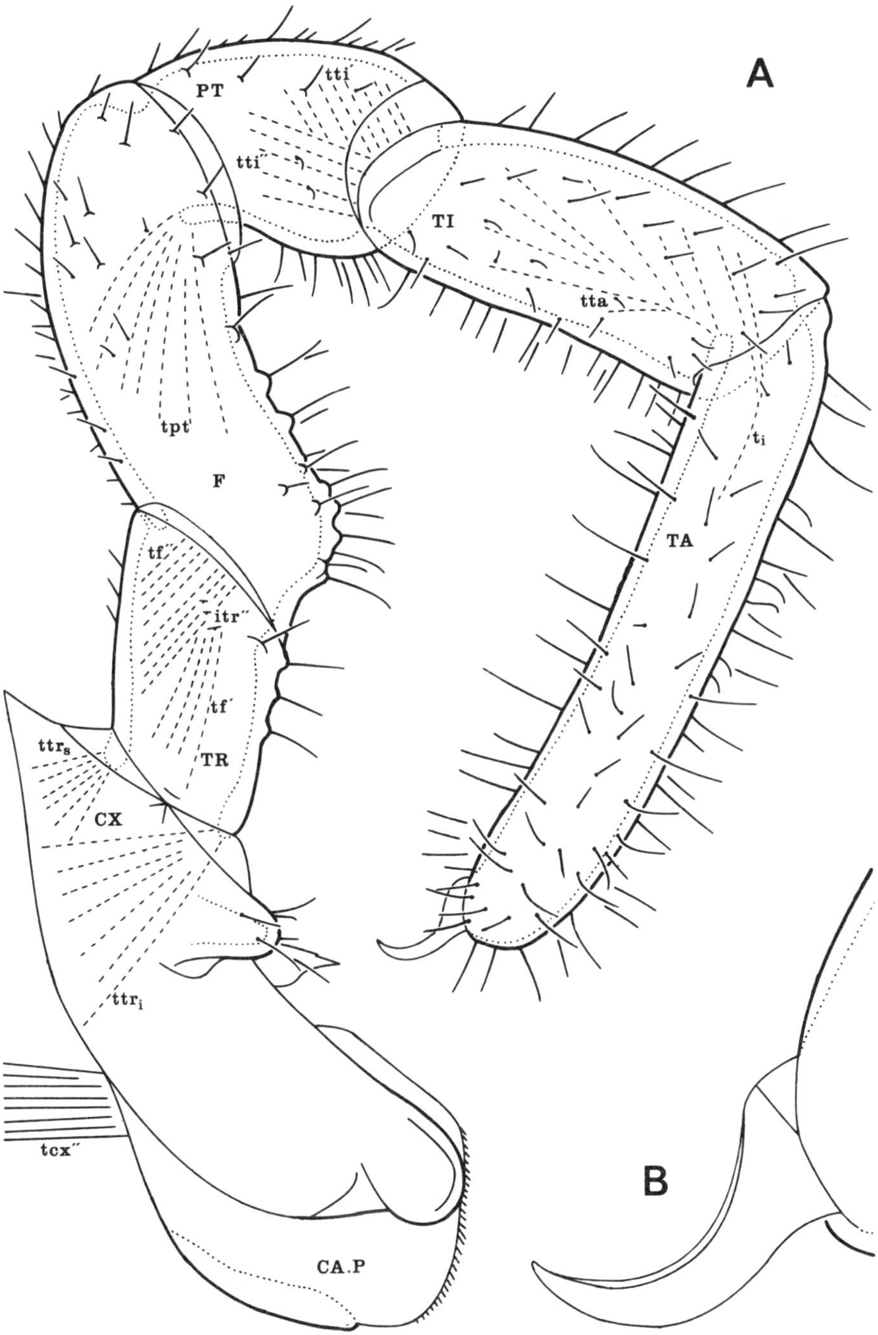

Fig. 288. For legend see page 530.

occurrence is probably related to the extension and retraction of these coxapophyses. The labrum has probably completely fused with the coxapophyses of the palp; it has apparently no free distal part.

The mouthparts of *Scotolemon lespesi* (see figs. 281, 282, 287A, 294A) are particularly characterized by: the membraneous labium; the large coxapophyses I and II, which are immovably connected with the coxae; and the labrum which has fused with the coxapophyses of the palp, and has no free distal part. The mouthparts of *Paroligolophus agrestis* (see figs. 267B, 268, 283, 288A, 296F) are particularly characterized by: the sclerotized labium; the large coxapophyses which can be rather soft and individually movable; and the superficial taenidia (called pseudotracheae by Kaestner, 1933), associated with the salivary glands in the palpal coxapophyses, which run up to the mouth.

Chelicera (figs. 282, 284, 285). – The Opilionid chelicera consists of three segments which are called trochanter, body of the chelicera (including the fixed jaw of the chela), and apotele (the movable jaw of the chela). The basal part of the trochanter is internal: the cheliceral frame and the arthrodial membrane are attached to the trochanter according to the line *acf*. The basal articulation point of the chelicera is constituted by a condyle (close to the line of attachment of the cheliceral frame to the cervix) and the antilateroventral part of the trochanter (close to the line *acf*). As mentioned above, in the section on the mouthparts, the chelicerae are operated by a large inferior depressor muscle (ttr_i) and a smaller superior levator muscle (ttr_s); there can also be additional muscles. The articulation between the trochanter and the body of the chelicera is constituted by a pivot joint with one or more superior levator muscles (tf_s, tf_s'') and one or more inferior depressor muscles (tf_i, tf_i', tf_i''). The joint between the body of the chelicera and the apotele is operated by the usual tendons of the claw segment (t_s and t_i). Owing to torsion, the articulation axis of the joint between the body of the chelicera and the apotele can have become more or less parallel to the sagittal plane of the body, instead of perpendicular to it; in these cases (*Scotolemon lespesi, Paroligolophus agrestis*; much less in *Siro rubens*), the position of the apotele is antiaxial instead of ventral (inferior).

The trochanter and the body of the chelicera present setae, especially in the dorsal regions. In *Scotolemon lespesi* and *Paroligolophus agrestis*, the setae are most numerous on the body of the chelicera. Lyrifissures can be present in the trochanter (*Scotolemon lespesi*, fig. 282: *itr''*) and in the body of the chelicera near the chela (in *Scotolemon lespesi*, see fig. 282, a single lyrifissure *ibc*; in *Paroligolophus agrestis*, see fig. 285C, a group of lyrifissures *ibc*). Cheliceral glands in the anterior prosomatic region, with an orifice in the dorso-

←

Fig. 288. Paroligolophus agrestis (Meade), right palp of adult male, lateral (antiaxial) view; A, coxa, trochanter, femur, patella, tibia, tarsus and apotele; B, terminal part of tarsus and apotele (setae and aciculate microsculpture omitted; the secondary sexual characters are not visible in this orientation); A, × 110; B, × 368.

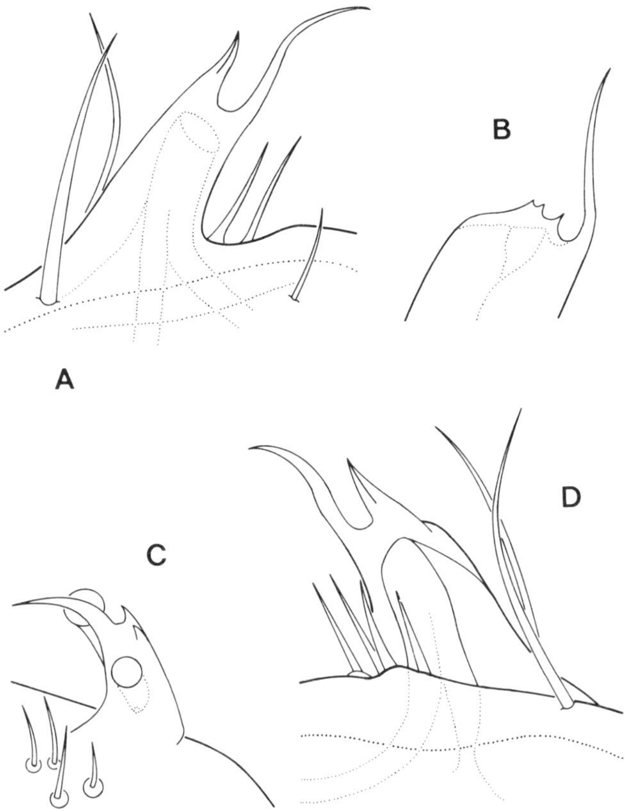

Fig. 289. Siro rubens Latreille, adult male, telotarsal organ of right leg IV; A, lateral (posterior) view; B, posterodorsal view; C, anterodorsal view; D, lateral (anterior) view; A–D, × 688.

distal part of the cheliceral trochanter, were described by Dumitrescu (1974).

Palp (figs. 286–288). – The Opilionid palp consists of coxa, trochanter, femur, patella, tibia, tarsus and apotele (or claw segment; it is absent in many cases); there is no separate basitarsus, and there are no other tarsal subdivisions. The coxa presents a coxapophysis associated with ingestion (the coxa-body articulation is dealt with in the paragraph on the mouthparts). The coxa-trochanter articulation is a pivot joint with superior levator and inferior depressor muscles. The trochanter-femur articulation is a rocking joint with ventral condyle and laterally inserted promotor and remotor muscles (*tf'*, *tf''*). The femur-patella articulation is a hinge joint with dorsal condyle and ventral muscle (*tpt*). The patella-tibia articulation is a rocking joint with one condyle (originally dorsal) and two lateral muscles (*tti'* and *tti''*). The tibia-tarsus articulation is a hinge joint with dorsal condyle and ventral muscle (*tta*). The tarsus-apotele articulation, when present, is a pivot joint with lateral condyles, and two tendons, viz., the superior tendon t_s, and the inferior tendon t_i.

All palpal segments present setae (the seta of the apotele is represented by the unguis); the setae can be inserted on large tubercles or apophyses (as in *Scotolemon lespesi*; see fig. 287). Solenidia can be present on the tarsus (in *Siro rubens* in the distal part, see fig. 286B; in *Paroligolophus agrestis* on the dorsal face, see fig. 288A). Lyrifissures can be present in various segments; in *Scotelemon lespesi* (fig. 287B), the femur presents two dorsal lyrifissures, if_1 and if_2; in *Paroligolophus agrestis* (fig. 288A), the trochanter presents a group of three lateral (antiaxial) lyrifissures *itr''*.

Apart from the coxapophysis (which is associated with ingestion), the function of the palp is probably either predominantly sensory, or predominantly prehensile (as in *Scotolemon lespesi*).

Legs (figs. 289–297). – The Opilionid legs consist of seven eudesmatic segments: coxa, trochanter, femur, patella, tibia, tarsus and apotele. The femur can be incompletely subdivided, by a basifemoral ring (figs. 294A, 296A), into two adesmatic segments, viz., basi- and telofemur. The tarsus is always subdivided into two adesmatic segments, viz., basi- and telotarsus, which can both present further subidivisions (basitarsus 1, 2, etc.; telotarsus 1, 2, etc.; see figs. 294A, 295E–F, 297A–C).

The coxae of the anterior legs are movable (abduction and adduction) in connection with ingestion, but not in connection with locomotion; the other coxae are fixed. The coxa-trochanter articulation is a bicondylar pivot joint with superior levator (ttr_s) and inferior depressor (ttr_i) muscles. The trochanter-femur articulation is a rocking joint with inferior condyle and laterally inserted muscles *tf'* and *tf''*. The femur-patella articulation is a hinge joint with superior condyle and inferior muscle *tpt*. The patella-tibia articulation is a rocking joint with superior condyle and laterally inserted muscles *tti'* and *tti''*. The tibia-tarsus articulation is a hinge joint with superior condyle and inferior muscle *tta*. The tarsus-apotele articulation is a pivot joint with lateral condyles and two tendons, viz., a superior tendon t_s and an inferior tendon t_i (which continues in the tibia). The adesmatic segments of basi- and telotarsus can move under the influence of t_s and t_i.

Setae can be present in all segments of the legs (the ungues of the apotele are homologous with setae); they can be inserted on tubercles or apophyses. Solenidia can be present on the tarsus only.

Lyrifissures (see: Barth & Stagl, 1976) are generally associated with articulations, particularly those of the proximal part of the legs (figs. 276A, 294A–C, 296A, C–F), but in Phalangiidae also associated with the basitarsus-telotarsus articulation (fig. 297A, F). They occur as solitary slits, or they are arranged in groups or lyriform organs.

Accessory stigmata and tracheae are present (proximally and distally) in the tibiae of the legs in the family Phalangiidae.

In *Siro rubens* (figs. 289–293), femur, basitarsus and telotarsus are not subdivided. Lyrifissures are not discernable (although this can be connected with the granulate microsculpture of the tegument). Telotarsus I presents, in my

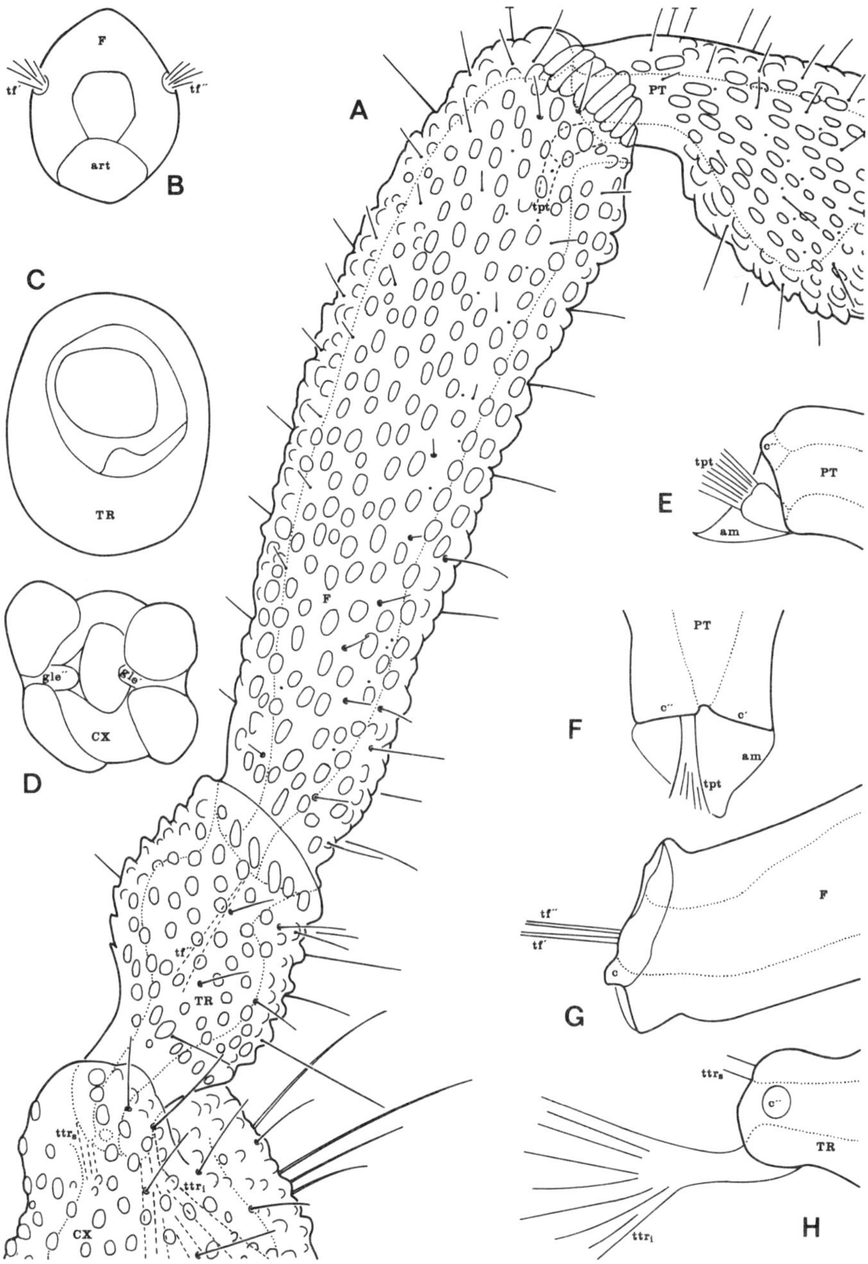

Fig. 290. Siro rubens Latreille, right leg I of adult male; A, distal part of coxa, trochanter, femur, and proximal part of patella, lateral (posterior) view; B, proximal view of femur; C, distal view of trochanter; D, distal view of coxa; E, lateral (posterior) view of patella; F, dorsal view of distal part of patella; G, lateral (posterior) view of proximal part of femur; H, lateral (posterior) view of proximal part of trochanter; A–C, E–H, × 246; D, × 187.

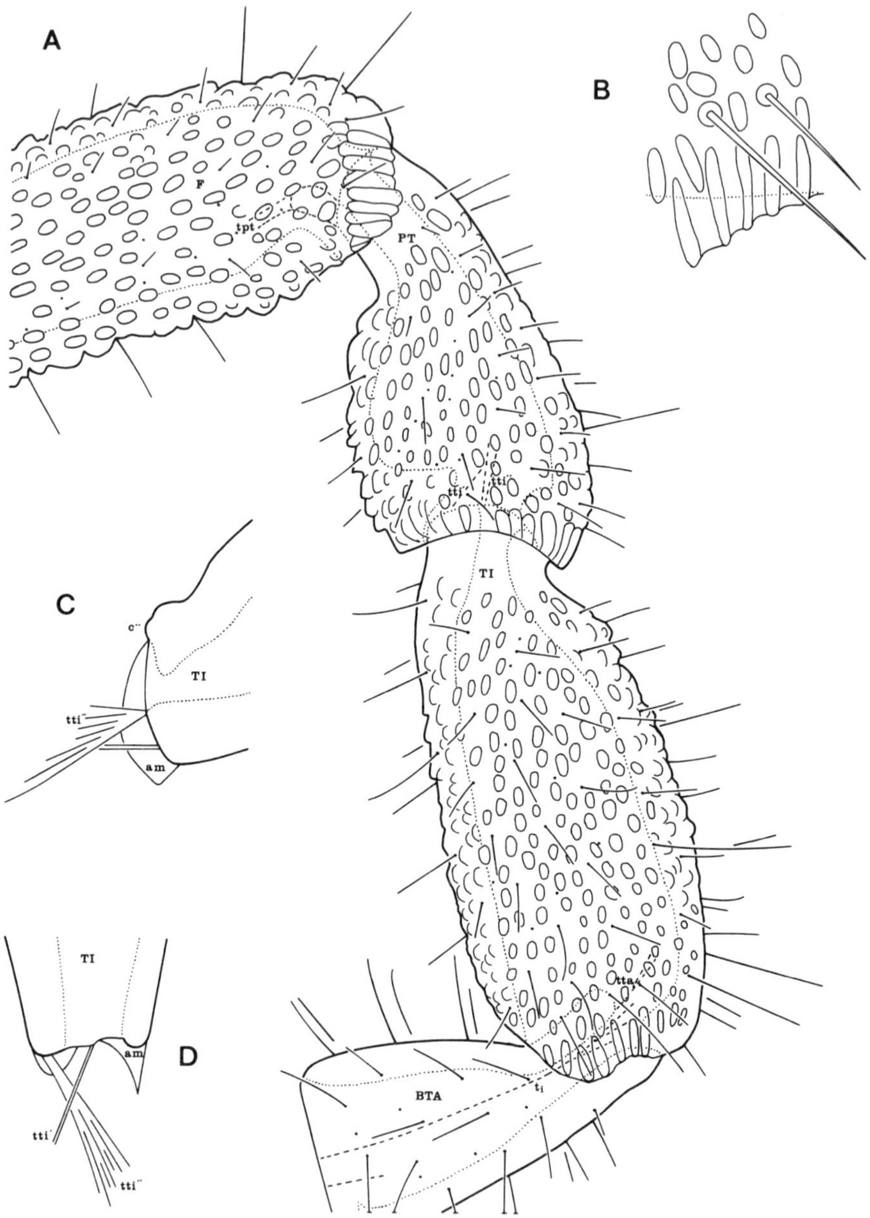

Fig. 291. Siro rubens Latreille, right leg I of adult male; A, lateral (posterior) view of distal part of femur, patella, tibia and basitarsus; B, tuberculate microsculpture of distal part of tibia, lateral (posterior) view; C, D, basal part of tibia (separated from patella); C, lateral (posterior) view; D, ventral view; A, C, D, × 246; B, × 458.

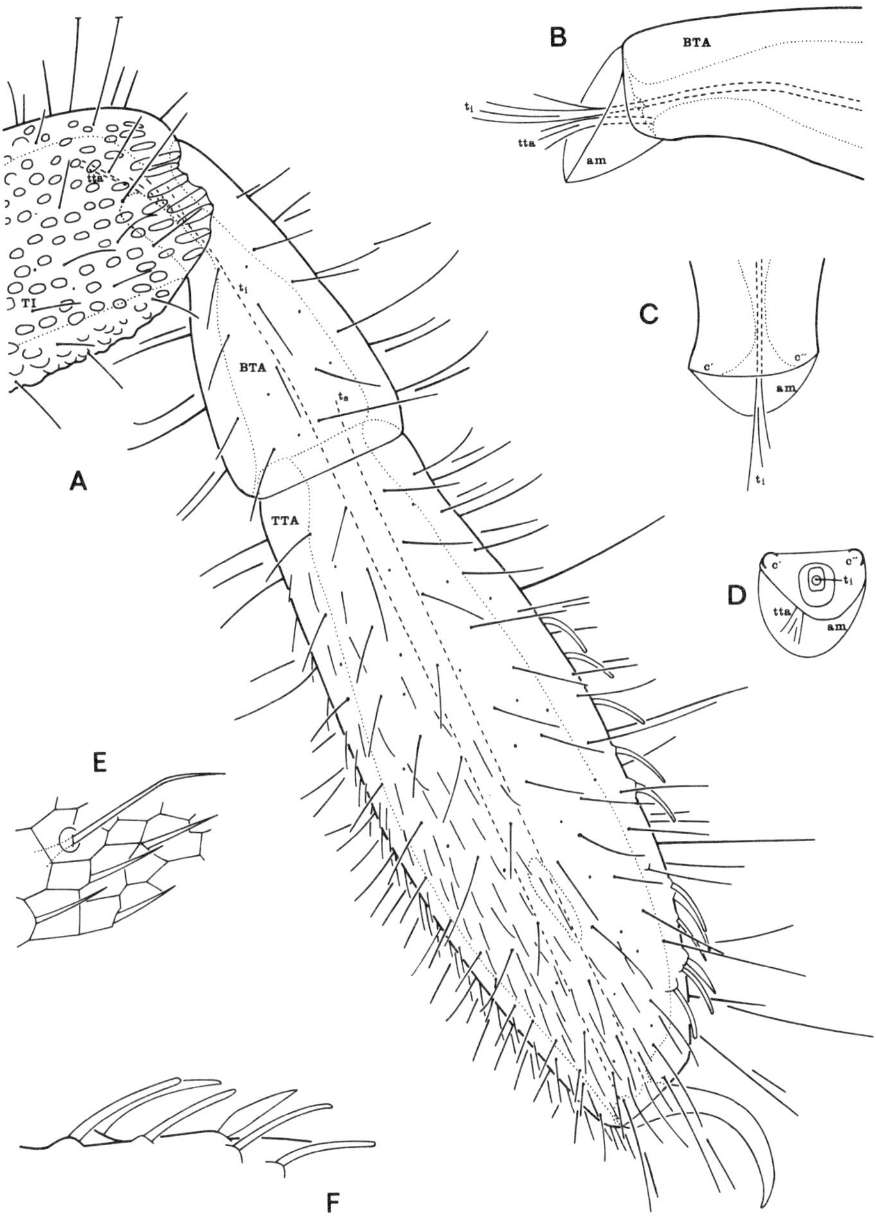

Fig. 292. Siro rubens Latreille, right leg I of adult male; A, lateral (posterior) view of distal part of tibia, basitarsus, telotarsus and apotele; B–D, basal part of basitarsus; B, lateral (posterior) view; C, dorsal view; D, proximal view; E, reticulate aciculate microsculpture of part of telotarsus, lateral (posterior) view; F, lateral (posterior) view of distal solenidia of telotarsus; A–D, × 246; E, F, × 458.

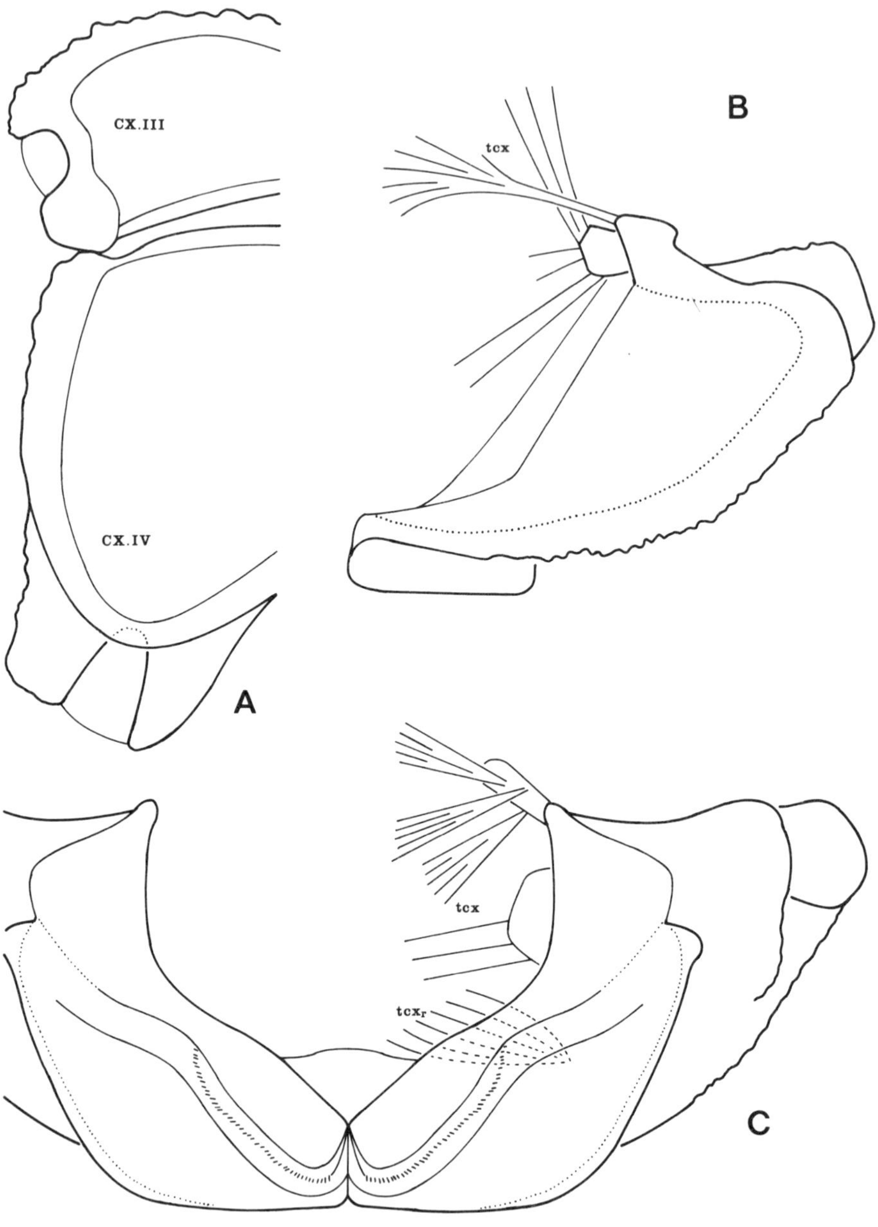

Fig. 293. Siro rubens Latreille, coxae of adult male; A, dorsal view of coxae III–IV, after removal of the dorsal body wall; B, anterior view of separated left coxa II; C, anterior view of separated coxae I; A–C, × 186.

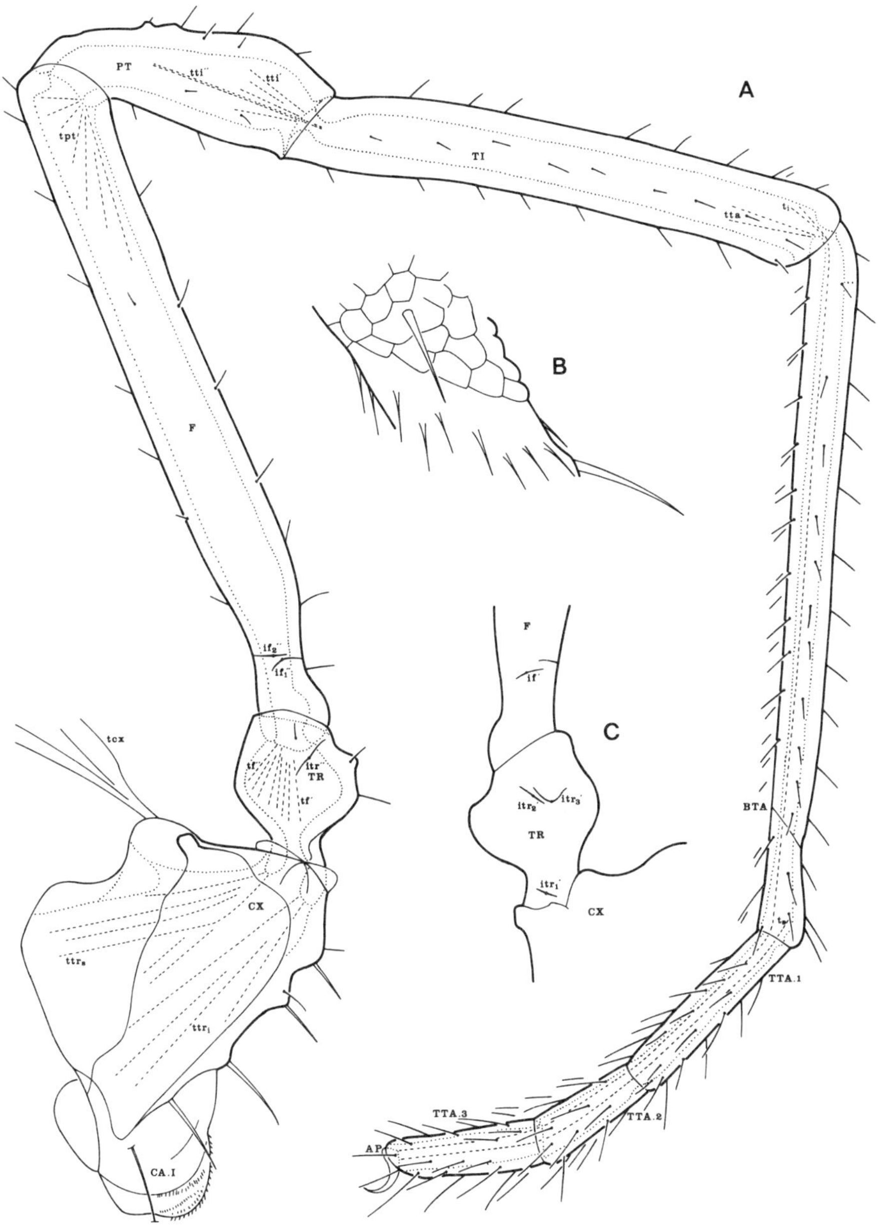

Fig. 294. Scotolemon lespesi Lucas, right leg I of adult male; A, lateral (posterior) view; B, transition from reticulate to aciculate microsculpture in the distal part of the basitarsus, lateral (posterior) view; C, lateral (anterior) view of distal part of coxa, trochanter, and proximal part of femur (setae omitted); A, C, × 99; B, × 385.

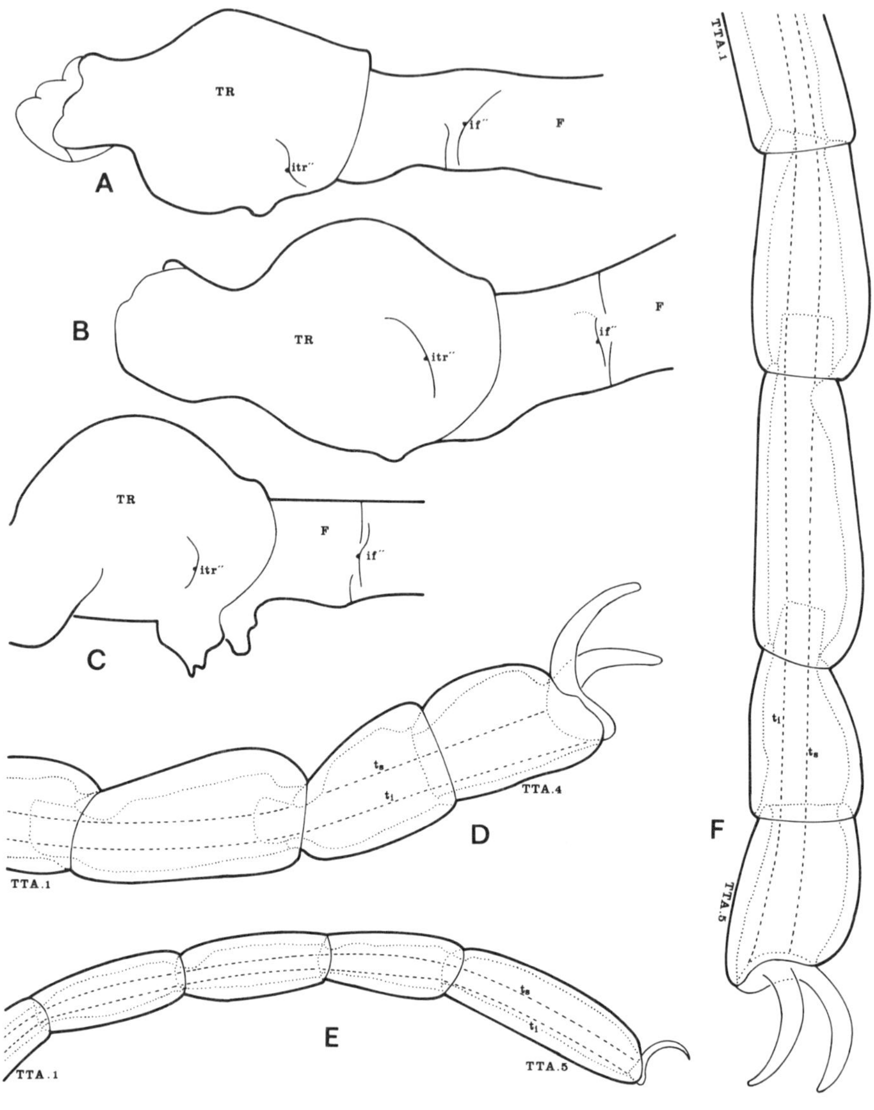

Fig. 295. Scotolemon lespesi Lucas, adult male; A, trochanter and proximal part of femur of right leg II, lateral (posterior) view; B, trochanter and proximal part of femur of right leg III, lateral (posterior) view; C, trochanter (basal part omitted) and proximal part of femur of right leg IV, lateral (posterior) view; D, terminal part of right leg III, lateral (posterior) view; E, terminal part of right leg II, lateral (posterior) view; F, terminal part of right leg IV, lateral (posterior) view; A, B, × 155; C, E, × 118; D, F, × 246.

specimens, a dorsal row of eleven solenidia; solenidia are also present on telotarsus II (but not on telotarsi III and IV). All legs are monodactyl. Leg IV of the male presents a telotarsal organ, consisting of a dorsal apophysis whch contains the orifice of the so-called tarsal gland (fig. 289; see also: Rafalski, 1958: 535–539, figs. 10–13; Juberthie, 1967: 162–63, fig. 6B).

In *Scotolemon lespesi* (figs. 294, 295), the femur of the legs is incompletely subdivided by a basifemoral ring (an arrangement of lyrifissures) into a small basifemur and a much larger telofemur. The telotarsi I–IV present three (leg I), four (leg III) or five (legs II, IV) adesmatic segments. Legs I and II are monodactyl, legs III and IV bidactyl. Lyrifissures are present in trochanter and femur of legs I–IV (figs. 294A, C, 295A–C).

In *Paroligolophus agrestis* (figs. 296, 297) the femur is subdivided, by a basifemoral ring (associated with lyrifissures), into a small basifemur and a much larger telofemur. The tarsus is subdivided into numerous adesmatic segments; in the case of leg I, the basitarsus presents an incomplete subdivision, whilst the telotarsus of the figured specimen consists of 26 adesmatic segments (telotarsi 1–26). The curved phaneres, in the distal part of the tarsus, apparently represent solenidia. Lyrifissures are present in coxa, trochanter, femur and basitarsus. The coxal lyrifissure *icx''* of leg I is close to the articulation point of coxa and trochanter. The posterior lyrifissures of the trochanter of leg I constitute a lyriform organ *itr*, situated close to the laterodistal border of the segment. Femur I presents, posteriorly, a single lyrifissure if_a (close to the articulation with the patella), and a group of six lyrifissures *if''* (close to the basifemoral ring). The posterior lyrifissure of basitarsus I (*ita*) is close to the distal border of the basitarsus. The basal accessory stigma of tibia I is represented in fig. 296B.

Internal anatomy. – Important summaries of Opilionid internal anatomy were published by Kaestner (1935–1937) and Berland (1949); Janczyk (1956) studied the internal anatomy of *Siro duricorius* Joseph, in comparison with other Opilionida. The present paragraph constitutes a compendious survey.

The alimentary canal consists of mouth, pharynx (with six longitudinal folds), oesophagus, midgut and large caeca, hindgut and anus (see: Dumitrescu, 1974a, 1975a, 1975b, 1976, 1980). The excretory system comprises the coxal glands (orifice near coxa III; according to Moritz, 1959, the gland belongs embryologically to segment III), nephrocytes and perineural organs (for storage of waste products); there are no Malpighian tubes. The endosternite (Firstman, 1973: 8–15, figs. 5–8) presents two or four pairs of dorsal suspensor muscles and two pairs of ventrals (transverse suspensors are lacking); in Cyphophthalmi the endosternite is vestigial, with a single pair of dorsoventral suspensors. The circulatory system consists of a heart and an anterior aorta. The respiratory system consists of the pair of stigmata, posterior to coxae IV (in segment VIII), and tracheae; Phalangiidae present, moreover, tibial stigmata and pedal tracheae. The central nervous system consists of a supraoesophageal ganglionic mass, a suboesophageal ganglionic mass, and a small opis-

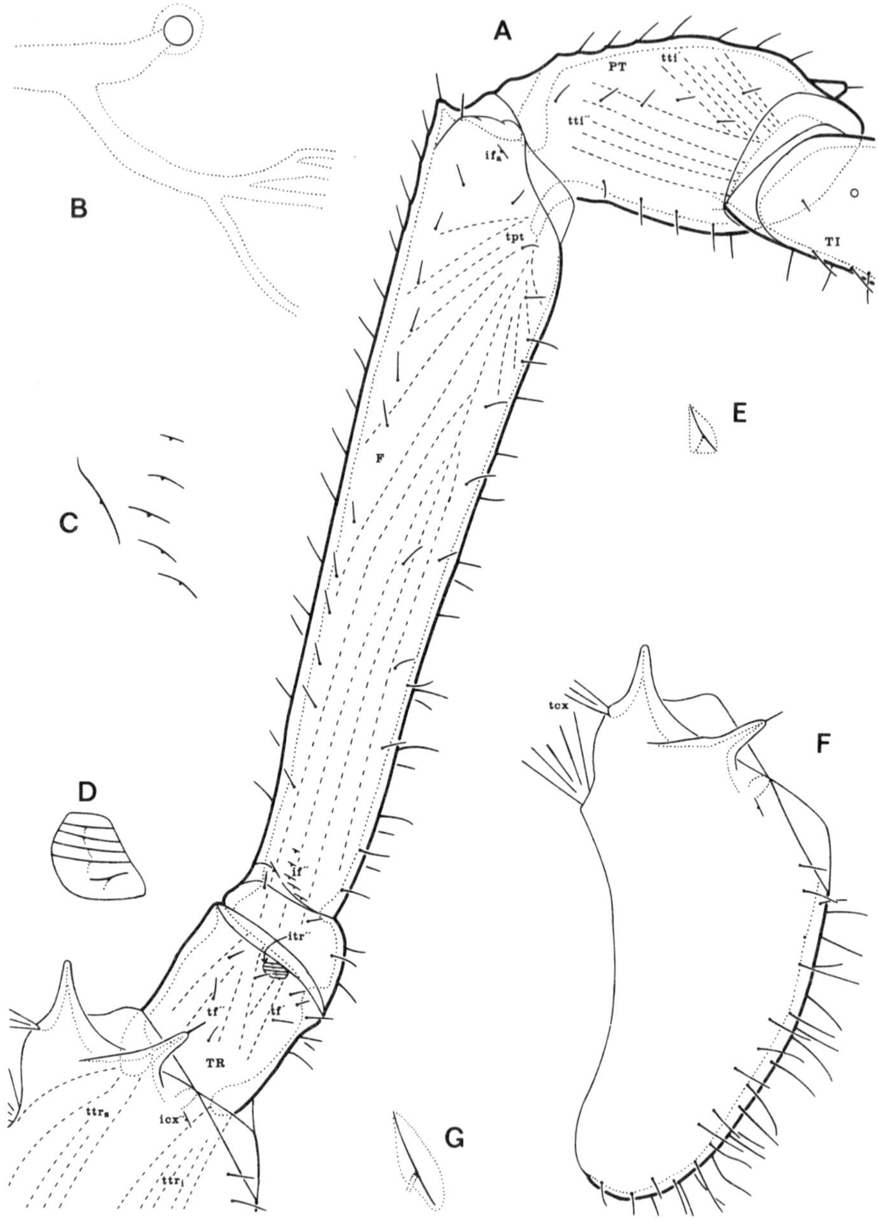

Fig. 296. Paroligolophus agrestis (Meade), proximal part of right leg I of adult male, lateral (posterior) view; A, distal part of coxa, trochanter, femur, patella, and proximal part of tibia; B, tibial stigma and tracheae; C, posterolateral lyrifissures associated with basifemoral ring; D, posterolateral lyrifissures of trochanter; E, distal posterolateral lyrifissure of femur; F, coxa; G, posterolateral lyrifissure of coxa (B–G have the same orientation as A); A, F, × 73; B–E, G, × 246.

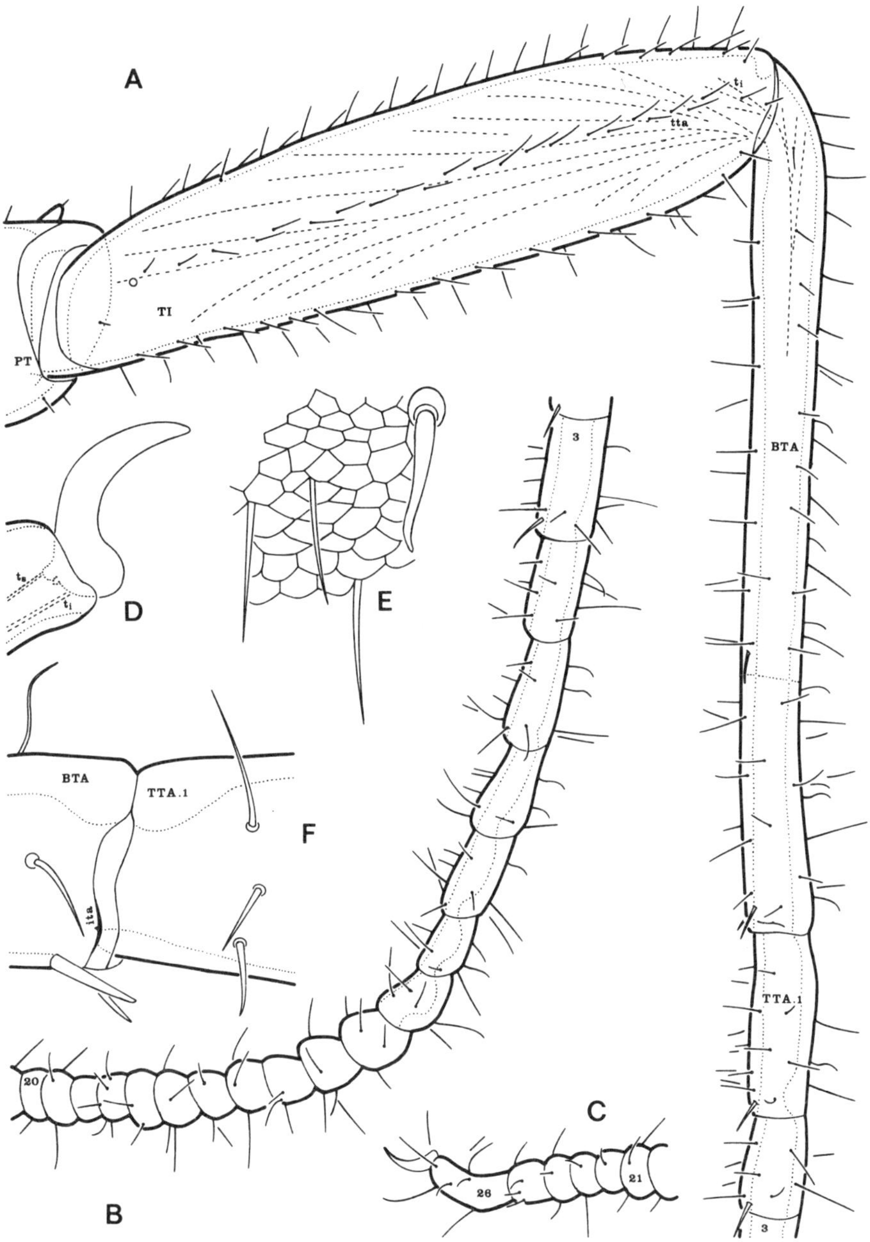

Fig. 297. Paroligolophus agrestis (Meade), distal part of right leg I of adult male, lateral (posterior) view; A, distal part of patella, tibia, basitarsus, and telotarsi 1–3; B, telotarsi 2–21; C, telotarsi 20–26 and apotele; D, terminal part of telotarsus 26 and apotele; E, reticulate aciculate microsculpture and seta of proximal part of basitarsus; F, articulation between basitarsus and telotarsus 1

→

thosomatic ganglionic mass (there is no true opistosomatic nervecord). The sensory organs include, among others, one pair of eyes (absent in several cases), stress-sensitive organs (lyrifissures), and sensory phaneres (setae and solenidia; Opilionida do not present trichobothria). The male reproductive system consists of testes, vasa deferentia, vesicula seminalis and penis (or spermatopositor); Opilionid sperm is completely aflagellate (with the exception of Cyphophthalmi, in which group a rudiment of a cilium is present during spermatogenesis; see Juberthie, Manier & Boissin, 1976). Opilionid sperm transfer is direct (by copulation; not known from Cyphophthalmi, where a spermatophore is produced which is transported to the female reproductive system in an unknown way). The female reproductive system consists of ovaria, oviducts, uterus and ovipositor (with receptacula seminis). The glandular system includes the prosomatic odoriferous glands (see Juberthie, 1961), salivary glands (see Dumitrescu, 1975), and cheliceral glands (see Dumitrescu, 1974). The opisthosomatic musculature includes dorsoventral and longitudinal muscles; the last-mentioned muscles can have disappeared in the case of fusion of sclerites.

Reproduction. – Opilionid reproduction is bisexual; parthenogenesis has been described from very few species. Sperm transfer is direct, by copulation (not known from Cyphophthalmi, where a spermatophore is produced): sperm is transferred by the male penis to the female genital system, and fertilization is internal. Opilionida are oviparous; eggs are laid in a suitable place by means of an ovipositor.

Embryonic development. – The Opilionid embryonic development was studied, among others, by Winkler (1957), Moritz (1957, 1959) and Juberthie (1964: 65–75, figs. 20–25). According to Weygoldt & Paulus (1979: 114), Chelicerate embryology furnishes hardly any characters which can be used in the study of Chelicerate phylogeny. According to Anderson (1973), the base of the pattern of embryonic development in Opilionida (and Mites, Araneida, Uropygi, Amblypygi, Solifugae and Pseudoscorpionida) could lie in that of the Xiphosura, but not in that of the Scorpionida; among these Chelicerate groups, Opilionida (and Mites) present a more generalized pattern than Araneida, Uropygi, Amblypygi, Solifugae and Pseudoscorpionida (which present more modified patterns with forwardly flexed opisthosoma).

Postembryonic development. – The Opilionid postembryonic development was studied, among others, by Pabst (1953), Immel (1954, 1955), Winkler (1957) and Juberthie (1964, 1965); a general study of the evolution of the Chelicerate life-cycle was published by Van der Hammen (1978). There are three phases: A larval phase, a nymphal phase and an adult phase. The larval phase consists of one instar (a stase), viz., the larva, which is an elattostase

---

(in figs. A–C, the phanerotaxy is represented schematically; in figs. A–D, F, the aciculate microsculpture is omitted); A–C, × 43; D, F, × 246; E, × 458.

(Cyphophthalmi) or a calyptostase (Laniatores, Palpatores). Consequently, the first level of postembryonic development has been subject to regression (as in other groups of Chelicerata; the evolutionary phenomenon is called protelattosis). The nymphal phase comprises a variable number of instars (4–6, mostly 5 or 6); the number of instars can even vary within one species, but a smaller number of moults is generally associated with adults of smaller size (these adults are probably neotenous). A number of four nymphs (rarely met with, but mentioned for *Anelasmocephalus cambridgei* (Westwood) and *Lophopilio palpinalis* (Herbst) corresponds with a so-called protostasic life-cycle of six instars (larva, four nymphs and adult). A higher number of nymphal instars (mentioned for the majority of Opilionida of which the life-cycle was studied) could either refer to a secondary increase in the protostasic number of instars, or to a more primitive condition; in both cases, the nymphal instars represent nymphoids. Judging from the data in literature, the various nymphal instars are not the result of growing moults or repetition moults (they are different from the preceding instars by distinct discontinuous characters), and consequently do not represent isophena. As far as is known, the Opilionid adults (which, in hypogeous species, can live for several years) do not present growing moults.

3. Additional remarks

There are important differences in the general occurrence of the three main groups of Opilionida (Cyphophthalmi, Laniatores, Palpatores). Some fifty species of Cyphophthalmi are known, particularly from subtropical and tropical regions. Some 1500 species of Laniatores are known, particularly from America, tropical Asia, southern Africa, and Australia. The distribution of Palpatores (some 700 species) is world-wide. Opilionida are found in open places as well as in forests, under stones and in caves, in the soil (sometimes rather deep), on the soil and between low vegetation, on bushes, on the trunks of trees, and even rather high in the top of trees. Extinct species of Opilionida are known since the Carboniferous.

List of the notations of figs. 266–297

acf line of attachment of cheliceral frame to cheliceral trochanter.
am arthrodial membrane.
AP apotele.
ap.m apodeme of myliosoma.
art articulation point.
at line of attachment of cheliceral frame to cervix.

b mouth.

BF	basifemur.
BTA	basitarsus.
c, c′, c″	condyles.
CA.P	coxapophysis of palp.
CA.I	coxapophysis of leg I.
CA.II	coxapophysis of leg II.
CE	cervix.
CH, CH′	chelicerae.
CX	coxa.
CX.P	coxa of palp.
CX.I	coxa of leg I.
CX.II	coxa of leg II.
CX.III	coxa of leg III.
CX.IV	coxa of leg IV.
dgm	ducts of male postero-opisthosomatic glands.
F	femur.
gle′, gle″	glenoid cavities.
GEN	genital region.
gen	genital opening.
ibc	lyrifissure of body of chelicera.
ica.II	lyrifissures of coxapophyses of leg II.
icx, icx'', icx_1, icx_2	lyrifissures of coxa.
$if', if'', if_1, if_1'', if_2, if_2''$	lyrifissures of femur.
if_a	distal lyrifissure of femur.
ita	lyrifissure of tarsus.
itr'', itr_{1-3}'	lyrifissures of trochanter.
LI	labium.
LS	labrum.
md	dorsal muscle of penis (spermatopositor).
ml, ml′	lateral muscles of penis (spermatopositor).
mv	ventral muscle of ovipositor.
OC	eye.
od	oviduct.
OE	oesophagus.
og	eugenital opening.
ogc	orifice of coxal gland.
oglo	orifice of odoriferous gland.
ogm	orifices of male postero-opisthosomatic glands.
op.an	anal cover (operculum anale).
opg	progenital opening.
op.ge	genital cover (operculum genitale).
PA	palp.
PH	pharynx.

PN	penis (spermatopositor).
PT	patella.
rs	receptaculum seminis.
SA.III	sternapophysis of the segment bearing leg III.
sc.d	dorsal sclerite of penis.
sc.l'	lateral sclerite of penis.
st	stigma.
t''	lateral tendon or muscle.
t_i	inferior tendon of apotele.
t_s	superior tendon of apotele.
TA	tarsus.
tap	lateral tendon of myliosomatic apodeme.
tcx, *tcx'*, *tcx''*	tendons of coxa.
tcx_i	inferior tendon of coxa.
tcx_r	retractor muscle of coxa.
tcx.P	tendon of palpal coxa.
tcx.I', *tcx.I''*	tendons of coxa I.
tf', *tf''*	lateral tendons of femur.
tf_i, tf_i', tf_i''	inferior tendons of body of chelicera.
tf_s, tf_s''	superior tendons of body of chelicera.
TI	tibia.
tm	median tendon of myliosomatic apodeme.
tpe	tendon of glans penis.
tpt	tendon of patella.
TR	trochanter.
tta	tendon of tarsus.
TTA	telotarsus.
TTA.1–5	telotarsi 1–5.
tti', *tti''*	lateral tendons of tibia.
ttr_i, ttr_i''	inferior tendons of trochanter.
ttr_s, ttr_s'	superior tendons of trochanter.
x	base of median tendon of myliosomatic apodeme.
V–XVI	segments V–XVI.
3–26	telotarsi 3–26.
' (prime)	anterior face of an appendage; the right one of two symmetrical structures.
'' (double prime)	posterior face of an appendage.

XI. SCORPIONIDA

In the fourth part of my Comparative Studies in Chelicerata (Van der Hammen, 1986a: 12, 17, 21, 26, 29, 43, figs. 3B, 6A, 13, 22), I have dealt with some parts of Scorpionid morphology, viz., segmentation of the body, respiratory system, orifices of coxal glands, mouthparts, and segmentation and articulation of the legs. Previously, I had mentioned Scorpionida only in some general papers dealing with Chelicerate morphology, postembryonic development and classification (Van der Hammen, 1963b: 445; 1974: 385, 386, 387; 1977a: 312, 313, 317, tab. 1–3; 1978: 45, 49, 52, 54; 1985: 52, 55, fig. 34; 1985a: 137, 139, 141, 142, 143, 144, 145, figs. 3, 7). Scorpionida are also mentioned in my final survey of Chelicerate relationships (Van der Hammen, 1986c: 225, fig. 3). The results of these studies are summarized in the present chapter, supplemented with data from literature. The chapter is subdivided into three sections: an introduction, a descriptive part, and some additional remarks.

1. Introduction

Data with reference to the comparative morphology of Scorpionida have, for instance, been summarized by Millot & Vachon (1949) and Kaestner (1940; 1956: 540–546, figs. 538–541). Snodgrass (1948: 39–43, figs. 13–14) studied the mouthparts, Firstman (1973: 19–21, figs. 13–14) the endosternite and the circulatory system, and Couzijn (1976) the functional anatomy of the walking-legs.

The original part of the present chapter is based on a study of specimens of *Euscorpius carpathicus* (Linnaeus), collected in Yugoslavia.

2. Descriptive part

Habitus. – Scorpionida are relatively large Chelicerata of which the segmented opisthosoma is subdivided into a mesosoma and a much narrower metasoma or tail (with terminal sting). The palp is characterized by the presence of a well-developed chela.

Divisions of the body, tagmata and segmentation. – Data with reference to the constitution and segmentation of the Scorpionid body have been summarized by Kaestner (1940: 119–125, figs. 91–93) and Millot & Vachon (1949: 387–391, figs. 158–160). A species of Scorpionida is represented, in lateral view, in fig. 5B. The Scorpionid prodorsum is constituted by a single prodorsal shield. The ventral surface of the prosoma is occupied by the coxae of the appendages. The opisthosoma consists of the segments VII–XIX and a 'telson' (the sting). The sternite between the coxae of legs IV is attributed here to segment VII; the tergite of segment VII is no more recognizable (segment VII is present in the embryo). The segments VIII–XIV each present a separate tergite, segments X–XIV moreover a separate sternite; tergites and sternites are connected by a pleural region of soft skin. Ventrally, segment VIII is represented by the genital operculum which is supposed to be homologous with the opisthosomatic appendages of this segment. Segment IX is represented ventrally by a plate on which the combs or pectines (homologous with the opisthosomatic appendages of segment IX) are inserted. Segments VII–XIV constitute the so-called mesosoma, a pseudotagma. Segments XV–XIX (which are distinctly narrowed) and the 'telson' constitute the so-called metasoma, another pseudotagma; because of the absence of pleural soft skin, these segments do not present separate tergites and sternites. The Scorpionid 'telson' is represented by the sting. The anal opening has a terminal position in segment XIX, ventral to the 'telson'.

Tegument. – The tegument is for the greater part heavily sclerotized. Generally, it presents a characteristic microsculpture. Various types of phaneres (spines, ordinary setae, trichobothria) and lyrifissures can be present.

Aspidosoma. – The aspidosoma is constituted by a single prodorsal shield. Generally, it presents a pair of relatively large median eyes and, on both sides, 2–5 small lateral eyes.

Dorsal and laterodorsal region of opisthosoma. – The slender, segmented metasoma or tail is sharply set off from the segmented mesosoma which is much broader. At the end of the tail, a 'telson' is present, of which the position is more or less dorsal; it has the shape of a bulb drawn out into a curved sting (the bulb contains the paired poison glands).

Ventral and lateroventral region of opisthosoma. – The sternite of segment VII is supposed to be represented by the triangular sclerite between coxae IV. The (secondary) genital opening is constituted by a longitudinal slit flanked by the genital opercula (homologues of opisthosomatic appendages); the eugenital opening is hidden by the opercula. The paired combs or pectines (homologues of opisthosomatic appendages) are inserted on a plate (the basal piece of the pectines). Each pecten consists of marginal and median lamellae, fulcra and teeth. There are four pairs of stigmata which occupy a ventral position in the segments X–XIII (fig. 298A); they are associated with booklungs which are regarded as homologues of opisthosomatic appendages. The anus has a terminal position in segment XIX, ventral to the base of the telson.

Ventral aspect of prosoma. – The ventral surface of the prosoma is oc-

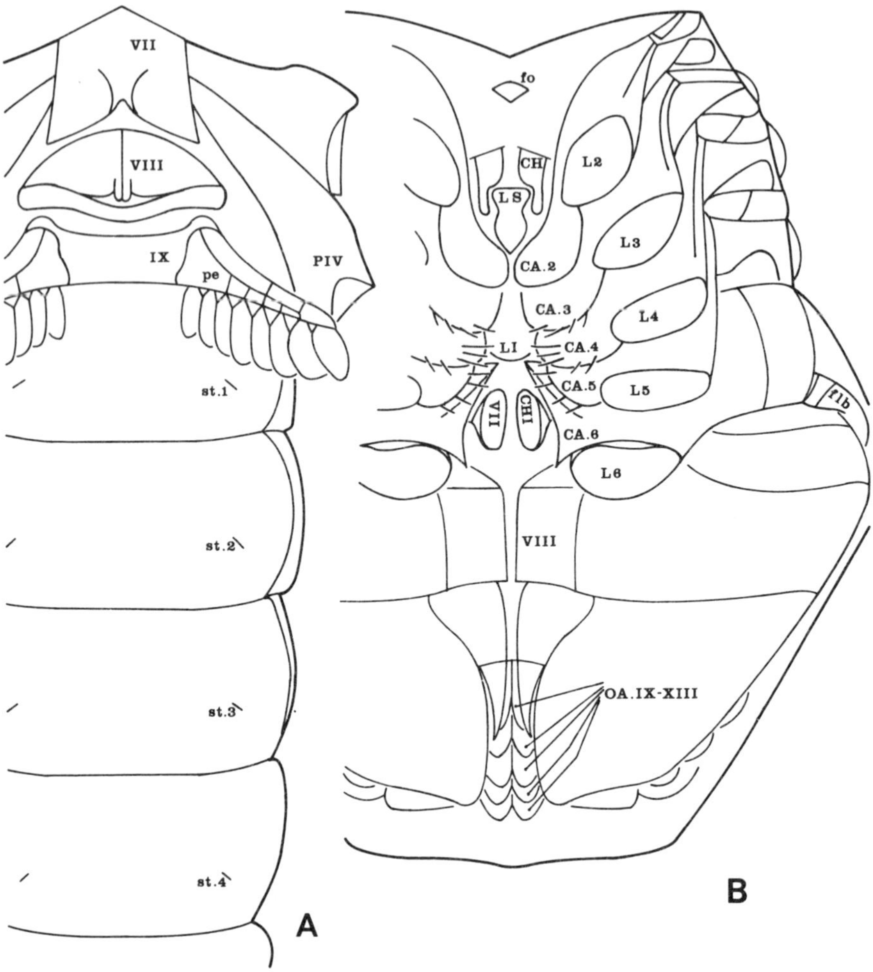

Fig. 298. Representatives of Scorpionida and Xiphosura in ventral view; A, *Euscorpius carpathicus* (Linnaeus), adult male, ventral view of posterior part of prosoma and anterior part of opisthosoma; B, *Limulus polyphemus* (Linnaeus), juvenile specimen, ventral view of prosoma and anterior part of opisthosoma; A, × 13.5; B, × 5.1.

cupied by the coxae of the legs. The Scorpionid coxal gland belongs to segment V; the orifice is paraxial, at the base of coxa III (fig. 299C). Coxae I and II are characterized by the presence of large coxapophyses.

Lateral aspect of prosoma. – This region is occupied by the antiaxial part of the coxae, and by soft skin.

Mouthparts. – The Scorpionid mouthparts are particularly characterized by the ventral position of the mouth (as a result of backward translocation) and by the participation of the coxapophyses of legs I and II in ingestion. This

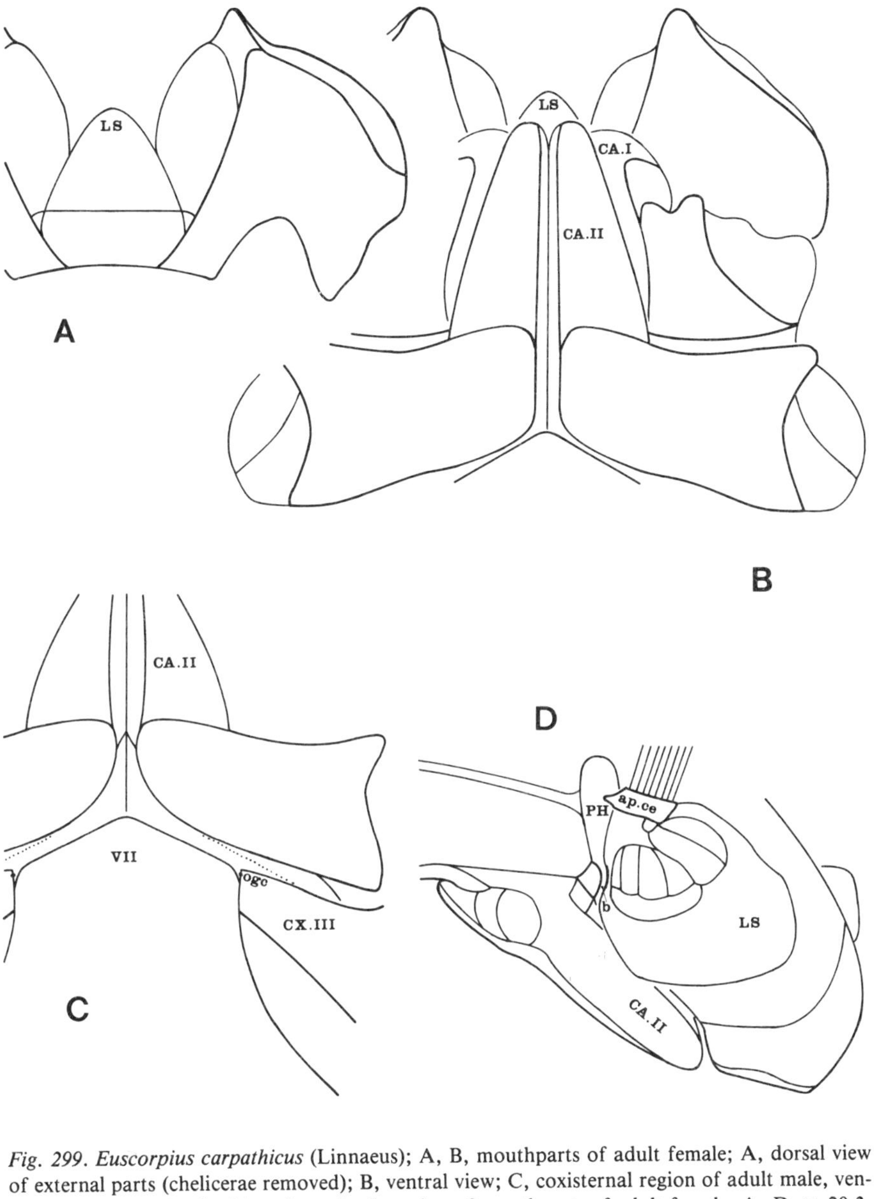

Fig. 299. Euscorpius carpathicus (Linnaeus); A, B, mouthparts of adult female; A, dorsal view of external parts (chelicerae removed); B, ventral view; C, coxisternal region of adult male, ventral view; D, longitudinal (nearly sagittal) section of mouthparts of adult female; A–D, × 20.3.

type of mouthparts (it is also found in Opilionida and Xiphosura) is called myliosoma; it is associated with coxisternal food-intake. Scorpionid mouthparts are represented in dorsal view in fig. 299A, in ventral view in fig. 299B, and in longitudinal (nearly sagittal) section in fig. 299D. The mouthparts include (besides the chelicerae): labrum, cervix, palpal coxae, labium (a prosomatic sternite) and the coxapophyses of legs I and II (there are no palpal coxapophyses). The pharynx is bent in ventral direction, in connection with the ventral position of the mouth. The orifices of the coxal glands, at the base of coxae III, could be associated with the mouthparts.

Chelicera. – The chelicerae are small and three-segmented; they consist of trochanter, body of the chelicera and apotele. A terminal chela is constituted by the apotele (movable jaw) and an extension of the body of the chelicera (the fixed jaw).

Palp. – The palp consists of six segments: coxa, trochanter, femur, patellotibia, tarsus and apotele. The distal part of the tarsus and the apotele constitute a powerful chela. The function of the palp is prehensile.

Legs. – In Scorpionida, the legs consist of coxa, trochanter, femur, patella, tibia, basitarsus, telotarsus and apotele. Leg IV of *Euscorpius carpathicus* is represented in fig. 300. With the exception of leg I, the Scorpionid coxae (fig. 300C) are not movably connected with the body (coxa I is movable because its coxapophysis is associated with food-intake); the coxae occupy the ventral surface of the prosoma, and extend laterally beyond the outline of the prosoma. The coxa-trochanter joint presents an anterior condyle (distal to the coxal apodeme) and, in the case of leg IV, also a posterodorsal arthrodial sclerite (fig. 300A, B, D); it allows of rocking movements. The trochanter-femur articulation is constituted by a bicondylar pivot joint with levator and depressor muscles. The femur-patella articulation is constituted by a bicondylar hinge-joint with inferior muscles. The patella-tibia articulation is a pivot joint with two superior condyles and flexor and extensor muscles. The tibia-basitarsus articulation is constituted by a hinge joint with two superior condyles and an inferior flexor; the inferior tendon of the apotele continues in the tibia. The articulation between basi- and telotarsus is constituted by an adesmatic joint operated by the tendons of the apotele. The articulation between telotarsus and apotele is constituted by a bicondylar pivot joint with the usual two tendons (superior and inferior). The apotele bears two ungues and a pseudonychium.

Internal anatomy. – The alimentary canal consists of mouth, pharynx, oesophagus, midgut with two pairs of prosomatic and five pairs of opisthosomatic caeca, short hindgut and anus. The excretory system includes coxal glands, Malpighian tubes and nephrocytes. The Scorpionid coxal glands belong to segment V; the orifice is at the base of coxa III. The prosomatic endosternite consists of a pair of rods which join each other posteriorly (at the place

---

→

Fig. 300. Euscorpius carpathicus (Linnaeus), right leg IV of adult male; A, lateral (posterior) view of distal part of coxa, trochanter, femur, patella, tibia, basitarsus, telotarsus and apotele; B, posterodorsal view of the coxa-trochanter articulation; C, lateral (posterior) view of coxa; D, lateral (anterior) view of the coxa-trochanter articulation; A–D, × 31.4.

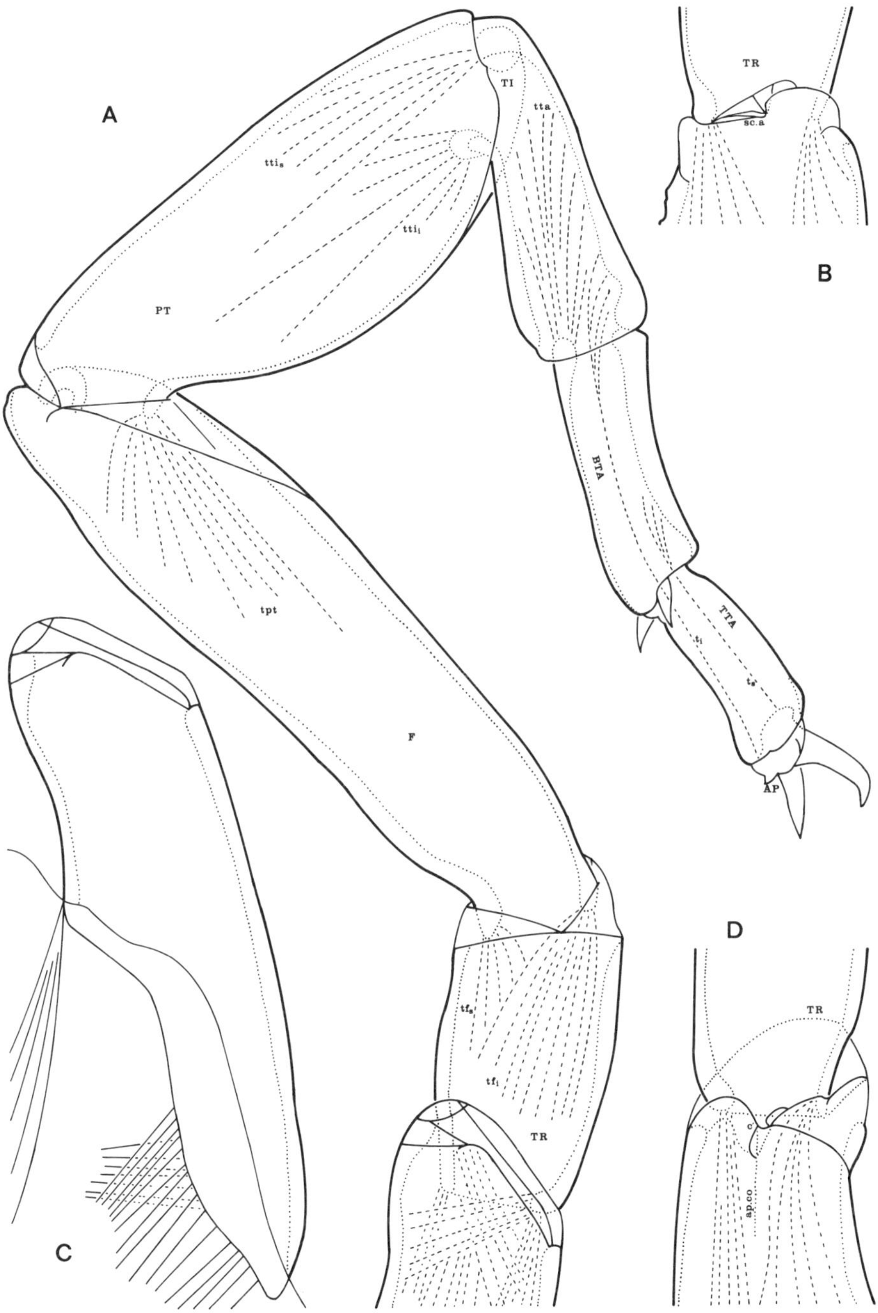
A
B
C
D
TR
sc.a
TI
tta
ttis
ttii
PT
BTA
TTA
ti
ts
AP
tpt
F
tfs
tfi
TR
TR
c
ap.co

of the diaphragm, a transverse muscular partition); there are three pairs of dorsal suspensor muscles, and three pairs of transverse suspensors. The circulatory system includes a mesosomatic heart with seven pairs of ostia, an arterial system (anterior aorta, prosomatic sinus, lateral arteries and posterior aorta) and a lacunar system. The respiratory system consists of four pairs of book-lungs and the associated four pairs of stigmata with ventral position in the segments X–XIII. The central nervous system comprises the supraoesophageal syncerebrum (which includes the cheliceral ganglia), a suboesophageal ganglionic mass (including the ganglia of the segments II–XI) and an opisthosomatic nerve-cord with the free ganglia of the segments XII–XIX (the ganglia of segments XVIII and XIX have fused); of all Chelicerata, Scorpionida present the central nervous system which is richest in free ganglia. The male reproductive system consists of the paired testes (with transverse branches), the paired vasa deferentia and vesicula seminalis, the genital atrium, the so-called paraxial organ (associated with spermatophore formation) and accessory glands. Scorpionid sperm is filiform with long flagellum. The female reproductive system comprises paired ovaria (with transverse branches), paired oviducts, a genital atrium and receptacula seminis. The glandular system includes the poison glands (associated with the sting). The opisthosomatic musculature includes dorsal and ventral longitudinal muscles, and four pairs of mesosomatic and one pair of metasomatic dorso-ventral muscles.

Reproduction. – Scorpionid reproduction is generally bisexual. Thelytokous parthenogenesis is known from a Brazilian species of *Tityus*. Sperm transfer is indirect (spermatophory) and includes courtship behaviour. Scorpionida are ovoviviparous.

Postembryonic development. – The Scorpionid postembryonic development can include an elattostasic larva, a variable number of nymphoids, an adult, and isophena of the adult (in cases of adult growing or repetition moults); neoteny can occur in the case of males.

3. Additional remarks

Some six hundred species of Scorpionida are now known, all with similar characteristic habitus. The geographic distribution is particularly tropical and subtropical, although some species have penetrated into the temporal regions. Scorpions are, generally, inhabitants of the soil; they are often found under rocks, but can also occur under bark. They are also found in houses, but rarely inhabit caves. They generally live in colonies.

Species of the extinct suborder Protoscorpionina (characterized by the presence of the tergite of segment VII) are known from the Silurian. Extinct species of the suborder Euscorpionina (in which tergite VII is absent) are known since the Devonian, whilst several families were already present in the Carboniferous.

List of the notations of figs. 298–300

AP	apotele.
ap.ce	cervical apodeme.
ap.co	coxal apodeme.
b	mouth.
BTA	basitarsus.
c'	condyle.
CA.I	coxapophysis of leg I.
CA.II	coxapophysis of leg II.
CA.2–6	coxapophyses of limbs 2–6.
CH	chelicera.
CHI	chilarium.
CX.III	coxa of leg III.
F	femur.
flb	flabellum.
fo	frontal organ.
LI	labium.
LS	labrum.
L2–6	limbs 2–6.
OA.IX–XIII	opisthosomatic appendages of segments IX–XIII.
ogc	orifice of coxal gland.
pe	comb (pecten).
PH	pharynx.
PT	patella.
P.IV	leg IV.
sc.a	arthrodial sclerite.
st.1–4	stigmata 1–4.
t_i	inferior tendon of apotele.
t_s	superior tendon of apotele.
tf_i	inferior tendons and muscles of femur.
tf_s	superior tendons and muscles of femur.
TI	tibia.
tpt	tendons and muscles of patella.
TR	trochanter.
tta	tendons and muscles of tarsus.
TTA	telotarsus.
tti_i	inferior tendons and muscles of tibia.
tti_s	superior tendons and muscles of tibia.
VII–IX	segments VII–IX.

XII. XIPHOSURA

In the fourth part of my Comparative Studies in Chelicerata (Van der Hammen, 1986a: 12–14, 17, 21, 26, 43–45, figs. 3A, 6B, 14, 23) I have also dealt with some parts of Xiphosurid morphology, viz., segmentation of the body, respiratory system, orifices of coxal glands, mouthparts, and segmentation and articulation of the legs. Previously, I had only exceptionally mentioned Xiphosura in some general papers dealing with Chelicerate morphology, postembryonic development and classification (Van der Hammen, 1977a: 310–312, 313, 314, 317, 318, tab. 1–3; 1978: 49; 1985: 52, 55, fig. 34; 1986a: 137, 139, 141, 142, 143, 144, 145, figs. 3, 7). Xiphosura are also mentioned in my final survey of Chelicerate relationships (Van der Hammen, 1986c: 225, fig. 3). The results of these studies are summarized in the present chapter, supplemented with data from literature. The chapter is subdivided into three sections: an introduction, a descriptive part, and some additional remarks.

1. Introduction

Data with reference to the comparative morphology of Xiphosura have been summarized, among others, by Gerhardt (1932–1935), Fage (1949) and Kaestner (1956: 513–523, figs. 518–521). Classical studies of Xiphosurid morphology include those by Van der Hoeven (1838), Lankester (1881), Lankester, Benham & Beck (1885) and Versluys & Demoll (1921, 1922). The mouthparts of Xiphosura were studied by Snodgrass (1948: 16–19, fig. 3A), the prosomatic appendages by Vachon (1945a) and the endosternite and the circulatory system by Firstman (1973: 29–32, figs. 22–23).

The original part of the present chapter is based on a study of *Limulus polyphemus* (Linnaeus); the material of this species, studied by me, originated from Woods Hole, Massachusetts, U.S.A.

2. Descriptive part

Habitus. – Xiphosura are large Chelicerata with indistinct segmentation and heavily sclerotized body. The prodorsal shield is large and convex, and horseshoe-shaped. The opisthosoma presents, on both sides, a lateral row of six movable spines. The 'telson' or tail is styliform and about as long as the body, with triangular transverse section.

Divisions of the body, tagmata and segmentation. – Data with reference to the constitution of the Xiphosurid body have been summarized, among others, by Gerhardt (1935: 50–53, figs. 32–33) and Fage (1949: 220–222, figs. 1–3). A species of Xiphosura is represented, in lateral view, in fig. 5A. The Xiphosurid prodorsum is constituted by a single, large prodorsal shield. The ventral surface of the prosoma is constituted by the coxae of the appendages and a narrow sternum (posterior to the mouth). The ventral region of the opisthosoma consists of: an anterior part with homologues of seven pairs of opisthosomatic appendages, viz., the chilaria (segment VII, with advanced position), the genital operculum (segment VIII) and five pairs of branchial appendages (segments IX–XIII); and a posterior part with a large 'telson' or tail. The dorsal part of segment VII is represented by a small area of soft skin. The segments VIII–XIII are dorsally recognizable by the presence of furrows (and muscle attachments) in the median part, and by the presence of six pairs of lateral spines. Because the posterior part of the opisthosoma of some extinct Xiphosura consists of three segments and a 'telson', the segments XIV–XVI of extant species are supposed to have been incorporated in the posterior part of the opisthosoma, withouth leaving any trace of segmentation (see Størmer, 1955: 14–23, figs. 11–16). The Xiphosurid anal opening is terminal, ventral to the base of the 'telson'.

Tegument. – The cuticle is heavily sclerotized and presents two large dorsal sclerites. Opisthosoma, palp and legs are characterized by the presence of some spines.

Aspidosoma. – The large, convex, horseshoe-shaped prodorsal shield presents three longitudinal ridges (separated by a pair of furrows), a pair of median eyes, and a pair of compound lateral eyes. The lateral and posterior extensions of the prodorsal shield protect the appendages.

Dorsal and laterodorsal region of opisthosoma. – The opisthosoma presents a median ridge and a pair of longitudinal furrows (in a line with those of the prodorsum). The central area of soft skin, separating the sclerotized parts of prosoma and opisthosoma, is supposed to represent the dorsal part of segment VII. Traces of opisthosomatic segmentation are represented by six pairs of small depressions (corresponding with muscle attachments) and six pairs of movable lateral spines. The styliform 'telson' or tail has a triangular transverse section and ends in a point.

Ventral region of opisthosoma (fig. 298B). – The ventral surface of the opisthosoma is characterized by the presence of homologues of seven pairs of opisthosomatic appendages. The appendages of segment VII are constituted by the so-called chilaria: short, unsegmented appendages (function unknown), provided with spines and setae. The appendages of segment VIII constitute the genital operculum; a segmented antiaxial branch and a smaller segmented

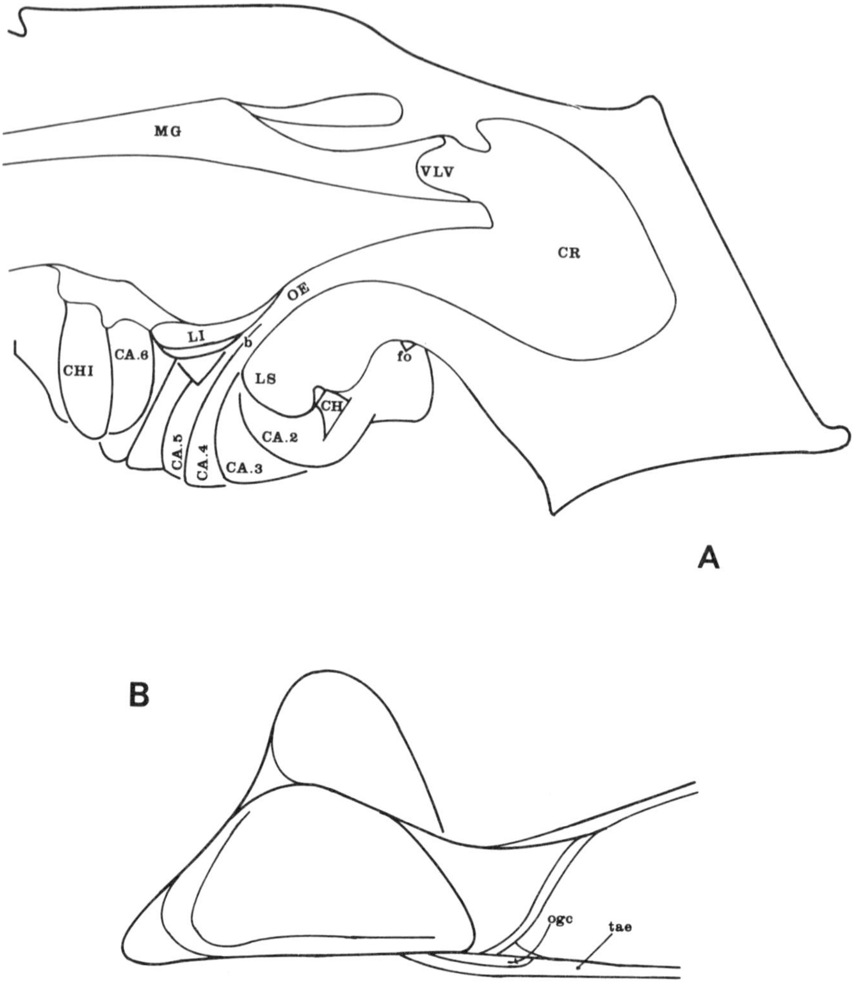

Fig. 301. Limulus polyphemus (Linnaeus), juvenile specimen; A, longitudinal (nearly sagittal) section of prosoma; B, posteroventral view of the coxal base of right limb 5, with the orifice of the coxal glands; A, × 5.1; B, × 20.3.

paraxial branch ('endite') are distinctly recognizable. The appendages of segments IX–XIII are branchial appendages; they consist also of two branches and carry the gills at the posterior face of the outer branch. The anus has a terminal position, ventral to the base of the tail.

Ventral aspect of prosoma. – The ventral surface of the prosoma is occupied by the mouthparts (dealt with below) and the coxae and coxapophyses of limbs 2–6 (dealt with in the paragraph on the limbs). A small frontal organ with a pair of eyes (functional in the nymphs, vestigial in the adult) occupies a ventral precheliceral position. The orifice of the Xiphosurid coxal gland (fig. 301B) is at the posterior border of the coxa of limb 5 (homologous with coxa III of other Chelicerata); it is associated with a taenidium. The orifice was first described (but not figured) by Tower (1895).

Mouthparts. – Xiphosura are characterized by the presence of a myliosoma, and by the participation of the coxapophyses of limbs 2–6 in food-intake (the coxapophyses of limbs 2–5 are associated with the mouth). A ventral view of the Xiphosurid mouthparts is represented in fig. 298B, a longitudinal (nearly sagittal) section in fig. 301A. Xiphosura differ from all other groups of Chelicerata by the absence of a pharynx; the foregut consists of oesophagus and crop, and is separated by a valve from the midgut (fig. 311A). The oesophagus is curved back in such a way that the labrum is ventral to the mouth, the labium dorsal; the supra- and suboesophageal ganglionic masses have been subject to a similar translocation. The chelicerae articulate with the labrum; there is no cheliceral frame (a cheliceral frame is present in all other Chelicerata). The Xiphosurid labium represents the fused prosomatic sternites.

Chelicera. – The chelicerae are short and three-segmented. They consist of trochanter, body of the chelicera, and apotele; the terminal part of the body of the chelicera (the fixed jaw) and the apotele (movable jaw) constitute a chela. The chelicerae are inserted on either side of the labrum.

Limbs 2–6. – In Xiphosura, limbs 2–6 function as ambulatory appendages (limb 1 is the chelicera, limb 2 is homologous with the palp of other Chelicerata). Limbs 2 and 6 of *Limulus polyphemus* are represented in fig. 302. Xiphosurid limbs 2–5 (see fig. 302C) consist of coxa, trochanter, femur, patella, tibiotarsus and apotele; limb 6 (see fig. 302A, B) consists of coxa, trochanter, femur, patella, tibia, tarsus and apotele. The coxae are extended in antiaxial (laterodorsal) direction and include pleural sclerites; they are also extended in paraxial direction (well-developed coxapophyses). The coxa of limb 6 presents an 'exite' (the so-called flabellum). The body-coxa articulation is associated with promotor-remotor movements (see Manton, 1977: 450, figs. 10.2a, 10.4f); this association is not known from other Chelicerata (it is supposed here that the body-coxa articulation was first associated with ingestion only, as in Opilionida and Scorpionida, and subsequently became also associated with locomotion). The coxa-trochanter articulation is constituted by a bicondylar pivot joint with levator and depressor muscles. The trochanter-femur articulation is also constituted by a bicondylar pivot joint with levator and depressor muscles; it is remarkable because the superior muscle is in the femur (its direction is opposite to the normal one). The femur-patella articulation is constituted by a hinge joint with two superior condyles and inferior flexors. In limbs 2–5, the patella-tibiotarsus articulation is constituted by a hinge joint with superior articulation, an inferior flexor and a superior extensor muscle. In limb 6, the patella-tibia articulation is similar to the patella-tibiotarsus articulation in limbs 2–5. The tibia-tarsus articulation of limb 6 is also characterized by the presence of flexor and extensor muscles. The tibiotarsus (or tarsus)-apotele

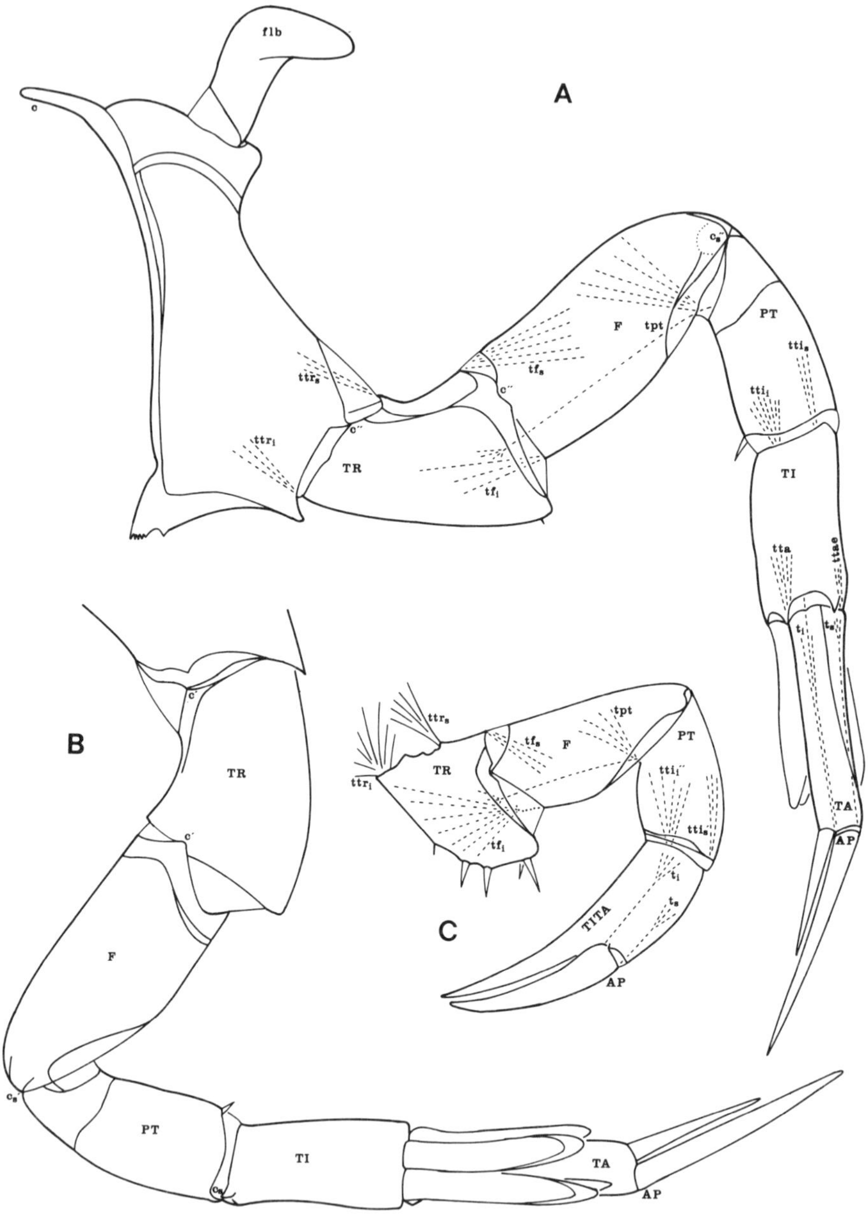

Fig. 302. Limulus polyphemus (Linnaeus), juvenile specimen; A, B, limb 6; A, lateral (posterior) view of entire limb; B, lateral (anterior) view of distal part of coxa, trochanter, femur, patella, tibia, tarsus and apotele; C, lateral (posterior) view of trochanter, femur, patella, tibiotarsus and apotele of limb 2; A–C, × 7.

articulation presents the usual two tendons (superior and inferior). It may be remarked here that, in literature, the patella is generally regarded as a patellotibia (because of the presence of a suture in limbs 3–6; as a consequence of this, tibia and tarsus of limb 6 were interpreted as basi-and telotarsus (this is, evidently, not correct because of the presence of a eudesmatic joint between these segments). Limbs 2–5 have a distinct terminal chela (constituted by the apotele and a projection of the tibiotarsus). In limb 6, a terminal spine of the tarsus is associated with the apotele. Limb 2 or limbs 2 and 3 of the Xiphosurid male present secondary sexual characters: the chela is replaced by a simple strong claw.

Internal anatomy. – The alimentary canal consists of the mouth (characterized by its backward translocation), a recurved oesophagus (there is no pharynx), a crop (separated by a valve from the midgut), the midgut with a pair of large digestive glands, a short hindgut and the anus. The excretory system is represented by a pair of coxal glands (see Tower, 1895; Patten & Hazen, 1900), originating from the fusion of the glands of segments II–V; the orifice is at the posterior border of the coxa of limb 5 (fig. 301B). The endosternite is located dorsal to the central nervous system (possibly as a result of the backward translocation of the anterior part of the prosoma); there are three pairs of dorsal suspensors, one pair of ventral suspensors and two pairs of lateral suspensors. The circulatory system comprises a heart (its posterior end is blind) with eight pairs of ostia, an arterial system (which includes an anterior aorta, dividing into three arteries, and a ventral arterial sinus), a veinous system and a lacunar system. The respiratory system is constituted by five pairs of gills carried by the opisthosomatic appendages of the segments IX–XIII. The central nervous system is enclosed in the ventral sinus; the brain (protocerebrum) lies on the same level with the suboesophageal ganglionic mass (probably as a result of the backward translocation of the anterior part of the prosoma); it is composed of eight neuromeres, the first of which corresponds with the tritocerebrum. The ventral nerve-cord presents four pairs of ganglia and a ganglionic mass (composed of three pairs). The male reproductive system comprises the originally paired testes, which constitute a network, and the paired vasa deferentia. Xiphosurid sperm is flagellate and represents the primitive type (associated with external fertilization). The female reproductive system comprises the originally paired ovaria, which constitute a network, and the paired oviducts. The opisthosomatic musculature includes six pairs of transverse muscles, attached to the opisthosomatic appendages (segments VIII–XIII); they are tendonified in the median part and constitute opisthosomatic endosternites which receive the insertion of paired dorsal suspensors.

Reproduction. – Xiphosurid reproduction is bisexual. Fertilization is external; males sit on the back of the female while she lays the eggs in a hole on the beach.

Postembryonic development. – The postembryonic development comprises calyptostasic prelarvae (which remain inside the egg; there are two prelarval moults), an elattostasic larva (the so-called trilobitic larva), a series of nymphs or nymphoids (the so-called prestwickianella larvae) and the adult. It is possible that the nymphal phase includes isophena (as a result of growing or repetition moults); it is also possible that the nymphal phase really comprises two phases (a prenymphal and a nymphal phase).

3. Additional remarks

Extant Xiphosura are classified with three genera and include four or five species. They live on the Atlantic coast of North America and on Southeast Asian coasts (from southern Japan to the Moluccas). Extinct species, supposed to be related to extant Xiphosura, are known since the Cambrium. Xiphosura, undoubtedly related to extant species, are known since the Silurian.

List of the notations of figs. 301–302

AP	apotele.
b	mouth.
c, c''	condyles.
c_s, c_s', c_s''	superior condyles.
CA.2–6	coxapophyses of limbs 2–6.
CH	chelicera.
CHI	chilarium.
CX	coxa.
F	femur.
flb	flabellum.
fo	frontal organ.
LI	labium.
LS	labrum.
MG	midgut.
OE	oesophagus.
ogc	orifice of coxal gland.
PT	patella.
t_i	inferior tendon of apotele.
t_s	superior tendon of apotele.
TA	tarsus.
tae	taenidium.
tf_i	inferior tendons and muscles of femur.
tf_s	superior tendons and muscles of femur.

TI	tibia.
TITA	tibiotarsus.
tpt	tendons and muscles of patella.
TR	trochanter.
tta	tendons and muscles of tarsus.
ttae	extensor muscles of tarsus.
tti_i	inferior tendons and muscles of tibia.
tti_s	superior tendons and muscles of tibia.
ttr_i	inferior tendons and muscles of trochanter.
ttr_s	superior tendons and muscles of trochanter.
VLV	valve (entrance to midgut).
′ (prime)	anterior (pertaining to the anterior face of an appendage).
″ (double prime)	posterior (pertaining to the posterior face of an appendage).

REFERENCES

Akimov, I.A., 1979. Morphological and functional characteristics of the mouthparts of the Acaridae mites (Acaridae Ewing et Nesbitt, 1942). – Proc. 4th Int. Congr. Acar. Saarfelden 1974: 569–574, figs. 1–6.

Alberti, G., 1979. Zur Feinstruktur der Spermien und Spermiocytogenese von Prokoenenia wheeleri (Rucker, 1901) (Palpigradi, Arachnida). – Zoomorphologie 94: 111–120, figs. 1–4.

Alberti, G., 1979a. Licht- und elektronenmikroskopische Untersuchungen an Coxaldrüsen von Walzenspinnen (Arachnida: Solifugae). – Zool. Anz. 203: 48–64, figs. 1–10.

Alberti, G., 1980. Zur Feinstruktur der Spermien und Spermiocytogenese der Milben (Acari) I. Anactinotrichida. – Zool. Jahrb. Anat. 104: 77–138, figs. 1–39.

Alberti, G., 1980a. Zur Feinstruktur der Spermien und Spermiocytogenese der Milben (Acari) II. Actinotrichida. – Zool. Jahrb. Anat. 104: 144–203, figs. 1–40.

Alberti, G., 1980b. Zur Feinstruktur des Hodenepithels und der Spermien von Eusimonia mirabilis Roewer 1934 (Solifugae, Arachnida). – Zool. Anz. 204: 345–352, figs. 1–4.

Alberti, G. & J.G. Palacios-Vargas, 1984. Fine structure of spermatogenesis and mature spermatozoa in Cryptocellus boneti Bolivar y Pieltain, 1941 (Arachnida, Ricinulei). – Journ. Ultrastruct. Res. 87: 1–12, figs. 1–23.

Alberti, G. & J.G. Palacios-Vargas, 1987. Fine structure of spermatozoa and spermatogenesis of Schizomus palaciosi, Riddell and Cokendolpher, 1986 (Arachnida: Uropygi, Schizomida). – Protoplasma 137: 1–14, figs. 1–30.

Alberti, G. & V. Storch, 1974. Über Bau und Funktion der Prosoma-Drüsen von Spinnmilben (Tetranychidae, Trombidiformes). – Zeitschr. Morph. Tiere 79: 133–153, figs. 1–13.

Alberti, G. & C. Weinmann, 1985. Fine structure of spermatozoa of some Labidognath spiders (Filistatidae, Segestriidae, Dysderidae, Oonopidae, Scytodidae, Pholcidae; Araneae; Arachnida). With remarks on spermiogenesis. – Journ. Morph. 185: 1–35, figs. 1–15.

Anderson, D.T., 1973. Embryology and phylogeny in Annelids and Arthropods. – Oxford (Pergamon Press), Int. Ser. Monogr. Pure Appl. Biol., Zool. Div. 50: xiv + 495, figs. 1–164, tab. 1–4.

André, M., 1949. Ordre des Acariens (Acari, Nitzsch, 1818). – In: P.-P. Grassé, Traité de Zoologie 6: 794–892, figs. 598–665, 2 pls.

Arthur, D.R., 1946. The feeding mechanism of Ixodes ricinus L. – Parasitology 37: 154–162, figs. 1–5.

Arthur, D.R., 1951. The capitulum and feeding mechanism of Ixodes hexagonus Leach. – Parasitology 41: 66–81, figs. 1–16, tab. 1.

Arthur, D.R., 1953. The morphology of the British Prostriata with particular reference to Ixodes hexagonus Leach. II. – Parasitology 42: 161–186, figs. 1–40, tab. 1–6.

Arthur, D.R., 1953a. The capitulum and feeding mechanism of Ixodes hexagonus Leach. II. – Parasitology 42: 187–191, figs. 1–7.

Arthur, D.R., 1956. The morphology of the British Prostriata with particular reference to Ixodes hexagonus Leach III. – Parasitology 46: 261–307, figs. 1–69, tab. 1–2.

Arthur, D.R., 1956a. The morphology of Haller's organ in Ixodes ticks. – Proc. XIV Int. Congr. Zool., Copenhagen 1953: 491–492.

Arthur, D.R., 1957. The capitulum and feeding mechanism of Dermacentor parumapertus Neumann 1901. – Parasitology 47: 169–184, figs. 1–34.
Arthur, D.R., 1962. Ticks and disease. – Oxford, etc. (Pergamon Press): xvi + 445 pp., pls. 1–7, figs. 1–95, tab. 1–15.
Athias-Henriot, C., 1969. Notes sur la morphologie externe des Gamasides (Acariens anactinotriches). – Acarologia 11: 609–625, figs. 1–60, tab. 1–4.
Athias-Henriot, C., 1969a. Les organes cuticulaires sensoriels et glandulaires des Gamasides. Poroïdotaxie et adénotaxie. – Bull. Soc. Zool. France 94: 485–492, fig. 1, tab. 1–3.
Babos, S., 1964. Die Zeckenfauna Mitteleuropas. – Budapest: 410 pp., 304 figs., 7 pls.
Baccetti, B., 1979. Ultrastructure of sperm and its bearing on Arthropod phylogeny. – In: A.P. Gupta (ed.), Arthropod Phylogeny: 609–644, figs. 11.1–11.29.
Baker, G.T. & G.W. Krantz, 1985. Structure of the male and female reproductive and digestive systems of Rhizoglyhus robini Claparède (Acari, Acaridae). – Acarologia 26: 55–65, figs. 1–23.
Barth, F.G. & J. Stagl, 1976. The slit sense organs of Arachnids. A comparative study of their topography on the walking legs (Chelicerata, Arachnida). – Zoomorphologie 86: 1–23, figs. 1–8, tab. 1.
Beier, M., 1932. 5. Ordnung der Arachnida: Pseudoscorpionidea = Afterscorpione. – In: W. Kükenthal & Th. Krumbach, Handbuch der Zoologie, Dritter Band, Zweite Hälfte, Erster Teil, Lief. 5(2): 117–192, figs. 158–267.
Bell, G., 1982. The masterpiece of nature. The evolution and genetics of sexuality. – London & Canberra (Croom Helm): 635 pp., 43 figs., 54 tab.
Berland, L., 1949. Ordre des Opilions (Opiliones, Sundeval; Phalangidae, auct.). – In: P.-P. Grassé, Traité de Zoologie 6: 761–793, figs. 553–597.
Bertram, D.S., 1939. The structure of the capitulum in Ornithodoros: A contribution to the study of the feeding mechanism in ticks. – Ann. Trop. Med. Parasitol. 33: 229–258, figs. 1–22.
Börner, C., 1901. Zur äusseren Morphologie von Koenenia mirabilis Grassi. – Zool. Anz. 24: 537–556, figs. 1–10.
Börner, C., 1902. Koenenia mirabilis und andere Pedipalpen. – Verh. Deutsch. Zool. Ges. 12: 214–215.
Börner, C., 1904. Beiträge zur Morphologie der Arthropoden. I. Ein Beitrag zur Kenntnis der Pedipalpen. – Zoologica, Stuttgart, vol. 17, Lief. 5, Heft 42: 1–174, pls. 1–7, figs. 1–114.
Brignoli, P.M., 1973. On some Ricinulei of Mexico with notes of the morphology of the female genital apparatus (Arachnida, Ricinulei). – Quad. Accad. Naz. Lincei 171: 153–174, figs. 1–3.
Camicas, J.L. & P.C. Morel, 1977. Position systématique et classification des tiques (Acarida: Ixodida). – Acarologia 18: 410–420.
Chamberlin, J.C., 1931. The Arachnid order Chelonethida. – Stanford Univ. Publ., Univ. Ser., Biol. Sci 7(1): 1–284, figs. 1–71, tab. 1–3.
Chamberlin, R.V. & S. Mulaik, 1942. On a new family in the Notostigmata. – Proc. Biol. Soc. Wash. 55: 125–132, pls. 1–2.
Clarke, J., 1984. On the relationship between structure and function in the leg joints of Heteropoda venatoria (L.) (Araneae: Eusparassidae). – Bull. Brit. Arachnol. Soc. 6(5): 181–192, figs. 1–14.
Coineau, Y., 1971. Eléments pour une monographie morphologique, écologique et biologique des Caeculidae (Acariens). – Paris (thèse), 2 vols.: 519 pp., 138 figs.
Coineau, Y., 1974. Eléments pour une monographie morphologique, écologique et biologique des Caeculidae (Acariens). – Mém. Mus. Nat. Hist. Natur. (N.S.), sér. A, Zool., 81: i–vi, 1–299, figs. 1–115, pls. 1–24.
Coineau, Y., 1974a. Nouvelles techniques pour l'étude de la morphologie des formations chitineuses des Acariens. – Acarologia 16: 4–10.
Coineau, Y., 1977. La première prélarve élattostasique connue chez les Acariens. – Acarologia 19: 46–54, figs. 1–4.
Coineau, Y., A. Fize & C. Delamare Deboutteville, 1967. Découverte en France des Acariens Nematalycidae Strenzke à l'occasion de travaux d'aménagement du Languedoc-Rousillon. – C.R. Séanc. Acad. Sci. Paris 265D: 685–688.

Coineau, Y. & L. van der Hammen, 1979. The postembryonic development of Opilioacarida, with notes on new taxa and on a general model for the evolution. – Proc. 4th Int. Congr. Acarol. Saalfelden 1974: 437–441, figs. 1–2.
Condé, B., 1965. Présence de Palpigrades dans le milieu interstitiel littoral. – C.R. Séanc. Acad. Sci. Paris 261: 1898–1900, 1 fig.
Cooke, J.A.L., 1967. Observations on the biology of Ricinulei (Arachnida) with descriptions of two new species of Cryptocellus. – Journ. Zool., Lond. 151: 31–42, pls. 1–5, figs. 1–5.
Couzijn, H.W.C., 1976. Functional anatomy of the walking-legs of Scorpionida with remarks on terminology and homologization of leg segments. – Netherl. Journ. Zool. 26: 453–501, figs. 1–10, tab. 1.
Cross, E.A., 1965. The generic relationships of the family Pyemotidae (Acarina: Trombidiformes). – Univ. Kans. Sci. Bull. 45: 29–275, figs. 1–100.
Cunliffe, F., 1955. A proposed classification of the Trombidiforme mites (Acarina). – Proc. Ent. Soc. Wash. 57: 209–218.
Dawydoff, C., 1949. Développement embryonnaire des Arachnides. – In: P.-P. Grassé, Traité de Zoologie 6: 320–385, figs. 97–157.
Dumitresco, M. & I. Juvara-Bals, 1973. Cryptocellus cubanicus n.sp. (Arachnida-Ricinulei). Premier représentant de la fam. Ricinuleidae de Cuba. – Résultats des expéditions biospéologiques Cubano-Roumaines à Cuba 1: 259–275, figs. 1–9.
Dumitresco, M. & I. Juvara-Bals, 1976. Position systématique de Heteroricinoides bordoni n.g. n.sp. dans la famille Ricinuleididae (Arachnida). – Bol. Soc. Venezol. Espel. 7(14): 147–180, figs. 1–13, tab. 1–2.
Dumitresco, M. & I. Juvara-Bals, 1977. Importance de la structure des organs copulateurs femelles dans la systématique des Ricinulei. – Trav. Inst. Spéol. 'Emile Racovitza' 16: 259–261, figs. 1–5.
Dumitresco, M. & I. Juvara-Bals, 1977a. Quelques détails de morphologie tégumentaire chez Cryptocellus cubanicus Dumitresco, Juvara. – Résultats des Expéditions Biospéologiques Cubano-Roumaines à Cuba 2: 145–146, figs. 1–2.
Dumitrescu, D., 1970. Contributions à l'étude de la morphologie des organes génitaux externes chez quelques Opilions de la famille Phalangiidae de Roumanie. – Trav. Mus. Hist. Natur. 'Gr. Antipa' 10: 79–90, pls. 1–4.
Dumitrescu, D., 1974. Les glandes chélicériennes chez les Opilions (Arachnida). – Trav. Mus. Hist. Natur. 'Gr. Antipa' 14: 109–113, pls. 1–2.
Dumitrescu, D., 1974a. Contribution à l'étude de l'appareil digestif (intestin moyen) des Troguli-dae (Arachnida, Opilionidea). – Trav. Mus. Hist. Natur. 'Gr. Antipa' 15: 57–67, pls. 1–3, tab. 1.
Dumitrescu, D., 1975. Glandes salivaires gnathocoxales chez les Opilions (Arachnida). – Trav. Mus. Hist. Natur. 'Gr. Antipa' 16: 121–126, pls. 1–2.
Dumitrescu, D., 1975a. Observations concernant l'appareil digestif (intestin moyen) des Opilions appartenant aux familles des Sironidae, Caddidae et Neopilionidae (Arachnida). – Trav. Mus. Hist. Natur. 'Gr. Antipa' 16: 115–119, figs. 1–9.
Dumitrescu, D., 1975b. Contribution à l'étude morphologique de l'appareil digestif (intestin moyen) des Opilions. – Proc. 6th Int. Arachn. Congr. 1974: 150–155, figs. 1–2.
Dumitrescu, D., 1976. Recherches morphologiques sur l'appareil digestif (intestin moyen) des Gonyleptomorphi (Arachnida, Opilionida). – Trav. Mus. Hist. Natur. 'Gr. Antipa' 17: 17–30, figs. 1–19.
Dumitrescu, D., 1980. Recherches sur la morphologie de l'appareil digestif (intestin moyen) des Opilions (Arachnida). – Trav. Mus. Hist. Natur. 'Gr. Antipa' 21: 43–50, figs. 1–13.
Edwards, M.A. & G.O. Evans, 1967. Some observations on the chaetotaxy of the legs of larval Ixodidae (Acari: Metastigmata). – Journ. Nat. Hist. 1: 595–601, figs. 1–3.
Engelmann, W., 1846. Verzeichniss der Bücher über Naturgeschichte welche in Deutschland, Scandinavien, Holland, England, Frankreich, Italien und Spanien in den Jahren 1700–1846 erschienen sind. – Leipzig (Engelmann), vol. 1: viii + 786 pp.
Evans, G.O., 1957. An introduction to the British Mesostigmata (Acarina) with keys to families

and genera. – Journ. Linn. Soc. London, Zool. 43: 203–259, figs. 1–92.

Evans, G.O. & E. Browning, 1956. British mites of the subfamily Macrochelinae Trägårdh (Gamasina-Macrochelidae). – Bull. Brit. Mus. (Nat. Hist.), Zool. 4: 1–55, figs. 1–85, pls. 1–4.

Ewing, H.E., 1929. A synopsis of the American Arachnids of the primitive order Ricinulei. – Ann. Ent. Soc. Amer. 22: 583–600, figs. 1–9.

Fage, L., 1949. Classe des Mérostomacés (Merostomata, Woodward 1866). – In: P.-P. Grassé, Traité de Zoologie 6: 219–262, figs. 1–46.

Filipponi, A. & F. Pegazzano, 1960. Acari del genere Glyptholaspis nom. nov. pro Macrocheles (Macrocheles) Berl. 1918 (Mesostigmata, Macrochelidae). – Redia 45: 133–171, figs. 1–9, pls. 3–11.

Firstman, B., 1973. The relationship of the Chelicerate arterial system to the evolution of the endosternite. – Journ. Arachnol. 1: 1–54, figs. 1–35.

Gerhardt, U., 1932–1935. Erste Klasse der Chelicerata. Merostoma. Merostomata Woodward. Synonyma: Poecilopoda McCay, Palaeostraca Grobben. – In: W. Kükenthal & Th. Krumbach, Handbuch der Zoologie, Dritter Band, Zweite Hälfte, Erster Teil. Lief. 3, 8: (1) 10–96, figs. 8–74. [Lief. 3 (pp. 10–32): 1932. Lief. 8 (pp. 33–96): 1935.]

Gerhardt, U. & A. Kaestner, 1937–1938. 8. Ordnung der Arachnida: Araneae = Echte Spinnen = Webspinnen. – In: W. Kükenthal & Th. Krumbach, Handbuch der Zoologie, Dritter Band, Zweite Hälfte, Erster Teil, Lief. 11, 12: (2) 394–656, figs. 484–854. [Lief. 11 (pp. 385–496): 1937. Lief. 12 (pp. 497–656): 1938.]

Gorirossi, F.E., 1955. The gnathosoma of the Celaenopsina (Acarina-Mesostigmata). – Amer. Midl. Natur. 54: 153–167, pls. 1–3.

Gorirossi, F. & G.W. Wharton, 1953. The anatomy of the feeding apparatus of Megisthanus floridanus Banks, 1904, a large Mesostigmatid mite. – Amer. Midl. Natur. 50: 433–447, fig. 1, pl. 1.

Grandjean, F., 1934. Observations sur les Oribates (6e série). – Bull. Mus. Nat. Hist. Natur. (2), 6: 353–360, figs. 1–3. [= Œuvres I(22).][1]

Grandjean, F., 1935. Les poils et les organes sensitifs portés par les pattes et le palpe chez les Oribates. – Bull. Soc. Zool. France 60: 6–39, figs. 1–8. [= Œuvres I(26).]

Grandjean, F., 1935a. Observations sur les Acariens (1re série). – Bull. Mus. Nat. Hist. Natur. (2), 7: 119–126, figs. 1–3. [= Œuvres I(27).]

Grandjean, F., 1936. Observations sur les Acariens (3e série). – Bull. Mus. Nat. Hist. Natur. (2), 8: 84–91, figs. 1–2. [= Œuvres I(31).]

Grandjean, F., 1936a. Le genre Pachygnathus Dugès (Alycus Koch) (Acariens). 1re partie. – Bull. Mus. Nat. Hist. Natur. (2), 8: 398–405. [= Œuvres I(36).]

Grandjean, F., 1936b. Un Acarien synthétique: Opilioacarus segmentatus With. – Bull. Soc. Hist. Natur. Afr. Nord 27: 413–444, figs. 1–5. [= Œuvres I(37).]

Grandjean, F., 1937. Le genre Pachygnathus Dugès (Alycus Koch) (Acariens). (Deuxième partie). – Bull. Mus. Nat. Hist. Natur. (2), 9: 56–61, figs. 1–3. [= Œuvres II(38).]

Grandjean, F., 1937a. Otodectes cynotis (Hering) et les prétendues trachées des Acaridiae. – Bull. Soc. Zool. France 62: 280–290, figs. 1–2. [= Œuvres II(40).].

Grandjean, F., 1937b. Sur quelques caractères des Acaridiae libres. – Bull. Soc. Zool. France 62: 388–398, figs. 1–3. [= Œuvres II(44).]

Grandjean, F., 1938. Observations sur les Bdelles (Acariens). – Ann. Soc. Ent. France 107: 1–24, figs. 1–4. [= Œuvres II(48).]

Grandjean, F., 1938a. Observations sur les Acaridiae (1re série). – Bull. Soc. Zool. France 63: 214–224, figs. 1–2. [= Œuvres II(52).]

Grandjean, F., 1938b. Observations sur les Acaridiae (2e série). – Bull. Soc. Zool. France 63: 278–288, figs. A–F. [= Œuvres II(53).]

Grandjean, F., 1939. La chaetotaxie des pattes chez les Acaridiae. – Bull. Soc. Zool. France 64: 50–60, figs. 1–2. [= Œuvres II(58).]

Grandjean, F., 1939a. Quelques genres d'Acariens appartenant au groupe des Endeostigmata. – Ann. Sci. Natur., Zool. (11), 2: 1–122, figs. 1–25. [= Œuvres II(60).]

---

[1] The numbers between square brackets are those of the Œuvres Acarologiques Complètes (Complete Acarological Works), vols. I–VII (The Hague, Lochem, 1972–1976).

Grandjean, F., 1939b. Observations sur les Oribates (12e série). – Bull. Mus. Nat. Hist. Natur. (2), 11: 300–307, figs. A–G. [= Œuvres II(61).]
Grandjean, F., 1942. La chaetotaxie comparée des pattes chez les Oribates (2e série). – Bull. Soc. Zool. France 67: 40–53, figs. 1–2. [= Œuvres II(78).]
Grandjean, F., 1943. Le développement postlarvaire d'Anystis (Acarien). – Mém. Mus. Nat. Hist. Natur. (N.S.), 18: 33–77, figs. 1–13. [= Œuvres II(87).]
Grandjean, F., 1944. Certaines trachées des Acariens ont-elles une origine néphridiale? – C.R. Séanc. Soc. Phys. Hist. Natur. Genève 61: 167–171. [= Œuvres II(93).]
Grandjean, F., 1947. Les Enarthronota (Acariens). Première série. – Ann. Sci. Natur., Zool. (11), 8: 213–248, figs. 1–6. [= Œuvres III(103).]
Grandjean, F., 1947a. Étude sur les Smarisidae et quelques autres Érythroides (Acariens). – Arch. Zool. Exp. Gén. 85: 1–126, figs. 1–11. [= Œuvres III(104).]
Grandjean, F., 1947b. L'origine de la pince mandibulaire chez les Acariens actinochitineux. – Arch. Sci. Phys. Natur. Genève (5), 29: 305–355, figs. 1–4. [= Œuvres III(113).]
Grandjean, F., 1948. Les Enarthronota (Acariens). (2e série). – Ann. Sci. Natur., Zool. (11), 10: 29–58, figs. 1–7. [= Œuvres III(119).]
Grandjean, F., 1949. Au sujet des variations individuelles et des polygones de fréquence. – C.R. Séanc. Acad. Sci. 229: 801–804 (sep. pp. 1–4). [= Œuvres IV(123).]
Grandjean, F., 1950. Observations sur les Oribates (20e série). – Bull. Mus. Nat. Hist. Natur. (2), 22: 73–80, figs. 1–2. [= Œuvres IV(126).]
Grandjean, F., 1952. Au sujet de l'ectosquelette du podosoma chez les Oribates supérieurs et de sa terminologie. – Bull. Soc. Zool. France 77: 13–36, figs. 1–2. [= Œuvres IV(140).]
Grandjean, F., 1952a. Le morcellement secondaire des tarses de Tarsolarkus sp. (Acarien). – Arch. Zool. Exp. Gén. 89, Notes et Rev. 3: 113–123, figs. A–H. [= Œuvres IV(146).]
Grandjean, F., 1953. La coalescence fémorogénuale chez Fusacarus (Acaridié, Acarien). – Bull. Mus. Nat. Hist. Natur. (2) 25: 387–394, figs. 1–2. [= Œuvres IV(151).]
Grandjean, F., 1954. Essai de classification des Oribates (Acariens). – Bull. Soc. Zool. France 78: 421–446. [= Œuvres IV(154).]
Grandjean, F., 1954a. Étude sur les Palaeacaroïdes (Acariens, Oribates). – Mém. Mus. Nat. Hist. Natur. (N.S.), sér. A, Zool., 7: 179–274, figs. 1–25. [= Œuvres IV(156).]
Grandjean, F., 1954b. Sur les nombres d'articles aux appendices des Acariens actinochitineux. – Arch. Sci., Genève 7: 335–362. [= Œuvres V(162).]
Grandjean, F., 1957. Les stases du développement ontogénétique chez Balaustium florale (Acarien, Erythroïde). Première partie. – Ann. Soc. Ent. France 125: 135–152, figs. 1–5. [= Œuvres V(177).]
Grandjean, F., 1957a. Observations sur les Oribates (37e série). – Bull. Mus. Nat. Hist. Natur. (2), 29: 88–95, figs. 1–2. [= Œuvres V(178).]
Grandjean, F., 1957b. L'infracapitulum et la manducation chez les Oribates et d'autres Acariens. – Ann. Sci. Natur., Zool. (11), 19: 233–281, figs. 1–7. [= Œuvres V(182).]
Grandjean, F., 1958. Au sujet du naso et de son œil infère chez les Oribates et les Endeostigmata (Acariens). – Bull. Mus. Nat. Hist. Natur. (2), 30: 427–435, figs. 1–2. [= Œuvres V(192).]
Grandjean, F., 1959. Observations sur les Oribates (40e série). – Bull. Mus. Nat. Hist. Natur. (2), 31: 359–366, figs. A–C. [= Œuvres VI(197).]
Grandjean, F., 1959a. Les stases du développement ontogénétique chez Balaustium florale (Acarien, Erythroïde). Deuxième partie. – Ann. Soc. Ent. France 128: 159–177, figs. 1–5. [= Œuvres VI(198).]
Grandjean, F., 1962. Au sujet des Hermanniellidae (Oribates). Première partie. – Acarologia 4: 237–273, figs. 1–11. [= Œuvres VI(209).]
Grandjean, F., 1962a. Nouvelles observations sur les Oribates (2e série). – Acarologia 4: 396–422, figs. 1–4. [= Œuvres VI(211).]
Grandjean, F., 1965. Oribates mexicains (2e série). Stelechobates megalotrichus n.g., n.sp. – Acarologia 7: 532–563, figs. 1–5. [= Œuvres VII(224).]

Grandjean, F., 1965a. Complément à mon travail de 1953 sur la classification des Oribates. – Acarologia 7: 713–734, figs. 1–4. [= Œuvres VII(225).]

Grandjean, F., 1967. Les Staurobatidae n. fam. (Oribates). – Acarologia 8: 696–727, figs. 1–8. [= Œuvres VII(230).]

Grandjean, F., 1969. Considérations sur le classement des Oribates. Leur division en 6 groupes majeurs. – Acarologia 11: 127–153. [= Œuvres VII(234).]

Grandjean, F., 1970. Stases. Actinopiline. Rappel de ma classification des Acariens en 3 groupes majeurs. Terminologie en soma. – Acarologia 11: 796–827, 1 fig. [= Œuvres VII(236).]

Grassi, B., 1885. Intorno ad un nuovo Aracnide Artrogastro (Koenenia mirabilis) che crediamo rappresentante d'un nuovo ordine (Microteliphonida). – Natural. Sicil. 4: 127–133, 162–169.

Grassi, B., 1886. I progenitori dei Miriapodi e degli Insetti. Memoria V. Intorno ad un nuovo Arachnide Artrogastro (Koenenia mirabilis) rappresentante di un nuovo ordine (Microthelyphonida. – Bull. Soc. Ent. Ital. 18: 153–172, pls. 9–10.

Griffiths, D.A., 1977. A new family of astigmatid mites from the Iles Crozet, sub-Antarctica; introducing a new concept relating to ontogenetic development of idiosomal setae. – Journ. Zool., London 182: 291–308, figs. 1–12, pls. 1–3, tab. 1–4.

Haase, E., 1890. Bemerkungen zur Palaeontologie der Insekten. – Neues Jahrb. Mineral. Geol. Palaeont. 2: 1–33, pl. 1.

Haase, E., 1890a. Beiträge zur Kenntniss der fossilen Arachniden. – Zeitschr. Deutsch. Geol. Ges. 42: 629–657, pls. 30–31.

Hammen, L. van der, 1953. Notes on the Oribatei (Acari) of Dutch New Guinea I. Allonothrus schuilingi nov. gen., nov. spec. – Proc. Kon. Ned. Akad. Wet. (C), 56: 244–250, figs. 1–3.

Hammen, L. van der, 1955. Notes on the Oribatei (Acari) of Dutch New Guinea II. A redescription of Archegozetes magna (Sellnick). – Proc. Kon. Ned. Akad. Wet. (C), 58: 90–97, figs. 1–4.

Hammen, L. van der, 1955a. Notes on the Oribatei (Acari) of Dutch New Guinea III. The development of Archegozetes magna (Sellnick) and Allonothrus schuilingi (Van der Hammen). – Proc. Kon. Ned. Akad. Wet. (C), 58: 195–205, figs. 1–8.

Hammen, L. van der, 1956. Notes on the Oribatei (Acari) of Dutch New Guinea IV. Description of Trhypochthonius montanus nov. spec., and preliminary revision of the genus Trhypochthonius. – Proc. Kon. Ned. Akad. Wet. (C), 59: 398–404, figs. 1–2.

Hammen, L. van der, 1960. Fortuynia marina nov. gen., nov. spec., an Oribatid mite from the intertidal zone in Netherlands New Guinea. – Zool. Meded. Leiden 37: 1–9, figs. 1–5.

Hammen, L. van der, 1961. Description of Holothyrus grandjeani nov. spec., and notes on the classification of the mites. – Nova Guinea, Zool., 9: 173–194, figs. 1–9, pl. 6.

Hammen, L. van der, 1963. Description of Fortuynia yunkeri nov. spec., and notes on the Fortuyniidae nov. fam. (Acarida, Oribatei). – Acarologia 5: 152–167, figs. 1–7.

Hammen, L. van der, 1963a. The Oribatid family Phthiracaridae I. Introduction and redescription of Hoplophthiracarus pavidus (Berlese). – Acarologia 5: 306–317, figs. 1–8.

Hammen, L. van der, 1963b. The addition of segments during the post-embryonic ontogenesis of the Actinotrichida (Acarida) and its importance for the recognition of the primary subdivision of the body and the original segmentation. – Acarologia 5: 443–454.

Hammen, L. van der, 1963c. The Oribatid family Phthiracaridae II. Redescription of Phthiracarus laevigatus (C.L. Koch). – Acarologia 5: 704–715, figs. 1–6.

Hammen, L. van der, 1964. The morphology of the palp in two families of ticks (Acarida: Ixodina). A contribution to the study of the Anactinotrichida. – Zool. Meded. Leiden 39: 147–152, figs. 1–3.

Hammen, L. van der, 1964a. The morphology of Glyptholaspis confusa (Foà, 1900) (Acarida, Gamasina). – Zool. Verh. Leiden 71: 1–56, figs. 1–21.

Hammen, L. van der, 1964b. The Oribatid family Phthiracaridae III. Redescription of Phthiracarus nitens (Nicolet). – Acarologia 6: 400–411, figs. 1–6.

Hammen, L. van der, 1965. Further notes on the Holothyrina (Acarida) I. Supplementary description of Holothyrus coccinella Gervais. – Zool. Meded. Leiden 40: 253–276, figs. 1–9.

Hammen, L. van der, 1965a. The Oribatid family Phthiracaridae IV. The leg chaetotaxy of Phthiracarus anonymum Grandjean. – Acarologia 7: 376–381, figs. 1–2.

Hammen, L. van der, 1966. Studies on Opilioacarida (Arachnida) I. Description of Opilioacarus texanus (Chamberlin & Mulaik) and revised classification of the genera. – Zool. Verh. Leiden 86: 1–80, figs. 1–21.
Hammen, L. van der, 1968. The gnathosoma of Hermannia convexa (C.L. Koch) (Acarida: Oribatina) and comparative remarks on its morphology in other mites. – Zool. Verh. Leiden 94: 1–45, figs. 1–12.
Hammen, L. van der, 1968a. Stray notes on Acarida (Arachnida) I. – Zool. Meded. Leiden 42: 261–280, figs. 1–3.
Hammen, L. van der, 1968b. Studies of Opilioacarida (Arachnida) II. – Redescription of Paracarus hexophthalmus (Redikorzev). – Zool. Meded. Leiden 43: 57–76, figs. 1–4.
Hammen, L. van der, 1968c. Introduction générale à la classification, la terminologie morphologique, l'ontogénèse et l'évolution des Acariens. – Acarologia 10: 401–412.
Hammen, L. van der, 1969. Notes on the morphology of Alycus roseus C.L. Koch. – Zool. Meded. Leiden 43: 177–202, figs. 1–7.
Hammen, L. van der, 1969a. Notes on the mouthparts of Eukoenenia mirabilis (Grassi) (Arachnidea: Palpigradida). – Zool. Meded. Leiden 44: 41–45, fig. 1.
Hammen, L. van der, 1969b. Studies on Opilioacarida (Arachnida) III. Opilioacarus platensis Silvestri, and Adenacarus arabicus (With). – Zool. Meded. Leiden 44: 113–131, figs. 1–5.
Hammen, L. van der, 1970. Tarsonemoides limbatus nov. spec., and the systematic position of the Tarsonemida (Acarida). – Zool. Verh. Leiden 108: 1–35, figs. 1–10.
Hammen, L. van der, 1970a. La segmentation primitive des Acariens. – Acarologia 12: 3–10, fig. 1, tab. 1.
Hammen, L. van der, 1970b. La segmentation des appendices chez les Acariens. – Acarologia 12: 11–15, fig. 1, tab. 1–2.
Hammen, L. van der, 1970c. Remarques générales sur la structure fondamentale du gnathosoma. – Acarologia 12: 16–22, fig. 1.
Hammen, L. van der, 1971. La phylogénèse des Opilioacarides, et leurs affinités avec les autres Acariens. – Acarologia 12: 465–473, figs. 1–5.
Hammen, L. van der, 1972. Spinachtigen – Arachnida IV. Mijten – Acarida. Algemene inleiding in de acarologie. – Wetensch. Meded. K.N.N.V. 91: 1–71, figs. 1–37, tab. 1–5.
Hammen, L. van der, 1972a. A revised classification of the mites (Arachnidea, Acarida) with diagnoses, a key, and notes on phylogeny. – Zool. Meded. Leiden 47: 273–292, fig. 1.
Hammen, L. van der, 1973. Classification and phylogeny of mites. – Proc. 3rd Int. Congr. Acarol. Prague 1971: 275–282, fig. 1.
Hammen, L. van der, 1974. L'évolution du cycle vital chez les Acariens et les autres groupes d'Arachnides. – Acarologia 15: 384–390.
Hammen, L. van der, 1975. L'évolution des Acariens, et les modèles de l'évolution des Arachnides. – Acarologia 16: 377–381, figs. 1–2.
Hammen, L. van der, 1976. Glossaire de la terminologie acarologique 2. Opilioacarida. – The Hague (Dr. W. Junk bv Publishers): viii + 137 pp., tab. 1–3, figs. 1–31, pls. 1–5.
Hammen, L. van der, 1977. Studies on Opilioacarida (Arachnidea) IV. The genera Panchaetes Naudo and Salfacarus nov. gen. – Zool. Meded. Leiden 51: 43–78, figs. 1–17.
Hammen, L. van der, 1977a. A new classification of Chelicerata. – Zool. Meded. Leiden 51: 307–319, fig. 1, tab. 1–3.
Hammen, L. van der, 1977b. The evolution of the coxa in mites and other groups of Chelicerata. – Acarologia 19: 12–19, figs. 1–2.
Hammen, L. van der, 1978. The evolution of the Chelicerate life-cycle. – Acta Biotheoretica 27: 44–60, figs. 1–6.
Hammen, L. van der, 1979. Evolution in mites, and the patterns of evolution in Arachnidea. – Proc. 4th Int. Congr. Acarol. Saalfelden 1974: 425–430, figs. 1–2.
Hammen, L. van der, 1979a. Comparative studies in Chelicerata I. The Cryptognomae (Ricinulei), Architarbi and Anactinotrichida). – Zool. Verh. Leiden 174: 1–62, figs. 1–31.
Hammen, L. van der, 1980. Glossary of acarological terminology 1. General terminology. – The Hague (Dr. W. Junk bv Publishers): viii + 244 pp.

Hammen, L. van der, 1981. Type-concept, higher classification and evolution. – Acta Biotheoretica 30: 3–48, figs. 1–7.
Hammen, L. van der, 1981a. Numerical changes and evolution in Actinotrichid mites (Chelicerata). – Zool. Verh. Leiden 182: 1–47, figs. 1–13.
Hammen, L. van der, 1982. Morphology and postembryonic development of Rhizoglyphus echinopus (Fumouze & Robin) (Chelicerata, Actinotrichida). – Zool. Meded. Leiden 56: 237–258, figs. 1–12.
Hammen, L. van der, 1982a. Comparative studies in Chelicerata II. Epimerata (Palpigradi and Actinotrichida). – Zool. Verh. Leiden 196: 1–70, figs. 1–31.
Hammen, L. van der, 1983. Unfoldment and manifestation: The natural philosophy of evolution. – Acta Biotheoretica 32: 179–193.
Hammen, L. van der, 1983a. Notes on the comparative morphology of ticks (Anactinotrichida: Ixodida). – Zool. Meded. Leiden 57: 209–242, figs. 1–23.
Hammen, L. van der, 1983b. New notes on Holothyrida (Anactinotrichid mites). – Zool. Verh. Leiden 207: 1–48, figs. 1–28.
Hammen, L. van der, 1985. Comparative studies in Chelicerata III. Opilionida. – Zool. Verh. Leiden 220: 1–60, figs. 1–34.
Hammen, L. van der, 1985a. Functional morphology and affinities of extant Chelicerata in evolutionary perspective. – Trans. Roy. Soc. Edinburgh, Earth Sciences 76: 137–146, figs. 1–8.
Hammen, L. van der, 1985b. A structuralist approach in the study of evolution and classification. –Zool. Meded. Leiden 59: 391–409, figs. 1–5.
Hammen, L. van der, 1986. On some aspects of parallel evolution in Chelicerata. – Acta Biotheoretica 35: 15–37, figs. 1–9.
Hammen, L. van der, 1986a. Comparative studies in Chelicerata IV. Apatellata, Arachnida, Scorpionida, Xiphosura. – Zool. Verh. Leiden 226: 1–52, figs. 1–23.
Hammen, L. van der, 1986b. The morphology of Epilohmannia zwarti spec. nov., an Oribatid mite from New Guinea. – Zool. Meded. Leiden 60: 71–85, figs. 1–5.
Hammen, L. van der, 1986c. Acarological and arachnological notes. – Zool. Meded. Leiden 60: 217–230, figs. 1–3.
Hammen, L. van der, 1986d. Some notes on taxonomic methodology. – Zool. Meded. Leiden 60: 231–256.
Hammen, L. van der, 1987. Structuralism in evolutionary biology and systematics. – Riv. di Biol., Biol. Forum 80: 216–220, 304–307, 377–408, figs. 1–7.
Hammen, L. van der, 1988. Unfoldment and manifestation. Seven essays on evolution and classification. – The Hague (SPB Academic Publishing): xi + 181 pp., 33 figs.
Hansen, H.J., 1901. On six species of Koenenia, with remarks on the order Palpigradi. – Entom. Tidskr. 22: 193–240, pls. 2–4.
Hansen, H.J., 1926. Palpigradi (Deuxième série). – Arch. Zool. Exp. Gén. 65: 167–180, pls. 2–3.
Hansen, H.J., 1930. Studies on Arthropoda 3. – Copenhagen: 376 pp., pls. 1–16.
Hansen, H.J. & W. Sørensen, 1897. The order Palpigradi Thor. (Koenenia mirabilis Grassi) and its relationship to the other Arachnida. – Entom. Tidskr. 18: 223–240, pl. 4.
Hansen, H.J. & W. Sørensen, 1904. On two orders of Arachnida. Opiliones, especially the suborder Cyphophthalmi, and Ricinulei, namely the family Cryptostemmatoidae. – Cambridge: xii + 182, pls. 1–9.
Heymons, R., 1901. Die Entwicklungsgeschichte der Scolopender. – Zoologica 13(33), Lief. 2–5: 244 pp., 8 pls.
Hirschmann, W., 1959. Gangsystematik der Parasitiformes. Teil 2. Mundwerkzeuge und Hypostombestimmungstafeln. – Acarologie, Schriftenr. Vergl. Milbenk. 2: 1–23, i–ii, figs. 1–91.
Hirst, S., 1923. On some Arachnid remains from the Old Red Sandstone (Rhynie Chert Bed, Aberdeenshire). – Ann. Mag. Nat. Hist. (9), 12: 455–474, figs. 1–13, pls. 11–15.
Hoeven, J. van der, 1838. Recherches sur l'histoire naturelle et l'anatomie des Limules. – Leiden: viii + 48 pp., pls. 1–7.
Hughes, A.M., 1948. The mites associated with stored food products. – Min. Agr. Fish. London: iv + 168 pp., figs. 1–250.

Hughes, A.M., 1961. The mites of stored food. – Min. Agr. Fish. Food, Techn. Bull. 9: vi + 287, figs. 1–385.
Hughes, T.E., 1949. The functional morphology of the mouthparts of Liponyssus bacoti. – Ann. Trop. Med. Parasitol. 43: 349–360, figs. 1–12, tab.
Hughes, T.E., 1959. Mites, or the Acari. – London: viii + 225 pp., pls. 1–52.
Immel, V., 1954. Zur Biologie und Physiologie von Nemastoma quadripunctatum (Opiliones, Dyspnoi). – Zool. Jahrb. Syst. 83: 129–184, figs. 1–47.
Immel, V., 1955. Einige Bemerkungen zur Biologie von Platybunus bucephalus (Opiliones, Eupnoi). – Zool. Jahrb. Syst. 83: 475–484, figs. 1–4.
Janczyk, F.S.W., 1956. Anatomie von Siro duricorius Joseph im Vergleich mit anderen Opilioniden. – S.B. Österr. Akad. Wiss., Math.-naturw. Kl., Abt. I, 165: 475–522, figs. 1–28.
Juberthie, C., 1961, Structure et fonction des glandes odorantes chez quelques Opilions (Arachnida). – Verh. Deutsch. Zool. Ges. Saarbrücken 1961: 533–537, figs. 1–4.
Juberthie, C., 1964. Recherches sur la biologie des Opilions. – Ann. Spéléol. 19: 5–238, pls. 1–4, figs. 1–75, tab. 1–25.
Juberthie, C., 1965. Données sur l'écologie, le développement et la reproduction des Opilions. – Rev. Ecol. Biol. Sol 2: 377–396, figs. 1–15, tab. 1–2.
Juberthie, C., 1967. Documents faunistiques et écologiques. Siro rubens (Opilion, Cyphophthalme). – Rev. Écol. Biol. Sol 4: 155–171, figs. 1–8.
Juberthie, C., 1967a. Caractères sexuels secondaires des Opilions: les glandes anales de Siro rubens. – Rev. Ecol. Biol. Sol 4: 489–496, figs. 1–4.
Juberthie, C., 1970. Sur Suzukielus sauteri (Roewer, 1916) Opilion Cyphophthalme du Japon. – Rev. Ecol. Biol. Sol 7: 563–569, figs. 1–4.
Juberthie, C., J.-F. Manier & L. Boissin, 1976. Étude ultrastructurale de la double spermiogenèse chez l'opilion cyphophthalme Siro rubens Latreille. – Journ. Microsc. Biol. Cell. 25: 137–148, figs. 1–7.
Kaestner, A., 1932. 2. Ordnung der Arachnida: Pedipalpi Latreille = Geissel-Scorpione. – In: W. Kükenthal & Th. Krumbach, Handbuch der Zoologie, Dritter Band, Zweite Hälfte, Erster Teil, Lief. 4: 1–76, figs. 1–94.
Kaestner, A., 1932a. 3. Ordnung der Arachnida: Palpigradi Thorell. – In: W. Kükenthal & Th. Krumbach, Handbuch der Zoologie, Dritter Band, Zweite Hälfte, Erster Teil: (2) 77–98, figs. 95–129.
Kaestner, A., 1932b. 4. Ordnung der Arachnida: Ricinulei Thorell. – In: W. Kükenthal & Th. Krumbach, Handbuch der Zoologie, Dritter Band, Zweite Hälfte, Erster Teil, Lief. 5: (2) 99–116, figs. 130–156.
Kaestner, A., 1933. Verdauungs- und Atemorgane der Weberknechte Opilio parietinus De Geer und Phalangium opilio L. – Zeitschr. Morph. Ökol. Tiere 27: 587–623, figs. 1–27.
Kaestner, A., 1933a. 6. Ordnung der Arachnida: Solifugae Sundevall. Walzenspinnen. – In: W. Kükenthal & Th. Krumbach, Handbuch der Zoologie, Dritter Band, Zweite Hälfte, Erster Teil, Lief. 7: (2) 193–209, figs. 268–362.
Kaestner, A., 1935–1937. 7. Ordnung der Arachnida: Opiliones Sundevall = Weberknechte. – In: W. Kükenthal & Th. Krumbach, Handbuch der Zoologie, Dritter Band, Zweite Hälfte, Erster Teil: (2) 300–384, figs. 363–476 (1935), 385–393, figs. 477–483 (1937).
Kaestner, A., 1940. 1. Ordnung der Arachnida: Scorpiones. – In: W. Kükenthal & Th. Krumbach, Handbuch der Zoologie, Dritter Band, Zweite Hälfte, Erster Teil, Lief. 14: (1) 117–240, figs. 88–222.
Kaestner, A., 1955. Spinnentiere (Trilobitomorphen, Fühlerlose) und Tausendfüssler. – Das Tierreich (Sammlung Göschen) 4(2): 1–96, figs. 1–55.
Kaestner, A., 1956. Lehrbuch der Speziellen Zoologie 1(3): 485–658, figs. 492–629. – Jena.
Kennaugh, J.H., 1968. An examination of the cuticle of three species of Ricinulei (Arachnida). – Journ. Zool., Londen 156: 393–404, pls. 1–6, fig. 1.
Knülle, W., 1958. Die Mehlmilbe (Acarus siro L.) und ihre Entwicklungsstadien. – Zeitschr. Angew. Ent. 43: 439–444, figs. 1–8.
Knülle, W., 1959. Morphologische und entwicklungsgeschichtliche Untersuchungen zum phyloge-

netischen System der Acari: Acariformes Zachv. II. Acaridiae: Acaridae. Grundlagen der Getreidemilben-Systematik. – Mitt. Zool. Mus. Berlin 55: 347–417, figs. 1–47.

Kramer, P., 1882. Ueber die Segmentierung bei den Milben. – Arch. Naturg. 48(1): 178–182, pl. 13, figs. 1–4.

Krantz, G.W., 1962. A review of the genera of the family Macrochelidae Vitzthum 1930 (Acarina: Mesostigmata). – Acarologia 4: 143–173, figs. 1–37.

Krantz, G.W., 1978. A manual of acarology. – Corvallis (second edition): viii + 509 pp., pls. 1–163.

Lamarck, J.B., 1801. Système des animaux sans vertèbres. – Paris.

Lankester, E.R., 1881. Limulus an Arachnid. – Quart. Journ. Micr. Sci. (N.S.) 21: 504–548, 609–649, figs. 1–20.

Lankester, E.R., 1902. Arthopoda. – Encyclopaedia Britannica (tenth edition) 25: 689–701, 11 figs.

Lankester, E.R., 1904. The structure and classification of the Arthropoda. – Quart. Journ. Micr. Sci. (N.S.) 47: 523–582, pl. 42.

Lankester, E.R., W.B.S. Benham & E.J. Beck, 1885. On the muscular and endoskeletal systems of Limulus and Scorpio; with some notes on the anatomy and generic characters of scorpions. – Trans. Zool. Soc. London 11: 311–384, pls. 72–83.

Latreille, P., 1829. Les Crustacés, les Arachnides et les Insectes. – Paris (Ballière). [= G. Cuvier, Le règne animal (second edition), vols. 4 and 5.]

Lawrence, R.F., 1947. Some observations on the eggs and newly hatched embryos of Solpuga hostilis White (Arachnida). – Proc. Zool. Soc. London 117: 429–434, figs. 1–2.

Legendre, R., 1958. Contributions à l'étude du système nerveux des Aranéides. – Année Biol. (3), 34: 193–223, figs. 1–11.

Legendre, R., 1959. Contribution à l'étude du système nerveux des Aranéides. – Ann. Sci. Nat., Zool. (12), 1: 339–474, pls. 1–4, figs. 1–51.

Legendre, R., 1967. Morphologie und Entwicklungsgeschichte der Cheliceraten. Entwicklung, Anatomie und Histologie der Milben (Acari). – Fortschr. Zool. 18: 207–222, figs. 1–9.

Legendre, R., 1968. La nomenclature anatomique chez les acariens. – Acarologia 10: 413–417.

Legg, G., 1973. Spermatophore formation in the pseudoscorpion Chthonius ischnocheles (Chthoniidae). – Journ. Zool., London 170: 367–394, figs. 1–11, tab. 1.

Legg, G., 1975. The possible significance of spermathecae in pseudoscorpions (Arachnida). – Bull. Brit. Arachn. Soc. 3(4): 91–95, figs. 1–12.

Legg, G., 1976. The external morphology of a new species of ricinuleid (Arachnida) from Sierra Leone. – Zool. Journ. Linn. Soc. 59: 1–58, pls. 1–6, figs. 1–43.

Lehtinen, P., 1981. New Holothyrina (Arachnida, Anactinotrichida) from New Guinea and South America. – Acarologia 22: 3–13, tab. 1, figs. 1–29.

Lindquist, E.E., 1976. Transfer of the Tarsocheylidae to the Heterostigmata, and reassignment of the Tarsonemina and Heterostigmata to lower hierarchic status in the Prostigmata (Acari). – Canad. Entomol. 108: 23–48, figs. 1–18, tab. 1–2.

Lindquist, E.E., 1977. Homology of dorsal opisthosomal plates, setae, and cupules of Heterostigmatic mites with those of other Eleutherengone Prostigmata (Acari). – Acarologia 19: 97–104, figs. 1–10.

Lindquist, E.E., 1984. Current theories on the evolution of major groups of Acari and on their relationships with other groups of Arachnida, with consequent implications for their classification. – In: D.A. Griffiths & C.E. Bowman, Acarology 6, vol. 1: 28–62.

Lindquist, E.E., 1985. Diagnosis and phylogenetic relationships. – In: W. Helle & M.W. Sabelis, Spider Mites. Their Biology, Natural Enemies and Control, vol. 1A: 63–74, fig. 1.1.3.1.

Luxton, M., 1967. Two new mites from the rocky shore of North Auckland. With a note on zoogeographical relations of the littoral Cryptostigmata (Acari) of New Zealand. – New Zealand Journ. Mar. Freshw. Res. 1: 76–87, figs. 1–5.

Luxton, M., 1986. A new species of Fortuynia (Acari: Cryptostigmata) from the marine littoral of Kenya. – Journ. Nat. Hist. 20: 65–69, figs. 1–2.

Manton, S.M., 1977. The Arthropoda. Habits, functional morphology, and evolution. – Oxford (Clarendon Press): xxii + 527 pp., 186 figs., 8 pls.

Martens, J., 1976. Genitalmorphologie, System und Phylogenie der Weberknechte (Arachnida: Opiliones). – Ent. Germ. 3: 51–68, figs. 1–46, tab. 1–2.
Martens, J., 1978. Spinnentiere, Arachnida. Weberknechte, Opiliones. – Die Tierwelt Deutschlands 64: 464 pp., 815 figs.
Martens, J., U. Hoheisel & M. Götze, 1981. Vergleichende Anatomie der Legeröhren der Opiliones als Beitrag zur Phylogenie der Ordnung (Arachnida). – Zool. Jahrb., Anat. 105: 13–76, figs. 1–44.
Millot, J., 1942. Sur l'anatomie et l'histophysiologie de Koenenia mirabilis Grassi (Arachnida Palpigradi). – Rev. Franç. Ent. 9: 33–51, figs. 1–10.
Millot, J., 1943. Notes complémentaires sur l'anatomie, l'histologie et la répartition géographique en France de Koenenia mirabilis Grassi (Arachnida Palpigradi). – Rev. Franç. Ent. 9: 127–135, figs. 1–6.
Millot, J., 1945. La constitution de l'abdomen des Ricinulei (Arachnides). – Bull. Soc. Ent. France 50: 72–74, figs. 1–3.
Millot, J., 1946. Les Ricinulei ne sont pas des Arachnides primitifs. – Bull. Soc. Zool. France 70: 106–108.
Millot, J., 1947. L'anatomie interne des Ricinulei (Arachnides). – Ann. Sci. Natur., Zool. (11), 7: 1–29, figs. 1–34.
Millot, J., 1949. Classe des Arachnides (Arachnida) I. – Morphologie générale et anatomie interne. – In: P.-P. Grassé, Traité de Zoologie 6: 263–319, figs. 47–96.
Millot, J., 1949a. Ordre des Palpigrades. Palpigradi, Thorell, 1800 [= 1888], Microtelyphonida, Grassi et Calandruccio, 1885. – In: P.-P. Grassé, Traité de Zoologie 6: 520–532, figs. 274–286.
Millot, J., 1949b. Ordre des Uropyges (Uropygi Thorell). – In: P.-P. Grassé, Traité de Zoologie 6: 533–562, figs. 287–317.
Millot, J., 1949c. Ordre des Amblypyges. Amblypygi, Thorell, 1900. –In: P.-P. Grassé, Traité de Zoologie 6: 563–588, figs. 318–348.
Millot, J., 1949d. Ordre des Aranéides (Araneae). – In: P.-P. Grassé, Traité de Zoologie 6: 589–743, figs. 349–528.
Millot, J., 1949e. Ordre des Ricinuléides (Ricinulei Thorell, 1892, Podogonata Cook, 1800 [= 1899]). – In: P.-P. Grassé, Traité de Zoologie 6: 744–760, figs. 529–552.
Millot, J. & M. Vachon, 1949. Ordre des Scorpions. Scorpionides Latreille, 1810. Scorpiones Hemprich et Ehrenberg, 1810. Scorpiides C.L. Koch, 1837. – In: P.-P. Grassé, Traité de Zoologie 6: 386–436, figs. 158–195.
Millot, J. & M. Vachon, 1949a. Ordre des Solifuges. Solifugae Sundevall 1833 – Solpugae C.L. Koch 1842 – Solpugides Gervais 1844. – In: P.-P. Grassé, Traité de Zoologie 6: 482–519, figs. 242–273.
Monniot, F., 1966. Un Palpigrade interstitiel: Leptokoenenia scurra n.sp. – Rev. Ecol. Biol. Sol 3: 41–64, figs. 1–6.
Monniot, F., 1970. Quelques caractères morphologiques de Eukoenenia mirabilis (Arachnide, Palpigrade) observé au stereoscan. – Rev. Ecol. Biol. Sol 7: 559–562, figs. A–G.
Moritz, M., 1957. Zur Embryonalentwicklung der Phalangiiden (Opiliones, Palpatores) unter besonderer Berücksichtigung der äusseren Morphologie, der Bildung des Mitteldarmes und der Genitalanlage. – Zool. Jahrb., Anat. 76: 331–370, figs. 1–30.
Moritz, M., 1959. Zur Embryonalentwicklung der Phalangiiden (Opiliones, Palpatores) II. Die Anlage und Entwicklung der Coxaldrüse bei Phalangium opilio L. – Zool. Jahrb., Anat. 77: 229–240, figs. 1–8.
Muma, M.H., 1966. The life cycle of Eremobates durangonus (Arachnida: Solpugida). – Florida Ent. 49: 233–242, figs. 1–6, tab. 1–2.
Nathanson, M.E., 1967. Comparative fine structure of sclerotized and unsclerotized integument of the rabbit tick, Haemaphysalis leporispalustris (Acari: Ixodides: Ixodidae). – Ann. Ent. Soc. Am. 60: 1125–1135, figs. 1–15.
Nordenskiöld, E., 1908. Zur Anatomie und Histologie von Ixodes reduvius. – Zool. Jahrb., Anat. 25: 637–674, figs. A–B, pls. 26–28.
Norton, R.A., 1973. Macrocheles ontariensis n.sp., a new mite (Acarina: Macrochelidae) from

fresh-water beach debris in New York. – Journ. Kansas Ent. Soc. 46: 101–107, figs. 1–13.
Norton, R.A., L.J. Metz & G.D. Sharma 1978. Observations on Epilohmannoides Jacot, 1936 (Acarina: Oribatei), with the description of a new species. – Journ. Georgia Ent. Soc. 13: 134–148, figs. 1–4.
Norton, R.A., B.M. OConnor & D.E. Johnston, 1983. Systematic relationship of the Pediculochelidae (Acari: Acariformes). – Proc. Ent. Soc. Wash. 85: 493–512, figs. 1–11.
Nuttall, G.H.F., W.F. Cooper & L.E. Robinson, 1908. The structure and biology of Haemaphysalis punctata, Canestrini and Fanzago, I. – Parasitology 1: 152–181, figs. 1–9, pls. 12–16.
Nuttall, G.H.F., W.F. Cooper & L.E. Robinson, 1908a. On the structure of 'Haller's organ' in the Ixodoidea. – Parasitology 1: 238–242, pl. 18, fig. 1.
Nuttall, G.H.F., W.F. Cooper & L.E. Robinson, 1908b. On the structure of the spiracle of a tick, Haemaphysalis punctata Can. et Fanz. – Parasitology 1: 347–351.
Oliver, Jr., J.H., 1971. Parthenogenesis in mites and ticks (Arachnida: Acari). – Am. Zoologist 11: 283–299, fig. 1.
Oudemans, A.C., 1906. Über Genitaltracheeen bei Chernetiden und Acari. – Zool. Anz. 30: 135–140, 1 fig.
Oudemans, A.C., 1928. Acarologische aanteekeningen XCIII. – Ent. Ber. 7: 346–348.
Oudemans, A.C., 1931. Acarologische aanteekeningen CXI. – Ent. Ber. 8: 312–331.
Oudemans, A.C., 1936. Kritisch historisch overzicht der acarologie. Derde gedeelte. 1805–1850, vol. A. – Leiden (E.J. Brill): xx + 430 pp., 179 figs.
Oudemans, A.C., 1940. Die Stellung von Koenenia und Stenarthron im System und die Auflösung der Gruppe 'Acari'. – Zool. Anz. 131: 287–296, figs. 1–10.
Pabst, W., 1953. Zur Biologie der mitteleuropäischen Troguliden. – Zool. Jahrb., Syst. 82: 1–46, pls. 1–5, figs. 1–10.
Pagenstecher, H.A., 1861. Beiträge zur Anatomie der Milben. Heft II. Ixodes ricinus. – Leipzig (Wilhelm Engelmann): iv + 45 pp., pls. 1–2.
Patten, W. & A.P. Hazen, 1900. The development of the coxal gland, branchial cartilages, and genital ducts of Limulus polyphemus. – Journ. Morphol. 16: 459–502, figs. 1–3, pls. 22–28.
Petrunkevitch, A., 1955. Arachnida. – In: R.C. Moore, Treatise on Invertebrate Paleontology, Part P, Arthropoda 2: 42–162, figs. 31–116.
Pittard, K. & R.W. Mitchell, 1972. Comparative morphology of the life stages of Cryptocellus pelaezi (Arachnida, Ricinulei). – Grad. Stud. Texas Tech. Univ. 1: 3–77, figs. 1–130.
Platnick, N.I. & N. Paz S., 1979. On the Cryptocellus magnus group (Arachnida, Ricinulei). – Amer. Mus. Novit. 2677: 1–9 figs. 1–11.
Platnick, N.I. & M.U. Shadab, 1976. On Colombian Cryptocellus (Arachnida, Ricinulei). – Amer. Mus. Novit. 2605: 1–8, figs. 1–17.
Platnick, N.I. & M.U. Shadab, 1977. On Amazonian Cryptocellus (Arachnida, Ricinulei). – Amer. Mus. Novit. 2633: 1–17, figs. 1–58.
Pocock, R.P., 1893. On some points in the morphology of the Arachnida (s.s.), with notes on the classification of the group. – Ann. Mag. Nat. Hist. (6), 11: 1–19, pls. 1–2.
Rafalski, J., 1958. A description of Siro carpathicus sp.n. with remarks on the morphology and systematics of the Cyphophthalmi (Opiliones). – Act. Zool. Cracov. 2: 523–556, figs. 1–21.
Rémy, P.A., 1950. Palpigrades de Madagascar. – Mém. Inst. Sci. Madagascar (A), 4: 135–164, figs. 1–13.
Robaux, P., 1974. Recherches sur le développement et la biologie des Acariens 'Thrombidiidae'. – Mém. Mus. Nat. Hist. Natur. (N.S.), sér. A, Zool., 85: 1–186, figs. 1–58.
Robinson, L.E. & J. Davidson, 1913. The anatomy of Argas persicus (Oken 1818). Part I. – Parasitology 6: 20–48, figs. 1–2, pls. 1–6.
Robinson, L.E. & J. Davidson, 1913a. The anatomy of Argas persicus (Oken 1818). Part II. – Parasitology 6: 217–256, figs. 1–8, pls. 14–17.
Robinson, L.E. & J. Davidson, 1914. The anatomy of Argas persicus (Oken 1818). Part III. – Parasitology 6: 382–424, figs. 1–8, pls. 25–28.
Roewer, C.F., 1932–1934. 1. Ordnung. Solifuga oder Walzenspinnen. – In: Bronns Klassen und

Ordnungen des Tierreichs, Fünfter Band, Vierte Abteilung, Viertes Buch, Lief. 1–5: 9–637, figs. 13–362. [Lief. 1 (pp. 9–160): 1932. Lief. 2 (pp. 161–320): 1933. Lief. 3–5 (pp. 321–637): 1934.]
Roewer, C.Fr., 1934a. Palpigradi. – In: Bronns Klassen und Ordnungen des Tierreichs, Fünfter Band, Vierte Abteilung, Viertes Buch, Lief. 5: 640–707, figs. 363–403.
Roth, V.D. & B.M. Roth, 1984. A review of appendotomy in spiders and other arachnids. – Bull. Brit. Arachn. Soc. 6: 137–146, figs. 1–8.
Rowland, J.M. & W.D. Sissom, 1980. Report on a fossil Palpigrade from the Tertiary of Arizona, and a review of the morphology and systematics of the order (Arachnida, Palpigradida). – Journ. Arachnol. 8: 69–86, figs. 1–7.
Rucker, A., 1901. The Texas Koenenia. – The American Naturalist, Boston 35: 615–630.
Rucker, A., 1903. Further observations on Koenenia. – Zool. Jahrb., Syst. 14: 401–434, pls. 21–23.
Rucker, A., 1903a. A new Koenenia from Texas. – Quart. Journ. Microscop. Sci. (N.S.) 47: 215–231, pl. 18.
Ruser, M., 1933. Beiträge zur Kenntnis des Chitins und der Muskulatur der Zecken (Ixodidae). – Zeitschr. Morph. Ökol. Tiere 27: 199–261, figs. 1–39.
Savory, T.H., 1935. The Arachnida. – London: xii +218 pp., figs. 1–99, pls. 1–8.
Savory, T.H., 1964. Arachnida. – London, New York (Academic Press): xii + 291 pp., figs. 1–109.
Savory, T.H., 1971. Evolution in the Arachnida. – Watford, Herts.: vi + 42 pp., figs. 1–8.
Savory, T.H., 1974. On the arachnid order Palpigradi. – Journ. Arachnol. 2: 43–45.
Savory, T.H., 1977. Arachnida. – London, etc. (Academic Press) (second edition). xii + 360 pp., figs. 1–107.
Schaarschmidt, L., 1959. Systematik und Ökologie der Tarsonemiden. – In: H.J. Stammer, Beiträge zur Systematik und Ökologie Mittel-europäischer Acarina 1(2): 713–823, figs. 1–55.
Schaller, F., 1965. Mating behaviour of lower terrestrial arthropods from the phylogenetic point of view. – Proc. 12th Int. Congr. Ent. London 1964: 297–298.
Schaller, F., 1971. Indirect sperm transfer by soil arthropods. – Ann. Rev. Ent. 16: 407–446, figs. 1–34, tab. 1.
Schaller, F., 1979. Significance of sperm transfer and formation of spermatophores in Arthropod phylogeny. – In: A.P. Gupta, Arthropod Phylogeny (New York): 587–608, figs. 10.1–10.3, pls. 10.I–10.IV.
Schulze, P., 1932. Über die Körpergliederung der Zecken, die Zusammensetzung des Gnathosoma und die Beziehungen der Ixodoidea zu den fossilen Anthracomarti. – S.B. Abh. Naturforsch. Ges. Rostock (3), 3: 104–126, figs. 1–21.
Schulze, P., 1935. Zur vergleichenden Anatomie der Zecken. Das Sternale, die Mundwerkzeuge, Analfurchen und Analbeschilderung und ihre Bedeutung, Ursprünglichkeit und Luxurieren. – Zeitschr. Morph. Ökol. Tiere 30: 1–40, figs. 1–37.
Schulze, P., 1941. Das Geruchsorgan der Zecken. Untersuchungen über die Abwandlungen eines Sinnesorgans und seine stammesgeschichtliche Bedeutung. – Zeitschr. Morph. Ökol. Tiere 37: 491–564, figs. 1–90.
Schulze, P., 1942. Über die Hautsinnesorgane der Zecken, besonders über eine bisher unbekannte Art von Arthropoden-Sinnesorgane, die Krobylophoren. – Zeitschr. Morph. Ökol. Tiere 38: 379–419, figs. 1–38.
Selden, P.A., 1981. Functional morphology of the prosoma of Baltoeurypterus tetragonophthalmus (Fischer) (Chelicerata: Eurypterida). – Trans. Roy. Soc. Edinburgh: Earth Sciences 72: 9–48, figs. 1–32.
Selden, P.A., 1983. The fossil Ricinuleids. – 9th Int. Arachnol. Congr. Panama (Abstract): 1 p.
Shear, W.A., P.M. Bonamo, J.D. Grierson, W.D.I. Rolfe, E.L. Smith & R.A. Norton, 1984. Early land animals in North America: Evidence from Devonian Age Arthropods from Gilboa, New York. – Science 224: 492–494, fig. 1, tab. 1.
Siebold, G.Th. von, 1845. Lehrbuch der vergleichenden Anatomie der Wirbellosen Thiere [= Von Siebold & Stannius, Lehrbuch der vergleichenden Anatomie, vol. 1], Heft 1. – Berlin (Veit & Comp.).

Snodgrass, R.E., 1948. The feeding organs of Arachnida, including mites and ticks. – Smithson. Misc. Coll. 110(10): 1–93, figs. 1–29.
Størmer, L., 1955. Merostomata. – In: R.C. Moore, Treatise on Invertebrate Paleontology, Part P, Arthropoda 2: 4–41, figs. 3–30.
Suski, Z.W., 1966. Nomenclature of leg setation in the mite family Tarsonemidae (Acarina), (Heterostigmata). – Bull. Ac. Pol. Sci., Sér. Sci. Biol. 14: 635–638, figs. 1–2.
Suski, Z.W., 1967. Badania nad roztoczami z rodziny Tarsonemidae (Acarina, Heterostigmata) występującymi na jabłoniach w Polsce. – Skierniewice: 268 pp., 75 figs.
Suski, Z.W., 1968. Setation of legs I, II and III in some species of the mite family Tarsonemidae (Acarina, Heterostigmata). – Roczn. Nauk. Roln. 93–A4: 757–787, figs. 1–5.
Thon, K., 1905. Neue Luftorgane bei Milben. – Zool. Anz. 28: 585–594, figs. 1–3.
Thorell, T., 1888. Pedipalpi e Scorpioni dell' Arcipelago Malese conservati nel Museo Civico di Storia Naturale di Genova. – Ann. Mus. Civ. Stor. Natur. Genova (2), 6 [= 26]: 327–428.
Tower, R.W., 1895. The external opening of the 'Brick-Red' gland in Limulus polyphemus. – Zool. Anz. 18: 471–472.
Travé, J., 1982. Premières observations sur le comportement de Thonius braueri (Thon, 1906) (Holothyrida). – Acarologia 23: 199–206, fig. 1.
Travé, J., 1983. Observations morphologiques sur les Holothyrides. Ire partie. Le corps de Thonius Braueri (Thon, 1906). – Acarologia 24: 333–341, figs. 1–3.
Tuxen, S.L., 1974. The African genus Ricinoides (Arachnida, Ricinulei). – Journ. Arachnol. 1: 85–106, figs. 1–44.
Vachon, M., 1938. Recherches anatomiques et biologiques sur la reproduction et le développement des Pseudoscorpions. – Ann. Sci. Natur., Zool. Biol. Anim. (1), 1: 1–207, figs. 1–85.
Vachon, M., 1945. L'appendice arachnidien et son évolution. Note préliminaire. – Bull. Soc. Zool. France 69: 172–177.
Vachon, M., 1945a. Remarques sur les appendices du prosoma des Limules et leur arthrogenèse. – Arch. Zool. Exp. Gén. 84(6): 271–300, figs. 1–24, tab. 1–2.
Vachon, M., 1949. Ordre des Pseudoscorpions. Faux-scorpion, Latreille 1817. Chelifères, Gervais 1844. Chernetidae, Menge 1855. Pseudoscorpiones, Pavesi 1880. Chelonethi, Thorell 1882. Chelonethida, Chamberlin 1929. Pseudoscorpionidea, Beier 1932. – In: P.-P. Grassé, Traité de Zoologie 6: 437–481, figs. 196–241.
Vercammen-Grandjean, P.H., 1965. On the true nature of the 'striate' solenidion in chiggers (Trombiculidae: Acarina). – Acarologia 7: 318–320, figs. 1–5.
Versluys, J. & R. Demoll, 1921. Die Verwandtschaft der Merostomata mit den Arachnida und den anderen Abteilungen der Arthropoda. – Proc. Kon. Akad. Wet. Amsterdam 23: 739–765, figs. 1–6.
Versluys, J. & R. Demoll, 1922. Das Limulus-Problem. Die Verwandtschaftsbeziehungen der Merostomen und Arachnoideen unter sich und mit anderen Arthropoden. – Ergebn. Fortschr. Zool. 5: 67–388, figs. 1–123.
Vitzthum, H. Graf, 1931. 9. Ordnung der Arachniden: Acari = Milben. – In: W. Kükenthal & Th. Krumbach, Handbuch der Zoologie, Dritter Band, Zweite Hälfte, Erster Teil: (3) 1–160, figs. 1–161.
Vitzthum, H. Graf, 1940. Acarina. – In: Bronns Klassen und Ordnungen des Tierreichs, Fünfter Band, Vierte Abteilung, Fünftes Buch, Lief. 1–3: 1–480, figs. 1–428.
Vitzthum, H. Graf, 1941. Acarina. – In: Bronns Klassen und Ordnungen des Tierreichs, Fünfter Band, Vierte Abteilung, Fünftes Buch, Lief. 4: 481–640, figs. 429–448.
Vitzthum, H. Graf, 1941a. 9. und letzte Ordnung der Arachnida: Acari = Milben. – In: W. Kükenthal & Th. Krumbach, Handbuch der Zoologie, Dritter Band, Zweite Hälfte, Erster Teil: (3) 215–220, fig. 102.
Vitzthum, H. Graf, 1942. Acarina. – In: Bronns Klassen und Ordnungen des Tierreichs, Fünfter Band, Vierte Abteilung, Fünftes Buch, Lief. 5–6: 641–912, figs. 449–522.
Vitzthum, H. Graf, 1943. Acarina. – In: Bronns Klassen und Ordnungen des Tierreichs, Fünfter Band, Vierte Abteilung, Fünftes Buch, Lief. 7: 913–1011.
Weygoldt, P., 1968. Vergleichend-embryologische Untersuchungen an Pseudoscorpionen IV.

Die Entwicklung von Chthonius tetrachelatus Preyssl., Chthonius ischnocheles Hermann (Chthoniinea, Chthoniidae) und Verrucaditha spinosa Banks (Chthoniinea, Tridenchthoniidae). – Zeitschr. Morph. Tiere 63: 111–154, figs. 1–31.

Weygoldt, P., 1969. Beobachtungen zur Fortpflanzungsbiologie und zum Verhalten der Geisselspinne Tarantula marginemaculata C.L. Koch (Chelicerata, Amblypygi). – Zeitschr. Morph. Tiere 64: 338–360, figs. 1–8.

Weygoldt, P., 1970. Lebenszyklus und postembryonale Entwicklung der Geisselspinne Tarantula marginemaculata C.L. Koch (Chelicerata, Amblypygi) im Laboratorium. – Zeitschr. Morph. Tiere 67: 58–85, figs. 1–15, tab. 1–4.

Weygoldt, P., 1970a. Vergleichende Untersuchungen zur Fortpflanzungsbiologie der Pseudoscorpione II. – Zeitschr. Zool. Syst. u. Evolutionsforsch. 8: 241–259, figs. 1–17, tab. 1.

Weygoldt, P., 1970b. Courtship behaviour and sperm transfer in the Giant Whip Scorpion, Mastigoproctus giganteus (Lucas) (Uropygi, Thelyphonidae). – Behaviour 36: 1–8, figs. 1–4.

Weygoldt, P., 1971. Notes on the life history and reproductive biology of the Giant Whip Scorpion, Mastigoproctus giganteus (Uropygi, Thelyphonidae) from Florida. – Journ. Zool. London 164: 137–147, figs. 1–6, tab. 1, pls. I–III.

Weygoldt, P., 1980. Towards a cladistic classification of the Chelicerata. – Proc. 8. Int. Arachn.-Kongr. Wien 1980: 331–334.

Weygoldt, P. & H.F. Paulus, 1979. Untersuchungen zur Morphologie, Taxonomie und Phylogenie der Chelicerata I. Morphologische Untersuchungen. – Zeitschr. Zool. Syst. u. Evolutionsforsch. 17: 85–116, figs. 1–7.

Weygoldt, P. & H.F. Paulus, 1979a. Untersuchungen zur Morphologie, Taxonomie und Phylogenie der Chelicerata II. Cladogramme und die Entfaltung der Chelicerata. – Zeitschr. Zool. Syst. u. Evolutionsforsch. 17: 177–200, figs. 1–5.

Weygoldt, P., A. Weisemann & K. Weisemann, 1972. Morphologisch-histologische Untersuchungen an den Geslechtsorganen der Amblypygi unter besonderer Berücksichtigung von Tarantula marginemaculata C.L. Koch (Arachnida). – Zeitschr. Morph. Tiere 73: 209–247, figs. 1–21.

Wheeler, W., 1900. A singular arachnid (Koenenia mirabilis Grassi) occurring in Texas. – Amer. Natural., Boston 34: 837–850.

Winkler, D., 1957. Die Entwicklung der äusseren Körpergestalt bei den Phalangiidae (Opiliones). – Mitt. Zool. Mus. Berlin 33: 355–389, figs. 1–46.

With, C.J., 1903. The Notostigmata, a new suborder of Acari. – Vidensk. Medd. Naturh. Foren. Kjøbenhavn 1904: 137–192, pls. 4–6.

Woodring, J.P., 1973. Comparative morphology, functions, and homologies of the coxal glands in Oribatid mites (Arachnida: Acari). – Journ. Morph. 139: 407–429, figs. 1–38, graph 1, tab. 1.

Zebrowsky, G., 1926. A preliminary report on the morphology of the American dog tick. – Trans. Amer. Ent. Soc. 51: 331–369, pls. 12–14.